Springer Tracts in Advanced Robotics
Volume 118

Editors: Bruno Siciliano · Oussama Khatib

Peter Corke

Robotics, Vision and Control

Fundamental Algorithms in MATLAB®

Second, completely revised, extended and updated edition

With 492 Images

Additional material is provided at www.petercorke.com/RVC

 Springer

Professor Bruno Siciliano

Dipartimento di Ingegneria Elettrica e Tecnologie dell'Informazione,
Università di Napoli Federico II, Via Claudio 21, 80125 Napoli, Italy,
e-mail: siciliano@unina.it

Professor Oussama Khatib

Artificial Intelligence Laboratory, Department of Computer Science,
Stanford University, Stanford, CA 94305-9010, USA,
e-mail: khatib@cs.stanford.edu

Author
Peter Corke

School of Electrical Engineering and Computer Science
Queensland University of Technology (QUT), Brisbane QLD 4000, Australia
e-mail: rvc@petercorke.com

ISSN 1610-7438   ISSN 1610-742X (electronic)
Springer Tracts in Advanced Robotics

ISBN 978-3-319-54412-0   ISBN 978-3-319-54413-7 (eBook)
DOI 10.1007/978-3-319-54413-7

Library of Congress Control Number: 2017934638

1st ed. 2011 © Springer-Verlag Berlin Heidelberg 2011
© Springer International Publishing AG 2017

Production: Armin Stasch and Scientific Publishing Services Pvt. Ltd. Chennai, India
Typesetting and layout: Stasch · Bayreuth (stasch@stasch.com)

Printed on acid-free paper

This Springer imprint is published by Springer Nature
The registered company is Springer International Publishing AG
The registered company address is: Gewerbestrasse 11, 6330 Cham, Switzerland

More information about this series at http://www.springer.com/series/5208

To my family Phillipa, Lucy and Madeline for their indulgence and support;
my parents Margaret and David for kindling my curiosity;
and to Lou Paul who planted the seed that became this book.

Foreword

Once upon a time, a very thick document of a dissertation from a faraway land came to me for evaluation. *Visual robot control* was the thesis theme and *Peter Corke* was its author. Here, I am reminded of an excerpt of my comments, which reads, *this is a masterful document, a quality of thesis one would like all of one's students to strive for, knowing very few could attain – very well considered and executed.*

The connection between robotics and vision has been, for over two decades, the central thread of Peter Corke's productive investigations and successful developments and implementations. This rare experience is bearing fruit in this second edition of his book on *Robotics, Vision, and Control.* In its melding of theory and application, this second edition has considerably benefited from the author's unique mix of academic and real-world application influences through his many years of work in robotic mining, flying, underwater, and field robotics.

There have been numerous textbooks in robotics and vision, but few have reached the level of integration, analysis, dissection, and practical illustrations evidenced in this book. The discussion is thorough, the narrative is remarkably informative and accessible, and the overall impression is of a significant contribution for researchers and future investigators in our field. Most every element that could be considered as relevant to the task seems to have been analyzed and incorporated, and the effective use of Toolbox software echoes this thoroughness.

The reader is taken on a realistic walkthrough the fundamentals of mobile robots, navigation, localization, manipulator-arm kinematics, dynamics, and joint-level control, as well as camera modeling, image processing, feature extraction, and multi-view geometry. These areas are finally brought together through extensive discussion of visual servo system. In the process, the author provides insights into how complex problems can be decomposed and solved using powerful numerical tools and effective software.

The *Springer Tracts in Advanced Robotics (STAR)* is devoted to bringing to the research community the latest advances in the robotics field on the basis of their significance and quality. Through a wide and timely dissemination of critical research developments in robotics, our objective with this series is to promote more exchanges and collaborations among the researchers in the community and contribute to further advancements in this rapidly growing field.

Peter Corke brings a great addition to our STAR series with an authoritative book, reaching across fields, thoughtfully conceived and brilliantly accomplished.

Oussama Khatib
Stanford, California
October 2016

Preface

These are exciting times for robotics. Since the first edition of this book was published we have seen much progress: the rise of the self-driving car, the Mars science laboratory rover making profound discoveries on Mars, the Philae comet landing attempt, and the DARPA Robotics Challenge. We have witnessed the drone revolution – flying machines that were once the domain of the aerospace giants can now be bought for just tens of dollars. All this has been powered by the continuous and relentless improvement in computer power and tremendous advances in low-cost inertial sensors and cameras – driven largely by consumer demand for better mobile phones and gaming experiences. It's getting easier for individuals to create robots – 3D printing is now very affordable, the Robot Operating System (ROS) is both capable and widely used, and powerful hobby technologies such as the Arduino, Raspberry Pi, Dynamixel servo motors and Lego's EV3 brick are available at low cost. This in turn has contributed to the rapid growth of the global maker community – ordinary people creating at home what would once have been done by a major corporation. We have also witnessed an explosion of commercial interest in robotics and computer vision – many startups and a lot of acquisitions by big players in the field. Robotics even featured on the front cover of the Economist magazine in 2014!

So how does a robot work? Robots are data-driven machines. They acquire data, process it and take action based on it. The data comes from sensors measuring the velocity of a wheel, the angle of a robot arm's joint or the intensities of millions of pixels that comprise an image of the world that the robot is observing. For many robotic applications the amount of data that needs to be processed, in real-time, is massive. For a vision sensor it can be of the order of tens to hundreds of megabytes per second.

Progress in robots and machine vision has been, and continues to be, driven by more effective ways to process data. This is achieved through new and more efficient algorithms, and the dramatic increase in computational power that follows Moore's law. ◄ When I started in robotics and vision in the mid 1980s, see Fig. 0.1, the IBM PC had been recently released – it had a 4.77 MHz 16-bit microprocessor and 16 kbytes (expandable to 256 k) of memory. Over the intervening 30 years computing power has perhaps doubled 20 times which is an increase by a factor of one million.

"Computers in the future may weigh no more than 1.5 tons." Popular Mechanics, forecasting the relentless march of science, 1949

Over the fairly recent history of robotics and machine vision a very large body of algorithms has been developed to efficiently solve large-scale problems in perception, planning, control and localization – a significant, tangible, and collective achievement of the research community. However its sheer size and complexity presents a very real barrier to somebody new entering the field. Given so many algorithms from which to choose, a real and important question is:

What is the right algorithm for this particular problem?

One strategy would be to try a few different algorithms and see which works best for the problem at hand, but this is not trivial and leads to the next question:

How can I evaluate algorithm X on my own data without spending days coding and debugging it from the original research papers?

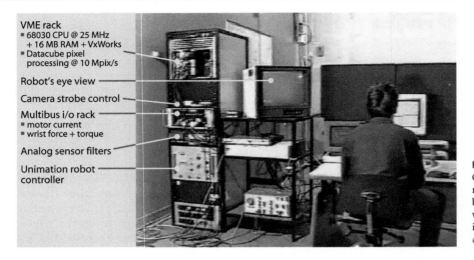

VME rack
- 68030 CPU @ 25 MHz
  + 16 MB RAM + VxWorks
- Datacube pixel
  processing @ 10 Mpix/s

Robot's eye view

Camera strobe control

Multibus i/o rack
- motor current
- wrist force + torque

Analog sensor filters

Unimation robot
controller

Fig. 0.1.
Once upon a time a lot of equipment was needed to do vision-based robot control. The author with a large rack full of real-time image processing and robot control equipment (1992)

Two developments come to our aid. The first is the availability of general purpose mathematical software which it makes it easy to prototype algorithms. There are commercial packages such as MATLAB®, Mathematica®, Maple® and MathCad®▸, as well as open source projects include SciLab, Octave, and PyLab. All these tools deal naturally and effortlessly with vectors and matrices, can create complex and beautiful graphics, and can be used interactively or as a programming environment. The second is the open-source movement. Many algorithms developed by researchers are available in open-source form. They might be coded in one of the general purpose mathematical languages just mentioned, or written in a mainstream language like C, C++, Java or Python.

Respectively the trademarks of The Math-Works Inc., Wolfram Research, MapleSoft and PTC.

For more than twenty years I have been part of the open-source community and maintained two open-source MATLAB Toolboxes: one for robotics and one for machine vision▸. They date back to my own Ph.D. work and have evolved since then, growing features and tracking changes to the MATLAB language. The Robotics Toolbox has also been translated into a number of different languages such as Python, SciLab and LabView. More recently some of its functionality is finding its way into the MATLAB Robotics System Toolbox™ published by The MathWorks.

These Toolboxes have some important virtues. Firstly, they have been around for a long time and used by many people for many different problems so the code can be accorded some level of trust. New algorithms, or even the same algorithms coded in new languages or executing in new environments, can be compared against implementations in the Toolbox.

The term machine vision is uncommon today, but it implied the use of real-time computer vision techniques in an industrial setting for some monitoring or control purpose. For robotics the real-time aspect is critical but today the interesting challenges are in nonindustrial applications such as outdoor robotics. The term robotic vision is gaining currency and is perhaps a modern take on machine vision.

» allow the user to work with real problems, not just trivial examples

Secondly, they allow the user to work with real problems, not just trivial examples. For real robots, those with more than two links, or real images with millions of pixels the computation required is beyond unaided human ability. Thirdly, they allow us to gain insight which can otherwise get lost in the complexity. We can rapidly and easily experiment, play *what if* games, and depict the results graphically using the powerful 2D and 3D graphical display tools of MATLAB. Fourthly, the Toolbox code makes many common algorithms tangible and accessible. You can read the code, you can apply it to your own problems, and you can extend it or rewrite it. It gives you a "leg up" as you begin your journey into robotics.

» a narrative that covers robotics and computer vision – both separately and together

The genesis of the book lies in the tutorials and reference material that originally shipped with the Toolboxes from the early 1990s, and a conference paper describing the Robotics Toolbox that was published in 1995. After a false start in 2004, the first edition of this book was written in 2009–2010. The book takes a conversational approach, weaving text, mathematics and examples into a narrative that covers robotics and computer vision – both separately and together. I wanted to show how complex problems can be decomposed and solved using just a few simple lines of code. More formally this is an inductive learning approach, going from specific and concrete examples to the more general.

» show how complex problems can be decomposed and solved

The topics covered in this book are based on my own interests but also guided by real problems that I observed over many years as a practitioner of both robotics and computer vision. I want to give the reader a flavor of what robotics and vision is about and what it can do – consider it a grand tasting menu. I hope that by the end of this book you will share my enthusiasm for these topics.

» consider it a grand tasting menu

I was particularly motivated to present a solid introduction to computer vision for roboticists. The treatment of vision in robotics textbooks tends to concentrate on simple binary vision techniques. In this book we will cover a broad range of topics including color vision, advanced segmentation techniques, image warping, stereo vision, motion estimation, bundle adjustment, visual odometry and image retrieval. We also cover nonperspective imaging using fisheye lenses, catadioptric optics and the emerging area of light-field cameras. These topics are growing in importance for robotics but are not commonly covered. Vision is a powerful sensor, and roboticists should have a solid grounding in modern fundamentals. The last part of the book shows how vision can be used as the primary sensor for robot control.

This book is unlike other text books, and deliberately so. Firstly, there are already a number of excellent text books that cover robotics and computer vision separately and in depth, but few that cover both in an integrated fashion. Achieving such integration is a principal goal of the book.

» software is a first-class citizen in this book

Secondly, software is a first-class citizen in this book. Software is a tangible instantiation of the algorithms described – it can be read and it can be pulled apart, modified and put back together again. There are a number of classic books that use software in an illustrative fashion and have influenced my approach, for example *LaTeX: A document preparation system* (Lamport 1994), *Numerical Recipes in C* (Press et al. 2007), *The Little Lisper* (Friedman et al. 1987) and *Structure and Interpretation of Classical Mechanics* (Sussman et al. 2001). Over 1 000 examples in this book illustrate how the Toolbox software can be used and generally provide *instant gratification* in just a couple of lines of MATLAB code.

» instant gratification in just a couple of lines of MATLAB code

Thirdly, building the book around MATLAB and the Toolboxes means that we are able to tackle more realistic and more complex problems than other books.

» this book provides a complementary approach

The emphasis on software and examples does not mean that rigor and theory are unimportant – they are very important, but this book provides a complementary approach. It is best read in conjunction with standard texts which do offer rigor and theoretical nourishment. The end of each chapter has a section on further reading and provides pointers to relevant textbooks and key papers. I try hard to use the least amount of mathematical notation required, if you seek deep mathematical rigor this may not be the book for you.

Writing this book provided the impetus to revise and extend the Toolboxes and to include some great open-source software. I am grateful to the following for code that has been either incorporated into the Toolboxes or which has been wrapped into the Toolboxes. Robotics Toolbox contributions include: mobile robot localization and mapping by Paul Newman; a quadrotor simulator by Paul Pounds; a Symbolic Manipulator Toolbox by Jörn Malzahn; pose-graph SLAM code by Giorgio Grisetti and 3D robot models from the ARTE Robotics Toolbox by Arturo Gil. Machine Vision Toolbox contributions include: RANSAC code by Peter Kovesi; pose estimation by Francesco Moreno-Noguer, Vincent Lepetit, and Pascal Fua; color space conversions by Pascal Getreuer; numerical routines for geometric vision by various members of the Visual Geometry Group at Oxford (from the web site of the Hartley and Zisserman book; Hartley and Zisserman 2003); k-means, SIFT and MSER algorithms from the wonderful VLFeat suite (vlfeat.org); graph-based image segmentation software by Pedro Felzenszwalb; and the OpenSURF feature detector by Dirk-Jan Kroon. The Camera Calibration Toolbox by Jean-Yves Bouguet is used unmodified.

Along the way I became fascinated by the mathematicians, scientists and engineers whose work, hundreds of years ago, underpins the science of robotic and computer vision today. Some of their names have become adjectives like Coriolis, Gaussian, Laplacian or Cartesian; nouns like Jacobian, or units like Newton and Coulomb. They are interesting characters from a distant era when science was a hobby and their day jobs were as doctors, alchemists, gamblers, astrologers, philosophers or mercenaries. In order to know whose shoulders we are standing on I have included small vignettes about the lives of some of these people – a smattering of history as a backstory.

In my own career I have had the good fortune to work with many wonderful people who have inspired and guided me. Long ago at the University of Melbourne John Anderson fired my interest in control and Graham Holmes tried with mixed success to have me "think before I code". Early on I spent a life-direction-changing ten months working with Richard (Lou) Paul in the GRASP laboratory at the University of Pennsylvania in the period 1988–1989. The genesis of the Toolboxes was my Ph.D. research (1991–1994) and my advisors Malcolm Good (University of Melbourne) and Paul Dunn (CSIRO) asked me good questions and guided my research. Laszlo Nemes (CSIRO) provided great wisdom about life and the ways of organizations, and encouraged me to publish and to open-source my software. Much of my career was spent at CSIRO where I had the privilege and opportunity to work on a diverse range of real robotics projects and to work with a truly talented set of colleagues and friends. Part way through writing the first edition I joined the Queensland University of Technology which made time available to complete that work, and in 2015 sabbatical leave to complete the second.

Many people have helped me in my endeavor and I thank them. I was generously hosted for periods of productive writing at Oxford (both editions) by Paul Newman, and at MIT (first edition) by Daniela Rus. Daniela, Paul and Cédric Pradalier made constructive suggestions and comments on early drafts of that edition. For the second edition I was helped by comments on draft chapters by: Tim Barfoot, Dmitry Bratanov, Duncan Campbell, Donald Dansereau, Tom Drummond, Malcolm Good, Peter Kujala, Obadiah Lam, Jörn Malzahn, Felipe Nascimento Martins, Ajay Pandey, Cédric Pradalier, Dan Richards, Daniela Rus, Sareh Shirazi, Surya Singh, Ryan Smith, Ben Talbot, Dorian Tsai and Ben Upcroft; and assisted with wisdom and content by: François Chaumette, Donald Dansereau, Kevin Lynch, Robert Mahony and Frank Park.

I have tried my hardest to eliminate errors but inevitably some will remain. Please email bug reports to me at rvc@petercorke.com as well as suggestions for improvements and extensions.

Writing the second edition was financially supported by EPSRC Platform Grant EP/M019918/1, QUT Science & Engineering Faculty sabbatical grant, QUT Vice Chancellor's Excellence Award, QUT Robotics and Autonomous Systems discipline and the ARC Centre of Excellence for Robotic Vision (grant CE140100016).

Over both editions I have enjoyed wonderful support from MathWorks, through their author program, and from Springer. My editor Thomas Ditzinger has been a great supporter of this project and Armin Stasch, with enormous patience and dedication in layout and typesetting, has transformed my untidy ideas into a thing of beauty.

Finally, my deepest thanks are to Phillipa who has supported me and "the book" with grace and patience for a very long time and in many different places – without her this book could never have been written.

Peter Corke
Brisbane,
Queensland
October 2016

Note on the Second Edition

It seems only yesterday that I turned in the manuscript for the first edition of this book, but it was in fact December 2010, the end of 20 months of writing. So the oldest parts of the book are over 6 years old – it's time for an update!

The revision principle was to keep the good (narrative style, code as a first-class citizen, soft plastic cover) and eliminate the bad (errors and missing topics). I started with the collected errata for the first edition and pencilled markup from a battered copy of the first edition that I've carried around for years. There were more errors than I would have liked and I thank everybody who submitted errata and suggested improvements.

The first edition was written before I taught in the university classroom or created the MOOCs, which is the inverse of the way books are normally developed. Preparing for teaching gave me insights into better ways to present some topics, particularly around pose representation, robot kinematics and dynamics so the presentation has been adjusted accordingly.

New content includes matrix exponential notation; the basics of screw theory and Lie algebra; inertial navigation; differential steer and omnidirectional mobile robots; a deeper treatment of SLAM systems including scan matching and pose graphs; greater use of MATLAB computer algebra; operational space control; deeper treatment of manipulator dynamics and control; visual SLAM and visual odometry; structured light; bundle adjustment; and light-field cameras.

In the first edition I shied away from Lie algebra, matrix exponentials and twists but I think it's important to cover them. The topic is deeply mathematical and I've tried to steer a middle ground between hardcore algebraic topology and the homogenous transformation only approach of most other texts, while also staying true to the overall approach of this book.

All MATLAB generated figures have been regenerated to reflect recent improvements to MATLAB graphics and all code examples have been updated as required and tested, and are available as MATLAB Live Scripts.

The second edition of the book is matched by new major releases of my Toolboxes: Robotics Toolbox (release 10) and the Machine Vision Toolbox (release 4). These newer versions of the toolboxes have some minor incompatibilities with previous releases of the toolboxes, and therefore also with the code examples in the first edition of the book.

Contents

Nomenclature

The notation used in robotics and computer vision varies considerably across books and research papers. The symbols used in this book, and their units where appropriate, are listed below. Some symbols have multiple meanings and their context must be used to disambiguate them.

Notation	Description
x^*	desired value of x
x^+	predicted value of x
$x^\#$	measured, or observed, value of x
$\hat{x}$	estimated value of x
$\bar{x}$	mean of x or relative value
$x\langle k \rangle$	k^{th} element of a time series
v	a vector
$\hat{v}$	a unit-vector parallel to v
$\tilde{v}$	homogeneous representation of vector v
$v[i]$	i^{th} element of vector v
v_x	a component of a vector
A	a matrix
$A[i,j]$	the element (i,j) of A
$A_{i,j}$	the element (i,j) of A
$f(x)$	a function of x
$F_x(x)$	the derivative $\partial f/\partial x$
$F_{xy}(x,y)$	the derivative $\partial^2 f/\partial x \partial y$
$\mathring{q}$	unit quaternion, $\mathring{q} \in \mathbb{S}^3$
$\mathbf{0}_{m \times n}$	an $m \times n$ matrix of zeros
$\mathbf{1}_{m \times n}$	an $m \times n$ matrix of ones

Symbol	Description	Unit
B	viscous friction coefficient	$\mathrm{N\,m\,s\,rad^{-1}}$
B	magnetic field intensity (or magnetic flux density)	T
C	camera matrix, $C \in \mathbb{R}^{3 \times 4}$	
$C(q, \dot{q})$	manipulator centripetal and Coriolis term	$\mathrm{kg\,m^2\,s^{-1}}$
$\mathcal{C}$	configuration space of a robot with N joints: $\mathcal{C} \subset \mathbb{R}^N$	
E	illuminance (lux)	lx
f	focal length	m
f	force	N
$\boldsymbol{f}$	vector of image features	
$F(\dot{q})$	friction torque	Nm
$G(q)$	manipulator gravity loading term	Nm
$\mathbb{H}$	the set of all quaternions (H for Hamilton)	
$I_{n \times n}$	$n \times n$ identity matrix	
J	inertia	$\mathrm{kg\,m^2}$
J	inertia tensor, $J \in \mathbb{R}^{3 \times 3}$	$\mathrm{kg\,m^2}$
J	Jacobian matrix	
$^{A}J_{B}$	Jacobian transforming velocities in frame B to frame A	
k, K	constant	
K	camera calibration matrix	
K_i	amplifier gain (transconductance)	$\mathrm{A\,V^{-1}}$
K_m	motor torque constant	$\mathrm{Nm\,A^{-1}}$
L	luminance (nit)	nt
m	mass	kg
$\mathbf{M}(q)$	manipulator inertia matrix	$\mathrm{kg\,m^2}$
$N(\mu, \sigma^2)$	a normal (Gaussian) distribution with mean μ and standard deviation σ	
$\mathbf{p}$	an image plane point, $\mathbf{p} \in \mathbb{R}^2$	
$\mathbf{P}$	a world point, $\mathbf{P} \in \mathbb{R}^3$	
$\mathbb{P}^2$	projective space of all 2-D points, a 3-tuple	
$\mathbb{P}^3$	projective space of all 3-D points, a 4-tuple	
q	generalized coordinates, configuration $q \in \mathcal{C}$	m, rad
Q	generalized force $Q \in \mathbb{R}^N$	N, Nm
R	an orthonormal rotation matrix, $R \in \mathbf{SO}(2)$ or $\mathbf{SO}(3)$	
$\mathbb{R}$	set of real numbers	
$\mathbb{R}^2$	set of all 2-D points	
$\mathbb{R}^3$	set of all 3-D points	
s	Laplace transform operator	
$\mathbb{S}^1$	unit circle, set of angles $[0, 2\pi)$	
$\mathbb{S}^n$	unit sphere embedded in $\mathbb{R}^{n+1}$	
$\mathbf{se}(n)$	Lie algebra for $\mathbf{SE}(n)$, an $\mathbb{R}^{(n+1) \times (n+1)}$ augmented skew-symmetric matrix	
$\mathbf{so}(n)$	Lie algebra for $\mathbf{SO}(n)$, an $\mathbb{R}^{n \times n}$ skew-symmetric matrix	
$\mathbf{SE}(n)$	special Euclidean group, the set of all poses in n dimensions, represented by an $\mathbb{R}^{(n+1) \times (n+1)}$ homogeneous transformation matrix	
$\mathbf{SO}(n)$	special orthogonal group, the set of all orientations in n dimensions, represented by an $\mathbb{R}^{n \times n}$ orthogonal matrix	
S	twist in 3 dimensions, $S \in \mathbb{R}^6$	
t	time	s
$\mathcal{T}$	task space of robot: $\mathcal{T} \subset \mathbf{SE}(3)$	K

Symbol	Description	Unit
T	sample interval	s
T	temperature	K
T	optical transmission	m^{-1}
T	homogeneous transformation, $T \in \mathbf{SE}(2)$ or $\mathbf{SE}(3)$	
$^A T_B$	homogeneous transform representing frame $\{B\}$ with respect to frame $\{A\}$. If A is not given then assumed relative to world coordinate frame 0. Note that $^A T_B = (^B T_A)^{-1}$	
u, v	camera image plane coordinates	pixels
u_0, v_0	coordinates of the principal point	pixels
$\bar{u}, \bar{v}$	normalized image plane coordinates, relative to the principal point	m
v	velocity	$m\ s^{-1}$
$\boldsymbol{v}$	velocity vector	$m\ s^{-1}$
$\boldsymbol{W}$	wrench, a vector of forces and moments $(f_x, f_y, f_z, m_x, m_y, m_z)$	N, Nm
X, Y, Z	Cartesian coordinates	
$\bar{x}, \bar{y}$	normalized image-plane coordinates	
$\mathbb{Z}$	set of all integers	
$\mathbb{Z}^+$	the set of all integers greater than zero	
ϕ	luminous flux (lumens)	lm
γ	robot steering angle	rad
$\boldsymbol{\Gamma}$	3-angle representation of rotation, $\boldsymbol{\Gamma} \in \mathbb{R}^3$	rad
$\boldsymbol{\Gamma}$	body torque, $\boldsymbol{\Gamma} \in \mathbb{R}^3$	Nm
θ	angle	rad
$\theta_r, \theta_p, \theta_y$	roll pitch yaw angles	rad
λ	wavelength	m
λ	an eigenvalue	
$\boldsymbol{\nu}$	innovation	
ν	spatial velocity, $\nu = (v_x, v_y, v_z, \omega_x, \omega_y, \omega_z) \in \mathbb{R}^6$	$m\ s^{-1}$, $rad\ s^{-1}$
ξ	abstract representation of Cartesian pose (pronounced ksi)	
$^A \xi_B$	abstract representation of relative pose, frame $\{B\}$ with respect to frame $\{A\}$ or rigid-body motion from frame $\{A\}$ to $\{B\}$	
π	mathematic constant	
$\boldsymbol{\pi}$	a plane	
ρ_w, ρ_h	pixel width and height	m
σ	standard deviation	
σ	robot joint type, $\sigma = $ R for revolute and $\sigma = $ P for prismatic	
$\boldsymbol{\Sigma}$	Lie algebra $\boldsymbol{\Sigma} = [\cdot] \in \mathbf{se}(3)$	
τ	torque	N m
τ_C	Coulomb friction torque	N m
ω	rotational rate	$rad\ s^{-1}$
$\boldsymbol{\omega}$	angular velocity vector	$rad\ s^{-1}$
ϖ	rotational speed of a motor or propellor	$rad\ s^{-1}$
$\boldsymbol{\Omega}$	Lie algebra $\boldsymbol{\Omega} = [\cdot]_\times \in \mathbf{so}(3)$	

Operator	Description	MATLAB
$\|\cdot\|$	norm, or length, of vector: $\mathbb{R}^n \mapsto \mathbb{R}$	norm, .norm
$v_1 \cdot v_2$	dot, or inner, product, also $v_1^T v_2$: $\mathbb{R}^n \times \mathbb{R}^n \mapsto \mathbb{R}$	dot
$v_1 \times v_2$	cross, or vector, product: $\mathbb{R}^n \times \mathbb{R}^n \mapsto \mathbb{R}^n$	cross
A^{-1}	inverse of A: $\mathbb{R}^{n \times n} \mapsto \mathbb{R}^{n \times n}$	inv
A^+	pseudo-inverse of A: $\mathbb{R}^{n \times m} \mapsto \mathbb{R}^{m \times n}$	pinv
A^*	adjugate of $A \mapsto \det(A)A^{-1}$, $\mathbb{R}^{n \times n} \mapsto \mathbb{R}^{n \times n}$	
A^T	transpose of A: $\mathbb{R}^{n \times m} \mapsto \mathbb{R}^{m \times n}$	'
A^{-T}	transpose of inverse $A \mapsto (A^T)^{-1} = (A^{-1})^T$, $\mathbb{R}^{n \times n} \mapsto \mathbb{R}^{n \times n}$	
$\bullet$	transform a point (coordinate vector) by a relative pose: $\mathbf{SE}(n) \times \mathbb{R}^n \mapsto \mathbb{R}^n$	*
$\oplus$	composition: $S_E^O(n) \times S_E^O(n) \mapsto S_E^O(n)$	*
$\ominus$	composition with inverse: $S_E^O(n) \times S_E^O(n) \mapsto S_E^O(n)$	/
$\ominus$	unary inverse: $S_E^O(n) \mapsto S_E^O(n)$	.inv
$\Delta(\cdot)$	maps incremental pose change to differential motion: $\mathbf{SE}(3) \mapsto \mathbb{R}^6$	tr2delta
$\Delta^{-1}(\cdot)$	maps differential motion to incremental pose change: $\mathbb{R}^6 \mapsto \mathbf{SE}(3)$	delta2tr
$\mathscr{R}_i(\theta)$	pure rotation about axis i: $\mathbb{R} \mapsto \mathbf{SE}(3)$	SE3.rotx|y|z
$\mathscr{R}(\omega)$	pure rotation by $\|\omega\|$ about ω: $\mathbb{R}^3 \mapsto \mathbf{SE}(3)$	SE3.angvec
$\mathscr{T}_i(d)$	pure translation along axis i: $\mathbb{R} \mapsto \mathbf{SE}(2), \mathbf{SE}(3)$	SE2, SE3
$\mathscr{T}(t)$	pure translation by vector: $\mathbb{R}^n \mapsto \mathbf{SE}(n)$	SE2, SE3
$[\cdot]_t$	translational component of pose: $\mathbf{SE}(n) \mapsto \mathbb{R}^n$	.t
$[\cdot]_R$	rotational component of pose: $\mathbf{SE}(n) \mapsto \mathbb{R}^{n \times n}$	.R
$[\cdot]_\times$	skew-symmetric matrix: $\mathbb{R} \mapsto \mathbf{so}(2)$, $\mathbb{R}^3 \mapsto \mathbf{so}(3)$	skew
$\vee_\times(\cdot)$	*unpack* skew-symmetric matrix: $\mathbf{so}(2) \mapsto \mathbb{R}$, $\mathbf{so}(3) \mapsto \mathbb{R}^3$	vex
$[\cdot]$	augmented skew-symmetric matrix: $\mathbb{R}^3 \mapsto \mathbf{se}(2)$, $\mathbb{R}^6 \mapsto \mathbf{se}(3)$	skewa
$\vee(\cdot)$	*unpack* augmented skew-symmetric matrix: $\mathbf{se}(2) \mapsto \mathbb{R}^3$, $\mathbf{so}(3) \mapsto \mathbb{R}^6$	vexa
$\mathrm{Ad}(\cdot)$	adjoint representation: $\mathbf{SE}(3) \mapsto \mathbb{R}^{6 \times 6}$	.Ad
$\mathrm{ad}(\cdot)$	logarithm of adjoint representation: $\mathbf{SE}(3) \mapsto \mathbb{R}^{6 \times 6}$	.ad
$\circ$	quaternion (Hamiltonian) multiplication: $\mathbb{H} \times \mathbb{H} \mapsto \mathbb{H}$	*
$\mathring{v}$	pure quaternion: $\mathbb{R}^3 \mapsto \mathbb{H}$	Quaternion.pure
$\sim$	equivalence of representations	
$\simeq$	homogeneous coordinate equivalence	
$\ominus$	smallest angular difference between two angles on a circle: $\mathbb{S}^1 \times \mathbb{S}^1 \mapsto \mathbb{R}$	angdiff
$\mathcal{K}(\cdot)$	forward kinematics: $\mathcal{C} \mapsto \mathcal{T}$	fkine
$\mathcal{K}^{-1}(\cdot)$	inverse kinematics: $\mathcal{T} \mapsto \mathcal{C}$	ikine
$\mathcal{D}^{-1}(\cdot)$	manipulator inverse dynamics function: $\mathcal{C}, \mathbb{R}^N, \mathbb{R}^N \mapsto \mathbb{R}^N$	rne
$\mathcal{P}(\cdot)$	camera projection function: $\mathbb{R}^3 \mapsto \mathbb{R}^2$	.project
$*$	convolution	iconv
$\otimes$	correlation	
$\equiv$	colormetric equivalence	
$\oplus$	morphological dilation	
$\ominus$	morphological erosion	
$\circ$	morphological opening	
$\bullet$	morphological closing	
$\{F\}$	coordinate frame F	
$[a, b]$	interval a to b inclusive	
(a, b)	interval a to b exclusive, not including a or b	
$[a, b)$	interval a to b, not including b	
$(a, b]$	interval a to b, not including a	

MATLAB® Toolbox Conventions

- A Cartesian coordinate, a point, is expressed as a column vector.
- A set of points is expressed as a matrix with columns representing the coordinates of individual points.
- A rectangular region by two opposite corners $[x_{min} \, x_{max}; \, y_{min} \, y_{max}]$.
- A robot configuration, a set of joint angles, is expressed as a row vector.
- Time series data is expressed as a matrix with rows representing time steps.
- A MATLAB matrix has subscripts (i, j) which represent row and column respectively. Image coordinates are written (u, v) so an image represented by a matrix I is indexed as $I(v, u)$.
- Matrices with three or more dimensions are frequently used:
  - A color image has 3 dimensions: row, column, color plane.
  - A greyscale image sequence has 3 dimensions: row, column, index.
  - A color image sequence has 4 dimensions: row, column, color plane, index.

Common Abbreviations

2D	2-dimensional
3D	3-dimensional
DOF	Degrees of freedom
n-tuple	A group of n numbers, it can represent a point of a vector

1 Introduction

The term robot means different things to different people. Science fiction books and movies have strongly influenced what many people expect a robot to be or what it can do. Sadly the practice of robotics is far behind this popular conception. One thing is certain though – robotics will be an important technology in this century. Products such as vacuum cleaning robots have already been with us for over a decade and self-driving cars are coming. These are the vanguard of a wave of smart machines that will appear in our homes and workplaces in the near to medium future.

In the eighteenth century the people of Europe were fascinated by automata such as Vaucanson's duck shown in Fig. 1.1a. These machines, complex by the standards of the day, demonstrated what then seemed *life-like* behavior. The duck used a cam mechanism to sequence its movements and Vaucanson went on to explore mechanization of silk weaving. Jacquard extended these ideas and developed a loom, shown in Fig. 1.1b, that was essentially a programmable weaving machine. The pattern to be woven was encoded as a series of holes on punched cards. This machine has many hallmarks of a modern robot: it performed a physical task and was reprogrammable.

The term robot first appeared in a 1920 Czech science fiction play "Rossum's Universal Robots" by Karel Čapek (pronounced Chapek). The term was coined by his brother Josef, and in the Czech language means serf labor but colloquially means hardwork or drudgery. The robots in the play were artificial people or androids and as in so many robot stories that follow this one, the robots rebel and it ends badly for humanity. Isaac Asimov's robot series, comprising many books and short stories written between 1950 and 1985, explored issues of human and robot interaction and morality. The robots in these stories are equipped with "positronic brains" in which the "Three laws of robotics" are encoded. These stories have influenced subsequent books and movies which in turn have shaped the public perception of what robots are. The mid twentieth century also saw the advent of the field of *cybernetics* – an uncommon term today but then an exciting science at the frontiers of understanding life and creating intelligent machines.

The first patent for what we would now consider a robot was filed in 1954 by George C. Devol and issued in 1961. The device comprised a mechanical arm with

Fig. 1.1.
Early programmable machines. **a** Vaucanson's duck (1739) was an automaton that could flap its wings, eat grain and defecate. It was driven by a clockwork mechanism and executed a single program; **b** The Jacquard loom (1801) was a reprogrammable machine and the program was held on punched cards (photograph by George P. Landow from www.victorianweb.org)

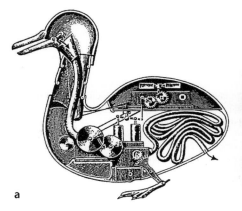

a gripper that was mounted on a track and the sequence of motions was encoded as magnetic patterns stored on a rotating drum. The first robotics company, Unimation, was founded by Devol and Joseph Engelberger in 1956 and their first industrial robot shown in Fig. 1.2 was installed in 1961. The original vision of Devol and Engelberger for robotic automation has become a reality and many millions of arm-type robots such as shown in Fig. 1.3 have been built and put to work at tasks such as welding, painting, machine loading and unloading, electronic assembly, packaging and palletizing. The use of robots has led to increased productivity and improved product quality. Today many products we buy have been assembled or handled by a robot.

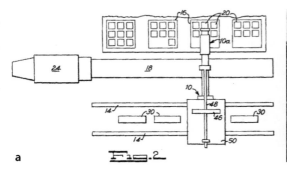

a   b

Fig. 1.2.
Universal automation. **a** A plan view of the machine from Devol's patent; **b** the first Unimation robot working at a General Motors factory (photo courtesy of George C. Devol)

Unimation Inc. (1956–1982). Devol sought financing to develop his unimation technology and at a cocktail party in 1954 he met Joseph Engelberger who was then an engineer with Manning, Maxwell and Moore. In 1956 they jointly established Unimation, the first robotics company, in Danbury Connecticut. The company was acquired by Consolidated Diesel Corp. (Condec) and became Unimate Inc. a division of Condec. Their first robot went to work in 1961 at a General Motors die-casting plant in New Jersey. In 1968 they licensed technology to Kawasaki Heavy Industries which produced the first Japanese industrial robot. Engelberger served as chief executive until it was acquired by Westinghouse in 1982. People and technologies from this company have gone on to be very influential on the whole field of robotics.

George C. Devol, Jr. (1912–2011) was a prolific American inventor. He was born in Louisville, Kentucky, and in 1932 founded United Cinephone Corp. which manufactured phonograph arms and amplifiers, registration controls for printing presses and packaging machines. In 1954, he applied for US patent 2,988,237 for Programmed Article Transfer which introduced the concept of Universal Automation or "Unimation". Specifically it described a track-mounted polar-coordinate arm mechanism with a gripper and a programmable controller – the precursor of all modern robots.

In 2011 he was inducted into the National Inventors Hall of Fame. (Photo on the right: courtesy of George C. Devol)

Joseph F. Engelberger (1925–2015) was an American engineer and entrepreneur who is often referred to as the "Father of Robotics". He received his B.S. and M.S. degrees in physics from Columbia University, in 1946 and 1949, respectively. Engelberger has been a tireless promoter of robotics. In 1966, he appeared on *The Tonight Show Starring Johnny Carson* with a Unimate robot which poured a beer, putted a golf ball, and directed the band. He promoted robotics heavily in Japan, which led to strong investment and development of robotic technology in that country.

Engelberger served as chief executive of Unimation until 1982, and in 1984 founded Transitions Research Corporation which became HelpMate Robotics Inc., an early entrant in the hospital service robot sector. He was elected to the National Academy of Engineering, received the Beckman Award and the Japan Prize, and has written two books: Robotics in Practice (1980) and Robotics in Service (1989). Each year the Robotics Industries Association presents an award in his honor to "persons who have contributed outstandingly to the furtherance of the science and practice of robotics."

These first generation robots are fixed in place and cannot move about the factory – they are not mobile. By contrast mobile robots as shown in Figs. 1.4 and 1.5 can move through the world using various forms of mobility. They can locomote over the ground using wheels or legs, fly through the air using fixed wings or multiple rotors, move through the water or sail over it. An alternative taxonomy is based on the function that the robot performs. *Manufacturing* robots operate in factories and are the technological descendents of the first generation robots. *Service robots* supply services to people such as cleaning, personal care, medical rehabilitation or fetching and carrying as shown in Fig. 1.5b. *Field robots*, such as those shown in Fig. 1.4, work outdoors on tasks such as environmental monitoring, agriculture, mining, construction and forestry. *Humanoid robots* such as shown in Fig. 1.6 have the physical form of a human being – they are both mobile robots and service robots. ◄

In practice the categorization of robots is not very consistently applied.

Fig. 1.3.
Manufacturing robots, technological descendants of the Unimate shown in Fig. 1.2. **a** A modern six-axis robot designed for high accuracy and throughput (image courtesy ABB robotics); **b** Baxter two-armed robot with built in vision capability and programmable by demonstration, designed for moderate throughput piece work (image courtesy Rethink Robotics)

Rossum's Universal Robots (RUR). In the introductory scene Helena Glory is visiting Harry Domin the director general of Rossum's Universal Robots and his robotic secretary Sulla.

Domin Sulla, let Miss Glory have a look at you.
Helena (stands and offers her hand) Pleased to meet you. It must be very hard for you out here, cut off from the rest of the world [the factory is on an island]
Sulla I do not know the rest of the world Miss Glory. Please sit down.
Helena (sits) Where are you from?
Sulla From here, the factory
Helena Oh, you were born here.
Sulla Yes I was made here.
Helena (startled) What?
Domin (laughing) Sulla isn't a person, Miss Glory, she's a robot.
Helena Oh, please forgive me …

The full play can be found at http://ebooks.adelaide.edu.au/c/capek/karel/rur. (Image on the left: Library of Congress item 96524672)

A manufacturing robot is typically an arm-type manipulator on a fixed base such as Fig. 1.3a that performs repetitive tasks within a local work cell. Parts are presented to the robot in an orderly fashion which maximizes the advantage of the robot's high speed and precision. High-speed robots are hazardous and safety is achieved by excluding people from robotic work places, typically placing the robot inside a cage. In contrast the Baxter robot shown in Fig. 1.3b is human safe, it operates at low speed and stops moving if it encounters an obstruction.

Field and service robots face specific and significant challenges. The first challenge is that the robot must operate and move in a complex, cluttered and changing environment. A delivery robot in a hospital must operate despite crowds of people and a time-varying configuration of parked carts and trolleys. A Mars rover as shown in Fig. 1.5a must navigate rocks and small craters despite not having an accurate local map in advance of its travel. Robotic, or self-driving cars, such as shown in Fig. 1.5c, must follow roads, avoid obstacles and obey traffic signals and the rules of the road. The second challenge for these types of robots is that they must operate safely in the presence of people. The hospital delivery robot operates among people, the robotic car contains people and a robotic surgical device operates *inside* people.

Fig. 1.4. Non-land-based mobile robots. **a** Small autonomous underwater vehicle (Todd Walsh © 2013 MBARI); **b** Global Hawk unmanned aerial vehicle (UAV) (photo courtesy of NASA)

Cybernetics, artificial intelligence and robotics. Cybernetics flourished as a research field from the 1930s until the 1960s and was fueled by a heady mix of new ideas and results from neurology, control theory and information theory. Research in neurology had shown that the brain was an electrical network of neurons. Harold Black, Henrik Bode and Harry Nyquist at Bell Labs were researching negative feedback and the stability of electrical networks, Claude Shannon's information theory described digital signals, and Alan Turing was exploring the fundamentals of computation. Walter Pitts and Warren McCulloch proposed an artificial neuron in 1943 and showed how it might perform simple logical functions. In 1951 Marvin Minsky built SNARC (from a B24 autopilot and comprising 3 000 vacuum tubes) which was perhaps the first neural-network-based learning machine as his graduate project. William Grey Walter's robotic tortoises showed life-like behavior. Maybe an electronic brain could be built!

An important early book was Norbert Wiener's *Cybernetics or Control and Communication in the Animal and the Machine* (Wiener 1965). A characteristic of a cybernetic system is the use of feedback which is common in engineering and biological systems. The ideas were later applied to evolutionary biology, psychology and economics.

In 1956 a watershed conference was hosted by John McCarthy at Dartmouth College and attended by Minsky, Shannon, Herbert Simon, Allen Newell and others. This meeting defined the term artificial intelligence (AI) as we know it today with an emphasis on digital computers and symbolic manipulation and led to new research in robotics, vision, natural language, semantics and reasoning. McCarthy and Minsky formed the AI group at MIT, and McCarthy left in 1962 to form the Stanford AI Laboratory. Minsky focused on artificially simple "blocks world". Simon, and his student Newell, were influential in AI research at Carnegie-Mellon University from which the Robotics Institute was spawned in 1979. These AI groups were to be very influential in the development of robotics and computer vision in the USA. Societies and publications focusing on cybernetics are still active today.

Fig. 1.5. Mobile robots. **a** Mars Science Lander, Curiosity, self portrait taken at "John Klein". The mast contains many cameras including two stereo camera pairs from which the robot can compute the 3-dimensional structure of its environment (image courtesy of NASA/JPL-Caltech/MSSS); **b** Savioke Relay delivery robot (image courtesy Savioke); **c** self driving car (image courtesy Dept. Information Engineering, Oxford Univ.); **d** Cheetah legged robot (image courtesy Boston Dynamics)

So what is a robot? There are many definitions and not all of them are particularly helpful. A definition that will serve us well in this book is

a goal oriented machine that can sense, plan and act.

A robot *senses* its environment and uses that information, together with a goal, to *plan* some *action*. The action might be to move the tool of an arm-robot to grasp an object or it might be to drive a mobile robot to some place.

Sensing is critical to robots. Proprioceptive sensors measure the state of the robot itself: the angle of the joints on a robot arm, the number of wheel revolutions on a mobile robot or the current drawn by an electric motor. Exteroceptive sensors measure the state of the world with respect to the robot. The sensor might be a simple bump sensor on a robot vacuum cleaner to detect collision. It might be a GPS receiver that measures distances to an orbiting satellite constellation, or a compass that measures the direction of the Earth's magnetic field vector relative to the robot. It might also be an active sensor that emits acoustic, optical or radio pulses in order to measure the distance to points in the world based on the time taken for a reflection to return to the sensor.

A camera is a passive device that captures patterns of optical energy reflected from the scene. Our own experience is that eyes are a very effective sensor for recognition, navigation, obstacle avoidance and manipulation so vision has long been of interest to robotics researchers. An important limitation of a single camera, or a single eye, is that the 3-dimensional structure of the scene is lost in the resulting 2-dimensional image. Despite this, humans are particularly good at inferring the 3-dimensional nature of a scene using a number of visual cues. Robots are currently not as well developed. Figure 1.7 shows some very early work on reconstructing a 3-dimensional wireframe model from a single

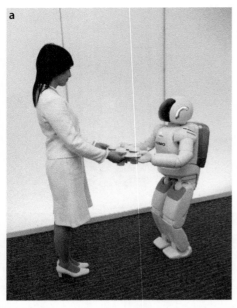

Fig. 1.6.
Humanoid robots. **a** Honda's Asimo humanoid robot (image courtesy Honda Motor Co. Japan); **b** Hubo robot that won the DARPA Robotics Challenge in 2015 (image courtesy KAIST, Korea)

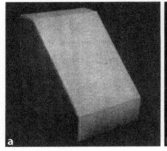 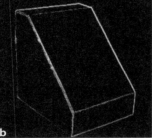 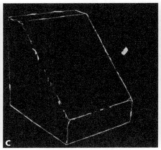 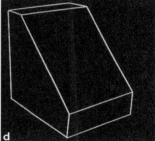

2-dimensional image and gives some idea of the difficulties involved. Another approach is stereo vision where information from two cameras is combined to estimate the 3-dimensional structure of the scene – this is a technique used by humans and robots, for example, the Mars rover shown in Fig. 1.5a has a stereo camera on its mast.

In this book we focus on the use of cameras as sensors for robots. Machine vision, discussed in Part IV, is the use of computers to process images from one or more cameras and to extract numerical features. For example determining the coordinate of a round red object in the scene, or how far a robot has moved based on how the world appears to have moved relative to the robot.

If the robot's environment is unchanging it can make do with an accurate map and have little need to sense the state of the world, apart from determining where it is. Imagine driving a car with the front window covered over and just looking at the GPS navigation system. If you had the road to yourself you could probably drive from A to B quite successfully albeit slowly. However if there were other cars, pedestrians, traffic signals or roadworks then you would be in some difficulty. To deal with this you need to look outwards – to sense the world and plan your actions accordingly. For humans this is easy, done without conscious thought, but it is not yet easy to program a machine to do the same – this is the challenge of *robotic vision*.

Telerobots are robot-like machines that are remotely controlled by a human operator. Perhaps the earliest was a radio controlled boat demonstrated by Nikola Tesla in 1898 and which he called a teleautomaton. According to the definition above these are not robots but they were an important precursor to robots and are still important today for many tasks where people cannot work but which are too complex for a machine to per-

Fig. 1.7. Early results in computer vision for estimating the shape and pose of objects, from the Ph.D. work of L.G. Roberts at MIT Lincoln Lab in 1963 (Roberts 1963). **a** Original picture; **b** gradient image; **c** connected feature points; **d** reconstructed line drawing

The Manhattan Project in World War 2 (WW II) developed the first nuclear weapons and this required handling of radioactive material. Remotely controlled arms were developed by Ray Goertz at Argonne National Laboratory to exploit the manual dexterity of human operators while keeping them away from the hazards of the material they were handling. The operators viewed the work space through thick lead-glass windows or via a television link and manipulated the master arm (on the left). The slave arm (on the right) followed the motion, and forces felt by the slave arm were reflected back to the master arm, allowing the operator to feel weight and interference force. Telerobotics is still important today for many tasks where people cannot work but which are too complex for a machine to perform by itself, for instance the underwater robots that surveyed the wreck of the Titanic. (Photo on the left: Courtesy Argonne National Laboratory)

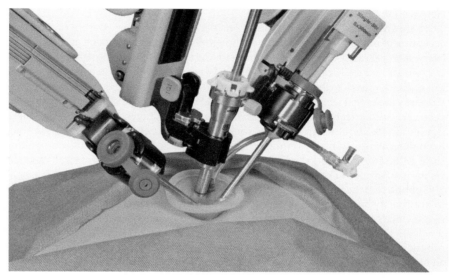

Fig. 1.8.
The working end of a surgical robot, multiple tools working through a single small incision (image © 2015 Intuitive Surgical, Inc)

form by itself. For example the underwater robots that surveyed the wreck of the Titanic were technically remotely operated vehicles (ROVs). A modern surgical robot as shown in Fig. 1.8 is also teleoperated – the motion of the small tools are remotely controlled by the surgeon and this makes it possible to use much smaller incisions than the old-fashioned approach where the surgeon works inside the body with their hands.

The various Mars rovers autonomously navigate the surface of Mars but human operators provide the high-level goals. That is, the operators tell the robot where to go and the robot itself determines the details of the route. Local decision making on Mars is essential given that the communications delay is several minutes. Some robots are hybrids and the control task is shared or traded with a human operator. In traded control, the control function is passed back and forth between the human operator and the computer. For example an aircraft pilot can pass control to an autopilot and take control back. In shared control, the control function is performed by the human operator and the computer working together. For example an autonomous passenger car might have the computer keeping the car safely in the lane while the human driver just controls speed.

1.1 Robots, Jobs and Ethics

A number of ethical issues arise from the advent of robotics. Perhaps the greatest concern to the wider public is "robots taking jobs from people". This is a complex issue but we cannot shy away from the fact that many jobs now done by people will, in the future, be performed by robots. Clearly there are dangerous jobs which people should not do, for example handling hazardous substances or working in dangerous environments. There are many low-skilled jobs where human labor is increasingly hard to source, for instance in jobs like fruit picking. In many developed countries people no longer aspire to hard physical outdoor work in remote locations. What are the alternatives if people don't want to do the work? In areas like manufacturing, particularly car manufacturing, the adoption of robotic automation has been critical in raising productivity which has allowed that industry to be economically viable in high-wage countries like Europe, Japan and the USA. Without robots these industries could not exist; they would not employ any people, not pay any taxes, and not consume products and services from other parts of the economy. Automated industry might employ fewer people but it still makes an important contribution to society. Rather than taking jobs we could argue that robotics and automation has helped to keep manufacturing industries viable in high-labor cost countries. How do we balance the good of the society with the good of the individual?

There are other issues besides jobs. Consider self-driving cars. We are surprisingly accepting of manually driven cars even though they kill more than one million people every year, yet many are uncomfortable with the idea of self-driving cars even though they will dramatically reduce this loss of life. We worry about who to blame if a robotic car makes a mistake while the carnage caused by human drivers continues. Similar concerns are raised when talking about robotic healthcare and surgery – human surgeons are not perfect but robots are seemingly held to a much higher account. There is a lot of talk about using robots to look after elderly people, but does this detract from their quality of life by removing human contact, conversation and companionship? Should we use robots to look after our children, and even teach them? What do we think of armies of robots fighting and killing human beings?

Robotic cars, health care, elder care and child care might bring economic benefits to our society but is it the right thing to do? Is it a direction that we want our society to go? Once again how do we balance the good of the society with the good of the individual? These are deep ethical questions that cannot and should not be decided by roboticists alone. But neither should roboticists ignore them. This is a discussion for all of society and roboticists have a duty to be active participants in this debate.

1.2   About the Book

This book is about robotics and computer vision – separately, and together as robotic vision. These are big topics and the combined coverage is necessarily broad. The intent is not to be shallow but rather to give the reader a flavor of what robotics and vision is about and what it can do – consider it a grand tasting menu.

The goals of the book are:

- to provide a broad and solid base of understanding through theory and examples;
- to make abstract concepts tangible
- to tackle more complex problems than other more specialized textbooks by virtue of the powerful numerical tools and software that underpins it;
- to provide instant gratification by solving complex problems with relatively little code;
- to complement the many excellent texts in robotics and computer vision;
- to encourage intuition through hands on numerical experimentation; and
- to limit the number of equations presented to those cases where (in my judgment) they add value or clarity.

The approach used is to present background, theory and examples in an integrated fashion. Code and examples are first-class citizens in this book and are not relegated to the end of the chapter or an associated web site. The examples are woven into the discussion like this

```
>> p = transl(Ts);
>> plot(t, p);
```

where the MATLAB® code illuminates the topic being discussed and generally results in a crisp numerical result or a graph in a figure that is then discussed. The examples illustrate how to use the associated Toolboxes and that knowledge can then be applied to other problems. Most of the figures in this book have been generated by the code examples provided and they are available from the book's website as described in Appendix A.

1.2.1 MATLAB Software and the Toolboxes

To do good work, one must first have good tools.
Chinese proverb

The computational foundation of this book is MATLAB®, a software package developed by The MathWorks Inc. MATLAB is an interactive mathematical software environment that makes linear algebra, data analysis and high-quality graphics a breeze. MATLAB is a popular package and one that is very likely to be familiar to engineering students as well as researchers. It also supports a programming language which allows the creation of complex algorithms.

A strength of MATLAB is its support for Toolboxes which are collections of functions targeted at particular topics. Toolboxes are available from MathWorks, third party companies and individuals. Some Toolboxes are products to be purchased while others are open-source and generally free to use. This book is based on two open-source Toolboxes written by the author: the Robotics Toolbox for MATLAB and the Machine Vision Toolbox for MATLAB. These Toolboxes, with MATLAB, turn a personal computer into a powerful and convenient environment for investigating complex problems in robotics, machine vision and vision-based control. The Toolboxes are free to use and distributed under the GNU Lesser General Public License (GNU LGPL).

The *Robotics Toolbox* (RTB) provides a diverse range of functions for simulating mobile and arm-type robots. The Toolbox supports a very general method of representing the structure of serial-link manipulators using MATLAB objects and provides functions for forward and inverse kinematics and dynamics. The Toolbox includes functions for manipulating and converting between datatypes such as vectors, homogeneous transformations, 3-angle representations, twists and unit-quaternions which are necessary to represent 3-dimensional position and orientation. The Toolbox also includes functionality for simulating mobile robots and includes models of wheeled vehicles and quadrotors and controllers for these vehicles. It also provides standard algorithms for robot path planning, localization, map making and SLAM.

The *Machine Vision Toolbox* (MVTB) provides a rich collection of functions for camera modeling, image processing, image feature extraction, multi-view geometry and vision-based control. The MVTB also contains functions for image acquisition and

The MATLAB software we use today has a long history. It starts with the LINPACK and EISPACK projects run by the Argonne National Laboratory in the 1970s to produce high quality, tested and portable mathematical software. LINPACK is a collection of routines for linear algebra and EISPACK is a library of numerical algorithms for computing eigenvalues and eigenvectors of matrices. These packages were written in Fortran which was then the language of choice for large-scale numerical problems.

Cleve Moler, then at the University of New Mexico, contributed to both projects and wrote the first version of MATLAB in the late 1970s. It allowed interactive use of LINPACK and EISPACK for problem solving without having to write and compile Fortran code. MATLAB quickly spread to other universities and found a strong audience within the applied mathematics and engineering community. In 1984 Cleve Moler and Jack Little founded The MathWorks Inc. which exploited the newly released IBM PC – the first widely available desktop computer.

Cleve Moler received his bachelor's degree from Caltech in 1961, and a Ph.D. from Stanford University. He was a professor of mathematics and computer science at universities including University of Michigan, Stanford University, and the University of New Mexico. He has served as president of the Society for Industrial and Applied Mathematics (SIAM) and was elected to the National Academy of Engineering in 1997.

See also http://www.mathworks.com/company/aboutus/founders/clevemoler.html which includes a video of Cleve Moler and also http://history.siam.org/pdfs2/Moler_final.pdf.

display; filtering; blob, point and line feature extraction; mathematical morphology; image warping; stereo vision; homography and fundamental matrix estimation; robust estimation; bundle adjustment; visual Jacobians; geometric camera models; camera calibration and color space operations. For modest image sizes on a modern computer the processing rate can be sufficiently "real-time" to allow for closed-loop control.

If you're starting out in robotics or vision then the Toolboxes are a significant initial base of code on which to build your project. The Toolboxes are provided in source code form. The bulk of the code is written in the MATLAB M-language but a few functions are written in C▶ or Java for increased computational efficiency. In general the Toolbox code is written in a straightforward manner to facilitate understanding, perhaps at the expense of computational efficiency. Appendix A provides details of how to obtain the Toolboxes and pointers to online resources including discussion groups.

This book provides examples of how to use many Toolbox functions in the context of solving specific problems but it is not a reference manual. Comprehensive documentation of all Toolbox functions is available through the MATLAB builtin help mechanism or the PDF format manual that is distributed with each Toolbox.

These are implemented as MEX files, which are written in C in a very specific way that allows them to be invoked from MATLAB just like a function written in M-language.

1.2.2   Notation, Conventions and Organization

The mathematical notation used in the book is summarized in the Nomenclature section on page xxv. Since the coverage of the book is broad there are just not enough good symbols to go around, so it is unavoidable that some symbols have different meanings in different parts of the book.

There is a lot of MATLAB code in the book and this is indicated in blue fixed-width font such as

```
>> a = 2 + 2
a =
    4
```

The MATLAB command prompt is >> and what follows is the command issued to MATLAB by the user. Subsequent lines, without the prompt, are MATLAB's response. All functions, classes and methods mentioned in the text or in code segments are cross-referenced and have their own indexes at the end of the book allowing you to find different ways that particular functions can be used.

Colored boxes are used to indicate different types of material. Orange informational boxes highlight material that is particularly important while red and orange warning boxes highlight points that are often traps for those starting out. Blue boxes provide technical, historical or biographical information that augment the main text but they are not critical to its understanding.

As an author there is a tension between completeness, clarity and conciseness. For this reason a lot of detail has been pushed into notes▶ and blue boxes and on a first reading these can be skipped. Some chapters have an Advanced Topics section at the end that can also be skipped on a first reading. However if you are trying to understand a particular algorithm and apply it to your own problem then understanding the details and nuances can be important and the notes or advanced topics are for you.

They are placed as marginal notes near the corresponding marker.

Each chapter ends with a *Wrapping Up* section that summarizes the important lessons from the chapter, discusses some suggested further reading, and provides some exercises. For clarity, references are cited sparingly in the text of each chapter. The *Further Reading* subsection discusses prior work and references that provide more rigor or more complete description of the algorithms. *Resources* provides links to relevant online code and datasets. *MATLAB Notes* provides additional details about the author's toolboxes and those with similar functionality from MathWorks. *Exercises* extend the concepts discussed within the chapter and are generally related to specific code examples discussed in the chapter. The exercises vary in difficulty from straightforward extension of the code examples to more challenging problems.

1.2.3   Audience and Prerequisites

The book is intended primarily for third or fourth year engineering undergraduate students, Masters students and first year Ph.D. students. For undergraduates the book will serve as a companion text for a robotics or computer vision course or to support a major project in robotics or vision. Students should study Part I and the appendices for foundational concepts, and then the relevant part of the book: mobile robotics, arm robots, computer vision or vision-based control. The Toolboxes provide a solid set of tools for problem solving, and the exercises at the end of each chapter provide additional problems beyond the worked examples in the book.

For students commencing graduate study in robotics, and who have previously studied engineering or computer science, the book will help fill the gaps between what you learned as an undergraduate and what will be required to underpin your deeper study of robotics and computer vision. The book's working code base can help bootstrap your research, enabling you to get started quickly and working productively on your own problems and ideas. Since the source code is available you can reshape it to suit your need, and when the time comes (as it usually does) to code your algorithms in some other language then the Toolboxes can be used to cross-check your implementation.

For those who are no longer students, the researcher or industry practitioner, the book will serve as a useful companion for your own reference to a wide range of topics in robotics and computer vision, as well as a handbook and guide for the Toolboxes.

The book assumes undergraduate-level knowledge of linear algebra (matrices, vectors, eigenvalues), basic set theory, basic graph theory, probability, dynamics (forces, torques, inertia) and control theory. Some of these topics will likely be more familiar to engineering students than computer science students. Computer science students may struggle with some concepts in Chap. 4 and 9 such as the Laplace transform, transfer functions, linear control (proportional control, proportional-derivative control, proportional-integral control) and block diagram notation. This material could be skimmed over on a first reading and Albertos and Mareels (2010) may be a useful introduction to some of these topics. The book also assumes the reader is familiar with using and programming in MATLAB and also familiar with object-oriented programming techniques (perhaps C++, Java or Python). Familiarity with Simulink®, the graphical block-diagram modeling tool integrated with MATLAB will be helpful but not essential.

1.2.4   Learning with the Book

The best way to learn is by doing. Although the book shows the MATLAB commands and the response there is something special about doing it for yourself. Consider the book as an invitation to tinker. By running the commands yourself you can look at the results in ways that you prefer, plot the results in a different way, or try the algorithm on different data or with different parameters. The book is especially designed to stay open which enables you to type in commands as you read. You can also look at the online documentation for the Toolbox functions, discover additional features and options, and experiment with those, or read the code to see how it really works and perhaps modify it.

Most of the commands are quite short so typing them in to MATLAB is not too onerous. However the book's web site, see Appendix A, includes all the MATLAB commands shown in the book (more than 1 600 lines) and these can be cut and pasted into MATLAB or downloaded and used to create your own scripts.

In 2015 two open online courses (MOOCs) were released – based on the content and approach of this book. Introduction to Robotics covers most of Parts I and III, while Robotic Vision covers some of Parts IV and V. Each MOOC is six weeks long and comprises 12 hours of video lecture material plus quizzes, assignments and an optional project. They can be reached via http://petercorke.com/moocs.

1.2.5   Teaching with the Book

The book can be used in support of courses in robotics, mechatronics and computer vision. All courses should include the introduction to coordinate frames and their composition which is discussed in Chap. 2. For a mobile robotics or image processing course it is sufficient to teach only the 2-dimensional case. For robotic manipulators or multi-view geometry the 2- and 3-dimensional cases should be taught.

Most figures (MATLAB-generated and line drawings) in this book are available as PDF format files from the book's web site and you are free to use them with attribution in any course material that you prepare. All the code in this book can be downloaded from the web site and used as the basis for demonstrations in lectures or tutorials. See Appendix A for details.

The exercises at the end of each chapter can be used as the basis of assignments, or as examples to be worked in class or in tutorials. Most of the questions are rather open ended in order to encourage exploration and discovery of the effects of parameters and the limits of performance of algorithms. This exploration should be supported by discussion and debate about performance measures and what *best* means. True understanding of algorithms involves an appreciation of the effects of parameters, how algorithms fail and under what circumstances.

The teaching approach could also be inverted, by diving headfirst into a particular problem and then teaching the appropriate prerequisite material. Suitable problems could be chosen from the Application sections of Chap. 7, 14 or 16, or from any of the exercises. Particularly challenging exercises are so marked.

If you wanted to consider a flipped learning approach then the two MOOCs mentioned on page 11 could be used in conjunction with your class. Students would watch the videos and undertake some formative assessment out of the classroom, and you could use classroom time to work through problem sets.

For graduate level teaching the papers and textbooks mentioned in the *Further Reading* could form the basis of a student's reading list. They could also serve as candidate papers for a reading group or journal club.

1.2.6   Outline

I promised a book with instant gratification but before we can get started in robotics there are some fundamental concepts that we absolutely need to understand, and understand well. Part I introduces the concepts of pose and coordinate frames – how we represent the position and orientation of a robot, a camera or the objects that the robot needs to work with. We discuss how motion between two poses can be *decomposed* into a sequence of elementary translations and rotations, and how elementary motions can be *composed* into more complex motions. Chapter 2 discusses how pose can be represented in a computer, and Chap. 3 discusses the relationship between velocity and the derivative of pose, estimating motion from sensors and generating a sequence of poses that smoothly follow some path in space and time.

With these formalities out of the way we move on to the first main event – robots. There are two important classes of robot: mobile robots and manipulator arms and these are covered in Parts II and III respectively.▶

Part II begins, in Chap. 4, with motion models for several types of wheeled vehicles and a multi-rotor flying vehicle. Various control laws are discussed for wheeled vehicles such as moving to a point, following a path and moving to a specific pose. Chapter 5 is concerned with navigation, that is, how a robot finds a path between points A and B in the world. Two important cases, with and without a map, are discussed. Most navigation techniques require knowledge of the robot's position and Chap. 6 discusses various approaches to this problem based on dead-reckoning, or

Although robot arms came first chronologically, mobile robotics is mostly a 2-dimensional problem and easier to understand than the 3-dimensional arm-robot case.

landmark observation and a map. We also show how a robot can make a map, and even determine its location while simultaneously mapping an unknown region.

Part III is concerned with arm-type robots, or more precisely serial-link manipulators. Manipulator arms are used for tasks such as assembly, welding, material handling and even surgery. Chapter 7 introduces the topic of kinematics which relates the angles of the robot's joints to the 3-dimensional pose of the robot's tool. Techniques to generate smooth paths for the tool are discussed and two examples show how an arm-robot can draw a letter on a surface and how multiple arms (acting as legs) can be used to create a model for a simple walking robot. Chapter 8 discusses the relationships between the rates of change of joint angles and tool pose. It introduces the Jacobian matrix and concepts such as singularities, manipulability, null-space motion, and resolved-rate motion control. It also discusses under- and over-actuated robots and the general numerical solution to inverse kinematics. Chapter 9 introduces the design of joint control systems, the dynamic equations of motion for a serial-link manipulator, and the relationship between joint forces and joint motion. It discusses important topics such as variation in inertia, the effect of payload, flexible transmissions and independent joint versus nonlinear control strategies.

Computer vision is a large field concerned with processing images in order to enhance them for human benefit, interpret the contents of the scene or create a 3D model corresponding to the scene. Part IV is concerned with machine vision, a subset of computer vision, and defined here as the extraction of numerical features from images to provide input for control of a robot. The discussion starts in Chap. 10 with the fundamentals of light, illumination and color. Chapter 11 describes the geometric model of perspective image creation using lenses and discusses topics such as camera calibration and pose estimation. We introduce nonperspective imaging using wide-angle lenses and mirror systems, camera arrays and light-field cameras. Chapter 12 discusses *image processing* which is a domain of 2-dimensional signal processing that transforms one image into another image. The discussion starts with acquiring real-world images and then covers various arithmetic and logical operations that can be performed on images. We then introduce spatial operators such as convolution, segmentation, morphological filtering and finally image shape and size changing. These operations underpin the discussion in Chap. 13 which describe how numerical features are extracted from images. The features describe homogeneous regions (blobs), lines or distinct points in the scene and are the basis for vision-based robot control. Chapter 14 is concerned with estimating the underlying three-dimensional geometry of a scene using classical methods such as structured lighting and also combining features found in different views of the same scene to provide information about the geometry and the spatial relationship between the camera views which is encoded in fundamental, essential and homography matrices. This leads to the topic of bundle adjustment and structure from motion and applications including perspective correction, mosaicing, image retrieval and visual odometry.

Part V discusses how visual features extracted from the camera's view can be used to control arm-type and mobile robots – an approach known as vision-based control or visual servoing. This part pulls together concepts introduced in the earlier parts of the book. Chapter 15 introduces the classical approaches to visual servoing known as position-based and image-based visual servoing and discusses their respective limitations. Chapter 16 discusses more recent approaches that address these limitations and also covers the use of nonperspective cameras, under-actuated robots and mobile robots.

This is a big book but any one of the parts can be read standalone, with more or less frequent visits to the required earlier material. Chapter 2 is the only mandatory material. Parts II, III or IV could be used respectively for an introduction to mobile robots, arm robots or computer vision class. An alternative approach, following the instant gratification theme, is to jump straight into any chapter and start exploring – visiting the earlier material as required.

Further Reading

The Handbook of Robotics (Siciliano and Khatib 2016) provides encyclopedic coverage of the field of robotics today, covering theory, technology and the different types of robot such as telerobots, service robots, field robots, flying robots, underwater robots and so on. The classic work by Sheridan (2003) discusses the spectrum of autonomy from remote control, through shared and traded control to full autonomy.

A comprehensive coverage of computer vision is the book by Szeliski (2011), and a solid introduction to artificial intelligence is the text by Russell and Norvig (2009).

A number of recent books discuss the future impacts of robotics and artificial intelligence on society, for example Ford (2015), Brynjolfsson and McAfee (2014), Bostrom (2016) and Neilson (2011). The YouTube video Grey (2014) makes some powerful points about the future of work and is always a great discussion starter.

Part I Foundations

2 Representing Position and Orientation

Numbers are an important part of mathematics. We use numbers for counting: *there are 2 apples*. We use *denominate numbers*, a number plus a unit, to specify distance: *the object is 2 m away*. We also call this single number a *scalar*. We use a vector, a denominate number plus a direction, to specify a location: *the object is 2 m due north*. We may also want to know the orientation of the object: *the object is 2 m due north and facing west*. The combination of position and orientation we call *pose*.

A point in space is a familiar concept from mathematics and can be described by a coordinate vector, as shown in Fig. 2.1a. The vector represents the displacement of the point with respect to some reference coordinate frame – we call this a bound vector since it cannot be freely moved. A coordinate frame, or Cartesian coordinate system, is a set of orthogonal axes which intersect at a point known as the origin. A vector can be described in terms of its components, a linear combination of unit vectors which are parallel to the axes of the coordinate frame. Note that points and vectors are different types of mathematical objects even though each can be described by a tuple of numbers. We can add vectors but adding points makes no sense. The difference of two points is a vector, and we can add a vector to a point to obtain another point.

A point is an interesting mathematical abstraction, but a real object comprises infinitely many points. An object, unlike a point, also has an orientation. If we attach a coordinate frame to an object, as shown in Fig. 2.1b, we can describe every point within the object as a constant vector with respect to that frame. ◄ Now we can describe the position and orientation – the pose – of that coordinate frame with respect to the reference coordinate frame. To distinguish the different frames we label them and in this case the object coordinate frame is labeled {B} and its axes are labeled x_B and y_B, adopting the frame's label as their subscript.

To completely describe the pose of a rigid object in a 3-dimensional world we need 6 not 3 dimensions: 3 to describe its position and 3 to describe its orientation. These dimensions behave quite differently. If we increase the value of one of the position dimensions the object will move continuously in a straight line, but if we increase the value of one of the orientation dimensions the object will rotate in some way and soon get back to its original orientation – this dimension is curved. We clearly need to treat the position and orientation dimensions quite differently.

We assume that the object is rigid, that is, the points do not move with respect to each other.

Fig. 2.1.
a The point **P** is described by a coordinate vector with respect to an absolute coordinate frame. **b** The points are described with respect to the object's coordinate frame {B} which in turn is described by a relative pose ξ_B. Axes are denoted by thick lines with an open arrow, vectors by thin lines with a swept arrow head and a pose by a thick line with a solid head

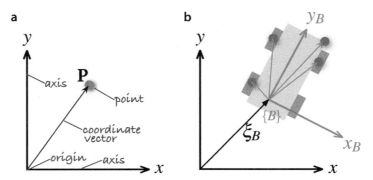

The pose of the coordinate frame is denoted by the symbol ξ – pronounced ksi. Figure 2.2 shows two frames $\{A\}$ and $\{B\}$ and the relative pose ${}^A\xi_B$ which describes $\{B\}$ with respect to $\{A\}$. The leading superscript denotes the reference coordinate frame and the subscript denotes the frame being described. We could also think of ${}^A\xi_B$ as describing some motion – imagine picking up $\{A\}$ and applying a displacement and a rotation so that it is transformed to $\{B\}$. If the initial superscript is missing we assume that the change in pose is relative to the world coordinate frame which is generally denoted $\{O\}$.

The point **P** in Fig. 2.2 can be described with respect to *either* coordinate frame by the vectors ${}^A\boldsymbol{p}$ or ${}^B\boldsymbol{p}$ respectively. Formally they are related by

$$
{}^A\boldsymbol{p} = {}^A\xi_B \cdot {}^B\boldsymbol{p} \tag{2.1}
$$

where the right-hand side expresses the motion from $\{A\}$ to $\{B\}$ and then to **P**. The operator $\cdot$ *transforms* the vector, resulting in a new vector that describes the same point but with respect to a different coordinate frame.

An important characteristic of relative poses is that they can be *composed* or *compounded*. Consider the case shown in Fig. 2.3. If one frame can be described in terms of another by a relative pose then they can be applied sequentially

$$
{}^A\xi_C = {}^A\xi_B \oplus {}^B\xi_C
$$

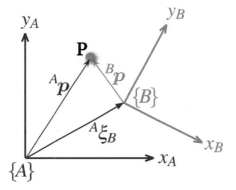

Fig. 2.2.
The point **P** can be described by coordinate vectors relative to either frame $\{A\}$ or $\{B\}$. The pose of $\{B\}$ relative to $\{A\}$ is ${}^A\xi_B$

In relative pose composition we can check that we have our reference frames correct by ensuring that the subscript and superscript on each side of the $\oplus$ operator are matched. We can then *cancel out* the intermediate subscripts and superscripts

$$
{}^X\xi_Z = {}^X\xi_Y \oplus {}^Y\xi_Z
$$

leaving just the end most subscript and superscript which are shown highlighted.

Euclid of Alexandria (ca. 325 BCE–265 BCE) was a Greek mathematician, who was born and lived in Alexandria Egypt, and is considered the "father of geometry". His great work *Elements* comprising 13 books, captured and systematized much early knowledge about geometry and numbers. It deduces the properties of planar and solid geometric shapes from a set of 5 axioms and 5 postulates.

Elements is probably the most successful book in the history of mathematics. It describes plane geometry and is the basis for most people's first introduction to geometry and formal proof, and is the basis of what we now call Euclidean geometry. Euclidean distance is simply the distance between two points on a plane. Euclid also wrote *Optics* which describes geometric vision and perspective.

which says, in words, that the pose of {C} relative to {A} can be obtained by compounding the relative poses from {A} to {B} and {B} to {C}. We use the operator ⊕ to indicate *composition* of relative poses.

For this case the point **P** can be described by

$$^A\boldsymbol{p} = \left(^A\xi_B \oplus \,^B\xi_C\right) \cdot {}^C\boldsymbol{p}$$

Later in this chapter we will convert these abstract notions of ξ, • and ⊕ into standard mathematical objects and operators that we can implement in MATLAB®.

In the examples so far we have shown 2-dimensional coordinate frames. This is appropriate for a large class of robotics problems, particularly for mobile robots which operate in a planar world. For other problems we require 3-dimensional coordinate frames to describe objects in our 3-dimensional world such as the pose of a flying or underwater robot or the end of a tool carried by a robot arm.

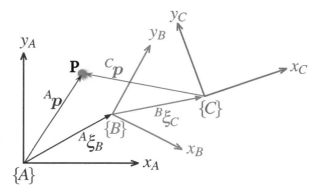

Fig. 2.3.
The point **P** can be described by coordinate vectors relative to either frame {A}, {B} or {C}. The frames are described by relative poses

Euclidean versus Cartesian geometry. Euclidean geometry is concerned with points and lines in the Euclidean plane (2D) or Euclidean space (3D). It is entirely based on a set of axioms and makes no use of arithmetic. Descartes added a coordinate system (2D or 3D) and was then able to describe points, lines and other curves in terms of algebraic equations. The study of such equations is called analytic geometry and is the basis of all modern geometry. The Cartesian plane (or space) is the Euclidean plane (or space) with all its axioms and postulates *plus* the extra facilities afforded by the added coordinate system. The term Euclidean geometry is often used to mean that Euclid's fifth postulate (parallel lines never intersect) holds, which is the case for a planar surface but not for a curved surface.

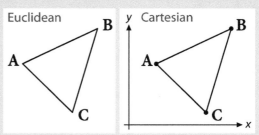

René Descartes (1596–1650) was a French philosopher, mathematician and part-time mercenary. He is famous for the philosophical statement "*Cogito, ergo sum*" or "*I am thinking, therefore I exist*" or "*I think, therefore I am*". He was a sickly child and developed a life-long habit of lying in bed and thinking until late morning. A possibly apocryphal story is that during one such morning he was watching a fly walk across the ceiling and realized that he could describe its position in terms of its distance from the two edges of the ceiling. This is the basis of the *Cartesian* coordinate system and modern (analytic) geometry, which he described in his 1637 book La Géométrie. For the first time mathematics and geometry were connected, and modern calculus was built on this foundation by Newton and Leibniz. In Sweden at the invitation of Queen Christina he was obliged to rise at 5 A.M., breaking his lifetime habit – he caught pneumonia and died. His remains were later moved to Paris, and are now lost apart from his skull which is in the Musée de l'Homme. After his death, the Roman Catholic Church placed his works on the Index of Prohibited Books.

Figure 2.4 shows a more complex 3-dimensional example in a graphical form where we have attached 3D coordinate frames to the various entities and indicated some relative poses. The fixed camera observes the object from its fixed viewpoint and estimates the object's pose $^F\xi_B$ relative to itself. The other camera is not fixed, it is attached to the robot at some constant relative pose and estimates the object's pose $^C\xi_B$ relative to itself.

An alternative representation of the spatial relationships is a directed graph (see Appendix I) which is shown in Fig. 2.5.► Each node in the graph represents a pose and each edge represents a relative pose. An arrow from X to Y is denoted $^X\xi_Y$ and describes the pose of Y relative to X. Recalling that we can compose relative poses using the $\oplus$ operator we can write some spatial relationships

It is quite possible that a pose graph can be inconsistent, that is, two paths through the graph give different results. In robotics these poses are only ever derived from noisy sensor data.

$$\xi_F \oplus {}^F\xi_B = \xi_R \oplus {}^R\xi_C \oplus {}^C\xi_B$$
$$\xi_F \oplus {}^F\xi_R = \xi_R$$

and each equation represents a loop in the graph with each side of the equation starting and ending at the same node. Each side of the first equation represents a path through the network from {0} to {B}, a sequence of edges (arrows) written in order.

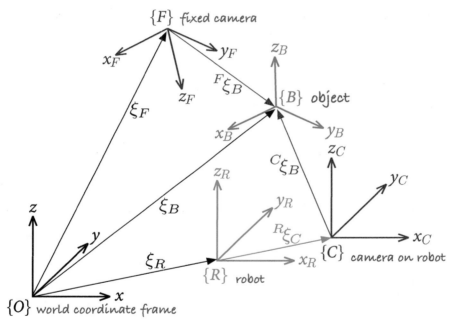

Fig. 2.4.
Multiple 3-dimensional coordinate frames and relative poses

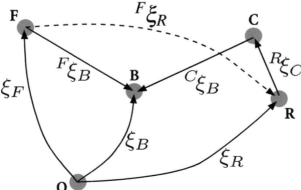

Fig. 2.5.
Spatial example of Fig. 2.4 expressed as a directed graph

In mathematical terms poses constitute a group – a set of objects that supports an associative binary operator (composition) whose result belongs to the group, an inverse operation and an identity element. In this case the group is the special Euclidean group in either 2 or 3 dimensions which are commonly referred to as **SE**(2) or **SE**(3) respectively.

There are just a few algebraic rules:◄

$$\xi \oplus 0 = \xi, \ \xi \ominus 0 = \xi$$
$$\xi \ominus \xi = 0, \ \ominus \xi \oplus \xi = 0$$

where 0 represents a zero relative pose. A pose has an inverse

$$\ominus^X\xi_Y = {}^Y\xi_X$$

which is represented graphically by an arrow from {Y} to {X}. Relative poses can also be composed or compounded

$${}^X\xi_Y \oplus {}^Y\xi_Z = {}^X\xi_Z$$

It is important to note that the algebraic rules for poses are different to normal algebra and that composition is *not* commutative

$$\xi_1 \oplus \xi_2 \neq \xi_2 \oplus \xi_1$$

with the exception being the case where $\xi_1 \oplus \xi_2 = 0$. A relative pose can transform a point expressed as a vector relative to one frame to a vector relative to another

$${}^X\boldsymbol{p} = {}^X\xi_Y \cdot {}^Y\boldsymbol{p}$$

A very useful property of poses is the ability to perform algebra. The second loop equation says, in words, that the pose of the robot is the same as composing two relative poses: from the world frame to the fixed camera and from the fixed camera to the robot. We can subtract ξ_F from both sides of the equation◄ by adding the inverse of ξ_F which we denote as $\ominus \xi_F$ and this gives

Order is important here, and we add $\ominus \xi_F$ to the left on each side of the equation.

$$\ominus \xi_F \oplus \xi_F \oplus {}^F\xi_R = \ominus \xi_F \oplus \xi_R$$
$${}^F\xi_R = \ominus \xi_F \oplus \xi_R$$

which is the pose of the robot relative to the fixed camera, shown as a dashed line in Fig. 2.5.

We can write these expressions quickly by inspection. To find the pose of node X with respect to node Y:

- find a path from Y to X and write down the relative poses on the edges in a left to right order;
- if you traverse the edge in the direction of its arrow precede it with the $\oplus$ operator, otherwise use $\ominus$.

So what is ξ? It can be any mathematical object that supports the algebra described above and is suited to the problem at hand. It will depend on whether we are considering a 2- or 3-dimensional problem. Some of the objects that we will discuss in the rest of this chapter will be familiar to us, for example vectors, but others will be more exotic mathematical objects such as homogeneous transformations, orthonormal rotation matrices, twists and quaternions. Fortunately all these mathematical objects are well suited to the mathematical programming environment of MATLAB.

To recap:

1. A point is described by a bound coordinate vector that represents its displacement from the origin of a reference coordinate system.
2. Points and vectors are different things even though they are each described by a tuple of numbers. We can add vectors but not points. The difference between two points is a vector.
3. A set of points that represent a rigid object can be described by a single coordinate frame, and its constituent points are described by constant vectors relative to that coordinate frame.
4. The position and orientation of an object's coordinate frame is referred to as its pose.
5. A relative pose describes the pose of one coordinate frame with respect to another and is denoted by an algebraic variable ξ.
6. A coordinate vector describing a point can be represented with respect to a different coordinate frame by applying the relative pose to the vector using the · operator.
7. We can perform algebraic manipulation of expressions written in terms of relative poses and the operators $\oplus$ and $\ominus$.

The remainder of this chapter discusses concrete representations of ξ for various common cases that we will encounter in robotics and computer vision. We start by considering the two-dimensional case which is comparatively straightforward and then extend those concepts to three dimensions. In each case we consider rotation first, and then add translation to create a description of pose.

2.1   Working in Two Dimensions (2D)

A 2-dimensional world, or plane, is familiar to us from high-school Euclidean geometry. We use a right-handed► Cartesian coordinate system or coordinate frame with orthogonal axes denoted x and y and typically drawn with the x-axis horizontal and the y-axis vertical. The point of intersection is called the origin. Unit-vectors parallel to the axes are denoted $\hat{x}$ and $\hat{y}$. A point is represented by its x- and y-coordinates (x, y) or as a bound vector

The relative orientation of the x- and y-axes obey the right-hand rule as shown on page 31.

$$p = x\hat{x} + y\hat{y} \tag{2.2}$$

Figure 2.6 shows a red coordinate frame $\{B\}$ that we wish to describe with respect to the blue reference frame $\{A\}$. We can see clearly that the origin of $\{B\}$ has been displaced by the vector $t = (x, y)$ and then rotated counter-clockwise by an angle θ.

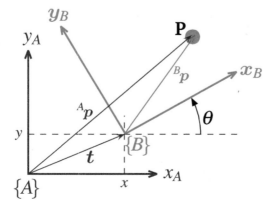

Fig. 2.6.
Two 2D coordinate frames $\{A\}$ and $\{B\}$ and a world point **P**. $\{B\}$ is rotated and translated with respect to $\{A\}$

A concrete representation of pose is therefore the 3-vector $^A\xi_B \sim (x, y, \theta)$, and we use the symbol $\sim$ to denote that the two representations are equivalent. Unfortunately this *representation* is not convenient for compounding since

$$(x_1, y_1, \theta_1) \oplus (x_2, y_2, \theta_2)$$

is a complex trigonometric function of both poses. Instead we will look for a different way to represent rotation and pose. We will consider the problem in two parts: rotation and then translation.

2.1.1 Orientation in 2-Dimensions

2.1.1.1 Orthonormal Rotation Matrix

Consider an arbitrary point **P** which we can express with respect to each of the coordinate frames shown in Fig. 2.6. We create a new frame $\{V\}$ whose axes are parallel to those of $\{A\}$ but whose origin is the same as $\{B\}$, see Fig. 2.7. According to Eq. 2.2 we can express the point **P** with respect to $\{V\}$ in terms of the unit-vectors that define the axes of the frame

$$
\begin{aligned}
^Vp &= {}^Vx\hat{x}_V + {}^Vy\hat{y}_V \\
&= (\hat{x}_V \quad \hat{y}_V)\begin{pmatrix} ^Vx \\ ^Vy \end{pmatrix}
\end{aligned}
\tag{2.3}
$$

which we have written as the product of a row and a column vector.

The coordinate frame $\{B\}$ is completely described by its two orthogonal axes which we represent by two unit vectors

$$
\begin{aligned}
\hat{x}_B &= \cos\theta\hat{x}_V + \sin\theta\hat{y}_V \\
\hat{y}_B &= -\sin\theta\hat{x}_V + \cos\theta\hat{y}_V
\end{aligned}
$$

which can be factorized into matrix form as

$$
(\hat{x}_B \quad \hat{y}_B) = (\hat{x}_V \quad \hat{y}_V)\begin{pmatrix} \cos\theta & -\sin\theta \\ \sin\theta & \cos\theta \end{pmatrix}
\tag{2.4}
$$

Using Eq. 2.2 we can represent the point **P** with respect to $\{B\}$ as

$$
\begin{aligned}
^Bp &= {}^Bx\hat{x}_B + {}^By\hat{y}_B \\
&= (\hat{x}_B \quad \hat{y}_B)\begin{pmatrix} ^Bx \\ ^By \end{pmatrix}
\end{aligned}
$$

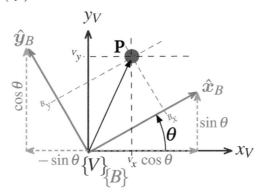

Fig. 2.7.
Rotated coordinate frames in 2D. The point **P** can be considered with respect to the red or blue coordinate frame

and substituting Eq. 2.4 we write

$$^B p = (\hat{x}_V \quad \hat{y}_V)\begin{pmatrix} \cos\theta & -\sin\theta \\ \sin\theta & \cos\theta \end{pmatrix}\begin{pmatrix} ^B x \\ ^B y \end{pmatrix} \tag{2.5}$$

Now by equating the coefficients of the right-hand sides of Eq. 2.3 and Eq. 2.5 we write

$$\begin{pmatrix} ^V x \\ ^V y \end{pmatrix} = \begin{pmatrix} \cos\theta & -\sin\theta \\ \sin\theta & \cos\theta \end{pmatrix}\begin{pmatrix} ^B x \\ ^B y \end{pmatrix}$$

which describes how points are transformed from frame $\{B\}$ to frame $\{V\}$ when the frame is rotated. This type of matrix is known as a rotation matrix since it transforms a point from frame $\{V\}$ to $\{B\}$ and is denoted $^V R_B$

$$\begin{pmatrix} ^V x \\ ^V y \end{pmatrix} = {}^V R_B \begin{pmatrix} ^B x \\ ^B y \end{pmatrix} \tag{2.6}$$

$$^X R_Y(\theta) = \begin{pmatrix} \cos\theta & -\sin\theta \\ \sin\theta & \cos\theta \end{pmatrix}$$

is a 2-dimensional rotation matrix with some special properties:

- it is *orthonormal* (also called *orthogonal*) since each of its columns is a unit vector and the columns are orthogonal. ▶
- the columns are the unit vectors that define the axes of the rotated frame Y with respect to X and are by definition both unit-length and orthogonal.
- it belongs to the special orthogonal group of dimension 2 or $R \in SO(2) \subset \mathbb{R}^{2\times2}$. This means that the product of any two matrices belongs to the group, as does its inverse.
- its *determinant* is $+1$, which means that the length of a vector is unchanged after transformation, that is, $\|^Y p\| = \|^X p\|, \forall\theta$.
- the inverse is the same as the transpose, that is, $R^{-1} = R^T$.

See Appendix B which provides a refresher on vectors, matrices and linear algebra.

We can rearrange Eq. 2.6 as

$$\begin{pmatrix} ^B x \\ ^B y \end{pmatrix} = \left(^V R_B\right)^{-1}\begin{pmatrix} ^V x \\ ^V y \end{pmatrix} = \left(^V R_B\right)^T\begin{pmatrix} ^V x \\ ^V y \end{pmatrix} = {}^B R_V\begin{pmatrix} ^V x \\ ^V y \end{pmatrix}$$

Note that inverting the matrix is the same as swapping the superscript and subscript, which leads to the identity $R(-\theta) = R(\theta)^T$.

It is interesting to observe that instead of representing an angle, which is a scalar, we have used a 2×2 matrix that comprises four elements, however these elements are not independent. Each column has a unit magnitude which provides two constraints. The columns are orthogonal which provides another constraint. Four elements and three constraints are effectively one independent value. The rotation matrix is an example of a nonminimum representation and the disadvantages such as the increased memory it requires are outweighed, as we shall see, by its advantages such as composability.

The Toolbox allows easy creation of these rotation matrices

```
>> R = rot2(0.2)
R =
    0.9801   -0.1987
    0.1987    0.9801
```

where the angle is specified in radians. We can observe some of the properties such as

```
>> det(R)
ans =
     1
```

and the product of two rotation matrices is also a rotation matrix

```
>> det(R*R)
ans =
     1
```

You will need to have the MATLAB Symbolic Math Toolbox™ installed.

The Toolbox also supports symbolic mathematics◄ for example

```
>> syms theta
>> R = rot2(theta)
R =
[ cos(theta), -sin(theta)]
[ sin(theta),  cos(theta)]
>> simplify(R*R)
ans =
[ cos(2*theta), -sin(2*theta)]
[ sin(2*theta),  cos(2*theta)]
>> simplify(det(R))
ans =
1
```

2.1.1.2   Matrix Exponential

Consider a pure rotation of 0.3 radians expressed as a rotation matrix

```
>> R = rot2(0.3)
ans =
     0.9553    -0.2955
     0.2955     0.9553
```

We can compute the logarithm of this matrix using the MATLAB builtin function

logm is different to the builtin function log which computes the logarithm of each element of the matrix. A logarithm can be computed using a power series, with a matrix rather than scalar argument. For a matrix the logarithm is not unique and logm computes the principal logarithm of the matrix.

logm◄

```
>> S = logm(R)
S =
     0.0000    -0.3000
     0.3000     0.0000
```

and the result is a simple matrix with two elements having a magnitude of 0.3, which intriguingly is the original rotation angle. There is something deep and interesting going on here – we are on the fringes of Lie group theory which we will encounter throughout this chapter.

In 2 dimensions the skew-symmetric matrix is

$$[\omega]_\times = \begin{pmatrix} 0 & -\omega \\ \omega & 0 \end{pmatrix} \tag{2.7}$$

which has clear structure and only one unique element $\omega \in \mathbb{R}$. A simple example of Toolbox support for skew-symmetric matrices is

```
>> skew(2)
ans =
     0    -2
     2     0
```

and the inverse operation is performed using the Toolbox function vex

```
>> vex(ans)
ans =
     2
```

This matrix has a zero diagonal and is an example of a 2×2 skew-symmetric matrix. The matrix has only one unique element and we can unpack it using the Toolbox function `vex`

```
>> vex(S)
ans =
    0.3000
```

to recover the rotation angle.

The inverse of a logarithm is exponentiation and using the builtin MATLAB matrix exponential function `expm`▶

```
>> expm(S)
ans =
    0.9553    -0.2955
    0.2955     0.9553
```

the result is, as expected, our original rotation matrix. In fact the command

```
>> R = rot2(0.3);
```

is equivalent to

```
>> R = expm( skew(0.3) );
```

Formally we can write

$$R = e^{[\theta]_\times} \in \mathbf{SO}(2)$$

where θ is the rotation angle, and the notation $[\cdot]_\times : \mathbb{R} \mapsto \mathbb{R}^{2 \times 2}$ indicates a mapping from a scalar to a skew-symmetric matrix.

2.1.2   Pose in 2-Dimensions

2.1.2.1   Homogeneous Transformation Matrix

Now we need to account for the translation between the origins of the frames shown in Fig. 2.6. Since the axes $\{V\}$ and $\{A\}$ are parallel, as shown in Figs. 2.6 and 2.7, this is simply vectorial addition

$$\begin{pmatrix} ^Ax \\ ^Ay \end{pmatrix} = \begin{pmatrix} ^Vx \\ ^Vy \end{pmatrix} + \begin{pmatrix} x \\ y \end{pmatrix} \tag{2.8}$$

$$= \begin{pmatrix} \cos\theta & -\sin\theta \\ \sin\theta & \cos\theta \end{pmatrix} \begin{pmatrix} ^Bx \\ ^By \end{pmatrix} + \begin{pmatrix} x \\ y \end{pmatrix} \tag{2.9}$$

$$= \begin{pmatrix} \cos\theta & -\sin\theta & x \\ \sin\theta & \cos\theta & y \end{pmatrix} \begin{pmatrix} ^Bx \\ ^By \\ 1 \end{pmatrix} \tag{2.10}$$

or more compactly as

$$\begin{pmatrix} ^Ax \\ ^Ay \\ 1 \end{pmatrix} = \begin{pmatrix} ^AR_B & t \\ 0_{1\times2} & 1 \end{pmatrix} \begin{pmatrix} ^Bx \\ ^By \\ 1 \end{pmatrix} \tag{2.11}$$

where $t = (x, y)$ is the translation of the frame and the orientation is ^AR_B. Note that $^AR_B = {}^VR_B$ since the axes of frames $\{A\}$ and $\{V\}$ are parallel. The coordinate vectors for point P are now expressed in homogeneous form and we write

A vector $p = (x, y)$ is written in homogeneous form as $\tilde{p} \in \mathbb{P}^2$, $\tilde{p} = (x_1, x_2, x_3)$ where $x = x_1/x_3$, $y = x_2/x_3$ and $x_3 \neq 0$. The dimension has been increased by one and a point on a plane is now represented by a 3-vector. To convert a point to homogeneous form we typically append an element equal to one $\tilde{p} = (x, y, 1)$. The tilde indicates the vector is homogeneous.

Homogeneous vectors have the important property that $\tilde{p}$ is equivalent to $\lambda \tilde{p}$ for all $\lambda \neq 0$ which we write as $\tilde{p} \simeq \lambda \tilde{p}$. That is $\tilde{p}$ represents the same point in the plane irrespective of the overall scaling factor. Homogeneous representation is important for computer vision that we discuss in Part IV. Additional details are provided in Sect. C.2.

$$
^A\tilde{p} = \begin{pmatrix} ^AR_B & t \\ 0_{1\times 2} & 1 \end{pmatrix} {}^B\tilde{p}
$$
$$
= {}^AT_B\, {}^B\tilde{p}
$$

and ^AT_B is referred to as a homogeneous transformation. The matrix has a very specific structure and belongs to the special Euclidean group of dimension 2 or $T \in \mathbf{SE}(2) \subset \mathbb{R}^{3\times 3}$.

By comparison with Eq. 2.1 it is clear that ^AT_B represents translation and orientation or relative pose. This is often referred to as a *rigid-body motion*.

$$
T = \begin{pmatrix} \cos\theta & -\sin\theta & x \\ \sin\theta & \cos\theta & y \\ 0 & 0 & 1 \end{pmatrix}
$$

A concrete representation of relative pose ξ is $\xi \sim T \in \mathbf{SE}(2)$ and $T_1 \oplus T_2 \mapsto T_1 T_2$ which is standard matrix multiplication

$$
T_1 T_2 = \begin{pmatrix} R_1 & t_1 \\ 0_{1\times 2} & 1 \end{pmatrix}\begin{pmatrix} R_2 & t_2 \\ 0_{1\times 2} & 1 \end{pmatrix} = \begin{pmatrix} R_1 R_2 & t_1 + R_1 t_2 \\ 0_{1\times 2} & 1 \end{pmatrix}
$$

One of the algebraic rules from page 21 is $\xi \oplus 0 = \xi$. For matrices we know that $TI = T$, where I is the identify matrix, so for pose $0 \mapsto I$ the identity matrix. Another rule was that $\xi \ominus \xi = 0$. We know for matrices that $TT^{-1} = I$ which implies that $\ominus T \mapsto T^{-1}$

$$
T^{-1} = \begin{pmatrix} R & t \\ 0_{1\times 2} & 1 \end{pmatrix}^{-1} = \begin{pmatrix} R^T & -R^T t \\ 0_{1\times 2} & 1 \end{pmatrix}
$$

For a point described by $\tilde{p} \in \mathbb{P}^2$ then $T \cdot \tilde{p} \mapsto T\tilde{p}$ which is a standard matrix-vector product.

To make this more tangible we will show some numerical examples using MATLAB and the Toolbox. We create a homogeneous transformation which represents a translation of $(1, 2)$ followed by a rotation of 30°

```
>> T1 = transl2(1, 2) * trot2(30, 'deg')
T1 =
    0.8660   -0.5000    1.0000
    0.5000    0.8660    2.0000
         0         0    1.0000
```

The function `transl2` creates a relative pose with a finite translation but zero rotation, while `trot2` creates a relative pose with a finite rotation but zero translation. ◀ We can plot this, relative to the world coordinate frame, by

Many Toolbox functions have variants that return orthonormal rotation matrices or homogeneous transformations, for example, `rot2` and `trot2`.

```
>> plotvol([0 5 0 5]);
>> trplot2(T1, 'frame', '1', 'color', 'b')
```

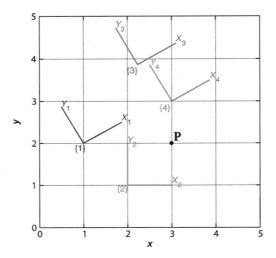

Fig. 2.8.
Coordinate frames drawn using
the Toolbox function `trplot2`

The options specify that the label for the frame is {1} and it is colored blue and this is shown in Fig. 2.8. We create another relative pose which is a displacement of (2, 1) and zero rotation

```
>> T2 = transl2(2, 1)
T2 =
     1     0     2
     0     1     1
     0     0     1
```

which we plot in red

```
>> trplot2(T2, 'frame', '2', 'color', 'r');
```

Now we can compose the two relative poses

```
>> T3 = T1*T2
T3 =
    0.8660   -0.5000    2.2321
    0.5000    0.8660    3.8660
         0         0    1.0000
```

and plot it, in green, as

```
>> trplot2(T3, 'frame', '3', 'color', 'g');
```

We see that the displacement of (2, 1) has been applied with respect to frame {1}. It is important to note that our final displacement is not (3, 3) because the displacement is with respect to the rotated coordinate frame. The noncommutativity of composition is clearly demonstrated by

```
>> T4 = T2*T1;
>> trplot2(T4, 'frame', '4', 'color', 'c');
```

and we see that frame {4} is different to frame {3}.

Now we define a point (3, 2) relative to the world frame

```
>> P = [3 ; 2 ];
```

which is a column vector and add it to the plot

```
>> plot_point(P, 'label', 'P', 'solid', 'ko');
```

To determine the coordinate of the point with respect to {1} we use Eq. 2.1 and write down

$$^{0}\!p = {}^{0}\xi_1 \cdot {}^{1}\!p$$

and then rearrange as

$$^{1}\boldsymbol{p} = {}^{1}\xi_{0} \cdot {}^{0}\boldsymbol{p}$$

$$= \left({}^{0}\xi_{1}\right)^{-1} \cdot {}^{0}\boldsymbol{p}$$

Substituting numerical values

```
>> P1 = inv(T1) * [P; 1]
P1 =
    1.7321
   -1.0000
    1.0000
```

where we first converted the Euclidean point coordinates to *homogeneous form* by appending a one. The result is also in homogeneous form and has a negative *y*-coordinate in frame {1}. Using the Toolbox we could also have expressed this as

```
>> h2e( inv(T1) * e2h(P) )
ans =
    1.7321
   -1.0000
```

where the result is in Euclidean coordinates. The helper function `e2h` converts Euclidean coordinates to homogeneous and `h2e` performs the inverse conversion.

2.1.2.2   Centers of Rotation

We will explore the noncommutativity property in more depth and illustrate with the example of a pure rotation. First we create and plot a reference coordinate frame {0} and a target frame {*X*}

```
>> plotvol([-5 4 -1 5]);
>> T0 = eye(3,3);
>> trplot2(T0, 'frame', '0');
>> X = transl2(2, 3);
>> trplot2(X, 'frame', 'X');
```

and create a rotation of 2 radians (approximately 115°)

```
>> R = trot2(2);
```

and plot the effect of the two possible orders of composition

```
>> trplot2(R*X, 'framelabel', 'RX', 'color', 'r');
>> trplot2(X*R, 'framelabel', 'XR', 'color', 'r');
```

    The results are shown as red coordinate frames in Fig. 2.9. We see that the frame {*RX*} has been rotated about the origin, while frame {*XR*} has been rotated about the origin of {*X*}.

    What if we wished to rotate a coordinate frame about an arbitrary point? First of all we will establish a new point **C** and display it

```
>> C = [1 2]';
>> plot_point(C, 'label', ' C', 'solid', 'ko')
```

and then compute a transform to rotate about point **C**

```
>> RC = transl2(C) * R * transl2(-C)
RC =
   -0.4161   -0.9093    3.2347
    0.9093   -0.4161    1.9230
         0         0    1.0000
```

and applying this

```
>> trplot2(RC*X, 'framelabel', 'XC', 'color', 'r');
```

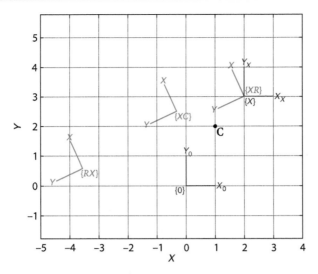

Fig. 2.9.
The frame {X} is rotated by
2 radians about {0} to give
frame {RX}, about {X} to
give {XR}, and about point **C**
to give frame {XC}

we see that the frame has indeed been rotated about point **C**. Creating the required transform was somewhat cumbersome and not immediately obvious. Reading from right to left▶ we first apply an origin shift, a translation from **C** to the origin of the reference frame, apply the rotation about that origin, and then apply the inverse origin shift, a translation from the reference frame origin back to **C**. A more descriptive way to achieve this is using twists.

RC left multiplies X, therefore the first transform applied to X is `transl(-C)`, then R, then `transl(C)`.

2.1.2.3 Twists in 2D

The corollary to what we showed in the last section is that, given any two frames we can find a rotational center that will *rotate* the first frame into the second. For the case of pure translational motion the rotational center will be at infinity. This is the key concept behind what is called a twist.

We can create a rotational twist about the point specified by the coordinate vector C

```
>> tw = Twist('R', C)
tw =
( 2 -1; 1 )
```

and the result is a `Twist` object that encodes a twist vector with two components: a 2-vector *moment* and a 1-vector *rotation*. The first argument `'R'` indicates a rotational twist is to be computed. This particular twist is a *unit twist* since the magnitude of the rotation, the last element of the twist, is equal to one.

To create an **SE**(2) transformation for a rotation about this unit twist by 2 radians we use the T method

```
>> tw.T(2)
ans =
   -0.4161   -0.9093    3.2347
    0.9093   -0.4161    1.9230
         0         0    1.0000
```

which is the same as that computed in the previous section, but more concisely specified in terms of the center of rotation. The center is also called the pole of the transformation and is encoded in the twist

```
>> tw.pole'
ans =
     1        2
```

If we wish to perform translational motion in the direction (1, 1) the relevant unit twist is◄

For a unit-translational twist the rotation is zero and the moment is a unit vector.

```
>> tw = Twist('T', [1 1])
tw =
( 0.70711 0.70711; 0 )
```

and for a displacement of $\sqrt{2}$ in the direction defined by this twist the **SE**(2) transformation is

```
>> tw.T(sqrt(2))
ans =
    1    0    1
    0    1    1
    0    0    1
```

which we see has a null rotation and a translation of 1 in the x- and y-directions.

For an arbitrary planar transform such as

```
>> T = transl2(2, 3) * trot2(0.5)
T =
    0.8776   -0.4794    2.0000
    0.4794    0.8776    3.0000
         0         0    1.0000
```

we can compute the twist vector

```
>> tw = Twist(T)
tw =
( 2.7082 2.4372; 0.5 )
```

and we note that the last element, the rotation, is not equal to one but is the required rotation angle of 0.5 radians. This is a nonunit twist. Therefore when we convert this to an **SE**(2) transform we don't need to provide a second argument since it is implicit in the twist

```
>> tw.T
ans =
    0.8776   -0.4794    2.0000
    0.4794    0.8776    3.0000
         0         0    1.0000
```

and we have regenerated our original homogeneous transformation.

2.2 Working in Three Dimensions (3D)

The 3-dimensional case is an extension of the 2-dimensional case discussed in the previous section. We add an extra coordinate axis, typically denoted by z, that is orthogonal to both the x- and y-axes. The direction of the z-axis obeys the *right-hand rule* and forms a *right-handed coordinate frame*. Unit vectors parallel to the axes are denoted $\hat{x}$, $\hat{y}$ and $\hat{z}$ such that◄

In all these identities, the symbols from left to right (across the equals sign) are a cyclic rotation of the sequence *xyz*.

$$\hat{z} = \hat{x} \times \hat{y}, \quad \hat{x} = \hat{y} \times \hat{z}; \quad \hat{y} = \hat{z} \times \hat{x} \qquad (2.12)$$

A point **P** is represented by its x-, y- and z-coordinates (x, y, z) or as a bound vector

$$p = x\hat{x} + y\hat{y} + z\hat{z}$$

Figure 2.10 shows a red coordinate frame {*B*} that we wish to describe with respect to the blue reference frame {*A*}. We can see clearly that the origin of {*B*} has been

Right-hand rule. A right-handed coordinate frame is defined by the first three fingers of your right hand which indicate the relative directions of the x-, y- and z-axes respectively.

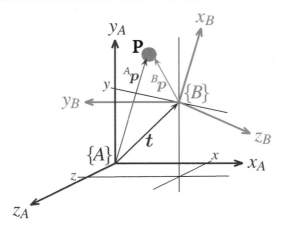

Fig. 2.10.
Two 3D coordinate frames {A} and {B}. {B} is rotated and translated with respect to {A}

displaced by the vector $t = (x, y, z)$ and then rotated in some complex fashion. Just as for the 2-dimensional case the way we represent orientation is very important.

Our approach is to again consider an arbitrary point **P** with respect to each of the coordinate frames and to determine the relationship between $^A\boldsymbol{p}$ and $^B\boldsymbol{p}$. We will again consider the problem in two parts: rotation and then translation. Rotation is surprisingly complex for the 3-dimensional case and we devote all of the next section to it.

2.2.1   Orientation in 3-Dimensions

> *Any two independent orthonormal coordinate frames*
> *can be related by a sequence of rotations (not more than **three**)*
> *about coordinate axes, where no two successive rotations may be about the same axis.*
> Euler's rotation theorem (Kuipers 1999).

Figure 2.10 shows a pair of right-handed coordinate frames with very different orientations, and we would like some way to describe the orientation of one with respect to the other. We can imagine picking up frame {A} in our hand and rotating it until it looked just like frame {B}. *Euler's rotation theorem* states that any rotation can be considered as a sequence of rotations about different coordinate axes.

We start by considering rotation about a single coordinate axis. Figure 2.11 shows a right-handed coordinate frame, and that same frame after it has been rotated by various angles about different coordinate axes.

The issue of rotation has some subtleties which are illustrated in Fig. 2.12. This shows a sequence of two rotations applied in different orders. We see that the final orientation depends on the order in which the rotations are applied. This is a deep and confounding characteristic of the 3-dimensional world which has intrigued mathematicians for a long time. There are implication for the pose algebra we have used in this chapter:

> In 3-dimensions rotation is not commutative – the order in which rotations are applied makes a difference to the result.

Mathematicians have developed many ways to represent rotation and we will discuss several of them in the remainder of this section: orthonormal rotation matrices, Euler and Cardan angles, rotation axis and angle, exponential coordinates, and unit quaternions. All can be represented as vectors or matrices, the natural datatypes of MATLAB or as a Toolbox defined class. The Toolbox provides many function to convert between these representations and these are shown in Tables 2.1 and 2.2 (pages 57, 58).

Rotation about a vector. Wrap your right hand around the vector with your thumb (your *x*-finger) in the direction of the arrow. The curl of your fingers indicates the direction of increasing angle.

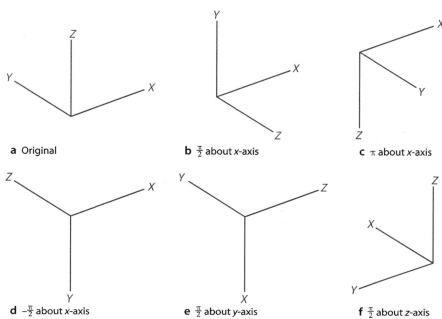

a Original   **b** $\frac{\pi}{2}$ about *x*-axis   **c** π about *x*-axis

Fig. 2.11.
Rotation of a 3D coordinate frame. **a** The original coordinate frame, **b–f** frame **a** after various rotations as indicated

d $-\frac{\pi}{2}$ about *x*-axis   **e** $\frac{\pi}{2}$ about *y*-axis   **f** $\frac{\pi}{2}$ about *z*-axis

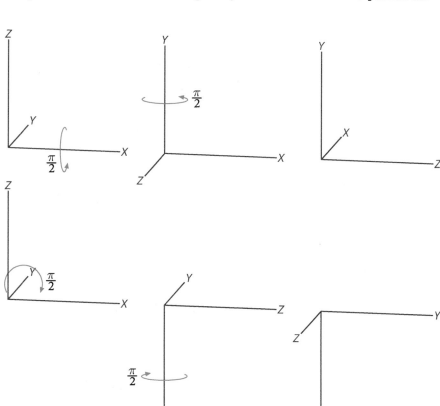

Fig. 2.12.
Example showing the noncommutativity of rotation. In the top row the coordinate frame is rotated by $\frac{\pi}{2}$ about the *x*-axis and then $\frac{\pi}{2}$ about the *y*-axis. In the bottom row the order of rotations has been reversed. The results are clearly different

2.2.1.1   Orthonormal Rotation Matrix

Just as for the 2-dimensional case we can represent the orientation of a coordinate frame by its unit vectors expressed in terms of the reference coordinate frame. Each unit vector has three elements and they form the columns of a 3×3 *orthonormal matrix* $^A R_B$

$$
\begin{pmatrix} ^A x \\ ^A y \\ ^A z \end{pmatrix} = {^A R_B} \begin{pmatrix} ^B x \\ ^B y \\ ^B z \end{pmatrix}
\tag{2.13}
$$

which transforms the description of a vector defined with respect to frame $\{B\}$ to a vector with respect to $\{A\}$.

A 3-dimensional rotation matrix $^X R_Y$ has some special properties:

- it is *orthonormal* (also called *orthogonal*) since each of its columns is a unit vector and the columns are orthogonal. ►
- the columns are the unit vectors that define the axes of the rotated frame Y with respect to X and are by definition both unit-length and orthogonal.
- it belongs to the special orthogonal group of dimension 3 or $R \in \mathbf{SO}(3) \subset \mathbb{R}^{3 \times 3}$. This means that the product of any two matrices within the group also belongs to the group, as does its inverse.
- its determinant is $+1$, which means that the length of a vector is unchanged after transformation, that is, $\|^Y p\| = \|^X p\|, \forall \theta$.
- the inverse is the same as the transpose, that is, $R^{-1} = R^T$.

See Appendix B which provides a refresher on vectors, matrices and linear algebra.

The orthonormal rotation matrices for rotation of θ about the x-, y- and z-axes are

$$
R_x(\theta) = \begin{pmatrix} 1 & 0 & 0 \\ 0 & \cos\theta & -\sin\theta \\ 0 & \sin\theta & \cos\theta \end{pmatrix}
$$

$$
R_y(\theta) = \begin{pmatrix} \cos\theta & 0 & \sin\theta \\ 0 & 1 & 0 \\ -\sin\theta & 0 & \cos\theta \end{pmatrix}
$$

$$
R_z(\theta) = \begin{pmatrix} \cos\theta & -\sin\theta & 0 \\ \sin\theta & \cos\theta & 0 \\ 0 & 0 & 1 \end{pmatrix}
$$

The Toolbox provides functions to compute these elementary rotation matrices, for example $R_x(\theta)$ is

```
>> R = rotx(pi/2)
R =
    1.0000         0         0
         0    0.0000   -1.0000
         0    1.0000    0.0000
```

and its effect on a reference coordinate frame is shown graphically in Fig. 2.11b. The functions `roty` and `rotz` compute $R_y(\theta)$ and $R_z(\theta)$ respectively.

If we consider that the rotation matrix represents a pose then the corresponding coordinate frame can be displayed graphically

```
>> trplot(R)
```

which is shown in Fig. 2.13a. We can visualize a rotation more powerfully using the Toolbox function `tranimate` which animates a rotation

```
>> tranimate(R)
```

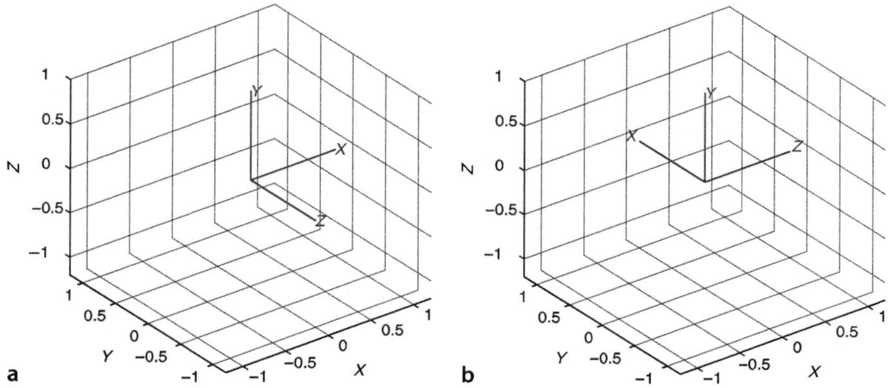

Fig. 2.13.
Coordinate frames displayed using `trplot`. **a** Reference frame rotated by $\frac{\pi}{2}$ about the *x*-axis, **b** frame **a** rotated by $\frac{\pi}{2}$ about the *y*-axis

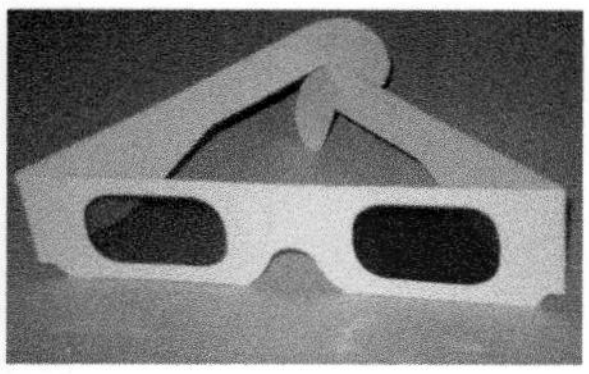

showing the world frame rotating into the specified coordinate frame. If you have a pair of anaglyph stereo glasses◄ you can see this in more realistic 3D by

```
>> tranimate(R, '3d')
```

To illustrate compounding of rotations we will rotate the frame of Fig. 2.13a again, this time around its *y*-axis

```
>> R = rotx(pi/2) * roty(pi/2)
R =
     0.0000          0    1.0000
     1.0000     0.0000   -0.0000
    -0.0000     1.0000    0.0000
>> trplot(R)
```

to give the frame shown in Fig. 2.13b. In this frame the *x*-axis now points in the direction of the world *y*-axis.

The noncommutativity of rotation can be shown by reversing the order of the rotations above

```
>> roty(pi/2)*rotx(pi/2)
ans =
     0.0000     1.0000     0.0000
          0     0.0000    -1.0000
    -1.0000     0.0000     0.0000
```

which has a very different value.

We recall that Euler's rotation theorem states that *any* rotation can be represented by *not more than three* rotations about coordinate axes. This means that in general an arbitrary rotation between frames can be decomposed into a sequence of three rotation angles and associated rotation axes – this is discussed in the next section.

The orthonormal rotation matrix has nine elements but they are not independent. The columns have unit magnitude which provides three constraints. The columns are orthogonal to each other which provides another three constraints.◄ Nine elements and six constraints is effectively three independent values.

If the column vectors are $c_i, i \in 1 \cdots 3$ then $c_1 \cdot c_2 = c_2 \cdot c_3 = c_3 \cdot c_1 = 0$ and $\|c_i\| = 1$.

Reading an orthonormal *rotation matrix*, the columns from left to right tell us the directions of the new frame's axes in terms of the current coordinate frame. For example if

```
R =
     1.0000          0          0
          0     0.0000    -1.0000
          0     1.0000     0.0000
```

the new frame has its *x*-axis in the old *x*-direction $(1, 0, 0)$, its *y*-axis in the old *z*-direction $(0, 0, 1)$, and the new *z*-axis in the old negative *y*-direction $(0, -1, 0)$. In this case the *x*-axis was unchanged since this is the axis around which the rotation occurred. The rows are the converse – the current frame axes in terms of the new frame axes.

2.2.1.2   Three-Angle Representations

Euler's rotation theorem requires successive rotation about three axes such that no two successive rotations are about the same axis. There are two classes of rotation sequence: Eulerian and Cardanian, named after Euler and Cardano respectively.

The Eulerian type involves repetition, but not successive, of rotations about one particular axis: XYX, XZX, YXY, YZY, ZXZ, or ZYZ. The Cardanian type is characterized by rotations about all three axes: XYZ, XZY, YZX, YXZ, ZXY, or ZYX.

> It is common practice to refer to all 3-angle representations as Euler angles but this is underspecified since there are twelve different types to choose from. The particular angle sequence is often a convention within a particular technological field.

The ZYZ sequence

$$R = R_z(\phi)R_y(\theta)R_z(\psi) \tag{2.14}$$

is commonly used in aeronautics and mechanical dynamics, and is used in the Toolbox. The Euler angles are the 3-vector $\varGamma = (\phi, \theta, \psi)$.

For example, to compute the equivalent rotation matrix for $\varGamma = (0.1, 0.2, 0.3)$ we write

```
>> R = rotz(0.1) * roty(0.2) * rotz(0.3);
```

or more conveniently

```
>> R = eul2r(0.1, 0.2, 0.3)
R =
    0.9021   -0.3836    0.1977
    0.3875    0.9216    0.0198
   -0.1898    0.0587    0.9801
```

The inverse problem is finding the Euler angles that correspond to a given rotation matrix

```
>> gamma = tr2eul(R)
gamma =
    0.1000    0.2000    0.3000
```

However if θ is negative

```
>> R = eul2r(0.1 , -0.2, 0.3)
R =
    0.9021   -0.3836   -0.1977
    0.3875    0.9216   -0.0198
    0.1898   -0.0587    0.9801
```

the inverse function

```
>> tr2eul(R)
ans =
   -3.0416    0.2000   -2.8416
```

returns a positive value for θ and quite different values for ϕ and ψ. However the corresponding rotation matrix

Leonhard Euler (1707–1783) was a Swiss mathematician and physicist who dominated eighteenth century mathematics. He was a student of Johann Bernoulli and applied new mathematical techniques such as calculus to many problems in mechanics and optics. He also developed the functional notation, $y = f(x)$, that we use today. In robotics we use his rotation theorem and his equations of motion in rotational dynamics.

He was prolific and his collected works fill 75 volumes. Almost half of this was produced during the last seventeen years of his life when he was completely blind.

```
>> eul2r(ans)
ans =
      0.9021    -0.3836    -0.1977
      0.3875     0.9216    -0.0198
      0.1898    -0.0587     0.9801
```

is the same – the two different sets of Euler angles correspond to the one rotation matrix. The mapping from a rotation matrix to Euler angles is not unique and the Toolbox *always* returns a positive angle for θ.

For the case where $\theta = 0$

```
>> R = eul2r(0.1, 0, 0.3)
R =
      0.9211    -0.3894          0
      0.3894     0.9211          0
           0          0     1.0000
```

the inverse function returns

```
>> tr2eul(R)
ans =
           0          0     0.4000
```

which is clearly quite different but the result is the same rotation matrix. The explanation is that if $\theta = 0$ then $\mathbf{R}_y = \mathbf{I}$ and Eq. 2.14 becomes

$$\mathbf{R} = \mathbf{R}_z(\phi)\,\mathbf{R}_z(\psi) = \mathbf{R}_z(\phi + \psi)$$

which is a function of the sum $\phi + \psi$. Therefore the inverse operation can do no more than determine this sum, and by convention we choose $\phi = 0$. The case $\theta = 0$ is a singularity and will be discussed in more detail in the next section.

Another widely used convention are the Cardan angles: roll, pitch and yaw. Confusingly there are two different versions in common use. Text books seem to define the roll-pitch-yaw sequence as ZYX or XYZ depending on whether they have a mobile robot or robot arm focus.◄ When describing the attitude of vehicles such as ships, aircraft and cars the convention is that the x-axis points in the forward direction and the z-axis points either up or down. It is intuitive to apply the rotations in the sequence: yaw (direction of travel), pitch (elevation of the front with respect to horizontal) and then finally roll (rotation about the forward axis of the vehicle). This leads to the ZYX angle sequence

> Well known texts such as Siciliano et al. (2008), Spong et al. (2006) and Paul (1981) use the XYZ sequence. The Toolbox supports both formats by means of the `'xyz'` and `'zyx'` options. The ZYX order is default for Release 10, but for Release 9 the default was XYZ.

$$\mathbf{R} = \mathbf{R}_z\!\left(\theta_y\right)\mathbf{R}_y\!\left(\theta_p\right)\mathbf{R}_x\!\left(\theta_r\right) \tag{2.15}$$

> Named after Peter Tait a Scottish physicist and quaternion supporter, and George Bryan an early Welsh aerodynamicist.

Roll-pitch-yaw angles are also known as Tait-Bryan angles◄ or nautical angles, and for aeronautical applications they can be called bank, attitude and heading angles respectively.

Gerolamo Cardano (1501–1576) was an Italian Renaissance mathematician, physician, astrologer, and gambler. He was born in Pavia, Italy, the illegitimate child of a mathematically gifted lawyer. He studied medicine at the University of Padua and later was the first to describe typhoid fever. He partly supported himself through gambling and his book about games of chance *Liber de ludo aleae* contains the first systematic treatment of probability as well as effective cheating methods. His family life was problematic: his eldest son was executed for poisoning his wife, and his daughter was a prostitute who died from syphilis (about which he wrote a treatise). He computed and published the horoscope of Jesus, was accused of heresy, and spent time in prison until he abjured and gave up his professorship.

He published the solutions to the cubic and quartic equations in his book *Ars magna* in 1545, and also invented the combination lock, the gimbal consisting of three concentric rings allowing a compass or gyroscope to rotate freely (see Fig. 2.15), and the Cardan shaft with universal joints – the drive shaft used in motor vehicles today.

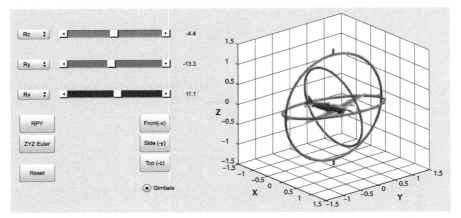

Fig. 2.14.
The Toolbox application `tripleangle` allows you to experiment with Euler angles and roll-pitch-yaw angles and see how the attitude of a body changes

When describing the attitude of a robot gripper, as shown in Fig. 2.16, the convention is that the z-axis points forward and the x-axis is either up or down. This leads to the XYZ angle sequence

$$R = R_x(\theta_y)R_y(\theta_p)R_z(\theta_r) \tag{2.16}$$

The Toolbox defaults to the ZYX sequence but can be overridden using the `'xyz'` option. For example

```
>> R = rpy2r(0.1, 0.2, 0.3)
R =
    0.9363   -0.2751    0.2184
    0.2896    0.9564   -0.0370
   -0.1987    0.0978    0.9752
```

and the inverse is

```
>> gamma = tr2rpy(R)
gamma =
    0.1000    0.2000    0.3000
```

The roll-pitch-yaw sequence allows all angles to have arbitrary sign and it has a singularity when $\theta_p = \pm\frac{\pi}{2}$ which is fortunately outside the range of feasible attitudes for most vehicles.

The Toolbox includes an interactive graphical tool

```
>> tripleangle
```

that allows you to experiment with Euler angles or roll-pitch-yaw angles and see their effect on the orientation of a body as shown in Fig. 2.14.

2.2.1.3   Singularities and Gimbal Lock

A fundamental problem with all the three-angle representations just described is singularity. This is also known as gimbal lock, a term made famous in the movie Apollo 13. This occurs when the rotational axis of the middle term in the sequence becomes parallel to the rotation axis of the first or third term.

A mechanical gyroscope used for spacecraft navigation is shown in Fig. 2.15. The innermost assembly is the *stable member* which has three orthogonal gyroscopes that hold it at a constant orientation with respect to the universe. It is mechanically connected to the spacecraft via a gimbal mechanism which allows the spacecraft to move around the stable platform without exerting any torque on it. The attitude of the spacecraft is determined directly by measuring the angles of the gimbal axes with respect to the stable platform – giving a direct indication of roll-pitch-yaw angles which in this design are a Cardanian YZX sequence.▶

"The LM Body coordinate system is right-handed, with the +X axis pointing up through the thrust axis, the +Y axis pointing right when facing forward which is along the +Z axis. The rotational transformation matrix is constructed by a 2-3-1 Euler sequence, that is: Pitch about Y, then Roll about Z and, finally, Yaw about X. Positive rotations are pitch up, roll right, yaw left." (Hoag 1963).

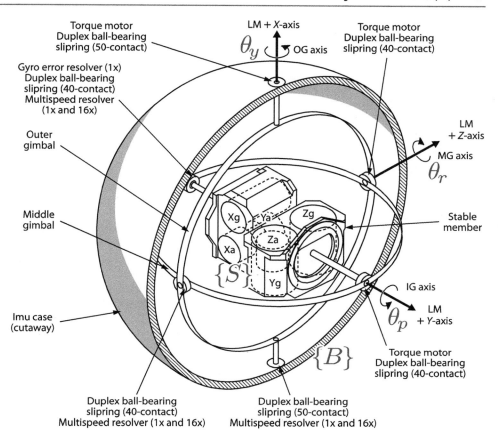

Fig. 2.15.
Schematic of Apollo Lunar Module (LM) inertial measurement unit (IMU). The vehicle's coordinate system has the x-axis pointing up through the thrust axis, the z-axis forward, and the y-axis pointing right. Starting at the stable platform {S} and working outwards toward the spacecraft's body frame {B} the rotation angle sequence is YZX. The components labeled X_g, Y_g and Z_g are the x-, y- and z-axis gyroscopes and those labeled X_a, Y_a and Z_a are the x-, y- and z-axis accelerometers (redrawn after Apollo Operations Handbook, LMA790-3-LM)

Operationally this was a significant limiting factor with this particular gyroscope (Hoag 1963) and could have been alleviated by adding a fourth gimbal, as was used on other spacecraft. It was omitted on the Lunar Module for reasons of weight and space.

Rotations obey the cyclic rotation rules
$Rx(\frac{\pi}{2}) Ry(\theta) Rx(\frac{\pi}{2})^T \equiv Rz(\theta)$
$Ry(\frac{\pi}{2}) Rz(\theta) Ry(\frac{\pi}{2})^T \equiv Rx(\theta)$
$Rz(\frac{\pi}{2}) Rx(\theta) Rz(\frac{\pi}{2})^T \equiv Ry(\theta)$
and anti-cyclic rotation rules
$Ry(\frac{\pi}{2})^T Rx(\theta) Ry(\frac{\pi}{2}) \equiv Rz(\theta)$
$Rz(\frac{\pi}{2})^T Ry(\theta) Rz(\frac{\pi}{2}) \equiv Rx(\theta)$.

Consider the situation when the rotation angle of the middle gimbal (rotation about the spacecraft's z-axis) is 90° – the axes of the inner and outer gimbals are aligned and they share the *same* rotation axis. Instead of the original three rotational axes, since two are parallel, there are now only two effective rotational axes – we say that one degree of freedom has been lost. ◄

In mathematical, rather than mechanical, terms this problem can be seen using the definition of the Lunar module's coordinate system where the rotation of the spacecraft's body-fixed frame {B} with respect to the stable platform frame {S} is

$$^{S}R_B = R_y\left(\theta_p\right)R_z\left(\theta_r\right)R_x\left(\theta_y\right)$$

For the case when $\theta_r = \frac{\pi}{2}$ we can apply the identity ◄

$$R_y(\theta)R_z\left(\tfrac{\pi}{2}\right) \equiv R_z\left(\tfrac{\pi}{2}\right)R_x(\theta)$$

leading to

$$^{S}R_B = R_z\left(\tfrac{\pi}{2}\right)R_x\left(\theta_p\right)R_x\left(\theta_y\right) = R_z\left(\tfrac{\pi}{2}\right)R_x\left(\theta_p + \theta_y\right)$$

which is unable to represent any rotation about the y-axis. This is not a good thing because spacecraft rotation about the y-axis would rotate the stable element and thus ruin its precise alignment with the stars: hence the anxiety on Apollo 13.

The loss of a degree of freedom means that mathematically we cannot invert the transformation, we can only establish a linear relationship between two of the angles. In this case the best we can do is determine the sum of the pitch and yaw angles. We observed a similar phenomena with the Euler angle singularity earlier.

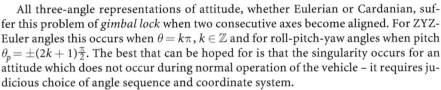

Apollo 13 mission clock: 02 08 12 47

- **Flight:** "Go, Guidance."
- **Guido:** "He's getting close to <u>gimbal lock</u> there."
- **Flight:** "Roger. CapCom, recommend he bring up C3, C4, B3, B4, C1 and C2 thrusters, and advise he's getting close to <u>gimbal lock</u>."
- **CapCom:** "Roger."

Apollo 13, mission control communications loop (1970) (Lovell and Kluger 1994, p 131; NASA 1970).

All three-angle representations of attitude, whether Eulerian or Cardanian, suffer this problem of *gimbal lock* when two consecutive axes become aligned. For ZYZ-Euler angles this occurs when $\theta = k\pi$, $k \in \mathbb{Z}$ and for roll-pitch-yaw angles when pitch $\theta_p = \pm(2k + 1)\frac{\pi}{2}$. The best that can be hoped for is that the singularity occurs for an attitude which does not occur during normal operation of the vehicle – it requires judicious choice of angle sequence and coordinate system.

Singularities are an unfortunate consequence of using a minimal representation. To eliminate this problem we need to adopt different representations of orientation. Many in the Apollo LM team would have preferred a four gimbal system and the clue to success, as we shall see shortly in Sect. 2.2.1.7, is to introduce a fourth parameter.

2.2.1.4   Two Vector Representation

For arm-type robots it is useful to consider a coordinate frame $\{E\}$ attached to the end-effector as shown in Fig. 2.16. By convention the axis of the tool is associated with the z-axis and is called the *approach vector* and denoted $\hat{a} = (a_x, a_y, a_z)$. For some applications it is more convenient to specify the approach vector than to specify Euler or roll-pitch-yaw angles.

However specifying the direction of the z-axis is insufficient to describe the coordinate frame – we also need to specify the direction of the x- and y-axes. An orthogonal vector that provides orientation, perhaps between the two fingers of the robot's gripper is called the *orientation vector*, $\hat{o} = (o_x, o_y, o_z)$. These two unit vectors are sufficient to completely define the rotation matrix

$$R = \begin{pmatrix} n_x & o_x & a_x \\ n_y & o_y & a_y \\ n_z & o_z & a_z \end{pmatrix} \tag{2.17}$$

since the remaining column, the normal vector, can be computed using Eq. 2.12 as $\hat{n} = \hat{o} \times \hat{a}$. Consider an example where the gripper's approach and orientation vectors are parallel to the world x- and y-directions respectively. Using the Toolbox this is implemented by

```
>> a = [1 0 0]';
>> o = [0 1 0]';
>> R = oa2r(o, a)
R =
     0     0     1
     0     1     0
    -1     0     0
```

Any two nonparallel vectors are sufficient to define a coordinate frame. Even if the two vectors $\hat{a}$ and $\hat{o}$ are not orthogonal they still define a plane and the computed $\hat{n}$ is normal to that plane. In this case we need to compute a new value for $\hat{o}' = \hat{a} \times \hat{n}$ which lies in the plane but is orthogonal to each of $\hat{a}$ and $\hat{n}$.

For a camera we might use the optical axis, by convention the z-axis, and the left side of the camera which is by convention the x-axis. For a mobile robot we might use

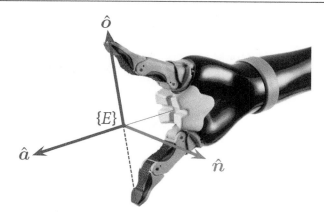

Fig. 2.16.
Robot end-effector coordinate system defines the pose in terms of an *approach* vector $\hat{a}$ and an *orientation* vector $\hat{o}$, from which $\hat{n}$ can be computed. $\hat{n}$, $\hat{o}$ and $\hat{a}$ vectors correspond to the *x*-, *y*- and *z*-axes respectively of the end-effector coordinate frame. (courtesy of Kinova Robotics)

the gravitational acceleration vector (measured with accelerometers) which is by convention the *z*-axis and the heading direction (measured with an electronic compass) which is by convention the *x*-axis.

2.2.1.5 Rotation about an Arbitrary Vector

Two coordinate frames of arbitrary orientation are related by a *single* rotation about some axis in space. For the example rotation used earlier

```
>> R = rpy2r(0.1 , 0.2, 0.3);
```

we can determine such an angle and vector by

```
>> [theta, v] = tr2angvec(R)
th =
    0.3655
v =
    0.1886    0.5834    0.7900
```

where `theta` is the angle of rotation and `v` is the vector◄ around which the rotation occurs.

This information is encoded in the eigenvalues and eigenvectors of **R**. Using the built-in MATLAB function `eig`

```
>> [x,e] = eig(R)
x =
   -0.6944 + 0.0000i   -0.6944 + 0.0000i    0.1886 + 0.0000i
    0.0792 + 0.5688i    0.0792 - 0.5688i    0.5834 + 0.0000i
    0.1073 - 0.4200i    0.1073 + 0.4200i    0.7900 + 0.0000i
e =
    0.9339 + 0.3574i    0.0000 + 0.0000i    0.0000 + 0.0000i
    0.0000 + 0.0000i    0.9339 - 0.3574i    0.0000 + 0.0000i
    0.0000 + 0.0000i    0.0000 + 0.0000i    1.0000 + 0.0000i
```

the eigenvalues are returned on the diagonal of the matrix `e` and the corresponding eigenvectors are the corresponding columns of `x`.◄

From the definition of eigenvalues and eigenvectors we recall that

$$Rv = \lambda v$$

where v is the eigenvector corresponding to the eigenvalue λ. For the case $\lambda = 1$

$$Rv = v$$

which implies that the corresponding eigenvector v is *unchanged* by the rotation. There is only one such vector and that is the one *about which* the rotation occurs. In the example the third eigenvalue is equal to one, so the rotation axis is the third column of `x`.

This is not unique. A rotation of $-\texttt{theta}$ about the vector $-\texttt{v}$ results in the same orientation.

Both matrices are complex, but some elements are real (zero imaginary part).

Olinde Rodrigues (1795–1850) was a French banker and mathematician who wrote extensively on politics, social reform and banking. He received his doctorate in mathematics in 1816 from the University of Paris, for work on his first well known formula which is related to Legendre polynomials. His eponymous rotation formula was published in 1840 and is perhaps the first time the representation of a rotation as a scalar and a vector was articulated. His formula is sometimes, and inappropriately, referred to as the Euler-Rodrigues formula. He is buried in the Pere-Lachaise cemetery in Paris.

An orthonormal rotation matrix will always have one real eigenvalue at $\lambda = 1$ and in general a complex pair $\lambda = \cos\theta \pm i\sin\theta$ where θ is the rotation angle. The angle of rotation◂ in this case is

```
>> theta = angle(e(1,1))
theta =
    0.3655
```

The inverse problem, converting from angle and vector to a rotation matrix, is achieved using Rodrigues' rotation formula

It can also be shown that the trace of a rotation matrix $tr(R) = 1 + 2\cos\theta$ from which we can compute the magnitude of θ but not its sign.

$$R = I_{3\times3} + \sin\theta[\hat{v}]_\times + (1 - \cos\theta)[\hat{v}]_\times^2 \qquad (2.18)$$

where $[\hat{v}]_\times$ is a skew-symmetric matrix. We can use this formula to determine the rotation of $\frac{\pi}{2}$ about the x-axis

```
>> R = angvec2r(pi/2, [1 0 0])
R =
    1.0000         0         0
         0    0.0000   -1.0000
         0    1.0000    0.0000
```

It is interesting to note that this representation of an arbitrary rotation is parameterized by four numbers: three for the rotation axis, and one for the angle of rotation. This is far fewer than the nine numbers required by a rotation matrix. However the direction can be represented by a unit vector which has only two parameters▸ and the angle can be encoded in the length to give a 3-parameter representation such as $\hat{v}\theta$, $\hat{v}\sin(\theta/2)$, $\hat{v}\tan(\theta)$ or the Rodrigues' vector $\hat{v}\tan(\theta/2)$. While these forms are minimal and efficient in terms of data storage they are analytically problematic and ill-defined when $\theta = 0$.

Imagine a unit-sphere. All possible unit vectors from the center can be described by the latitude and longitude of the point at which they touch the surface of the sphere.

2.2.1.6   Matrix Exponentials

Consider an x-axis rotation expressed as a rotation matrix

```
>> R = rotx(0.3)
R =
    1.0000         0         0
         0    0.9553   -0.2955
         0    0.2955    0.9553
```

As we did for the 2-dimensional case we can compute the logarithm of this matrix using the MATLAB builtin function `logm`▸

```
>> S = logm(R)
S =
         0         0         0
         0    0.0000   -0.3000
         0    0.3000    0.0000
```

`logm` is different to the builtin function `log` which computes the logarithm of each element of the matrix. A logarithm can be computed using a power series, with a matrix rather than scalar argument. For a matrix the logarithm is not unique and `logm` computes the principal logarithm of the matrix.

and the result is a sparse matrix with two elements that have a magnitude of 0.3, which is the original rotation angle. This matrix has a zero diagonal and is another example of a skew-symmetric matrix, in this case 3×3.

Applying `vex` to the skew-symmetric matrix gives

```
>> vex(S)'
ans =
    0.3000         0         0
```

and we find the original rotation angle is in the first element, corresponding to the *x*-axis about which the rotation occurred. For the 3-dimensional case the Toolbox function `trlog` is equivalent◄

`trlog` uses a more efficient closed-form solution as well as being able to return the angle and axis information separately.

```
>> [th,w] = trlog(R)
th =
    0.3000
w =
    1.0000
         0
         0
```

The inverse of a logarithm is exponentiation and applying the builtin MATLAB matrix exponential function `expm`◄

`expm` is different to the builtin function `exp` which computes the exponential of each element of the matrix:
$expm(A) = I + A + A^2/2! + A^3/3! + \cdots$

```
>> expm(S)
ans =
    1.0000         0         0
         0    0.9553   -0.2955
         0    0.2955    0.9553
```

we have regenerated our original rotation matrix. In fact the command

```
>> R = rotx(0.3);
```

is equivalent to

```
>> R = expm( skew([1 0 0]) * 0.3 );
```

where we have specified the rotation in terms of a rotation angle and a rotation axis (as a unit-vector). This generalizes to rotation about *any* axis and formally we can write

$$R = e^{[\hat{\omega}]_\times \theta} \in \mathbf{SO}(3)$$

where θ is the rotation angle, $\hat{\omega}$ is a unit-vector parallel to the rotation axis, and the notation $[\cdot]_\times : \mathbb{R}^3 \mapsto \mathbb{R}^{3\times3}$ indicates a mapping from a vector to a skew-symmetric matrix. Since $[\omega]_\times \theta = [\omega\theta]_\times$ we can treat $\omega\theta \in \mathbb{R}^3$ as a rotational parameter called exponential coordinates. For the 3-dimensional case, Rodrigues' rotation formula (Eq. 2.18) is a computationally efficient means of computing the matrix exponential for the special case where the argument is a skew-symmetric matrix, and this is used by the Toolbox function `trexp` which is equivalent to `expm`.

In 3-dimensions the skew-symmetric matrix has the form

$$[\omega]_\times = \begin{pmatrix} 0 & -\omega_z & \omega_y \\ \omega_z & 0 & -\omega_x \\ -\omega_y & \omega_x & 0 \end{pmatrix} \tag{2.19}$$

which has clear structure and only three unique elements $\omega \in \mathbb{R}^3$. The matrix can be used to implement the vector cross product $v_1 \times v_2 = [v_1]_\times v_2$. A simple example of Toolbox support for skew-symmetric matrices is

```
>> skew([1 2 3])
ans =
         0        -3         2
         3         0        -1
        -2         1         0
```

and the inverse operation is performed using the Toolbox function `vex`

```
>> vex(ans)'
ans =
         1         2         3
```

Both functions work for the 3D case, shown here, and the 2D case where the vector is a 1-vector.

2.2.1.7   Unit Quaternions

Quaternions came from Hamilton after his really good work had been done;
and, though beautifully ingenious, have been an unmixed evil to those
who have touched them in any way, including Clark Maxwell.
Lord Kelvin, 1892

Quaternions were discovered by Sir William Hamilton over 150 years ago and, while initially controversial, have great utility for robotics. The quaternion is an extension of the complex number – a hypercomplex number – and is written as a scalar plus a vector

$$q = s + v$$
$$= s + v_1 i + v_2 j + v_3 k$$

(2.20)

where $s \in \mathbb{R}$, $v \in \mathbb{R}^3$ and the orthogonal complex numbers i, j and k are defined such that

$$i^2 = j^2 = k^2 = ijk = -1$$

(2.21)

and we denote a quaternion as

$$q = s < v_1, v_2, v_3 >$$

In the Toolbox quaternions are implemented by the `Quaternion` class. Quaternions support addition and subtraction, performed element-wise, multiplication by a scalar and multiplication

$$q_1 \circ q_2 = s_1 s_2 - v_1 \cdot v_2 < s_1 v_2 + s_2 v_1 + v_1 \times v_2 >$$

which is known as the quaternion or Hamilton product.▶

One early objection to quaternions was that multiplication was not commutative but as we have seen above this is exactly what we require for rotations. Despite the initial controversy quaternions are elegant, powerful and computationally straightforward and they are *widely* used for robotics, computer vision, computer graphics and aerospace navigation systems.

To represent rotations we use unit-quaternions denoted by $\mathring{q}$. These are quaternions of unit magnitude; that is, those for which $\|q\| = s^2 + v_1^2 + v_2^2 + v_3^2 = 1$. They can be considered as a rotation of θ about the unit vector $\hat{v}$ which are related to the quaternion components by▶

If we write the quaternion as a 4-vector (s, v_1, v_2, v_2) then multiplication can be expressed as a matrix-vector product where

$$\mathring{q} \circ \mathring{q}' = \begin{pmatrix} s & -v_1 & -v_2 & -v_3 \\ v_1 & s & -v_3 & v_2 \\ v_2 & v_3 & s & -v_1 \\ v_3 & -v_2 & v_1 & s \end{pmatrix} \begin{pmatrix} s' \\ v_1' \\ v_2' \\ v_3' \end{pmatrix}$$

As for the angle-vector representation this is not unique. A rotation of θ about the vector $-v$ results in the same orientation. This is referred to as a double mapping or double cover.

Sir William Rowan Hamilton (1805–1865) was an Irish mathematician, physicist, and astronomer. He was a child prodigy with a gift for languages and by age thirteen knew classical and modern European languages as well as Persian, Arabic, Hindustani, Sanskrit, and Malay. Hamilton taught himself mathematics at age 17, and discovered an error in Laplace's Celestial Mechanics. He spent his life at Trinity College, Dublin, and was appointed Professor of Astronomy and Royal Astronomer of Ireland while still an undergraduate. In addition to quaternions he contributed to the development of optics, dynamics, and algebra. He also wrote poetry and corresponded with Wordsworth who advised him to devote his energy to mathematics.

According to legend the key quaternion equation, Eq. 2.21, occured to Hamilton in 1843 while walking along the Royal Canal in Dublin with his wife, and this is commemorated by a plaque on Broome bridge:

Here as he walked by on the 16th of October 1843 Sir William Rowan Hamilton in a flash of genius discovered the fundamental formula for quaternion multiplication $i^2 = j^2 = k^2 = ijk = -1$ & cut it on a stone of this bridge.

His original carving is no longer visible, but the bridge is a pilgrimage site for mathematicians and physicists.

For the case of unit quaternions our generalized pose is a rotation $\xi \sim \mathring{q} \in \mathbb{S}^3$ and

$$\mathring{q}_1 \oplus \mathring{q}_2 \mapsto \mathring{q}_1 \circ \mathring{q}_2$$

and

$$\ominus \mathring{q} \mapsto \mathring{q}^{-1} = s <-v>$$

which is the quaternion conjugate. The zero rotation $0 \mapsto 1 <0, 0, 0>$ which is the identity quaternion. A vector $v \in \mathbb{R}^3$ is rotated by

$$\mathring{q} \cdot v \mapsto \mathring{q} \circ \mathring{v} \circ \mathring{q}^{-1}$$

where $\mathring{v} = 0 <v>$ is known as a pure quaternion.

$$\mathring{q} = \cos\tfrac{\theta}{2} < \hat{v}\sin\tfrac{\theta}{2}> \tag{2.22}$$

and has similarities to the angle-axis representation of Sect. 2.2.1.5.

In the Toolbox these are implemented by the `UnitQuaternion` class and the constructor converts a passed argument such as a rotation matrix to a unit quaternion, for example

```
>> q = UnitQuaternion( rpy2tr(0.1, 0.2, 0.3)  )
q =
0.98335 < 0.034271, 0.10602, 0.14357 >
```

This class overloads a number of standard methods and functions. Quaternion multiplication◄ is invoked through the overloaded multiplication operator

Compounding two orthonormal rotation matrices requires 27 multiplications and 18 additions. The quaternion form requires 16 multiplications and 12 additions. This saving can be particularly important for embedded systems.

```
>> q = q * q;
```

and inversion, the conjugate of a unit quaternion, is

```
>> inv(q)
ans =
0.93394 < -0.0674, -0.20851, -0.28236 >
```

Multiplying a quaternion by its inverse yields the identity quaternion

```
>> q*inv(q)
ans =
1 < 0, 0, 0 >
```

which represents a null rotation, or more succinctly

```
>> q/q
ans =
1 < 0, 0, 0 >
```

The quaternion can be converted to an orthonormal rotation matrix by

```
>> q.R
ans =
    0.7536   -0.4993    0.4275
    0.5555    0.8315   -0.0081
   -0.3514    0.2436    0.9040
```

and we can also plot the orientation represented by a quaternion

```
>> q.plot()
```

which produces a result similar in style to that shown in Fig. 2.13. A vector is rotated by a quaternion using the overloaded multiplication operator

```
>> q*[1 0 0]'
ans =
    0.7536
    0.5555
   -0.3514
```

The Toolbox implementation is quite complete and the `UnitQuaternion` class has many methods and properties which are described fully in the online documentation.

2.2.2   Pose in 3-Dimensions

We return now to representing relative pose in three dimensions – the position and orientation change between the two coordinate frames as shown in Fig. 2.10. This is often referred to as a rigid-body displacement or rigid-body motion.

We have discussed several different representations of orientation, and we need to combine one of these with translation, to create a tangible representation of relative pose.

2.2.2.1   Homogeneous Transformation Matrix

The derivation for the homogeneous transformation matrix is similar to the 2D case of Eq. 2.11 but extended to account for the z-dimension. $t \in \mathbb{R}^3$ is a vector defining the origin of frame $\{B\}$ with respect to frame $\{A\}$, and R is the 3×3 orthonormal matrix which describes the orientation of the axes of frame $\{B\}$ with respect to frame $\{A\}$.

$$\begin{pmatrix} {}^A x \\ {}^A y \\ {}^A z \\ 1 \end{pmatrix} = \begin{pmatrix} {}^A R_B & t \\ 0_{1\times3} & 1 \end{pmatrix} \begin{pmatrix} {}^B x \\ {}^B y \\ {}^B z \\ 1 \end{pmatrix}$$

If points are represented by homogeneous coordinate vectors then

$$\begin{aligned} {}^A \tilde{p} &= \begin{pmatrix} {}^A R_B & t \\ 0_{1\times3} & 1 \end{pmatrix} {}^B \tilde{p} \\ &= {}^A T_B \, {}^B \tilde{p} \end{aligned} \tag{2.23}$$

and ${}^A T_B$ is a 4×4 homogeneous transformation matrix. This matrix has a very specific structure and belongs to the special Euclidean group of dimension 3 or $T \in \mathrm{SE}(3) \subset \mathbb{R}^{4\times4}$.

A concrete representation of relative pose is $\xi \sim T \in \mathrm{SE}(3)$ and $T_1 \oplus T_2 \mapsto T_1 T_2$ which is standard matrix multiplication.

$$T_1 T_2 = \begin{pmatrix} R_1 & t_1 \\ 0_{1\times3} & 1 \end{pmatrix} \begin{pmatrix} R_2 & t_2 \\ 0_{1\times3} & 1 \end{pmatrix} = \begin{pmatrix} R_1 R_2 & t_1 + R_1 t_2 \\ 0_{1\times3} & 1 \end{pmatrix} \tag{2.24}$$

One of the rules of pose algebra from page 21 is $\xi \oplus 0 = \xi$. For matrices we know that $TI = T$, where I is the identify matrix, so for pose $0 \mapsto I$ the identity matrix. Another rule of pose algebra was that $\xi \ominus \xi = 0$. We know for matrices that $TT^{-1} = I$ which implies that $\ominus T \mapsto T^{-1}$

$$T^{-1} = \begin{pmatrix} R & t \\ 0_{1\times3} & 1 \end{pmatrix}^{-1} = \begin{pmatrix} R^T & -R^T t \\ 0_{1\times3} & 1 \end{pmatrix} \tag{2.25}$$

The 4×4 homogeneous transformation is very commonly used in robotics, computer graphics and computer vision. It is supported by the Toolbox and will be used throughout this book as a concrete representation of 3-dimensional pose.

The Toolbox has many functions to create homogeneous transformations. For example we can demonstrate composition of transforms by

```
>> T = transl(1, 0, 0) * trotx(pi/2) * transl(0, 1, 0)
T =
    1.0000         0         0    1.0000
         0    0.0000   -1.0000    0.0000
         0    1.0000    0.0000    1.0000
         0         0         0    1.0000
```

The function `transl` creates a relative pose with a finite translation but no rotation, while `trotx` creates a relative pose corresponding to a rotation of $\frac{\pi}{2}$ about the x-axis with zero translation. We can think of this expression as representing a walk along the x-axis for 1 unit, then a rotation by 90° about the x-axis and then a walk of 1 unit along the new y-axis which was the previous z-axis. The result, as shown in the last column of the resulting matrix is a translation of 1 unit along the original x-axis and 1 unit along the original z-axis. The orientation of the final pose shows the effect of the rotation about the x-axis. We can plot the corresponding coordinate frame by

Many Toolbox functions have variants that return orthonormal rotation matrices or homogeneous transformations, for example, `rotx` and `trotx`, `rpy2r` and `rpy2tr` etc. Some Toolbox functions accept an orthonormal rotation matrix or a homogeneous transformation and ignore the translational component, for example, `tr2rpy`.

```
>> trplot(T)
```

The rotation matrix component of T is

```
>> t2r(T)
ans =
    1.0000         0         0
         0    0.0000   -1.0000
         0    1.0000    0.0000
```

and the translation component is a column vector

```
>> transl(T)'
ans =
    1.0000    0.0000    1.0000
```

2.2.2.2 Vector-Quaternion Pair

A compact and practical representation is the vector and unit quaternion pair. It represents pose using just 7 numbers, is easy to compound, and singularity free.

This representation is not implemented in the Toolbox.

> For the vector-quaternion case $\xi \sim (t, \mathring{q})$ where $t \in \mathbb{R}^3$ is a vector defining the frame's origin with respect to the reference coordinate frame, and $\mathring{q} \in \mathbb{S}^3$ is the frame's orientation with respect to the reference frame.
>   Composition is defined by
> $$\xi_1 \oplus \xi_2 = (t_1 + \mathring{q}_1 \cdot t_2, \mathring{q}_1 \circ \mathring{q}_2)$$
> and negation is
> $$\ominus \xi = \left(-\mathring{q}^{-1} \cdot t, \mathring{q}^{-1}\right)$$
> and a point coordinate vector is transformed to a coordinate frame by
> $$^X p = {}^X \xi_Y \cdot {}^Y p = \mathring{q} \cdot {}^Y p + t$$

2.2.2.3 Twists

In Sect. 2.1.2.3 we introduced twists for the 2D case. Any rigid-body motion in 3D space is equivalent to a screw motion – motion about and along some line in space. We represent a screw as a pair of 3-vectors $s = (v, \omega) \in \mathbb{R}^6$.

Pure translation can be considered as rotation about a point at infinity.

The ω component of the twist vector is the direction of the screw axis. The v component is called the moment and encodes the position of the line of the twist axis in space and also the pitch of the screw. The pitch is the ratio of the distance along the screw axis to the rotation about the screw axis.

Consider the example of a rotation of 0.3 radians about the *x*-axis. We first specify a unit twist▸ with an axis that is parallel to the *x*-axis and passes through the origin

A rotational unit twist has $\|\omega\| = 1$.

```
>> tw = Twist('R', [1 0 0], [0 0 0])
tw =
( -0 -0 -0; 1 0 0 )
```

which we convert, for the required rotation angle, to an **SE**(3)-homogeneous transformation

```
>>  tw.T(0.3)
ans =
    1.0000         0         0         0
         0    0.9553   -0.2955         0
         0    0.2955    0.9553         0
         0         0         0    1.0000
```

and has the same value we would obtain using `trotx(0.3)`.

For pure translation in the *y*-direction the unit twist▸ would be

A translational unit twist has $\|v\| = 1$ and $\omega = 0$.

```
>> tw = Twist('T', [0 1 0])
tw =
( 0 1 0; 0 0 0 )
```

which we convert, for the required translation distance, to an **SE**(3)-homogeneous transformation.

```
>> tw.T(2)
ans =
    1    0    0    0
    0    1    0    2
    0    0    1    0
    0    0    0    1
```

which is, as expected, an identity matrix rotational component (no rotation) and a translational component of 2 in the *y*-direction.

To illustrate the underlying screw model we define a coordinate frame {*X*}

```
>> X = transl(3, 4, -4);
```

which we will rotate by a range of angles

```
>> angles = [0:0.3:15];
```

around a screw axis parallel to the *z*-axis, direction (0, 0, 1), through the point (2, 3, 2) and with a pitch of 0.5

```
>> tw = Twist('R', [0 0 1], [2 3 2], 0.5);
```

The next line packs a lot of functionality. For values of θ drawn successively from the vector `angles` we use an anonymous function to evaluate the twist for each value of θ and apply it to the frame {*X*}. This sequence is animated and each frame in the sequence is retained

```
>> tranimate( @(theta) tw.T(theta) * X, angles, ...
        'length', 0.5, 'retain', 'rgb', 'notext');
```

and the result is shown in Fig. 2.17. We can clearly see the screw motion in the successive poses of the displaced reference frame as it is rotated about the screw axis.

The screw axis is the line

```
>> L = tw.line
L =
{ 3  -2  0; 0  0  1 }
```

which is described in terms of its Plücker coordinates which we can plot

```
>> L.plot('k:', 'LineWidth', 2)
```

Finally we can convert an arbitrary homogeneous transformation to a nonunit twist

```
>> T = transl(1, 2, 3) * eul2tr(0.3, 0.4, 0.5);
>> tw = Twist(T)
tw =
( 1.1204 1.6446 3.1778; 0.041006 0.4087 0.78907 )
```

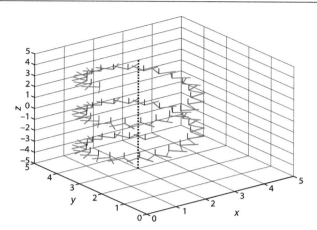

Fig. 2.17.
A coordinate frame $\{X\}$ displayed for different values of θ about a screw parallel to the z-axis and passing through the point $(2, 3, 2)$. The x-, y- and z-axes are indicated by red, green and blue lines respectively

which has a pitch of

```
>> tw.pitch
ans =
    3.2256
```

and the rotation about the axis is

```
>> tw.theta
ans =
    0.8896
```

and a point lying on the twist axis is

```
>> tw.pole'
ans =
    0.0011    0.8473    -0.4389
```

2.3   Advanced Topics

2.3.1   Normalization

The IEEE standard for double precision floating point, the standard MATLAB numeric format, has around 16 decimal digits of precision.

Floating-point arithmetic has finite precision◄ and consecutive operations will accumulate error. A rotation matrix has by definition, a determinant of one

```
>> R = eye(3,3);
>> det(R) - 1
ans =
    0
```

but if we repeatedly multiply by a valid rotation matrix the result

```
>> for i=1:100
     R = R * rpy2r(0.2, 0.3, 0.4);
   end
>> det(R) - 1
ans =
  4.4409e-15
```

indicates a small error – the determinant is no longer equal to one and the matrix is no longer a proper orthonormal rotation matrix. To fix this we need to normalize the matrix, a process which enforces the constraints on the columns c_i of an orthonormal matrix $R = [c_1, c_2, c_3]$. We need to assume that one column has the correct direction

$$c_3' = c_3$$

then the first column is made orthogonal to the last two

$$c_1' = c_2 \times c_3'$$

However the last two columns may not have been orthogonal so

$$c_2' = c_1' \times c_3'$$

Finally the columns are all normalized to unit magnitude

$$c_i'' = \frac{c_1'}{\|c_1'\|}, \quad i = 1 \cdots 3$$

In the Toolbox normalization is implemented by

```
>> R = trnorm(R);
```

and the determinant is now much closer to one▶

```
>> det(R) - 1
ans =
   -2.2204e-16
```

This error is now at the limit of double precision arithmetic which is 2.2204×10^{-16} and given by the MATLAB function eps.

A similar issue arises for unit quaternions when the norm, or magnitude, of the unit quaternion is no longer equal to one. However this is much easier to fix since normalizing the quaternion simply involves dividing all elements by the norm

$$\mathring{q}' = \frac{\mathring{q}}{\|\mathring{q}\|}$$

which is implemented by the unit method

```
>> q = q.unit();
```

The UnitQuaternion class also supports a variant of multiplication

```
>> q = q .* q2;
```

which performs an explicit normalization after the multiplication.

Normalization does not need to be done after every multiplication since it is an expensive operation. However for situations like the example above where one transform is being repeatedly updated it is advisable.

2.3.2   Understanding the Exponential Mapping

In this chapter we have glimpsed some connection between rotation matrices, skew-symmetric matrices and matrix exponentiation. The basis for this lies in the mathematics of Lie groups which are covered in text books on algebraic geometry and algebraic topology. These require substantial knowledge of advanced mathematics and many people starting out in robotics will find their content quite inaccessible. An introduction to the essentials of this topic is given in Appendix D. In this section we will use an intuitive approach, based on undergraduate engineering mathematics, to shed some light on these relationships.

Consider a point **P**, defined by a coordinate vector p, being rotated with an angular velocity ω which is a vector whose direction defines the axis of rotation and whose magnitude $\|\omega\|$ specifies the rate of rotation about the axis which we assume passes through the origin.▶ We wish to rotate the point by an angle θ about this axis and the velocity of the point is known from mechanics to be

Angular velocity will be properly introduced in the next chapter.

$$\dot{p} = \omega \times p$$

and we replace the cross product with a skew-symmetric matrix giving a matrix-vector product

$$\dot{p} = [\omega]_\times p \tag{2.26}$$

We can find the solution to this first-order differential equation by analogy to the simple scalar case

$$\dot{x} = ax$$

whose solution is

$$x(t) = e^{at}x(0)$$

This implies that the solution to Eq. 2.26 is

$$p(t) = e^{[\omega]_\times t}p(0)$$

If $\|\omega\| = 1$ then after t seconds the vector will have rotated by t radians. We require a rotation by θ so we can set $t = \theta$ to give

$$p(\theta) = e^{[\hat{\omega}]_\times \theta}p(0)$$

which describes the vector $p(0)$ being rotated to $p(\theta)$. A matrix that rotates a vector is a rotation matrix, and this implies that our matrix exponential is a rotation matrix

$$R(\theta, \hat{\omega}) = e^{[\hat{\omega}]_\times \theta} \in \mathbf{SO}(3)$$

Now consider the more general case of rotational and translational motion. We can write

$$\dot{p} = [\omega]_\times p + v$$

and rearranging into matrix form

$$\begin{pmatrix} \dot{p} \\ 0 \end{pmatrix} = \begin{pmatrix} [\omega]_\times & v \\ 0 & 0 \end{pmatrix} \begin{pmatrix} p \\ 1 \end{pmatrix}$$

and introducing homogeneous coordinates this becomes

$$\dot{\tilde{p}} = \begin{pmatrix} [\omega]_\times & v \\ 0 & 0 \end{pmatrix} \tilde{p}$$
$$= \Sigma \tilde{p}$$

where Σ is a 4×4 augmented skew-symmetric matrix. Again, by analogy with the scalar case we can write the solution as

$$\tilde{p}(\theta) = e^{\Sigma \theta} \tilde{p}(0)$$

A matrix that rotates and translates a point in homogeneous coordinates is a homogeneous transformation matrix, and this implies that our matrix exponential is a homogeneous transformation matrix

$$T(\theta, \hat{\omega}, v) = e^{\begin{pmatrix} [\hat{\omega}]_\times & v \\ 0 & 0 \end{pmatrix} \theta} \in \mathbf{SE}(3)$$

where $[\hat{\omega}]_\times \theta$ defines the magnitude and axis of rotation and $v\theta$ is the translation.

The exponential of a scalar can be computed using a power series, and the matrix case is analogous and relatively straightforward to compute. The MATLAB function `expm` uses a polynomial approximation for the general matrix case. If A is skew-symmetric or augmented-skew-symmetric then an efficient closed-form solution for a rotation matrix – the Rodrigues' rotation formula (Eq. 2.18) – can be used and this is implemented by the Toolbox function `trexp`.

2.3.3   More About Twists

In this chapter we introduced and applied twists and here we will more formally define them. We also highlight the very close relationship between twists and homogeneous transformation matrices via the exponential mapping.

The key concept comes from Chasle's theorem: "*any displacement of a body in space can be accomplished by means of a rotation of the body about a unique line in space accompanied by a translation of the body parallel to that line*". Such a line is called a screw axis and is illustrated in Fig. 2.18. The mathematics of screw theory was developed by Sir Robert Ball in the late 19th century for the analysis of mechanisms. At the core of screw theory are pairs of vectors: angular and linear velocity; forces and moments; and Plücker coordinates (see Sect. C.1.2.2).

The general displacement of a rigid body in 3D can be represented by a twist vector

$$S = (v, \omega) \in \mathbb{R}^6$$

where $v \in \mathbb{R}^3$ is referred to as the moment and encodes the position of the action line in space and the pitch of the screw and $\omega \in \mathbb{R}^3$ is the direction of the screw axis.

For rotational motion where the screw axis is parallel to the vector $\hat{a}$, passes through a point Q defined by its coordinate vector q, and the screw pitch p is the ratio of the distance along the screw axis to the rotation about the axis, the twist elements are

$$S = (q \times \hat{a} + p\hat{a}, \hat{a})$$

and the pitch can be recovered by

$$p = \hat{w}^T v$$

For the case of pure rotation the pitch of the screw is zero and the unit twist is

$$S = (q \times \hat{a}, \hat{a})$$

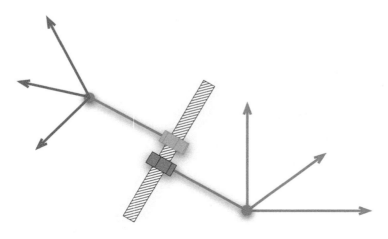

Fig. 2.18.
Conceptual depiction of a screw. A coordinate frame is attached to a nut by a rigid rod and rotated around the screw thread. The pose changes from the red frame to the blue frame. The corollary is that given any two frames we can determine a screw axis to rotate one into the other

Michel Chasles (1793–1880) was a French mathematician born at Épernon. He studied at the École Polytechnique in Paris under Poisson and in 1814 was drafted to defend Paris in the War of the Sixth Coalition. In 1837 he published a work on the origin and development of methods in geometry, which gained him considerable fame and he was appointed as professor at the École Polytechnique in 1841, and at the Sorbonne in 1846.

He was an avid collector and purchased over 27 000 forged letters purporting to be from Newton, Pascal and other historical figures – all written in French! One from Pascal claimed he had discovered the laws of gravity before Newton, and in 1867 Chasles took this to the French Academy of Science but scholars recognized the fraud. Eventually Chasles admitted he had been deceived and revealed he had spent nearly 150 000 francs on the letters. He is buried in Cimetière du Père Lachaise in Paris.

For purely translational motion in the direction parallel to the vector a, the pitch is infinite which leads to a zero rotational component and the unit twist is

$$S = (\hat{a}, 0)$$

A twist is related to the rigid-body displacement in SE(3) by the exponential mapping already discussed.

$$T(\theta, S) = e^{[S]\theta} \in \mathbf{SE}(3)$$

where the augmented skew-symmetric matrix

$$[S] = \left(\begin{array}{ccc|c} 0 & -\omega_3 & \omega_2 & v_1 \\ \omega_3 & 0 & -\omega_1 & v_2 \\ -\omega_2 & \omega_1 & 0 & v_3 \\ \hline 0 & 0 & 0 & 0 \end{array} \right) \in \mathbf{se}(3)$$

belongs to the Lie algebra **se**(3) and is the *generator* of the rigid-body displacement. The matrix exponential has an efficient closed-form

$$T(\theta, S) = \begin{pmatrix} R(\theta, \hat{\omega}) & \left(I_{3\times3}\theta + (1-\cos\theta)[\hat{\omega}]_{\times} + (\theta - \sin\theta)[\hat{\omega}]_{\times}^2 \right)v \\ 0 & 1 \end{pmatrix}$$

where $R(\theta, \hat{\omega})$ is computed using Rodrigues' rotation formula (Eq. 2.18). For a nonunit rotational twist, that is $\|\omega\| \neq 1$, then $\theta = \|\omega\|$.

> For real numbers, if $x = \log X$ and $y = \log Y$ then
>
> $$Z = XY = e^x e^y = e^{x+y}$$
>
> but for the matrix case this is *only* true if the matrices commute, and rotation matrices do not, therefore
>
> $$Z = XY = e^x e^y \neq e^{x+y} \quad \text{if } x, y \in \mathbf{so}(n) \text{ or } \mathbf{se}(n)$$
>
> The bottom line is that there is no shortcut to compounding rotations, we must compute $z = \log(e^x e^y)$ not $z = x + y$.

The Toolbox provides many ways to create twists and to convert them to rigid-body displacements expressed as homogeneous transformations. Now that we understand more about the exponential mapping we will revisit the example from page 48

```
>> tw = Twist('R', [1 0 0], [0 0 0])
tw =
( -0 -0 -0; 1 0 0 )
```

A unit twist describes a *family* of motions that have a single parameter, either a rotation and translation about and along some screw axis, or a pure translation in some direction. We can visualize it as a mechanical screw in space, or represent it as a 6-vector $S = (v, \omega)$ where $\|\omega\| = 1$ for a rotational twist and $\|v\| = 1, \omega = 0$ for a translational twist.

A *particular* rigid-body motion is described by a unit-twist s and a motion parameter θ which is a scalar specifying the amount of rotation or translation. The motion is described by the twist $S\theta$ which is in general not a unit-twist. The exponential of this in 4×4 matrix format is the 4×4 homogeneous transformation matrix describing that particular rigid-body motion in $\mathbf{SE}(3)$.

which is a unit twist that describes rotation about the *x*-axis in $\mathbf{SE}(3)$. The `Twist` has a number of properties

```
>> tw.S'
ans =
     0      0      0      1      0      0
>> tw.v'
ans =
     0      0      0
>> tw.w'
ans =
     1      0      0
```

as well as various methods. We can create the **se**(3) Lie algebra using the `se` method of this class

```
>> tw.se
ans =
     0      0      0      0
     0      0     -1      0
     0      1      0      0
     0      0      0      0
```

which is the augmented skew-symmetric version of S. The method `T` performs the exponentiation▸ of this to create an **SE**(3) homogeneous transformation for the specified rotation about the unit twist

The `expm` method is synonomous and both invoke the Toolbox function `trexp`.

```
>>   tw.T(0.3)
ans =
   1.0000          0          0          0
        0     0.9553    -0.2955          0
        0     0.2955     0.9553          0
        0          0          0     1.0000
```

The Toolbox functions `trexp` and `trlog` are respectively closed-form alternatives to `expm` and `logm` when the arguments are in **so**(3)/**se**(3) or **SO**(3)/**SE**(3).

The `line` method returns a `Plucker` object that represents the line of the screw in Plücker coordinates

```
>> tw.line
ans =
{ 0  0  0; 1  0  0 }
```

Finally, the overloaded multiplication operator for the `Twist` class will compound two twists.

```
>> t2 = tw * tw
t2 =
( -0  -0  -0; 2  0  0 )
>> tr2angvec(t2.T)
Rotation: 2.000000 rad x [1.000000 0.000000 0.000000]
```

and the result in this case is a nonunit twist of two units, or 2 rad, about the *x*-axis.

2.3.4 Dual Quaternions

Quaternions were developed by William Hamilton in 1843 and we have already seen their utility for representing orientation, but using them to represent pose proved more difficult. One early approach was Hamilton's bi-quaternion where the quaternion coefficients were complex numbers. Somewhat later William Clifford◀ developed the dual number, defined as an ordered pair $d = (x, y)$ which can be written as $d = x + y\varepsilon$ where $\varepsilon^2 = 0$ and for which specific addition and multiplication rules exist. Clifford created a quaternion dual number with $x, y \in \mathbb{H}$ which he also called a bi-quaternion but is today called a dual quaternion

$$\mathring{q}_\xi = r + \tfrac{1}{2}\varepsilon \mathring{t} \circ \mathring{r}$$

where $\mathring{r} \in \mathbb{H}$ is a unit quaternion representing the rotational part of the pose and $\mathring{t} \in \mathbb{H}$ is a pure quaternion representing translation. This type of mathematical object has been largely eclipsed by modern matrix and vector approaches, but there seems to be a recent resurgence of interest in alternative approaches. The dual quaternion is quite compact, requiring just 8 numbers; it is easy to compound using a special multiplication table; and it is easy to renormalize to eliminate the effect of imprecise arithmetic. However it has no real useful computational advantage over matrix methods.

2.3.5 Configuration Space

We have so far considered the pose of objects in terms of the position and orientation of a coordinate frame affixed to them. For an arm-type robot we might affix a coordinate frame to its end-effector, while for a mobile robot we might affix a frame to its body – its body-fixed frame. This is sufficient to describe the state of the robot in the familiar 2D or 3D Euclidean space which is referred to as the task space or operational space since it is where the robot performs tasks or operates.

An alternative way of thinking about this comes from classical mechanics and is referred to as the *configuration* of a system. The configuration is the smallest set of parameters, called generalized coordinates, that are required to fully describe the position of *every* particle in the system. This is not as daunting as it may appear since in general a robot comprises one or more rigid elements, and in each of these the particles maintain a constant relative offset to each other.

If the *system* is a train moving along a track then all the particles comprising the train move together and we need only a single generalized coordinate q, the distance along the track from some datum, to describe their location. A robot arm with a fixed base and two rigid links, connected by two rotational joints has a configuration that is completely described by two generalized coordinates – the two joint angles (q_1, q_2). The generalized coordinates can, as their name implies, represent displacements or rotations.

1845–1879, an English mathematician and geometer.

Sir Robert Ball (1840–1913) was an Irish astronomer born in Dublin. He became Professor of Applied Mathematics at the Royal College of Science in Dublin in 1867, and in 1874 became Royal Astronomer of Ireland and Andrews Professor of Astronomy at the University of Dublin. In 1892 he was appointed Lowndean Professor of Astronomy and Geometry at Cambridge University and became director of the Cambridge Observatory. He was a Fellow of the Royal Society and in 1900 became the first president of the Quaternion Society.

He is best known for his contributions to the science of kinematics described in his treatise "The Theory of Screws" (1876), but he also published "A Treatise on Spherical Astronomy" (1908) and a number of popular articles on astronomy. He is buried at the Parish of the Ascension Burial Ground in Cambridge.

The number of independent▶ generalized coordinates N is known as the number of degrees of freedom of the system. Any configuration of the system is represented by a point in its N-dimensional configuration space, or C-space, denoted by $\mathcal{C}$ and $q \in \mathcal{C}$. We can also say that dim $\mathcal{C} = N$. For the train example $\mathcal{C} \subset \mathbb{R}$ which says that the displacement is a bounded real number. For the 2-joint robot the generalized coordinates are both angles so $\mathcal{C} \subset \mathbb{S}^1 \times \mathbb{S}^1$.

That is, there are no holonomic constraints on the system.

Any point in the configuration space can be mapped to a point in the task space $q \in \mathcal{C} \mapsto \tau \in \mathcal{T}$ but the inverse is not necessarily true. This mapping depends on the task space that we choose and this, as its name suggests, is task specific.

Consider again the train moving along its rail. We might be interested to describe the train in terms of its position on a plane in which case the task space would be $\mathcal{T} \subset \mathbb{R}^2$, or in terms of its latitude and longitude, in which case the task space would be $\mathcal{T} \subset \mathbb{S}^1 \times \mathbb{S}^1$. We might choose a 3-dimensional task space $\mathcal{T} \subset \mathbf{SE}(3)$ to account for height changes as the train moves up and down hills and its orientation changes as it moves around curves. However in all these case the dimension of the task space exceeds the dimension of the configuration space dim $\mathcal{T} >$ dim $\mathcal{C}$ and this means that the train cannot *access* all points in the task space. While every point along the rail line can be mapped to the task space, most points in the task space will not map to a point on the rail line. The train is constrained by its fixed rails to move in a subset of the task space.

The simple 2-joint robot arm can access a subset of points in a plane so a useful task space might be $\mathcal{T} \subset \mathbb{R}^2$. The dimension of the task space equals the dimension of the configuration space dim $\mathcal{T} =$ dim $\mathcal{C}$ and this means that the mapping between task and configuration spaces is bi-directional but it is not necessarily unique – for this type of robot, in general, two different configurations map to a single point in task space. Points in the task space beyond the physical reach of the robot are not mapped to the configuration space. If we chose a task space with more dimensions such as $\mathbf{SE}(2)$ or $\mathbf{SE}(3)$ then dim $\mathcal{T} >$ dim $\mathcal{C}$ and the robot would only be able to access points within a subset of that space.

Now consider a snake-robot arm, such as shown in Fig. 8.9, with 20 joints and $\mathcal{C} \subset \mathbb{S}^1 \times \cdots \times \mathbb{S}^1$ and dim $\mathcal{T} <$ dim $\mathcal{C}$. In this case an infinite number of configurations in a $20 - 6 = 14$-dimensional subspace of the 20-dimensional configuration space will map to the same point in task space. This means that in addition to the task of positioning the robot's end-effector we can *simultaneously* perform motion in the configuration subspace to control the shape of the arm to avoid obstacles in the environment. Such a robot is referred to as over-actuated or redundant and this topic is covered in Sect. 8.4.2.

The body of a quadrotor, such as shown in Fig. 4.19d, is a single rigid-body whose configuration is completely described by six generalized coordinates, its position and orientation in 3D space $\mathcal{C} \subset \mathbb{R}^3 \times \mathbb{S}^1 \times \mathbb{S}^1 \times \mathbb{S}^1$ where the orientation is expressed in some three-angle representation. For such a robot the most logical task space would be $\mathbf{SE}(3)$ which is equivalent to the configuration space and dim $\mathcal{T} =$ dim $\mathcal{C}$. However the quadrotor has only four actuators which means it cannot *directly* access all the points in its configuration space and hence its task space. Such a robot is referred to as under-actuated and we will revisit this in Sect. 4.2.

2.4   Using the Toolbox

The Toolbox supports all the different representations discussed in this chapter as well as conversions between many of them. The representations and possible conversions are shown in tabular form in Tables 2.1 and 2.2 for the 2D and 3D cases respectively.

In this chapter we have mostly used native MATLAB matrices to represent rotations and homogeneous transformations▶ and historically this has been what the Toolbox supported – the Toolbox *classic* functions. From Toolbox release 10 there are classes that

Quaternions and twists are implemented as classes not native types, but in very old versions of the Toolbox quaternions were 1×4 vectors.

represent rotations and homogeneous transformations, named respectively SO2 and SE2 for 2 dimensions and SO3 and SE3 for 3 dimensions. These provide real advantages in terms of code readability and type safety and can be used in an almost identical fashion to the native matrix types. They are also polymorphic meaning they support many of the same operations which makes it very easy to switch between using say rotation matrices and quaternions or lifting a solution from 2- to 3-dimensions. A quick illustration of the new functionality is the example from page 27 which becomes

```
>> T1 = SE2(1, 2, 30, 'deg');
>> about T1
T1 [SE2] : 1x1 (176 bytes)
```

The size of the object in bytes, shown in parentheses, will vary between MATLAB. versions and computer types.

which results in an SE2 class object not a 3 × 3 matrix.◄ If we display it however it does look like a 3 × 3 matrix►

If you have the cprintf package from MATLAB File Exchange installed then the rotation submatrix will be colored red.

```
>> T1
T1 =
    0.8660   -0.5000        1
    0.5000    0.8660        2
         0         0        1
```

The matrix is encapsulated within the object and we can extract it readily if required

```
>> T1.T
ans =
    0.8660   -0.5000    1.0000
    0.5000    0.8660    2.0000
         0         0    1.0000
>> about ans
ans [double] : 3x3 (72 bytes)
```

Returning to that earlier example we can quite simply transform the vector

```
>> inv(T1) * P
ans =
    1.7321
   -1.0000
```

and the class handles the details of converting the vector between Euclidean and homogeneous forms.

This new functionality is also covered in Tables 2.1 and 2.2, and Table 2.3 is a map between the classic and new functionality to assist you in using the Toolbox. From here on the book will use a mixture of classic functions and the newer classes.

Table 2.1. Toolbox supported data types for representing 2D pose: constructors and conversions

Input type	Output type							
	t	θ	R	T	Twist vector	Twist	SO2	SE2
t (2-vector)				trans12		Twist('T')		SE2()
θ (scalar)			rot2	trot2		Twist('R')	SO2()	SE2()
R (2 × 2 matrix)				r2t			SO2()	SE2()
T (3 × 3 matrix)	trans12		t2r			Twist()		SE2()
Twist vector (1- or 3-vector)			trexp2	trexp2		Twist()	SO2.exp()	SE3.exp()
Twist				.T	.S			.SE
SO2		.theta	.R	.T	.log			.SE2
SE2	.t	.theta	.R	.T	.log	.Twist	.SO2	

Dark grey boxes are not possible conversions. Light grey boxes are possible conversions but the Toolbox has no direct conversion, you need to convert via an intermediate type. Red text indicates classical Robotics Toolbox functions that work with native MATLAB® vectors and matrices. **Bold text** indicates a Toolbox class. Class.type() indicates a static factory method that constructs a Class object from input of that type. Functions shown starting with a dot are a method on the class corresponding to that row.

Input type	Output type										
	t	Euler	RPY	θ, v	R	T	Twist vector	Twist	Unit-Quaternion	SO3	SE3
t (3-vector)						`transl`		`Twist('T')`			`SE3()`
Euler (3-vector)					`eul2r`	`eul2tr`			`UnitQuaternion.eul()`	`SO3.eul()`	`SE3.eul()`
RPY (3-vector)					`rpy2r`	`rpy2tr`			`UnitQuaternion.rpy()`	`SO3.rpy()`	`SE3.rpy()`
θ, v (scalar + 3-vector)					`angvec2r`	`angvec2tr`			`UnitQuaternion.angvec()`	`SO3.angvec()`	`SE3.angvec()`
R (3×3 matrix)		`tr2eul`	`tr2rpy`	`tr2angvec`		`r2t`	`trlog`		`UnitQuaternion()`	`SO3()`	`SE3()`
T (4×4 matrix)	`transl`	`tr2eul`	`tr2rpy`	`tr2angvec`	`t2r`		`trlog`	`Twist()`	`UnitQuaternion()`	`SO3()`	`SE3()`
Twist vector (3- or 6-vector)					`trexp`	`trexp`		`Twist()`		`SO3.exp()`	`SE3.exp()`
Twist						`.T`	`.S`				`.SE`
Unit-Quaternion		`.toeul`	`.torpy`	`.toangvec`	`.R`	`.T`				`.SO3`	`.SE3`
SO3		`.toeul`	`.torpy`	`.toangvec`	`.R`	`.T`	`.log`		`.UnitQuaternion`		`.SE3`
SE3	`.t`	`.toeul`	`.torpy`	`.toangvec`	`.R`	`.T`	`.log`	`.Twist`	`.UnitQuaternion`	`.SO3`	

Dark grey boxes are not possible conversions. Light grey boxes are possible conversions but the Toolbox has no direct conversion, you need to convert via an intermediate type. Red text indicates classical Robotics Toolbox functions that work with native MATLAB® vectors and matrices. Class.type() indicates a static factory method that constructs a Class object from input of that type. Functions shown starting with a dot are a method on the class corresponding to that row.

Table 2.2. Toolbox supported data types for representing 3D pose: constructors and conversions

2.5   Wrapping Up

In this chapter we learned how to represent points and poses in 2- and 3-dimensional worlds. Points are represented by coordinate vectors relative to a coordinate frame. A set of points that belong to a rigid object can be described by a coordinate frame, and its constituent points are described by constant vectors in the object's coordinate frame. The position and orientation of any coordinate frame can be described relative to another coordinate frame by its relative pose ξ. We can think of a relative pose as a motion – a rigid-body motion – and these motions can be applied sequentially (composed or compounded). It is important to remember that composition is noncommutative – the order in which relative poses are applied is important.

We have shown how relative poses can be expressed as a pose graph or manipulated algebraically. We can also use a relative pose to transform a vector from one coordinate frame to another. A simple graphical summary of key concepts is given in Fig. 2.19.

We have discussed a variety of mathematical objects to tangibly represent pose. We have used orthonormal rotation matrices for the 2- and 3-dimensional case to represent orientation and shown how it can rotate a points' coordinate vector from one coordinate frame to another. Its extension, the homogeneous transformation matrix, can be used to represent both orientation and translation and we have shown how it can rotate and translate a point expressed in homogeneous coordinates from one frame

Orientation		Pose	
Classic	**New**	**Classic**	**New**
rot2	SO2	trot2	SE2
		trans12	SE2
trplot2	.plot	trplot2	.plot
rotx, roty, rotz	SO3.Rx, SO3.Ry, SO3.Rz	trotx, troty, trotz	SE3.Rx, SE3.Ry, SE3.Rz
		T = transl(v)	SE3(v)
eul2r, rpy2r	SO3.eul, SO3.rpy	eul2tr, rpy2tr	SE3.eul, SE3.rpy
angvec2r	SO3.angvec	angvec2tr	SE3.angvec
oa2r	SO3.oa	oa2tr	SE3.oa
		v = transl(T)	.t, .transl
tr2eul, tr2rpy	.toeul, .torpy	tr2eul, tr2rpy	.toeul, .torpy
tr2angvec	.toangvec	tr2angvec	.toangvec
trexp	SO3.exp	trexp	SE3.exp
trlog	.log	trlog	.log
trplot	.plot	trplot	.plot

Functions starting with dot are methods on the new objects. You can use them in functional form `toeul(R)` or in dot form `R.toeul()` or `R.toeul`. It's a personal preference. The trailing parentheses are not required if no arguments are passed, but it is a useful convention and reminder that you that you are invoking a method not reading a property. The old function `transl` appears twice since it maps a vector to a matrix as well as the inverse.

Table 2.3. Table of subsitutions from classic Toolbox functions that operate on and return a matrix, to the corresponding new classes and methods

to another. Rotation in 3-dimensions has subtlety and complexity and we have looked at various parameterizations such as Euler angles, roll-pitch-yaw angles and unit quaternions. Using Lie group theory we showed that rotation matrices, from the group **SO**(2) or **SO**(3), are the result of exponentiating skew-symmetric generator matrices. Similarly, homogeneous transformation matrices, from the group **SE**(2) or **SE**(3), are the result of exponentiating augmented skew-symmetric generator matrices. We have also introduced twists as a concise way of describing relative pose in terms of rotation around a screw axis, a notion that comes to us from screw theory and these twists are the unique elements of the generator matrices.

There are two important lessons from this chapter. The first is that there are *many* mathematical objects that can be used to represent pose and these are summarized in Table 2.4. There is no right or wrong – each has strengths and weaknesses and we typically choose the representation to suit the problem at hand. Sometimes we wish for a vectorial representation, perhaps for interpolation, in which case (x, y, θ) or (x, y, z, Γ) might be appropriate, but this representation cannot be easily compounded. Sometime we may only need to describe 3D rotation in which case Γ or $\overset{\circ}{q}$ is appropriate. Converting between representations is easy as shown in Tables 2.1 and 2.2.

The second lesson is that coordinate frames are your friend. The essential first step in many vision and robotics problems is to assign coordinate frames to all objects of interest, indicate the relative poses as a directed graph, and write down equations for the loops. Figure 2.20 shows you how to build a coordinate frame out of paper that you can pick up and rotate – making these ideas more tangible. Don't be shy, embrace the coordinate frame.

We now have solid foundations for moving forward. The notation has been defined and illustrated, and we have started our hands-on work with MATLAB. The next chapter discusses motion and coordinate frames that change with time, and after that we are ready to move on and discuss robots.

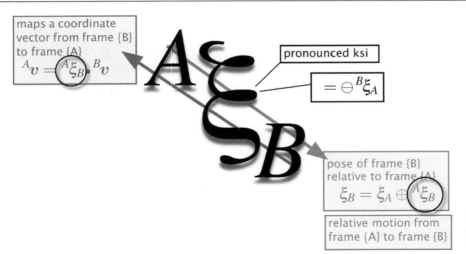

maps a coordinate vector from frame {B} to frame {A}

$$^A v = {}^A_B \xi \cdot {}^B v$$

pronounced ksi

$$= \ominus\, {}^B \xi_A$$

pose of frame {B} relative to frame {A}

$$\xi_B = \xi_A \oplus {}^A \xi_B$$

relative motion from frame {A} to frame {B}

Fig. 2.19.
Everything you need to know about pose

	2D		Composition	3D	Composition
Position	2-vector		+	3-vector	+
Orientation	Angle 3 × 3 rotation matrix		+ *	3 angles $\boldsymbol{\Gamma}$: Euler, RPY, etc. 2 vectors: OA (angle, vector) UnitQuaternion $\mathring{q}$ 3 × 3 rotation matrix	⊗ ⊗ ⊗ * *
Pose	(angle, 2-vector) 4 × 4 transformation matrix		⊗ *	(3 angles, 3-vector) (3-vector, UnitQuaternion) 4 × 4 transformation matrix	⊗ ☺ *

Toolbox composition operators are shown in blue. Composition operators shown in red are ⊗ difficult to implement, ☺ less difficult to implement.

Table 2.4. Summary of the various concrete representations of pose ξ introduced in this chapter

Further Reading

The treatment in this chapter is a hybrid mathematical and graphical approach that covers the 2D and 3D cases by means of abstract representations and operators which are later made tangible. The standard robotics textbooks such as Kelly (2013), Siciliano et al. (2009), Spong et al. (2006), Craig (2005), and Paul (1981) all introduce homogeneous transformation matrices for the 3-dimensional case but differ in their approach. These books also provide good discussion of the other representations such as angle-vector and 3-angle representations. Spong et al. (2006, sect. 2.5.1) have a good discussion of singularities. The book Lynch and Park (2017) covers the standard matrix approaches but also introduces twists and screws. Siegwart et al. (2011) explicitly cover the 2D case in the context of mobile robotics.

Quaternions are discussed in Kelly (2013) and briefly in Siciliano et al. (2009). The book by Kuipers (1999) is a very readable and comprehensive introduction to quaternions. Quaternion interpolation is widely used in computer graphics and animation and the classic paper by Shoemake (1985) is very readable introduction to this topic. The first publication about quaternions for robotics is probably Taylor (1979), and followed up in subsequent work by Funda (1990).

You will encounter a wide variety of different notation for rotations and transformations in textbooks and research articles. This book uses $^A T_B$ to denote a transform giving the pose of frame {B} with respect to frame {A}. A common alternative notation is T_B^A or even $_B^A T$. To denote points this book uses $^A p_B$ to denote a vector from the origin of frame {A} to the point **B** whereas others use p_B^A, or even $^C p_B^A$ to denote a vector

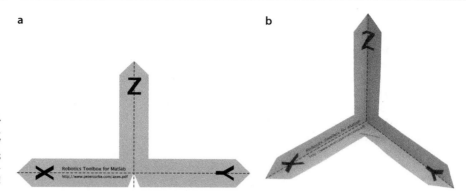

a

b

Fig. 2.20.
Build your own coordinate frame. **a** Get the PDF file from http://www.petercorke.com/axes.pdf; **b** cut it out, fold along the dotted lines and add a staple. Voila!

from the origin of frame $\{A\}$ to the point **B** but with respect to coordinate frame $\{C\}$. Twists can be written as either (v, ω) as in this book, or as (ω, v).

Historical and general. Hamilton and his supporters, including Peter Tait, were vigorous in defending Hamilton's precedence in inventing quaternions, and for opposing the concept of vectors which were then beginning to be understood and used. Rodrigues developed his eponymous formula in 1840 although Gauss discovered it in 1819 but, as usual, did not publish it. It was published in 1900. Quaternions had a tempestuous beginning. The paper by Altmann (1989) is an interesting description on this tussle of ideas, and quaternions have even been woven into fiction (Pynchon 2006).

Exercises

1. Explore the many options associated with `trplot`.
2. Animate a rotating cube
    a) Write a function to plot the edges of a cube centered at the origin.
    b) Modify the function to accept an argument which is a homogeneous transformation which is applied to the cube vertices before plotting.
    c) Animate rotation about the x-axis.
    d) Animate rotation about all axes.
3. Create a vector-quaternion class to describe pose and which supports composition, inverse and point transformation.
4. Create a 2D rotation matrix. Visualize the rotation using `trplot2`. Use it to transform a vector. Invert it and multiply it by the original matrix; what is the result? Reverse the order of multiplication; what is the result? What is the determinant of the matrix and its inverse?
5. Create a 3D rotation matrix. Visualize the rotation using `trplot` or `tranimate`. Use it to transform a vector. Invert it and multiply it by the original matrix; what is the result? Reverse the order of multiplication; what is the result? What is the determinant of the matrix and its inverse?
6. Compute the matrix exponential using the power series. How many terms are required to match the result shown to standard MATLAB precision?
7. Generate the sequence of plots shown in Fig. 2.12.
8. For the 3-dimensional rotation about the vector $[2, 3, 4]$ by 0.5 rad compute an **SO**(3) rotation matrix using: the matrix exponential functions `expm` and `trexp`, Rodrigues' rotation formula (code this yourself), and the Toolbox function `angvec2tr`. Compute the equivalent unit quaternion.
9. Create two different rotation matrices, in 2D or 3D, representing frames $\{A\}$ and $\{B\}$. Determine the rotation matrix $^A R_B$ and $^B R_A$. Express these as a rotation axis and angle, and compare the results. Express these as a twist.

10. Create a 2D or 3D homogeneous transformation matrix. Visualize the rigid-body displacement using `tranimate`. Use it to transform a vector. Invert it and multiply it by the original matrix, what is the result? Reverse the order of multiplication; what happens?

11. Create two different rotation matrices, in 2D or 3D, representing frames {A} and {B}. Determine the rotation matrix $^A R_B$ and $^B R_A$. Express these as a rotation axis and angle and compare the results. Express these as a twist.

12. Create three symbolic variables to represent roll, pitch and yaw angles, then use these to compute a rotation matrix using `rpy2r`. You may want to use the `simplify` function on the result. Use this to transform a unit vector in the z-direction. Looking at the elements of the rotation matrix devise an algorithm to determine the roll, pitch and yaw angles. Hint – find the pitch angle first.

13. Experiment with the `tripleangle` application in the Toolbox. Explore roll, pitch and yaw motions about the nominal attitude and at singularities.

14. If you have an iPhone or iPad download from the App Store the free "Euler Angles" app by École de Technologie Supérieure and experiment with it.

15. Using Eq. 2.24 show that $TT^{-1} = I$.

16. Is the inverse of a homogeneous transformation matrix equal to its transpose?

17. In Sect. 2.1.2.2 we rotated a frame about an arbitrary point. Derive the expression for computing RC that was given.

18. Explore the effect of negative roll, pitch or yaw angles. Does transforming from RPY angles to a rotation matrix then back to RPY angles give a different result to the starting value as it does for Euler angles?

19. From page 53 show that $e^x e^y \neq e^{x+y}$ for the case of matrices. Hint – expand the first few terms of the exponential series.

20. A camera has its z-axis parallel to the vector $[0, 1, 0]$ in the world frame, and its y-axis parallel to the vector $[0, 0, -1]$. What is the attitude of the camera with respect to the world frame expressed as a rotation matrix and as a unit quaternion?

3

Time and Motion

*The only reason for time is
so that everything doesn't happen at once*
Albert Einstein

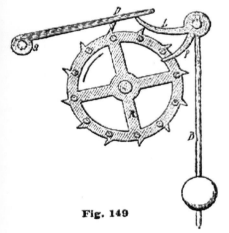

Fig. 149

In the previous chapter we learned how to describe the pose of objects in 2- or 3-dimensional space. This chapter extends those concepts to poses that change as a function of time. Section 3.1 introduces the derivative of time-varying position, orientation and pose and relates that to concepts from mechanics such as velocity and angular velocity. Discrete-time approximations to the derivatives are covered which are useful for computer implementation of algorithms such as inertial navigation. Section 3.2 is a brief introduction to the dynamics of objects moving under the influence of forces and torques and discusses the important difference between inertial and noninertial reference frames.

Section 3.3 discusses how to generate a temporal sequence of poses, a trajectory, that smoothly changes from an initial pose to a final pose. For robots this could be the path of a robot gripper moving to grasp an object or the flight path of a flying robot. Section 3.4 brings many of these topics together for the important application of inertial navigation. We introduce three common types of inertial sensor and learn how to how to use their measurements to update the estimate of pose for a moving object such as a robot.

3.1 Time-Varying Pose

In this section we discuss how to describe the rate of change of pose which has both a translational and rotational velocity component. The translational velocity is straightforward: it is the rate of change of the position of the origin of the coordinate frame. Rotational velocity is a little more complex.

3.1.1 Derivative of Pose

There are many ways to represent the orientation of a coordinate frame but most convenient for present purposes is the exponential form

$$^A\boldsymbol{R}_B(t) = e^{\left[^A\hat{\omega}(t)\right]_\times \theta(t)} \in \mathbf{SO}(3)$$

where the rotation is described by a rotational axis $^A\hat{\omega}(t)$ defined with respect to frame $\{A\}$ and a rotational angle $\theta(t)$, and where $[\cdot]_\times$ is a skew-symmetric matrix.

At an instant in time t we will assume that the axis has a fixed direction and the frame is rotating around the axis. The derivative with respect to time is

$$^A\dot{\boldsymbol{R}}_B(t) = \left[^A\hat{\boldsymbol{\omega}}(t)\right]_\times \dot{\theta} \, e^{\left[^A\hat{\omega}(t)\right]_\times \theta(t)} \in \mathbb{R}^{3\times3}$$

$$= \left[^A\hat{\boldsymbol{\omega}}(t)\right]_\times \dot{\theta} \, ^A\boldsymbol{R}_B(t)$$

which we write succinctly as

$$
{}^{A}\dot{\boldsymbol{R}}_{B} = \left[{}^{A}\boldsymbol{\omega}\right]_{\times} {}^{A}\boldsymbol{R}_{B} \in \mathbb{R}^{3\times3} \tag{3.1}
$$

where ${}^{A}\boldsymbol{\omega} = {}^{A}\hat{\boldsymbol{\omega}}\dot{\theta}$ is the angular velocity in frame $\{A\}$. This is a vector quantity ${}^{A}\boldsymbol{\omega} = (\omega_{x}, \omega_{y}, \omega_{z})$ that defines the *instantaneous* axis and rate of rotation. The direction of ${}^{A}\boldsymbol{\omega}$ is parallel to the axis about which the coordinate frame is rotating at a particular instant of time, and the magnitude $\|{}^{A}\boldsymbol{\omega}\|$ is the rate of rotation about that axis.▸ Note that the derivative of a rotation matrix is not a rotation matrix, it is a general 3×3 matrix.

> For a tumbling object the axis of rotation changes with time.

Consider now that angular velocity is expressed in frame $\{B\}$ and we know that

$$
{}^{A}\boldsymbol{\omega} = {}^{A}\boldsymbol{R}_{B}{}^{B}\boldsymbol{\omega}
$$

and using the identity $[\boldsymbol{Av}]_{\times} = \boldsymbol{A}[\boldsymbol{v}]_{\times}\boldsymbol{A}^{T}$ it follows that

$$
{}^{A}\dot{\boldsymbol{R}}_{B} = {}^{A}\boldsymbol{R}_{B}\left[{}^{B}\boldsymbol{\omega}\right]_{\times} \in \mathbb{R}^{3\times3} \tag{3.2}
$$

The derivative of a unit quaternion, the quaternion equivalent of Eq. 3.1, is defined as

$$
{}^{A}\dot{\boldsymbol{q}}_{B} = \tfrac{1}{2}{}^{A}\mathring{\boldsymbol{\omega}} \circ {}^{A}\mathring{\boldsymbol{q}}_{B} = \tfrac{1}{2}{}^{A}\mathring{\boldsymbol{q}}_{B} \circ {}^{B}\mathring{\boldsymbol{\omega}} \in \mathbb{H} \tag{3.3}
$$

where $\mathring{\omega}$ is a pure quaternion formed from the angular velocity vector. These are implemented by the Toolbox methods `dot` and `dotb` respectively. The derivative of a unit-quaternion is not a unit-quaternion, it is a regular quaternion which can also be considered as a 4-vector.

The derivative of pose can be determined by expressing pose as a homogeneous transformation matrix

$$
\xi \sim {}^{A}\boldsymbol{T}_{B} = \begin{pmatrix} {}^{A}\boldsymbol{R}_{B} & {}^{A}\boldsymbol{t}_{B} \\ 0_{1\times3} & 1 \end{pmatrix}
$$

and taking the derivative with respect to time and substituting Eq. 3.1 gives

$$
\dot{\xi} \sim {}^{A}\dot{\boldsymbol{T}}_{B} = \begin{pmatrix} {}^{A}\dot{\boldsymbol{R}}_{B} & {}^{A}\dot{\boldsymbol{t}}_{B} \\ 0_{1\times3} & 0 \end{pmatrix} = \begin{pmatrix} \left[{}^{A}\boldsymbol{\omega}\right]_{\times}{}^{A}\boldsymbol{R}_{B} & {}^{A}\dot{\boldsymbol{t}}_{B} \\ 0_{1\times3} & 0 \end{pmatrix}
$$

The rate of change can be described in terms of the current orientation ${}^{A}\boldsymbol{R}_{B}$ and *two* velocities. The linear or translational velocity $\boldsymbol{v} = {}^{A}\dot{\boldsymbol{t}}_{B}$ is the velocity of the origin of $\{B\}$ with respect to $\{A\}$. The angular velocity ${}^{A}\boldsymbol{\omega}_{B}$ we have already introduced. We can combine these two velocity vectors to create the spatial velocity vector

$$
{}^{A}\boldsymbol{\nu}_{B} = \left({}^{A}\boldsymbol{v}_{B}, {}^{A}\boldsymbol{\omega}_{B}\right) \in \mathbb{R}^{6}
$$

which is the instantaneous velocity of frame $\{B\}$ with respect to $\{A\}$.

Every point in the body has the same angular velocity. Knowing that, plus the translational velocity vector of any point is enough to fully describe the instantaneous motion of a rigid body. It is common to place $\{B\}$ at the body's center of mass.

3.1.2   Transforming Spatial Velocities

The velocity of a moving body can be expressed with respect to a world reference frame $\{A\}$ or the moving body frame $\{B\}$ as shown in Fig. 3.1. The spatial velocities are linearly related by

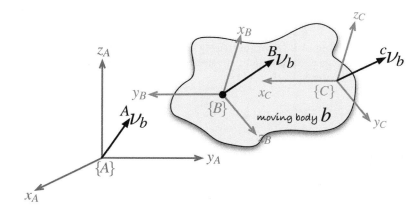

Fig. 3.1.
Representing the spatial velocity of a moving body b with respect to various coordinate frames. Note that ν is a 6-dimensional vector

$$
{}^{A}\nu = \begin{pmatrix} {}^{A}\boldsymbol{R}_{B} & \boldsymbol{0}_{3\times3} \\ \boldsymbol{0}_{3\times3} & {}^{A}\boldsymbol{R}_{B} \end{pmatrix} {}^{B}\nu = {}^{A}\boldsymbol{J}_{B}\!\left({}^{A}\xi_{B}\right){}^{B}\nu \tag{3.4}
$$

where ${}^{A}\xi_{B} \sim ({}^{A}\boldsymbol{R}_{B}, {}^{A}\boldsymbol{t}_{B})$ and ${}^{A}\boldsymbol{J}_{B}(\cdot)$ is a Jacobian or interaction matrix. For example, we can define a body-fixed frame and a spatial velocity in that frame

```
>> TB = SE3(1, 2, 0) * SE3.Rz(pi/2);
>> vb = [0.2 0.3 0 0 0 0.5]';
```

and the spatial velocity in the world frame is

```
>> va = TB.velxform * vb;
>> va'
ans =
    0.2000    0.0000    0.3000         0   -0.5000    0.0000
```

For the case where frame $\{C\}$ is also on the moving body the transformation becomes

$$
{}^{C}\nu = \begin{pmatrix} {}^{C}\boldsymbol{R}_{B} & \left[{}^{C}\boldsymbol{t}_{B}\right]_{\times}{}^{C}\boldsymbol{R}_{B} \\ \boldsymbol{0}_{3\times3} & {}^{C}\boldsymbol{R}_{B} \end{pmatrix} {}^{B}\nu = \mathrm{Ad}\!\left({}^{C}\xi_{B}\right){}^{B}\nu
$$

and involves the adjoint matrix of the relative pose which is discussed in Appendix D. Continuing the example above we will define an additional frame $\{C\}$ relative to frame $\{B\}$

```
>> TBC = SE3(0, 0.4, 0);
```

To determine velocity at the origin of this frame we first compute ${}^{C}\xi_{B}$

```
>> TCB = inv(TBC);
```

and the velocity in frame $\{C\}$ is

```
>> vc = TBC.Ad * vb;
>> vc'
ans =
         0    0.3000         0         0         0    0.5000
```

which has zero velocity in the x_C-direction since the rotational and translational velocity components cancel out.

Lynch and Park (2017) use the term velocity twist while Murray et al. 1994 call this a spatial velocity.

The scalar product of a velocity twist and a wrench represents power.

Some texts introduce a *velocity twist* $\boldsymbol{V}$ which is different to the spatial velocity introduced above.◄ The velocity twist of a body-fixed frame $\{B\}$ is ${}^{B}\boldsymbol{V} = ({}^{B}\boldsymbol{v}, {}^{B}\boldsymbol{\omega})$ which has a translational and rotational velocity component but ${}^{B}\boldsymbol{v}$ is the body-frame velocity of an imaginary point rigidly attached to the body and located at the world frame origin. The body- and world-frame velocity twists are related by the adjoint matrix rather than Eq. 3.4. The velocity twist is the dual of the wrench described in Sect. 3.2.2.◄

3.1.3   Incremental Rotation

The physical meaning of $\dot{R}$ is not intuitively obvious – it is simply the way that the elements of R change with time. To gain some insight we consider a first-order approximation to the derivative▶

$$\dot{R} \approx \frac{R\langle t+\delta_t\rangle - R\langle t\rangle}{\delta_t} \in \mathbb{R}^{3\times3} \qquad (3.5)$$

The only valid operator for the group $\mathsf{SO}(n)$ is composition $\oplus$, so the result of subtraction cannot belong to the group. The result is a 3×3 matrix of element-wise differences. Groups are introduced in Appendix D.

Consider an object whose body frames $\{B\}$ at two consecutive timesteps are related by a small rotation $^B R_\Delta$ expressed in the body frame

$$R_B\langle t+\delta_t\rangle = R_B\langle t\rangle\, ^B R_\Delta$$

We substitute Eq. 3.2 into 3.5 and rearrange to obtain

$$^B R_\Delta \approx \delta_t \left[^B\omega\right]_\times + I_{3\times3} \qquad (3.6)$$

which says that an infinitesimally small rotation can be approximated by the sum of a skew-symmetric matrix and an identity matrix.▶ For example

This is the first two terms of the Rodrigues' rotation formula on, Eq. 2.18, when $\theta = \delta_t\omega$.

```
>> rotx(0.001)
ans =
    1.0000         0         0
         0    1.0000   -0.0010
         0    0.0010    1.0000
```

Equation 3.6 directly relates rotation between timesteps to the angular velocity. Rearranging it allows us to compute the approximate angular velocity vector

$$\omega \approx \frac{1}{\delta_t} \vee_\times \left(R_B\langle t\rangle^T R_B\langle t+\delta_t\rangle - I_{3\times3}\right)$$

from two consecutive rotation matrices where $\vee_\times(\cdot)$ is the inverse skew-symmetric matrix operator such that if $S = [v]_\times$ then $v = \vee_\times(S)$. Alternatively, if the angular velocity in the body frame is known we can approximately update the rotation matrix

$$R_B\langle t+\delta_t\rangle \approx R_B\langle t\rangle + \delta_t R_B\langle t\rangle[\omega]_\times \qquad (3.7)$$

which is cheap to compute, involves no trigonometric operations, and is key to inertial navigation systems which we discuss in Sect. 3.4.

The only valid operator for the group $\mathsf{SO}(n)$ is composition $\oplus$, so the result of addition cannot be within the group. The result is a general 3×3 matrix.

> Adding any nonzero matrix to a rotation matrix results in a matrix that is *not* a rotation matrix.◀ However if the increment is sufficiently small, that is the angular velocity and/or sample time is small,▶ the result will be close to orthonormal and we can *straighten it up*. The resulting matrix should be normalized, as discussed in Sect. 2.3.1, to make it a proper rotation matrix. This is a common approach when implementing inertial navigation systems on low-end computing hardware.

Which is why inertial navigation systems operate at a high sample rate and δ_t is small.

We can also approximate the quaternion derivative by a first-order difference▶

$$\dot{q} \approx \frac{\mathring{q}\langle k+1\rangle - \mathring{q}\langle k\rangle}{\delta_t} \in \mathbb{H}$$

which combined with Eq. 3.3 gives us the approximation

Similar to the case for $\mathsf{SO}(n)$, addition and subtraction are not operators for the unit-quaternion group $\mathbb{S}^3$ so the result will be a quaternion $q \in \mathbb{H}$ for which addition and substraction are permitted. The Toolbox supports this with overloaded operators + and − and appropriate object class conversions.

$$\mathring{q}\langle k+1 \rangle \approx \mathring{q}\langle k \rangle + \frac{\delta_t}{2} \mathring{\omega} \circ \mathring{q}\langle k \rangle \tag{3.8}$$

which is even cheaper to compute than the rotation matrix approach. Adding a nonzero vector to a unit-quaternion results in a nonunit quaternion but if the angular velocity and/or sample time is small then the approximation is reasonable. Normalizing the result to create a unit-quaternion is computationally cheaper than normalizing a rotation matrix, as discussed in Sect. 2.3.1.

3.1.4 Incremental Rigid-Body Motion

Consider two poses ξ_1 and ξ_2 which differ infinitesimally and are related by

$$\xi_2 = \xi_1 \oplus \xi_\Delta$$

where $\xi_\Delta = \ominus \xi_1 \oplus \xi_2$. In homogeneous transformation matrix form

$$\xi_\Delta \sim T_\Delta = \begin{pmatrix} R_\Delta & t_\Delta \\ 0_{1\times3} & 1 \end{pmatrix}$$

where t_Δ is an incremental displacement and R_Δ is an incremental rotation matrix which will be skew symmetric with only three unique elements $\vee_x(R_\Delta - I_{3\times3})$ plus an identity matrix. The incremental rigid-body motion can therefore be described by just six parameters

$$\Delta(\xi_1, \xi_2) \mapsto \Delta_\xi \in \mathbb{R}^6$$

This is useful in optimization procedures that seek to minimize the error between two poses: we can choose the cost function $e = \|\Delta(\xi_1, \xi_2)\|$ which is equal to zero when $\xi_1 \equiv \xi_2$. This is very approximate when the poses are significantly different, but becomes ever more accurate as $\xi_1 \to \xi_2$.

where $\Delta_\xi = (\Delta_t, \Delta_R)$ can be considered as a spatial displacement. ◄ A body with constant spatial velocity ν for δ_t seconds undergoes a spatial displacement of $\Delta_\xi = \delta_t \nu$.

The inverse operator

$$\Delta^{-1}(\Delta_\xi) \mapsto \xi_\Delta \in \mathbf{SE(3)}$$

is given by

$$\xi_\Delta \sim T_\Delta = \begin{pmatrix} [\Delta_R]_\times + I_{3\times3} & \Delta_t \\ 0_{1\times3} & 1 \end{pmatrix}$$

The spatial displacement operator and its inverse are implemented by the Toolbox functions `tr2delta` and `delta2tr` respectively. These functions assume that the displacements are infinitesimal and become increasingly approximate with displacement magnitude.

Sir Isaac Newton (1642–1727) was an English mathematician and alchemist. He was Lucasian professor of mathematics at Cambridge, Master of the Royal Mint, and the thirteenth president of the Royal Society. His achievements include the three laws of motion, the mathematics of gravitational attraction, the motion of celestial objects and the theory of light and color (see page 287), and building the first reflecting telescope.

Many of these results were published in 1687 in his great 3-volume work "The Philosophiae Naturalis Principia Mathematica" (Mathematical principles of natural philosophy). In 1704 he published "Opticks" which was a study of the nature of light and color and the phenomena of diffraction. The SI unit of force is named in his honor. He is buried in Westminster Abbey, London.

3.2   Accelerating Bodies and Reference Frames

So far we have considered only the first derivative, the velocity of coordinate frames. However all motion is ultimately caused by a force or a torque which leads to acceleration and the consideration of dynamics.

3.2.1   Dynamics of Moving Bodies

For translational motion Newton's second law describes, in the inertial frame, the acceleration of a particle with position x and mass m

$$m\,{}^0\ddot{\boldsymbol{x}} = {}^0\boldsymbol{f} \tag{3.9}$$

due to the applied force f.

Rotational motion in $\mathbf{SO}(3)$ is described by Euler's equations of motion which relates the angular acceleration of the body in the body frame

$$ {}^B\boldsymbol{J}\,{}^B\dot{\boldsymbol{\omega}} + {}^B\boldsymbol{\omega} \times \left({}^B\boldsymbol{J}\,{}^B\boldsymbol{\omega} \right) = {}^B\boldsymbol{\tau} \tag{3.10}$$

to the applied torque or moment τ and a positive-definite rotational inertia matrix ${}^B\boldsymbol{J} \in \mathbb{R}^{3\times3}$.► Nonzero angular acceleration implies that angular velocity, the axis and/or angle of rotation, evolves over time.◄

Consider the motion of a tumbling object which we can easily simulate. We define an inertia matrix◄

```
>> J = [2 -1 0;-1 4 0;0 0 3];
```

and initial conditions for orientation and angular velocity

```
>> attitude = UnitQuaternion();
>> w = 0.2*[1 2 2]';
```

The simulation loop computes angular acceleration with Eq. 3.10, uses rectangular integration to obtain angular velocity and attitude, and then updates a graphical coordinate frame

```
>> dt = 0.05;
>> h = attitude.plot();
>> for t=0:dt:10
        wd = -inv(J) * (cross(w, J*w));
        w = w + wd*dt; attitude = attitude .* UnitQuaternion.omega(wd*dt);
        attitude.plot('handle', h); pause(dt)
   end
```

Notice that inertia has an associated reference frame, it is a matrix and its elements depend on the choice of the coordinate frame.

In the absence of torque a body generally rotates with a time-varying angular velocity – this is quite different to the linear velocity case. It is angular momentum $h = J\omega$ in the inertial frame that is constant.

The matrix must be positive definite, that is symmetric and all its eigenvalues are positive.

The rotational inertia of a body that moves in $\mathbf{SE}(3)$ is represented by the 3×3 symmetric matrix

$$J = \begin{pmatrix} J_{xx} & J_{xy} & J_{xz} \\ J_{xy} & J_{yy} & J_{yz} \\ J_{xz} & J_{yz} & J_{zz} \end{pmatrix}$$

The diagonal elements are the positive moments of inertia, and the off-diagonal elements are products of inertia. Only six of these nine elements are unique: three moments and three products of inertia. The products of inertia are all zero if the object's mass distribution is symmetrical with respect to the coordinate frame.

3.2.2 Transforming Forces and Torques

The spatial velocity is a vector quantity that represents translational and rotational velocity. In a similar fashion we can combine translational force and rotational torque into a 6-vector that is called a wrench $W = (f_x, f_y, f_z, m_x, m_y, m_z) \in \mathbb{R}^6$. A wrench ^BW is defined with respect to the coordinate frame $\{B\}$ and applied at the origin of that frame.

The wrench ^CW is equivalent if it causes the same motion of the body when applied to the origin of coordinate frame $\{C\}$ and defined with respect to $\{C\}$. The wrenches are related by

$$^CW = \begin{pmatrix} ^B\boldsymbol{R}_C & \left[^B\boldsymbol{t}_C \right]_\times {}^B\boldsymbol{R}_C \\ \mathbf{0}_{3\times3} & ^B\boldsymbol{R}_C \end{pmatrix}^T {}^BW = \mathrm{Ad}\left(^B\xi_C \right)^T {}^BW \tag{3.11}$$

which is similar to the spatial velocity transform of Eq. 3.4 but uses the transpose of the adjoint of the *inverse* relative pose.

Continuing the MATLAB example from page 65 we define a wrench with respect to frame $\{B\}$ with forces of 3 and 4 Nm in the x- and y-directions respectively

```
>> WB = [3 4 0 0 0 0]';
```

The equivalent wrench in frame $\{C\}$ would be

```
>> WC = TBC.Ad' * WB;
>> WC'
ans =
    3.0000    4.0000         0         0         0    1.2000
```

which is the same forces as applied at $\{B\}$ *plus* a torque of 1.2 Nm about the z-axis to counter the moment due to the application of the x-axis force along a different line of action.

3.2.3 Inertial Reference Frame

The term *inertial reference frame* is frequently used in robotics and it is crisply defined as "*a reference frame that is not accelerating or rotating*".

Consider a particle P at rest with respect to a stationary reference frame $\{0\}$. Frame $\{B\}$ is moving with constant velocity ^0v_B relative to frame $\{0\}$. From the perspective of $\{B\}$ the particle would be moving at constant velocity, in fact $^Bv_P = -^0v_B$. The particle is not accelerating and obeys Newton's first law "*that in the absence of an applied force a particle moves at a constant velocity*". Frame $\{B\}$ is therefore also an inertial reference frame.

Now imagine that frame $\{B\}$ is accelerating at a constant acceleration ^0a_B with respect to $\{0\}$. From the perspective of $\{B\}$ the particle appear to be accelerating, in fact $^Ba_P = -^0a_B$ and this violates Newton's first law. An observer in frame $\{B\}$ who was aware of Newton's theories might invoke some magical force to explain what they observe. We call such a force a fictitious, apparent, pseudo, inertial or d'Alembert force – they only exist in an accelerating or noninertial reference frame. This accelerating

Gaspard-Gustave de Coriolis (1792–1843) was a French mathematician, mechanical engineer and scientist. Born in Paris, in 1816 he became a tutor at the École Polytechnique where he carried out experiments on friction and hydraulics and later became a professor at the École des Ponts and Chaussées (School of Bridges and Roads). He extended ideas about kinetic energy and work to rotating systems and in 1835 wrote the famous paper *Sur les équations du mouvement relatif des systèmes de corps* (On the equations of relative motion of a system of bodies) which dealt with the transfer of energy in rotating systems such as waterwheels. In the late 19th century his ideas were picked up by the meteorological community to incorporate effects due to the Earth's rotation. He is buried in Paris's Montparnasse Cemetery.

frame {B} is *not* an inertial reference frame. In Newtonian mechanics, gravity is considered a real body force *mg*– a free object will accelerate relative to the inertial frame.▶

An everyday example of a noninertial reference frame is an accelerating car or airplane. Inside an accelerating vehicle we observe fictitious forces pushing objects around in a way that is not explained by Newton's law in an inertial reference frame. We also experience real forces acting on our body which, in this case, are provided by the seat and the restraint.

For a rotating reference frame things are more complex still. Imagine that you and a friend are standing on a large rotating turntable, and throwing a ball back and forth. You will observe that the ball follows a curved path in space.▶ As a Newton-aware observer in this noninertial reference frame you would have to resort to invoking some magical force that explains why flying objects follow curved paths.

If the reference frame {B} is rotating with angular velocity ω about its origin then Newton's second law Eq. 3.9 becomes

$$m\left(\underbrace{{}^B\dot{v} + \omega \times \left(\omega \times {}^B p\right)}_{\text{centripetal}} + \underbrace{2\omega \times {}^B v}_{\text{Coriolis}} + \underbrace{\frac{d\omega}{dt} \times {}^B p}_{\text{Euler}}\right) = {}^0 f$$

with three *new* acceleration terms. Centripetal acceleration always acts inward toward the origin. If the point is moving then Coriolis acceleration will be normal to its velocity. If rotational velocity is time varying then Euler acceleration will be normal to the position vector. Frequently the centripetal term is moved to the right-hand side in which case it becomes a fictitious outward centrifugal force. This complexity is symptomatic of being in a noninertial reference frame, and another definition of an inertial frame is one in which the *"physical laws hold good in their simplest form"*.▶

In robotics the term inertial frame and world coordinate frame tend to be used loosely and interchangeably to indicate a frame fixed to some point on the Earth. This is to distinguish it from the body-frame attached to the robot or vehicle. The surface of the Earth is an approximation of an inertial reference frame – the effect of the Earth's rotation is a finite acceleration less than 0.04 m s^{-2} due to centripetal acceleration. From the perspective of an Earth-bound observer a moving body will experience Coriolis acceleration. Both effects are small,▶ dependent on latitude, and typically ignored.

3.3   Creating Time-Varying Pose

In robotics we often need to generate a time-varying pose that moves smoothly in translation and rotation. A path is a spatial construct – a locus in space that leads from an initial pose to a final pose. A trajectory is a path with specified timing. For example there is a path from A to B, but there is a trajectory from A to B in 10 s or at 2 m s^{-1}.

An important characteristic of a trajectory is that it is *smooth* – position and orientation vary smoothly with time. We start by discussing how to generate smooth trajectories in one dimension. We then extend that to the multi-dimensional case and then to piecewise-linear trajectories that visit a number of intermediate points without stopping.

3.3.1   Smooth One-Dimensional Trajectories

We start our discussion with a scalar function of time. Important characteristics of this function are that its initial and final value are specified and that it is *smooth*. Smoothness in this context means that its first few temporal derivatives are continuous. Typically velocity and acceleration are required to be continuous and sometimes also the derivative of acceleration or jerk.

Albert Einstein's equivalence principle is that *"we assume the complete physical equivalence of a gravitational field and a corresponding acceleration of the reference system"* – we are unable to distinguish between gravity and being on a rocket accelerating at 1 *g* far from the gravitational influence of any celestial object.

Of course if we look down onto the turntable from an inertial reference frame the ball is moving in a straight line.

Einstein, *"The foundation of the general theory of relativity"*.

Coriolis acceleration is significant for weather systems and meteorological prediction but below the sensitivity of low-cost sensors.

An obvious candidate for such a function is a polynomial function of time. Polynomials are simple to compute and can easily provide the required smoothness and boundary conditions. A quintic (fifth-order) polynomial is often used

$$s(t) = At^5 + Bt^4 + Ct^3 + Dt^2 + Et + F \qquad (3.12)$$

where time $t \in [0, T]$. The first- and second-derivatives are also smooth polynomials

$$\dot{s}(t) = 5At^4 + 4Bt^3 + 3Ct^2 + 2D + E \qquad (3.13)$$

$$\ddot{s}(t) = 20At^3 + 12Bt^2 + 6Ct + 2D \qquad (3.14)$$

Time	s	$\dot{s}$	$\ddot{s}$
$t = 0$	s_0	$\dot{s}_0$	$\ddot{s}_0$
$t = T$	s_T	$\dot{s}_T$	$\ddot{s}_T$

The trajectory has defined boundary conditions for position, velocity and acceleration◀ and frequently the velocity and acceleration boundary conditions are all zero.

Writing Eq. 3.12 to Eq. 3.14 for the boundary conditions $t = 0$ and $t = T$ gives six equations which we can write in matrix form as

$$\begin{pmatrix} s_0 \\ s_T \\ \dot{s}_0 \\ \dot{s}_T \\ \ddot{s}_0 \\ \ddot{s}_T \end{pmatrix} = \begin{pmatrix} 0 & 0 & 0 & 0 & 0 & 1 \\ T^5 & T^4 & T^3 & T^2 & T & 1 \\ 0 & 0 & 0 & 0 & 1 & 0 \\ 5T^4 & 4T^3 & 3T^2 & 2T & 1 & 0 \\ 0 & 0 & 0 & 2 & 0 & 0 \\ 20T^3 & 12T^2 & 6T & 2 & 0 & 0 \end{pmatrix} \begin{pmatrix} A \\ B \\ C \\ D \\ E \\ F \end{pmatrix}$$

This is the reason for choice of quintic polynomial. It has six coefficients that enable it to meet the six boundary conditions on initial and final position, velocity and acceleration.

Since the matrix is square◀ we can solve for the coefficient vector (A, B, C, D, E, F) using standard linear algebra methods such as the MATLAB \-operator. For a quintic polynomial acceleration will be a smooth cubic polynomial, and jerk will be a parabola.

The Toolbox function `tpoly` generates a quintic polynomial trajectory as described by Eq. 3.12. For example

```
>> tpoly(0, 1, 50);
```

Fig. 3.2. Quintic polynomial trajectory. From top to bottom is position, velocity and acceleration versus time step. **a** With zero-velocity boundary conditions, **b** initial velocity of 0.5 and a final velocity of 0. Note that velocity and acceleration are in units of timestep not seconds

generates a polynomial trajectory and plots it, along with the corresponding velocity and acceleration, as shown in Fig. 3.2a. We can get these values into the workspace by providing output arguments

```
>> [s,sd,sdd] = tpoly(0, 1, 50);
```

where `s`, `sd` and `sdd` are respectively the trajectory, velocity and acceleration – each a 50×1 column vector. We observe that the initial and final velocity and acceleration

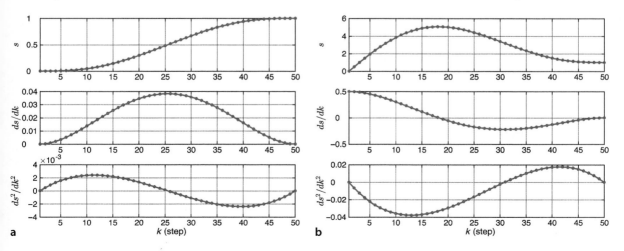

a   b

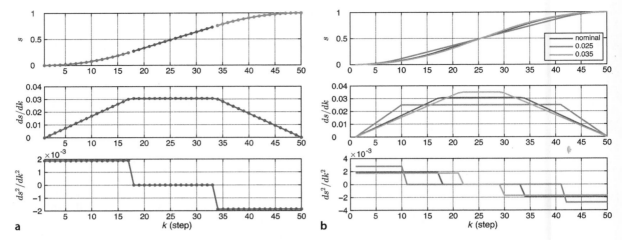

are all zero – the default value. The initial and final velocities can be set to nonzero values

```
>> tpoly(0, 1, 50, 0.5, 0);
```

in this case, an initial velocity of 0.5 and a final velocity of 0. The results shown in Fig. 3.2b illustrate an important problem with polynomials. The nonzero initial velocity causes the polynomial to overshoot the terminal value – it peaks at 5 on a trajectory from 0 to 1.

Another problem with polynomials, a very practical one, can be seen in the middle graph of Fig. 3.2a. The velocity peaks at $k = 25$ which means that for most of the time the velocity is far less than the maximum. The mean velocity

```
>> mean(sd) / max(sd)
ans =
    0.5231
```

is only 52% of the peak so we are not using the motor as fully as we could. A real robot joint has a well defined maximum velocity and for minimum-time motion we want to be operating at that maximum for as much of the time as possible. We would like the velocity curve to be *flatter* on top.

A well known alternative is a hybrid trajectory which has a constant velocity segment with polynomial segments for acceleration and deceleration. Revisiting our first example the hybrid trajectory is

```
>> lspb(0, 1, 50);
```

where the arguments have the same meaning as for `tpoly` and the trajectory is shown in Fig. 3.3a. The trajectory comprises a linear segment (constant velocity) with parabolic blends, hence the name `lspb`. The term blend is commonly used to refer to a trajectory segment that smoothly joins linear segments. As with `tpoly` we can also return the trajectory and its velocity and acceleration

```
>> [s,sd,sdd] = lspb(0, 1, 50);
```

This type of trajectory is also referred to as trapezoidal due to the shape of the velocity curve versus time, and is commonly used in industrial motor drives.▶

The function `lspb` has *chosen* the velocity of the linear segment to be

```
>> max(sd)
ans =
    0.0306
```

but this can be overridden by specifying it as a fourth input argument

Fig. 3.3. Linear segment with parabolic blend (LSPB) trajectory: **a** default velocity for linear segment; **b** specified linear segment velocity values

The trapezoidal trajectory is smooth in velocity, but not in acceleration.

```
>> s = lspb(0, 1, 50, 0.025);
>> s = lspb(0, 1, 50, 0.035);
```

The trajectories for these different cases are overlaid in Fig. 3.3b. We see that as the velocity of the linear segment increases its duration decreases and ultimately its duration would be zero. In fact the velocity cannot be chosen arbitrarily◄, too high or too low a value for the maximum velocity will result in an infeasible trajectory and the function returns an error.

The system has one design degree of freedom. There are six degrees of freedom (blend time, three parabolic coefficients and two linear coefficients) and five constraints (total time, initial and final position and velocity).

3.3.2 Multi-Dimensional Trajectories

Most useful robots have more than one axis of motion and it is quite straightforward to extend the smooth scalar trajectory to the vector case. In terms of configuration space (Sect. 2.3.5), these axes of motion correspond to the dimensions of the robot's configuration space – to its degrees of freedom. We represent the robot's configuration as a vector $q \in \mathbb{R}^N$ where N is the number of degrees of freedom. The configuration of a 3-joint robot would be its joint angles $q = (q_1, q_2, q_3)$. The configuration vector of wheeled mobile robot might be its position $q = (x, y)$ or its position and heading angle $q = (x, y, \theta)$. For a 3-dimensional body that had an orientation in $\mathbf{SO}(3)$ we would use a configuration vector $q = (\theta_r, \theta_p, \theta_y)$ or for a pose in $\mathbf{SE}(3)$ we would use $q = (x, y, z, \theta_r, \theta_p, \theta_y)$◄. In all these cases we would require smooth multi-dimensional motion from an initial configuration vector to a final configuration vector.

Or an equivalent 3-angle representation.

In the Toolbox this is achieved using the function `mtraj` and to move from configuration $(0, 2)$ to $(1, -1)$ in 50 steps we write

```
>> q = mtraj(@lspb, [0 2], [1 -1], 50);
```

which results in a 50 × 2 matrix `q` with one row per time step and one column per axis. The first argument is a handle to a function that generates a *scalar* trajectory, `@lspb` as in this case or `@tpoly`. The trajectory for the `@lspb` case

```
>> plot(q)
```

is shown in Fig. 3.4.

If we wished to create a trajectory for 3-dimensional pose we might consider converting a pose T to a 6-vector by a command like

```
q = [T1.t' T1.torpy]
```

though as we shall see later interpolation of 3-angle representations has some limitations.

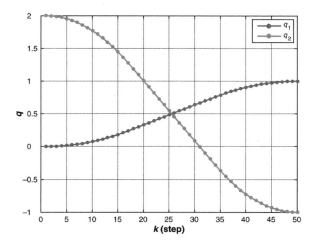

Fig. 3.4.
Multi-dimensional motion.
q_1 varies from $0 \rightarrow 1$ and
q_2 varies from $2 \rightarrow -1$

3.3.3   Multi-Segment Trajectories

In robotics applications there is often a need to move smoothly along a path through one or more intermediate or *via* points without stopping. This might be to avoid obstacles in the workplace, or to perform a task that involves following a piecewise continuous trajectory such as welding a seam or applying a bead of sealant in a manufacturing application.

To formalize the problem consider that the trajectory is defined by M configurations q_k, $k \in [1, M]$ and there are $M - 1$ motion segments. As in the previous section $q_k \in \mathbb{R}^N$ is a *vector* representation of configuration.

The robot starts from q_1 at rest and finishes at q_M at rest, but moves through (or close to) the intermediate configurations without stopping. The problem is over constrained and in order to attain continuous velocity we surrender the ability to reach each intermediate configuration. This is easiest to understand for the 1-dimensional case shown in Fig. 3.5. The motion comprises linear motion segments with polynomial blends, like `lspb`, but here we choose quintic polynomials because they are able to match boundary conditions on position, velocity and acceleration at their start and end points.

The first segment of the trajectory accelerates from the initial configuration q_1 and zero velocity, and joins the line heading toward the second configuration q_2. The blend time is set to be a constant t_{acc} and $t_{acc} / 2$ before reaching q_2 the trajectory executes a polynomial blend, of duration t_{acc}, onto the line from q_2 to q_3, and the process repeats. The constant velocity $\dot{q}_k$ can be specified for each segment. The average acceleration during the blend is

$$\ddot{q} = \frac{\dot{q}_{k+1} - \dot{q}_k}{t_{acc}}$$

If the maximum acceleration capability of the axis is known then the minimum blend time can be computed.▶

On a particular motion segment each axis will have a different distance to travel and traveling at its maximum speed there will be a minimum time before it can reach its goal. The first step in planning a segment is to determine which axis will be the slowest to complete the segment, based on the distance that each axis needs to travel for the segment and its maximum achievable velocity. From this the duration of the segment can be computed and then the required velocity of each axis. This ensures that all axes reach the next target q_k at the *same time*.

The Toolbox function `mstraj` generates a multi-segment multi-axis trajectory based on a matrix of via points. For example 2-axis motion via the corners of a rotated square can be generated by

```
>> via = SO2(30, 'deg') * [-1 1; 1 1; 1 -1; -1 -1]';
>> q0 = mstraj(via(:,[2 3 4 1])', [2,1], [], via(:,1)', 0.2, 0);
```

The first argument is the matrix of via points, each row is the coordinates of a point. The remaining arguments are respectively: a vector of maximum speeds per axis, a vector of

The real limit of the axis will be its peak, rather than average, acceleration. The peak acceleration for the blend can be determined from Eq. 3.14 once the quintic coefficients are known.

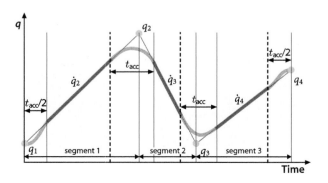

Fig. 3.5.
Notation for multi-segment trajectory showing four points and three motion segments. Blue indicates constant velocity motion, red indicates regions of acceleration

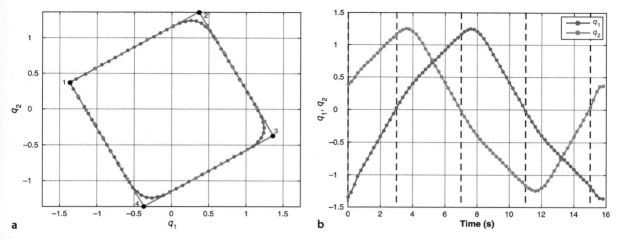

Fig. 3.6. Multi-segment multi-axis trajectories: **a** configuration of robot (tool position) for acceleration time of $t_{acc} = 0$ s (red) and $t_{acc} = 2$ s (blue), the via points are indicated by solid black markers; **b** configuration versus time with segment transitions ($t_{acc} = 2$ s) indicated by dashed black lines. The discrete-time points are indicated by dots

Only one of the maximum axis speed or time per segment can be specified, the other is set to MATLAB's empty matrix [ ] .

Acceleration time if given is rounded up internally to a multiple of the time step.

durations for each segment,➤ the initial configuration, the sample time step, and the acceleration time.➤ The function mstraj returns a matrix with one row per time step and the columns correspond to the axes. We can plot q_2 against q_1 to see the path of the robot

```
>> plot(q0(:,1), q0(:,2))
```

and is shown by the red path in Fig. 3.6a. If we increase the acceleration time

```
>> q2 = mstraj(via(:,[2 3 4 1])', [2,1], [], via(:,1)', 0.2, 2);
```

the trajectory becomes more rounded (blue path) as the polynomial blending functions do their work. The smoother trajectory also takes more time to complete.

```
>> [numrows(q0) numrows(q2)]
ans =
    28    80
```

The configuration variables as a function of time are shown in Fig. 3.6b. This function also accepts optional initial and final velocity arguments and t_{acc} can be a vector specifying different acceleration times for each of the N blends.

Keep in mind that this function simply interpolates pose represented as a vector. In this example the vector was assumed to be Cartesian coordinates, but this function could also be applied to Euler or roll-pitch-yaw angles but this is not an ideal way to interpolate rotation. This leads us nicely to the next section where we discuss interpolation of orientation.

3.3.4   Interpolation of Orientation in 3D

In robotics we often need to interpolate orientation, for example, we require the end-effector of a robot to smoothly change from orientation $\boldsymbol{\xi}_0$ to $\boldsymbol{\xi}_1$ in **SO**(3). We require some function $\boldsymbol{\xi}(s) = \sigma(\boldsymbol{\xi}_0, \boldsymbol{\xi}_1, s)$ where $s \in [0, 1]$ which has the boundary conditions $\sigma(\boldsymbol{\xi}_0, \boldsymbol{\xi}_1, 0) = \boldsymbol{\xi}_0$ and $\sigma(\boldsymbol{\xi}_0, \boldsymbol{\xi}_1, 1) = \boldsymbol{\xi}_1$ and where $\sigma(\boldsymbol{\xi}_0, \boldsymbol{\xi}_1, s)$ varies *smoothly* for intermediate values of s. How we implement this depends very much on our concrete representation of $\boldsymbol{\xi}$.

If pose is represented by an orthonormal rotation matrix, $\boldsymbol{\xi} \sim \boldsymbol{R} \in \mathbf{SO}(3)$, we might consider a simple linear interpolation $\sigma(\boldsymbol{R}_0, \boldsymbol{R}_1, s) = (1 - s)\boldsymbol{R}_0 + s\boldsymbol{R}_1$ but this would not, in general, be a valid orthonormal matrix which has strict column norm and inter-column orthogonality constraints.

A workable and commonly used approach is to consider a 3-angle representation such as Euler or roll-pitch-yaw angles, $\boldsymbol{\xi} \sim \boldsymbol{\Gamma} \in \mathbb{S}^1 \times \mathbb{S}^1 \times \mathbb{S}^1$ and use linear interpolation

$$\sigma(\boldsymbol{\Gamma}_0, \boldsymbol{\Gamma}_1, s) = (1 - s)\boldsymbol{\Gamma}_0 + s\boldsymbol{\Gamma}_1$$

and converting the interpolated angles back to a rotation matrix always results in a valid form. For example we define two orientations

```
>> R0 = SO3.Rz(-1) * SO3.Ry(-1);
>> R1 = SO3.Rz(1) * SO3.Ry(1);
```

and find the equivalent roll-pitch-yaw angles

```
>> rpy0 = R0.torpy();   rpy1 = R1.torpy();
```

and create a trajectory between them over 50 time steps

```
>> rpy = mtraj(@tpoly, rpy0, rpy1, 50);
```

which is most easily visualized as an animation►

```
>> SO3.rpy( rpy ).animate;
```

rpy is a 50 × 3 matrix and the result of SO3.rpy is a 1 × 50 vector of SO3 objects, and their animate method is then called.

For large orientation changes we see that the axis around which the coordinate frame rotates changes along the trajectory. The motion, while smooth, sometimes looks uncoordinated. There will also be problems if either ξ_0 or ξ_1 is close to a singularity in the particular 3-angle system being used. This particular trajectory passes very close to the singularity, at around steps 24 and 25, and a symptom of this is the very rapid rate of change of roll-pitch-yaw angles at this point. The frame is not rotating faster at this point – you can verify that in the animation – the rotational parameters are changing very quickly and this is consequence of the particular representation.

Interpolation of unit-quaternions is only a little more complex than for 3-angle vectors and produces a change in orientation that is a rotation around a *fixed* axis in space. Using the Toolbox we first find the two equivalent quaternions

```
>> q0 = R0.UnitQuaternion;   q1 = R1.UnitQuaternion;
```

and then interpolate them

```
>> q = interp(q0, q1, 50);
>> about(q)
q [UnitQuaternion] : 1x50 (1.7 kB)
```

which results in a vector of 50 UnitQuaternion objects which we can animate by

```
>> q.animate
```

Quaternion interpolation is achieved using spherical linear interpolation (*slerp*) in which the unit quaternions follow a great circle path on a 4-dimensional hypersphere. The result in 3-dimensions is rotation about a fixed axis in space.

3.3.4.1   Direction of Rotation

When traveling on a circle we can move clockwise or counter-clockwise to reach the goal – the result is the same but the distance traveled may be different. On a sphere or hypersphere the principle is the same but now we are traveling on a great circle►. In this example we animate a rotation about the z-axis, from an angle of -2 radians to $+2$ radians

```
>> q0 = UnitQuaternion.Rz(-2);   q1 = UnitQuaternion.Rz(2);
>> q = interp(q0, q1, 50);
>> q.animate()
```

but this is taking the long way around the circle, moving 4 radians when we could travel $2\pi - 4 \approx 2.28$ radians in the opposite direction. The 'shortest' option requests the rotational interpolation to select the shortest path

A great circle on a sphere is the intersection of the sphere and a plane that passes through the center. On Earth the equator and all lines of longitude are great circles. Ships and aircraft prefer to follow great circles because they represent the shortest path between two points on the surface of a sphere.

```
>> q = interp(q0, q1, 50, 'shortest');
>> q.animate()
```

and the animation clearly shows the difference.

3.3.5   Cartesian Motion in 3D

Another common requirement is a smooth path between two poses in **SE**(3) which involves change in position as well as in orientation. In robotics this is often referred to as Cartesian motion.

We represent the initial and final poses as homogeneous transformations

```
>> T0 = SE3([0.4, 0.2, 0]) * SE3.rpy(0, 0, 3);
>> T1 = SE3([-0.4, -0.2, 0.3]) * SE3.rpy(-pi/4, pi/4, -pi/2);
```

The SE3 object has a method interp that interpolates between two poses for normalized distance $s \in [0, 1]$ along the path, for example the midway pose between T0 and T1 is

```
>> interp(T0, T1, 0.5)
ans =
    0.0975   -0.7020    0.7055         0
    0.7020    0.5510    0.4512         0
   -0.7055    0.4512    0.5465      0.15
         0         0         0
```

where the translational component is linearly interpolated and the rotation is spherically interpolated using the unit-quaternion interpolation method interp.

A trajectory between the two poses in 50 steps is created by

```
>> Ts = interp(T0, T1, 50);
```

This could also be written as T0.interp(T1, 50).

where the arguments are the initial and final pose and the trajectory length.◄ The resulting trajectory Ts is a vector of SE3 objects

```
>> about(Ts)
Ts [SE3] : 1x50 (6.5 kB)
```

representing the pose at each time step. The homogeneous transformation for the first point on the path is

```
>> Ts(1)
ans =
   -0.9900   -0.1411         0       0.4
    0.1411   -0.9900         0       0.2
         0         0         1         0
         0         0         0         1
```

and once again the easiest way to visualize this is by animation

```
>> Ts.animate
```

which shows the coordinate frame moving and rotating from pose T0 to pose T1. The translational part of this trajectory is obtained by◄

```
>> P = Ts.transl;
```

The .t property applied to a vector of SE3 objects returns a MATLAB comma-separated list of translation vectors. The .transl method returns the translations in a more useful matrix form.

which returns the Cartesian position for the trajectory in matrix form

```
>> about(P)
P [double] : 50x3 (1.2 kB)
```

which has one row per time step that is the corresponding position vector. This is plotted

```
>> plot(P);
```

in Fig. 3.7 along with the orientation in roll-pitch-yaw format

```
>> rpy = Ts.torpy;
>> plot(rpy);
```

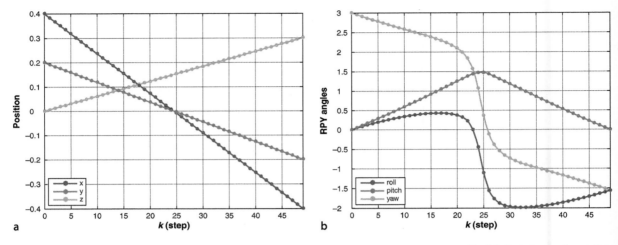

Fig. 3.7. Cartesian motion. **a** Cartesian position versus time, **b** roll-pitch-yaw angles versus time

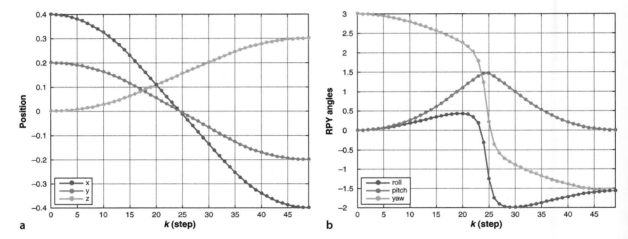

We see that the position coordinates vary smoothly and linearly with time and that orientation varies smoothly with time. ◢

However the motion has a velocity and acceleration *discontinuity* at the first and last points. While the path is smooth in space the distance *s* along the path is not smooth in time. Speed along the path jumps from zero to some finite value and then drops to zero at the end – there is no initial acceleration or final deceleration. The scalar functions `tpoly` and `lspb` discussed earlier can be used to generate *s* so that motion *along* the path is smooth. We can pass a vector of normalized distances along the path as the second argument to `interp`

```
>> Ts = T0.interp(T1, lspb(0, 1, 50) );
```

The trajectory is unchanged but the coordinate frame now accelerates to a constant speed along the path and then decelerates and this is reflected in smoother curves for the trajectory shown in Fig. 3.8. The Toolbox provides a convenient shorthand `ctraj` for the above

```
>> Ts = ctraj(T0, T1, 50);
```

where the arguments are the initial and final pose and the number of time steps.

Fig. 3.8. Cartesian motion with LSPB path distance profile. **a** Cartesian position versus time, **b** roll-pitch-yaw angles versus time

The roll-pitch-yaw angles do not vary linearly with time because they represent a nonlinear transformation of the linearly varying quaternion.

b

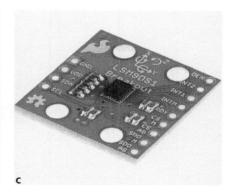

c

Fig. 3.9. a SPIRE (Space Inertial Reference Equipment) from 1953 was 1.5 m in diameter and weighed 1 200 kg. **b** A modern inertial navigation system the LORD MicroStrain 3DM-GX4-25 has triaxial gyroscopes, accelerometers and magnetometer, a pressure altimeter, is only 36 × 24 × 11 mm and weighs 16 g (image courtesy of LORD MicroStrain); **c** 9 Degrees of Freedom IMU Breakout (LSM9DS1-SEN-13284 from SparkFun Electronics), the chip itself is only 3.5 × 3 mm

As discussed in Sect. 3.2.3 the Earth's surface is not an inertial reference frame, but for most robots with nonmilitary grade sensors this is a valid assumption.

3.4 Application: Inertial Navigation

An inertial navigation system or INS is a "black box" that estimates its velocity, orientation and position by measuring accelerations and angular velocities and integrating them over time. Importantly it has no external inputs such as radio signals from satellites. This makes it well suited to applications such as submarine, spacecraft and missile guidance where it is not possible to communicate with radio navigation aids or which must be immune to radio jamming. These particular applications drove development of the technology during the cold war and space race of the 1950s and 1960s. Those early systems were large, see Fig. 3.9a, extremely expensive and the technical details were national secrets. Today INSs are considerably cheaper and smaller as shown in Fig. 3.9b; the sensor chips shown in Fig. 3.9c can cost as little as a few dollars and they are built into every smart phone.

An INS estimates its pose with respect to an inertial reference frame which is typically denoted {0} and fixed to some point on the Earth's surface – the world coordinate frame.◄ The frame typically has its z-axis upward or downward and the x- and y-axes establish a local tangent plane. Two common conventions have the x-, y- and z-axes respectively parallel to north-east-down (NED) or east-north-up (ENU) directions. The coordinate frame {B} is attached to the moving vehicle or robot and is known as the body- or body-fixed frame.

3.4.1 Gyroscopes

Any sensor that measures the rate of change of orientation is known, for historical reasons, as a gyroscope.

3.4.1.1 How Gyroscopes Work

The term gyroscope conjures up an image of a childhood toy – a spinning disk in a round frame that can balance on the end of a pencil. Gyroscopes are confounding devices – you try to turn them one way but they resist and turn (precess) in a different direction. This unruly behavior is described by a simplified version of Eq. 3.10

$$\tau = \boldsymbol{\omega} \times \boldsymbol{h} \tag{3.15}$$

where $\boldsymbol{h}$ is the angular momentum of the gyroscope, a vector parallel to the rotor's axis of spin and with magnitude $\|\boldsymbol{h}\| = J\varpi$, where J is the rotor's inertia and ϖ its rotational speed. It is the cross product in Eq. 3.15 that makes the gyroscope move in a contrary way.

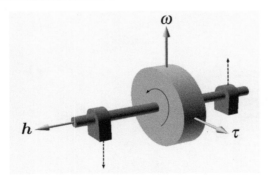

Fig. 3.10.
Gyroscope in strapdown configuration. Angular velocity ω
induces a torque τ which can be
sensed as forces at the bearings
shown in red

If no torque is applied to the gyroscope its angular momentum remains constant in
the inertial reference frame which implies that the axis will maintain a *constant direction* in that frame. Two gyroscopes with orthogonal axes form a stable platform that will
maintain a *constant orientation* with respect to the inertial reference frame – fixed with
respect to the universe. This was the principle of many early spacecraft navigation systems
such as that shown in Fig. 2.15 – the vehicle was able to rotate about the stable platform
and the spacecraft's orientation could be measured with respect to the platform.▶

The challenge was to create a mechanism
that allowed the vehicle to rotate around
the stable platform without exerting any
torque on the gyroscopes. This required
exquisitely engineered low-friction gimbals and bearing systems.

Alternatively we can fix the gyroscope to the vehicle in the strapdown configuration as shown in Fig. 3.10. If the vehicle rotates with an angular velocity ω the attached
gyroscope will *resist* and exert an orthogonal torque τ which can be measured.▶ If the
magnitude of h is high then this kind of sensor is very sensitive – a very small angular
velocity leads to an easily measurable torque.

Typically by strain gauges attached to the
bearings of the rotor shaft.

Over the last few decades this rotating disk technology has been eclipsed by sensors
based on optical principles such as the ring-laser gyroscope (RLG) and the fiber-optic
gyroscope (FOG). These are high quality sensors but expensive and bulky. The low-cost
sensors used in mobile phones and drones are based on micro-electro-mechanical systems (MEMS) fabricated on silicon chips. Details of the designs vary but all contain a mass
vibrating at high frequency▶ in a plane, and rotation about an axis normal to the plane
causes an orthogonal displacement within the plane that is measured capacitively.

Typically over 10 kHz.

Gyroscopic angular velocity sensors measure rotation about a single axis. Typically
three gyroscopes are packaged together and arranged so that their sensitive axes are
orthogonal. The three outputs of such a triaxial gyroscope are the components of the
angular velocity vector ${}^{B}\omega^{\#}$ measured in the body frame $\{B\}$, and we introduce the
superscript to explicitly indicate a sensor measurement.

Interestingly, nature has invented gyroscopic sensors. All vertebrates have angular velocity sensors as part of their vestibular system. In each inner ear we have three
semi-circular canals – fluid filled organs that measure angular velocity. They are arranged orthogonally, just like a triaxial gyroscope, with two measurement axes in a
vertical plane and one diagonally across the head.

3.4.1.2   Estimating Orientation

If we assume that ${}^{B}\omega$ is constant over a time interval δ_t the equivalent rotation at the
timestep k is

$$
{}^{B}\xi_{\Delta}\langle k\rangle \sim e^{\left[{}^{B}\omega^{\#}\right]_{\times}\delta_t}
\tag{3.16}
$$

If the orientation of the sensor frame is initially ξ_B then the evolution of estimated
pose can be written in discrete-time form as

$$
\hat{\xi}_B\langle k{+}1\rangle \leftarrow \hat{\xi}_B\langle k\rangle \oplus {}^{B}\xi_{\Delta}\langle k\rangle
\tag{3.17}
$$

Much important development was undertaken by the MIT Instrumentation Laboratory under the leadership of Charles Stark Draper. In 1953 the feasibility of inertial navigation for aircraft was demonstrated in a series of flight tests with a system called SPIRE (Space Inertial Reference Equipment) shown in Fig. 3.9a. It was 1.5 m in diameter and weighed 1 200 kg. SPIRE guided a B-29 bomber on a 12 hour trip from Massachusetts to Los Angeles without the aid of a pilot and with Draper aboard. In 1954 the first self-contained submarine navigation system (SINS) was introduced to service. The Instrumentation Lab also developed the Apollo Guidance Computer, a one-cubic-foot computer that guided the Apollo Lunar Module to the surface of the Moon in 1969.

Today high-performance inertial navigation systems based on fiber-optic gyroscopes are widely available and weigh around one 1 kg while low-cost systems based on MEMS technology can weigh just a few grams and cost a few dollars.

where we use the hat notation to explicitly indicate an estimate of pose and $k \in \mathbb{Z}^+$ is the index of the time step. In concrete terms we can compute this *update* using $SO(3)$ rotation matrices or unit-quaternions as discussed in Sect. 3.1.3 and taking care to normalize the rotation after each step.

We will demonstrate this integration using unit quaternions and simulated angular velocity data for a tumbling body. The script

```
>> ex_tumble
```

creates a matrix w whose columns represent consecutive body-frame angular velocity measurements with corresponding times given by elements of the vector t. We choose the initial pose to be the null rotation

```
>> attitude(1) = UnitQuaternion();
```

and then for each time step we update the orientation and keep the orientation history in a vector of quaternions

```
>> for k=1:numcols(w)-1
       attitude(k+1) = attitude(k) .* UnitQuaternion.omega( w(:,k)*dt );
   end
```

The technical term is an oblate spheroid, it bulges out at the equator because of centrifugal acceleration due to the Earth's rotation. The equatorial diameter is around 40 km greater than the polar diameter.

The .increment method of the UnitQuaternion class does this in a single call.

The omega method creates a unit-quaternion corresponding to a rotation angle and axis given by the magnitude and direction of its argument. The .* operator performs quaternion multiplication and normalizes the product, ensuring the result has a unit norm.◄ We can animate the changing orientation of the body frame

```
>> attitude.animate('time', t)
```

or view the roll-pitch-yaw angles as a function of time

```
>> mplot(t, attitude.torpy() )
```

3.4.2   Accelerometers

Accelerometers are sensors that measure acceleration. Even when not moving they sense the acceleration due to gravity which defines the direction we know as *downward*. Gravitational acceleration is a function of the material in the Earth beneath us and our distance from the Earth's center. The Earth is not a perfect sphere◄ and points in the equatorial region are further from the center. Gravitational acceleration can be approximated by

$$g \approx 9.780327\left(1 + 0.0053024\sin^2\theta - 0.0000058\sin^2 2\theta\right) - 0.000003086h$$

where θ is the angle of latitude and h is height above sea level. A map of gravity showing the effect of latitude and topography is shown in Fig. 3.11.

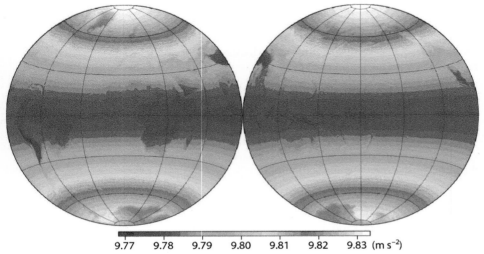

Fig. 3.11.
Variation in Earth's gravitational acceleration, continents and mountain ranges are visible. The hemispheres shown are centerd on the prime (left) and anti (right) meridian respectively (from Hirt et al. 2013)

3.4.2.1   How Accelerometers Work

An accelerometer is conceptually a very simple device comprising a mass, known as the proof mass, supported by a spring as shown in Fig. 3.12. In the inertial reference frame Newton's second law for the proof mass is

$$m\ddot{x}_m = F_s - mg \qquad (3.18)$$

and for a spring with natural length l_0 the relationship between force and extension d is

$$F_s = kd$$

The various displacements are related

$$x_b - (l_0 + d) = x_m$$

and taking the double derivative then substituting Eq. 3.18 gives

$$\ddot{x}_b - \ddot{d} = \tfrac{1}{m}(kd - mg)$$

The quantity we wish to measure is the acceleration of the accelerometer $a = \ddot{x}_b$ ▶ and the relative displacement of the proof mass

$$d = \frac{m}{k}(a + g)$$

We assume that $\ddot{d} = 0$ in steady state. Typically there would be a damping element to increase friction and stop the proof mass oscillating. This adds a term $-B\dot{x}_m$ to the right-hand side of Eq. 3.18.

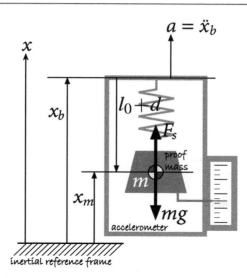

$$a = \ddot{x}_b$$

Fig. 3.12.
The essential elements of an accelerometer and notation

is linearly related to that acceleration. In an accelerometer the displacement is measured and scaled by k / m so that the output of the sensor is

$$a^\# = a + g \ \text{m s}^{-2}$$

If this accelerometer is stationary then $a = 0$ yet the measured acceleration would be $a^\# = 0 + g = g$ in the upward direction. This is because our model has included the Newtonian gravity force mg, as discussed in Sect. 3.2.3. Accelerometer output is sometimes referred to as specific, inertial or proper acceleration.

> The fact that a stationary accelerometer indicates an upward acceleration of 1 g is unintuitive since the accelerometer is clearly stationary and not accelerating. Intuition would suggest, that if anything, the acceleration should be in the downward direction where the device would accelerate if dropped. However the reality is that an accelerometer at rest in a gravity field reports upward acceleration. ◄

A number of iPhone sensor apps incorrectly report acceleration in the *downward* direction when the phone is stationary.

Accelerometers measure acceleration along a single axis. Typically three accelerometers are packaged together and arranged so that their sensitive axes are orthogonal. The three outputs of such a triaxial accelerometer are the components of the acceleration vector $^B a^\#$ measured in the body frame $\{B\}$.

Nature has also invented the accelerometer. All vertebrates have acceleration sensors called ampullae as part of their vestibular system. We have two in each inner ear: the saccule which measures vertical acceleration, and the utricle which measures front-to-back acceleration, and they help us maintain balance. ◄ The proof mass in the ampullae is a collection of calcium carbonate crystals called otoliths, literally ear stones, on a gelatinous substrate which serves as the spring and damper. Hair cells embedded in the substrate measure the displacement of the otoliths due to acceleration.

Inconsistency between motion sensed in our ears and motion perceived by our eyes is the root cause of motion sickness.

3.4.2.2   Estimating Pose and Body Acceleration

In frame $\{0\}$ with its z-axis vertically upward, the gravitational acceleration vector is

$$^0 a = \begin{pmatrix} 0 \\ 0 \\ g \end{pmatrix}$$

where g is the local gravitational acceleration from Fig. 3.11. In a body-fixed frame $\{B\}$ at an arbitrary orientation expressed in terms of ZYX roll-pitch-yaw angles ►

We could use any 3-angle sequence.

$$^0\xi_B = \mathscr{R}_z(\theta_y) \oplus \mathscr{R}_y(\theta_p) \oplus \mathscr{R}_x(\theta_r)$$

the gravitational acceleration will be

$$^B\boldsymbol{a} = \left(\ominus\,^0\xi_B\right) \cdot {}^0\boldsymbol{a} = \begin{pmatrix} -g\sin\theta_p \\ g\cos\theta_p\sin\theta_r \\ g\cos\theta_p\cos\theta_r \end{pmatrix} \qquad (3.19)$$

The *measured* acceleration vector from the sensor in frame $\{B\}$ is

$$^B\boldsymbol{a}^{\#} = \begin{pmatrix} a_x \\ a_y \\ a_z \end{pmatrix}$$

and equating this with Eq. 3.19 we can solve for the roll and pitch angles

$$\sin\hat{\theta}_p = \frac{-a_x}{g} \qquad (3.20)$$

$$\tan\hat{\theta}_r = \frac{a_y}{a_z}, \quad \theta_p \neq \pm\frac{\pi}{2} \qquad (3.21)$$

and we use the hat notation to indicate that these are estimates of the angles. ► Notice that there is no solution for the yaw angle and in fact θ_y does not even appear in Eq. 3.19. The gravity vector is parallel to the vertical axis and rotating around that axis, yaw rotation, will not change the measured value at all. ►

> We have made a very strong assumption that the measured acceleration $^B\boldsymbol{a}^{\#}$ is only due to gravity. On a robot the sensor will experience additional acceleration as the vehicle moves and this will introduce an error in the estimated orientation.

Frequently we want to estimate the motion of the vehicle in the inertial frame, and the total measured acceleration in $\{0\}$ is due to gravity *and* motion

$$^0\boldsymbol{a}^{\#} = {}^0\boldsymbol{g} + {}^0\boldsymbol{a}_v$$

We observe acceleration in the body frame so the vehicle acceleration in the world frame is

$$^0\hat{\boldsymbol{a}}_v = {}^0\hat{\boldsymbol{R}}_B\,{}^B\boldsymbol{a}^{\#} - {}^0\boldsymbol{g} \qquad (3.22)$$

and we assume that $^0\hat{\boldsymbol{R}}_B$ and $\boldsymbol{g}$ are both known. ► Integrating that with respect to time

$$^0\hat{\boldsymbol{v}}_v(t) = \int {}^0\hat{\boldsymbol{a}}_v(t)\,\mathrm{d}t \qquad (3.23)$$

gives the velocity of the vehicle, and integrating again

$$^0\hat{\boldsymbol{p}}_v(t) = \int {}^0\hat{\boldsymbol{v}}_v(t)\,\mathrm{d}t \qquad (3.24)$$

gives its position. Note that we can assume vehicle acceleration is zero and estimate attitude, or assume attitude and estimate vehicle acceleration. We cannot estimate both since there are more unknowns than measurements.

These angles are sufficient to determine whether a phone, tablet or camera is in portrait or landscape orientation.

Another way to consider this is that we are essentially measuring the direction of the gravity vector with respect to the frame $\{B\}$ and a vector provides only two unique *pieces* of directional information, since one component of a unit vector can be written in terms of the other two.

The first assumption is a strong one and problematic in practice. Any error in the rotation matrix results in incorrect cancellation of the gravity component of $\boldsymbol{a}^{\#}$ which leads to an error in the estimated body acceleration.

3.4.3   Magnetometers

The Earth is a massive but weak magnet. The poles of this geomagnet are the Earth's north and south magnetic poles which are constantly moving and located quite some distance from the planet's rotational axis.

At any point on the planet the magnetic flux lines can be considered a vector m whose magnitude and direction can be accurately predicted and mapped as shown in Fig. 3.13. We describe the vector's direction in terms of two angles: declination and inclination. A horizontal projection of the vector m points in the direction of magnetic north and the declination angle D is measured from true north◄ clockwise to that projection. The inclination angle I of the vector is measured in a vertical plane downward◄ from horizontal to m. The length of the vector, the magnetic field intensity, is measured by a magnetometer in units of Tesla (T) and for the Earth this varies from $25-65\ \mu\text{T}^{\blacktriangleright}$ as shown in Fig. 3.13a.

The direction of the Earth's north rotational pole, where the rotational axis intersects the surface of the northern hemisphere.

In the Northern hemisphere inclination is positive, that is, the vector points into the ground.

By comparison a modern MRI machine has a magnetic field strength of 4-8 T.

3.4.3.1   How Magnetometers Work

The key element of most modern magnetometers is a Hall-effect sensor, a semiconductor device which produces a voltage proportional to the magnetic field intensity in a direction normal to the current flow. Typically three Hall-effect sensors are packaged together and arranged so that their sensitive axes are orthogonal. The three outputs of such a triaxial magnetometer are the components of the Earth's magnetic field intensity vector $^{B}m^{\#}$ measured in the body frame $\{B\}$.

Yet again nature leads, and creatures from bacteria to turtles and birds are known to sense magnetic fields. The effect is particularly well known in pigeons and there is debate about whether or not humans have this sense. The actual biological sensing mechanism has not yet been discovered.

3.4.3.2   Estimating Heading

Consider an inertial coordinate frame $\{0\}$ with its z-axis vertically upward and its x-axis pointing toward *magnetic* north. The magnetic field intensity vector therefore lies in the xz-plane

$$^{0}m = B\begin{pmatrix}\cos I\\ 0\\ \sin I\end{pmatrix}$$

where B is the magnetic field intensity and I the inclination angle which are both known from Fig. 3.13. In a body-fixed frame $\{B\}$ at an arbitrary orientation expressed in terms of roll-pitch-yaw angles◄

We could use any 3-angle sequence.

$$^{0}\xi_{B} = \mathscr{R}_{z}\!\left(\theta_{y}\right) \oplus \mathscr{R}_{y}\!\left(\theta_{p}\right) \oplus \mathscr{R}_{x}\!\left(\theta_{r}\right)$$

Edwin Hall (1855–1938) was an American physicist born in Maine. His Ph.D. research in physics at the Johns Hopkins University in 1880 discovered that a magnetic field exerts a force on a current in a conductor. He passed current through thin gold leaf and in the presence of a magnetic field normal to the leaf was able to measure a very small potential difference between the sides of the leaf. This is now known as the Hall effect. While it was then known that a magnetic field exerted a force on a current carrying conductor it was believed the force acted on the conductor not the current itself – electrons were yet to be discovered. He was appointed as professor of physics at Harvard in 1895 where he worked on thermoelectric effects.

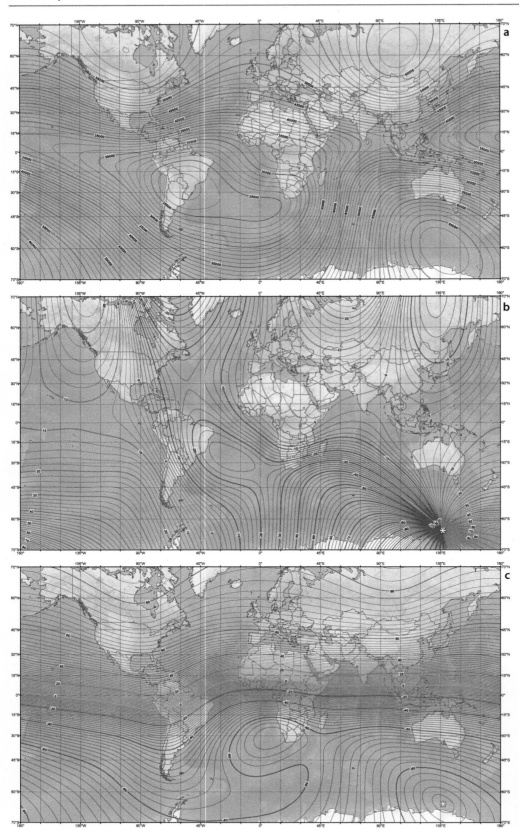

Fig. 3.13.
A predicted model of the Earth magnetic field parameters for 2015. **a** Magnetic field intensity (nT); **b** magnetic declination (degrees); **c** magnetic inclination (degrees). Magnetic poles indicated by *asterisk* (maps by NOAA/NGDC and CIRES http://ngdc.noaa.gov/geomag/WMM, published Dec 2014)

the magnetic field intensity will be

$$^{B}m = \left(\ominus^{0}\xi_{B}\right)^{0} \bullet m \tag{3.25}$$

The measured magnetic field intensity vector from the sensor in frame $\{B\}$ is

$$^{B}m^{\#} = \begin{pmatrix} m_{x} \\ m_{y} \\ m_{z} \end{pmatrix}$$

and equating this with Eq. 3.25 we can solve for the yaw angle

$$\hat{\theta}_{y} = \tan^{-1}\frac{\cos\theta_{p}\left(m_{z}\sin\theta_{r} - m_{y}\cos\theta_{r}\right)}{m_{x} + B\sin I\,\sin\theta_{p}}$$

Many triaxial Hall-effect sensor chips also include a triaxial accelerometer for just this purpose.

assuming that the roll and pitch angles have been determined, perhaps using measured acceleration and Eq. 3.21.◄

We defined yaw angle as the orientation of the frame $\{B\}$ x-axis▲ with respect to *magnetic* north. To obtain the heading angle with respect to true-north we subtract the local declination angle

Typically in vehicle navigation the x-axis points forward and the yaw angle is also called the heading angle.

$$^{\mathrm{tn}}\hat{\theta}_{y} = \hat{\theta}_{y} - D$$

Magnetometers are great in theory but problematic in practice. Firstly, our modern world is full of magnets and electromagnets. Buildings contain electrical wiring and robots themselves are full of electric motors, batteries and electronics. These all add to, or overwhelm, the local geomagnetic field. Secondly, many objects in our world contain ferromagnetic materials such as the reinforcing steel in buildings or the steel bodies of cars or ships. These distort the geomagnetic field leading to local changes in its direction. These effects are referred to respectively as hard- and soft-iron distortion of the magnetic field.◄

These can be calibrated out but the process requires that the sensor is rotated by 360 degrees.

3.4.4   Sensor Fusion

An inertial navigation system uses the devices we have just discussed to determine the pose of a vehicle – its position and its orientation. Early inertial navigation systems, such as shown in Fig. 2.15, used mechanical gimbals to keep the accelerometers at a constant attitude with respect to the stars using a gyro-stabilized platform. The acceleration measured on this platform is by definition referred to the inertial frame and simply needs to be integrated to obtain the velocity of the platform, and integrated again to obtain its position. In order to achieve accurate position estimates over periods of hours or days the gimbals and gyroscopes had to be of extremely high quality so that the stable platform did not drift, and the acceleration sensors needed to be extremely accurate.

The modern strapdown inertial measurement configuration uses no gimbals. The angular velocity, acceleration and magnetic field sensors are rigidly attached to the vehicle. The collection of inertial sensors is referred to as an inertial measurement unit or IMU. A 6-DOF IMU comprises triaxial gyroscopes and accelerometers while a 9-DOF IMU comprises triaxial gyroscopes, accelerometers and magnetometers.◄ A system that only determines attitude is called an attitude and heading reference system or AHRS.

Increasingly these sensor packages also include a barometric pressure sensor to measure changes in altitude.

The sensors we use, particularly the low-cost ones in phones and drones, are far from perfect. Consider any sensor value – gyroscope, accelerometer or magnetometer – the measured signal

$$x^{\#} = sx + b + \varepsilon$$

is related▶ to the unknown true value x by a scale factor s, offset or bias b and random noise ε. s is usually specified by the manufacturer to some tolerance, perhaps $\pm 1\%$, and for a particular sensor this can be determined by some calibration procedure. Bias b is ideally equal to zero but will vary from device to device. Bias that varies over time is often called sensor drift. Scale factor and bias are typically both a function of temperature.▶

In practice bias is the biggest problem because it varies with time and temperature and has a very deleterious effect on the estimated pose and position. Consider a positive bias on the output of a gyroscopic sensor – the output is higher than it should be. At each time step in Eq. 3.17 the incremental rotation will be bigger than it should be, which means that the orientation error will grow linearly with time.▶

If we use Eq. 3.22 to estimate the vehicle's acceleration then the error in attitude means that the measured gravitation acceleration is incorrectly canceled out and will be indistinguishable from *actual* vehicle acceleration. This offset in acceleration becomes a linear time error in velocity and a quadratic time error in position. Given that the pose error is already linear in time we end up with a cubic time error in position, and this is ignoring the effects of accelerometer bias. Sensor bias is problematic! A rule of thumb is that gyroscopes with bias stability of $0.01 \deg h^{-1}$ will lead to position error growing at a rate of 1 nmi h^{-1} (1.85 km h^{-1}). Military grade systems have very impressive stability, for missiles $<0.00002 \deg h^{-1}$ which is in stark contrast to consumer grade devices which are in the range $0.01 - 0.2$ deg per *second*.

A simple approach to this problem is to estimate bias by leaving the IMU stationary for a few seconds and computing the average value of all the sensors.▶ This value is then subtracted from future sensor readings. This is really only valid over a short time period because the bias is not constant.

A more sophisticated approach is to estimate the bias online▶, but to do this we need to combine information from different sensors – an approach known as sensor fusion. We rely on the fact that different sensors have complementary characteristics. Bias on angular rate sensors causes the attitude estimate error to grow with time, but for accelerometers it will only cause an attitude offset. However accelerometers respond to motion of the vehicle while good gyroscopes do not. Magnetometers provide partial information about roll, pitch and yaw, are immune to acceleration, but do respond to stray magnetic fields and other distortions. There are many ways to achieve this kind of fusion. A common approach is to use an estimation tool called an extended Kalman filter described in Appendix H. Given a full nonlinear mathematical model that relates the sensor signals and their biases to the vehicle pose and knowledge about the noise (uncertainty) on the sensor signals, the filter gives an optimal estimate of the pose and bias that best explain the sensor signals.

Here we will consider a simple but still very effective alternative called the explicit complementary filter. The rotation update step is performed using Eq. 3.17 but compared to Eq. 3.16 the incremental rotation is more complex

$$^{B}\xi_{\Delta}\langle k\rangle = e^{\left[^{B}\omega^{\#}\langle k\rangle - \hat{\boldsymbol{b}}\langle k\rangle + k_{P}\sigma_{R}\langle k\rangle\right]_{\times}\delta_{t}} \tag{3.26}$$

The key differences are that the estimated bias $\hat{\boldsymbol{b}}$ is subtracted from the sensor measurement and a term based on the orientation error σ_{R} is added. The estimated bias changes with time according to

$$\hat{\boldsymbol{b}}\langle k+1\rangle \leftarrow \hat{\boldsymbol{b}}\langle k\rangle - k_{I}\sigma_{R}\langle k\rangle \tag{3.27}$$

and also depends on the orientation error σ_{R}. $k_{P} > 0$ and $k_{I} > 0$ are both well chosen constants.

The orientation error is derived from N *vector measurements* $^{0}\boldsymbol{v}_{i}^{\#}$

$$\sigma_{R}\langle k+1\rangle = \sum_{i=1}^{N} k_{i}\,^{0}\boldsymbol{v}_{i} \times \,^{0}\boldsymbol{v}_{i}^{\#}\langle k\rangle$$

We assume a linear relationship but check the fine print in a datasheet to understand what a sensor really does.

Some sensors also exhibit cross-sensitivity. They may give a weak response to a signal in an orthogonal direction or from a different mode, quite commonly low-cost gyroscopes respond to vibration and acceleration as well as rotation.

The effect of an attitude error is dangerous on something like a quadrotor. For example, if the estimated pitch angle is too high then the vehicle control system will pitch down by the same amount to keep the craft "level", and this will cause it to accelerate forward.

A lot of hobby drones do this just before they take off.

Our brain has an online mechanism to cancel out the bias in our vestibular gyroscopes. It uses the recent average rotation as the bias, based on the reasonable assumption that we do not undergo prolonged rotation. If we do, then that angular rate becomes the new normal so that when we stop rotating we perceive rotation in the opposite direction. We call this dizziness.

where ^0v_i is the known value of a vector signal in the inertial frame (for example gravitational acceleration) and

$$^0v_i^\#\langle k\rangle = {}^0\hat{\xi}_B\langle k\rangle \bullet {}^Bv^\#\langle k\rangle$$

is the value measured in the body-fixed frame and rotated into the inertial frame by the estimated orientation $^0\hat{\xi}_B$. Any error in direction between these vectors will yield a nonzero cross-product which is the axis around which to rotate one vector into the other. The filter uses this difference – the innovation – to improve the orientation estimate by feeding it back into Eq. 3.26. This filter allows an unlimited number of vectorial measurements ^0v_i to be fused together; for example we could add magnetic field or any other kind of direction data such as the altitude and azimuth of visual landmarks, stars or planets.

The script

```
>> ex_tumble
```

provides simulated "measured" gyroscope, accelerometer and magnetometer data organized as columns of the matrices wm, gm and mm respectively and all include a fixed bias. Corresponding times are given by elements of the vector t. Firstly we will repeat the example from page 81 but now with sensor bias

```
>> attitude(1) = UnitQuaternion();
>> for k=1:numcols(wm)-1
       attitude(k+1) = attitude(k) .* UnitQuaternion.omega( wm(:,k)*dt );
   end
```

To see the effect of bias on the estimated attitude we will compare it to the true attitude truth that was also computed by the script. As a measure of error we plot the angle between the corresponding unit quaternions in the sequence

```
>> plot(t, angle(attitude, truth), 'r' );
```

which is shown as the red line in Fig. 3.14a. We can clearly see growth in angular error over time. Now we implement the explicit complementary filter with just a few extra lines of code

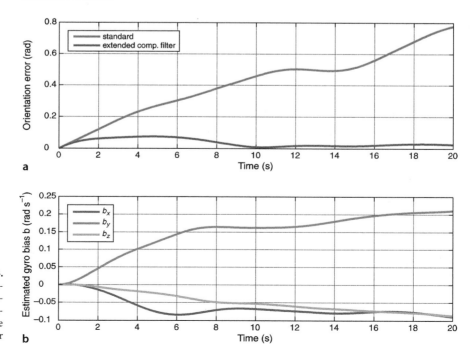

Fig. 3.14.
a Effect of gyroscope bias on naive INS (red) and explicit complementary filter (blue); **b** estimated gyroscope bias from the explicit complementary filter

```
>> kI = 0.2; kP = 1;
>> b = zeros(3, numcols(w));
>> attitude_ecf(1) = UnitQuaternion(); b = [0 0 0]';
>> for k=1:numcols(wm)-1
      invq = inv( attitude_ecf(k) );
      sigmaR = cross(gm(:,k), invq*g0) + cross(mm(:,k), invq*m0);
      wp = wm(:,k) - b(:,k) + kP*sigmaR;
      attitude_ecf(k+1) = attitude_ecf(k) .* UnitQuaternion.omega( wp*dt );
      b(:,k+1) = b(:,k) - kI*sigmaR*dt;
   end
```

and plot the angular difference between the estimated and the attitude as a blue line

```
>> plot(t, angle(attitude_ecf, truth), 'b' );
```

Bringing together information from multiple sensors has checked the growth in attitude error, despite all the sensors having a bias. The estimated gyroscope bias is shown in Fig. 3.14b and we can see the bias estimates converging on their true value.

3.5   Wrapping Up

In this chapter we have considered pose that varies as a function of time from several perspectives.

Firstly we took a calculus perspective and showed that the temporal derivative of an orthonormal rotation matrix or a quaternion is a function of the angular velocity of the body – a concept from mechanics. The skew-symmetric matrix appears in the rotation matrix case and we should no longer be surprised about this given its intimate connection to rotation via Lie group theory. We then looked at finite time differences as an approximation to the derivative and showed how these lead to computationally cheap methods to update rotation matrices and quaternions given knowledge of angular velocity. We also discussed the dynamics of moving bodies that translate and rotate under the influence of forces and torques, inertial and noninertial reference frames and the notion of fictitious forces.

The second perspective was to create motion – a sequence of poses, a trajectory, that a robot can follow. An important characteristic of a trajectory is that it is *smooth* – the position and orientation changes smoothly with time. We started by discussing how to generate smooth trajectories in one dimension and then extended that to the multi-dimensional case and then to piecewise-linear trajectories that visit a number of intermediate points. Smoothly varying rotation was achieved by interpolating roll-pitch-yaw angles and quaternions.

With all this under our belt we were then able to tackle an application, the important problem of inertial navigation. Given imperfect measurements from sensors on a moving body we are able to estimate the pose of that moving body.

Further Reading

The earliest work on manipulator Cartesian trajectory generation was by Paul (1972, 1979) and Taylor (1979). The multi-segment trajectory is discussed by Paul (1979, 1981) and the concept of segment transitions or blends is discussed by Lloyd and Hayward (1991). These early papers, and others, are included in the compilation on Robot Motion (Brady et al. 1982). Polynomial and LSPB trajectories are described in detail by Spong et al. (2006) and multi-segment trajectories are covered at length in Siciliano et al. (2009) and Craig (2005).

The book *Digital Apollo* (Mindell 2008) is a very readable story of the development of the inertial navigation system for the Apollo Moon landings. The article by Corke et al. (2007) describes the principles of inertial sensors and the functionally equivalent sensors located in the inner ear of mammals that play a key role in maintaining balance.

There is a lot of literature related to the theory and practice of inertial navigation systems. The thesis of Achtelik (2014) describes a sophisticated extended Kalman filter for estimating the pose, velocity and sensor bias for a small flying robot. The explicit complementary filter used in this chapter is described by Hua et al. (2014). The recently revised book Groves (2013) covers inertial and terrestrial radio and satellite navigation and has a good coverage of Kalman filter state estimation techniques. Titterton and Weston (2005) provides a clear and concise description of the principles underlying inertial navigation with a focus on the sensors themselves but is perhaps a little dated with respect to modern low-cost sensors. Data sheets on many low-cost inertial and magnetic field sensing chips can be found at https://www.sparkfun.com in the Sensors category.

Exercises

1. Express the incremental rotation $^{B}R_{\Delta}$ as an exponential series and verify Eq. 3.7.
2. Derive the unit-quaternion update equation Eq. 3.8.
3. Make a simulation with a particle moving at constant velocity and a rotating reference frame. Plot the position of the particle in the inertial and the rotating reference frame and observe how the motion changes as a function of the inertial frame velocity.
4. Redo the quaternion-based angular velocity integration on page 81 using rotation matrices.
5. Derive the expression for fictitious forces in a rotating reference frame from Sect. 3.2.3.
6. At your location determine the magnitude and direction of centripetal acceleration you would experience. If you drove at 100 km h^{-1} due east what is the magnitude and direction of the Coriolis acceleration you would experience? What about at 100 km h^{-1} due north? The vertical component is called the Eötvös effect, how much lighter does it make you?
7. For a `tpoly` trajectory from 0 to 1 in 50 steps explore the effects of different initial and final velocities, both positive and negative. Under what circumstances does the quintic polynomial overshoot and why?
8. For a `lspb` trajectory from 0 to 1 in 50 steps explore the effects of specifying the velocity for the constant velocity segment. What are the minimum and maximum bounds possible?
9. For a trajectory from 0 to 1 and given a maximum possible velocity of 0.025 compare how many time steps are required for each of the `tpoly` and `lspb` trajectories?
10. Use `animate` to compare rotational interpolation using quaternions, Euler angles and roll-pitch-yaw angles. Hint: use the quaternion `interp` method and `mtraj`.
11. Repeat the example of Fig. 3.7 for the case where:
    a) the interpolation does *not* pass through a singularity. Hint – change the start or goal pitch angle. What happens?
    b) the final orientation is at a singularity. What happens?
12. Develop a method to quantitatively compare the performance of the different orientation interpolation methods. Hint: plot the locus followed by $\hat{z}$ on a unit sphere.
13. For the `mstraj` example (page 75)
    a) Repeat with different initial and final velocity.
    b) Investigate the effect of increasing the acceleration time. Plot total time as a function of acceleration time.
14. Modify `mstraj` so that acceleration limits are taken into account when determining the segment time.
15. There are a number of iOS and Android apps that display sensor data from gyros, accelerometers and magnetometers. You could also use MATLAB, see http://mathworks.com/hardware-support/iphone-sensor.html. Run one of these and explore how the sensor signals change with orientation and movement. What happens when you throw the phone into the air?

16. Consider a gyroscope with a 20 mm diameter steel rotor that is 4 mm thick and rotating at 10 000 rpm. What is the magnitude of h? For an angular velocity of 5 deg s^{-1}, what is the generated torque?

17. Using Eq. 3.15 can you explain how a toy gyroscope is able to balance on a single point with its spin axis horizontal? What holds it up?

18. A triaxial accelerometer has fallen off the table, ignoring air resistance what value does it return as it falls?

19. Implement the algorithm to determine roll and pitch angles from accelerometer measurements.
    a) Devise an algorithm to determine if you are in portrait or landscape orientation?
    b) Create a trajectory for the accelerometer using `tpoly` to generate motion in either the x- or y-direction. What effect does the acceleration along the path have on the estimated angle?
    c) Calculate the orientation using quaternions rather than roll-pitch-yaw angles.

20. You are in an aircraft flying at 30 000 feet over your current location. How much lighter are you?

21. Determine the Euler angles as a function of the measured acceleration. If you have the Symbolic Math Toolbox™ you might like to use that.

22. Determine the magnetic field strength, declination and inclination at your location. Visit the website http://www.ngdc.noaa.gov/geomag-web.

23. Using the sensor reading app from above, orient the phone so that the magnetic field vector has only a z-axis component, where is the magnetic field vector with respect to your phone?

24. Using the sensor reading app from above log some inertial sensor data from a phone while moving it around. Use that data to estimate the changing attitude or full pose of the phone. Can you do this in real time?

25. Experiment with varying the parameters of the explicit complementary filter on page 90. Change the bias or add Gaussian noise to the simulated sensor readings.

Part II Mobile Robots

II

Mobile Robots

In this part we discuss mobile robots, a class of robots that are able to move through the environment. The figures show an assortment of mobile robots that can move over the ground, over the water, through the air, or through the water. This highlights the diversity of what is referred to as the *robotic platform* – the robot's physical embodiment and means of locomotion as shown in Figs. II.2 through II.4.

However these mobile robots are very similar in terms of what they do and how they do it. One of the most important functions of a mobile robot is to move to some place. That place might be specified in terms of some feature in the environment, for instance move to the light, or in terms of some geometric coordinate or map reference. In either case the robot will take some path to reach its destination and it faces challenges such as obstacles that might block its way or having an incomplete map, or no map at all.

One strategy is to have very simple sensing of the world and to react to what is sensed. For example *Elsie* the robotic tortoise, shown in Fig. II.1a, was built in the 1940s and *reacted* to her environment to seek out a light source without having any explicit plan or knowledge of the position of the light. An alternative to the reactive approach was embodied in the 1960s robot Shakey, shown in Fig. II.1b, which was capable of 3D perception and created a map of its environment and then reasoned about the map to plan a path to its destination.

These two approaches exemplify opposite ends of the spectrum for mobile robot navigation. Reactive systems can be fast and simple since sensation is connected directly to action – there is no need for resources to hold and maintain a representation of the world nor any capability to reason about that representation. In nature such

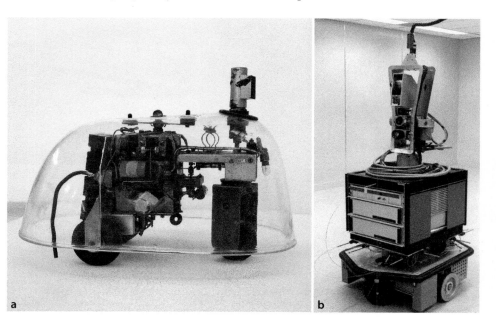

Fig. II.1.
a Elsie the tortoise. Burden Institute Bristol (1948). Now in the collection of the Smithsonian Institution but not on display (photo courtesy Reuben Hoggett collection). **b** Shakey. SRI International (1968). Now in the Computer Museum in Mountain View (photo courtesy SRI International)

a   b

strategies are used by simple organisms such as insects. Systems that make maps and reason about them require more resources but are capable of performing more complex tasks. In nature such strategies are used by more complex creatures such as mammals.

The first commercial applications of mobile robots came in the 1980s when automated guided vehicles (AGVs) were developed for transporting material around factories and these have since become a mature technology. Those early free-ranging mobile wheeled vehicles typically use fixed infrastructure for guidance, for example, a painted line on the floor, a buried cable that emits a radio-frequency signal, or wall-mounted bar codes. The last decade has seen significant achievements in mobile robotics that can operate without navigational infrastructure. Figure II.2a shows a robot vacuum cleaner which use reactive strategies to clean the floor, after the fashion of *Elsie*. Figure II.2b shows an early self-driving vehicle developed for the DARPA series of grand challenges for autonomous cars (Buehler et al. 2007, 2010). We see a multitude of sensors that provide the vehicle with awareness of its surroundings. Other examples are shown in Figs. 1.4 to 1.6. Mobile robots are not just limited to operations on the ground. Figure II.3 shows examples of unmanned aerial vehicles (UAVs), autonomous underwater vehicles (AUVs), and robotic boats which are known as autonomous surface vehicles (ASVs). Field robotic systems such as trucks in mines, container transport vehicles in shipping ports, and self-driving tractors for broad-acre agriculture are now commercially available for various applications are shown in Fig. II.4.

Fig. II.2.
Some mobile ground robots: **a** The Roomba robotic vacuum cleaner, 2008 (photo courtesy iRobot Corporation). **b** *Boss*, Tartan racing team's autonomous car that won the Darpa Urban Grand Challenge, 2007 (Carnegie-Mellon University)

Fig. II.3.
Some mobile air and water robots: **a** Yamaha RMAX helicopter with 3 m blade diameter (photo by Sanjiv Singh). **b** Fixed-wing robotic aircraft (photo of ScanEagle courtesy of Insitu). **c** DEPTHX: Deep Phreatic Thermal Explorer, a 6-thruster under-water robot. Stone Aerospace/CMU (2007) (photo by David Wettergreen, © Carnegie-Mellon University). **d** Autonomous Surface Vehicle (photo by Matthew Dunbabin)

Fig. II.4.
a Exploration: Mars Science Laboratory (MSL) rover, known as Curiosity, undergoing testing (image courtesy NASA/Frankie Martin). **b** Logistics: an automated straddle carrier that moves containers; Port of Brisbane, 2006 (photo courtesy of Port of Brisbane Pty Ltd). **c** Mining: autonomous haul truck (Copyright © 2015 Rio Tinto). **d** Agriculture: broad-acre weeding robot (image courtesy Owen Bawden)

The chapters in this part of the book cover the fundamentals of mobile robotics. Chapter 4 discusses the motion and control of two exemplar robot platforms: wheeled vehicles that operate on a planar surface, and flying robots that move in 3-dimensional space – specifically quadrotor flying robots. Chapter 5 is concerned with navigation. We will cover in some detail the reactive and plan-based approaches to guiding a robot through an environment that contains obstacles. Most navigation strategies require knowledge of the robot's position and this is the topic of Chap. 6 which examines techniques such dead reckoning and the use of maps along with observations of landmarks. We also show how a robot can make a map, and even determine its location while simultaneously mapping an unknown region.

4

Mobile Robot Vehicles

This chapter discusses how a robot platform moves, that is, how its pose changes with time as a function of its control inputs. There are many different types of robot platform as shown on pages 95–97 but in this chapter we will consider only four important exemplars. Section 4.1 covers three different types of wheeled vehicle that operate in a 2-dimensional world. They can be propelled forwards or backwards and their heading direction controlled by some steering mechanism. Section 4.2 describes a quadrotor, a flying vehicle, which is an example of a robot that moves in 3-dimensional space. Quadrotors are becoming increasing popular as a robot platform since they are low cost and can be easily modeled and controlled.

Section 4.3 revisits the concept of configuration space and dives more deeply into important issues of under-actuation and nonholonomy.

4.1 Wheeled Mobile Robots

Wheeled locomotion is one of humanity's great innovations. The wheel was invented around 3000 BCE and the two-wheeled cart around 2000 BCE. Today four-wheeled vehicles are ubiquitous and the total automobile population of the planet is over one billion. The effectiveness of cars, and our familiarity with them, makes them a natural choice for robot platforms that move across the ground.

We know from our everyday experience with cars that there are limitations on how they move. It is not possible to drive sideways, but with some practice we can learn to follow a path that results in the vehicle being to one side of its initial position – this is parallel parking. Neither can a car rotate on the spot, but we can follow a path that results in the vehicle being at the same position but rotated by 180° – a three-point turn. The necessity to perform such maneuvers is the hall mark of a system that is nonholonomic – an important concept which is discussed further in Sect. 4.3. Despite these minor limitations the car is the simplest and most effective means of moving in a planar world that we have yet found. The car's motion model and the challenges it raises for control will be discussed in Sect. 4.1.1.

In Sect. 4.1.2 we will introduce differentially-steered vehicles which are mechanically simpler than cars and do not have steered wheels. This is a common configuration for small mobile robots and also for larger machines like bulldozers. Section 4.1.3 introduces novel types of wheels that *are* capable of omnidirectional motion and then models a vehicle based on these wheels.

4.1.1 Car-Like Mobile Robots

Cars with steerable wheels are a very effective class of vehicle and the archetype for most ground robots such as those shown in Fig. II.4a–c. In this section we will create a model for a car-like vehicle and develop controllers that can drive the car to a point, along a line, follow an arbitrary trajectory, and finally, drive to a specific pose.

A commonly used model for the low-speed behavior of a four-wheeled car-like vehicle is the kinematic bicycle model► shown in Fig. 4.1. The bicycle has a rear wheel fixed to the body and the plane of the front wheel rotates about the vertical axis to steer the vehicle. We assume that the velocity of each wheel is in the plane of the wheel, and that the wheel rolls without slipping sideways

Often incorrectly called the Ackermann model.

$$^B\boldsymbol{v} = (v, 0)$$

The pose of the vehicle is represented by its body coordinate frame {B} shown in Fig. 4.1, with its x-axis in the vehicle's forward direction and its origin at the center of the rear axle. The *configuration* of the vehicle is represented by the generalized coordinates $\boldsymbol{q} = (x, y, \theta) \in \mathcal{C}$ where $\mathcal{C} \subset \mathbb{R}^2 \times \mathbb{S}^1$.

The dashed lines show the direction along which the wheels cannot move, the lines of no motion, and these intersect at a point known as the Instantaneous Center of Rotation (ICR). The reference point of the vehicle thus follows a circular path and its angular velocity is

$$\dot{\theta} = \frac{v}{R_B} \tag{4.1}$$

and by simple geometry the turning radius is $R_B = L / \tan \gamma$ where L is the length of the vehicle or *wheel base*. As we would expect the turning circle increases with vehicle length. The steering angle γ is typically limited mechanically and its maximum value dictates the minimum value of R_B.

> **Vehicle coordinate system.** The coordinate system that we will use, and a common one for vehicles of all sorts is that the x-axis is forward (longitudinal motion), the y-axis is to the left side (lateral motion) which implies that the z-axis is upward. For aerospace and underwater applications the z-axis is often downward and the x-axis is forward.

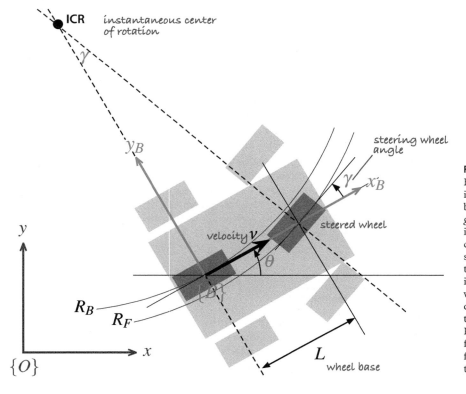

Fig. 4.1.
Bicycle model of a car. The car is shown in *light grey*, and the bicycle approximation is *dark grey*. The vehicle's body frame is shown in *red*, and the world coordinate frame in *blue*. The steering wheel angle is γ and the velocity of the back wheel, in the x-direction, is v. The two wheel axes are extended as dashed lines and intersect at the Instantaneous Center of Rotation (ICR) and the distance from the ICR to the back and front wheels is R_B and R_F respectively

Arcs with smoothly varying radius. Dubbins and Reeds-Shepp paths comprises constant radius circular arcs and straight line segments.

For a fixed steering wheel angle the car moves along a circular arc. For this reason curves on roads are circular arcs or clothoids◄ which makes life easier for the driver since constant or smoothly varying steering wheel angle allow the car to follow the road. Note that $R_F > R_B$ which means the front wheel must follow a longer path and therefore rotate more quickly than the back wheel. When a four-wheeled vehicle goes around a corner the two steered wheels follow circular paths of different radii and therefore the angles of the steered wheels γ_L and γ_R should be very slightly different. This is achieved by the commonly used Ackermann steering mechanism which results in lower wear and tear on the tyres. The driven wheels must rotate at different speeds on corners which is why a differential gearbox is required between the motor and the driven wheels.

The velocity of the robot in the world frame is $(v\cos\theta, v\sin\theta)$ and combined with Eq. 4.1 we write the equations of motion as

FIG. 124.

From Sharp 1896

$$\dot{x} = v\cos\theta$$
$$\dot{y} = v\sin\theta$$
$$\dot{\theta} = \frac{v}{L}\tan\gamma$$

(4.2)

This model is referred to as a kinematic model since it describes the velocities of the vehicle but not the forces or torques that cause the velocity. The rate of change of heading $\dot{\theta}$ is referred to as turn rate, heading rate or yaw rate and can be measured by a gyroscope. It can also be deduced from the angular velocity of the nondriven wheels on the left- and right-hand sides of the vehicle which follow arcs of different radius, and therefore rotate at different speeds.

Equation 4.2 captures some other important characteristics of a car-like vehicle. When $v = 0$ then $\dot{\theta} = 0$; that is, it is not possible to change the vehicle's orientation when it is not moving. As we know from driving, we must be moving in order to turn. When the steering angle $\gamma = \frac{\pi}{2}$ the front wheel is orthogonal to the back wheel, the vehicle cannot move forward and the model enters an undefined region.

In the world coordinate frame we can write an expression for velocity in the vehicle's y-direction

$$\dot{y}\cos\theta - \dot{x}\sin\theta \equiv 0$$

(4.3)

which is the called a nonholonomic constraint and will be discussed further in Sect. 4.3.1. This equation cannot be integrated to form a relationship between x, y and θ.

The Simulink® system

```
>> sl_lanechange
```

The model also includes a maximum velocity limit, a velocity rate limiter to model finite acceleration, and a limiter on the steering angle to model the finite range of the steered wheel. These can be accessed by double clicking the Bicycle block in Simulink.

shown in Fig. 4.2 uses the Toolbox `Bicycle` block which implements Eq. 4.2◄. The velocity input is a constant, and the steering wheel angle is a finite positive pulse followed by a negative pulse. Running the model simulates the motion of the vehicle and adds a new variable `out` to the workspace

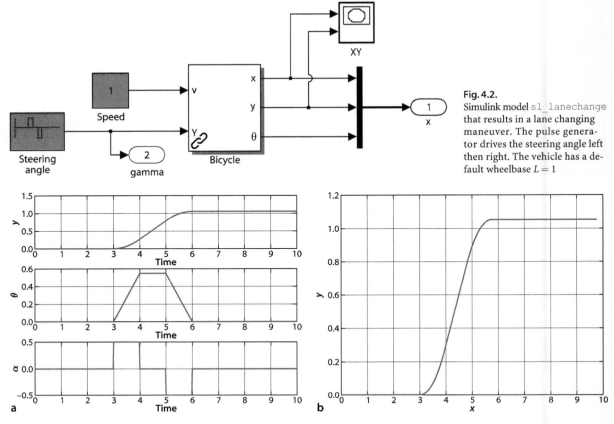

Fig. 4.2.
Simulink model `sl_lanechange` that results in a lane changing maneuver. The pulse generator drives the steering angle left then right. The vehicle has a default wheelbase $L = 1$

Fig. 4.3. Simple lane changing maneuver. **a** Vehicle response as a function of time, **b** motion in the xy-plane, the vehicle moves in the positive x-direction

```
>> out
Simulink.SimulationOutput:
    t: [504x1 double]
    y: [504x4 double]
```

from which we can retrieve the simulation time and other variables

```
>> t = out.get('t'); q = out.get('y');
```

Configuration is plotted against time

```
>> mplot(t, q)
```

in Fig. 4.3a and the result in the xy-plane

```
>> plot(q(:,1), q(:,2))
```

shown in Fig. 4.3b demonstrates a simple *lane-changing* trajectory.

4.1.1.1   Moving to a Point

Consider the problem of moving toward a goal point (x^*, y^*) in the plane. We will control the robot's velocity to be proportional to its distance from the goal

$$v^* = K_v \sqrt{\left(x^* - x\right)^2 + \left(y^* - y\right)^2}$$

and to steer toward the goal which is at the vehicle-relative angle▶ in the world frame of

$$\theta^* = \tan^{-1} \frac{y^* - y}{x^* - x}$$

This angle can be anywhere in the interval $[-\pi, \pi)$ and is computed using the `atan2` function.

using a proportional controller

$$\gamma = K_h\big(\theta^* \ominus \theta\big), \quad K_h > 0$$

which turns the steering wheel toward the target. Note the use of the operator $\ominus$ since θ^* and θ are angles $\in \mathbb{S}^1$ not real numbers◀. A Simulink model

The Toolbox function `angdiff` computes the difference between two angles and returns a difference in the interval $[-\pi, \pi]$. This is also the shortest distance around the circle, as discussed in Sect. 3.3.4.1. Also available in the Toolbox Simulink blockset `roblocks`.

```
>> sl_drivepoint
```

is shown in Fig. 4.4. We specify a goal coordinate

```
>> xg = [5 5];
```

and an initial pose

```
>> x0 = [8 5 pi/2];
```

and then simulate the motion

```
>> r = sim('sl_drivepoint');
```

The variable `r` is an object that contains the simulation results from which we extract the configuration as a function of time

```
>> q = r.find('y');
```

The vehicle's path in the plane is

```
>> plot(q(:,1), q(:,2));
```

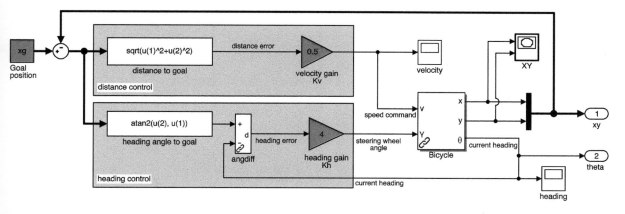

Fig. 4.4. `sl_drivepoint`, the Simulink model that drives the vehicle to a point. Red blocks have parameters that you can adjust to investigate the effect on performance

To run the Simulink model called `model` we first load it

```
>> model
```

and a new window is popped up that displays the model in block-diagram form. The simulation can be started by pressing the play button on the toolbar of the model's window. The model can also be run directly from the MATLAB command line

```
>> sim('model')
```

Many Toolbox models create additional figures to display robot animations or graphs as they run.

All models in this chapter have the simulation data export option set to create a MATLAB `SimulationOutput` object. All the unconnected output signals are concatenated, in port number order, to form a row vector and these are stacked to form a matrix `y` with one row per timestep. The corresponding time values form a vector `t`. These variables are packaged in a `SimulationOutput` object which is written to the workspace variable `out` or returned if the simulation is invoked from MATLAB

```
>> r = sim('model')
```

Displaying `r` or `out` lists the variables that it contains and their value is obtained using the find method, for example

```
>> t = r.find('t');
```

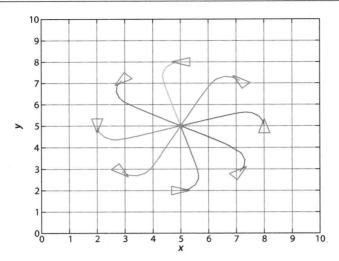

Fig. 4.5.
Simulation results for
`sl_drivepoint` for different
initial poses. The goal is (5, 5)

which is shown in Fig. 4.5 for a number of starting poses. In each case the vehicle has moved forward and turned onto a path toward the goal point. The final part of each path is a straight line and the final orientation therefore depends on the starting point.

4.1.1.2 · Following a Line

Another useful task for a mobile robot is to follow a line on the plane▸ defined by $ax + by + c = 0$. This requires two controllers to adjust steering. One controller

> 2-dimensional lines in homogeneous form are discussed in Sect. C.2.1.

$$\alpha_d = -K_d d, \; K_d > 0$$

turns the robot toward the line to minimize the robot's normal distance from the line

$$d = \frac{(a, b, c) \cdot (x, y, 1)}{\sqrt{a^2 + b^2}}$$

The second controller adjusts the heading angle, or orientation, of the vehicle to be parallel to the line

$$\theta^* = \tan^{-1} \frac{-a}{b}$$

using the proportional controller

$$\alpha_h = K_h \big(\theta^* \ominus \theta \big), \; K_h > 0$$

The combined control law

$$\gamma = -K_d d + K_h \big(\theta^* \ominus \theta \big)$$

turns the steering wheel so as to drive the robot toward the line and move along it.
The Simulink model

```
>> sl_driveline
```

is shown in Fig. 4.6. We specify the target line as a 3-vector (a, b, c)

```
>> L = [1 -2 4];
```

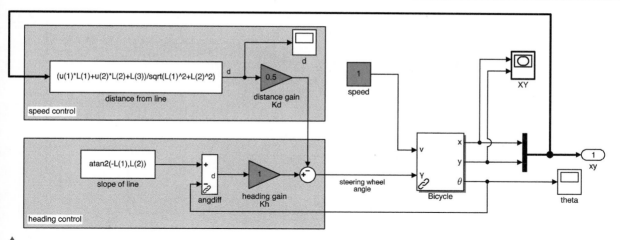

Fig. 4.6. The Simulink model `sl_driveline` drives the vehicle along a line. The line parameters (a, b, c) are set in the workspace variable `L`. Red blocks have parameters that you can adjust to investigate the effect on performance

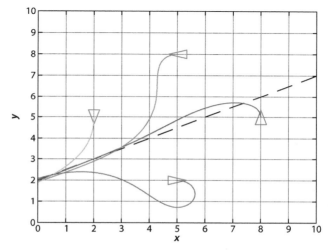

Fig. 4.7.
Simulation results from different initial poses for the line $(1, -2, 4)$

and an initial pose

```
>> x0 = [8 5 pi/2];
```

and then simulate the motion

```
>> r = sim('sl_driveline');
```

The vehicle's path for a number of different starting poses is shown in Fig. 4.7.

4.1.1.3   Following a Trajectory

Instead of a straight line we might wish to follow a trajectory that is a timed sequence of points on the xy-plane. This might come from a motion planner, such as discussed in Sect. 3.3 or 5.2, or in real-time based on the robot's sensors.

A simple and effective algorithm for trajectory following is pure pursuit in which the goal point $(x^*\langle t\rangle, y^*\langle t\rangle)$ moves along the trajectory, in its simplest form at constant speed. The vehicle always heads toward the goal – think carrot and donkey.

This problem is very similar to the control problem we tackled in Sect. 4.1.1.1, moving to a point, except this time the point is moving. The robot maintains a distance d^* behind the pursuit point and we formulate an error

$$e = \sqrt{\left(x^* - x\right)^2 + \left(y^* - y\right)^2} - d^*$$

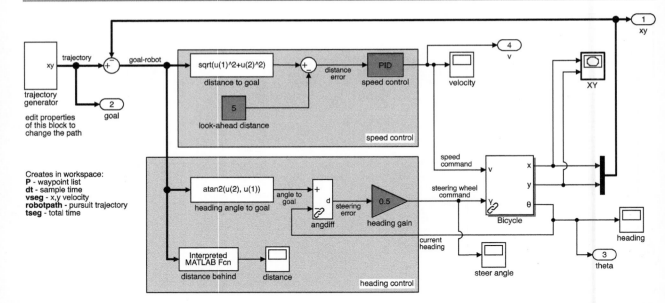

that we regulate to zero by controlling the robot's velocity using a proportional-integral (PI) controller

$$v^* = K_v e + K_i \int e\, \mathrm{d}t$$

The integral term is required to provide a nonzero velocity demand v^* when the following error is zero. The second controller steers the robot toward the target which is at the relative angle

$$\theta^* = \tan^{-1}\frac{y^* - y}{x^* - x}$$

and a simple proportional controller

$$\gamma = K_h\left(\theta^* \ominus \theta\right), \quad K_h > 0$$

turns the steering wheel so as to drive the robot toward the target.

The Simulink model

```
>> sl_pursuit
```

shown in Fig. 4.8 includes a target that moves at constant velocity along a piecewise linear path defined by a number of waypoints. It can be simulated

```
>> r = sim('sl_pursuit')
```

and the results are shown in Fig. 4.9a. The robot starts at the origin but catches up to, and follows, the moving goal. Figure 4.9b shows how the speed converges on a steady state value when following at the desired distance. Note the slow down at the end of each segment as the robot *short cuts* across the corner.

Fig. 4.8. The Simulink model `sl_pursuit` drives the vehicle along a piecewise linear trajectory. Red blocks have parameters that you can adjust to investigate the effect on performance

4.1.1.4   Moving to a Pose

The final control problem we discuss is driving to a specific pose (x^*, y^*, θ^*). The controller of Fig. 4.4 could drive the robot to a goal position but the final orientation depended on the starting position.

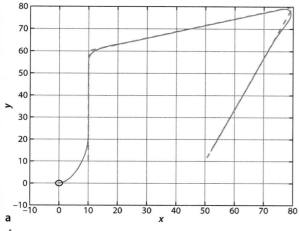

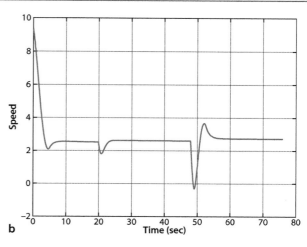

Fig. 4.9. Simulation results from pure pursuit. **a** Path of the robot in the *xy*-plane. The red dashed line is the path to be followed and the blue line in the path followed by the robot, which starts at the origin. **b** The speed of the robot versus time

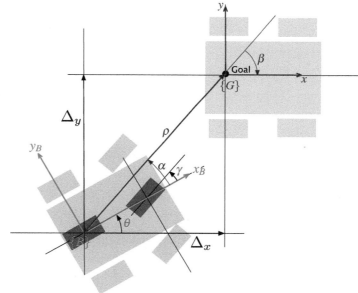

Fig. 4.10.
Polar coordinate notation for the bicycle model vehicle moving toward a goal pose: ρ is the distance to the goal, β is the angle of the goal vector with respect to the world frame, and α is the angle of the goal vector with respect to the vehicle frame

In order to control the final orientation we first rewrite Eq. 4.2 in matrix form

$$\begin{pmatrix} \dot{x} \\ \dot{\omega} \\ \dot{\theta} \end{pmatrix} = \begin{pmatrix} \cos\theta & 0 \\ \sin\theta & 0 \\ 0 & 1 \end{pmatrix} \begin{pmatrix} v \\ \omega \end{pmatrix}$$

where the inputs to the vehicle model are the speed v and the turning rate ω which can be achieved by applying the steering angle ◄

We have effectively converted the Bicycle kinematic model to a Unicycle model which we discuss in Sect. 4.1.2.

$$\gamma = \tan^{-1}\frac{\omega L}{v}$$

We then transform the equations into polar coordinate form using the notation shown in Fig. 4.10 and apply a change of variables

$$\rho = \sqrt{\Delta_x^2 + \Delta_y^2}$$

$$\alpha = \tan^{-1}\frac{\Delta_y}{\Delta_x} - \theta$$

$$\beta = -\theta - \alpha$$

which results in

$$
\begin{pmatrix} \dot{\rho} \\ \dot{\alpha} \\ \dot{\beta} \end{pmatrix} = \begin{pmatrix} -\cos\alpha & 0 \\ \dfrac{\sin\alpha}{\rho} & -1 \\ -\dfrac{\sin\alpha}{\rho} & 0 \end{pmatrix} \begin{pmatrix} v \\ \omega \end{pmatrix}, \ \text{if } \alpha \in \left(-\frac{\pi}{2}, \frac{\pi}{2} \right]
$$

and assumes the goal frame {G} is in front of the vehicle. The linear control law

$$v = k_\rho \rho$$
$$\omega = k_\alpha \alpha + k_\beta \beta$$

drives the robot to a unique equilibrium▶ at $(\rho, \alpha, \beta) = (0, 0, 0)$. The intuition behind this controller is that the terms $k_\rho \rho$ and $k_\alpha \alpha$ drive the robot along a line toward {G} while the term $k_\beta \beta$ rotates the line so that $\beta \to 0$. The closed-loop system

> The control law introduces a discontinuity at $\rho = 0$ which satisfies Brockett's theorem.

$$
\begin{pmatrix} \dot{\rho} \\ \dot{\alpha} \\ \dot{\beta} \end{pmatrix} = \begin{pmatrix} -k_\rho \cos\alpha \\ k_\rho \sin\alpha - k_\alpha \alpha - k_\beta \beta \\ -k_\rho \sin\alpha \end{pmatrix}
$$

is stable so long as

$$k_\rho > 0, \ k_\beta < 0, \ k_\alpha - k_\rho > 0$$

The distance and bearing to the goal (ρ, α) could be measured by a camera or laser range finder, and the angle β could be derived from α and vehicle orientation θ as measured by a compass.

For the case where the goal is behind the robot, that is $\alpha \notin (-\frac{\pi}{2}, \frac{\pi}{2}]$, we reverse the vehicle by negating v and γ in the control law. The velocity v always has a constant sign which depends on the initial value of α.

So far we have described a *regulator* that drives the vehicle to the pose $(0, 0, 0)$. To move the robot to an arbitrary pose (x^*, y^*, θ^*) we perform a change of coordinates

$$x' = x - x^*, \ y' = y - y^*, \ \theta' = \theta, \ \beta = \beta' + \theta^*$$

This pose controller is implemented by the Simulink model

```
>> sl_drivepose
```

shown in Fig. 4.11 and the transformation from Bicycle to Unicycle kinematics is clearly shown, mapping angular velocity ω to steering wheel angle γ. We specify a goal pose

Fig. 4.11. The Simulink model `sl_drivepose` drives the vehicle to a pose. The initial and final poses are set by the workspace variable `x0` and `xf` respectively. Red blocks have parameters that you can adjust to investigate the effect on performance

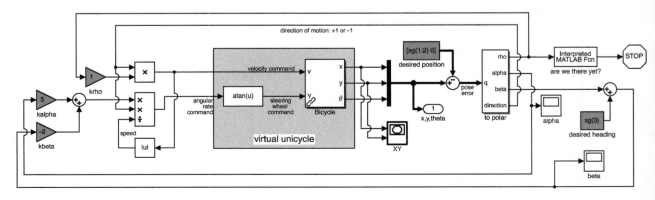

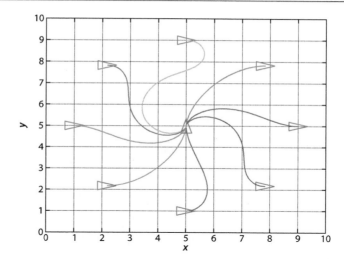

Fig. 4.12.
Simulation results from different initial poses to the final pose $(5, 5, \frac{\pi}{2})$. Note that in some cases the robot has *backed into* the final pose

```
>> xg = [5 5 pi/2];
```

and an initial pose

```
>> x0 = [9 5 0];
```

and then simulate the motion

```
>> r = sim('sl_drivepose');
```

As before, the simulation results are stored in `r` and can be plotted

```
>> q = r.find('y');
>> plot(q(:,1), q(:,2));
```

The controller is based on the bicycle model but the Simulink model `Bicycle` has additional hard nonlinearities including steering angle limits and velocity rate limiting. If those limits are violated the pose controller may fail.

to show the vehicle's path in the plane. The vehicle's path for a number of starting poses is shown in Fig. 4.12. The vehicle moves forwards or backward and takes a smooth path to the goal pose. ◄

4.1.2 Differentially-Steered Vehicle

Having steerable wheels as in a car-like vehicle is mechanically complex. Differential steering does away with this and steers by independently controlling the speed of the wheels on each side of the vehicle – if the speeds are not equal the vehicle will turn. Very simple differential steer robots have two driven wheels and a front and back castor to provide stability. Larger differential steer vehicles such as the one shown in Fig. 4.13 employ a pair of wheels on each side, with each pair sharing a drive motor via some mechanical transmission. Very large differential steer vehicles such as bulldozers and tanks sometimes employ caterpillar tracks instead of wheels. The vehicle's velocity is by definition v in the vehicle's x-direction, and zero in the y-direction since the wheels cannot slip sideways. In the vehicle frame $\{B\}$ this is

$$^{B}\!v = (v, 0)$$

The pose of the vehicle is represented by the body coordinate frame $\{B\}$ shown in Fig. 4.14, with its x-axis in the vehicle's forward direction and its origin at the centroid of the four wheels. The configuration of the vehicle is represented by the generalized coordinates $q = (x, y, \theta) \in \mathcal{C}$ where $\mathcal{C} \subset \mathbb{R}^2 \times \mathbb{S}^1$.

The vehicle follows a curved path centered on the Instantaneous Center of Rotation (ICR). The left-hand wheels move at a speed of v_L along an arc with a radius of R_L

Fig. 4.13.
Clearpath Husky robot with differential drive steering (photo by Tim Barfoot)

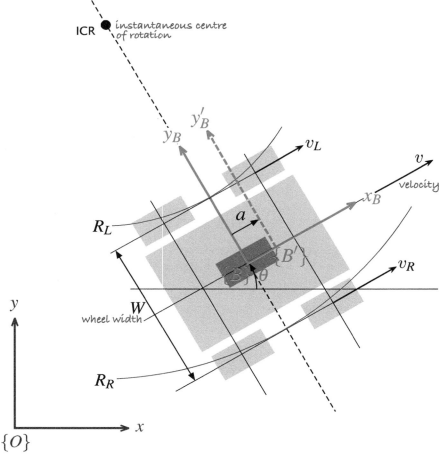

Fig. 4.14.
Differential drive robot is shown in *light grey*, and the unicycle approximation is *dark grey*. The vehicle's body coordinate frame is shown in *red*, and the world coordinate frame in *blue*. The vehicle follows a path around the Instantaneous Center of Rotation (ICR) and the distance from the ICR to the left and right wheels is R_L and R_R respectively. We can use the alternative body frame $\{B'\}$ for trajectory tracking control

while the right-hand wheels move at a speed of v_R along an arc with a radius of R_R. The angular velocity of $\{B\}$ is

$$\dot{\theta} = \frac{v_L}{R_L} = \frac{v_R}{R_R}$$

and since $R_R = R_L + W$ we can write the turn rate

$$\dot{\theta} = \frac{v_R - v_L}{W} \tag{4.4}$$

in terms of the differential velocity and wheel separation W. The equations of motion are therefore

$$\begin{aligned} \dot{x} &= v\cos\theta \\ \dot{y} &= v\sin\theta \\ \dot{\theta} &= \frac{v_\Delta}{W} \end{aligned} \tag{4.5}$$

where $v = \frac{1}{2}(v_R + v_L)$ and $v_\Delta = v_R - v_L$ are the average and differential velocities respectively. For a desired speed v and turn rate $\dot{\theta}$ we can solve for v_R and v_L. This kinematic model is often called the *unicycle model*.

There are similarities and differences to the bicycle model of Eq. 4.2. The turn rate for this vehicle is directly proportional to v_Δ and is independent of speed – the vehicle can turn even when not moving forward. For the 4-wheel case shown in Fig. 4.14 the axes of the wheels do not intersect the ICR, so when the vehicle is turning the wheel velocity vectors $\mathbf{v}_L$ and $\mathbf{v}_R$ are not tangential to the path – there is a component in the lateral direction which violates the no-slip constraint. This causes skidding or scuffing◄ which is extreme when the vehicle is turning on the spot – hence differential steering is also called skid steering. Similar to the car-like vehicle we can write an expression for velocity in the vehicle's y-direction expressed in the world coordinate frame

$$\dot{y}\cos\theta - \dot{x}\sin\theta \equiv 0 \tag{4.6}$$

which is the nonholonomic constraint. It is important to note that the ability to turn on the spot does not make the vehicle holonomic and is fundamentally different to the ability to move in an arbitrary direction which we will discuss next.

If we move the vehicle's reference frame to $\{B'\}$ and ignore orientation we can rewrite Eq. 4.5 in matrix form as

$$\begin{pmatrix} \dot{x} \\ \dot{y} \end{pmatrix} = \begin{pmatrix} \cos\theta & -a\sin\theta \\ \sin\theta & a\cos\theta \end{pmatrix} \begin{pmatrix} v \\ \omega \end{pmatrix}$$

and if $a \neq 0$ this can be be inverted

$$\begin{pmatrix} v \\ \omega \end{pmatrix} = \begin{pmatrix} \cos\theta & \sin\theta \\ -\frac{1}{a}\sin\theta & \frac{1}{a}\cos\theta \end{pmatrix} \begin{pmatrix} \dot{x} \\ \dot{y} \end{pmatrix} \tag{4.7}$$

to give the required forward speed and turn rate to achieve an arbitrary velocity $(\dot{x}, \dot{y})$ for the origin of frame $\{B'\}$.

The Toolbox Simulink block library `roblocks` contains a block called `Unicycle` to implement this model and the coordinate frame shift a is one of its parameters. It has the same outputs as the `Bicycle` model used in the last section. Equation 4.7 is implemented in the block called `Tracking Controller`.

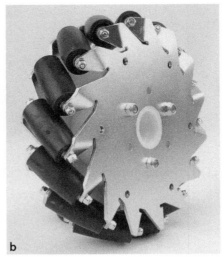

Fig. 4.15.
Two types of omnidirectional wheel, note the different roller orientation. **a** Allows the wheel to *roll* sideways (courtesy Vex Robotics); **b** allows the wheel to *drive* sideways (courtesy of Nexus Robotics)

4.1.3   Omnidirectional Vehicle

The vehicles we have discussed so far have a constraint on lateral motion, the nonholonomic constraint, which necessitates complex maneuvers in order to achieve some goal poses. Alternative wheel designs such as shown in Fig. 4.15 remove this constraint and allow omnidirectional motion. Even more radical is the spherical wheel shown in Fig. 4.16.

In this section we will discuss the mecanum or "Swedish" wheel▶ shown in Fig. 4.15b and schematically in Fig. 4.17. It comprises a number of rollers set around the circumference of the wheel with their axes at an angle of α relative to the axle of the wheel. The dark roller is the one on the bottom of the wheel and currently in contact with the ground. The rollers have a barrel shape so only one point on the roller is in contact with the ground at any time.

As shown in Fig. 4.17 we establish a wheel coordinate frame $\{W\}$ with its x-axis pointing in the direction of wheel motion. Rotation of the wheel will cause forward velocity of $R\varpi\hat{\boldsymbol{x}}_w$ where R is the wheel radius and ϖ is the wheel rotational rate. However because the roller is free to roll in the direction indicated by the green line, normal to the roller's axis, there is potentially arbitrary velocity in that direction. A desired velocity $\boldsymbol{v}$ can be resolved into two components, one parallel to the direction of wheel motion $\hat{\boldsymbol{x}}_w$ and one parallel to the rolling direction

$$\boldsymbol{v} = \underbrace{v_w\hat{\boldsymbol{x}}_w}_{\text{driven}} + \underbrace{v_r\left(\cos\alpha\hat{\boldsymbol{x}}_w + \sin\alpha\hat{\boldsymbol{y}}_w\right)}_{\text{rolling}} \tag{4.8}$$
$$= (v_w + v_r\cos\alpha)\hat{\boldsymbol{x}}_w + v_r\sin\alpha\hat{\boldsymbol{y}}_w$$

where v_w is the speed due to wheel rotation and v_r is the rolling speed. Expressing $\boldsymbol{v} = v_x\hat{\boldsymbol{x}}_w + v_y\hat{\boldsymbol{y}}_w$ in component form allows us to solve for the rolling speed $v_r = \boldsymbol{v}_y/\sin\alpha$ and substituting this into the first term we can solve for the required wheel velocity

$$v_w = \boldsymbol{v}_x - \boldsymbol{v}_y\cot\alpha \tag{4.9}$$

The required wheel rotation rate is then $\varpi = v_w/R$. If $\alpha = 0$ then v_w is undefined since the roller axes are parallel to the wheel axis and the wheel can provide no traction. If $\alpha = \frac{\pi}{2}$ as in Fig. 4.15a, the wheel allows sideways rolling but not sideways driving since there is zero coupling from v_w to $\boldsymbol{v}_y$.

Mecanum was a Swedish company where the wheel was invented by Bengt Ilon in 1973. It is described in US patent 3876255.

Fig. 4.16. The Rezero ballbot developed at ETH Zurich (photo by Péter Fankhauser)

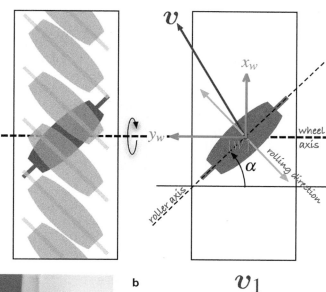

Fig. 4.17.
Schematic of a mecanum wheel in plan view. The *light rollers* are on top of the wheel, the *dark roller* is in contact with the ground. The *green arrow* indicates the rolling direction

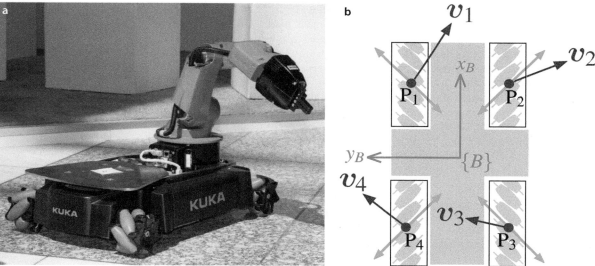

Fig. 4.18. a Kuka youBot, which has has four mecanum wheels (image courtesy youBot Store); **b** schematic of a vehicle with four mecanum wheels in the youBot configuration

A single mecanum wheel does not allow any control in the rolling direction but for three or more mecanum wheels, suitably arranged, the motion in the rolling direction of any one wheel will be driven by the other wheels. A vehicle with four mecanum wheels is shown in Fig. 4.18. Its pose is represented by the body frame {B} with its x-axis in the vehicle's forward direction and its origin at the centroid of the four wheels. The configuration of the vehicle is represented by the generalized coordinates $q = (x, y, \theta) \in \mathcal{C}$ where $\mathcal{C} \subset \mathbb{R}^2 \times \mathbb{S}^1$. The rolling axes of the wheels are orthogonal which means that when the wheels are not rotating the vehicle cannot roll in any direction or rotate.

The four wheel contact points indicated by *grey dots* have coordinate vectors $^B p_i$. For a desired body velocity $^B v_B$ and angular rate $^B \omega$ the velocity at each wheel contact point is

$$^B v_i = {}^B v_B + {}^B \omega \hat{z}_B \times {}^B p_i$$

and we then apply Eq. 4.8 and 4.9 to determine wheel rotational rates ϖ_i, while noting that α has the opposite sign for wheels 2 and 4 in Eq. 4.8.

4.2   Flying Robots

> *In order to fly, all one must do is simply miss the ground.*
> Douglas Adams

Flying robots or unmanned aerial vehicles (UAV) are becoming increasingly common and span a huge range of size and shape as shown in shown in Fig. 4.19. Applications include military operations, surveillance, meteorological observation, robotics research, commercial photography and increasingly hobbyist and personal use. A growing class of flying machines are known as micro air vehicles or MAVs which are smaller than 15 cm in all dimensions. Fixed wing UAVs are similar in principle to passenger aircraft with wings to provide lift, a propeller or jet to provide forward thrust and control surface for maneuvering. Rotorcraft UAVs have a variety of configurations that include conventional *helicopter* design with a main and tail rotor, a *coax* with counter-rotating coaxial rotors and *quadrotors*. Rotorcraft UAVs have the advantage of being able to take off vertically and to hover.

Flying robots differ from ground robots in some important ways. Firstly they have 6 degrees of freedom and their configuration $q \in \mathcal{C}$ where $\mathcal{C} \subset \mathbb{R}^3 \times \mathbb{S}^1 \times \mathbb{S}^1 \times \mathbb{S}^1$. Secondly they are actuated by forces; that is their motion model is expressed in terms of forces, torques and accelerations rather than velocities as was the case for the ground vehicle models – we use a dynamic rather than a kinematic model. Underwater robots have many similarities to flying robots and can be considered as vehicles that *fly through water* and there are underwater equivalents to fixed wing aircraft and rotorcraft. The principal differences underwater are an upward buoyancy force, drag forces that are much more significant than in air, and added mass.

In this section we will create a model for a quadrotor flying vehicle such as shown in Fig. 4.19d. Quadrotors are now widely available, both as commercial products and as open-source projects. Compared to fixed wing aircraft they are highly maneuverable and can be flown safely indoors which makes them well suited for laboratory or hobbyist use. Compared to conventional helicopters, with a large main rotor and tail rotor, the quadrotor is easier to fly, does not have the complex swash plate mechanism and is easier to model and control.

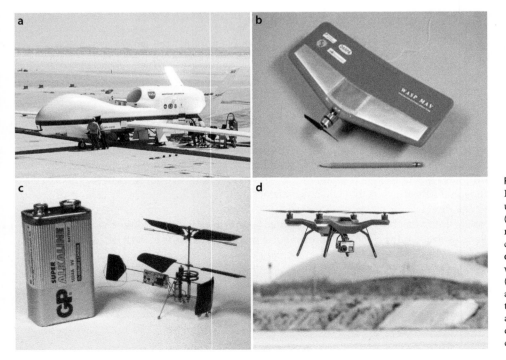

Fig. 4.19.
Flying robots. **a** Global Hawk unmanned aerial vehicle (UAV) (photo courtesy of NASA), **b** a micro air vehicle (MAV) (photo courtesy of AeroVironment, Inc.), **c** a 1 gram co-axial helicopter with 70 mm rotor diameter (photo courtesy of Petter Muren and Proxflyer AS), **d** a quadrotor which has four rotors and a block of sensing and control electronics in the middle (photo courtesy of 3DRobotics)

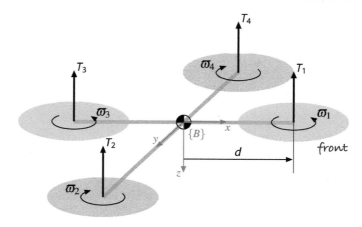

Fig. 4.20.
Quadrotor notation showing the four rotors, their thrust vectors and directions of rotation. The body frame {B} is attached to the vehicle and has its origin at the vehicle's center of mass. Rotors 1 and 3 (*blue*) rotate counter-clockwise (viewed from above) while rotors 2 and 4 (*red*) rotate clockwise

The notation for the quadrotor model is shown in Fig. 4.20. The body coordinate frame {B} has its z-axis downward following the aerospace convention. The quadrotor has four rotors, labeled 1 to 4, mounted at the end of each cross arm. Hex- and octo-rotors are also popular, with the extra rotors providing greater payload lift capability. The approach described here can be generalized to N rotors, where N is even.

The rotors are driven by electric motors powered by electronic speed controllers. Some low-cost quadrotors use small motors and reduction gearing to achieve sufficient torque. The rotor speed is ϖ_i and the thrust is an upward vector

$$T_i = b\varpi_i^2, \quad i = 1, 2, 3, 4 \tag{4.10}$$

in the vehicle's negative z-direction, where $b > 0$ is the lift constant that depends on the air density, the cube of the rotor blade radius, the number of blades, and the chord length of the blade. ◄

Close to the ground, height $<2d$, the vehicle experiences increased lift due to a cushion of air beneath it – this is ground effect.

The translational dynamics of the vehicle in world coordinates is given by Newton's second law

$$m\dot{v} = \begin{pmatrix} 0 \\ 0 \\ mg \end{pmatrix} - {}^0R_B \begin{pmatrix} 0 \\ 0 \\ T \end{pmatrix} - Bv \tag{4.11}$$

where v is the velocity of the vehicle's center of mass in the world frame, g is gravitational acceleration, m is the total mass of the vehicle, B is aerodynamic friction and $T = \Sigma T_i$ is the total upward thrust. The first term is the force of gravity which acts downward in the world frame, the second term is the total thrust in the vehicle frame rotated into the world coordinate frame and the third term is aerodynamic drag.

Pairwise differences in rotor thrusts cause the vehicle to rotate. The torque about the vehicle's x-axis, the *rolling* torque, is generated by the moments

$$\tau_x = dT_4 - dT_2$$

The propeller blades on a rotor craft have fascinating dynamics. When flying into the wind the blade tip coming forward experiences greater lift while the receding blade has less lift. This is equivalent to a torque about an axis pointing into the wind and the rotor blades behave like a gyroscope (see Sect. 3.4.1.1) so the net effect is that the rotor blade plane pitches up by an amount proportional to the apparent or nett wind speed, countered by the blade's bending stiffness and the change in lift as a function of blade bending. The pitched blade plane causes a component of the thrust vector to retard the vehicle's forward motion and this velocity dependent force acts like a friction force. This is known as blade flapping and is an important characteristic of blades on all types of rotorcraft.

where d is the distance from the rotor axis to the center of mass. We can write this in terms of rotor speeds by substituting Eq. 4.10

$$\tau_x = db\left(\varpi_4^2 - \varpi_2^2\right) \tag{4.12}$$

and similarly for the y-axis, the *pitching* torque is

$$\tau_y = db\left(\varpi_1^2 - \varpi_3^2\right) \tag{4.13}$$

The torque applied to each propeller by the motor is opposed by aerodynamic drag

$$Q_i = k\varpi_i^2$$

where k depends on the same factors as b. This torque exerts a reaction torque on the airframe which acts to rotate the airframe about the propeller shaft in the opposite direction to its rotation. The total reaction torque about the z-axis is

$$\begin{aligned}
\tau_z &= Q_1 - Q_2 + Q_3 - Q_4 \\
&= k\left(\varpi_1^2 + \varpi_3^2 - \varpi_2^2 - \varpi_4^2\right)
\end{aligned} \tag{4.14}$$

where the different signs are due to the different rotation directions of the rotors. A yaw torque can be created simply by appropriate coordinated control of all four rotor speeds.

The total torque applied to the airframe according to Eq. 4.12 to 4.14 is $\boldsymbol{\tau} = (\tau_x, \tau_y, \tau_z)^T$ and the rotational acceleration is given by Euler's equation of motion from Eq. 3.10

$$J\dot{\boldsymbol{\omega}} = -\boldsymbol{\omega} \times J\boldsymbol{\omega} + \boldsymbol{\tau} \tag{4.15}$$

where J is the 3×3 inertia matrix of the vehicle and $\boldsymbol{\omega}$ is the angular velocity vector.

The motion of the quadrotor is obtained by integrating the forward dynamics equations Eq. 4.11 and Eq. 4.15 where the forces and moments on the airframe

$$\begin{pmatrix} T \\ \boldsymbol{\tau} \end{pmatrix} = \begin{pmatrix} -b & -b & -b & -b \\ 0 & -db & 0 & db \\ db & 0 & -db & 0 \\ k & -k & k & -k \end{pmatrix} \begin{pmatrix} \varpi_1^2 \\ \varpi_2^2 \\ \varpi_3^2 \\ \varpi_4^2 \end{pmatrix} = A \begin{pmatrix} \varpi_1^2 \\ \varpi_2^2 \\ \varpi_3^2 \\ \varpi_4^2 \end{pmatrix} \tag{4.16}$$

are functions of the rotor speeds. The matrix A is constant, and full rank if $b, k, d > 0$ and can be inverted

$$\begin{pmatrix} \varpi_1^2 \\ \varpi_2^2 \\ \varpi_3^2 \\ \varpi_4^2 \end{pmatrix} = A^{-1} \begin{pmatrix} T \\ \tau_x \\ \tau_y \\ \tau_z \end{pmatrix} \tag{4.17}$$

to solve for the rotor speeds► required to apply a specified thrust T and moment $\boldsymbol{\tau}$ to the airframe.

To control the vehicle we will employ a nested control structure which we describe for pitch and x-translational motion. The innermost loop uses a proportional and derivative controller► to compute the required pitching torque on the airframe

$$\tau_y^* = K_{\tau,p}\left(\theta_p^* - \theta_p^{\#}\right) + K_{\tau,d}\left(\dot{\theta}_p^* - \dot{\theta}_p^{\#}\right) \tag{4.18}$$

based on the error between desired and actual pitch angle.► The gains $K_{\tau,p}$ and $K_{\tau,d}$ are determined by classical control design approaches based on an approximate dy-

The direction of rotation is as shown in Fig. 4.20. Control of motor velocity is discussed in Sect. 9.1.6.

The rotational dynamics has a second-order transfer function of $\Theta_y(s) / \tau_y(s) = 1 / (Js^2 + Bs)$ where J is rotational inertia and B is aerodynamic damping which is generally quite small. To regulate a second-order system requires a proportional-derivative controller.

The term $\dot{\theta}_p^*$ is commonly ignored.

namic model and then tuned to achieve good performance. The actual vehicle pitch angle $\theta_p^{\#}$ would be estimated by an inertial navigation system as discussed in Sect. 3.4 and $\dot\theta_p^{\#}$ would be derived from gyroscopic sensors. The required rotor speeds are then determined using Eq. 4.17.

Consider a coordinate frame $\{B'\}$ attached to the vehicle and with the same origin as $\{B\}$ but with its x- and y-axes in the horizontal plane and parallel to the ground. The thrust vector is parallel to the z-axis of frame $\{B\}$ and pitching the nose down, rotating about the y-axis by θ_p, generates a force

$$^{B'}\!f = \mathscr{R}_y\!\left(\theta_p\right) \cdot \begin{pmatrix} 0 \\ 0 \\ T \end{pmatrix} = \begin{pmatrix} T\sin\theta_p \\ 0 \\ T\cos\theta_p \end{pmatrix}$$

which has a component

$$^{B'}\!f_x = T\sin\theta_p \approx T\theta_p$$

that accelerates the vehicle in the $^{B'}\!x$-direction, and we have assumed that θ_p is small. We can control the velocity in this direction with a proportional control law

$$^{B'}\!f_x^* = mK_f\!\left(^{B'}\!v_x^* - {}^{B'}\!v_x^{\#}\right)$$

where $K_f > 0$ is a gain. Combining these two equations we obtain the desired pitch angle

$$\theta_p^* \approx \frac{m}{T}K_f\!\left(^{B'}\!v_x^* - {}^{B'}\!v_x^{\#}\right) \tag{4.19}$$

required to achieve the desired forward velocity. Using Eq. 4.18 we compute the required pitching torque, and then using Eq. 4.17 the required rotor speeds. For a vehicle in vertical equilibrium the total thrust equals the weight force so $m / T \approx 1 / g$.

The actual vehicle velocity $^{B}\!v_x$ would be estimated by an inertial navigation system as discussed in Sect. 3.4 or a GPS receiver. If the position of the vehicle in the xy-plane of the world frame is $p \in \mathbb{R}^2$ then the desired velocity is given by the proportional control law

$$^{0}\!v^* = K_p\!\left(^{0}\!p^* - {}^{0}\!p^{\#}\right) \tag{4.20}$$

based on the error between the desired and actual position. The desired velocity in the xy-plane of frame $\{B'\}$ is

$$^{B'}\!v = \ominus{}^{0}\mathscr{R}_{B'}\!\left(\theta_y\right) \cdot {}^{0}\!v, \quad \mathscr{R} \in \mathbf{SO}(2)$$

which is a function of the yaw angle θ_y

$$\begin{pmatrix} ^{B'}\!v_x \\ ^{B'}\!v_y \end{pmatrix} = \begin{pmatrix} \cos\theta_y & -\sin\theta_y \\ \sin\theta_y & \cos\theta_y \end{pmatrix}^{T} \begin{pmatrix} v_x \\ v_y \end{pmatrix}$$

Figure 4.21 shows a Simulink model of the complete control system for a quadrotor◀ which can be loaded and displayed by

```
>> sl_quadrotor
```

This model is hierarchical and organized in terms of subsystems. Click the down arrow on a subsystem (can be seen on-screen but not in the figure) to reveal the detail. Double-click on the subsystem box to modify its parameters.

Working our way left to right and starting at the top we have the desired position of the quadrotor in world coordinates. The position error is rotated from the world frame to the body frame and becomes the desired velocity. The velocity controller implements Eq. 4.19 and its equivalent for the roll axis and outputs the desired pitch and roll angles of the quadrotor. The attitude controller is a proportional-derivative controller that determines the appropriate pitch and roll torques to achieve these

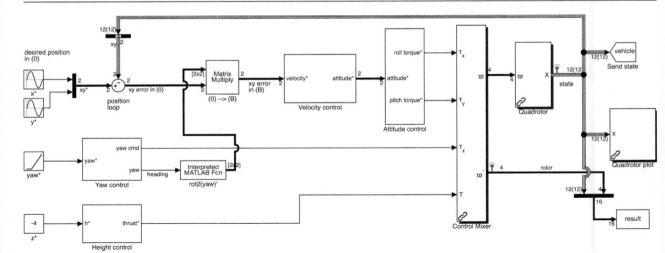

angles based on feedback of current attitude and attitude rate. ◂ The yaw control block determines the error in heading angle and implements a proportional-derivative controller to compute the required yaw torque which is achieved by speeding up one pair of rotors and slowing the other pair.

Altitude is controlled by a proportional-derivative controller

$$T = K_p\left(z^* - z^{\#}\right) + K_d\left(\dot{z}^* - \dot{z}^{\#}\right) + T_0$$

which determines the average rotor speed. $T_0 = mg$ is the weight of the vehicle and this is an example of feedforward control – used here to counter the effect of gravity which otherwise is a constant disturbance to the altitude control loop. The alternatives to feedforward control would be to have very high gain for the altitude loop which often leads to actuator saturation and instability, or a proportional-integral (PI) controller which might require a long time for the integral term to increase to a useful value and then lead to overshoot. We will revisit gravity compensation in Chap. 9 applied to arm-type robots.

The control mixer block combines the three torque demands and the vertical thrust demand and implements Eq. 4.17 to determine the appropriate rotor speeds. Rotor speed limits are applied here. These are input to the quadrotor block▸ which implements the forward dynamics integrating Eq. 4.16 to give the position, velocity, orientation and orientation rate. The output of this block is the state vector $x = (^0p, {}^0\Gamma, {}^Bp, {}^B\dot{\Gamma}) \in \mathbb{R}^{12}$. As is common in aerospace applications we represent orientation Γ and orientation rate $\dot{\Gamma}$ in terms of roll-pitch-yaw angles. Note that position and attitude are in the world frame while the rates are expressed in the body frame.

The parameters of a specific quadrotor can be loaded

```
>> mdl_quadrotor
```

which creates a structure called `quadrotor` in the workspace, and its elements are the various dynamic properties of the quadrotor. The simulation can be run using the Simulink menu or from the MATLAB command line

```
>> sim('sl_quadrotor');
```

and it displays an animation in a separate window.▸ The vehicle lifts off and flies around a circle while spinning slowly about its own z-axis. A snapshot is shown in Fig. 4.22. The simulation writes the results from each timestep into a matrix in the workspace

```
>> about result
result [double] : 2412x16 (308.7 kB)
```

Fig. 4.21. The Simulink® model `sl_quadrotor` which is a closed-loop simulation of the quadrotor. The vehicle takes off and flies in a circle at constant altitude. A Simulink bus is used for the 12-element state vector X output by the `Quadrotor` block. To reduce the number of lines in the diagram we have used `Goto` and `From` blocks to transmit and receive the state vector

Note that according to the coordinate conventions shown in Fig. 4.20 x-direction motion requires a negative rotation about the y-axis (pitch angle) and y-direction motion requires a positive rotation about the x-axis (roll angle) so the gains have different signs for the roll and pitch loops.

The Simulink library `roblocks` also includes a block for an N-rotor vehicle.

Loading and displaying the model using
`>> sl_quadrotor` automatically loads the default quadrotor model. This is done by the PreLoadFcn callback set from model's properties File+Model Properties+Model Properties+Callbacks+PreLoadFcn.

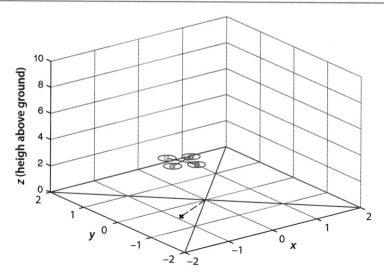

Fig. 4.22.
One frame from the quadrotor simulation. The marker on the ground plane is a projection of the vehicle's centroid

which has one row per timestep, and each row contains the time followed by the state vector (elements 2–13) and the commanded rotor speeds ω_i (elements 14–17). To plot x and y versus time is

```
>> plot(result(:,1), result(:,2:3));
```

To recap on control of the quadrotor. A position error results in a required translational velocity. To achieve this requires appropriate pitch and roll angles so that a component of the vehicle's thrust acts in the horizontal plane and generates a force to accelerate the vehicle.◀ As it approaches its goal the airframe must be rotated in the opposite direction so that a component of thrust decelerates the motion. To achieve the pitch and roll angles requires differential propeller thrust to create a moment that rotationally accelerates the airframe.

The total thrust must be increased so that the vertical thrust component still balances gravity.

This indirection from translational motion to rotational motion is a consequence of the vehicle being under-actuated – we have just four rotor speeds to adjust but the vehicle's configuration space is 6-dimensional. In the configuration space we cannot move in the x- or y-direction, but we can move in the pitch- or roll-direction which results in motion in the x- or y-direction. The cost of under actuation is once again a maneuver. The pitch and roll angles are a means to achieve translation control and cannot be independently set.

4.3   Advanced Topics

4.3.1   Nonholonomic and Under-Actuated Systems

We introduced the notion of configuration space in Sect. 2.3.5 and it is useful to revisit it now that we have discussed several different types of mobile robot platform. Common vehicles – as diverse as cars, hovercrafts, ships and aircraft – are all able to move forward effectively but are unable to instantaneously move sideways. This is a very sensible tradeoff that simplifies design and caters to the motion we most commonly require of the vehicle. Sideways motion for occasional tasks such as parking a car, docking a ship or landing an aircraft are possible, albeit with some complex maneuvering but humans can learn this skill.

Consider a hovercraft which moves over a planar surface. To fully describe all its constituent particles we need to specify three generalized coordinates: its position in the xy-plane and its rotation angle. It has three degrees of freedom and its configuration space is $\mathcal{C} \subset \mathbb{R}^2 \times \mathbb{S}^1$. This hovercraft has two propellers whose axes are parallel but not collinear. The sum of their thrusts provide a forward force and the difference in thrusts generates a yawing torque for steering. The number of actuators, two, is less than its degrees of freedom dim $\mathcal{C} = 3$ and we call this an under-actuated system. This imposes significant limitations on the way in which it can move. At any point in time we can control the forward (parallel to the thrust vectors) acceleration and the rotational acceleration of the hovercraft but there is zero sideways (or lateral) acceleration since it cannot generate any lateral thrust. Nevertheless with some clever maneuvering, like with a car, the hovercraft can follow a path that will take it to a place to one side of where it started. In the hovercraft's 3-dimensional configuration space this means that at any point there are certain directions in which *acceleration* is not possible. We can reach points in those direction but not directly, only by following some circuitous path.

All flying and underwater vehicles have a configuration that is completely described by six generalized coordinates – their position and orientation in 3D space. $\mathcal{C} \subset \mathbb{R}^3 \times \mathbb{S}^1 \times \mathbb{S}^1 \times \mathbb{S}^1$ where the orientation is expressed in some three-angle representation – since dim $\mathcal{C} = 6$ the vehicles have six degrees of freedom. A quadrotor has four actuators, four thrust-generating propellers, and this is fewer than its degrees of freedom making it under-actuated. Controlling the four propellers causes motion in the up/down, roll, pitch and yaw directions of the configuration space but not in the forward/backward or left/right directions. To access those degrees of freedom it is necessary to perform a *maneuver*: pitch down so that the thrust vector provides a horizontal force component, accelerate forward, pitch up so that the thrust vector provides a horizontal force component to decelerate, and then level out.

For a helicopter only four of the six degrees of freedom are practically useful: up/down, forward/backward, left/right and yaw. Therefore a helicopter requires a minimum of four actuators: the main rotor generates a thrust vector whose magnitude is controlled by the collective pitch and whose direction is controlled by the lateral and longitudinal cyclic pitch. The tail rotor provides a yawing moment. This leaves two degrees of freedom unactuated, roll and pitch angles, but clever design ensures that gravity actuates them and keeps them close to zero – without gravity a helicopter cannot work. A fixed-wing aircraft moves forward very efficiently and also has four actuators: engine thrust provides acceleration in the forward direction and the ailerons, elevator and rudder exert respectively roll, pitch and yaw moments on the aircraft.▸ To access the missing degrees of freedom such as up/down and left/right translation, the aircraft must pitch or yaw while moving forward.

> Some low-cost hobby aircraft have no rudder and rely only on ailerons to bank and turn the aircraft. Even cheaper hobby aircraft have no elevator and rely on engine speed to control height.

The advantage of under-actuation is having fewer actuators. In practice this means real savings in terms of cost, complexity and weight. The consequence is that at any point in its configuration space there are certain directions in which the vehicle cannot move. Full actuation is possible but not common, for example the DEPTHX underwater robot shown on page 96 has six degrees of freedom and six actuators. These can exert an arbitrary force and torque on the vehicle, allowing it to accelerate in any direction or about any axis.

A 4-wheeled car has many similarities to the hovercraft discussed above. It moves over a planar surface and its configuration can be fully described by its generalized coordinates: its position in the xy-plane and a rotation angle. It has three degrees of freedom and its configuration space is $\mathcal{C} \subset \mathbb{R}^2 \times \mathbb{S}^1$. A car has two actuators, one to move forwards or backwards and one to change the heading direction. A car, like a hovercraft, is under-actuated.

We know from our experience with cars that we cannot move directly in certain directions and sometimes needs to perform a maneuver to reach our goal. A differential- or skid-steered vehicle, such as a tank, is also under-actuated – it has only two actuators, one for each track. While this type of vehicle can turn on the spot it cannot move sideways. To do that it has to turn, proceed, stop then turn – this need to maneuver is the clear signature of an under-actuated system.

We might often wish for an ability to drive our car sideways but the standard wheel provides real benefit when cornering – lateral friction between the wheels and the

Table 4.1.
Summary of configuration space
characteristics for various robots.
A nonholonomic system is
under-actuated and/or has a
rolling constraint

	dim $\mathcal{C}$	Degrees of freedom	Number of actuators	Actuation	Rolling constraints	Holonomic
Train	1	1	1	full		✓
2-joint robot arm	2	2	2	full		✓
6-joint robot arm	6	6	6	full		✓
10-joint robot arm	10	10	10	over		✓
Hovercraft	3	3	2	under		
Car	3	2	2	under	✓	
Helicopter	6	6	4	under		
Fixed wing aircraft	6	6	4	under		
DEPTHX AUV	6	6	6	full		✓

road provides, for free, the centripetal force which would otherwise require an extra actuator to provide. The hovercraft has many similarities to a car but we can push a hovercraft sideways – we cannot do that with a car. This lateral friction is a distinguishing feature of the car.

The inability to slip sideways is a constraint, the *rolling* constraint, on the velocity◄ of the vehicle just as under-actuation is. A vehicle with one or more velocity constraints, due to under-actuation or a rolling constraint, is referred to as a nonholonomic system. A key characteristic of these systems is that they cannot move *directly* from one configuration to another – they must perform a maneuver or sequence of motions. A car has a velocity constraint due to its wheels and is also under-actuated.

A holonomic constraint restricts the possible configurations that the system can achieve – it can be expressed as an equation written in terms of the configuration variables.◄ A nonholonomic constraint such as Eq. 4.3 and 4.6 is one that restricts the *velocity* (or acceleration) of a system in configuration space – it can only be expressed in terms of the derivatives of the configuration variables.◄ The nonholonomic constraint does not restrict the possible configurations the system can achieve but it does preclude instantaneous velocity or acceleration in certain directions.

In control theoretic terms Brockett's theorem (Brockett 1983) states that nonholonomic systems are controllable but they cannot be stabilized to a desired state using a differentiable, or even continuous, pure state-feedback controller. A time-varying or nonlinear control strategy is required which means that the robot follows some generally nonlinear path. One exception is an under-actuated system moving in 3-dimensional space within a force field, for example a gravity field – gravity acts like an additional actuator and makes the system linearly controllable (but not holonomic), as we showed for the quadrotor example in Sect. 4.2.

Mobility parameters for the various robots that we have discussed here, and earlier in Sect. 2.3.5, are tabulated in Table 4.1. We will discuss under- and over-actuation in the context of arm robots in Chap. 8.

The hovercraft, aerial and underwater vehicles are controlled by forces so in this case the constraints are on vehicle acceleration in configuration space not velocity.

For example fixing the end of the 10-joint robot arm introduces six holonomic constraints (position and orientation) so the arm would have only 4 degrees of freedom.

The constraint cannot be integrated to a constraint in terms of configuration variables, so such systems are also known as nonintegrable systems.

4.4   Wrapping Up

In this chapter we have created and discussed models and controllers for a number of common, but quite different, robot platforms. We first discussed wheeled robots. For car-like vehicles we developed a kinematic model which we used to develop a number of different controllers in order that the platform could perform useful tasks such as driving to a point, driving along a line, following a trajectory or driving to a pose. We then discussed differentially steered vehicles on which many robots are based, and omnidirectional robots based on novel wheel types. Then we we discussed a simple but common flying vehicle, the quadrotor, and developed a dynamic model and a hierarchical control system that allowed the quadrotor to fly a circuit. This hierarchical or nested control approach is described in more detail in Sect. 9.1.7 in the context of robot arms.

We also extended our earlier discussion about configuration space to include the velocity constraints due to under actuation and rolling constraints from wheels.

The next chapters in this Part will discuss how to plan paths for robots through complex environments that contain obstacles and then how to determine the location of a robot.

Further Reading

Comprehensive modeling of mobile ground robots is provided in the book by Siegwart et al. (2011). In addition to the models covered here, it presents in-depth discussion of a variety of wheel configurations with different combinations of driven wheels, steered wheels and passive castors. The book by Kelly (2013) also covers vehicle modeling and control. Both books also provide a good introduction to perception, localization and navigation which we will discuss in the coming chapters.

The paper by Martins et al. (2008) discusses kinematics, dynamics and control of differential steer robots. The Handbook of Robotics (Siciliano and Khatib 2016, part E) covers modeling and control of various vehicle types including aerial and underwater. The theory of helicopters with an emphasis on robotics is provided by Mettler (2003) but the definitive reference for helicopter dynamics is the very large book by Prouty (2002). The book by Antonelli (2014) provides comprehensive coverage of modeling and control of underwater robots.

Some of the earliest papers on quadrotor modeling and control are by Pounds, Mahony and colleagues (Hamel et al. 2002; Pounds et al. 2004, 2006). The thesis by Pounds (2007) presents comprehensive aerodynamic modeling of a quadrotor with a particular focus on blade flapping, a phenomenon well known in conventional helicopters but largely ignored for quadrotors. A tutorial introduction to the control of multi-rotor flying robots is given by Mahony, Kumar, and Corke (2012). Quadrotors are now commercially available from many vendors at quite low cost. There are also a number of hardware kits and open-source software projects such as ArduCopter and Mikrokopter.

Mobile ground robots are now a mature technology for transporting parts around manufacturing plants. The research frontier is now for vehicles that operate autonomously in outdoor environments (Siciliano and Khatib 2016, part F). Research into the automation of passenger cars has been ongoing since the 1980s and a number of automative manufacturers are talking about commercial autonomous cars by 2020.

Historical and interesting. The Navlab project at Carnegie-Mellon University started in 1984 and a series of autonomous vehicles, Navlabs, were built and a large body of research has resulted. All vehicles made strong use of computer vision for navigation. In 1995 the supervised autonomous Navlab 5 made a 3 000-mile journey, dubbed "No Hands Across America" (Pomerleau and Jochem 1995, 1996). The vehicle steered itself 98% of the time largely by visual sensing of the white lines at the edge of the road.

In Europe, Ernst Dickmanns and his team at Universität der Bundeswehr München demonstrated autonomous control of vehicles. In 1988 the VaMoRs system, a 5 tonne Mercedes-Benz van, could drive itself at speeds over 90 km h^{-1} (Dickmanns and Graefe 1988b; Dickmanns and Zapp 1987; Dickmanns 2007). The European Prometheus Project ran from 1987–1995 and in 1994 the robot vehicles VaMP and VITA-2 drove more than 1 000 km on a Paris multi-lane highway in standard heavy traffic at speeds up to 130 km h^{-1}. They demonstrated autonomous driving in free lanes, convoy driving, automatic tracking of other vehicles, and lane changes with autonomous passing of other cars. In 1995 an autonomous S-Class Mercedes-Benz made a 1 600 km trip from Munich to Copenhagen and back. On the German Autobahn speeds exceeded 175 km h^{-1} and the vehicle executed traffic maneuvers such as overtaking. The mean time between human interventions was 9 km and it drove up to 158 km without any human intervention. The UK part of the project demonstrated autonomous driving of an XJ6 Jaguar with vision (Matthews et al. 1995) and radar-based sensing for lane keeping and collision avoidance. More recently, in the USA a series of Grand Challenges were run for autonomous cars. The 2005 desert and 2007 urban challenges are comprehensively described in compilations of papers from the various teams in Buehler et al. (2007, 2010). More recent demonstrations of self-driving vehicles are a journey along the fabled silk road described by Bertozzi et al. (2011) and a classic road trip through Germany by Ziegler et al. (2014).

Ackermann's magazine can be found online at http://smithandgosling.wordpress.com/2009/12/02/ackermanns-repository-of-arts and the carriage steering mechanism is published in the March and April issues of 1818. King-Hele (2002) provides a comprehensive discussion about the prior work on steering geometry and Darwin's earlier invention.

Toolbox and MATLAB Notes

In addition to the Simulink `Bicycle` model used in this chapter the Toolbox also provides a MATLAB class which implements these kinematic equations and which we will use in Chap. 6. For example we can create a vehicle model with steer angle and speed limits

```
>> veh = Bicycle('speedmax', 1, 'steermax', 30*pi/180);
```

and evaluate Eq. 4.2 for a particular state and set of control inputs (v, γ)

```
>> veh.deriv([], [0 0 0], [0.3, 0.2])
ans =
    0.3000         0    0.0608
```

The `Unicycle` class is used for a differentially-steered robot and has equivalent methods.

The Robotics System Toolbox™ from The MathWorks has support for differentially-steered mobile robots which can be created using the function `ExampleHelperRobotSimulator`. It also includes a class `robotics.PurePursuit` that implements pure pursuit for a differential steer robot. An example is given in the Toolbox RST folder.

Exercises

1. For a 4-wheel vehicle with $L = 2$ m and width between wheel centers of 1.5 m
   a) What steering wheel angle is needed for a turn rate of 10 deg s^{-1} at a forward speed of 20 km h^{-1}?
   b) compute the difference in wheel steer angle for Ackermann steering around curves of radius 10, 50 and 100 m.
   c) If the vehicle is moving at 80 km h^{-1} compute the difference in back wheel rotation rates for curves of radius 10, 50 and 100 m.
2. Write an expression for turn rate in terms of the angular rotation rate of the two back wheels. Investigate the effect of errors in wheel radius and vehicle width.
3. Consider a car and bus with $L = 4$ and 12 m respectively. To follow a curve with radius of 10, 20 and 50 m determine the respective steered wheel angles.
4. For a number of steered wheel angles in the range -45 to 45° and a velocity of 2 m s^{-1} overlay plots of the vehicle's trajectory in the xy-plane.

5. Implement the $\ominus$ operator used in Sect. 4.1.1.1 and check against the code for `angdiff`.
6. Moving to a point (page 103) plot x, y and θ against time.
7. Pure pursuit example (page 106)
   a) Investigate what happens if you vary the look-ahead distance, heading gain or proportional gain in the speed controller.
   b) Investigate what happens when the integral gain in the speed controller is zero.
   c) With integral set to zero, add a constant to the output of the controller. What should the value of the constant be?
   d) Add a velocity feedforward term.
   e) Modify the pure pursuit example so the robot follows a slalom course.
   f) Modify the pure pursuit example to follow a target moving back and forth along a line.
8. Moving to a pose (page 107)
   a) Repeat the example with a different initial orientation.
   b) Implement a parallel parking maneuver. Is the resulting path practical?
   c) Experiment with different control parameters.
9. Use the MATLAB GUI interface to make a simple steering wheel and velocity control, and use this to create a very simple driving simulator. Alternatively interface a gaming steering wheel and pedal to MATLAB.
10. Adapt the various controllers in Sect. 4.1.1 to the differentially steered robot.
11. Derive Eq. 4.4 from the preceding equation.
12. For constant forward velocity, plot v_L and v_R as a function of ICR position along the y-axis. Under what conditions do v_L and v_R have a different sign?
13. Using Simulink implement a controller using Eq. 4.7 that moves the robot in its y-direction. How does the robot's orientation change as it moves?
14. Sketch the design for a robot with three mecanum wheels. Ensure that it cannot roll freely and that it can drive in any direction. Write code to convert from desired vehicle translational and rotational velocity to wheel rotation rates.
15. For the 4-wheel omnidirectional robot of Sect. 4.1.3 write an algorithm that will allow it to move in a circle of radius 0.5 m around a point with its nose always pointed toward the center of the circle.
16. Quadrotor (page 115)
   a) At equilibrium, compute the speed of all the propellers.
   b) Experiment with different control gains. What happens if you reduce the damping gains to zero?
   c) Remove the gravity feedforward and experiment with large altitude gain or a PI controller.
   d) When the vehicle has nonzero roll and pitch angles, the magnitude of the vertical thrust is reduced and the vehicle will slowly descend. Add compensation to the vertical thrust to correct this.
   e) Simulate the quadrotor flying inverted, that is, its z-axis is pointing upwards.
   f) Program a ballistic motion. Have the quadrotor take off at 45 deg to horizontal then remove all thrust.
   g) Program a smooth landing.
   h) Program a barrel roll maneuver. Have the quadrotor fly horizontally in its x-direction and then increase the roll angle from 0 to 2π.
   i) Program a flip maneuver. Have the quadrotor fly horizontally in its x-direction and then increase the pitch angle from 0 to 2π.
   j) Add another four rotors.
   k) Use the function `mstraj` to create a trajectory through ten via points $(X_i, Y_i, Z_i, \theta_y)$ and modify the controller of Fig. 4.21 for smooth pursuit of this trajectory.
   l) Use the MATLAB GUI interface to make a simple joystick control, and use this to create a very simple flying simulator. Alternatively interface a gaming joystick to MATLAB.

5 Navigation

the process of directing a vehicle so as to reach the intended destination
IEEE Standard 172-1983

Robot navigation is the problem of guiding a robot towards a goal. The human approach to navigation is to make maps and erect signposts, and at first glance it seems obvious that robots should operate the same way. However many robotic tasks can be achieved without any map at all, using an approach referred to as *reactive navigation*. For example, navigating by heading towards a light, following a white line on the ground, moving through a maze by following a wall, or vacuuming a room by following a random path. The robot is reacting directly to its environment: the intensity of the light, the relative position of the white line or contact with a wall. Grey Walter's tortoise Elsie from page 95 demonstrated "life-like" behaviors – she *reacted* to her environment and could seek out a light source. Today tens of millions of robotic vacuum cleaners are cleaning floors and most of them do so without using any map of the rooms in which they work. Instead they do the job by making random moves and sensing only that they have made contact with an obstacle as shown in Fig. 5.1.

Human-style *map-based navigation* is used by more sophisticated robots and is also known as motion planning. This approach supports more complex tasks but is itself more complex. It imposes a number of requirements, not the least of which is having a map of the environment. It also requires that the robot's position is always known. In the next chapter we will discuss how robots can determine their position and create maps. The remainder of this chapter discusses the reactive and map-based approaches to robot navigation with a focus on wheeled robots operating in a planar environment.

Fig. 5.1.
Time lapse photograph of a Roomba robot cleaning a room (photo by Chris Bartlett)

Valentino Braitenberg (1926–2011) was an Italian-Austrian neuroscientist and cyberneticist, and former director at the Max Planck Institute for Biological Cybernetics in Tübingen, Germany. His 1986 book *"Vehicles: Experiments in Synthetic Psychology"* (image on right is the cover of this book, published by The MIT Press, ©MIT 1984) describes reactive goal-achieving vehicles, and such systems are now commonly known as Braitenberg Vehicles.

A Braitenberg vehicle is an automaton which combines sensors, actuators and their direct interconnection to produce goal-oriented behaviors. In the book these vehicles are described conceptually as analog circuits, but more recently small robots based on a digital realization of the same principles have been developed. Grey Walter's tortoise predates the use of this term but was nevertheless an example of such a vehicle.

VEHICLES
Experiments in Synthetic Psychology

Valentino Braitenberg

5.1   Reactive Navigation

Surprisingly complex tasks can be performed by a robot even if it has no map and no real *idea* about where it is. As already mentioned robotic vacuum cleaners use only random motion and information from contact sensors to perform a complex task as shown in Fig. 5.1. Insects such as ants and bees gather food and return it to their nest based on input from their senses, they have far too few neurons to create any kind of mental map of the world and plan paths through it. Even single-celled organisms such as flagellate protozoa exhibit goal-seeking behaviors. In this case we need to temporarily modify our earlier definition of a robot to

a goal oriented machine that can sense, ~~plan~~ and act.

Grey Walter's robotic tortoise demonstrated that it could moves toward a light source, a behavior known as phototaxis.▶ This was an important result in the then emerging scientific field of cybernetics.

More generally a *taxis* is the response of an organism to a stimulus gradient.

5.1.1   Braitenberg Vehicles

A very simple class of goal achieving robots are known as Braitenberg vehicles and are characterized by direct connection between sensors and motors. They have no explicit internal representation of the environment in which they operate and nor do they make explicit plans.▶

Consider the problem of a robot moving in two dimensions that is seeking the local maxima of a scalar field – the field could be light intensity or the concentration of some chemical.▶ The Simulink® model

This is a fine philosophical point, the plan could be considered to be implicit in the details of the connections between the motors and sensors.

This is similar to the problem of moving to a point discussed in Sect. 4.1.1.1.

```
>> sl_braitenberg
```

shown in Fig. 5.2 achieves this using a steering signal derived directly from the sensors.▶

This is similar to Braitenberg's Vehicle 4a.

William Grey Walter (1910–1977) was a neurophysiologist and pioneering cyberneticist born in Kansas City, Missouri and studied at King's College, Cambridge. Unable to obtain a research fellowship at Cambridge, he worked on neurophysiological research in hospitals in London and from 1939 at the Burden Neurological Institute in Bristol. He developed electro-encephalographic brain topography which used multiple electrodes on the scalp and a triangulation algorithm to determine the amplitude and location of brain activity.

Walter was influential in the then new field of cybernetics. He built robots to study how complex reflex behavior could arise from neural interconnections. His tortoise Elsie (of the species *Machina Speculatrix*) is shown, without its shell, on page 95. Built in 1948 Elsie was a three-wheeled robot capable of phototaxis that could also find its way to a recharging station. A second generation tortoise (from 1951) is in the collection of the Smithsonian Institution. He published popular articles in "Scientific American" (1950 and 1951) and a book "The Living Brain" (1953). He was badly injured in a car accident in 1970 from which he never fully recovered. (Image courtesy Reuben Hoggett collection)

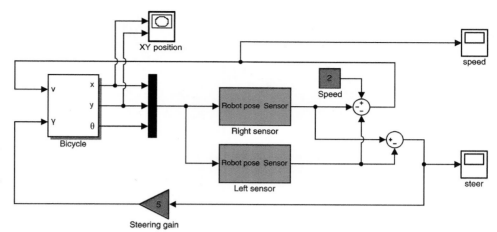

Fig. 5.2.
The Simulink® model
`sl_braitenberg` drives the
vehicle toward the maxima of
a provided scalar function. The
vehicle plus controller is an example of a Braitenberg vehicle

We can make the measurements simultaneously using two spatially separated
sensors or from one sensor over time as
the robot moves.

To ascend the gradient we need to estimate the gradient direction at the current
location and this requires at least two measurements of the field.◄ In this example we
use two sensors, bilateral sensing, with one on each side of the robot's body. The sensors are modeled by the green sensor blocks shown in Fig. 5.2 and are parameterized
by the position of the sensor with respect to the robot's body, and the sensing function.
In this example the sensors are at ±2 units in the vehicle's lateral or y-direction.

The field to be sensed is a simple inverse square field defined by

```
1   function sensor = sensorfield(x, y)
2       xc = 60; yc = 90;
3       sensor = 200./((x-xc).^2 + (y-yc).^2 + 200);
```

which returns the sensor value $s(x, y) \in [0, 1]$ which is a function of the sensor's position in the plane. This particular function has a peak value at the point $(60, 90)$.

The vehicle speed is

$$v = 2 - s_R - s_L$$

where s_R and s_L are the right and left sensor readings respectively. At the goal, where
$s_R = s_L = 1$ the velocity becomes zero.

Steering angle is based on the difference between the sensor readings

$$\gamma = k(s_R - s_L)$$

Similar strategies are used by moths
whose two antennae are exquisitely
sensitive odor detectors that are used
to steer a male moth toward a phero-
mone emitting female.

so when the field is equal in the left- and right-hand sensors the robot moves straight ahead.◄

We start the simulation from the Simulink menu or the command line

```
>> sim('sl_braitenberg');
```

and the path of the robot is shown in Fig. 5.3. The starting pose can be changed through
the parameters of the `Bicycle` block. We see that the robot turns toward the goal and
slows down as it approaches, asymptotically achieving the goal position.

This particular sensor-action control law results in a specific robotic *behavior*. We
could add additional logic to the robot to detect that it had arrived near the goal and
then switch to a stopping behavior. An obstacle would block this robot since its only
behavior is to steer toward the goal, but an additional behavior could be added to handle this case and drive around an obstacle. We could add another behavior to search
randomly for the source if none was visible. Grey Walter's tortoise had four behaviors
and switching was based on light level and a touch sensor.

Multiple behaviors and the ability to switch between them leads to an approach
known as behavior-based robotics. The subsumption architecture was proposed as a

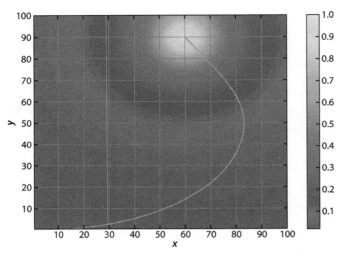

Fig. 5.3.
Path of the Braitenberg vehicle moving toward the maximum of a 2D scalar field whose magnitude is shown color coded

means to formalize the interaction between different behaviors. Complex, some might say *intelligent looking*, behaviors can be manifested by such systems. However as more behaviors are added the complexity of the system grows rapidly and interactions between behaviors become more complex to express and debug. Ultimately the penalty of not using a map becomes too great.

5.1.2   Simple Automata

Another class of reactive robots are known as *bugs* – simple automata that perform goal seeking in the presence of nondriveable areas or obstacles. There are a large number of *bug* algorithms and they share the ability to sense when they are in proximity to an obstacle. In this respect they are similar to the Braitenberg class vehicle, but the *bug* includes a state machine and other logic in between the sensor and the motors. The automata have memory which our earlier Braitenberg vehicle lacked.► In this section we will investigate a specific *bug* algorithm known as *bug2*.

We start by loading an obstacle field to challenge the robot

```
>> load house
>> about house
house [double] : 397x596 (1.9 MB)
```

which defines a matrix variable house in the workspace. The elements are zero or one representing free space or obstacle respectively and this is shown in Fig. 5.4. Tools to generate such maps are discussed on page 131. This matrix is an example of an occupancy grid which will be discussed further in the next section. This command also loads a list of named places within the house, as elements of a structure

```
>> place
place =
     kitchen: [320 190]
      garage: [500 150]
         br1: [50 220]
          .
          .
          .
```

At this point we state some assumptions. Firstly, the robot operates in a grid world and occupies one grid cell. Secondly, the robot is capable of omnidirectional motion and can move to any of its eight neighboring grid cells. Thirdly, it is able to determine its position on the plane which is a nontrivial problem that will be discussed in detail in Chap. 6. Finally, the robot can only sense its goal and whether adjacent cells are occupied.

Braitenberg's book describes a series of increasingly complex vehicles, some of which incorporate memory. However the term *Braitenberg vehicle* has become associated with the simplest vehicles he described.

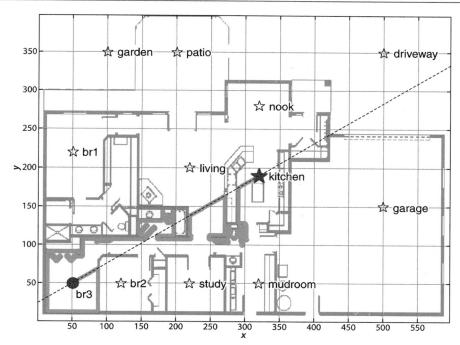

Fig. 5.4.
Obstacles are indicated by *red pixels*. Named places are indicated by *hollow black stars*. Approximate scale is 4.5 cm per cell. The start location is a *solid blue circle* and the goal is a *solid blue star*. The path taken by the bug2 algorithm is marked by a *green line*. The *black dashed line* is the m-line, the direct path from the start to the goal

We create an instance of the `bug2` class

```
>> bug = Bug2(house);
```

and pass in the occupancy grid. The *bug2* algorithm does not use the map to plan a path – the map is used by the simulator to provide sensory inputs to the robot. We can display the robot's environment by

```
>> bug.plot();
```

The simulation is run using the `query` method

```
>> bug.query(place.br3, place.kitchen, 'animate');
```

whose arguments are the start and goal positions of the robot within the house.

The method displays an animation of the robot moving toward the goal and the path is shown as a series of green dots in Fig. 5.4.

The strategy of the *bug2* algorithm is quite simple. It is given a straight line – the m-line – towards its goal. If it encounters an obstacle it turns right and continues until it encounters a point on the m-line that is closer to the goal than when it departed from the m-line.◄

If an output argument is specified

```
>> p = bug.query(place.br3, place.kitchen)
```

it returns the path as a matrix `p`

```
>> about p
p [double] : 1299x2 (20.8 kB)
```

It could be argued that the m-line represents an explicit plan. Thus *bug* algorithms occupy a position somewhere between Braitenberg vehicles and map-based planning systems in the spectrum of approaches to navigation.

which has one row per point, and comprises 1 299 points for this example. Invoking the function with an empty matrix

```
>> p = bug.query([], place.kitchen);
```

will prompt for the corresponding point to be selected by clicking on the plot.

In this example the *bug2* algorithm has reached the goal but it has taken a very suboptimal route, traversing the inside of a wardrobe, behind doors and visiting two bathrooms. It would perhaps have been quicker in this case to turn left, rather than right, at the first obstacle but that strategy might give a worse outcome somewhere else. Many variants of the *bug* algorithm have been developed, but while they improve the performance for one type of environment they can degrade performance in others. Fundamentally the robot is limited by not using a map. It cannot see the big picture and therefore takes paths that are locally, rather than globally, optimal.

5.2   Map-Based Planning

The key to achieving the *best* path between points A and B, as we know from everyday life, is to use a map. Typically best means the shortest distance but it may also include some penalty term or cost related to traversability which is how easy the terrain is to drive over – it might be quicker to travel further but faster over better roads. A more sophisticated planner might also consider the size of the robot, the kinematics and dynamics of the vehicle and avoid paths that involve turns that are tighter than the vehicle can execute. Recalling our earlier definition of a robot as a

> *goal oriented machine that can sense,* plan *and act,*

this section concentrates on planning.

There are many ways to represent a map and the position of the vehicle within the map. Graphs, as discussed in Appendix I, can be used to represent places and paths between them. Graphs can be efficiently searched to find a path that minimizes some measure or cost, most commonly the distance traveled. A simpler and very computer-friendly representation is the occupancy grid which is widely used in robotics.

An occupancy grid treats the world as a grid of cells and each cell is marked as occupied or unoccupied. We use zero to indicate an unoccupied cell or free space where the robot can drive. A value of one indicates an occupied or nondriveable cell. The size of the cell depends on the application. The memory required to hold the occupancy grid increases with the spatial area represented and inversely with the cell size. However for modern computers this representation is very feasible. For example a cell size 1×1 m requires▶ just 125 kbyte km^{-2}.

> Considering a single bit to represent each cell. The occupancy grid could be compressed or could be kept on a disk with only the local region in memory.

In the remainder of this section we use code examples to illustrate several different planners and all are based on the occupancy grid representation. To create uniformity the planners are all implemented as classes derived from the `Navigation` superclass which is briefly described on page 133. The *bug2* class we used previously was also an instance of this class so the remaining examples follow a familiar pattern.

Once again we state some assumptions. Firstly, the robot operates in a grid world and occupies one grid cell. Secondly, the robot does not have any nonholonomic constraints and can move to any neighboring grid cell. Thirdly, it is able to determine its position on the plane. Fourthly, the robot is able to use the map to compute the path it will take.

In all examples we will use the house map introduced in the last section and find paths from bedroom 3 to the kitchen. These parameters can be varied, and the occupancy grid changed using the tools described above.

5.2.1   Distance Transform

Consider a matrix of zeros with just a single nonzero element representing the goal. The distance transform of this matrix is another matrix, of the same size, but the value of each element is its distance▶ from the original nonzero pixel. For robot path planning we use the default Euclidean distance. The distance transform is actually an image processing technique and will be discussed further in Chap. 12.

> The distance between two points (x_1, y_1) and (x_2, y_2) where $\Delta_x = x_2 - x_1$ and $\Delta_y = y_2 - y_1$ can be Euclidean $\sqrt{\Delta_x^2 + \Delta_y^2}$ or CityBlock (also known as Manhattan) distance $|\Delta_x| + |\Delta_y|$.

Making a map. An occupancy grid is a matrix that corresponds to a region of 2-dimensional space. Elements containing zeros are free space where the robot can move, and those with ones are obstacles where the robot cannot move. We can use many approaches to create a map. For example we could create a matrix filled with zeros (representing all free space)

```
>> map = zeros(100, 100);
```

and use MATLAB operations such as

```
>> map(40:50,20:80) = 1;
```

or the MATLAB builtin matrix editor to create obstacles but this is quite cumbersome. Instead we can use the Toolbox map editor `makemap` to create more complex maps using an interactive editor

```
>> map = makemap(100)
```

that allows you to add rectangles, circles and polygons to an occupancy grid. In this example the grid is 100 × 100. See online help for details.

The occupancy grid in Fig. 5.4 was derived from a scanned image but online buildings plans and street maps could also be used.

Note that the occupancy grid is a matrix whose coordinates are conventionally expressed as (row, column) and the row is the vertical dimension of a matrix. We use the Cartesian convention of a horizontal *x*-coordinate first, followed by the *y*-coordinate therefore the matrix is always indexed as `y, x` in the code.

To use the distance transform for robot navigation we create a `DXform` object, which is derived from the `Navigation` class

```
>> dx = DXform(house);
```

and then create a plan to reach a specific goal

```
>> dx.plan(place.kitchen)
```

which can be visualized

```
>> dx.plot()
```

as shown in Fig. 5.5. We see the obstacle regions in red overlaid on the distance map whose grey level at any point indicates the distance from that point to the goal, in grid cells, taking into account travel *around* obstacles.

The hard work has been done and to find the shortest path from *any* point to the goal we simply consult or query the plan.◄ For example a path from the bedroom to the goal is

> For the bug2 algorithm there was no planning step so the query in that case was the simulated robot querying its proximity sensors.

```
>> dx.query(place.br3, 'animate');
```

which displays an animation of the robot moving toward the goal. The path is indicated by a series of green dots as shown in Fig. 5.5.◄

The plan is the distance map. Wherever the robot starts, it moves to the neighboring cell that has the smallest distance to the goal. The process is repeated until the robot reaches a cell with a distance value of zero which is the goal.

If the `path` method is called with an output argument the path

> By convention the plan is based on the goal location and we query for a start location, but we could base the plan on the start position and then query for a goal.

```
>> p = dx.query(place.br3);
```

is returned as a matrix, one row per point, which we can visualize overlaid on the occupancy grid and distance map

```
>> dx.plot(p)
```

The path comprises

```
>> numrows(p)
ans =
   336
```

points which is considerably shorter than the path found by *bug2*.

This navigation algorithm has exploited its global view of the world and has, through exhaustive computation, found the shortest possible path. In contrast, *bug2* without

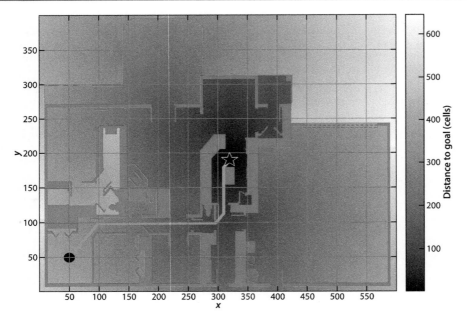

Fig. 5.5.
The distance transform path. Obstacles are indicated by *red cells*. The background grey intensity represents the cell's distance from the goal in units of cell size as indicated by the *scale* on the right-hand side

the global view has just bumped its way through the world. The penalty for achieving the optimal path is computational cost. This particular implementation of the distance transform is iterative. Each iteration has a cost of $O(N^2)$ and the number of iterations is at least $O(N)$, where N is the dimension of the map.

We can visualize the iterations of the distance transform by

```
>> dx.plan(place.kitchen, 'animate');
```

which shows the distance values propagating as a wavefront outward from the goal. The wavefront moves outward, spills through doorways into adjacent rooms and outside the house.▸ Although the plan is expensive to create, once it has been created it can be used to plan a path from *any* initial point to that goal.

We have converted a fairly complex planning problem into one that can now be handled by a Braitenberg-class robot that makes local decisions based on the distance to the goal. Effectively the robot is rolling *downhill* on the distance function which we can plot as a 3D surface

```
>> dx.plot3d(p)
```

shown in Fig. 5.6 with the robot's path and room locations overlaid.

For large occupancy grids this approach to planning will become impractical. The roadmap methods that we discuss later in this chapter provide an effective means to find paths in large maps at greatly reduced computational cost.

The scale associated with this occupancy grid is 4.5 cm per cell and we have assumed the robot occupies a single grid cell – this is a very small robot. The planner could therefore find paths that a larger real robot would be unable to fit through. A common solution to this problem is to *inflate* the occupancy grid – making the obstacles bigger is equivalent to leaving the obstacles unchanged and making the robot bigger. For example, if we inflate the obstacles by 5 cells

```
>> dx = DXform(house, 'inflate', 5);
>> dx.plan(place.kitchen);
>> p = dx.query(place.br3);
>> dx.plot(p)
```

the path shown in Fig. 5.7b now takes the wider corridors to reach its goal. To illustrate how this works we can overlay this new path on the inflated occupancy grid

```
>> dx.plot(p, 'inflated');
```

More efficient algorithms exist such as fast marching methods and Dijkstra's method, but the iterative wavefront method used here is easy to code and to visualize.

This is morphological dilation which is discussed in Sect. 12.6.

and this is shown in Fig. 5.7a. The inflation parameter of 5 has grown the obstacles by 5 grid cells in all directions, a bit like applying a very thick layer of paint. ◄ This is equivalent to growing the robot by 5 grid cells in all directions – the robot grows from a single grid cell to a disk with a diameter of 11 cells which is equivalent to around 50 cm.

Fig. 5.6.
The distance transform as a 3D function, where height is distance from the goal. Navigation is simply a downhill run. Note the discontinuity in the distance transform where the split wave-fronts met

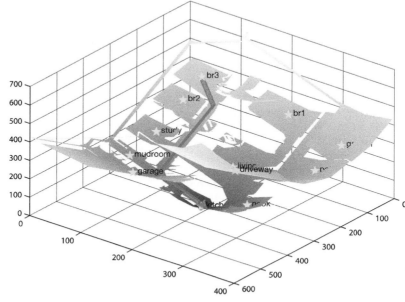

Fig. 5.7. Distance transform path with obstacles inflated by 5 cells. **a** Path shown with inflated obstacles; **b** path computed for inflated obstacles overlaid on original obstacle map, black regions are where no distance was computed ▼

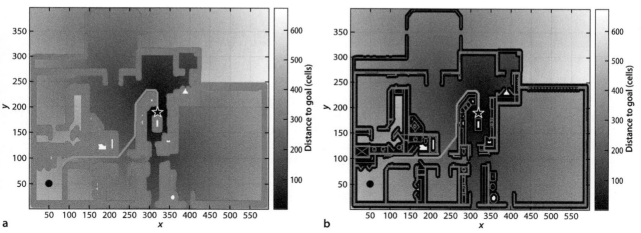

Navigation superclass. The examples in this chapter are all based on classes derived from the `Navigation` class which is designed for 2D grid-based navigation. Each example consists of essentially the following pattern. Firstly we create an instance of an object derived from the `Navigation` class by calling the class constructor.

```
nav = MyNavClass(map)
```

which is passed the occupancy grid. Then a plan is computed

```
nav.plan()
nav.plan(goal)
```

and depending on the planner the goal may or may not be required. A path from an initial position to the goal is computed by

```
p = nav.query(start, goal)
p = nav.query(start)
```

again depending on whether or not the planner requires a goal. The optional return value `p` is the path, a sequence of points from `start` to `goal`, one row per point, and each row comprises the x- and y-coordinate. If `start` or `goal` is given as [] the user is prompted to interactively click the point. The 'animate' option causes an animation of the robot's motion to be displayed.

The map and planning information can be visualized by

```
nav.plot()
```

or have a path overlaid

```
nav.plot(p)
```

5.2.2   D*

A popular algorithm for robot path planning is D* which finds the best path▸ through a graph, which it first computes, that corresponds to the input occupancy grid. D* has a number of features that are useful for real-world applications. Firstly, it generalizes the occupancy grid to a cost map which represents the cost $c \in \mathbb{R}$, $c > 0$ of traversing each cell in the horizontal or vertical direction. The cost of traversing the cell diagonally is $c\sqrt{2}$. For cells corresponding to obstacles $c = \infty$ (Inf in MATLAB).

Secondly, D* supports incremental replanning. This is important if, while we are moving, we discover that the world is different to our map. If we discover that a route has a higher than expected cost or is completely blocked we can incrementally replan to find a better path. The incremental replanning has a lower computational cost than completely replanning as would be required using the distance transform method just discussed.

D* finds the path which minimizes the total cost of travel. If we are interested in the shortest time to reach the goal then cost is the time to drive across the cell and is inversely related to traversability. If we are interested in minimizing damage to the vehicle or maximizing passenger comfort then cost might be related to the roughness of the terrain within the cell. The costs assigned to cells will also depend on the characteristics of the vehicle: a large 4-wheel drive vehicle may have a finite cost to cross a rough area whereas for a small car that cost might be infinite.

To implement the D* planner using the Toolbox we use a similar pattern and first create a D* navigation object

```
>> ds = Dstar(house);
```

The D* planner converts the passed occupancy grid map into a cost map which we can retrieve

```
>> c = ds.costmap();
```

where the elements of c will be 1 or ∞ representing free and occupied cells respectively.

A plan for moving to the goal is generated by

```
>> ds.plan(place.kitchen);
```

which creates a very dense directed graph (see Appendix I). Every cell is a graph vertex and has a cost, a distance to the goal, and a link to the neighboring cell that is closest to the goal. Each cell also has a state $t \in \{\text{NEW, OPEN, CLOSED}\}$. Initially every cell is in the NEW state, the cost of the goal cell is zero and its state is OPEN. We can consider the set of all cells in the OPEN state as a wavefront propagating outward from the goal.▸ The cost of

> D* is an extension of the A* algorithm for finding minimum cost paths through a graph, see Appendix I.

> The distance transform also evolves as a wavefront outward from the goal. However D* represents the frontier efficiently as a list of cells whereas the distance transform computes the frontier on a per-cell basis at every iteration – the frontier is implicitly where a cell with infinite cost (the initial value of all cells) is adjacent to a cell with finite cost.

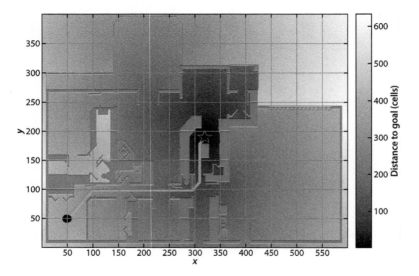

Fig. 5.8.
The D* planner path. Obstacles are indicated by *red cells* and all driveable cells have a cost of 1. The background grey intensity represents the cell's distance from the goal in units of cell size as indicated by the *scale* on the right-hand side

reaching cells that are neighbors of an OPEN cell is computed and these cells in turn are set to OPEN and the original cell is removed from the open list and becomes CLOSED. In MATLAB this initial planning phase is quite slow◄ and takes over a minute and

```
>> ds.niter
ans =
     245184
```

iterations of the planning loop.

The path from an arbitrary starting point to the goal

```
>> ds.query(place.br3);
```

is shown in Fig. 5.8. The robot has again taken a short and efficient path around the obstacles that is almost identical to that generated by the distance transform.

The real power of D* comes from being able to efficiently change the cost map during the mission. This is actually quite a common requirement in robotics since real sensors have a finite range and a robot discovers more of world as it proceeds. We inform D* about changes using the `modify_cost` method, for example to raise the cost of entering the kitchen via the bottom doorway

```
>> ds.modify_cost( [300,325; 115,125], 5 );
```

we have raised the cost to 5 for a small rectangular region across the doorway. This region is indicated by the yellow dashed rectangle in Fig. 5.9. The other driveable cells have a default cost of 1. The plan is updated by invoking the planning algorithm again

```
>> ds.plan();
```

and this time the number of iterations is only

```
>> ds.niter
ans =
     169580
```

which is 70% of that required to create the original plan.◄ The new path for the robot

```
>> ds.query(place.br3);
```

is shown in Fig. 5.9. The cost change is relatively small but we notice that the increased cost of driving within this region is indicated by a subtle brightening of those cells – in a cost sense these cells are now further from the goal. Compared to Fig. 5.8 the robot has taken a different route to the kitchen and avoided the bottom door. D* allows updates to the map to be made at any time while the robot is moving. After replanning the robot simply moves to the adjacent cell with the lowest cost which ensures continuity of motion even if the plan has changed.

D* is more efficient than the distance transform but it executes more slowly because it is implemented entirely in MATLAB code whereas the distance transform is a MEX-file written in C.

The cost increases with the number of cells modified and the effect those changes have on the distance map. It is possible that incremental replanning takes more time than planning from scratch.

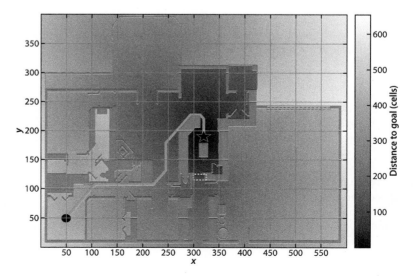

Fig. 5.9.
Path from D* planner with modified map. The higher-cost region is indicated by the *yellow dashed rectangle* and has changed the path compared to Fig. 5.7

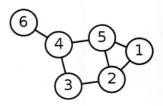

A graph is an abstract representation of a set of objects connected by links typically denoted $G(V, E)$ and depicted diagrammatically as shown to the right. The objects, V, are called vertices or nodes, and the links, E, that connect some pairs of vertices are called edges or arcs. Edges can be directed (arrows) or undirected as in this case. Edges can have an associated weight or cost associated with moving from one of its vertices to the other. A sequence of edges from one vertex to another is a path. Graphs can be used to represent transport or communications networks and even social relationships, and the branch of mathematics is graph theory. Minimum cost path between two nodes in the graph can be computed using well known algorithms such as Dijstrka's method or A*.

The navigation classes use a simple MATLAB graph class called `PGraph`, see Appendix I.

5.2.3   Introduction to Roadmap Methods

In robotic path planning the analysis of the map is referred to as the *planning phase*. The *query phase* uses the result of the planning phase to find a path from A to B. The two previous planning algorithms, distance transform and D*, require a significant amount of computation for the planning phase, but the query phase is very cheap. However the plan depends on the goal. If the goal changes the expensive planning phase must be re-executed. Even though D* allows the path to be recomputed as the costmap changes it does not support a changing goal.

The disparity in planning and query costs has led to the development of roadmap methods where the query can include both the start and goal positions. The planning phase provides analysis that supports changing starting points and changing goals. A good analogy is making a journey by train. We first find a local path to the nearest train station, travel through the train network, get off at the station closest to our goal, and then take a local path to the goal. The train network is invariant and planning a path through the train network is straightforward. Planning paths to and from the entry and exit stations respectively is also straightforward since they are, ideally, short paths. The robot navigation problem then becomes one of building a network of obstacle free paths through the environment which serve the function of the train network. In the literature such a network is referred to as a roadmap. The roadmap need only be computed once and can then be used like the train network to get us from any start location to any goal location.

We will illustrate the principles by creating a roadmap from the occupancy grid's free space using some image processing techniques. The essential steps in creating the roadmap are shown in Fig. 5.10. The first step is to find the free space in the map which is simply the complement of the occupied space

```
>> free = 1 - house
```

and is a matrix with nonzero elements where the robot is free to move. The boundary is also an obstacle so we mark the outermost cells as being not free

```
>> free(1,:) = 0; free(end,:) = 0;
>> free(:,1) = 0; free(:,end) = 0;
```

and this map is shown in Fig. 5.10a where free space is depicted as white.

The topological skeleton of the free space is computed by a morphological image processing algorithm known as thinning▸ applied to the free space of Fig. 5.10a

> Also known as skeletonization. We will cover this topic in Sect. 12.6.3.

```
>> skeleton = ithin(free);
```

and the result is shown in Fig. 5.10b. We see that the obstacles have grown and the free space, the white cells, have become a thin network of connected white cells which are equidistant from the boundaries of the original obstacles.

Figure 5.10c shows the free space network overlaid on the original map. We have created a network of paths that span the space and which can be used for obstacle-free travel around the map.▸ These paths are the edges of a generalized Voronoi

> The junctions in the roadmap are indicated by black dots. The junctions, or triple points, are identified using the morphological image processing function `triplepoint`.

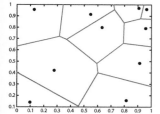

The Voronoi tessellation of a set of planar points, known as sites, is a set of Voronoi cells as shown to the left. Each cell corresponds to a site and consists of all points that are closer to its site than to any other site. The edges of the cells are the points that are equidistant to the two nearest sites. A generalized Voronoi diagram comprises cells defined by measuring distances to objects rather than points. In MATLAB we can generate a Voronoi diagram by

```
>> sites = rand(10,2)
>> voronoi(sites(:,1), sites(:,2))
```

Georgy Voronoi (1868–1908) was a Russian mathematician, born in what is now Ukraine. He studied at Saint Petersburg University and was a student of Andrey Markov. One of his students Boris Delaunay defined the eponymous triangulation which has dual properties with the Voronoi diagram.

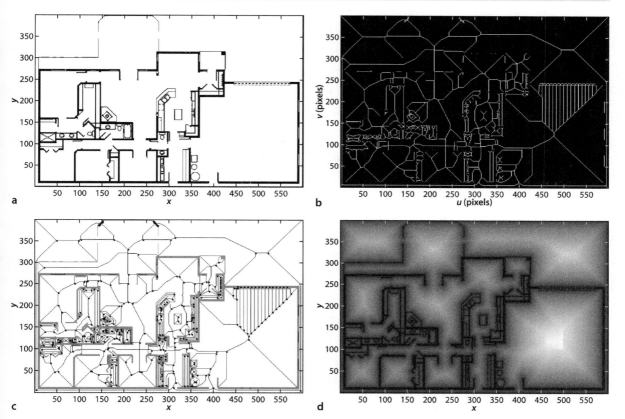

Fig. 5.10. Steps in the creation of a Voronoi roadmap. **a** Free space is indicated by *white cells*; **b** the skeleton of the free space is a network of adjacent cells no more than one cell thick; **c** the skeleton with the obstacles overlaid in red and road-map junction points indicated by *black dots*; **d** the distance transform of the obstacles, pixel values correspond to distance to the nearest obstacle

diagram. We could obtain a similar result by computing the distance transform of the obstacles, Fig. 5.10a, and this is shown in Fig. 5.10d. The value of each pixel is the distance to the nearest obstacle and the ridge lines correspond to the skeleton of Fig. 5.10b. Thinning or skeletonization, like the distance transform, is a computationally expensive iterative algorithm but it illustrates well the principles of finding paths through free space. In the next section we will examine a cheaper alternative.

5.2.4   Probabilistic Roadmap Method (PRM)

The high computational cost of the distance transform and skeletonization methods makes them infeasible for large maps and has led to the development of probabilistic methods. These methods sparsely sample the world map and the most well known of these methods is the probabilistic roadmap or PRM method.

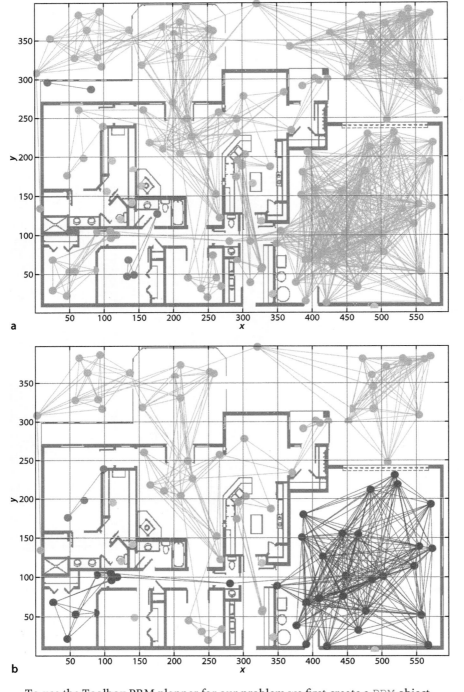

Fig. 5.11.
Probablistic roadmap (PRM)
planner and the random graphs
produced in the planning phase.
a Well connected network with
150 nodes; **b** poorly connected
network with 100 nodes

To use the Toolbox PRM planner for our problem we first create a PRM object

```
>> prm = PRM(house)
```

and then create the plan

```
>> prm.plan('npoints', 150)▶
```

To replicate the following result be sure to
initialize the random number generator
first using randinit. See page 139.

with 150 roadmap nodes. Note that we do not pass the goal as an argument since the
plan is independent of the goal. Creating the path is a two phase process: planning, and

query. The planning phase finds *N* random points, 150 in this case, that lie in free space. Each point is connected to its nearest neighbors by a straight line path that does not cross any obstacles, so as to create a network, or graph, with a minimal number of disjoint components and no cycles. The advantage of PRM is that relatively few points need to be tested to ascertain that the points and the paths between them are obstacle free. The resulting network is stored within the PRM object and a summary can be displayed

```
>> prm
prm =
PRM navigation class:
  occupancy grid: 397x596
  graph size: 150
  dist thresh: 178.8
  2 dimensions
  150 vertices
  1223 edges
  14 components
```

which indicates the number of edges and connected components in the graph. The graph can be visualized

This is derived automatically from the size of the occupancy grid.

```
>> prm.plot()
```

Random numbers. The MATLAB random number generator (used for rand and randn) generates a very long sequence of numbers that are an excellent approximation to a random sequence. The generator maintains an internal state which is effectively the position within the sequence. After startup MATLAB always generates the following random number sequence

```
>> rand
ans =
    0.8147
>> rand
ans =
    0.9058
>> rand
ans =
    0.1270
```

Many algorithms discussed in this book make use of random numbers and this means that the results can never be repeated. Before all such examples in this book is an invisible call to randinit which resets the random number generator to a known state

```
>> randinit
>> rand
ans =
    0.8147
>> rand
ans =
    0.9058
```

and we see that the random sequence has been restarted.

as shown in Fig. 5.11a. The dots represent the randomly selected points and the lines are obstacle-free paths between the points. Only paths less than 178.8 cells long are selected⌐ which is the distance threshold parameter of the PRM class. Each edge of the graph has an associated cost which is the distance between its two nodes. The color of the node indicates which component it belongs to and each component is assigned a unique color. In this case there are 14 components but the bulk of nodes belong to a single large component.

The query phase finds a path from the start point to the goal. This is simply a matter of moving to the closest node in the roadmap (the start node), following a minimum cost A* route through the roadmap, getting off at the node closest to the goal and then traveling to the goal. For our standard problem this is

```
>> prm.query(place.br3, place.kitchen)
>> prm.plot()
```

and the path followed is shown in Fig. 5.12. The path that has been found is quite efficient although there are two areas where the path doubles back on itself. Note that we provide the start and the goal position to the query phase. An advantage of this planner is that once the roadmap is created by the planning phase we can change the goal and starting points very cheaply, only the query phase needs to be repeated. The path taken is

```
>> p = prm.query(place.br3, place.kitchen);
>> about p
p [double] : 9x2 (144 bytes)
```

which is a list of the node coordinates that the robot passes through – via points. These could be passed to a trajectory following controller as discussed in Sect. 4.1.1.3.

There are some important tradeoffs in achieving this computational efficiency. Firstly, the underlying random sampling of the free space means that a different roadmap is created every time the planner is run, resulting in different paths and path lengths. Secondly, the planner can fail by creating a network consisting of disjoint components. The roadmap in Fig. 5.11b, with only 100 nodes has several large disconnected components and the nodes in the kitchen and bedrooms belong to different components. If the start and goal nodes are not connected by the roadmap, that is, they are close to different components the query method will report an error. The only solution is to rerun the planner and/or increase the number of nodes. Thirdly, long narrow gaps between obstacles such as corridors are unlikely to be exploited since the probability of randomly choosing points that lie along such spaces is very low.

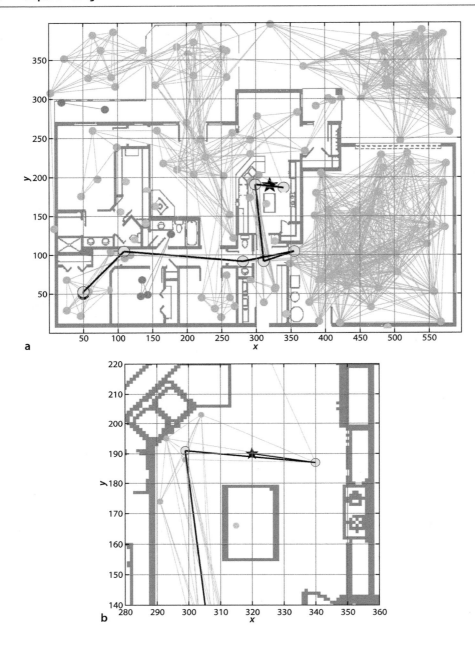

Fig. 5.12.
Probablistic roadmap (PRM) planner **a** showing the path taken by the robot via nodes of the roadmap which are highlighted in yellow; **b** closeup view of goal region where the short path from the final roadmap node to the goal can be seen

5.2.5   Lattice Planner

The planners discussed so far have generated paths independent of the motion that the vehicle can actually achieve, and we learned in Chap. 4 that wheeled vehicles have significant motion constraints. One common approach is to use the output of the planners we have discussed and move a point along the paths at constant velocity and then follow that point, using techniques such as the trajectory following controller described in Sect. 4.1.1.3.

An alternative is to *design* a path from the outset that we know the vehicle can follow. The next two planners that we introduce take into account the motion model of the vehicle, and relax the assumption we have so far made that the robot is capable of omnidirectional motion.

We consider that the robot is moving between discrete points in its 3-dimensional configuration space. The robot is initially at the origin and can drive forward to the three points shown in black in Fig. 5.13a.◄ Each path is an arc▲ which requires a constant steering wheel setting and the arc radius is chosen so that at the end of each arc the robot's heading direction is some multiple of $\frac{\pi}{2}$ radians.

At the end of each branch we can add the same set of three motions suitably rotated and translated, and this is shown in Fig. 5.13b. The graph now contains 13 nodes and represents 9 paths each 2 segments long. We can create this lattice by using the `Lattice` planner class

```
>> lp = Lattice();
>> lp.plan('iterations', 2)
13 nodes created
>> lp.plot()
```

which will generate a plot like Fig. 5.13b. Each node represents a configuration (x, y, θ), not just a position, and if we rotate the plot we can see in Fig. 5.14 that the paths lie in the 3-dimensional configuration space.

While the paths appear smooth and continuous the curvature is in fact discontinuous – at some nodes the steering wheel angle would have to change instantaneously from hard left to hard right for example.◄

The pitch of the grid is dictated by the turning radius of the vehicle.

Sometimes called Dubins curves.

A real robot would take a finite time to adjust its steering angle and this would introduce some error in the robot path. The steering control system could compensate for this by turning harder later in the segment so as to bring the robot to the end point with the correct orientation.

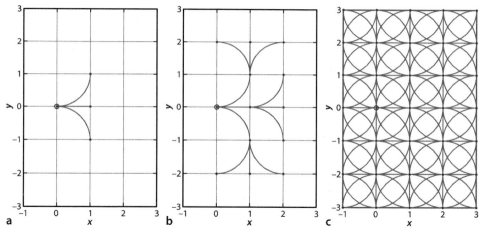

Fig. 5.13.
Lattice plan after 1, 2 and 8 iterations

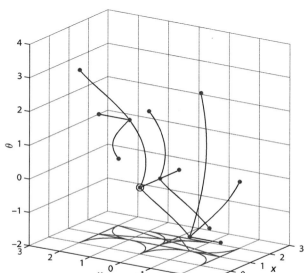

Fig. 5.14.
Lattice plan after 2 iterations shown in 3-dimensional configuration space

By increasing the number of iterations

```
>> lp.plan('iterations', 8)
780 nodes created
>> lp.plot()
```

we can fill in more possible paths as shown in Fig. 5.13c and the paths now extend well beyond the area shown.

Now that we have created the lattice we can compute a path between any two nodes using the query method

```
>> lp.query( [1 2 pi/2], [2 -2 0] );
A* path cost 6
```

where the start and goal are specified as configurations (x, y, θ) and the lowest cost path found by an A* search is reported.► We can overlay this on the vertices

```
>> lp.plot
```

and is shown in Fig. 5.15a. This is a path that takes into account the fact that the vehicle has an orientation and preferred directions of motion, as do most wheeled robot platforms. We can also access the configuration-space coordinates of the nodes

```
>> p = lp.query( [1 2 pi/2], [2 -2 0] )
A* path cost 6
>> about p
p [double] : 7x3 (168 bytes)
```

where each row represents the configuration-space coordinates (x, y, θ) of a node in the lattice along the path from start to goal configuration.

Implicit in our search for the lowest cost path is the cost of traversing each edge of the graph which by default gives equal cost to the three steering options: straight ahead, turn left and turn right. We can increase the cost associated with turning

```
>> lp.plan('cost', [1 10 10])
>> lp.query(start, goal);
A* path cost 35
>> lp.plot()
```

and now we now have the path shown in Fig. 5.15b which has only 3 turns compared to 5 previously. However the path is longer – having 8 rather than 6 segments.

Consider a more realistic scenario with obstacles in the environment. Specifically we want to find a path to move the robot 2 m in the lateral direction with its final heading angle the same as its initial heading angle

```
>> load road
>> lp = Lattice(road, 'grid', 5, 'root', [50 50 0])
>> lp.plan();
```

Every segment in the lattice has a default cost of 1 so the cost of 6 simply reflects the total number of segments in the path. A* search is introduced in Appendix I.

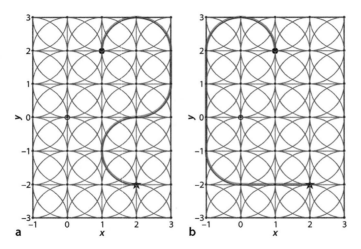

a   b

Fig. 5.15.
Paths over the lattice graph.
a With uniform cost; **b** with increased penalty for turns

where we have loaded an obstacle grid that represents a simple parallel-parking scenario and planned a lattice with a grid spacing of 5 units and the root node at a central obstacle-free configuration. In this case the planner continues to iterate until it can add no more nodes to the free space. We query for a path from the road to the parking spot

```
>> lp.query([30 45 0], [50 20 0])
```

and the result is shown in Fig. 5.16.

Paths generated by the lattice planner are inherently driveable by the robot but there are clearly problems driving along a diagonal with this simple lattice. The planner would generate a continual sequence of hard left and right turns which would cause undue wear and tear on a real vehicle and give a very uncomfortable ride. More sophisticated version of lattice planners are able to deal with this by using motion primitives with hundreds of arcs, such as shown in Fig. 5.17, instead of the three shown in these examples.

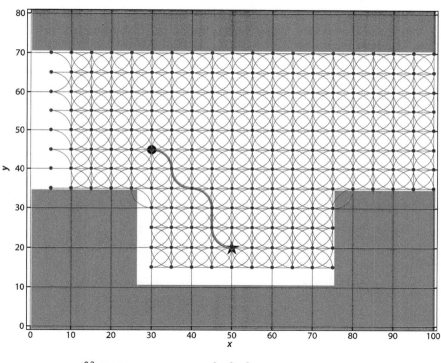

Fig. 5.16.
A simple parallel parking scenario based on the lattice planner with an occupancy grid (cells are 10 cm square)

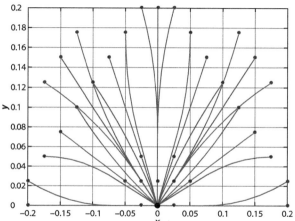

Fig. 5.17.
A more sophisticated lattice generated by the package **sbpl** with 43 paths based on the kinematic model of a unicycle

5.2.6   Rapidly-Exploring Random Tree (RRT)

The final planner that we introduce is also able to take into account the motion model of the vehicle. Unlike the lattice planner which plans over a regular grid, the RRT uses probabilistic methods like the PRM planner.

The underlying insight is similar to that for the lattice planner and Fig. 5.18 shows a family of paths that the bicycle model of Eq. 4.2 would follow in configuration space. The paths are computed over a fixed time interval for discrete values of velocity, forward or backward, and various steering angles. This demonstrates clearly the subset of all possible configurations that a nonholonomic vehicle can reach from a given initial configuration.

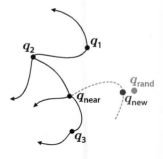

The main steps in creating an RRT are as follows, with the notation shown in the figure to the right. A graph of robot configurations is maintained and each node is a configuration $q \in \mathbb{R}^2 \times \mathbb{S}^1$ which is represented by a 3-vector $q \sim (x, y, \theta)$. The first, or root, node in the graph is the goal configuration of the robot. A random configuration q_{rand} is chosen, and the node with the closest configuration q_{near} is found – this configuration is near in terms of a cost function that includes distance and orientation.► A control is computed that moves the robot from q_{near} toward q_{rand} over a fixed path simulation time. The configuration that it reaches is q_{new} and this is added to the graph.

For any desired starting configuration we can find the closest configuration in the graph, and working backward toward the starting configuration we could determine the sequence of steering angles and velocities needed to move from the start to the goal configuration. This has some similarities to the roadmap methods discussed previously, but the limiting factor is the combinatoric explosion in the number of possible poses.

We first of all create a model to describe the vehicle kinematics

```
>> car = Bicycle('steermax', 0.5);
```

and here we have specified a car-like vehicle with a maximum steering angle of 0.5 rad. Following our familiar programming pattern we create an RRT object

```
>> rrt = RRT(car, 'npoints', 1000)
```

for an obstacle free environment which by default extends from –5 to +5 in the *x*- and *y*-directions and create a plan

```
>> rrt.plan();
>> rrt.plot();
```

> The distance measure must account for a difference in position and orientation and requires appropriate weighting of these quantities. From a consideration of units this is not quite proper since we are adding meters and radians.

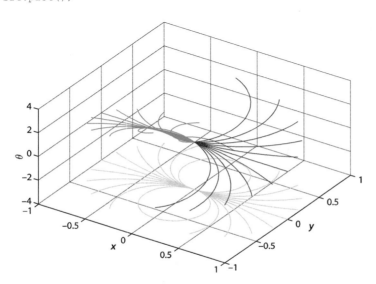

Fig. 5.18.
A set of possible paths that the bicycle model robot could follow from an initial configuration of $(0, 0, 0)$. For $v = \pm 1$, $\alpha \in [-1, 1]$ over a 2 s period. *Red lines* correspond to $v < 0$

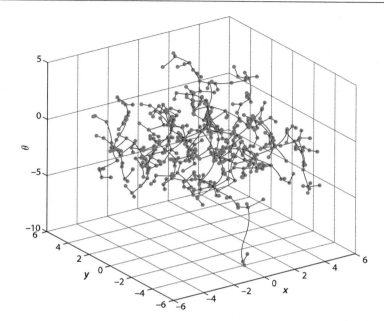

Fig. 5.19.
An RRT computed for the bicycle model with a velocity of ± 1 m s^{-1}, steering angle limits of ± 0.5 rad, integration period of 1 s, and initial configuration of $(0, 0, 0)$. Each node is indicated by a *green circle* in the 3-dimensional space of vehicle poses (x, y, θ)

The random tree is shown in Fig. 5.19 and we see that the paths have a good coverage of the configuration space, not just in the x- and y-directions but also in orientation, which is why the algorithm is known as *rapidly exploring*.

An important part of the RRT algorithm is computing the control input that moves the robot from an existing configuration in the graph to q_{rand}. From Sect. 4.1 we understand the difficulty of driving a nonholonomic vehicle to a specified configuration. Rather than the complex nonlinear controller of Sect. 4.1.1.4 we will use something simpler that fits with the randomized sampling strategy used in this class of planner. The controller randomly chooses whether to drive forwards or backwards and randomly chooses a steering angle within the limits ◄. It then simulates motion of the vehicle model for a fixed period of time, and computes the closest distance to q_{rand}. This is repeated multiple times and the control input with the best performance is chosen. The configuration on its path that was closest to q_{rand} is chosen as q_{near} and becomes a new node in the graph.

Handling obstacles with the RRT is quite straightforward. The configuration q_{rand} is discarded if it lies within an obstacle, and the point q_{near} will not be added to the graph if the path from q_{near} toward q_{rand} intersects an obstacle. The result is a set of paths, a roadmap, that is collision free and driveable by this nonholonomic vehicle. ◄

We will repeat the parallel parking example from the last section

```
>> rrt = RRT(car, road, 'npoints', 1000, 'root', [50 22 0], 'simtime', 4)
>> rrt.plan();
```

where we have specified the vehicle kinematic model, an occupancy grid, the number of sample points, the location of the first node, and that each random motion is simulated for 4 seconds. We can query the RRT plan for a path between two configurations

```
>> p = rrt.query([40 45 0], [50 22 0]);
```

and the result is a continuous path

```
>> about p
p [double] : 520x3 (12.5 kB)
```

which will take the vehicle from the street to the parking slot. We can overlay the path on the occupancy grid and RRT

```
>> rrt.plot(p)
```

Uniformly randomly distributed between the steering angle limits.

We have chosen the first node to be the goal configuration, and we search from here toward possible start configurations. However we could also make the first node the start configuration. Alternatively we could choose the start node to be neither the start or goal position, the planner will find a path through the RRT between configurations close to the start and goal.

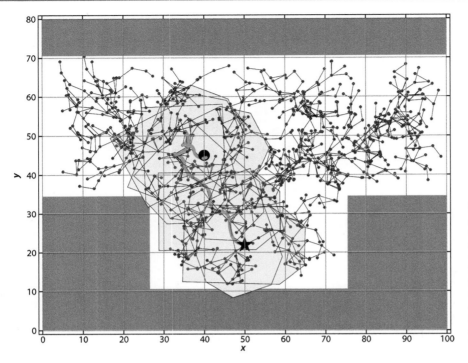

Fig. 5.20.
A simple parallel parking example based on the RRT planner with an occupancy grid (cells are 10 cm square). RRT nodes are shown in *blue*, the initial configuration is a *solid circle* and the goal is a *solid star*. The path through the RRT is shown in *green*, and a few snapshots of the vehicle configuration are overlaid in *pink*

and the result is shown in Fig. 5.20 with some vehicle configurations overlaid. We can also animate the motion along the path

```
>> plot_vehicle(p, 'box', 'size', [20 30], 'fill', 'r', 'alpha', 0.1)
```

where we have specified the vehicle be displayed as a red translucent shape of width 20 and length 30 units.

This example illustrates some important points about the RRT. Firstly, as for the PRM planner, there may be some distance (and orientation) between the start and goal configuration and the nearest node. Minimizing this requires tuning RRT parameters such as the number of nodes and path simulation time. Secondly, the path is feasible but not quite optimal. In this case the vehicle has changed direction twice before driving into the parking slot. This is due to the random choice of nodes – rerunning the planner and/or increasing the number of nodes may help. Finally, we can see that the vehicle body collides with the obstacle, and this is very apparent if you view the animation. This is actually not surprising since the collision check we did when adding a node only tested if the node's position lay in an obstacle – it should properly check if a finite-sized vehicle with that configuration intersects an obstacle. Alternatively, as discussed on page 132 we could inflate the obstacles by the radius of the smallest disk that contains the robot.

5.3   Wrapping Up

Robot navigation is the problem of guiding a robot towards a goal and we have covered a spectrum of approaches. The simplest was the purely reactive Braitenberg-type vehicle. Then we added limited memory to create state machine based automata such as *bug2* which can deal with obstacles, however the paths that it finds are far from optimal.

A number of different map-based planning algorithms were then introduced. The distance transform is a computationally intense approach that finds an optimal path to the goal. D* also finds an optimal path, but supports a more nuanced travel cost – individual cells have a continuous traversability measure rather than being considered as only free space or obstacle. D* also supports computationally cheap incremental replanning for small changes in the map. PRM reduces the computational burden significantly by probabilistic sampling but at the expense of somewhat less optimal paths. In particular it may not discover narrow routes between areas of free space. The lattice planner takes into account the motion constraints of a real vehicle to create paths which are feasible to drive, and can readily account for the orientation of the vehicle as well as its position. RRT is another random sampling method that also generates kinematically feasible paths. All the map-based approaches require a map and knowledge of the robot's location, and these are both topics that we will cover in the next chapter.

Further Reading

Comprehensive coverage of planning for robots is provided by two text books. Choset et al. (2005) covers geometric and probabilistic approaches to planning as well as the application to robots with dynamics and nonholonomic constraints. LaValle (2006) covers motion planning, planning under uncertainty, sensor-based planning, reinforcement learning, nonlinear systems, trajectory planning, nonholonomic planning, and is available online for free at http://planning.cs.uiuc.edu. In particular these books provide a much more sophisticated approach to representing obstacles in configuration space and cover potential-field planning methods which we have not discussed. The powerful planning techniques discussed in these books can be applied beyond robotics to very high order systems such as vehicles with trailers, robotic arms or even the shape of molecules. LaValle (2011a) and LaValle (2011b) provide a concise two-part tutorial introduction. More succinct coverage of planning is given by Kelly (2013), Siegwart et al. (2011), the Robotics Handbook (Siciliano and Khatib 2016, § 7), and also in Spong et al. (2006) and Siciliano et al. (2009).

The *bug1* and *bug2* algorithms were described by Lumelsky and Stepanov (1986). More recently eleven variations of Bug algorithm were implemented and compared for a number of different environments (Ng and Bräunl 2007). The distance transform is well described by Borgefors (1986) and its early application to robotic navigation was explored by Jarvis and Byrne (1988). Efficient approaches to implementing the distance transform include the two-pass method of Hirata (1996), fast marching methods or reframing it as a graph search problem which can be solved using Dijkstra's method; the last two approaches are compared by Alton and Mitchell (2006). The A* algorithm (Nilsson 1971) is an efficient method to find the shortest path through a graph, and we can always compute a graph that corresponds to an occupancy grid map. D* is an extension by Stentz (1994) which allows cheap replanning when the map changes and there have been many further extensions including, but not limited to, Field D* (Ferguson and Stentz 2006) and D* lite (Koenig and Likhachev 2002). D* is used in many real-world robot systems and many implementations exist including open source.

The ideas behind PRM started to emerge in the mid 1990s and it was first described by Kavraki et al. (1996). Geraerts and Overmars (2004) compare the efficacy of a number of subsequent variations that have been proposed to the basic PRM algorithm. Approaches to planning that incorporate the vehicle's dynamics include state-space sampling (Howard et al. 2008), and the RRT which is described in LaValle (1998, 2006) and related resources at http://msl.cs.uiuc.edu. More recently RRT* has been proposed by Karaman et al. (2011). Lattice planners are covered in Pivtoraiko, Knepper, and Kelly (2009).

Historical and interesting. The defining book in cybernetics was written by Wiener in 1948 and updated in 1965 (Wiener 1965). Grey Walter published a number of popular articles (1950, 1951) and a book (1953) based on his theories and experiments with robotic tortoises.

The definitive reference for Braitenberg vehicles is Braitenberg's own book (1986) which is a whimsical, almost poetic, set of thought experiments. Vehicles of increasing complexity (fourteen vehicle families in all) are developed, some including nonlinearities, memory and logic to which he attributes anthropomorphic characteristics such as love, fear, aggression and egotism. The second part of the book outlines the factual basis of these machines in the neural structure of animals.

Early behavior-based robots included the Johns Hopkins Beast, built in the 1960s, and Genghis (Brooks 1989) built in 1989. Behavior-based robotics are covered in the book by Arkin (1999) and the Robotics Handbook (Siciliano and Khatib 2016, § 13). Mataric's Robotics Primer (Mataric 2007) and associated comprehensive web-based resources is also an excellent introduction to reactive control, behavior based control and robot navigation. A rich collection of archival material about early cybernetic machines, including Grey-Walter's tortoise and the Johns Hopkins Beast can be found at the Cybernetic Zoo http://cyberneticzoo.com.

Resources

A very powerful set of motion planners exist in OMPL, the Open MotionPLanning Library (http://ompl.kavrakilab.org) written in C++. It has a Python-based app that provides a convenient means to explore planning problems. Steve LaValle's web site http://msl.cs.illinois.edu/~lavalle/code.html has many code resources (C++ and Python) related to motion planning. Lattice planners are included in the sbpl package from the Search-Based Planning Lab (http://sbpl.net) which has MATLAB tools for generating motion primitives and C++ code for planning over the lattice graphs.

MATLAB Notes

The Robotics System Toolbox™ from The MathWorks Inc. includes functions `Binary-OccupancyGrid` and `PRM` to create occupancy grids and plan paths using probabilistic roadmaps. Other functions support reading and writing ROS navigation and map messages. The Image Processing Toolbox™ function `bwdist` is an efficient implementation of the distance transform.

Exercises

1. Braitenberg vehicles (page 127)
   a) Experiment with different starting configurations and control gains.
   b) Modify the signs on the steering signal to make the vehicle light-phobic.
   c) Modify the `sensorfield` function so that the peak moves with time.
   d) The vehicle approaches the maxima asymptotically. Add a stopping rule so that the vehicle stops when the when either sensor detects a value greater than 0.95.
   e) Create a scalar field with two peaks. Can you create a starting pose where the robot gets confused?
2. Occupancy grids. Create some different occupancy grids and test them on the different planners discussed.
   a) Create an occupancy grid that contains a maze and test it with various planners. See http://rosettacode.org/wiki/Maze_generation.
   b) Create an occupancy grid from a downloaded floor plan.
   c) Create an occupancy grid from a city street map, perhaps apply color segmentation (Chap. 13) to segment roads from other features. Can you convert this to a cost map for D* where different roads or intersections have different costs?
   d) Experiment with obstacle inflation.
   e) At 1 m cell size how much memory is required to represent the surface of the Earth? How much memory is required to represent just the land area of Earth? What cell size is needed in order for a map of your country to fit in 1 Gbyte of memory?

3. Bug algorithms (page 128)
   a) Using the function `makemap` create a new map to challenge *bug2*. Try different starting points.
   b) Create an obstacle map that contains a maze. Can *bug2* solve the maze?
   c) Experiment with different start and goal locations.
   d) Create a bug trap. Make a hollow box, and start the bug inside a box with the goal outside. What happens?
   e) Modify *bug2* to change the direction it turns when it hits an obstacle.
   f) Implement other bug algorithms such as *bug1* and *tangent bug*. Do they perform better or worse?
4. Distance transform (page 132)
   a) Create an obstacle map that contains a maze and solve it using distance transform.
5. D* planner (page 134)
   a) Add a low cost region to the living room. Can you make the robot prefer to take this route to the kitchen?
   b) Block additional doorways to challenge the robot.
   c) Implement D* as a mex-file to speed it up.
6. PRM planner (page 138)
   a) Run the PRM planner 100 times and gather statistics on the resulting path length.
   b) Vary the value of the distance threshold parameter and observe the effect.
   c) Use the output of the PRM planner as input to a pure pursuit planner as discussed in Chap. 4.
   d) Implement a nongrid based version of PRM. The robot is represented by an arbitrary polygon as are the obstacles. You will need functions to determine if a polygon intersects or is contained by another polygon (see the Toolbox `Polygon` class). Test the algorithm on the piano movers problem.
7. Lattice planner (page 140)
   a) How many iterations are required to completely fill the region of interest shown in Fig. 5.13c?
   b) How does the number of nodes and the spatial extent of the lattice increase with the number of iterations?
   c) Given a car with a wheelbase of 4.5 m and maximum steering angles of ±30 deg what is the smallest possible grid size?
   d) Redo Fig. 5.15b to achieve a path that uses only right hand turns.
   e) Compute curvature as a function of path length for a path through the lattice such as the one shown in Fig. 5.15a.
   f) Design a controller in Simulink that will take a unicycle or bicycle model with a finite steering angle rate (there is a block parameter to specify this) that will drive the vehicle along the three paths shown in Fig. 5.13a.
8. RRT planner (page 144)
   a) Find a path to implement a 3-point turn.
   b) Experiment with RRT parameters such as the number of points, the vehicle steering angle limits, and the path integration time.
   c) Additional information in the node of each graph holds the control input that was computed to reach the node. Plot the steering angle and velocity sequence required to move from start to goal pose.
   d) Add a local planner to move from initial pose to the closest vertex, and from the final vertex to the goal pose.
   e) Determine a path through the graph that minimizes the number of reversals of direction.
   f) The collision test currently only checks that the center point of the robot does not lie in an occupied cell. Modify the collision test so that the robot is represented by a rectangular robot body and check that the entire body is obstacle free.

6

Localization

in order to get somewhere we need to know where we are

In our discussion of map-based navigation we assumed that the robot had a means of knowing its position. In this chapter we discuss some of the common techniques used to estimate the location of a robot in the world – a process known as localization.

Today GPS makes outdoor localization so easy that we often take this capability for granted. Unfortunately GPS is a far from perfect sensor since it relies on very weak radio signals received from distant orbiting satellites. This means that GPS cannot work where there is no *line of sight* radio reception, for instance indoors, underwater, underground, in urban canyons or in deep mining pits. GPS signals are also extremely weak and can be easily jammed and this is not acceptable for some applications.

GPS has only been in use since 1995 yet humankind has been navigating the planet and localizing for many thousands of years. In this chapter we will introduce the *classical* navigation principles such as dead reckoning and the use of landmarks on which modern robotic navigation is founded.

Dead reckoning is the estimation of location based on estimated speed, direction and time of travel with respect to a previous estimate. Figure 6.1 shows how a ship's position is updated on a chart. Given the average compass heading over the previous hour and a distance traveled the position at 3 P.M. can be found using elementary geometry from the position at 2 P.M. However the measurements on which the update is based are subject to both systematic and random error. Modern instruments are quite precise but 500 years ago clocks, compasses and speed measurement were primitive. The recursive nature of the process, each estimate is based on the previous

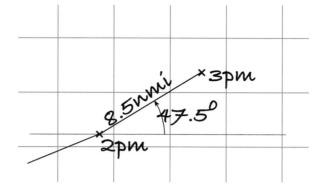

Fig. 6.1.
Location estimation by dead reckoning. The ship's position at 3 P.M. is based on its position at 2 P.M., the estimated distance traveled since, and the average compass heading

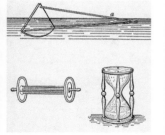

Measuring speed at sea. A ship's log is an instrument that provides an estimate of the distance traveled. The oldest method of determining the speed of a ship at sea was the Dutchman's log – a floating object was thrown into the water at the ship's bow and the time for it to pass the stern was measured using an hourglass. Later came the chip log, a flat quarter-circle of wood with a lead weight on the circular side causing it to float upright and resist towing. It was tossed overboard and a line with knots at 50 foot intervals was payed out. A special hourglass, called a log glass, ran for 30 s, and each knot on the line over that interval corresponds to approximately 1 nmi h^{-1} or 1 knot. A nautical mile (nmi) is now defined as 1.852 km. (Image modified from Text-Book of Seamanship, Commodore S. B. Luce 1891)

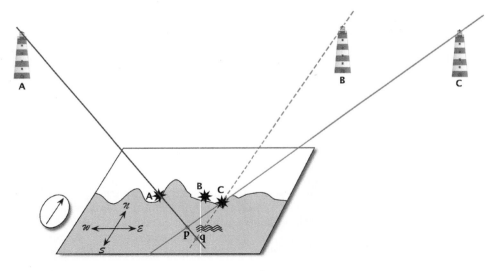

Fig. 6.2.
Location estimation using a map. Lines of sight from two light-houses, *A* and *C*, and their corresponding locations on the map provide an estimate *p* of our location. However if we mistake lighthouse *B* for *C* then we obtain an incorrect estimate *q*

one, means that errors will accumulate over time and for sea voyages of many-years this approach was quite inadequate.

The Phoenicians were navigating at sea more than 4 000 years ago and they did not even have a compass – that was developed 2 000 years later in China. The Phoenicians navigated with crude dead reckoning but wherever possible they used *additional information* to correct their position estimate – sightings of islands and headlands, primitive maps and observations of the Sun and the Pole Star.

A landmark is a visible feature in the environment whose location is known with respect to some coordinate frame. Figure 6.2 shows schematically a map and a number of lighthouse landmarks. We first of all use a compass to align the north axis of our map with the direction of the north pole. The direction of a single landmark constrains our position to lie along a line on the map. Sighting a second landmark places our position on another constraint line, and our position must be at their intersection – a process known as resectioning.▶ For example lighthouse **A** constrains us to lie along the blue line. Lighthouse **C** constrains us to lie along the red line and the intersection is our true position **p**.

However this process is critically reliant on correctly associating the observed landmark with the feature on the map. If we mistake one lighthouse for another, for example we see **B** but think it is **C** on the map, then the red dashed line leads to a

Resectioning is the estimation of position by measuring the bearing angles to known landmarks. Triangulation is the estimation of position by measuring the bearing angles to the unknown point from each of the landmarks.

Celestial navigation. The position of celestial bodies in the sky is a predictable function of the time and the observer's latitude and longitude. This information can be tabulated and is known as ephemeris (meaning daily) and such data has been published annually in Britain since 1767 as the "*The Nautical Almanac*" by HM Nautical Almanac Office. The elevation of a celestial body with respect to the horizon can be measured using a sextant, a handheld optical instrument.

Time and longitude are coupled, the star field one hour later is the same as the star field 15° to the east. However the northern Pole Star, *Polaris* or the *North Star*, is very close to the celestial pole and its elevation angle is independent of longitude and time, allowing latitude to be determined very conveniently from a single sextant measurememt.

Solving the longitude problem was the greatest scientific challenge to European governments in the eighteenth century since it was a significant impediment to global navigation and maritime supremacy. The British Longitude Act of 1714 created a prize of £20 000 which spurred the development of nautical chronometers, clocks that could maintain high accuracy onboard ships. More than fifty years later a suitable chronometer was developed by John Harrison, a copy of which was used by Captain James Cook on his second voyage of 1772–1775. After a three year journey the error in estimated longitude was just 13 km. With accurate knowledge of time, the elevation angle of stars could be used to estimate latitude and longitude. This technological advance enabled global exploration and trade. (Image courtesy archive.org)

THE

NAUTICAL ALMANAC

AND

ASTRONOMICAL EPHEMERIS,

FOR THE YEAR 1781.

Published by Order of the

COMMISSIONERS OF LONGITUDE.

LONDON

PRINTED BY WILLIAM RICHARDSON,
PRINTER;
AND SOLD BY
J. NOURSE, in the Strand, and Meff. MOUNT and PAGE
on Tower-Hill,
Bookfellers to the faid COMMISSIONERS.
M DCC LXXIX.
[Price Three Shillings and Six Pence.]

Radio-based localization. One of the earliest systems was LORAN, based on the British World War II GEE system. LORAN transmitters around the world emit synchronized radio pulses and a receiver measures the difference in arrival time between pulses from a pair of radio transmitters. Knowing the identity of two transmitters and the time difference (TD) constrains the receiver to lie along a hyperbolic curve shown on navigation charts as *TD lines*. Using a second pair of transmitters (which may include one of the first pair) gives another hyperbolic constraint curve, and the receiver must lie at the intersection of the two curves.

The Global Positioning System (GPS) was proposed in 1973 but did not become fully operational until 1995. It comprises around 30 active satellites orbiting the Earth in six planes at a distance of 20 200 km. A GPS receiver works by measuring the time of travel of radio signals from four or more satellites whose orbital position is encoded in the GPS signal. With four known points in space and four measured time delays it is possible to compute the (x, y, z) position of the receiver and the time. If the GPS signals are received after reflecting off some surface the dis-

tance traveled is longer and this will introduce an error in the position estimate. This effect is known as multi-pathing and is a common problem in large-scale industrial facilities.

Variations in the propagation speed of radio waves through the atmosphere is the main cause of error in the position estimate. However these errors vary slowly with time and are approximately constant over large areas. This allows the error to be measured at a reference station and transmitted as an augmentation to compatible nearby receivers which can offset the error – this is known as Differential GPS (DGPS). This information can be transmitted via the internet, via coastal radio networks to ships, or by satellite networks such as WAAS, EGNOS or OmniSTAR to aircraft or other users. RTK GPS achieves much higher precision in time measurement by using phase information from the carrier signal. The original GPS system deliberately added error, euphemistically termed selective availability, to reduce its utility to military opponents but this *feature* was disabled in May 2000. Other satellite navigation systems include the Russian GLONASS, the European Galileo, and the Chinese Beidou.

significant error in estimated position – we would believe we were at **q** instead of **p**. This belief would lead us to overestimate our distance from the coastline. If we decided to sail toward the coast we would run aground on rocks and be surprised since they were not where we expected them to be. This is unfortunately a very common error and countless ships have foundered because of this fundamental data association error. This is why lighthouses flash! In the eighteenth century technological advances enabled lighthouses to emit unique flashing patterns so that the identity of the particular lighthouse could be reliably determined and associated with a point on a navigation chart.

Of course for the earliest mariners there were no maps, or lighthouses or even compasses. They had to create maps as they navigated by incrementally adding new nonmanmade features to their maps just beyond the boundaries of what was already known. It is perhaps not surprising that so many early explorers came to grief◄ and that maps were tightly kept state secrets.

Magellan's 1519 expedition started with 237 men and 5 ships but most, including Magellan, were lost along the way. Only 18 men and 1 ship returned.

Robots operating today in environments without GPS face *exactly* the same problems as ancient navigators and, perhaps surprisingly, borrow heavily from navigational strategies that are centuries old. A robot's estimate of distance traveled will be imperfect and it may have no map, or perhaps an imperfect or incomplete map. Additional information from observed features in the world is critical to minimizing a robot's localization error but the possibility of data association error remains.

We can define the localization problem more formally where x is the true, but unknown, position of the robot and $\hat{x}$ is our best estimate of that position. We also wish to know the *uncertainty* of the estimate which we can consider in statistical terms as the standard deviation associated with the position estimate $\hat{x}$.

For robot pose (x, y, θ) the PDF is a 4-dimensional surface.

It is useful to describe the robot's estimated position in terms of a probability density function (PDF) over all possible positions of the robot.◄ Some example PDFs are shown in Fig. 6.3 where the magnitude of the function at any point is the relative likelihood of the vehicle being at that position. Commonly a Gaussian function is used which can be described succinctly in terms of its mean and standard deviation. The robot is most likely to be at the location of the peak (the mean) and increasingly less likely to be at positions further away from the peak. Figure 6.3a shows a peak with a small standard deviation which indicates that the vehicle's position is very well known. There is an almost zero probability that the vehicle is at the point indicated by the vertical black line. In contrast the peak in Fig. 6.3b has a large standard deviation which means that we are less certain about the location of the vehicle. There is a reasonable probability that the vehicle is at the point indicated by the vertical line.

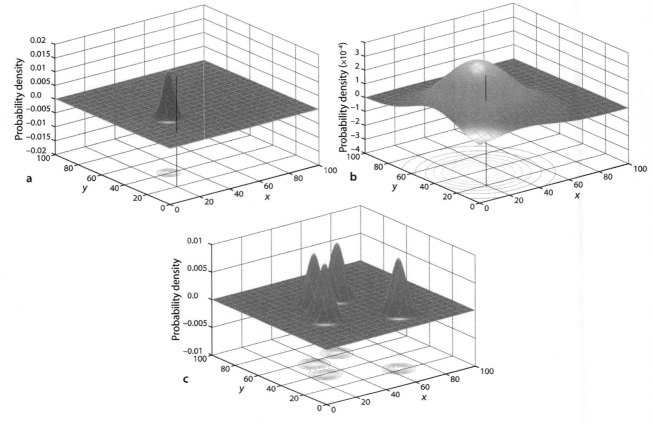

Using a PDF also allows for multiple hypotheses about the robot's position. For example the PDF of Fig. 6.3c describes a robot that is quite certain that it is at one of four places. This is more useful than it seems at face value. Consider an indoor robot that has observed a vending machine and there are four such machines marked on the map. In the absence of any other information the robot must be equally likely to be in the vicinity of *any* of the four vending machines. We will revisit this approach in Sect. 6.7.

Determining the PDF based on knowledge of how the vehicle moves and its observations of the world is a problem in estimation which we can define as:

the process of inferring the value of some quantity of interest, x, by processing data that is in some way dependent on x.

For example a ship's navigator or a surveyor estimates location by measuring the bearing angles to known landmarks or celestial objects, and a GPS receiver estimates latitude and longitude by observing the time delay from moving satellites whose locations are known.

For our robot localization problem the true and estimated state are vector quantities so uncertainty will be represented as a covariance matrix, see Appendix G. The diagonal elements represent uncertainty of the corresponding states, and the off-diagonal elements represent correlations between states.

Fig. 6.3. Notions of vehicle position and uncertainty in the *xy*-plane, where the vertical axis is the relative likelihood of the vehicle being at that position, sometimes referred to as belief or bel(*x*). Contour lines are displayed on the lower plane. **a** The vehicle has low position uncertainty, $\sigma = 1$; **b** the vehicle has much higher position uncertainty, $\sigma = 20$; **c** the vehicle has multiple hypotheses for its position, each $\sigma = 1$

> The value of a PDF is *not* the probability of being at that location. Consider a small region of the *xy*-plane, the volume under that region of the PDF is the probability of being in that region.

6.1   Dead Reckoning

Dead reckoning is the estimation of a robot's pose based on its estimated speed, direction and time of travel with respect to a previous estimate.

An odometer is a sensor that measures distance traveled and sometimes also change in heading direction. For wheeled vehicles this can be determined by measuring the angular rotation of the wheels. The direction of travel can be measured using an electronic compass, or the change in heading can be measured using a gyroscope or differential odometry.◀ These sensors are imperfect due to systematic errors such an incorrect wheel radius or gyroscope bias, and random errors such as slip between wheels and the ground. Robots without wheels, such as aerial and underwater robots, can use visual odometry – a computer vision approach based on observations of the world moving past the robot which is discussed in Sect. 14.7.4.

Measuring the difference in angular velocity of a left- and right-hand side wheel.

6.1.1   Modeling the Vehicle

The first step in estimating the robot's pose is to write a function, $f(\cdot)$, that describes how the vehicle's configuration changes from one time step to the next. A vehicle model such as Eq. 4.2 or 4.4 describes the evolution of the robot's configuration as a function of its control inputs, however for real robots we rarely have access to these control inputs. Most robotic platforms have proprietary motion control systems that accept motion commands from the user (speed and direction) and report odometry information.

Instead of using Eq. 4.2 or 4.4 directly we will write a discrete-time model for the evolution of configuration based on odometry where $\delta\langle k\rangle = (\delta_d, \delta_\theta)$ is the distance traveled and change in heading over the preceding interval, and k is the time step. The initial pose is represented in **SE(2)** as

$$\xi\langle k\rangle \sim \begin{pmatrix} \cos\theta\langle k\rangle & -\sin\theta\langle k\rangle & x\langle k\rangle \\ \sin\theta\langle k\rangle & \cos\theta\langle k\rangle & y\langle k\rangle \\ 0 & 0 & 1 \end{pmatrix}$$

We make a simplifying assumption that motion over the time interval is *small* so the order of applying the displacements is not significant. We choose to move forward in the vehicle x-direction by δ_d, and then rotate by δ_θ giving the new configuration

$$\xi\langle k+1\rangle \sim \begin{pmatrix} \cos\theta\langle k\rangle & -\sin\theta\langle k\rangle & x\langle k\rangle \\ \sin\theta\langle k\rangle & \cos\theta\langle k\rangle & y\langle k\rangle \\ 0 & 0 & 1 \end{pmatrix}\begin{pmatrix} 1 & 0 & \delta_d \\ 0 & 1 & 0 \\ 0 & 0 & 1 \end{pmatrix}\begin{pmatrix} \cos\delta_\theta & -\sin\delta_\theta & 0 \\ \sin\delta_\theta & \cos\delta_\theta & 0 \\ 0 & 0 & 1 \end{pmatrix}$$

$$\sim \begin{pmatrix} \cos(\theta\langle k\rangle + \delta_\theta) & -\sin(\theta\langle k\rangle + \delta_\theta) & x\langle k\rangle + \delta_d\cos\theta\langle k\rangle \\ \sin(\theta\langle k\rangle + \delta_\theta) & \cos(\theta\langle k\rangle + \delta_\theta) & y\langle k\rangle + \delta_d\sin\theta\langle k\rangle \\ 0 & 0 & 1 \end{pmatrix}$$

which we can represent concisely as a 3-vector $x = (x, y, \theta)$

$$x\langle k+1\rangle = \begin{pmatrix} x\langle k\rangle + \delta_d\cos\theta\langle k\rangle \\ y\langle k\rangle + \delta_d\sin\theta\langle k\rangle \\ \theta\langle k\rangle + \delta_\theta \end{pmatrix} \tag{6.1}$$

which gives the new configuration in terms of the previous configuration and the odometry.

In practice odometry is not perfect and we model the error by imagining a random number generator that corrupts the output of a perfect odometer. The measured output of the real odometer is the perfect, but unknown, odometry $(\delta_d, \delta_\theta)$ plus the output of the random number generator (v_d, v_θ). Such random errors are often referred to as noise, or more specifically as sensor noise. The random numbers are not known and cannot be measured, but we assume that we know the distribution from which they are drawn.

The robot's configuration at the next time step, including the odometry error, is

$$\boldsymbol{x}\langle k+1 \rangle = \boldsymbol{f}\big(\boldsymbol{x}\langle k \rangle, \, \delta\langle k \rangle, \, \boldsymbol{v}\langle k \rangle\big) = \begin{pmatrix} x\langle k \rangle + (\delta_d + v_d)\cos\theta\langle k \rangle \\ y\langle k \rangle + (\delta_d + v_d)\sin\theta\langle k \rangle \\ \theta\langle k \rangle + \delta_\theta + v_\theta \end{pmatrix} \qquad (6.2)$$

where k is the time step, $\delta\langle k \rangle = (\delta_d, \delta_\theta)^T \in \mathbb{R}^{2 \times 1}$ is the odometry measurement and $\boldsymbol{v}\langle k \rangle = (v_d, v_\theta)^T \in \mathbb{R}^{2 \times 1}$ is the random measurement noise over the preceding interval. ►

In the absence of any information to the contrary we model the odometry noise as $\boldsymbol{v} = (v_d, v_\theta)^T \sim N(0, \boldsymbol{V})$, a zero-mean multivariate Gaussian process ► with variance

$$\boldsymbol{V} = \begin{pmatrix} \sigma_d^2 & 0 \\ 0 & \sigma_\theta^2 \end{pmatrix}$$

This constant matrix, the covariance matrix, is diagonal which means that the errors in distance and heading are *independent*. ► Choosing a value for $\boldsymbol{V}$ is not always easy but we can conduct experiments or make some reasonable engineering assumptions. In the examples which follow, we choose $\sigma_d = 2$ cm and $\sigma_\theta = 0.5°$ per sample interval which leads to a covariance matrix of

```
>> V = diag([0.02, 0.5*pi/180].^2);
```

All objects of the Toolbox `Vehicle` superclass provide a method `f()` that implements the appropriate odometry update equation. For the case of a vehicle with bicycle kinematics that has the motion model of Eq. 4.2 and the odometric update Eq. 6.2, we create a `Bicycle` object

```
>> veh= Bicycle('covar', V)
veh =
Bicycle object
  L=1
  Superclass: Vehicle
    max speed=1, max steer input=0.5, dT=0.1, nhist=0
    V=(0.0004, 7.61544e-05)
    configuration: x=0, y=0, theta=0
```

which shows the default parameters such as the vehicle's length, speed, steering limit and the sample interval which defaults to 0.1 s. The object provides a method to simulate motion over one time step

```
>> odo = veh.step(1, 0.3)
odo =
    0.1108    0.0469
```

where we have specified a speed of 1 m s^{-1} and a steering angle of 0.3 rad. The function updates the robot's true configuration and returns a noise corrupted odometer reading. ► With a sample interval of 0.1 s the robot reports that is moving approximately 0.1 m each interval and changing its heading by approximately 0.03 rad. The robot's true (but 'unknown') configuration can be seen by

```
>> veh.x'
ans =
    0.1000         0    0.0309
```

Given the reported odometry we can estimate the configuration of the robot after one time step using Eq. 6.2 which is implemented by

```
>> veh.f([0 0 0], odo)
ans =
    0.1106    0.0052    0.0469
```

where the discrepancy with the exact value is due to the use of a noisy odometry measurement.

The odometry noise is *inside* the model of our process (vehicle motion) and is referred to as process noise.

A normal distribution of angles on a circle is actually not possible since $\theta \in \mathbb{S}^1 \not\subset \mathbb{R}$, that is angles wrap around 2π. However if the covariance for angular states is small this problem is minimal. A normal-like distribution of angles on a circle is the von Mises distribution.

In reality this is unlikely to be the case since odometry distance errors tend to be worse when change of heading is high.

We simulate the odometry noise using MATLAB generated random numbers that have zero-mean and a covariance given by the diagonal of $\vee$. The random noise means that repeated calls to this function will return different values.

For the scenarios that we want to investigate we require the simulated robot to drive for a long time period within a defined spatial region. The RandomPath class is a *driver* that steers the robot to randomly selected waypoints within a specified region. We create an instance of the driver object and connect it to the robot

```
>> veh.add_driver( RandomPath(10) )
```

where the argument to the RandomPath constructor specifies a working region that spans ±10 m in the *x*- and *y*-directions. We can display an animation of the robot with its driver by

```
>> veh.run()
```

> The number of history records is indicated by nhist= in the displayed value of the object. The hist property is an array of structures that hold the vehicle state at each time step.

which repeatedly calls the step method and maintains a history of the true state of the vehicle over the course of the simulation within the Bicycle object.◄ The RandomPath and Bicycle classes have many parameters and methods which are described in the online documentation.

6.1.2 Estimating Pose

The problem we face, just like the ship's navigator, is how to estimate our new pose given the previous pose and noisy odometry. We want the best estimate of where we are and how certain we are about that. The mathematical tool that we will use is the Kalman filter which is described more completely in Appendix H. This filter provides the optimal estimate of the system state, vehicle configuration in this case, assuming that the noise is zero-mean and Gaussian. The filter is a recursive algorithm that updates, at each time step, the optimal estimate of the unknown true configuration and the uncertainty associated with that estimate based on the previous estimate and noisy measurement data. The Kalman filter is formulated for linear systems but our model of the vehicle's motion Eq. 6.2 is nonlinear – the tool of choice is the extended Kalman filter (EKF).

For this problem the state vector is the vehicle's configuration

$$\boldsymbol{x} = \left(x_v, y_v, \theta_v\right)^T$$

> The Kalman filter, Fig. 6.6, has two steps: prediction based on the model and update based on sensor data. In this dead-reckoning case we use only the prediction equation.

and the prediction equations◄

$$\hat{\boldsymbol{x}}^+\langle k+1\rangle = \boldsymbol{f}\big(\hat{\boldsymbol{x}}\langle k\rangle, \hat{\boldsymbol{u}}\langle k\rangle\big) \tag{6.3}$$

$$\hat{\boldsymbol{P}}^+\langle k+1\rangle = \boldsymbol{F}_x\hat{\boldsymbol{P}}\langle k\rangle\boldsymbol{F}_x^T + \boldsymbol{F}_v\hat{\boldsymbol{V}}\boldsymbol{F}_v^T \tag{6.4}$$

describe how the state and covariance evolve with time. The term $\hat{\boldsymbol{x}}^+\langle k+1\rangle$ indicates an estimate of $\boldsymbol{x}$ at time $k + 1$ based on information up to, and including, time k. $\hat{\boldsymbol{u}}\langle k\rangle$ is the

Reverend Thomas Bayes (1702–1761) was a nonconformist Presbyterian minister. He studied logic and theology at the University of Edinburgh and lived and worked in Tunbridge-Wells in Kent. There, through his association with the 2nd Earl Stanhope he became interested in mathematics and was elected to the Royal Society in 1742. After his death his friend Richard Price edited and published his work in 1763 as *An Essay towards solving a Problem in the Doctrine of Chances* which contains a statement of a special case of Bayes' theorem. Bayes is buried in Bunhill Fields Cemetery in London.

Bayes' theorem shows the relation between a conditional probability and its inverse: the probability of a hypothesis given observed evidence and the probability of that evidence given the hypothesis. Consider the hypothesis that the robot is at location X and it makes a sensor observation S of a known landmark. The *posterior* probability that the robot is at X given the observation S is

$$P(X|S) = \frac{P(S|X)P(X)}{P(S)}$$

where $P(X)$ is the *prior* probability that the robot is at X (not accounting for any sensory information), $P(S|X)$ is the likelihood of the sensor observation S given that the robot is at X, and $P(S)$ is the prior probability of the observation S. The Kalman filter, and the Monte-Carlo estimator we discuss later in this chapter, are essentially two different approaches to solving this inverse problem.

Rudolf Kálmán (1930–2016) was a mathematical system theorist born in Budapest. He obtained his bachelors and masters degrees in electrical engineering from MIT, and Ph.D. in 1957 from Columbia University. He worked as a Research Mathematician at the Research Institute for Advanced Study, in Baltimore, from 1958–1964 where he developed his ideas on estimation. These were met with some skepticism among his peers and he chose a mechanical (rather than electrical) engineering journal for his paper *A new approach to linear filtering and prediction problems* because "When you fear stepping on hallowed ground with entrenched interests, it is best to go sideways". He has received many awards including the IEEE Medal of Honor, the Kyoto Prize and the Charles Stark Draper Prize.

Stanley F. Schmidt (1926–2015) was a research scientist who worked at NASA Ames Research Center and was an early advocate of the Kalman filter. He developed the first implementation as well as the nonlinear version now known as the extended Kalman filter. This led to its incorporation in the Apollo navigation computer for trajectory estimation. (Extract from Kálmán's famous paper (1960) on the right reprinted with permission of ASME)

input to the process, which in this case is the measured odometry, so $\hat{\boldsymbol{u}}\langle k\rangle = \delta\langle k\rangle$. $\hat{\boldsymbol{P}} \in \mathbb{R}^{3\times3}$ is a covariance matrix representing uncertainty in the estimated vehicle configuration. $\hat{\boldsymbol{V}}$ is our estimate of the covariance of the odometry noise which in reality we do not know.

F_x and F_v are Jacobian matrices – the vector version of a derivative. They are obtained by differentiating Eq. 6.2 and evaluating the result at $v = 0$ giving▸

Since the noise value cannot actually be measured we use the mean value which is zero.

$$F_x = \frac{\partial \boldsymbol{f}}{\partial \boldsymbol{x}}\bigg|_{v=0} = \begin{pmatrix} 1 & 0 & -\delta_d \sin\theta_v \\ 0 & 1 & \delta_d \cos\theta_v \\ 0 & 0 & 1 \end{pmatrix} \tag{6.5}$$

$$F_v = \frac{\partial \boldsymbol{f}}{\partial \boldsymbol{v}}\bigg|_{v=0} = \begin{pmatrix} \cos\theta_v & 0 \\ \sin\theta_v & 0 \\ 0 & 1 \end{pmatrix} \tag{6.6}$$

which are functions of the current state and odometry.▸ Jacobians are reviewed in Appendix E. All objects of the `Vehicle` superclass provide methods `Fx` and `Fv` to compute these Jacobians, for example

The time step notation $\langle k\rangle$ is dropped to reduce clutter.

```
>> veh.Fx( [0,0,0], [0.5, 0.1] )
ans =
    1.0000         0   -0.0499
         0    1.0000    0.4975
         0         0    1.0000
```

where the first argument is the state at which the Jacobian is computed and the second is the odometry.

To simulate the vehicle and the EKF using the Toolbox we define the initial covariance to be quite small since, we assume, we have a good idea of where we are to begin with

```
>> P0 = diag([0.005, 0.005, 0.001].^2);
```

and we pass this to the constructor for an `EKF` object

```
>> ekf = EKF(veh, V, P0);
```

Running the filter for 1 000 time steps

```
>> ekf.run(1000);
```

drives the robot as before, along a random path. At each time step the filter updates the state estimate using various methods provided by the `Vehicle` superclass.

We can plot the true path taken by the vehicle, stored within the `Vehicle` superclass object, by

```
>> veh.plot_xy()
```

and the filter's estimate of the path stored within the `EKF` object,

```
>> hold on
>> ekf.plot_xy('r')
```

These are shown in Fig. 6.4 and we see some divergence between the true and estimated robot path.

The covariance at the 700th time step is

```
>> P700 = ekf.history(700).P
P700 =
    1.8929   -0.5575   -0.1851
   -0.5575    3.4184    0.3400
   -0.1851    0.3400    0.0533
```

The matrix is symmetric and the diagonal elements are the estimated variance associated with the states, that is σ_x^2, σ_y^2 and σ_θ^2 respectively. The standard deviation σ_x of the PDF associated with the vehicle's x-coordinate is

```
>> sqrt(P700(1,1))
ans =
    1.3758
```

There is a 95% chance that the robot's x-coordinate is within the $\pm 2\sigma$ bound or ± 2.75 m in this case. We can compute uncertainty for y and θ similarly.

The off-diagonal terms are correlation coefficients and indicate that the uncertainties between the corresponding variables are related. For example the value $P_{1,3} = P_{3,1} = -0.5575$ indicates that the uncertainties in x and θ are related – error in heading angle causes error in x-position and vice versa. Conversely new information about θ can be used to correct θ as well as x. The uncertainty in position is described by the top-left 2×2 covariance submatrix of $\hat{P}$. This can be interpreted as an ellipse defining a confidence bound on position. We can overlay such ellipses on the plot by

```
>> ekf.plot_ellipse('g')
```

as shown in Fig. 6.4. These correspond to the default 95% confidence bound and are plotted by default every 20 time steps. The vehicle started at the origin and as it progresses we see that the ellipses become larger as the estimated uncertainty increases. The ellipses only show x- and y-position but uncertainty in θ also grows.

The total uncertainty,◀ position and heading, is given by $\sqrt{\det(\hat{P})}$ and is plotted as a function of time

```
>> ekf.plot_P();
```

The elements of P have different units: m² and rad². The uncertainty is therefore a mixture of spatial and angular uncertainty with an implicit weighting. If the range of the position variables $x, y \gg \pi$ then positional uncertainty dominates.

as shown in Fig. 6.5 and we observe that it never decreases. This is because the second term in Eq. 6.4 is positive definite which means that P, the position uncertainty, can never decrease.

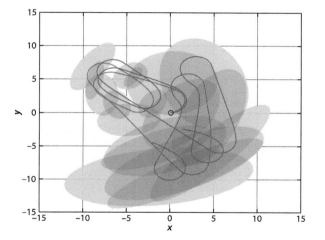

Fig. 6.4.
Deadreckoning using the EKF. The true path of the robot, *blue*, and the path estimated from odometry in *red*. 95% confidence ellipses are indicated in *green*. The robot starts at the origin

Error ellipses. We consider the PDF of the robot's position (ignoring orientation) such as shown in Fig. 6.3 to be a 2-dimensional Gaussian probability density function

$$p(\boldsymbol{x}) = \frac{1}{(2\pi)\det\left(\boldsymbol{P}_{xy}\right)^{1/2}} \exp\left\{ -\frac{1}{2}(\boldsymbol{x} - \mu_x)^T \boldsymbol{P}_{xy}^{-1}(\boldsymbol{x} - \mu_x) \right\}$$

where $\boldsymbol{x} = (x, y)^T$ is the position of the robot, $\mu_x = (\hat{x}, \hat{y})^T$ is the estimated mean position and $\boldsymbol{P}_{xy} \in \mathbb{R}^{2 \times 2}$ is the position covariance matrix, the top left of the covariance matrix $\boldsymbol{P}$ computed by the Kalman filter. A horizontal cross-section is a contour of constant probability which is an ellipse defined by the points $\boldsymbol{x}$ such that

$$(\boldsymbol{x} - \mu_x)^T \boldsymbol{P}_{xy}^{-1}(\boldsymbol{x} - \mu_x) = s$$

Such error ellipses are often used to represent positional uncertainty as shown in Fig. 6.4. A large ellipse corresponds to a wider PDF peak and less certainty about position. To obtain a particular confidence contour (eg. 99%) we choose s as the inverse of the χ^2 cumulative distribution function for 2 degrees of freedom, in MATLAB that is chi2inv(C, 2) where $C \in [0, 1]$ is the confidence value. Such confidence values can be passed to several EKF methods when specifying error ellipses.

A handy scalar measure of total position uncertainty is the area of the ellipse $\pi r_1 r_2$ where the radii $r_i = \sqrt{\lambda_i}$ and λ_i are the eigenvalues of $\boldsymbol{P}_{xy}$. Since $\det(\boldsymbol{P}_{xy}) = \Pi\lambda_i$ the ellipse area – the scalar uncertainty – is proportional to $\sqrt{\det(\boldsymbol{P}_{xy})}$. See also Appendices C.1.4 and G.

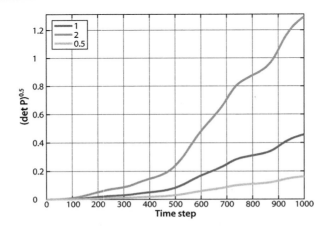

Fig. 6.5.
Overall uncertainty is given by $\sqrt{\det(\hat{\boldsymbol{P}})}$ which shows monotonically increasing uncertainty (*blue*). The effect of changing the magnitude of $\hat{\boldsymbol{V}}$ is to change the rate of uncertainty growth. Curves are shown for $\hat{\boldsymbol{V}} = \alpha\boldsymbol{V}$ where $\alpha = 1/2, 1, 2$

Note that we have used the odometry covariance matrix V twice. The first usage, in the `Vehicle` constructor, is the covariance *V* of the Gaussian noise source that is added to the true odometry to *simulate* odometry error in Eq. 6.2. In a real application this noise is generated by some physical process *hidden inside* the robot and we would not know its parameters. The second usage, in the EKF constructor, is $\hat{\boldsymbol{V}}$ which is our best *estimate* of the odometry covariance and is used in the filter's state covariance update equation Eq. 6.4.

The relative values of *V* and $\hat{\boldsymbol{V}}$ control the rate of uncertainty growth as shown in Fig. 6.5. If $\hat{\boldsymbol{V}} > \boldsymbol{V}$ then *P* will be larger than it should be and the filter is pessimistic – it overestimates uncertainty and is less certain than it should be. If $\hat{\boldsymbol{V}} < \boldsymbol{V}$ then *P* will be smaller than it should be and the filter will be *overconfident* of its estimate – the actual uncertainty is greater than the estimated uncertainty. In practice some experimentation is required to determine the appropriate value for the estimated covariance.

6.2   Localizing with a Map

We have seen how uncertainty in position grows without bound using dead-reckoning alone. The solution, as the Phoenicians worked out 4000 years ago, is to bring in additional information from observations of known features in the world. In the examples that follow we will use a map that contains *N* fixed but randomly located landmarks whose positions are known.

The Toolbox supports a LandmarkMap object

```
>> map = LandmarkMap(20, 10)
```

that in this case contains $N = 20$ landmarks uniformly randomly spread over a region spanning ± 10 m in the x- and y-directions and this can be displayed by

```
>> map.plot()
```

The robot is equipped with a sensor that provides observations of the landmarks *with respect to the robot* as described by

$$z = h(x, p_i) \tag{6.7}$$

where $x = (x_v, y_v, \theta_v)^T$ is the vehicle state, and $p_i = (x_i, y_i)^T$ is the *known* location of the ith landmark in the world frame.

To make this tangible we will consider a common type of sensor that measures the range and bearing angle to a landmark in the environment, for instance a radar or a scanning-laser rangefinder such as shown in Fig. 6.22a. The sensor is mounted on-board the robot so the observation of the ith landmark is

$$z = h(x, p_i) = \begin{pmatrix} \sqrt{(y_i - y_v)^2 + (x_i - x_v)^2} \\ \tan^{-1}(y_i - y_v)/(x_i - x_v) - \theta_v \end{pmatrix} + \begin{pmatrix} w_r \\ w_\beta \end{pmatrix} \tag{6.8}$$

where $z = (r, \beta)^T$ and r is the range, β the bearing angle, and $w = (w_r, w_\beta)^T$ is a zero-mean Gaussian random variable that models errors in the sensor

$$\begin{pmatrix} w_r \\ w_\beta \end{pmatrix} \sim N(0, W), \quad W = \begin{pmatrix} \sigma_r^2 & 0 \\ 0 & \sigma_\beta^2 \end{pmatrix}$$

It also indicates that covariance is independent of range but in reality covariance may increase with range since the strength of the return signal, laser or radar, drops rapidly ($1/r^4$) with distance (r) to the target.

The constant diagonal covariance matrix indicates that range and bearing errors are independent.◄

For this example we set the sensor uncertainty to be $\sigma_r = 0.1$ m and $\sigma_\beta = 1°$ giving a sensor covariance matrix

```
>> W = diag([0.1, 1*pi/180].^2);
```

A subclass of `Sensor`.

We model this type of sensor with a `RangeBearingSensor` object◄

```
>> sensor = RangeBearingSensor(veh, map, 'covar', W)
```

The landmark is chosen randomly from the set of visible landmarks, those that are within the field of view and the minimum and maximum range limits. If no landmark is visible i is assigned a value of 0.

which is connected to the vehicle and the map, and the sensor covariance matrix `W` is specified along with the maximum range and the bearing angle limits. The `reading` method provides the range and bearing to a randomly selected visible◄ landmark along with its identity, for example

```
>> [z,i] = sensor.reading()
z =
    9.0905
    1.0334
i =
    17
```

The identity is an integer $i \in [1, 20]$ since the map was created with 20 landmarks. We have avoided the data association problem by assuming that we know the identity of the sensed landmark. The position of landmark 17 can be looked up in the map

```
>> landmark(17)
   -4.4615
   -9.0766
```

Using Eq. 6.8 the robot can estimate the range and bearing angle to the landmark based on its own estimated position and the known position of the landmark from the map. Any difference between the observation $z^\#$ and the estimated observation indicates an error in the robot's pose estimate $\hat{x}$ – it isn't where it *thought* it was. However this *difference*

$$\nu = z^{\#}\langle k+1 \rangle - h\left(\hat{x}^{+}\langle k+1 \rangle, p_i\right) \tag{6.9}$$

has real value and is key to the operation of the Kalman filter. It is called the innovation since it represents *new* information. The Kalman filter uses the innovation to correct the state estimate and update the uncertainty estimate in an optimal way.

The predicted state computed earlier using Eq. 6.3 and Eq. 6.4 is updated by

$$\hat{x}\langle k+1 \rangle = \hat{x}^{+}\langle k+1 \rangle + K\nu \tag{6.10}$$

$$\hat{P}\langle k+1 \rangle = \hat{P}^{+}\langle k+1 \rangle - KH_x\hat{P}^{+}\langle k+1 \rangle \tag{6.11}$$

which are the Kalman filter update equations. These take the *predicted* values for the next time step denoted with the $^+$ and compute the optimal estimate by applying landmark measurements from time step $k+1$. The innovation is added to the estimated state after multiplying by the Kalman gain matrix K which is defined as

$$K = P^{+}\langle k+1 \rangle H_x^T S^{-1} \tag{6.12}$$

$$S = H_x P^{+}\langle k+1 \rangle H_x^T + H_w \hat{W} H_w^T \tag{6.13}$$

where $\hat{W}$ is the estimated covariance of the sensor noise and H_x and H_w are Jacobians obtained by differentiating Eq. 6.8 yielding

$$H_x = \left.\frac{\partial h}{\partial x}\right|_{w=0} = \begin{pmatrix} -\dfrac{x_i - x_v}{r} & -\dfrac{y_i - y_v}{r} & 0 \\ \dfrac{y_i - y_v}{r^2} & -\dfrac{x_i - x_v}{r^2} & -1 \end{pmatrix} \tag{6.14}$$

which is a function of landmark position, vehicle pose and landmark range; and

$$H_w = \left.\frac{\partial h}{\partial w}\right|_{w=0} = \begin{pmatrix} 1 & 0 \\ 0 & 1 \end{pmatrix} \tag{6.15}$$

The `RangeBearingSensor` object above includes methods h to implement Eq. 6.8 and Hx and Hw to compute these Jacobians respectively.

The Kalman gain matrix K in Eq. 6.10 *distributes* the innovation from the landmark observation, a 2-vector, to update every element of the state vector – the position and orientation of the vehicle. Note that the second term in Eq. 6.11 is *subtracted* from the estimated covariance and this provides a means for covariance to decrease which was

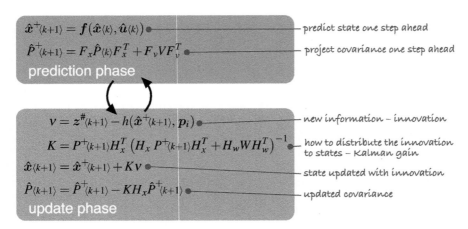

$\hat{x}^{+}\langle k+1 \rangle = f(\hat{x}\langle k \rangle, \hat{u}\langle k \rangle)$ —— predict state one step ahead

$\hat{P}^{+}\langle k+1 \rangle = F_x\hat{P}\langle k \rangle F_x^T + F_v V F_v^T$ —— project covariance one step ahead

prediction phase

$\nu = z^{\#}\langle k+1 \rangle - h(\hat{x}^{+}\langle k+1 \rangle, p_i)$ —— new information – innovation

$K = P^{+}\langle k+1 \rangle H_x^T \left(H_x P^{+}\langle k+1 \rangle H_x^T + H_w W H_w^T\right)^{-1}$ —— how to distribute the innovation to states – Kalman gain

$\hat{x}\langle k+1 \rangle = \hat{x}^{+}\langle k+1 \rangle + K\nu$ —— state updated with innovation

$\hat{P}\langle k+1 \rangle = \hat{P}^{+}\langle k+1 \rangle - KH_x\hat{P}^{+}\langle k+1 \rangle$ —— updated covariance

update phase

Fig. 6.6.
Summary of extended Kalman filter algorithm showing the prediction and update phases

not possible for the dead-reckoning case of Eq. 6.4. The EKF comprises two phases: prediction and update, and these are summarized in Fig. 6.6.

We now have all the piece to build an estimator that uses odometry and observations of map features. The Toolbox implementation is

```
>> map = LandmarkMap(20);
>> veh = Bicycle('covar', V);
>> veh.add_driver( RandomPath(map.dim) );
>> sensor = RangeBearingSensor(veh, map, 'covar', W, 'angle',↵
   [-pi/2 pi/2], 'range', 4, 'animate');
>> ekf = EKF(veh, V, P0, sensor, W, map);
```

The LandmarkMap constructor has a default map dimension of ±10 m which is accessed by its dim property.

Running the simulation for 1 000 time steps

```
>> ekf.run(1000);
```

shows an animation of the robot moving and observations being made to the landmarks. We plot the saved results

```
>> map.plot()
>> veh.plot_xy();
>> ekf.plot_xy('r');
>> ekf.plot_ellipse('k')
```

which are shown in Fig. 6.7a. The error ellipses are now much smaller and many can hardly be seen.

Figure 6.7b shows a zoomed view of the robot's actual and estimated path – the robot is moving from top to bottom. We can see the error ellipses growing as the robot moves and then shrinking, just after a jag in the estimated path. This corresponds to the observation of a landmark. New information, beyond odometry, has been used to correct the state in the Kalman filter update phase.

Figure 6.8a shows that the overall uncertainty is no longer growing monotonically. When the robot sees a landmark it is able to dramatically reduce its estimated covariance. Figure 6.8b shows the error associated with each component of pose and the pink background is the estimated 95% confidence bound (derived from the covariance matrix) and we see that the error is mostly within this envelope. Below this is plotted the landmark observations and we see that the confidence bounds are tight (indicating low uncertainty) while landmarks are being observed but that they start to grow once observations stop. However as soon as an observation is made the uncertainty rapidly decreases.

This EKF framework allows data from many and varied sensors to update the state which is why the estimation problem is also referred to as sensor fusion. For example heading angle from a compass, yaw rate from a gyroscope, target bearing angle from a camera, position

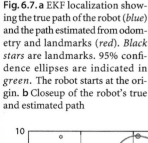

Fig. 6.7. a EKF localization showing the true path of the robot (*blue*) and the path estimated from odometry and landmarks (*red*). *Black stars* are landmarks. 95% confidence ellipses are indicated in *green*. The robot starts at the origin. **b** Closeup of the robot's true and estimated path

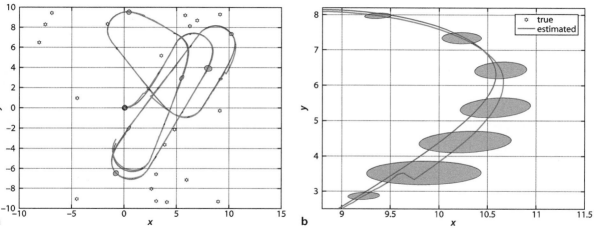

from GPS could all be used to update the state. For each sensor we need only to provide the observation function $h(\cdot)$, the Jacobians H_x and H_w and some estimate of the sensor covariance W. The function $h(\cdot)$ can be nonlinear and even noninvertible – the EKF will do the rest.

> As discussed earlier for V, we also use W twice. The first usage, in the constructor for the `RangeBearingSensor` object, is the covariance W of the Gaussian noise that is added to the computed range and bearing to *simulate* sensor error as in Eq. 6.8. In a real application this noise is generated by some physical process *hidden inside* the sensor and we would not know its parameters. The second usage, $\hat{W}$ is our best *estimate* of the sensor covariance which is used by the Kalman filter Eq. 6.12.

Fig. 6.8. **a** Covariance magnitude as a function of time. Overall uncertainty is given by $\sqrt{\det(P)}$ and shows that uncertainty does not continually increase with time. **b** Top: pose estimation error with 95% confidence bound shown in *pink*; bottom: observed landmarks the *bar* indicates which landmark is seen at each time step, 0 means no observation

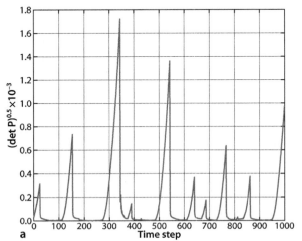

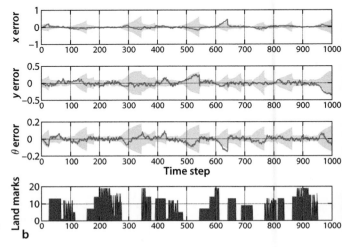

Data association. So far we have assumed that the observed landmark reveals its identity to us, but in reality this is rarely the case. Instead we compare our observation to the predicted position of all currently known landmarks and make a decision as to which landmark it is most likely to be, or whether it is a new landmark. This decision needs to take into account the uncertainty associated with the vehicle's pose, the sensor measurement and the landmarks in the map. This is the data association problem. Errors in this step are potentially catastrophic – incorrect innovation is coupled via the Kalman filter to the state of the vehicle and all the other landmarks which increases the chance of an incorrect data association on the next cycle. In practice, filters only use a landmark when there is a very high confidence in its estimated identity – a process that involves Mahalanobis distance and χ^2 confidence tests. If the situation is ambiguous it is best not to use the landmark – it can do more harm than good.

An alternative is to use a multi-hypothesis estimator, such as the particle filter that we will discuss in Sect. 6.7, which can model the possibility of observing landmark A or landmark B, and future observations will reinforce one hypothesis and weaken the others. The extended Kalman filter uses a Gaussian probability model, with just one peak, which limits it to holding only one hypothesis about the robot's pose. (Picture: the wreck of the Tararua, 1881)

Simple landmarks. For educational purposes it might be appropriate to use artificial landmarks that can be cheaply sensed by a camera. These need to be not only visually distinctive in the environment but also encode an identity. 2-dimensional bar codes such as QR codes or ARTags are well suited for this purpose. The Toolbox supports a variant called AprilTags, shown to the right, and

```
>> tags = apriltags(im);
```

returns a vector of `AprilTag` objects whose elements correspond to tags found in the image `im`. The centroid of the tag (`centre` property) can be used to determine relative bearing (see page 161), and the length of the edges (from the `corners` property) is a

A landmark might be some easily identifiable pattern such as this April tag (36h11) which can be detected in an image. Its position and size in the image encodes the bearing angle and range. The pattern itself encodes a number between 0 and 586 which could be used to uniquely identify the landmark in a map.

function of distance. The tag object also includes an homography (see Sect. 14.2.4) (`H` property) which encodes information about the orientation of the plane of the April tag. More details about April tags can be found at http://april.eecs.umich.edu.

6.3   Creating a Map

So far we have taken the existence of the map for granted, an understandable mindset given that maps today are common and available for free via the internet. Nevertheless somebody, or something, has to create the maps we will use. Our next example considers the problem of a robot moving in an environment with landmarks and creating a map of their locations.

As before we have a range and bearing sensor mounted on the robot which measures, imperfectly, the position of landmarks with respect to the robot. There are a total of N landmarks in the environment and as for the previous example we assume that the sensor can determine the identity of each observed landmark. However for this case we assume that the robot knows its own location perfectly – it has ideal localization. This is unrealistic but this scenario is an important stepping stone to the next section. ◀

Since the vehicle pose is known perfectly we do not need to estimate it, but we do need to estimate the coordinates of the landmarks. For this problem the state vector comprises the estimated coordinates of the M landmarks that have been observed so far

A close and realistic approximation would be a high-end RTK GPS+INS system operating in an environment with no buildings or hills to obscure satellites.

$$\hat{x} = \left(x_1, y_1, x_2, y_2, \cdots x_M, y_M\right)^T \in \mathbb{R}^{2M \times 1}$$

The corresponding estimated covariance $\hat{P}$ will be a $2M \times 2M$ matrix. The state vector has a variable length since we do not know in advance how many landmarks exist in the environment. Initially $M = 0$ and is incremented every time a previously unseen landmark is observed.

The prediction equation is straightforward in this case since the landmarks are assumed to be stationary

$$\hat{x}^+\langle k{+}1\rangle = \hat{x}\langle k\rangle \tag{6.16}$$

$$\hat{P}^+\langle k{+}1\rangle = \hat{P}\langle k\rangle \tag{6.17}$$

We introduce the function $g(\cdot)$ which is the inverse of $h(\cdot)$ and gives the coordinates of the observed landmark based on the known vehicle pose and the sensor observation

$$g(x, z) = \begin{pmatrix} x_v + r\cos(\theta_v + \beta) \\ y_v + r\sin(\theta_v + \beta) \end{pmatrix}$$

Since $\hat{x}$ has a variable length we need to extend the state vector and the covariance matrix whenever we encounter a landmark we have not previously seen. The state vector is extended by the function $y(\cdot)$

$$x\langle k\rangle' = y\big(x\langle k\rangle, z\langle k\rangle, x_v\langle k\rangle\big) \tag{6.18}$$

$$= \begin{pmatrix} x\langle k\rangle \\ g\big(x_v\langle k\rangle, z\langle k\rangle\big) \end{pmatrix} \tag{6.19}$$

which appends the sensor-based estimate of the new landmark's coordinates to those already in the map. The order of feature coordinates within $\hat{x}$ therefore depends on the order in which they are observed.

The covariance matrix also needs to be extended when a new landmark is observed and this is achieved by

$$\hat{P}\langle k\rangle' = Y_z \begin{pmatrix} \hat{P}\langle k\rangle & 0 \\ 0 & \hat{W} \end{pmatrix} Y_z^T$$

where Y_z is the insertion Jacobian

$$Y_z = \frac{\partial y}{\partial z} = \begin{pmatrix} I_{n \times n} & & 0_{n \times 2} \\ G_x & 0_{2 \times n-3} & G_z \end{pmatrix} \tag{6.20}$$

that relates the rate of change of the extended state vector to the new observation. n is the dimension of $\hat{P}$ prior to it being extended and

$$G_x = \frac{\partial g}{\partial x} = \begin{pmatrix} 0 & 0 & 0 \\ 0 & 0 & 0 \end{pmatrix} \qquad (6.21)$$

$$G_z = \frac{\partial g}{\partial z} = \begin{pmatrix} \cos(\theta_v + \beta) & -r\sin(\theta_v + \beta) \\ \sin(\theta_v + \beta) & r\cos(\theta_v + \beta) \end{pmatrix} \qquad (6.22)$$

G_x is zero since $g(\cdot)$ is independent of the map in x. An additional Jacobian for $h(\cdot)$ is

$$H_{p_i} = \frac{\partial h}{\partial p_i} = \begin{pmatrix} \frac{x_i - x_v}{r} & \frac{y_i - y_v}{r} \\ -\frac{y_i - y_v}{r^2} & \frac{x_i - x_v}{r^2} \end{pmatrix} \qquad (6.23)$$

which describes how the landmark observation changes with respect to landmark position for a particular robot pose, and is implemented by the method Hp.

For the mapping case the Jacobian H_x used in Eq. 6.11 describes how the landmark observation changes with respect to the full state vector. However the observation depends only on the position of that landmark so this Jacobian is mostly zeros

$$H_x = \frac{\partial h}{\partial x}\bigg|_{w=0} = \left(0 \cdots H_{p_i} \cdots 0 \right) \in \mathbb{R}^{2 \times 2M} \qquad (6.24)$$

where H_{p_i} is at the location in the vector corresponding to the state p_i. This structure represents the fact that observing a particular landmark provides information to estimate the position of that landmark, but no others.

The Toolbox implementation is

```
>> map = LandmarkMap(20);
>> veh = Bicycle();   % error free vehicle
>> veh.add_driver( RandomPath(map.dim) );
>> W = diag([0.1, 1*pi/180].^2);
>> sensor = RangeBearingSensor(veh, map, 'covar', W);
>> ekf = EKF(veh, [], [], sensor, W, []);
```

the empty matrices passed to EKF indicate respectively that there is no estimated odometry covariance for the vehicle (the estimate is perfect), no initial vehicle state covariance, and the map is unknown. We run the simulation for 1 000 time steps

```
>> ekf.run(1000);
```

Fig. 6.9. EKF mapping results. **a** The estimated landmarks are indicated by *black dots* with 95% confidence ellipses (*green*), the true location (*black ✿-marker*) and the robot's path (*blue*). The landmark estimates have not fully converged on their true values and the estimated covariance ellipses can only be seen by zooming; **b** the nonzero elements of the final covariance matrix

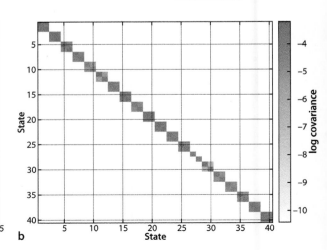

and see an animation of the robot moving and the covariance ellipses associated with the map features evolving over time. The estimated landmark positions

```
>> map.plot();
>> ekf.plot_map('g');
>> veh.plot_xy('b');
```

are shown in Fig. 6.9a as 95% confidence ellipses along with the true landmark positions and the path taken by the robot. The covariance matrix has a block diagonal structure which is shown graphically in Fig. 6.9b. The off-diagonal elements are zero, which implies that the landmark estimates are uncorrelated or independent. This is to be expected since observing one landmark provides no new information about any other landmark.

Internally the EKF object maintains a table to relate the landmark's identity, returned by the RangeBearingSensor, to the position of that landmark's coordinates in the state vector. For example the landmark with identity 6

```
>> ekf.landmarks(:,6)
ans =
      19
      71
```

was seen a total of 71 times during the simulation and comprises elements 19 and 20 of $\hat{x}$

```
>> ekf.x_est(19:20)'
ans =
   -6.4803    9.6233
```

which is its estimated location. Its estimated covariance is a submatrix within $\hat{P}$

```
>> ekf.P_est(19:20,19:20)
ans =
   1.0e-03 *
      0.2913      0.1814
      0.1814      0.3960
```

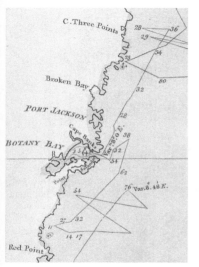

Fig. 6.10. Map of the New Holland coast (now eastern Australia) by Captain James Cook in 1770. The path of the ship and the map of the coast were determined at the same time. *Numbers* indicate depth in fathoms (1.83 m) (National Library of Australia, MAP NK 5557 A)

6.4 Localization and Mapping

Finally we tackle the problem of determining our position and creating a map at the same time. This is an old problem in marine navigation and cartography – incrementally extending maps while also using the map for navigation. Figure 6.10 shows what can be done without GPS from a moving ship with poor odometry and infrequent celestial position "fixes". In robotics this problem is known as simultaneous localization and mapping (SLAM) or concurrent mapping and localization (CML). This is often considered to be a "chicken and egg" problem – we need a map to localize and we need to localize to make the map. However based on what we have learned in the previous sections this problem is now quite straightforward to solve.

The state vector comprises the vehicle configuration *and* the coordinates of the *M* landmarks that have been observed so far

$$\hat{x} = \left(x_v, y_v, \theta_v, x_1, y_1, x_2, y_2, \cdots x_M, y_M\right)^T \in \mathbb{R}^{2M+3 \times 1}$$

The estimated covariance is a $(2M + 3) \times (2M + 3)$ matrix and has the structure

$$\hat{P} = \begin{pmatrix} \hat{P}_{vv} & \hat{P}_{vm} \\ \hat{P}_{vm}^T & \hat{P}_{mm} \end{pmatrix}$$

where $\hat{P}_{vv}$ is the covariance of the vehicle pose, $\hat{P}_{mm}$ the covariance of the map landmark positions, and $\hat{P}_{vm}$ is the correlation between vehicle and landmark states.

The predicted vehicle state and covariance are given by Eq. 6.3 and Eq. 6.4 and the sensor-based update is given by Eq. 6.10 to 6.15. When a new feature is observed the state vector is updated using the insertion Jacobian Y_z given by Eq. 6.20 but in this case G_x is nonzero

$$G_x = \frac{\partial g}{\partial x} = \begin{pmatrix} 1 & 0 & -r\sin(\theta_v + \beta) \\ 0 & 1 & r\cos(\theta_v + \beta) \end{pmatrix} \tag{6.25}$$

since the estimate of the new landmark depends on the state vector which now contains the vehicle's pose.

For the SLAM case the Jacobian H_x used in Eq. 6.11 describes how the landmark observation changes with respect to the state vector. The observation will depend on the position of the vehicle and on the position of the observed landmark and is

$$H_x = \left. \frac{\partial h}{\partial x} \right|_{w=0} = \left(H_{x_v} \cdots 0 \cdots H_{p_i} \cdots 0 \right) \in \mathbb{R}^{2\times(2M+3)} \tag{6.26}$$

where H_{p_i} is at the location corresponding to the landmark p_i. This is similar to Eq. 6.24 but with an extra nonzero block H_{x_v} given by Eq. 6.14.

The Kalman gain matrix K *distributes* innovation from the landmark observation, a 2-vector, to update *every* element of the state vector – the pose of the vehicle *and* the position of *every* landmark in the map.

The Toolbox implementation is by now quite familiar

```
>> P0 = diag([.01, .01, 0.005].^2);
>> map = LandmarkMap(20);
>> veh = Bicycle('covar', V);
>> veh.add_driver( RandomPath(map.dim) );
>> sensor = RangeBearingSensor(veh, map, 'covar', W);
>> ekf = EKF(veh, V, P0, sensor, W, []);
```

and the empty matrix passed to EKF indicates that the map is unknown. P0 is the initial 3×3 covariance for the vehicle state.

We run the simulation for 1 000 time steps

```
>> ekf.run(1000);
```

and as usual an animation is shown of the vehicle moving. We also see the covariance ellipses associated with the map features evolving over time. We can plot the results

```
>> map.plot();
>> ekf.plot_map('g');
>> ekf.plot_xy('r');
>> veh.plot_xy('b');
```

which are shown in Fig. 6.11.

Figure 6.12a shows that uncertainty is decreasing over time. The final covariance matrix is shown graphically in Fig. 6.12b and we see a complex structure. Unlike the mapping case af Fig. 6.9 $\hat{P}_{mm}$ is not block diagonal, and the finite off-diagonal terms

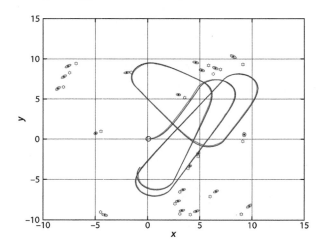

Fig. 6.11.
Simultaneous localization and mapping showing the true (*blue*) and estimated (*red*) robot path superimposed on the true map (*black* ✿-*marker*). The estimated map features are indicated by *black dots* with 95% confidence ellipses (*green*)

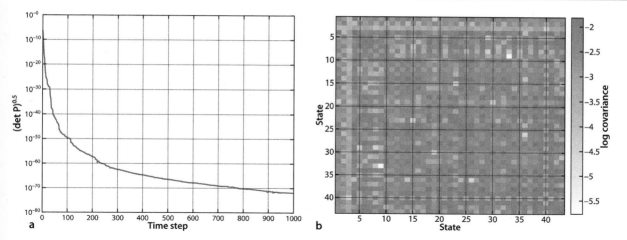

Fig. 6.12. Simultaneous localization and mapping. **a** Covariance versus time; **b** the final covariance matrix

represent correlation *between* the landmarks in the map. The landmark uncertainties never increase, the position prediction model is that they do not move, but they also never drop below the initial uncertainty of the vehicle which was set in P_0. The block $\hat{P}_{vm}$ is the correlation between errors in the vehicle pose and the landmark locations. A landmark's location estimate is a function of the vehicle's location and errors in the vehicle location appear as errors in the landmark location – and vice versa.

The correlations are used by the Kalman filter to connect the observation of any landmark to an improvement in the estimate of every other landmark in the map as well as the vehicle pose. Conceptually it is as if all the states were connected by springs and the movement of any one affects all the others.

The extended Kalman filter introduced here has a number of drawbacks. Firstly the size of the matrices involved increase with the number of landmarks and can lead to memory and computational bottlenecks as well as numerical problems. The underlying assumption of the Kalman filter is that all errors are Gaussian and this is far from true for sensors like laser rangefinders which we will discuss later in this chapter. We also need good estimates of covariance of the noise sources which in practice is challenging.

6.5 Rao-Blackwellized SLAM

We will briefly and informally introduce the underlying principle of Rao-Blackwellized SLAM of which FastSLAM is a popular and well known instance. The approach is motivated by the fact that the size of the covariance matrix for EKF SLAM is quadratic in the number of landmarks, and for large-scale environments becomes computationally intractable.

If we compare the covariance matrices shown in Fig. 6.9b and 6.12b we notice a stark difference. In both cases we were creating a map of unknown landmarks but Fig. 6.9b is mostly zero with a finite block diagonal structure whereas Fig. 6.12b has no zero values at all. The difference is that for Fig. 6.9b we assumed the robot trajectory was known exactly and that makes the landmark estimates *independent* – observing one landmark provides information about *only* that landmark. The landmarks are *uncorrelated*, hence all the zeros in the covariance matrix. If the robot trajectory is not known, the case for Fig. 6.12b, then the landmark estimates are correlated – error in one landmark position is related to errors in robot pose and other landmark positions. The Kalman filter uses the correlation information so that a measurement of any one landmark provides information to improve the estimate of all the other landmarks and the robot's pose.

In practice we don't know the true pose of the robot but imagine a multi-hypothesis estimator◄ where every hypothesis represents a robot trajectory that we *assume* is correct. This means that the covariance matrix will be block diagonal like Fig. 6.9b – rather than a filter with a $2N \times 2N$ covariance matrix we can have N simple filters which are

Such as the particle filter that we will discuss in Sect. 6.7.

each *independently* estimating the position of a single landmark and have a 2×2 covariance matrix. Independent estimation leads to a considerable saving in both memory and computation. Importantly though, we are only able to do this because we *assumed* that the robot's estimated trajectory is correct.

Each hypothesis also holds an estimate of the robot's trajectory to date. Those hypotheses that best explain the landmark measurements are retained and propagated while those that don't are removed and recycled. If there are M hypotheses the overall computational burden falls from $O(N^2)$ for the EKF SLAM case to $O(M \log N)$ and in practice works well for M in the order of tens to hundreds but can work for a value as low as $M = 1$.

6.6   Pose Graph SLAM

An alternative approach to the SLAM problem is to separate it into two components: a front end and a back end, connected by a pose graph as shown in Fig. 6.13. The robot's path is considered to be a sequence of distinct poses and the task is to estimate those poses. Constraints between the unknown poses are based on measurements from a variety of sensors including odometry, laser scanners and cameras. The problem is formulated as a directed graph as shown in Fig. 6.14. A node corresponds to a robot pose or a landmark position. An edge between two nodes represents a spatial constraint between the nodes derived from some sensor data.

As the robot progresses it compounds an increasing number of uncertain relative poses so that the cumulative error in the pose of the nodes will increase – the problem with dead reckoning we discussed earlier. This is shown in exaggerated fashion in Fig. 6.14 where the robot is traveling around a square. By the time the robot reaches node 4 the error is significant. However when it makes a measurement of node 1 a constraint is added – the dashed edge – indicating that the nodes are closer than the estimated relative pose based on the chain of relative poses from odometry: ${}^1\xi_2^\# \oplus {}^2\xi_3^\# \oplus {}^3\xi_4^\#$. The back-end algorithm will then *pull* all the nodes closer to their correct pose.

The front end adds new nodes as the robot travels▶ as well as edges that define constraints between poses. For example, when moving from one place to another wheel odometry gives an estimate of distance and change in orientation which is a constraint. In addition the robot's exteroceptive sensors may observe the relative position of a landmark and this also adds a constraint. Every measurement adds a constraint – an edge in the graph. There is no limit to the number of edges entering or leaving a node.

> Typically a new place is declared every meter or so of travel, or after a sharp turn.

The back end adjusts the poses of the nodes▶ so that the constraints are satisfied as well as possible, that is, that the sensor observations are best explained.

> Also the positions of landmarks as we discuss later in this section.

Figure 6.15 shows the notation associated with two poses in the graph. Coordinate frames $\{i\}$ and $\{j\}$ are associated with robot poses i and j respectively and we seek to estimate ${}^0\xi_i$ and ${}^0\xi_j$ in the world coordinate frame. The robot makes a measurement of the relative pose ${}^i\xi_j^\#$ which will, in general, be different to the relative pose ${}^i\xi_j$ inferred from the poses ${}^0\xi_i$ and ${}^0\xi_j$. This difference, or innovation, is caused by error in the sensor measurement ${}^i\xi_j^\#$ and/or the node poses ${}^0\xi_i$ and ${}^0\xi_j$ and we use it to adjust the poses of the nodes. However there is insufficient information to determine where the error lies so naively adjusting ${}^0\xi_i$ and ${}^0\xi_j$ to better explain the measurement might increase

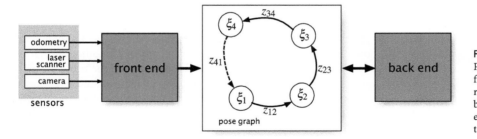

Fig. 6.13.
Pose-graph SLAM system. The front end creates nodes as the robot travels, and creates edges based on sensor data. The back end adjusts the node positions to minimize total error

the error in another part of the graph – we need to minimize the error consistently over the whole graph.

The first step is to express the error associated with the graph edge in terms of the sensor measurement and our best estimates of the node poses with respect to the world frame◄

We have used our pose notation here but in the literature measurements are typically denoted by z, error by e and pose or position by x.

$$\xi_\varepsilon = \ominus {}^i\xi_j^\# \ominus {}^0\hat{\xi}_i \oplus {}^0\hat{\xi}_j \in \mathbf{SE}(2) \tag{6.27}$$

which is ideally zero.

We can formulate this as a minimization problem and attempt to find the poses of all the nodes $x = \{\xi_1, \xi_2 \cdots \xi_N\}$ that minimizes the error across all the edges

$$x^* = \arg\min_x \sum_k F_k(x) \tag{6.28}$$

where x is the state of the pose graph and contains the pose of every node, and $F_k(x)$ is a nonnegative scalar cost associated with the edge k connecting node i to node j.

We convert the edge pose error in Eq. 6.27 to a vector representation $\xi_\varepsilon \sim (x, y, \theta)$ which is a function $f_k(x) \in \mathbb{R}^3$ of the state. The scalar cost can be obtained from a quadratic expression

In practice this matrix is diagonal reflecting confidence in the x-, y- and θ-directions. The "bigger" (in a matrix sense) Ω is, the more the edge *matters* in the optimization procedure. Different sensors have different accuracy and this must be taken into account. Information from a high-quality sensor should be given more weight than information from a low-quality sensor.

$$F_k(x) = f_k^T(x)\Omega_k f_k(x) \tag{6.29}$$

where Ω_k is a positive-definite information matrix used as a weighting term.◄ Although Eq. 6.29 is written as a function of all poses x, it in fact depends only on the pose of its two vertices ξ_i and ξ_j and the measurement ${}^i\xi_j^\#$. Solving Eq. 6.28 is a complex optimization problem which does not have a closed-form solution, but this kind of nonlinear least squares problem can be solved numerically if we have a good initial estimate of x. Specifically this is a sparse nonlinear least squares problem which is discussed in Sect. F.2.4.

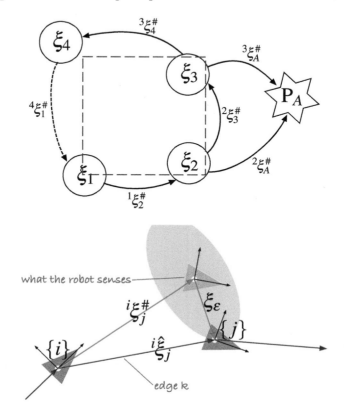

Fig. 6.14.
Pose-graph SLAM example. Places are shown as *circular nodes* and have an associated pose. Landmarks are shown as *star-shaped nodes* and have an associated position. Edges represent a measurement of a relative pose or position with respect to the node at the tail of the arrow

Fig. 6.15.
Pose graph notation. The *light grey robot* is the estimated pose of {j} based on the sensor measurement ${}^i\xi_j^\#$. The *yellow ellipse* indicates uncertainty associated with that measurement

The edge error $f_k(x)$ can be linearized about the current state x_0 of the pose graph

$$f'_k(\Delta) \approx f_{0,k} + J_k\Delta$$

where $f_{0,k} = f_k(x_0)$ and

$$J_k = \frac{\partial f_k(x)}{\partial x} \in \mathbb{R}^{3\times 3N}$$

is a Jacobian matrix which depends only on the pose of its two vertices ξ_i and ξ_j so it is mostly zeros

$$J_k = \begin{pmatrix} 0\cdots A_i \cdots B_j \cdots 0 \end{pmatrix}, \quad \text{where } A_i = \frac{\partial f_k(x)}{\partial \xi_i} \in \mathbb{R}^{3\times 3}, B_j = \frac{\partial f_k(x)}{\partial \xi_j} \in \mathbb{R}^{3\times 3}$$

and more details are provided in Appendix E.

There are many ways to compute the Jacobians but here will demonstrate use of the MATLAB Symbolic Math Toolbox™

```
>> syms xi yi ti xj yj tj xm ym tm assume real
>> xi_e = inv( SE2(xm, ym, tm) ) * inv( SE2(xi, yi, ti) ) * SE2(xj, yj, tj);
>> fk = simplify(xi_e.xyt);
```

and the Jacobian which describes how the function f_k varies with respect to ξ_i is

```
>> jacobian( fk, [xi yi ti] );
>> Ai = simplify(ans)
Ai =
[ -cos(ti+tm), -sin(ti+tm), yj*cos(ti+tm)-yi*cos(ti+tm)+xi*sin(ti+tm)-xj*sin(ti+tm)]
[  sin(ti+tm), -cos(ti+tm), xi*cos(ti+tm)-xj*cos(ti+tm)+yi*sin(ti+tm)-yj*sin(ti+tm)]
[          0,           0,                                                       -1]
```

and we follow a similar procedure for B_j.

It is quite straightforward to solve this type of pose-graph problem with the Toolbox. We load a simple pose graph, similar to Fig. 6.14, from a data file►

```
>> pg = PoseGraph('pg1.g2o')
loaded g2o format file: 4 nodes, 4 edges in 0.00 sec
```

The file format is one used by the popular posegraph optimization package g²o which you can find at http://openslam.org.

which returns a Toolbox `PoseGraph` object that describes the pose graph. We can visualize this by►

The nodes have an orientation which is in the z-direction, rotate the graph to see this.

```
>> pg.plot()
```

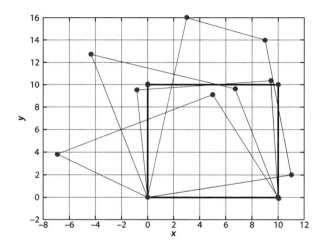

Fig. 6.16.
Pose graph optimization showing the result over consecutive iterations, the final configuration is the *square* shown in **bold**

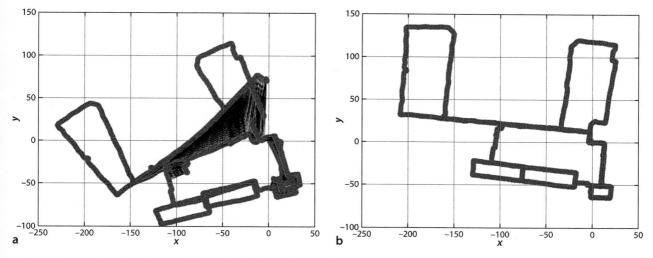

Fig. 6.17. Pose graph with 1 941 nodes and 3 995 edges from the MIT Killian Court dataset. **a** Initial configuration; **b** final configuration after optimization

The optimization reduces the error in the network while animating the changing pose of the nodes

```
>> pg.optimize('animate')
solving....done in 0.075 sec.   Total cost 316.88
solving....done in 0.0033 sec.  Total cost 47.2186
  .

solving....done in 0.0023 sec.  Total cost 3.14139e-11
```

The displayed text indicates that the total cost is decreasing while the graphics show the nodes moving into a configuration that minimizes the overall error in the network. The pose graph configurations are overlaid and shown in Fig. 6.16.

Now let's look a much larger example based on real robot data

```
>> pg = PoseGraph('killian-small.toro');
loaded TORO/LAGO format file: 1941 nodes, 3995 edges in 0.68 sec
```

There are a lot of nodes and this takes a few seconds.

which we can plot◄

```
>> pg.plot()
```

and this is shown in Fig. 6.17a. Note the mass of edges in the center of the graph, and if you zoom in you can see these in detail. We optimize the pose graph by

```
>> pg.optimize()
solving....done in 0.91 sec.   Total cost 1.78135e+06
  .

solving....done in 1.1 sec.  Total cost 5.44567
```

and the final configuration is shown in Fig. 6.17b. The original pose graph had severe pose errors from accumulated odometry error which meant that two trips along the corridor were initially very poorly aligned.

The pose graph can also include landmarks as shown in Fig. 6.18. Landmarks have a position $P_j \in \mathbb{R}^2$ not a pose, and therefore differ from the nodes discussed so far. To accomodate this we redefine the state vector to be $x = \{\xi_1, \xi_2 \cdots \xi_N \,|\, P_1, P_2 \cdots P_M\}$ which includes N robot poses and M landmark positions. The robot at pose i observes landmark j at range and bearing $z^\# = (r^\#, \beta^\#)$ which is converted to Cartesian form in frame $\{i\}$

$$^iP_j^\# = \left(r^\# \cos\beta^\#, r^\# \sin\beta^\#\right) \in \mathbb{R}^2$$

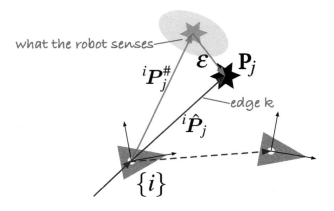

Fig. 6.18.
Notation for a pose graph with a landmark indicated by the *star-shaped symbol*. The measured position of landmark j with respect to robot pose i is ${}^iP_j^\#$. The *yellow ellipse* indicates uncertainty associated with that measurement

The estimated position of the landmark in frame $\{i\}$ is

$${}^i\hat{P}_j = \left(\ominus{}^0\xi_i\right)\bullet\hat{P}_j \in \mathbb{R}^2$$

and the error vector is

$$f_k(x) = \varepsilon = {}^i\hat{P}_j - {}^iP_j^\# \in \mathbb{R}^2$$

We follow a similar approach as earlier but the Jacobian matrix is now

$$J_k = \frac{\partial f_k(x)}{\partial x} \in \mathbb{R}^{2\times(3N+2M)}$$

which again is mostly zero but the two nonzero blocks now have different widths

$$A_i = \frac{\partial f_k(x)}{\partial \xi_i} \in \mathbb{R}^{2\times3}, \; B_j = \frac{\partial f_k(x)}{\partial P_j} \in \mathbb{R}^{2\times2}$$

and the solution can be achieved as before, see Sect. F.2.3 for more details.

Pose graph optimization results in a graph that has optimal relative poses and positions between the nodes but the absolute poses and positions are not necessarily correct. To remedy this we can fix or *anchor* one or more nodes (poses or landmarks) and not update them during the optimization, and this is discussed in Sect. F.2.4.

In practice the front and back ends can operate asynchronously. The graph is continually extended by the front end while the back end runs periodically to opti-

> **Monte Carlo methods** are a class of computational algorithms that rely on repeated random sampling to compute their results. An early example of this idea is Buffon's needle problem posed in the eighteenth century by Georges-Louis Leclerc (1707–1788), Comte de Buffon: *Suppose we have a floor made of parallel strips of wood of equal width t, and a needle of length l is dropped onto the floor. What is the probability that the needle will lie across a line between the strips?* If n needles are dropped and h cross the lines, the probability can be shown to be $h/n = 2l/\pi t$ and in 1901 an Italian mathematician Mario Lazzarini performed the experiment, tossing a needle 3 408 times, and obtained the estimate $\pi \approx 355/113$ (3.14159292).
>
> Monte Carlo methods are often used when simulating systems with a large number of coupled degrees of freedom with significant uncertainty in inputs. Monte Carlo methods tend to be used when it is infeasible or impossible to compute an exact result with a deterministic algorithm. Their reliance on repeated computation and random or pseudo-random numbers make them well suited to calculation by a computer. The method was developed at Los Alamos as part of the Manhattan project during WW II by the mathematicians John von Neumann, Stanislaw Ulam and Nicholas Metropolis. The name Monte Carlo alludes to games of chance and was the code name for the secret project.

mize the pose graph. Since the graph is only ever extended in a local region it is possible to optimize just a local subset of the pose graph and less frequently optimize the entire graph. If nodes are found to be equivalent after optimization they can be merged. The parallel tracking and mapping system (PTAM) is a vision-based SLAM system that has two parallel computational threads. One is the map builder which performs the front- and back-end tasks, adding landmarks to the pose graph based on estimated camera (vehicle) pose and performing graph optimization. The other thread is the localizer which matches observed landmarks to the estimated map to estimate the camera pose.

6.7   Sequential Monte-Carlo Localization

The estimation examples so far have assumed that the error in sensors such as odometry and landmark range and bearing have a Gaussian probability density function. In practice we might find that a sensor has a one sided distribution (like a Poisson distribution) or a multi-modal distribution with several peaks. The functions we used in the Kalman filter such as Eq. 6.2 and Eq. 6.7 are strongly nonlinear which means that sensor noise with a Gaussian distribution will not result in a Gaussian error distribution on the value of the function – this is discussed further in Appendix H. The probability density function associated with a robot's configuration may have multiple peaks to reflect several hypotheses that equally well explain the data from the sensors as shown for example in Fig. 6.3c.

The Monte-Carlo estimator that we discuss in this section makes no assumptions about the distribution of errors. It can also handle multiple hypotheses for the state of the system. The basic idea is disarmingly simple. We maintain many *different* values of the vehicle's configuration or state vector. When a new measurement is available we score how well each particular value of the state explains what the sensor just observed. We keep the best fitting states and randomly sample from the prediction distribution to form a new generation of states. Collectively these many possible states and their scores form a discrete approximation of the probability density function of the state we are trying to estimate. There is never any assumption about Gaussian distributions nor any need to linearize the system. While computationally expensive it is quite feasible to use this technique with today's standard computers. If we plot these state vectors as points in the state space we have a cloud of particles hence this type of estimator is often referred to as a particle filter.

We will apply Monte-Carlo estimation to the problem of localization using odometry and a map. Estimating only three states $x = (x, y, \theta)$ is computationally tractable to solve with straightforward MATLAB code. The estimator is initialized by creating N particles x_i, $i \in [1, N]$ distributed randomly over the configuration space of the vehicle. All particles have the same initial weight or likelihood $w_i = 1 / N$. The steps in the main iteration of the algorithm are:

1. Apply the state update to each particle

$$x_i^+ \langle k{+}1 \rangle = f\big(x_i \langle k \rangle, u \langle k \rangle + q \langle k \rangle\big)$$

where $\hat{u} \langle k \rangle$ is the input to the system or the measured odometry $\hat{u} \langle k \rangle = \delta \langle k \rangle$. We also add a random vector $q \langle k \rangle$ which represents uncertainty in the model or the odometry. Often q is drawn from a Gaussian random variable with covariance Q but any physically meaningful distribution can be used. The state update is often simplified to

$$x_i^+ \langle k{+}1 \rangle = f\big(x_i \langle k \rangle, u \langle k \rangle\big) + q \langle k \rangle$$

where $q \langle k \rangle$ represents uncertainty in the pose of the vehicle.

2. We make an observation $z^{\#}$ of landmark j which has, according to the map, coordinate p_j. For each particle we compute the innovation

$$\nu_i = h\left(x_i^+, p_j\right) - z^{\#}$$

which is the error between the predicted and actual landmark observation. A likelihood function provides a scalar measure of how well the particular particle explains this observation. In this example we choose a likelihood function

$$w_i = e^{-\nu_i^T L^{-1} \nu_i} + w_0$$

where w is referred to as the *importance* or *weight* of the particle, L is a covariance-like matrix, and $w_0 > 0$ ensures that there is a finite probability of a particle being retained despite sensor error. We use a quadratic exponential function only for convenience, the function does not need to be smooth or invertible but only to adequately describe the likelihood of an observation.▶

In this bootstrap type filter the weight is computed at each step, with no dependence on previous values.

3. Select the particles that best explain the observation, a process known as resampling▶ or importance sampling. A common scheme is to randomly select particles according to their weight. Given N particles x_i with corresponding weights w_i we first normalize the weights $w_i' = w_i / \Sigma_{i=1}^N w_i$ and construct a cumulative histogram $c_j = \Sigma_{i=1}^j w_i'$. We then draw a uniform random number $r \in [0, 1]$ and find

$$\arg \min_i r < c_i$$

where particle i is selected for the next generation. The process is repeated N times.

There are many resampling strategies for particle filters, both the resampling algorithm and the resampling frequency. Here we use the simplest strategy known variously as multinomial resampling, simple random resampling or select with replacement, at every time step. This is sometimes referred to as bootstrap particle filtering or condensation.

Particles with a large weight will correspond to a larger fraction of the vertical span of the cumulative histogram and therefore be more likely to be chosen. The result will have the same number of particles, some will have been copied▶ multiple times, others not at all. Resampling is a critical component of the particle filter without which the filter would quickly produce a degenerate set of particles where a few have high weights and the bulk have almost zero weight.

Step 1 of the next iteration will spread out these copies through the addition of $q_{(k)}$.

These steps are summarized in Fig. 6.19. The Toolbox implementation is broadly similar to the previous examples. We create a map

```
>> map = LandmarkMap(20);
```

and a robot with noisy odometry and an initial condition

```
>> W = diag([0.1, 1*pi/180].^2);
>> veh = Bicycle('covar', V);
>> veh.add_driver( RandomPath(10) );
```

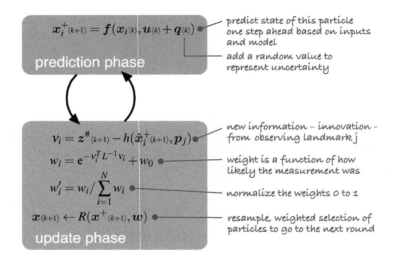

predict state of this particle one step ahead based on inputs and model

add a random value to represent uncertainty

new information – innovation – from observing landmark j

weight is a function of how likely the measurement was

normalize the weights 0 to 1

resample, weighted selection of particles to go to the next round

Fig. 6.19.
The particle filter estimator showing the prediction and update phases

and then a sensor with noisy readings

```
>> V = diag([0.005, 0.5*pi/180].^2);
>> sensor = RangeBearingSensor(veh, map, 'covar', W);
```

For the particle filter we need to define two covariance matrices. The first is the covariance of the random noise added to the particle states at each iteration to represent uncertainty in configuration. We choose the covariance values to be comparable with those of W

```
>> Q = diag([0.1, 0.1, 1*pi/180]).^2;
```

and the covariance of the likelihood function applied to innovation

```
>> L = diag([0.1 0.1]);
```

Finally we construct a `ParticleFilter` estimator

```
>> pf = ParticleFilter(veh, sensor, Q, L, 1000);
```

which is configured with 1 000 particles. The particles are initially uniformly distributed over the 3-dimensional configuration space.

We run the simulation for 1 000 time steps

```
>> pf.run(1000);
```

and watch the animation, two snapshots of which are shown in Fig. 6.20. We see the particles move about as their states are updated by odometry and random perturbation. The initially randomly distributed particles begin to aggregate around those regions of the configuration space that best *explain* the sensor observations that are made. In Darwinian fashion these particles become more highly weighted and survive the resampling step while the lower weight particles are extinguished.

The particles approximate the probability density function of the robot's configuration. The most likely configuration is the expected value or mean of all the particles. A measure of uncertainty of the estimate is the spread of the particle cloud or its standard deviation. The `ParticleFilter` object keeps the history of the mean and standard deviation of the particle state at each time step, taking into account the particle weighting[▸]. As usual we plot the results of the simulation

```
>> map.plot();
>> veh.plot_xy('b');
```

and overlay the mean of the particle cloud

```
>> pf.plot_xy('r');
```

Here we take statistics over all particles. Other strategies are to estimate the kernel density at every particle – the sum of the weights of all neighbors within a fixed radius – and take the particle with the largest value.

Fig. 6.20. Particle filter results showing the evolution of the particle cloud (*green dots*) over time. The vehicle is shown as a *blue triangle*. The *red diamond* is a waypoint, or temporary goal. When the simulation is running this is actually a 3D plot with orientation plotted in the *z*-direction, rotate the plot to see this dimension

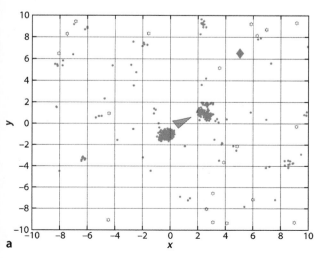

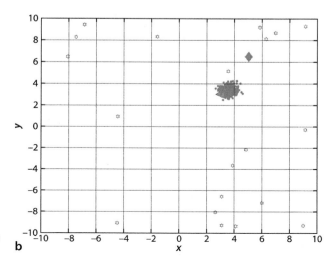

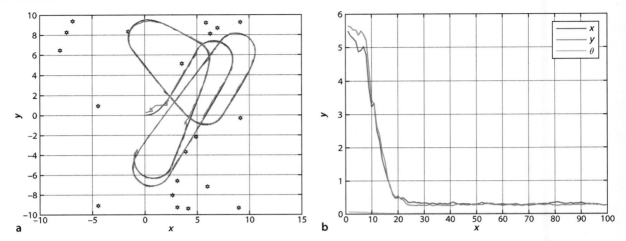

Fig. 6.21. Particle filter results. **a** True (*blue*) and estimated (*red*) robot path; **b** standard deviation of the particles versus time

which is shown in Fig. 6.21. The initial part of the estimated path has quite high standard deviation since the particles have not converged on the true configuration. We can plot the standard deviation against time

```
>> plot(pf.std(1:100,:))
```

and this is shown in Fig. 6.21b. We can see the sudden drop between timesteps 10–20 as the particles that are distant from the true solution are eliminated. As mentioned at the outset the particles are a sampled approximation to the PDF and we can display this as

```
>> pf.plot_pdf()
```

The problem we have just solved is known in robotics as the kidnapped robot problem where a robot is placed in the world with no idea of its initial location. To represent this large uncertainty we uniformly distribute the particles over the 3-dimensional configuration space and their sparsity can cause the particle filter to take a long time to converge unless a very large number of particles is used. It is debatable whether this is a realistic problem. Typically we have some approximate initial pose of the robot and the particles would be initialized to that part of the configuration space. For example, if we know the robot is in a corridor then the particles would be placed in those areas of the map that are corridors, or if we know the robot is pointing north then set all particles to have that orientation.

Setting the parameters of the particle filter requires a little experience and the best way to learn is to experiment. For the kidnapped robot problem we set Q and the number of particles high so that the particles explore the configuration space but once the filter has converged lower values could be used. There are many variations on the particle filter in the shape of the likelihood function and the resampling strategy.

6.8   Application: Scanning Laser Rangefinder

As we have seen, robot localization is informed by measurements of range and bearing to landmarks. Sensors that measure range can be based on many principles such as laser rangefinding (Fig. 6.22a, 6.22b), ultrasonic ranging (Fig. 6.22c), computer vision or radar.

A laser rangefinder emits short pulses of infra-red laser light and measures how long it takes for the reflected pulse to return. Operating range can be up to 50 m with an accuracy of the order of centimeters.

A *scanning* laser rangefinder, as shown in Fig. 6.22a, contains a rotating laser rangefinder and typically emits a pulse every quarter, half or one degree over an angular range of 180 or 270 degrees and returns a planar cross-section of the world in polar coordinate form $\{(r_i, \theta_i), i \in 1 \cdots N\}$. Some scanning laser rangefinders also measure

Fig. 6.22.
Robot rangefinders. **a** A scanning laser rangefinder with a maximum range of 30 m, an angular range of 270 deg in 0.25 deg intervals at 40 scans per second (courtesy of Hokuyo Automatic Co. Ltd.); **b** a low-cost time-of-flight rangefinder with maximum range of 20 cm at 10 measurements per second (VL6180 courtesy of SparkFun Electronics); **c** a low-cost ultrasonic rangefinder with maximum range of 6.5 m at 20 measurements per second (LV-MaxSonar-EZ1 courtesy of SparkFun Electronics)

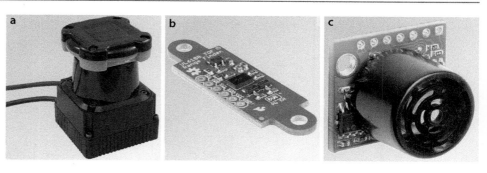

the return signal strength, remission, which is a function of the infra-red reflectivity of the surface. The rangefinder is typically configured to scan in a plane parallel to, and slightly above, the ground.

Laser rangefinders have advantages and disadvantages compared to cameras and computer vision which we discuss in Parts IV and V of this book. On the positive side laser scanners provide metric data, that is, the actual range to points in the world in units of meters, and they can work in the dark. However laser rangefinders work less well than cameras outdoors since the returning laser pulse is overwhelmed by infra-red light from the sun. Other disadvantages include providing only a linear cross section of the world, rather than an area as a camera does; inability to discern fine texture or color; having moving parts; as well as being bulky, power hungry and expensive compared to cameras.

Laser Odometry

A common application of scanning laser rangefinders is laser odometry, estimating the change in robot pose using laser scan data rather than wheel encoder data. We will illustrate this with laser scan data from a real robot

```
>> pg = PoseGraph('killian.g2o', 'laser');
loaded g2o format file: 3873 nodes, 4987 edges in 1.78 sec
   3873 laser scans: 180 beams, fov -90 to 90 deg, max range 50
```

and each scan is associated with a vertex of this already optimized pose graph. The range and bearing data for the scan at node 2 580 is

```
>> [r, theta] = pg.scan(2580);
>> about r theta
r [double] : 1x180 (1.4 kB)
theta [double] : 1x180 (1.4 kB)
```

represented by two vectors each of 180 elements. We can plot these in polar form

```
>> polar(theta, r)
```

or convert them to Cartesian coordinates and plot them

```
>> [x,y] = pol2cart(theta, r);
>> plot(x, y, '.')
```

The method `scanxy` is a simpler way to perform these operations. We load scans from two closely spaced nodes

```
>> p2580 = pg.scanxy(2580);
>> p2581 = pg.scanxy(2581);
>> about p2580
p2580 [double] : 2x180 (2.9 kB)
```

Note that the points close to the laser, at coordinate (0,0) in this sensor reference frame are much more tightly clustered and this is a characteristic of laser scanners where the points are equally spaced in angle not over an area.

which creates two matrices whose columns are Cartesian point coordinates and these are overlaid in Fig. 6.23a.◄

To determine the change in pose of the robot between the two scans we need to align these two sets of points and this can be achieved with iterated closest-point-matching

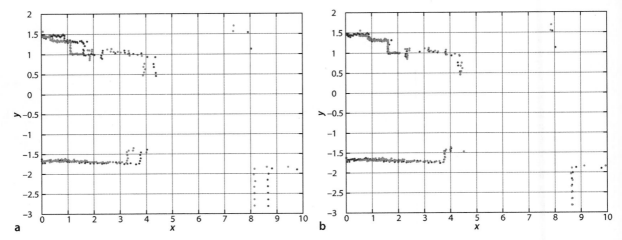

a   **b**

or ICP. This is implemented by the Toolbox function `icp`◢ and we pass in the second and first set of points, each organized as a $2 \times N$ matrix

```
>> T = icp( p2581, p2580, 'verbose' , 'T0', transl2(0.5, 0), 'distthresh', 3)
[1]: n=132/180, d=   0.466, t = (   0.499   -0.006), th = (  -0.0) deg
[2]: n=130/180, d=   0.429, t = (   0.500   -0.009), th = (   0.0) deg
     .
     .
     .
[6]: n=130/180, d=   0.425, t = (   0.503   -0.011), th = (   0.0) deg

T =
     1.0000   -0.0002    0.5032
     0.0002    1.0000   -0.0113
          0         0    1.0000
```

and the algorithm converges after a few iterations with an estimate of $T \sim {}^{2580}\xi_{2581}$ $\in \mathbf{SE}(2)$.◢ This transform maps points from the second scan so that they are as close as possible to the points in the first scan. Figure 6.23b shows the first set of points transformed and overlaid on the second set and we see good alignment. The translational part of this transform is an estimate of the robot's motion between scans – around 0.50 m in the x-direction. The nodes of the graph also hold time stamp information and these two scans were captured

```
>> pg.time(2581)-pg.time(2580)
ans =
    1.7600
```

seconds apart which indicates that the robot is moving quite slowly – a bit under 0.3 m s^{-1}.

At each iteration ICP assigns each point in the second set to the closest point in the first set and then computes a transform that minimizes the sum of distances between all corresponding points. Some points may not actually be corresponding but as long as enough are, the algorithm will converge. The `'verbose'` option causes data about each iteration to be displayed and `d` is the total distance between corresponding points which is decreasing but does not reach zero. This is due to many factors. The beams from the laser at the two different poses will not strike the walls at the same location so ICP's assumption about point correspondence is not actually valid.◢

In practice there are additional challenges. Some laser pulses will not return to the sensor if they fall on a surface with low reflectivity or on an oblique polished surface that specularly reflects the pulse away from the sensor – in these cases the sensor typically reports its maximum value. People moving through the environment change the shape of the world and temporarily cause a shorter range to be reported. In very large spaces all the walls may be beyond the maximum range of the sensor. Outdoors the beams can be reflected from rain drops, absorbed by fog or smoke and the return pulse can be overwhelmed by ambient sunlight. Finally the laser rangefinder, like all sensors, has measurement noise.

Fig. 6.23. Laser scan matching. **a** Laser scans from location 2 580 (*blue*) and 2 581 (*red*); **b** location 2 580 points (*blue*) and transformed points from location 2 581 (*red*)

The ICP algorithm is described more fully for the **SE**(3) case in Sect. 14.5.2.

We demonstrate the principle using ICP but in practice more robust algorithms are used. Here we provide an initial estimate of the translation between frames, based on odometry, so as to avoid getting stuck in a local minimum. ICP works poorly in plain corridors where the points lie along lines – this example was deliberately chosen because it has wall segments in orthogonal directions.

To remove invalid correspondences we pass the `'distthresh'` option to `icp()`. This causes any correspondences that involve a distance more than three times the median distance between all corresponding points to be dropped. In the `icp()` output the notation `132/180` means that 132 out of 180 possible correspondences met this test, 48 were rejected.

Laser-Based Map Building

If the robot pose is sufficiently well known, through some localization process, then we can transform all the laser scans to a global coordinate frame and build a map. Various map representations are possible but here we will outline how to build an occupancy grid as discussed in Chap. 5.

For a robot at a given pose, each beam in the scan is a ray and tells us several things. From the range measurement we can determine the coordinates of a cell that contains an obstacle but we can tell nothing about cells further along the ray. It is also implicit that all the cells between the sensor and the obstacle must be obstacle free. A maximum distance value, 50 m in this case, is the sensor's way of indicating that there was no returning laser pulse so we ignore all such measurements. We create the occupancy grid as a matrix and use the Bresenham algorithm to find all the cells along the ray based on the robot's pose and the laser range and bearing measurement, then a simple voting scheme to determine whether cells are free or occupied

```
>> pg.scanmap()
>> pg.plot_occgrid()
```

and the result is shown in Fig. 6.24. More sophisticated approaches treat the beam as a wedge of finite angular width and employ a probabilistic model of sensor return versus range. The principle can be extended to creating 3-dimensional point clouds from a scanning laser rangefinder on a moving vehicle as shown in Fig. 6.25.

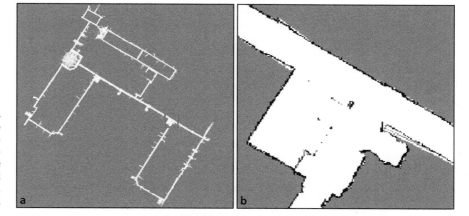

Fig. 6.24.
a Laser scans rendered into an occupancy grid, the area enclosed in the *green square* is displayed in **b**. *White cells* are free space, *black cells* are occupied and *grey cells* are unknown. Grid cell size is 10 cm

Fig. 6.25.
3D point cloud created by integrating multiple scans from a vehicle-mounted scanning laser rangefinder, where the scans are in a vertical plane normal to the vehicle's forward axis. This is sometimes called a "2.5D" representation since only the front surfaces of objects are described – note the range shadows on the walls behind cars. Note also that the density of laser points is not constant across the map, for example the point density on the road surface is much greater than it is high on the walls of buildings (image courtesy Alex Stewart; Stewart 2014)

Laser-Based Localization

We have mentioned landmarks a number of times in this chapter but avoided concrete examples of what they are. They could be distinctive visual features as discussed in Sect. 13.3 or artificial markers as discussed on page 164. If we consider a laser scan such as shown in Fig. 6.23a or 6.24b we see a fairly distinctive arrangement of points – a geometric signature – which we can use as a landmark. In many cases the signature will be ambiguous and of little value, for example a long corridor where all the points are collinear, but some signatures will be highly unique and can serve as a useful landmark. Naively we could match the current laser scan against all others and if the fit is good (the ICP error is low) we could add another constraint to the pose graph. However this strategy would be expensive with a large number of scans so typically only scans in the vicinity of the robot's estimated position are checked, and this once again raises the data association problem.

6.9   Wrapping Up

In this chapter we learned about two very different ways of estimating a robot's position: by dead reckoning, and by observing landmarks whose true position is known from a map. Dead reckoning is based on the integration of odometry information, the distance traveled and the change in heading angle. Over time errors accumulate leading to increased uncertainty about the pose of the robot.

We modeled the error in odometry by adding noise to the sensor outputs. The noise values are drawn from some distribution that describes the errors of that particular sensor. For our simulations we used zero-mean Gaussian noise with a specified covariance, but only because we had no other information about the specific sensor. The most realistic noise model available should be used. We then introduced the Kalman filter which provides an optimal estimate, in the least-squares sense, of the true configuration of the robot based on noisy measurements. The Kalman filter is however only optimal for the case of zero–mean Gaussian noise and a linear model. The model that describes how the robot's configuration evolves with time can be nonlinear in which case we approximate it with a linear model which included some partial derivatives expressed as Jacobian matrices – an approach known as extended Kalman filtering.

The Kalman filter also estimates uncertainty associated with the pose estimate and we see that the magnitude can never decrease and typically grows without bound. Only additional sources of information can reduce this growth and we looked at how observations of landmarks, with known locations, relative to the robot can be used. Once again we use the Kalman filter but in this case we use both the prediction and the update phases of the filter. We see that in this case the uncertainty can be decreased by a landmark observation, and that over the longer term the uncertainty does not grow. We then applied the Kalman filter to the problem of estimating the positions of the landmarks given that we knew the precise position of the vehicle. In this case, the state vector of the filter was the coordinates of the landmarks themselves.

Next we brought all this together and estimated the vehicle's position, the position of the landmarks and their uncertainties – simultaneous localization and mapping. The state vector in this case contained the configuration of the robot and the coordinates of the landmarks.

An important problem when using landmarks is data association, being able to determine which landmark has been known or observed by the sensor so that its position can be looked up in a map or in a table of known or estimated landmark positions. If the wrong landmark is looked up then an error will be introduced in the robot's position.

The Kalman filter scales poorly with an increasing number of landmarks and we introduced two alternative approaches: Rao-Blackwellized SLAM and pose-graph SLAM. The latter involves solving a large but sparse nonlinear least squares problem, turning the problem from one of (Kalman) filtering to one of optimization.

We finished our discussion of localization methods with Monte-Carlo estimation and introduced the particle filter. This technique is computationally intensive but makes no assumptions about the distribution of errors from the sensor or the linearity of the vehicle model, and supports multiple hypotheses. Particles filters can be considered as providing an approximate solution to the true system model, whereas a Kalman filter provides an exact solution to an approximate system model.

Finally we introduced laser rangefinders and showed how they can be applied to robot navigation, odometry and creating detailed floor plan maps.

Further Reading

Localization and SLAM. The tutorials by Bailey and Durrant-Whyte (2006) and Durrant-Whyte and Bailey (2006) are a good introduction to this topic, while the textbook *Probabilistic Robotics* (Thrun et al. 2005) is a readable and comprehensive coverage of all the material touched on in this chapter.

The book by Siegwart et al. (2011) also has a good treatment of robot localization. FastSLAM (Montemerlo et al. 2003; Montemerlo and Thrun 2007) is a state-of-the-art algorithm for Rao-Blackwellized SLAM.

Particle filters are described by Thrun et al. (2005), Stachniss and Burgard (2014) and the tutorial introduction by Rekleitis (2004). There are many variations such as fixed or adaptive number of particles and when and how to resample – and Li et al. (2015) provide a comprehensive review of resampling strategies. Determining the most likely pose was demonstrated by taking the weighted mean of the particles but many more approaches have been used. The kernel density approach takes the particle with the highest weight of neighboring particles within a fixed-size surrounding hypersphere.

Pose graph optimization, also known as GraphSLAM, has a long history starting with Lu and Milios (1997). There has been significant recent interest with many publications and open-source tools including g^{2}o (Kümmerle et al. 2011), $\sqrt{\text{SAM}}$ (Dellaert and Kaess 2006), iSAM (Kaess et al. 2007) and factor graphs. Agarwal et al. (2014) provides a gentle introduction to pose-graph SLAM and discusses the connection to land-based geodetic survey which is centuries old. Parallel Tracking and Mapping (PTAM) was described in Klein and Murray (2007), the code is available on github and there is also a blog.

There are many online resources related to SLAM. A collection of open-source SLAM implementations such as gmapping and iSam is available from OpenSLAM at http://www.openslam.org. An implementation of smoothing and mapping using factor graphs is available at https://bitbucket.org/gtborg/gtsam and has C++ and MATLAB bindings. MATLAB implementations include a 6DOF SLAM system at http://www.iri.upc.edu/people/jsola/JoanSola/eng/toolbox.html and the now dated CAS Robot Navigation Toolbox for planar SLAM at http://www.cas.kth.se/toolbox. Tim Bailey's website http://www-personal.acfr.usyd.edu.au/tbailey has MATLAB implementations of various SLAM and scan matching algorithms.

Many of the SLAM summer schools have websites that host excellent online resources such as lecture notes and practicals. Great teaching resources available online include Giorgio Grisetti's site http://www.dis.uniroma1.it/~grisetti and Paul Newman's *C4B Mobile Robots and Estimation Resources* ebook at https://www.free-ebooks.net/ebook/C4B-Mobile-Robotics.

Scan matching and map making. Many versions and variants of the ICP algorithm exist and it is discussed further in Chap. 14. Improved convergence and accuracy can be obtained using the normal distribution transform (NDT), originally proposed for 2D by Biber and Straßer (2003), extended to 3D by Magnusson et al. (2007) and implementations are available at pointclouds.org. A comparison of ICP and NDT for a field robotic application is described by Magnusson et al. (2009). A fast and popular approach to laser scan matching is that of Censi (2008).

When attempting to match a local geometric signature in a large point cloud (2D or 3D) to determine loop closure we often wish to limit our search to a local spatial region. An efficient way to achieve this is to organize the data using a kd-tree which is provided in MATLAB's Statistics and Machine Learning Toolbox™ and various contributions on File Exchange. FLANN (Muja and Lowe 2009) is a fast approximation which is available on github and has a MATLAB binding, and is also included in the VLFeat package.

For creating a map from robotic laser scan data in Sect. 6.8 we used a naive approach – a more sophisticated technique is the beam model or likelihood field as described in Thrun et al. (2005).

Kalman filtering. There are many published and online resources for Kalman filtering. Kálmán's original paper, Kálmán (1960), over 50 years old, is quite readable. The book by Zarchan and Musoff (2005) is a very clear and readable introduction to Kalman filtering. I have always found the classic book, recently republished, Jazwinski (2007) to be very readable. Bar-Shalom et al. (2001) provide comprehensive coverage of estimation theory and also the use of GPS. Groves (2013) also covers Kalman filtering. Welch and Bishop's online resources at http://www.cs.unc.edu/~welch/kalman have pointers to papers, courses, software and links to other relevant web sites.

A significant limitation of the EKF is its first-order linearization, particularly for processes with strong nonlinearity. Alternatives include the iterated EKF described by Jazwinski (2007) or the Unscented Kalman Filter (UKF) (Julier and Uhlmann 2004) which uses discrete sample points (sigma points) to approximate the PDF. Some of these topics are covered in the Handbook (Siciliano and Khatib 2016, §5 and §35). The information filter is an equivalent filter that maintains an inverse covariance matrix which has some useful properties, and is discussed in Thrun et al. (2005) as the sparse extended information filter.

Data association. SLAM techniques are critically dependent on accurate data association between observations and mapped landmarks, and a review of data association techniques is given by Neira and Tardós (2001). FastSLAM (Montemerlo and Thrun 2007) is capable of estimating data association as well as landmark position. The April tag which can be used as an artificial landmark is described in Olson (2011) and is supported by the Toolbox function `apriltags`. Mobile robots can uniquely identify places based on their visual appearance using tools such as OpenFABMAP (Glover et al. 2012).

Data association for Kalman filtering is covered in the Robotics Handbook (Siciliano and Khatib 2016). Data association in the tracking context is covered in considerable detail in, the now very old, book by Bar-Shalom and Fortmann (1988).

Sensors. The book by Kelly (2013) has a good coverage of sensors particularly laser range finders. For flying and underwater vehicles, odometry cannot be determined from wheel motion and an alternative, also suitable for wheeled vehicles, is visual odometry (VO). This is introduced in the tutorials by Fraundorfer and Scaramuzza (2012) and Scaramuzza and Fraundorfer (2011) and will be covered in Chap. 14. The Robotics Handbook (Siciliano and Khatib 2016) has good coverage of a wide range of robotic sensors. The principles of GPS and other radio-based localization systems are covered in some detail in the book by Groves (2013), and a number of links to GPS technical data are provided from this book's web site. The SLAM problem can be formulated in terms of bearing-only or range-only measurements. A camera is effectively a bearing-only sensor, giving the direction to a feature in the world. A VSLAM system is one that performs SLAM using bearing-only visual information, just a camera, and an introduction to the topic is given by Neira et al. (2008) and the associated special issue. Interestingly the robotic VSLAM problem is the same as the bundle adjustment problem known to the computer vision community and which will be discussed in Chap. 14.

The book by Borenstein et al. (1996) although dated has an excellent discussion of robotic sensors in general and odometry in particular. It is out of print but can be found

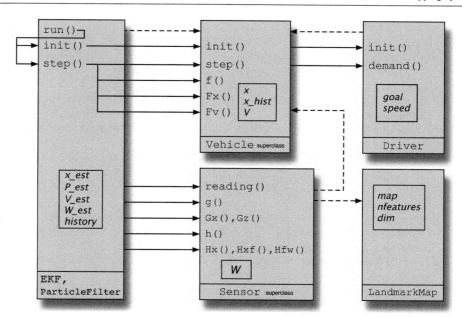

Fig. 6.26.
Toolbox class relationship for localization and mapping. Each class is shown as a *rectangle*, method calls are shown as *arrows* from caller to callee, properties are *boxed*, and *dashed lines* represent object references

online. The book by Everett (1995) covers odometry, range and bearing sensors, as well as radio, ultrasonic and optical localization systems. Unfortunately the discussion of range and bearing sensors is now quite dated since this technology has evolved rapidly over the last decade.

General interest. Bray (2014) gives a very readable account of the history of techniques to determine our location on the planet. If you ever wondered how to navigate by the stars or use a sextant Blewitt (2011) is a slim book that provides an easy to understand introduction.

The book *Longitude* (Sobel 1996) is a very readable account of the longitude problem and John Harrison's quest to build a marine chronometer.

Toolbox and MATLAB Notes

This chapter has introduced a number of Toolbox classes to solve mapping and localization problems. The principle was to decompose the problem into clear functional subsystems and implement these as a set of cooperating classes, and this allows quite complex problems to be expressed in very few lines of code.

The relationships between the objects and their methods and properties are shown in Fig. 6.26. As always more documentation is available through the online help system or comments in the code. `Vehicle` is a superclass and concrete subclasses include `Unicycle` and `Bicycle`.

The MATLAB Computer Vision System Toolbox™ includes a fast version of ICP called `pcregrigid`. The Robotics System Toolbox™ contains a generic particle filter class `ParticleFilter` and a particle filter based localizer class `MonteCarloLocalization`.

Exercises

1. What is the value of the Longitude Prize in today's currency?
2. Implement a driver object (page 157) that drives the robot around inside a circle with specified center and radius.
3. Derive an equation for heading change in terms of the rotational rate of the left and right wheels for the car-like and differential-steer vehicle models.

4. Dead-reckoning (page 156)
   a) Experiment with different values of P_0, V and $\hat{V}$.
   b) Figure 6.4 compares the actual and estimated position. Plot the actual and esti-
      mated heading angle.
   c) Compare the variance associated with heading to the variance associated with
      position. How do these change with increasing levels of range and bearing angle
      variance in the sensor?
   d) Derive the Jacobians in Eq. 6.5 and 6.6 for the case of a differential steer robot.
5. Using a map (page 163)
   a) Vary the characteristics of the sensor (covariance, sample rate, range limits and
      bearing angle limits) and investigate the effect on performance
   b) Vary W and $\hat{W}$ and investigate what happens to estimation error and final co-
      variance.
   c) Modify the `RangeBearingSensor` to create a bearing-only sensor, that is, as
      a sensor that returns angle but not range. The implementation includes all the
      Jacobians. Investigate performance.
   d) Modify the sensor model to return occasional errors (specify the error rate) such
      as incorrect range or beacon identity. What happens?
   e) Modify the EKF to perform data association instead of using the landmark iden-
      tity returned by the sensor.
   f) Figure 6.7 compares the actual and estimated position. Plot the actual and esti-
      mated heading angle.
   g) Compare the variance associated with heading to the variance associated with
      position. How do these change with increasing levels of range and bearing angle
      variance in the sensor?
6. Making a map (page 166)
   a) Vary the characteristics of the sensor (covariance, sample rate, range limits and
      bearing angle limits) and investigate the effect on performance.
   b) Use the bearing-only sensor from above and investigate performance relative to
      using a range and bearing sensor.
   c) Modify the EKF to perform data association instead of using identity returned
      by the sensor.
7. Simultaneous localization and mapping (page 168)
   a) Vary the characteristics of the sensor (covariance, sample rate, range limits and
      bearing angle limits) and investigate the effect on performance.
   b) Use the bearing-only sensor from above and investigate performance relative to
      using a range and bearing sensor.
   c) Modify the EKF to perform data association instead of using the landmark iden-
      tity returned by the sensor.
   d) Figure 6.11 compares the actual and estimated position. Plot the actual and es-
      timated heading angle.
   e) Compare the variance associated with heading to the variance associated with
      position. How do these change with increasing levels of range and bearing angle
      variance in the sensor?
8. Modify the pose-graph optimizer and test using the simple graph `pg1.g2o`
   a) anchor one node at a particular pose.
   b) add one or more landmarks. You will need to derive the relevant Jacobians
      first and add the landmark positions, constraints and information matrix to
      the data file.
9. Create a simulator for Buffon's needle problem, and estimate π for 10, 100, 1 000
   and 10 000 needle throws. How does convergence change with needle length?
10. Particle filter (page 176)
    a) Run the filter numerous times. Does it always converge?
    b) Vary the parameters Q, L, w_0 and N and understand their effect on convergence
       speed and final standard deviation.

c) Investigate variations to the kidnapped robot problem. Place the initial particles around the initial pose. Place the particles uniformly over the *xy*-plane but set their orientation to its actual value.

d) Use a different type of likelihood function, perhaps inverse distance, and compare performance.

11. Experiment with April tags. Print some tags and extract them from images using the `apriltags` function. Check out Sect. 12.1 on how to acquire images using MATLAB.

12. Implement a laser odometer and test it over the entire path saved in `killian.g2o`. Compare your odometer with the relative pose changes in the file.

13. In order to measure distance using laser rangefinding what timing accuracy is required to achieve 1 cm depth resolution?

14. Reformulate the localization, mapping and SLAM problems using a bearing-only landmark sensor.

15. Implement a localization or SLAM system using an external simulator such as V-REP or Gazebo. Obtain range measurements from the simulated robot, do laser odometry and landmark recognition, and send motion commands to the robot. You can communicate with these simulators from MATLAB using the ROS protocol if you have the Robotics System Toolbox. Alternatively you can communicate with V-REP using the Toolbox `VREP` class, see the documentation.

Part III | Arm-Type Robots

III Arm-Type Robots

Arm-type robots or robot manipulators are a very common and familiar type of robot. We are used to seeing pictures or video of them at work in factories doing jobs such as assembly, welding and handling tasks, or even in operating rooms doing surgery. The first robot manipulators started work nearly 60 years ago and have been enormously successful in practice – many millions of robot manipulators are working in the world today. Many products we buy have been assembled, packed or handled by a robot.

Unlike the mobile robots we discussed in the previous part, robot manipulators do not move through the world. They have a static base and therefore operate within a limited workspace. Many different types of robot manipulator have been created and Fig. III.1 shows some of the diversity. The most common is the 6DOF arm-type of robot comprising a series of rigid-links and actuated joints. The SCARA (Selective Compliance Assembly Robot Arm) is rigid in the vertical direction and compliant in the horizontal plane which is an advantage for planar tasks such as electronic circuit board assembly. A gantry robot has one or two degrees of freedom of motion along overhead rails which gives it

Fig. III.1.
a A 6DOF serial-link manipulator. General purpose industrial manipulator (source: ABB).
b SCARA robot which has 4DOF, typically used for electronic assembly (photo of Adept Cobra s600 SCARA robot courtesy of Adept Technology, Inc.).
c A gantry robot; the arm moves along an overhead rail (image courtesy of Güdel AG Switzerland | Mario Rothenbühler | www.gudel.com). **d** A parallel-link manipulator, the end-effector is driven by 6 parallel links (source: ABB)

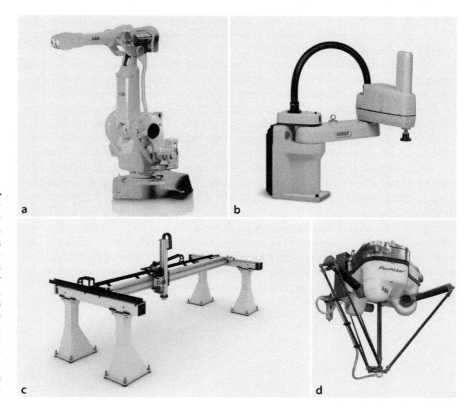

Fig. III.2.
Robot end-effectors. **a** A vacuum gripper holds a sheet of glass. **b** A human-like robotic hand (© Shadow Robot Company 2008)

a very large working volume. A parallel-link manipulator has its links connected in parallel to the tool which brings a number of advantages such as having all the motors on the base and providing a very stiff structure. The focus of this part is serial-link arm-type robot manipulators.

These nonmobile robots allow some significant simplifications to problems such as perception and safety. The work environment for a factory robot can be made very orderly so the robot can be fast and precise and *assume* the location of objects that it is working with. The safety problem is simplified since the robot has a limited working volume – it is straightforward to just exclude people from the robot's work space using safety barriers or even cages.

A robot manipulates objects using its end-effector or tool as shown in Fig. III.2. End-effectors range in complexity from simple 2-finger or parallel-jaw grippers to complex human-like hands with multiple actuated finger joints and an opposable thumb.

The chapters in this part cover the fundamentals of serial-link manipulators. Chapter 7 is concerned with the kinematics of serial-link manipulators. This is the geometric relationship between the angles of the robot's joints and the pose of its end-effector. We discuss the creation of smooth paths that the robot can follow and present an example of a robot drawing a letter on a plane and a 4-legged walking robot. Chapter 8 introduces the relationship between the rate of change of joint coordinates and the end-effector velocity which is described by the manipulator Jacobian matrix. It also covers alternative methods of generating paths in Cartesian space and introduces the relationship between forces on the end-effector and torques at the joints. Chapter 9 discusses independent joint control and some performance limiting factors such as gravity load and varying inertia. This leads to a discussion of the full nonlinear dynamics of serial-link manipulators – effects such as inertia, gyroscopic forces, friction and gravity – and more sophisticated model-based control approaches.

7

Robot Arm Kinematics

Take to kinematics. It will repay you.
It is more fecund than geometry; it adds a fourth dimension to space.
Chebyshev to Sylvester 1873

Kinematics⬆ is the branch of mechanics that studies the motion of a body, or a system of bodies, without considering its mass or the forces acting on it.

A robot arm, more formally a serial-link manipulator, comprises a chain of rigid links and joints. Each joint has one degree of freedom, either translational (a sliding or prismatic joint) or rotational (a revolute joint). Motion of the joint changes the relative pose of the links that it connects. One end of the chain, the base, is generally fixed and the other end is free to move in space and holds the tool or end-effector that does the useful work.

Figure 7.1 shows two modern arm-type robots that have six and seven joints respectively. Clearly the pose of the end-effector will be a complex function of the state of each joint and Sect. 7.1 describes how to compute the pose of the end-effector. Section 7.2 discusses the inverse problem, how to compute the position of each joint given the end-effector pose. Section 7.3 describes methods for generating smooth paths for the end-effector. The remainder of the chapter covers advanced topics and two complex applications: writing on a plane surface and a four-legged walking robot whose legs are simple robotic arms.

From the Greek word for motion.

7.1 Forward Kinematics

Forward kinematics is the mapping from joint coordinates, or robot configuration, to end-effector pose. We start in Sect. 7.1.1 with conceptually simple robot arms that move in 2-dimensions in order to illustrate the principles, and in Sect. 7.1.2 extend this to more useful robot arms that move in 3-dimensions.

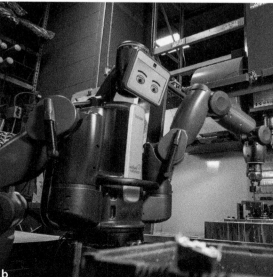

Fig. 7.1.
a Mico 6-joint robot with 3-fingered hand (courtesy of Kinova Robotics). **b** Baxter 2-armed robotic coworker, each arm has 7 joints (courtesy of Rethink Robotics)

a

b

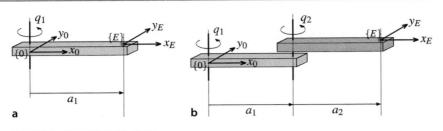

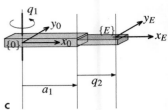

Fig. 7.2. Some simple planar robotic arms. **a** Planar arm with one rotational joint; **b** planar arm with two rotational joints; **c** planar arm with two joints: one rotational and one prismatic. The base {0} and end-effector {E} coordinate frames are shown. The joint variables, angle or prismatic extension, are generalized coordinates and denoted by q_j

We use the symbols $\mathscr{R}, \mathscr{T}_x, \mathscr{T}_y$ to denote relative poses in SE(2) that are respectively pure rotation and pure translation in the x- and y-directions.

7.1.1 2-Dimensional (Planar) Robotic Arms

Consider the simple robot arm shown in Fig. 7.2a which has a single rotational joint. We can describe the pose of its end-effector – frame {E} – by a sequence of relative poses: a rotation about the joint axis and then a translation by a_1 along the rotated x-axis◄

$$\xi_E(q) = \mathscr{R}_z(q_1) \oplus \mathscr{T}_x(a_1)$$

The Toolbox allows us to express this, for the case $a_1 = 1$, by

```
>> import ETS2.*
>> a1 = 1;
>> E = Rz('q1') * Tx(a1)
```

which is a sequence of ETS2 class objects. The argument to Rz is a string which indicates that its parameter is a joint variable whereas the argument to Tx is a constant numeric robot dimension.

The forward kinematics for a *particular* value of $q_1 = 30$ deg

```
>> E.fkine( 30, 'deg')
ans =
    0.8660   -0.5000    0.866
    0.5000    0.8660    0.5
         0         0      1
```

is an **SE**(2) homogeneous transformation matrix representing the pose of the end-effector – coordinate frame {E}.

An easy and intuitive way to understand how this simple robot behaves is interactively

```
>> E.teach
```

which generates a graphical representation of the robot arm as shown in Fig. 7.3. The rotational joint is indicated by a grey vertical cylinder and the link by a red horizontal pipe. You can adjust the joint angle q_1 using the slider and the arm pose and the displayed end-effector position and orientation will be updated. Clearly this is not a very useful robot arm since its end-effector can only reach points that lie on a circle.

Consider now a robot arm with two joints as shown in Fig. 7.2b. The pose of the end-effector is

$$\xi_E(q) = \underbrace{\mathscr{R}_z(q_1)}_{\text{joint 1}} \oplus \underbrace{\mathscr{T}_x(a_1)}_{\text{link 1}} \oplus \underbrace{\mathscr{R}_z(q_2)}_{\text{joint 2}} \oplus \underbrace{\mathscr{T}_x(a_2)}_{\text{link 2}} \tag{7.1}$$

We can represent this using the Toolbox as

```
>> a1 = 1; a2 = 1;
>> E = Rz('q1') * Tx(a1) * Rz('q2') * Tx(a2)
```

When computing the forward kinematics the joint angles are now specified by a vector

```
>> E.fkine( [30, 40], 'deg')
ans =
    0.3420   -0.9397    1.208
    0.9397    0.3420    1.44
         0         0      1
```

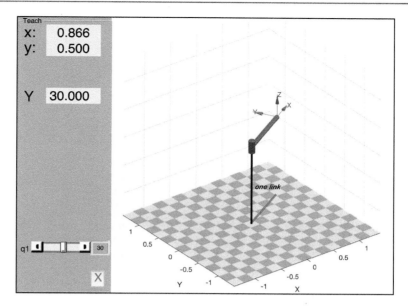

Fig. 7.3.
Toolbox depiction of 1-joint planar robot using the `teach` method. The *blue panel* contains the joint angle slider and displays the position and orientation (yaw angle) of the end-effector (in degrees)

and the result is the end-effector pose when $q_1 = 30$ and $q_2 = 40$ deg. We could display the robot interactively as in the previous example, or noninteractively by

```
>> E.plot( [30, 40], 'deg')
```

The joint structure of a robot is often referred to by a shorthand comprising the letters R (for revolute) or P (for prismatic) to indicate the number and types of its joints. For this robot

```
>> E.structure
ans =
RR
```

indicates a revolute-revolute sequence of joints. The notation underneath the terms in Eq. 7.1 describes them in the context of a physical robot manipulator which comprises a series of joints and links.

You may have noticed a few characteristics of this simple planar robot arm. Firstly, most end-effector positions can be reached with two *different* joint angle vectors. Secondly, the robot can position the end-effector at any point within its reach but we cannot specify an arbitrary orientation. This robot has 2 degrees of freedom and its configuration space is $\mathcal{C} = \mathbb{S}^1 \times \mathbb{S}^1$. This is sufficient to achieve positions in the task space $\mathcal{T} \subset \mathbb{R}^2$ since dim $\mathcal{C} = $ dim $\mathcal{T}$. However if our task space includes orientation $\mathcal{T} \subset \mathbf{SE}(2)$ then it is under-actuated since dim $\mathcal{C} < $ dim $\mathcal{T}$ and the robot can access only a subset of the task space.

So far we have only considered revolute joints but we could use a prismatic joint instead as shown in Fig. 7.2c. The end-effector pose is

$$\xi_E(\boldsymbol{q}) = \underbrace{\mathcal{R}_z(q_1)}_{\text{joint 1}} \oplus \underbrace{\mathcal{T}_x(a_1)}_{\text{link 1}} \oplus \underbrace{\mathcal{T}_x(q_2)}_{\text{joint 2}}$$

Prismatic joints. Robot joints are commonly revolute (rotational) but can also be prismatic (linear, sliding, telescopic, etc.). The SCARA robot of Fig. III.1b has a prismatic third joint while the gantry robot of Fig. III.1c has three prismatic joints for motion in the x-, y- and z-directions.

The Stanford arm shown here has a prismatic third joint. It was developed at the Stanford AI Lab in 1972 by robotics pioneer Victor Scheinman who went on to design the PUMA robot arms. This type of arm supported a lot of important early research work in robotics and one can be seen in the Smithsonian Museum of American History, Washington DC. (Photo courtesy Oussama Khatib)

and the Toolbox representation follows a familiar pattern

```
>> a1 = 1;
>> E = Rz('q1') * Tx(a1) * Tz('q2')
```

and the arm structure is now

```
>> E.structure
ans =
RP
```

which is commonly called a polar-coordinate robot arm.

We can easily add a third joint

$$\xi_e(q) = \underbrace{\mathcal{R}_z(q_1)}_{\text{joint 1}} \oplus \underbrace{\mathcal{T}_x(a_1)}_{\text{link 1}} \oplus \underbrace{\mathcal{R}_z(q_2)}_{\text{joint 2}} \oplus \underbrace{\mathcal{T}_x(a_2)}_{\text{link 2}} \oplus \underbrace{\mathcal{R}_z(q_3)}_{\text{joint 3}} \oplus \underbrace{\mathcal{T}_x(a_3)}_{\text{link 3}}$$

and use the now familiar Toolbox functionality to represent and work with this arm. This robot has 3 degrees of freedom and is able to access all points in the task space $\mathcal{T} \subset \mathbf{SE}(2)$, that is, achieve any pose in the plane (limited by reach).

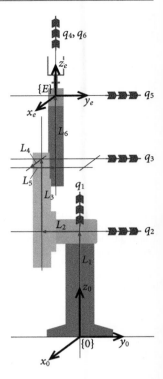

Fig. 7.4. Puma robot in the zero-joint-angle configuration showing dimensions and joint axes (indicated by *blue triple arrows*) (after Corke 2007)

We use the symbols $\mathcal{R}_i, \mathcal{T}_i, i \in \{x, y, z\}$ to denote relative poses in SE(3) that are respectively pure rotation about, or pure translation along, the *i*-axis.

7.1.2   3-Dimensional Robotic Arms

Truly useful robots have a task space $\mathcal{T} \subset \mathbf{SE}(3)$ enabling arbitrary position and orientation of the end-effector. This requires a robot with a configuration space dim $\mathcal{C} \geq$ dim $\mathcal{T}$ which can be achieved by a robot with six or more joints. In this section we will use the Puma 560 as an exemplar of the class of all-revolute six-axis robot manipulators with $\mathcal{C} \subset (\mathbb{S}^1)^6$.

We can extend the technique from the previous section for a robot like the Puma 560 whose dimensions are shown in Fig. 7.4. Starting with the world frame {0} we move up, rotate about the waist axis (q_1), rotate about the shoulder axis (q_2), move to the left, move up and so on. As we go, we write down the transform expression◄

$$\xi_E = \mathcal{T}_z(L_1) \oplus \mathcal{R}_z(q_1) \oplus \mathcal{R}_y(q_2) \oplus \mathcal{T}_y(L_2) \oplus \mathcal{T}_z(L_3) \oplus \mathcal{R}_y(q_3)$$
$$\oplus \mathcal{T}_x(L_4) \oplus \mathcal{T}_y(L_5) \oplus \mathcal{T}_z(L_6) \oplus \underbrace{\mathcal{R}_z(q_4) \oplus \mathcal{R}_y(q_5) \oplus \mathcal{R}_z(q_6)}_{\text{wrist}}$$

The marked term represents the kinematics of the robot's wrist and should be familiar to us as a *ZYZ* Euler angle sequence from Sect. 2.2.1.2 – it provides an arbitrary orientation but is subject to a singularity when the middle angle $q_5 = 0$.

We can represent this using the 3-dimensional version of the Toolbox class we used previously

```
>> import ETS3.*
>> L1 = 0; L2 = -0.2337; L3 = 0.4318; L4 = 0.0203; L5 = 0.0837; L6 = 0.4318;
>> E3 = Tz(L1) * Rz('q1') * Ry('q2') * Ty(L2) * Tz(L3) * Ry('q3') ↵
   * Tx(L4) * Ty(L5) * Tz(L6) * Rz('q4') * Ry('q5') * Rz('q6');
```

We can use the interactive teach facility or compute the forward kinematics

```
>> E3.fkine([0 0 0 0 0 0])
ans =
         1        0        0    0.0203
         0        1        0     -0.15
         0        0        1    0.8636
         0        0        0        1
```

While this notation is intuitive it does becomes cumbersome as the number of robot joints increases. A number of approaches have been developed to more concisely describe a serial-link robotic arm: Denavit-Hartenberg notation and product of exponentials.

Joint angle	θ_j	the angle between the x_{j-1} and x_j axes about the z_{j-1} axis	revolute joint variable
Link offset	d_j	the distance from the origin of frame $j-1$ to the x_j axis along the z_{j-1} axis	prismatic joint variable
Link length	a_j	the distance between the z_{j-1} and z_j axes along the x_j axis; for intersecting axes is parallel to $\hat{z}_{j-1} \times \hat{z}_j$	constant
Link twist	α_j	the angle from the z_{j-1} axis to the z_j axis about the x_j axis	constant
Joint type	σ_j	$\sigma = R$ for a revolute joint, $\sigma = P$ for a prismatic joint	constant

Table 7.1.
Denavit-Hartenberg parameters: their physical meaning, symbol and formal definition

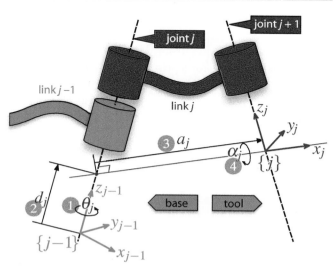

Fig. 7.5.
Definition of standard Denavit and Hartenberg link parameters. The colors red and blue denote all things associated with links $j-1$ and j respectively. The numbers in circles represent the order in which the elementary transforms are applied. x_j is parallel to $z_{j-1} \times z_j$ and if those two axes are parallel then d_j can be arbitrarily chosen

7.1.2.1 Denavit-Hartenberg Parameters

One systematic way of describing the geometry of a serial chain of links and joints is Denavit-Hartenberg notation.

> For a manipulator with N joints numbered from 1 to N, there are $N+1$ links, numbered from 0 to N. Joint j connects link $j-1$ to link j and moves them relative to each other. It follows that link ℓ connects joint ℓ to joint $\ell+1$. Link 0 is the base of the robot, typically fixed and link N, the last link of the robot, carries the end-effector or tool.

In Denavit-Hartenberg, notation a link defines the spatial relationship between two neighboring joint axes as shown in Fig. 7.5. A link is specified by four parameters. The relationship between two link coordinate frames would ordinarily entail six parameters, three each for translation and rotation. For Denavit-Hartenberg notation there are only four parameters but there are also two constraints: axis x_j intersects z_{j-1} and axis x_j is perpendicular to z_{j-1}. One consequence of these constraints is that sometimes the link coordinate frames are not actually located on the physical links of the robot. Another consequence is that the robot must be placed into a particular configuration – the zero-angle configuration – which is discussed further in Sect. 7.4.1. The Denavit-Hartenberg parameters are summarized in Table 7.1.

The coordinate frame $\{j\}$ is attached to the far (distal) end of link j. The z-axis of frame $\{j\}$ is aligned with the axis of joint $j+1$.

The transformation from link coordinate frame $\{j-1\}$ to frame $\{j\}$ is defined in terms of elementary rotations and translations as

$$^{j-1}\xi_j\left(\theta_j, d_j, a_j, \alpha_j\right) = \mathcal{R}_z\left(\theta_j\right) \oplus \mathcal{T}_z\left(d_j\right) \oplus \mathcal{T}_x\left(a_j\right) \oplus \mathcal{R}_x\left(\alpha_j\right) \tag{7.2}$$

which can be expanded in homogeneous matrix form as

$$^{j-1}A_j = \begin{pmatrix} \cos\theta_j & -\sin\theta_j\cos\alpha_j & \sin\theta_j\sin\alpha_j & a_j\cos\theta_j \\ \sin\theta_j & \cos\theta_j\cos\alpha_j & -\cos\theta_j\sin\alpha_j & a_j\sin\theta_j \\ 0 & \sin\alpha_j & \cos\alpha_j & d_j \\ 0 & 0 & 0 & 1 \end{pmatrix} \tag{7.3}$$

The parameters α_j and a_j are always constant. For a revolute joint, θ_j is the joint variable and d_j is constant, while for a prismatic joint, d_j is variable, θ_j is constant and $\alpha_j = 0$. In many of the formulations that follow, we use generalized joint coordinates q_j

$$\text{if } \sigma_j = \begin{cases} R: & \theta_j \leftarrow q_j \\ P: & d_j \leftarrow q_j \end{cases}$$

For an N-axis robot, the generalized joint coordinates $q \in \mathcal{C}$ where $\mathcal{C} \subset \mathbb{R}^N$ is called the joint space or configuration space.► For the common case of an all-revolute robot $\mathcal{C} \subset (\mathbb{S}^1)^N$ the joint coordinates are referred to as joint angles. The joint coordinates are also referred to as the *pose of the manipulator* which is different to the *pose of the end-effector* which is a Cartesian pose $\xi \in \mathbf{SE}(3)$. The term *configuration* is shorthand for *kinematic configuration* which will be discussed in Sect. 7.2.2.1.

This is the same concept as was introduced for mobile robots in Sect. 2.3.5.

Within the Toolbox a robot revolute joint and link can be created by

```
>> L = Revolute('a', 1)
L =
Revolute(std): theta=q, d=0, a=1, alpha=0, offset=0
```

which is a revolute-joint object of type `Revolute` which is a subclass of the generic `Link` object. The displayed value of the object shows the kinematic parameters (most of which have defaulted to zero), the joint type and that standard Denavit-Hartenberg convention is used (the tag `std`).►

A slightly different notation, *modifed Denavit-Hartenberg* notation is discussed in Sect. 7.4.3.

A `Link` object has many parameters and methods which are described in the online documentation, but the most common ones are illustrated by the following examples. The link transform Eq. 7.3 for $q = 0.5$ rad is an `SE3` object

```
>> L.A(0.5)
ans =
   0.8776   -0.4794        0    0.8776
   0.4794    0.8776       -0    0.4794
        0         0        1         0
        0         0        0         1
```

representing the homogeneous transformation due to this robot link with the particular value of θ. Various link parameters can be read or altered, for example

```
>> L.type
ans =
   R
```

indicates that the link is revolute and

```
>> L.a
ans =
   1.0000
```

returns the kinematic parameter a. Finally a link can contain an offset

```
>> L.offset = 0.5;
>> L.A(0)
ans =
   0.8776   -0.4794      · 0    0.8776
   0.4794    0.8776       -0    0.4794
        0         0        1         0
        0         0        0         1
```

which is added to the joint variable before computing the link transform and will be discussed in more detail in Sect. 7.4.1.

The forward kinematics is a function of the joint coordinates and is simply the composition of the relative pose due to each link

$$^{0}\xi_N = \mathcal{K}(q; \theta, d, a, \alpha, \sigma) = {}^{0}\xi_1 \oplus {}^{1}\xi_2 \cdots {}^{N-1}\xi_N \tag{7.4}$$

In this notation link 0 is the base of the robot and commonly for the first link $d_1 = 0$ but we could set $d_1 > 0$ to represent the height of the first joint above the world coordinate frame. The final link, link N, carries the tool – the parameters d_N, a_N and α_N provide a limited means to describe the tool-tip pose with respect to the $\{N\}$ frame. By convention the robot's tool points in the z-direction as shown in Fig. 2.16.

More generally we add two extra transforms to the chain◀

We have used W to denote the world frame in this case since 0 designates link 0, the base link.

$$^{W}\xi_E = \underbrace{{}^{W}\xi_0}_{\xi_B} \oplus {}^{0}\xi_1 \oplus {}^{1}\xi_2 \cdots {}^{N-1}\xi_N \oplus \underbrace{{}^{N}\xi_E}_{\xi_T}$$

The base transform ξ_B puts the base of the robot arm at an arbitrary pose within the world coordinate frame. In a manufacturing system the base is usually fixed to the environment but it could be mounted on a mobile ground, aerial or underwater robot, a truck, or even a space shuttle.

The frame $\{N\}$ is often defined as the center of the spherical wrist mechanism, and the tool transform ξ_T describes the pose of the tool tip with respect to that. In practice ξ_T might consist of several components. Firstly, a transform to a tool-mounting flange on the physical end of the robot. Secondly, a transform from the flange to the end of the tool that is bolted to it, where the tool might be a gripper, screwdriver or welding torch.

In the Toolbox we connect `Link` class objects in series using the `SerialLink` class

```
>> robot = SerialLink( [ Revolute('a', 1) Revolute('a', 1) ],↵
  'name', 'my robot')
robot =
my robot:: 2 axis, RR, stdDH
+---+-----------+-----------+-----------+-----------+-----------+
| j |     theta |         d |         a |     alpha |    offset |
+---+-----------+-----------+-----------+-----------+-----------+
|  1|      q1|0 |        |1 |        |0 |        |0 |           |
|  2|      q2|0 |        |1 |        |0 |        |0 |           |
+---+-----------+-----------+-----------+-----------+-----------+
```

We have just recreated the 2-robot robot we looked at earlier, but now it is embedded in **SE**(3). The forward kinematics are

```
>> robot.fkine([30 40], 'deg')
ans =
    0.3420   -0.9397        0    1.208
    0.9397    0.3420        0    1.44
         0         0        1        0
         0         0        0        1
```

The Toolbox contains a large number of robot arm models defined in this way and they can be listed by

```
>> models
ABB, IRB140, 6DOF, standard_DH (mdl_irb140)
Aldebaran, NAO, humanoid, 4DOF, standard_DH (mdl_nao)
Baxter, Rethink Robotics, 7DOF, standard_DH (mdl_baxter)
  . . .
```

where the name of the Toolbox script to load the model is given in parentheses at the end of each line, for example

```
>> mdl_irb140
```

The `models` function also supports searching by keywords and robot arm type. You can adjust the parameters of any model using the editing method, for example

```
>> robot.edit
```

Determining the Denavit-Hartenberg parameters for a particular robot is described in more detail in Sect. 7.4.2.

7.1.2.2 · Product of Exponentials

In Chap. 2 we introduced twists. A twist is defined by a screw axis direction and pitch, and a point that the screw axis passes through. In matrix form the twist $S \in \mathbb{R}^6$

$$T' = e^{[S]\theta} T$$

rotates the coordinate frame described by the pose T about the screw axis by an angle θ.► This is exactly the case of the single-joint robot of Fig. 7.2a, where the screw axis *is* the joint axis and T is the pose of the end-effector when $q_1 = 0$. We can therefore write the forward kinematics as

For a prismatic twist, the motion is a displacement of θ along the screw axis. Here we are working in the plane so $T \in$ SE(2) and $S \in \mathbb{R}^3$.

$$T_E = e^{[S_1]q_1} T_E(0)$$

where $T_E(0)$ is the end-effector pose in the zero-angle joint configuration: $q_1 = 0$.
For the 2-joint robot of Fig. 7.2b we would write

$$T_E = e^{[S_1]q_1} \underbrace{e^{[S_2]q_2} T_E(0)}$$

where S_1 and S_2 are the screws defined by the joint axes and $T_E(0)$ is the end-effector pose in the zero-angle joint configuration: $q_1 = q_2 = 0$. The indicated term is similar to the single-joint robot above, and the first twist rotates that joint and link about S_1. In MATLAB we define the link lengths and compute $T_E(0)$

```
>> a1 = 1; a2 = 1;
>> TE0 = SE2(a1+a2, 0, 0);
```

define the two twists, in **SE**(2), for this example

```
>> S1 = Twist( 'R', [0 0] );
>> S2 = Twist( 'R', [a1 0] );
```

and apply them to $T_E(0)$

```
>> TE = S1.T(30, 'deg') * S2.T(40, 'deg') * TE0
TE =
     0.3420    -0.9397      1.208
     0.9397     0.3420       1.44
          0          0          1
```

For a general robot that moves in 3-dimensions we can write the forward kinematics in product of exponential (PoE) form as

$$\xi_E \sim {}^0T_E = e^{[S_1]q_1} \cdots e^{[S_N]q_N}\, {}^0T_E(0)$$

where ${}^0T_E(0)$ is the end-effector pose when the joint coordinates are all zero and S_j is the twist for joint j expressed in the world frame.◄ This can also be written as

$$\xi_E \sim {}^0T_E = {}^0T_E(0)e^{[{}^ES_1]q_1} \cdots e^{[{}^ES_N]q_N}$$

and ${}^ES_j$ is the twist for joint j expressed in the end-effector frame which is related to the twists above by ${}^ES_j = \mathrm{Ad}({}^E\xi_0)S_j$.

A serial-link manipulator can be succinctly described by a table listing the 6 screw parameters for each joint as well as the zero-joint-coordinate end-effector pose.

The tool and base transform are effectively included in ${}^0T_E(0)$, but an explicit base transform could be added if the screw axes are defined with respect to the robot's base rather than the world coordinate frame, or use the adjoint matrix to transform the screw axes from base to world coordinates.

7.1.2.3   6-Axis Industrial Robot

Truly useful robots have a task space $\mathcal{T} \subset$ **SE**(3) enabling arbitrary position and attitude of the end-effector – the task space has six spatial degrees of freedom: three translational and three rotational. This requires a robot with a configuration space $\mathcal{C} \subset \mathbb{R}^6$ which can be achieved by a robot with six joints. In this section we will use the Puma 560 as an example of the class of all-revolute six-axis robot manipulators. We define an instance of a Puma 560 robot using the script

```
>> mdl_puma560
```

which creates a `SerialLink` object, `p560`, in the workspace. Displaying the variable shows the table of its Denavit-Hartenberg parameters

```
>> p560
Puma 560 [Unimation]:: 6 axis, RRRRRR, stdDH, slowRNE
 - viscous friction; params of 8/95;
+---+-----------+-----------+-----------+-----------+-----------+
| j |   theta   |     d     |     a     |   alpha   |  offset   |
+---+-----------+-----------+-----------+-----------+-----------+
| 1 |       q1  |        0  |        0  |    1.571  |        0  |
| 2 |       q2  |        0  |   0.4318  |        0  |        0  |
| 3 |       q3  |     0.15  |   0.0203  |   -1.571  |        0  |
| 4 |       q4  |   0.4318  |        0  |    1.571  |        0  |
| 5 |       q5  |        0  |        0  |   -1.571  |        0  |
| 6 |       q6  |        0  |        0  |        0  |        0  |
+---+-----------+-----------+-----------+-----------+-----------+
```

The **Puma 560 robot** (Programmable Universal Manipulator for Assembly) released in 1978 was the first modern industrial robot and became enormously popular. It featured an anthropomorphic design, electric motors and a spherical wrist – the archetype of all that followed. It can be seen in the Smithsonian Museum of American History, Washington DC.

The Puma 560 catalyzed robotics research in the 1980s and it was a very common laboratory robot. Today it is obsolete and rare but in homage to its important role in robotics research we use it here. For our purposes the advantages of this robot are that it has been well studied and its parameters are very well known – it has been described as the "white rat" of robotics research.

Most modern 6-axis industrial robots are very similar in structure and can be accomodated simply by changing the Denavit-Hartenberg parameters. The Toolbox has kinematic models for a number of common industrial robots from manufacturers such as Rethink, Kinova, Motoman, Fanuc and ABB. (Puma photo courtesy Oussama Khatib)

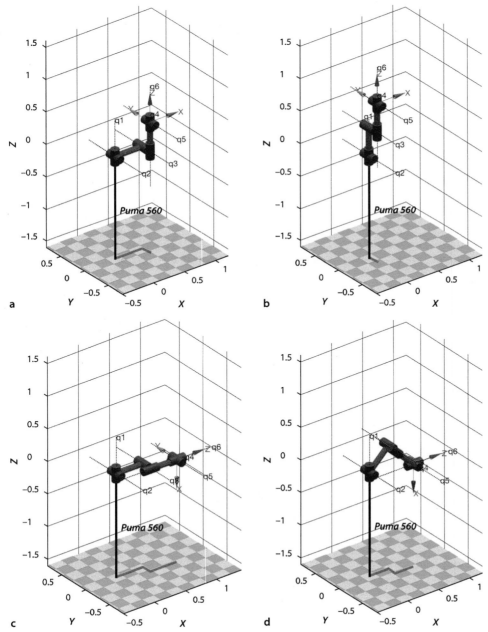

Fig. 7.6.
The Puma robot in 4 different poses. **a** *Zero angle*; **b** *ready pose*; **c** *stretch*; **d** *nominal*

> **Anthropomorphic** means having human-like characteristics. The Puma 560 robot was designed to have approximately the dimensions and reach of a human worker. It also had a spherical joint at the wrist just as humans have.
>
> Roboticists also tend to use anthropomorphic terms when describing robots. We use words like waist, shoulder, elbow and wrist when describing serial link manipulators. For the Puma these terms correspond respectively to joint 1, 2, 3 and 4–6.

Note that a_j and d_j are in SI units which means that the translational part of the forward kinematics will also have SI units.

The script `mdl_puma560` also creates a number of joint coordinate vectors in the workspace which represent the robot in some canonic configurations:

qz	$(0, 0, 0, 0, 0, 0)$	*zero angle*
qr	$(0, \frac{\pi}{2}, -\frac{\pi}{2}, 0, 0, 0)$	*ready*, the arm is straight and vertical
qs	$(0, 0, -\frac{\pi}{2}, 0, 0, 0)$	*stretch*, the arm is straight and horizontal
qn	$(0, \frac{\pi}{4}, -\pi, 0, \frac{\pi}{4}, 0)$	*nominal*, the arm is in a dextrous working pose ◄

Well away from singularities, which will be discussed in Sect. 7.3.4.

and these are shown graphically in Fig. 7.6. These plots are generated using the `plot` method, for example

```
>> p560.plot(qz)
```

which shows a skeleton of the robot with pipes that connect the link coordinate frames as defined by the Denavit-Hartenberg parameters. The `plot` method has many options for showing the joint axes, wrist coordinate frame, shadows and so on. More realistic-looking plots such as shown in Fig. 7.7 can be created by the `plot3d` method for a limited set of Toolbox robot models.

Forward kinematics can be computed as before

```
>> TE = p560.fkine(qz)
TE =
    1.0000         0         0    0.4521
         0    1.0000         0   -0.1500
         0         0    1.0000    0.4318
         0         0         0    1.0000
```

By the Denavit-Hartenberg parameters of the model in the `mdl_puma560` script.

where the joint coordinates are given as a row vector. This returns the homogeneous transformation corresponding to the end-effector pose. The origin of this frame, the link-6 coordinate frame {6}, is defined◄ as the point of intersection of the axes of the last 3 joints – physically this point is inside the robot's wrist mechanism. We can define a tool transform, from the T_6 frame to the actual tool tip by

```
>> p560.tool = SE3(0, 0, 0.2);
```

Alternatively we could change the kinematic parameter d_6. The tool transform approach is more general since the final link kinematic parameters only allow setting of d_6, a_6 and α_6 which provide z-axis translation, x-axis translation and x-axis rotation respectively.

in this case a 200 mm extension in the T_6 z-direction.◄ The pose of the tool tip, often referred to as the tool center point or TCP, is now

```
>> p560.fkine(qz)
ans =
    1.0000         0         0    0.4521
         0    1.0000         0   -0.1500
         0         0    1.0000    0.6318
         0         0         0    1.0000
```

The kinematic definition we have used considers that the base of the robot is the intersection point of the waist and shoulder axes which is a point inside the structure of the robot. The Puma 560 robot includes a "30-inch" tall pedestal. We can shift the origin of the robot from the point inside the robot to the base of the pedestal using a base transform

```
>> p560.base = SE3(0, 0, 30*0.0254);
```

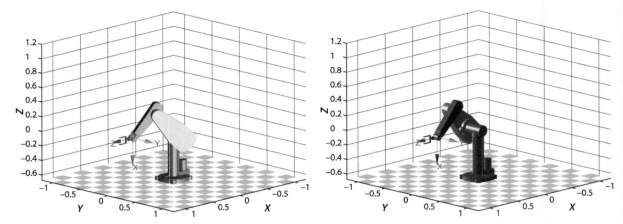

where for consistency we have converted the pedestal height to SI units. Now, with both base and tool transform, the forward kinematics are

```
>> p560.fkine(qz)
ans =
    1.0000         0         0    0.4521
         0    1.0000         0   -0.1500
         0         0    1.0000    1.3938
         0         0         0    1.0000
```

Fig. 7.7. These two different robot configurations result in the same end-effector pose. They are called the left- and right-handed configurations, respectively. These graphics, produced using the plot3d method, are available for a limited subset of robot models

and we can see that the z-coordinate of the tool is now greater than before.
   We can also do more interesting things, for example

```
>> p560.base = SE3(0,0,3) * SE3.Rx(pi);
>> p560.fkine(qz)
ans =
    1.0000         0         0    0.4521
         0   -1.0000   -0.0000    0.1500
         0    0.0000   -1.0000    2.3682
         0         0         0    1.0000
```

which positions the robot's origin 3 m above the world origin with its coordinate frame rotated by 180° about the x-axis. This robot is now hanging from the ceiling!
   The Toolbox supports joint angle time series, or trajectories, such as

```
>> q
q =
         0         0         0         0         0         0
         0    0.0365   -0.0365         0         0         0
         0    0.2273   -0.2273         0         0         0
         0    0.5779   -0.5779         0         0         0
         0    0.9929   -0.9929         0         0         0
         0    1.3435   -1.3435         0         0         0
         0    1.5343   -1.5343         0         0         0
         0    1.5708   -1.5708         0         0         0
```

where each row represents the joint coordinates at a different timestep and the columns represent the joints.▶ In this case the method fkine

```
>> T = p560.fkine(q);
```

Generated by the jtraj function, which is discussed in Sect. 7.3.1.

returns an array of SE3 objects

```
>> about T
T [SE3] : 1x8 (1.0 kB)
```

one per timestep. The homogeneous transform corresponding to the joint coordinates in the fourth row of q is

```
>> T(4)
ans =
    1.0000         0         0     0.382
         0        -1         0      0.15
         0         0   -1.0000     2.132
         0         0         0         1
```

Creating trajectories will be covered in Sect. 7.3.

7.2   Inverse Kinematics

We have shown how to determine the pose of the end-effector given the joint coordinates and optional tool and base transforms. A problem of real practical interest is the inverse problem: given the desired pose of the end-effector ξ_E what are the required joint coordinates? For example, if we know the Cartesian pose of an object, what joint coordinates does the robot need in order to reach it? This is the inverse kinematics problem which is written in functional form as

$$q = \mathcal{K}^{-1}(\xi) \tag{7.5}$$

and in general this function is not unique, that is, several joint coordinate vectors q will result in the same end-effector pose.

Two approaches can be used to determine the inverse kinematics. Firstly, a closed-form or analytic solution can be determined using geometric or algebraic approaches. However this becomes increasingly challenging as the number or robot joints increases and for some serial-link manipulators no closed-form solution exists. Secondly, an iterative numerical solution can be used. In Sect. 7.2.1 we again use the simple 2-dimensional case to illustrate the principles and then in Sect. 7.2.2 extend these to robot arms that move in 3-dimensions.

7.2.1   2-Dimensional (Planar) Robotic Arms

We will illustrate inverse kinematics for the 2-joint robot of Fig. 7.2b in two ways: algebraic closed-form and numerical.

7.2.1.1   Closed-Form Solution

We start by computing the forward kinematics algebraically as a function of joint angles. We can do this easily, and in a familiar way

```
>> import ETS2.*
>> a1 = 1; a2 = 1;
>> E = Rz('q1') * Tx(a1) * Rz('q2') * Tx(a2)
```

but now using the MATLAB Symbolic Math Toolbox™ we define some real-valued symbolic variables to represent the joint angles

```
>> syms q1 q2 real
```

Flange rotation
(joint 6)

Wrist
bend
(joint 5)

Wrist
rotation
(joint 4)

Spherical wrists are a key component of almost all modern arm-type robots. They have three axes of rotation that are orthogonal and intersect at a common point. This is a gimbal-like mechanism, and as discussed in Sect. 2.2.1.3 and will have a singularity.

The robot end-effector pose, position and an orientation, is defined at the center of the wrist. Since the wrist axes intersect at a common point they cause zero translation, therefore the position of the end-effector is a function only of the first three joints. This is a critical simplification that makes it possible to find closed-form inverse kinematic solutions for 6-axis industrial robots. An arbitrary end-effector orientation is achieved independently by means of the three wrist joints.

and then compute the forward kinematics

```
>> TE = E.fkine( [q1, q2] )
TE =
[ cos(q1 + q2), -sin(q1 + q2), cos(q1 + q2) + cos(q1)]
[ sin(q1 + q2),  cos(q1 + q2), sin(q1 + q2) + sin(q1)]
[            0,             0,                       1]
```

which is an algebraic representation of the robot's forward kinematics – the end-effector pose as a function of the joint variables.

We can define two more symbolic variables to represent the desired end-effector position (x, y)

```
>> syms x y real
```

and equate them with the results of the forward kinematics▶

With the MATLAB Symbolic Math Toolbox™ the == operator denotes equality, as opposed to = which denotes assignment.

```
>> e1 =  x == TE.t(1)
e1 =
x == cos(q1 + q2) + cos(q1)
>> e2 =  y == TE.t(2)
e2 =
y == sin(q1 + q2) + sin(q1)
```

which gives two scalar equations that we can solve simultaneously

```
>> [s1,s2] = solve( [e1 e2], [q1 q2] )
```

where the arguments are respectively the set of equations and the set of unknowns to solve for. The outputs are the solutions for q_1 and q_2 respectively. We observed in Sect. 7.1.1 that two different sets of joint angles give the same end-effector position, and this means that the inverse kinematics does not have a unique solution. Here MATLAB has returned

```
>> length(s2)
ans =
     2
```

indicating two solutions. One solution for q_2 is

```
>> s2(1)
ans =
-2*atan((-(x^2 + y^2)*(x^2 + y^2 - 4))^(1/2)/(x^2 + y^2))
```

and would be used in conjunction with the corresponding element of the solution vector for q_1 which is s1(1).

As mentioned earlier the complexity of algebraic solution increases dramatically with the number of joints and more sophisticated symbolic solution approaches need to be used. The SerialLink class has a method ikine_sym that generates symbolic inverse kinematics solutions for a limited class of robot manipulators.

7.2.1.2   Numerical Solution

We can think of the inverse kinematics problem as one of adjusting the joint coordinates until the forward kinematics matches the desired pose. More formally this is an optimization problem – to minimize the error between the forward kinematic solution and the desired pose ξ^*

$$q^* = \arg\min_{q} \left\| \mathcal{K}(q) \ominus \xi^* \right\|$$

For our simple 2-link example the error function comprises only the error in the end-effector position, not its orientation

$$E(q) = \left\| [\mathcal{K}(q)]_t - \left(x^*\ y^* \right)^T \right\|$$

We can solve this using the builtin MATLAB multi-variable minimization function `fminsearch`

```
>> pstar = [0.6; 0.7];
>> q = fminsearch( @(q) norm( E.fkine(q).t - pstar ), [0 0] )
q =
    -0.2295    2.1833
```

where the first argument is the error function, expressed here as a MATLAB anonymous function, that incorporates the desired end-effector position; and the second argument is the initial guess at the joint coordinates. The computed joint angles indeed give the desired end-effector position

```
>> E.fkine(q).print
t = (0.6, 0.7), theta = 111.9 deg
```

As already discussed there are two solutions for q but the solution that is found using this approach depends on the initial choice of q.

7.2.2   3-Dimensional Robotic Arms

7.2.2.1   Closed-Form Solution

Closed-form solutions have been developed for most common types of 6-axis industrial robots and many are included in the Toolbox. A necessary condition for a closed-form solution of a 6-axis robot is a spherical wrist mechanism. We will illustrate closed-form inverse kinematics using the Denavit-Hartenberg model for the Puma robot

```
>> mdl_puma560
```

At the *nominal* joint coordinates shown in Fig. 7.6d

```
>> qn
qn =
         0    0.7854    3.1416         0    0.7854         0
```

the end-effector pose is

```
>> T = p560.fkine(qn)
T =
   -0.0000    0.0000    1.0000    0.5963
   -0.0000    1.0000   -0.0000   -0.1501
   -1.0000   -0.0000   -0.0000   -0.0144
         0         0         0    1.0000
```

Since the Puma 560 is a 6-axis robot arm with a spherical wrist we use the method `ikine6s` to compute the inverse kinematics using a closed-form solution. ◄ The required joint coordinates to achieve the pose `T` are

The method `ikine6s` checks the Denavit-Hartenberg parameters to determine if the robot meets these criteria.

```
>> qi = p560.ikine6s(T)
qi =
    2.6486   -3.9270    0.0940    2.5326    0.9743    0.3734
```

Surprisingly, these are quite different to the joint coordinates we started with. However if we investigate a little further

```
>> p560.fkine(qi)
ans =
   -0.0000    0.0000    1.0000    0.5963
    0.0000    1.0000   -0.0000   -0.1500
   -1.0000    0.0000   -0.0000   -0.0144
         0         0         0    1.0000
```

we see that these two different sets of joint coordinates result in the *same* end-effector pose and these are shown in Fig. 7.7. The shoulder of the Puma robot is horizontally offset from the waist, so in one solution the arm is to the left of the waist, in the other

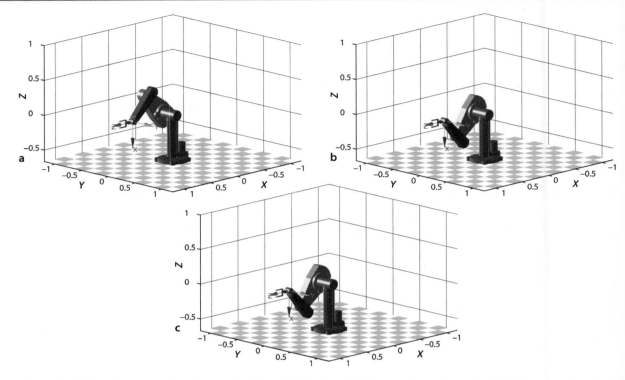

Fig. 7.8. Different configurations of the Puma 560 robot. **a** Right-up-noflip; **b** right-down-noflip; **c** right-down-flip

it is to the right. These are referred to as the left- and right-handed kinematic configurations respectively. In general there are eight sets of joint coordinates that give the same end-effector pose – as mentioned earlier the inverse solution is not unique.

We can *force* the right-handed solution

```
>> qi = p560.ikine6s(T, 'ru')
qi =
   -0.0000    0.7854    3.1416    0.0000    0.7854   -0.0000
```

which gives the original set of joint angles by specifying a *right handed* configuration with the elbow *up*.

In addition to the left- and right-handed solutions, there are solutions with the elbow either up or down,▸ and with the wrist flipped or not flipped. For the Puma 560 robot the wrist joint, θ_4, has a large rotational range and can adopt one of two angles that differ by π radians.

More precisely the elbow is above or below the shoulder.

Some different various kinematic configurations are shown in Fig. 7.8. The kinematic configuration returned by ikine6s is controlled by one or more of the options:

left or right handed	'l','r'
elbow up or down	'u','d'
wrist flipped or not flipped	'f','n'

Due to mechanical limits on joint angles and possible collisions between links not all eight solutions are physically achievable. It is also possible that no solution can be achieved. For example

```
>> p560.ikine6s( SE3(3, 0, 0) )
Warning: point not reachable
ans =
    NaN    NaN    NaN    NaN    NaN    NaN
```

has failed because the arm is simply not long enough to reach this pose.

A pose may also be unachievable due to singularity where the alignment of axes reduces the effective degrees of freedom (the gimbal lock problem again). The Puma 560 has a wrist singularity when q_5 is equal to zero and the axes of joints 4 and 6 become aligned. In this case the best that ikine6s can do is to constrain $q_4 + q_6$ but their individual values are arbitrary. For example consider the configuration

```
>> q = [0 pi/4 pi 0.1 0 0.2];
```

for which $q_4 + q_6 = 0.3$. The inverse kinematic solution is

```
>> p560.ikine6s(p560.fkine(q), 'ru')
ans =
   -0.0000    0.7854    3.1416   -3.0409    0.0000   -2.9423
```

which has quite different values for q_4 and q_6 but their sum

```
>> q(4)+q(6)
ans =
    0.3000
```

remains the same.

7.2.2.2   Numerical Solution

For the case of robots which do not have six joints and a spherical wrist we need to use an iterative numerical solution. Continuing with the example of the previous section we use the method ikine to compute the general inverse kinematic solution

```
>> T = p560.fkine(qn)
ans =
   -0.0000    0.0000    1.0000    0.5963
   -0.0000    1.0000   -0.0000   -0.1501
   -1.0000   -0.0000   -0.0000   -0.0144
         0         0         0    1.0000
>> qi = p560.ikine(T)
qi =
    0.0000   -0.8335    0.0940   -0.0000   -0.8312    0.0000
```

which is different to the original value

```
>> qn
qn =
         0    0.7854    3.1416         0    0.7854         0
```

but does result in the correct tool pose

```
>> p560.fkine(qi)
ans =
   -0.0000    0.0000    1.0000    0.5963
   -0.0000    1.0000   -0.0000   -0.1501
   -1.0000   -0.0000   -0.0000   -0.0144
         0         0         0    1.0000
```

Plotting the pose

```
>> p560.plot(qi)
```

shows clearly that ikine has found the elbow-down configuration.

A limitation of this general numeric approach is that it does not provide explicit control over the arm's kinematic configuration as did the analytic approach – the only control is implicit via the initial value of joint coordinates (which defaults to zero). If we specify the initial joint coordinates

```
>> qi = p560.ikine(T, 'q0', [0 0 3 0 0 0])
qi =
    0.0000    0.7854    3.1416    0.0000    0.7854   -0.0000
```

When solving for a trajectory as on p. 204 the inverse kinematic solution for one point is used to initialize the solution for the next point on the path.

we have forced the solution to converge on the elbow-up configuration. ◄

As would be expected the general numerical approach of ikine is considerably slower than the analytic approach of ikine6s. However it has the great advantage of being able to work with manipulators at singularities and manipulators with less than six or more than six joints. Details of the principle behind ikine is provided in Sect. 8.6.

7.2.2.3   Under-Actuated Manipulator

An under-actuated manipulator is one that has fewer than six joints, and is therefore limited in the end-effector poses that it can attain. SCARA robots such as shown on page 191 are a common example. They typically have an x-y-z-θ task space, $\mathcal{T} \subset \mathbb{R}^3 \times \mathbb{S}^1$ and a configuration space $\mathcal{C} \subset (\mathbb{S}^1)^3 \times \mathbb{R}$.

We will load a model of SCARA robot

```
>> mdl_cobra600
>> c600
c600 =
Cobra600 [Adept]:: 4 axis, RRPR, stdDH
+---+-----------+-----------+-----------+-----------+-----------+
| j |   theta   |     d     |     a     |   alpha   |   offset  |
+---+-----------+-----------+-----------+-----------+-----------+
| 1 |       q1  |    0.387  |    0.325  |        0  |        0  |
| 2 |       q2  |        0  |    0.275  |    3.142  |        0  |
| 3 |        0  |       q3  |        0  |        0  |        0  |
| 4 |       q4  |        0  |        0  |        0  |        0  |
+---+-----------+-----------+-----------+-----------+-----------+
```

and then define a desired end-effector pose

```
>> T = SE3(0.4, -0.3, 0.2) * SE3.rpy(30, 40, 160, 'deg')
```

where the end-effector approach vector is pointing downward but is not vertically aligned. This *pose* is over-constrained for the 4-joint SCARA robot – the tool physically cannot meet the orientation requirement for an approach vector that is not vertically aligned. Therefore we require the `ikine` method to not consider rotation about the x- and y-axes when computing the end-effector pose error. We achieve this by specifying a mask vector as the fourth argument

```
>> q = c600.ikine(T, 'mask', [1 1 1 0 0 1])
q =
    -0.1110    -1.1760     0.1870    -0.8916
```

The elements of the mask vector correspond respectively to the three translations and three orientations: $t_x, t_y, t_z, r_x, r_y, r_z$ in the end-effector coordinate frame. In this example we specified that rotation about the x- and y-axes are to be ignored (the zero elements). The resulting joint angles correspond to an achievable end-effector pose

```
>> Ta = c600.fkine(q);
>> Ta.print('xyz')
t = (0.4, -0.3, 0.2), RPY/xyz = (22.7, 0, 180) deg
```

which has the desired translation and yaw angle, but the roll and pitch angles are incorrect, as *we allowed* them to be. They are what the robot mechanism actually permits. We can also compare the desired and achievable poses graphically

```
>> trplot(T, 'color', 'b')
>> hold on
>> trplot(Ta, 'color', 'r')
```

7.2.2.4   Redundant Manipulator

A redundant manipulator is a robot with more than six joints. As mentioned previously, six joints is theoretically sufficient to achieve any desired pose in a Cartesian taskspace $\mathcal{T} \subset \mathbf{SE}(3)$. However practical issues such as joint limits and singularities mean that not all poses within the robot's reachable space can be achieved. Adding additional joints is one way to overcome this problem but results in an infinite number of joint-coordinate solutions. To find a single solution we need to introduce constraints – a common one is the minimum-norm constraint which returns a solution where the joint-coordinate vector elements have the smallest magnitude.

We will illustrate this with the Baxter robot shown in Fig. 7.1b. This is a two armed robot, and each arm has 7 joints. We load the Toolbox model

```
>> mdl_baxter
```

which defines two SerialLink objects in the workspace, one for each arm. We will work with the left arm

```
>> left
left =
Baxter LEFT [Rethink Robotics]:: 7 axis, RRRRRRR, stdDH
+---+-----------+-----------+-----------+-----------+-----------+
| j |    theta  |       d   |       a   |     alpha |    offset |
+---+-----------+-----------+-----------+-----------+-----------+
|  1|        q1 |      0.27 |     0.069 |    -1.571 |         0 |
|  2|        q2 |         0 |         0 |     1.571 |     1.571 |
|  3|        q3 |     0.364 |     0.069 |    -1.571 |         0 |
|  4|        q4 |         0 |         0 |     1.571 |         0 |
|  5|        q5 |     0.374 |      0.01 |    -1.571 |         0 |
|  6|        q6 |         0 |         0 |     1.571 |         0 |
|  7|        q7 |      0.28 |         0 |         0 |         0 |
+---+-----------+-----------+-----------+-----------+-----------+
  base:    t = (0.064614,0.25858,0.119), RPY/xyz = (0, 0, 45) deg
```

which we can see has a base offset that reflects where the arm is attached to Baxter's torso. We want the robot to move to this pose

```
>> TE = SE3(0.8, 0.2, -0.2) * SE3.Ry(pi);
```

which has its approach vector downward. The required joint angles are obtained using the numerical inverse kinematic solution and

```
>> q = left.ikine(TE)
q =
    0.0895   -0.0464   -0.4259    0.6980   -0.4248    1.0179    0.2998
```

is the joint-angle vector with the smallest norm that results in the desired end-effector pose. We can verify this by computing the forward kinematics or plotting

```
>> left.fkine(q).print('xyz')
t = (0.8, 0.2, -0.2), RPY/xyz = (180, 180, 180) deg
>> left.plot(q)
```

7.3   Trajectories

One of the most common requirements in robotics is to move the end-effector smoothly from pose A to pose B. Building on what we learned in Sect. 3.3 we will discuss two approaches to generating such trajectories: straight lines in configuration space and straight lines in task space. These are known respectively as joint-space and Cartesian motion.

7.3.1   Joint-Space Motion

In this robot configuration, similar to Fig. 7.6d, we specify the pose to include a rotation so that the end-effector z-axis is not pointing straight up in the world z-direction. For the Puma 560 robot this would be physically impossible to achieve in the elbow-up configuration.

Consider the end-effector moving between two Cartesian poses◄

```
>> T1 = SE3(0.4,  0.2, 0) * SE3.Rx(pi);
>> T2 = SE3(0.4, -0.2, 0) * SE3.Rx(pi/2);
```

which describe points in the xy-plane with different end-effector orientations. The joint coordinate vectors associated with these poses are

```
>> q1 = p560.ikine6s(T1);
>> q2 = p560.ikine6s(T2);
```

and we require the motion to occur over a time period of 2 seconds in 50 ms time steps

```
>> t = [0:0.05:2]';
```

A joint-space trajectory is formed by smoothly interpolating between the joint configurations q1 and q2. The scalar interpolation functions `tpoly` or `lspb` from Sect. 3.3.1 can be used in conjunction with the multi-axis *driver* function `mtraj`

```
>> q = mtraj(@tpoly, q1, q2, t);
```

or

```
>> q = mtraj(@lspb, q1, q2, t);
```

which each result in a 50 × 6 matrix q with one row per time step and one column per joint. From here on we will use the equivalent `jtraj` convenience function

```
>> q = jtraj(q1, q2, t);▶
```

This is equivalent to `mtraj` with `tpoly` interpolation but optimized for the multi-axis case and also allowing initial and final velocity to be set using additional arguments.

For `mtraj` and `jtraj` the final argument can be a time vector, as here, or an integer specifying the number of time steps.

We can obtain the joint velocity and acceleration vectors, as a function of time, through optional output arguments

```
>> [q,qd,qdd] = jtraj(q1, q2, t);
```

An even more concise way to achieve the above steps is provided by the `jtraj` method of the `SerialLink` class

```
>> q = p560.jtraj(T1, T2, t)
```

Fig. 7.9. Joint-space motion. **a** Joint coordinates versus time; **b** Cartesian position versus time; **c** Cartesian position locus in the *xy*-plane **d** roll-pitch-yaw angles versus time

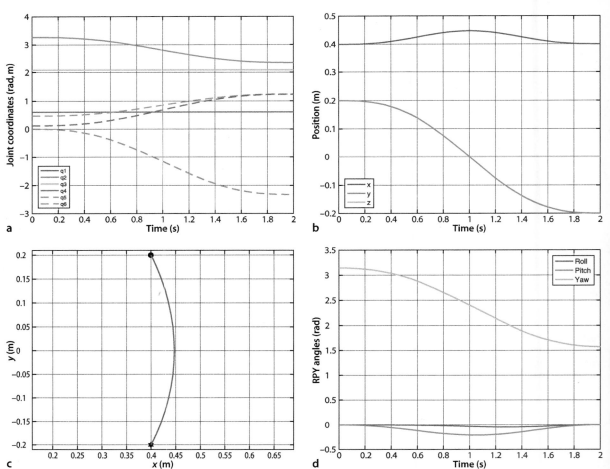

The trajectory is best viewed as an animation

```
>> p560.plot(q)
```

but we can also plot the joint angle, for instance q_2, versus time

```
>> plot(t, q(:,2))
```

or all the angles versus time

```
>> qplot(t, q);
```

as shown in Fig. 7.9a. The joint coordinate trajectory is smooth but we do not know how the robot's end-effector will move in Cartesian space. However we can easily determine this by applying forward kinematics to the joint coordinate trajectory

```
>> T = p560.fkine(q);
```

which results in an array of SE3 objects. The translational part of this trajectory is

```
>> p = T.transl;
```

which is in matrix form

Fig. 7.10. Cartesian motion. **a** Joint coordinates versus time; **b** Cartesian position versus time; **c** Cartesian position locus in the xy-plane; **d** roll-pitch-yaw angles versus time

```
>> about(p)
p [double] : 41x3 (984 bytes)
```

and has one column per time step, and each column is the end-effector position vector. This is plotted against time in Fig. 7.9b. The path of the end-effector in the xy-plane

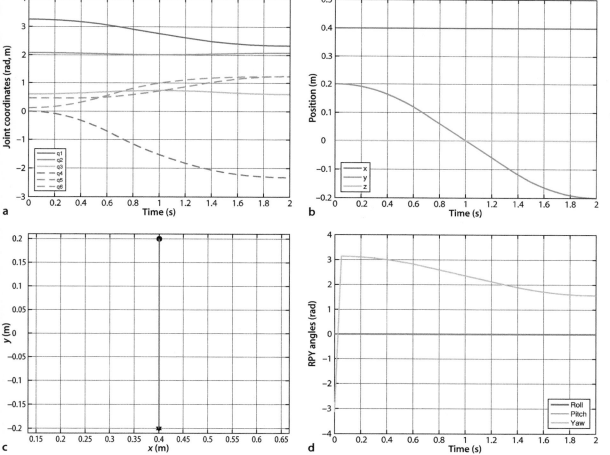
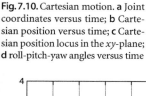

```
>> plot(p(1,:), p(2,:))
```

is shown in Fig. 7.9c and it is clear that the path is not a straight line. This is to be expected since we only specified the Cartesian coordinates of the end-points. As the robot rotates about its waist joint during the motion the end-effector will naturally follow a circular arc. In practice this could lead to collisions between the robot and nearby objects even if they do not lie on the path between poses A and B. The orientation of the end-effector, in XYZ roll-pitch-yaw angle form, can also be plotted against time

```
>> plot(t, T.torpy('xyz'))
```

as shown in Fig. 7.9d. Note that the yaw angle▶ varies from 0 to $\frac{\pi}{2}$ radians as we specified. However while the roll and pitch angles have met their boundary conditions they have varied along the path.

▶ Rotation about *x*-axis for a robot end-effector from Sect. 2.2.1.2.

7.3.2   Cartesian Motion

For many applications we require straight-line motion in Cartesian space which is known as Cartesian motion. This is implemented using the Toolbox function `ctraj` which was introduced in Sect. 3.3.5. Its usage is very similar to `jtraj`

```
>> Ts = ctraj(T1, T2, length(t));
```

where the arguments are the initial and final pose and the *number of* time steps and it returns the trajectory as an array of `SE3` objects.

As for the previous joint-space example we will extract and plot the translation

```
>> plot(t, Ts.transl);
```

and orientation components

```
>> plot(t, Ts.torpy('xyz'));
```

of this motion which is shown in Fig. 7.10 along with the path of the end-effector in the *xy*-plane. Compared to Fig. 7.9 we note some important differences. Firstly the end-effector follows a straight line in the *xy*-plane as shown in Fig. 7.10c. Secondly the roll and pitch angles shown in Fig. 7.10d are constant at zero along the path.

The corresponding joint-space trajectory is obtained by applying the inverse kinematics

```
>> qc = p560.ikine6s(Ts);
```

and is shown in Fig. 7.10a. While broadly similar to Fig. 7.9a the minor differences are what result in the straight line Cartesian motion.

7.3.3   Kinematics in Simulink

We can also implement this example in Simulink®

```
>> sl_jspace
```

and the block diagram model is shown in Fig. 7.11. The parameters of the `jtraj` block are the initial and final values for the joint coordinates and the time duration of the motion segment. The smoothly varying joint angles are wired to a `plot` block which will animate a robot in a separate window, and to an `fkine` block to compute the forward kinematics. Both the `plot` and `fkine` blocks have a parameter which is a `SerialLink` object, in this case `p560`. The Cartesian position of the end-effector pose is extracted using the `T2xyz` block which is analogous to the Toolbox function `transl`. The `XY Graph` block plots `y` against `x`.

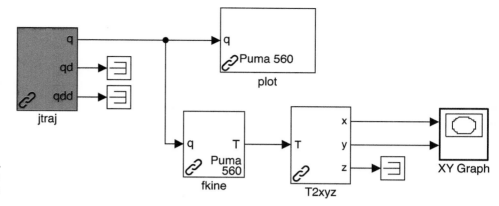

Fig. 7.11.
Simulink model sl_jspace for joint-space motion

7.3.4   Motion through a Singularity

We have already briefly touched on the topic of singularities (page 209) and we will revisit it again in the next chapter. In the next example we deliberately choose a trajectory that moves through a robot wrist singularity. We change the Cartesian endpoints of the previous example to

```
>> T1 = SE3(0.5,  0.3, 0.44) * SE3.Ry(pi/2);
>> T2 = SE3(0.5, -0.3, 0.44) * SE3.Ry(pi/2);
```

which results in motion in the y-direction with the end-effector z-axis pointing in the world x-direction. The Cartesian path is

```
>> Ts = ctraj(T1, T2, length(t));
```

which we convert to joint coordinates

```
>> qc = p560.ikine6s(Ts)
```

q_6 has increased rapidly, while q_4 has decreased rapidly and wrapped around from $-\pi$ to π. This counter-rotational motion of the two joints means that the gripper does not rotate but the two motors are working hard.

and is shown in Fig. 7.12a. At time $t \approx 0.7$ s we observe that the rate of change of the wrist joint angles q_4 and q_6 has become very high.◄ The cause is that q_5 has become almost zero which means that the q_4 and q_6 rotational axes are almost aligned – another gimbal lock situation or singularity.

The joint axis alignment means that the robot has lost one degree of freedom and is now effectively a 5-axis robot. Kinematically we can only solve for the sum $q_4 + q_6$ and there are an infinite number of solutions for q_4 and q_6 that would have the same sum. From Fig. 7.12b we observe that the generalized inverse kinematics method ikine handles the singularity with far less unnecessary joint motion. This is a consequence of the minimum-norm solution which has returned the smallest magnitude q_4 and q_6 which have the correct sum. The joint-space motion between the two poses, Fig. 7.12c, is immune to this problem since it is does not involve inverse kinematics. However it will not maintain the orientation of the tool in the x-direction for the whole path – only at the two end points.

The dexterity of a manipulator, its ability to move easily in any direction, is referred to as its manipulability. It is a scalar measure, high is good, and can be computed for each point along the trajectory

```
>> m = p560.maniplty(qc);
```

and is plotted in Fig. 7.12d. This shows that manipulability was almost zero around the time of the rapid wrist joint motion. Manipulability and the generalized inverse kinematics function are based on the manipulator's Jacobian matrix which is the topic of the next chapter.

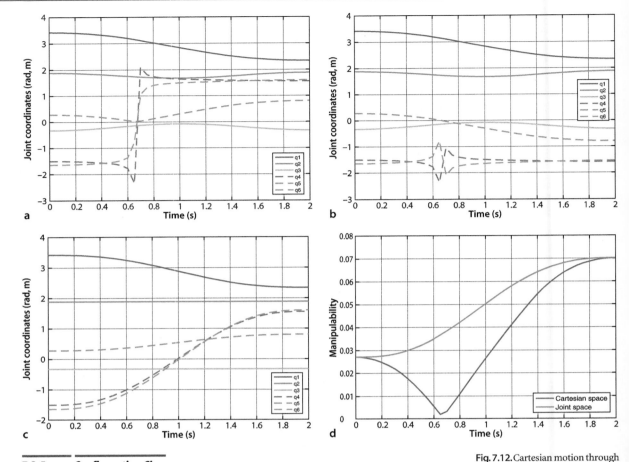

Fig. 7.12. Cartesian motion through a wrist singularity. **a** Joint coordinates computed by inverse kinematics (`ikine6s`); **b** joint coordinates computed by numerical inverse kinematics (`ikine`); **c** joint coordinates for joint-space motion; **d** manipulability

7.3.5   Configuration Change

Earlier (page 208) we discussed the kinematic configuration of the manipulator arm and how it can work in a left- or right-handed manner and with the elbow up or down. Consider the problem of a robot that is working for a while left-handed at one work station, then working right-handed at another. Movement from one configuration to another ultimately results in no change in the end-effector pose since both configuration have the same forward kinematic solution – therefore we *cannot* create a trajectory in Cartesian space. Instead we must use joint-space motion.

For example to move the robot arm from the right- to left-handed configuration we first define some end-effector pose

```
>> T = SE3(0.4, 0.2, 0) * SE3.Rx(pi);
```

and then determine the joint coordinates for the right- and left-handed elbow-up configurations

```
>> qr = p560.ikine6s(T, 'ru');
>> ql = p560.ikine6s(T, 'lu');
```

and then create a joint-space trajectory between these two joint coordinate vectors

```
>> q = jtraj(qr, ql, t);
```

Although the initial and final end-effector pose is the same, the robot makes some quite significant joint space motion as shown in Fig. 7.13 – in the real world you need to be careful the robot doesn't hit something. Once again, the best way to visualize this is in animation

```
>> p560.plot(q)
```

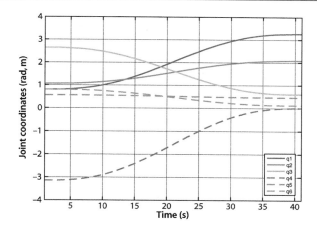

Fig. 7.13.
Joint space motions for configuration change from right-handed
to left-handed

7.4   Advanced Topics

7.4.1   Joint Angle Offsets

The pose of the robot with zero joint angles is an arbitrary decision of the robot
designer and might even be a mechanically unachievable pose. For the Puma robot
the zero-angle pose is a nonobvious *L-shape* with the upper arm horizontal and the
lower arm vertically upward as shown in Fig. 7.6a. This is a consequence of constraints imposed by the Denavit-Hartenberg formalism.

It is actually implemented within the
`Link` object.

The joint coordinate offset provides a mechanism to set an arbitrary configuration for the zero joint coordinate case. The offset vector, q_0, is added to the user
specified joint angles before any kinematic or dynamic function is invoked, ◄ for
example

$$\xi_E = \mathcal{K}(q + q_0) \tag{7.6}$$

Similarly it is subtracted after an operation such as inverse kinematics

$$q = \mathcal{K}^{-1}(\xi_E) - q_0 \tag{7.7}$$

The offset is set by assigning the `offset` property of the `Link` object, or giving the
`'offset'` option to the `SerialLink` constructor.

7.4.2   Determining Denavit-Hartenberg Parameters

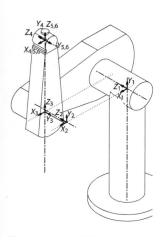

Fig. 7.14. Puma 560 robot coordinate frames. Standard Denavit-
Hartenberg link coordinate frames
for Puma in the zeroangle pose
(Corke 1996b)

The classical method of determining Denavit-Hartenberg parameters is to system-
atically assign a coordinate frame to each link. The link frames for the Puma robot
using the standard Denavit-Hartenberg formalism are shown in Fig. 7.14. However
there are strong constraints on placing each frame since joint rotation must be
about the *z*-axis and the link displacement must be in the *x*-direction. This in turn
imposes constraints on the placement of the coordinate frames for the base and the
end-effector, and ultimately dictates the zero-angle pose just discussed. Determining
the Denavit-Hartenberg parameters and link coordinate frames for a completely
new mechanism is therefore more difficult than it should be – even for an experienced roboticist.

An alternative approach, supported by the Toolbox, is to simply describe the manipulator as a series of elementary translations and rotations from the base to the
tip of the end-effector as we discussed in Sect. 7.1.2. Some of the elementary operations are constants such as translations that represent link lengths or offsets, and

some are functions of the generalized joint coordinates q_j. Unlike the conventional approach we impose no constraints on the axes about which these rotations or translations can occur.

For the Puma robot shown in Fig. 7.4 we first define a convenient coordinate frame at the base and then write down the sequence of translations and rotations, from "toe to tip", in a string▶

```
>> s = 'Tz(L1) Rz(q1) Ry(q2) Ty(L2) Tz(L3) Ry(q3) Tx(L4) Ty(L5)
        Tz(L6) Rz(q4) Ry(q5) Rz(q6)'
```

All lengths must start with *L* and negative signs cannot be used in the string, but the values themselves can be negative. You can generate this string from an ETS3 sequence (page 196) using its `string` method.

Note that we have described the second joint as `Ry(q2)`, a rotation about the *y*-axis, which is not permissible using the Denavit-Hartenberg formalism.

This string is input to a symbolic algebra function▶

```
>> dh = DHFactor(s);
```

Written in Java, the MATLAB® Symbolic Math Toolbox™ is not required.

which returns a `DHFactor` object▶ that holds the kinematic structure of the robot that has been factorized into Denavit-Hartenberg parameters. We can display this in a human-readable form

Actually a Java object.

```
>> dh
dh =
DH(q1, L1, 0, -90).DH(q2+90, 0, -L3, 0).DH(q3-90, L2+L5, L4, 90).
DH(q4, L6, 0, -90).DH(q5, 0, 0, 90).DH(q6, 0, 0, 0)
```

which shows the Denavit-Hartenberg parameters for each joint in the order θ, d, a and α. Joint angle offsets (the constants added to or subtracted from joint angle variables such as `q2` and `q3`) are generated automatically, as are base and tool transformations. The object can generate a string that is a complete Toolbox command to create the robot named "puma"

```
>> cmd = dh.command('puma')
cmd =
SerialLink([0, L1, 0, -pi/2, 0; 0, 0, -L3, 0, 0; 0, L2+L5, L4,↵
pi/2, 0; 0, L6, 0, -pi/2, 0; 0, 0, 0, pi/2, 0; 0, 0, 0, 0, 0; ], ...
  'name', 'puma', ...
  'base', eye(4,4), 'tool', eye(4,4), ...
  'offset', [0 pi/2 -pi/2 0 0 0 ])
```

which can be executed

```
>> robot = eval(cmd)
```

to create a workspace variable called `robot` that is a `SerialLink` object.▶

The length parameters `L1` to `L6` must be defined in the workspace first.

7.4.3   Modified Denavit-Hartenberg Parameters

The Denavit-Hartenberg notation introduced in this chapter is commonly used and described in many robotics textbooks. Craig (1986) first introduced the modified Denavit-Hartenberg parameters where the link coordinate frames shown in Fig. 7.15 are attached to the near (proximal), rather than the far (distal) end of each link. This modified notation is in some ways clearer and tidier and is also now commonly used. However its introduction has increased the scope for confusion, particularly for those who are new to robot kinematics. The root of the problem is that the algorithms for kinematics, Jacobians and dynamics depend on the kinematic conventions used. According to Craig's convention the link transform matrix is

$$^{j-1}\xi_j\big(\alpha_{j-1}, a_{j-1}, d_j, \theta_j\big) = \mathscr{R}_x\big(\alpha_{j-1}\big) \oplus \mathscr{T}_x\big(a_{j-1}\big) \oplus \mathscr{T}_z\big(d_j\big) \oplus \mathscr{R}_z\big(\theta_j\big) \tag{7.8}$$

denoted in that book as $^{j-1}_jA$. This has the same terms as Eq. 7.2 but in a different order – remember rotations are not commutative – and this is the nub of the problem. a_j is

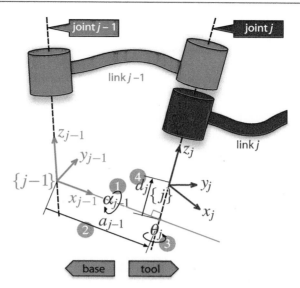

Fig. 7.15.
Definition of modified Denavit and Hartenberg link parameters. The colors *red* and *blue* denote all things associated with links $j-1$ and j respectively. The *numbers in circles* represent the order in which the elementary transforms are applied

always the length of link j, but it is the displacement between the origins of frame $\{j\}$ and frame $\{j+1\}$ in one convention, and frame $\{j-1\}$ and frame $\{j\}$ in the other.

> If you intend to build a Toolbox robot model from a table of kinematic parameters provided in a research paper it is really important to know which convention the author of the table used. Too often this important fact is not mentioned. An important clue lies in the column headings. If they all have the same subscript, i.e. θ_j, d_j, a_j and α_j then this is standard Denavit-Hartenberg notation. If half the subscripts are different, i.e. θ_j, d_j, a_{j-1} and α_{j-1} then you are dealing with modified Denavit-Hartenberg notation. In short, you must know which kinematic convention your Denavit-Hartenberg parameters conform to.
>
> You can also help the field when publishing by stating clearly which kinematic convention is used for your parameters.

The Toolbox can handle either form, it only needs to be specified, and this is achieved using variant classes when creating a link object

```
>> L1 = RevoluteMDH('d', 1)
L1 =
Revolute(mod): theta=q, d=1, a=0, alpha=0, offset=0
```

Everything else from here on, creating the robot object, kinematic and dynamic functions works as previously described.

The two forms can be interchanged by considering the link transform as a string of elementary rotations and translations as in Eq. 7.2 or Eq. 7.8. Consider the transformation chain for standard Denavit-Hartenberg notation

$$\underbrace{\mathscr{R}_z(\theta_1) \oplus \mathscr{T}_z(d_1) \oplus \mathscr{T}_x(a_1) \oplus \mathscr{R}_x(\alpha_1)}_{\text{DH}_1} \oplus \underbrace{\mathscr{R}_z(\theta_2) \oplus \mathscr{T}_z(d_2) \oplus \mathscr{T}_x(a_2) \oplus \mathscr{R}_x(\alpha_2)}_{\text{DH2}} \cdots$$

which we can regroup as

$$\underbrace{\mathscr{R}_z(\theta_1) \oplus \mathscr{T}_z(d_1)}_{\text{base}} \oplus \underbrace{\mathscr{T}_x(a_1)\mathscr{R}_x(\alpha_1) \oplus \mathscr{R}_z(\theta_2) \oplus \mathscr{T}_z(d_2)}_{\text{MDH}_1} \oplus \underbrace{\mathscr{T}_x(a_2) \oplus \mathscr{R}_x(\alpha_2)}_{\text{MDH}_2} \cdots$$

where the terms marked as MDH$_j$ have the form of Eq. 7.8 taking into account that translation along, and rotation about the same axis *is* commutative, that is, $\mathscr{R}_i(\theta) \oplus \mathscr{T}_i(d) = \mathscr{T}_i(d) \oplus \mathscr{R}_i(\theta)$ for $i \in \{x, y, z\}$.

7.5   Applications

7.5.1   Writing on a Surface [examples/drawing.m]

Our goal is to create a trajectory that will allow a robot to draw a letter. The Toolbox comes with a preprocessed version of the Hershey font▶

```
>> load hershey
```

Developed by Dr. Allen V. Hershey at the Naval Weapons Laboratory in 1967, data from http://paulbourke.net/data-formats/hershey.

as a cell array of character descriptors. For an upper-case 'B'

```
>> B = hershey{'B'}
B =
    stroke: [2x23 double]
     width: 0.8400
       top: 0.8400
    bottom: 0
```

the structure describes the dimensions of the character, vertically from 0 to 0.84 and horizontally from 0 to 0.84▶. The path to be drawn is

This is a variable-width font, and all characters fit within a unit-height rectangle.

```
>> B.stroke
ans =
  Columns 1 through 11
    0.1600    0.1600       NaN    0.1600    0.5200    0.6400    ...
    0.8400         0       NaN    0.8400    0.8400    0.8000    ...
```

where the rows are the x- and y-coordinates respectively, and a column of NaNs indicates the end of a segment – the pen is lifted and placed down again at the beginning of the next segment. We perform some processing

```
>> path = [ 0.25*B.stroke; zeros(1,numcols(B.stroke))];
>> k = find(isnan(path(1,:)));
>> path(:,k) = path(:,k-1); path(3,k) = 0.2;
```

to scale the path by 0.25 so that the character is around 20 cm tall, append a row of zeros (add z-coordinates to this 2-dimensional path), find the columns that contain NaNs and replace them with the preceding column but with the z-coordinate set to 0.2 in order to lift the pen off the surface.

Next we convert this to a continuous trajectory

```
>> traj = mstraj(path(:,2:end)', [0.5 0.5 0.5], [], path(:,1)',↵
   0.02, 0.2);
```

where the second argument is the maximum speed in the x-, y- and z-directions, the fourth argument is the initial coordinate followed by the sample interval and the acceleration time. The number of steps in the interpolated path is

```
>> about(traj)
traj [double] : 487x3 (11.7 kB)
```

and will take

```
>> numrows(traj) * 0.02
ans =
   9.7400
```

seconds to execute at the 20 ms sample interval. The trajectory can be plotted

```
>> plot3(traj(:,1), traj(:,2), traj(:,3))
```

as shown in Fig. 7.16.

We now have a sequence of 3-dimensional points but the robot end-effector has a pose, not just a position, so we need to attach a coordinate frame to every point. We assume that the robot is writing on a horizontal surface so these frames must have their approach vector pointing downward, that is, $a = [0, 0, -1]$, with the gripper ar-

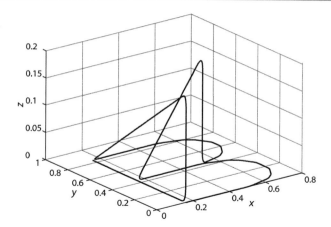

Fig. 7.16.
The end-effector path drawing
the letter 'B'

bitrarily oriented in the y-direction with $o = [0, 1, 0]$. The character is also placed at
$(0.6, 0, 0)$ in the workspace, and all this is achieved by

```
>> Tp = SE3(0.6, 0, 0) * SE3(traj) * SE3.oa( [0 1 0], [0 0 -1]);
```

Now we can apply inverse kinematics

```
>> q = p560.ikine6s(Tp);
```

to determine the joint coordinates and then animate it

```
>> p560.plot(q)
```

We have not considered the force that
the robot-held pen exerts on the paper,
we cover force control in Chap. 9. In a
real implementation of this example it
would be prudent to use a spring to push
the pen against the paper with sufficient
force to allow it to write.

The Puma is drawing the letter 'B', and lifting its pen in between strokes! The approach is quite general and we could easily change the size of the letter, write whole
words and sentences, write on an arbitrary plane or use a robot with quite different
kinematics. ◀

7.5.2   A Simple Walking Robot [examples/walking.m]

Four legs good, two legs bad!
Snowball the pig, Animal Farm by George Orwell

Our goal is to create a four-legged walking robot. We start by creating a 3-axis robot
arm that we use as a leg, plan a trajectory for the leg that is suitable for walking, and
then instantiate four instances of the leg to create the walking robot.

Kinematics

Kinematically a robot leg is much like a robot arm. For this application a three joint
serial-link manipulator is sufficient since the foot has point contact with the ground
and orientation is not important. Determining the Denavit-Hartenberg parameters,
even for a simple robot like this, is an involved procedure and the zero-angle offsets
need to be determined in a separate step. Therefore we will use the procedure introduced in Sect. 7.4.2.

As always we start by defining our coordinate frame. This is shown in Fig. 7.17
along with the robot leg in its zero-angle pose. We have chosen the aerospace coordinate convention which has the x-axis forward and the z-axis downward, constrain-
ing the y-axis to point to the right-hand side. The first joint will be hip motion, forward and backward, which is rotation about the z-axis or $R_z(q1)$. The second joint
is hip motion up and down, which is rotation about the x-axis, $R_x(q_2)$. These form a

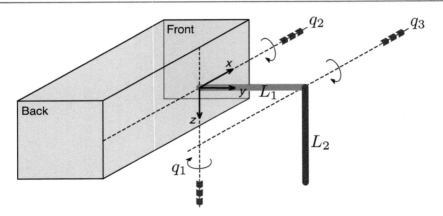

Fig. 7.17.
The coordinate frame and axis
rotations for the simple leg. The
leg is shown in its zero angle pose

spherical hip joint since the axes of rotation intersect. The knee is translated by L_1 in the y-direction or $T_y(L_1)$. The third joint is knee motion, toward and away from the body, which is $R_x(q_3)$. The foot is translated by L_2 in the z-direction or $T_z(L_2)$. The transform sequence of this robot, from hip to toe, is therefore $R_z(q1)R_x(q_2)T_y(L_1)R_x(q_3)T_z(L_2)$.

Using the technique of Sect. 7.4.2 we write this sequence as the string

```
>> s = 'Rz(q1) Rx(q2) Ty(L1) Rx(q3) Tz(L2)';
```

The string can be automatically manipulated into Denavit-Hartenberg factors

```
>> dh = DHFactor(s)
DH(q1+90, 0, 0, 90).DH(q2, 0, L1, 0).DH(q3-90, 0, -L2, 0)↵
.Rz(+90).Rx(-90).Rz(-90)
```

The last three terms in this factorized sequence is a tool transform

```
>> dh.tool
ans =
trotz(pi/2)*trotx(-pi/2)*trotz(-pi/2)
```

that changes the orientation of the frame at the foot. However for this problem the foot is simply a point that contacts the ground so we are not concerned about its orientation. The method `dh.command` generates a string that is the Toolbox command to create a `SerialLink` object

```
>> dh.command('leg')
ans =
SerialLink([0, 0, 0, pi/2, 0; 0, 0, L1, 0, 0; 0, 0, -L2, 0, 0; ],↵
  'name', 'leg', 'base', eye(4,4),↵
  'tool', trotz(pi/2)*trotx(-pi/2)*trotz(-pi/2),↵
  'offset', [pi/2 0 -pi/2 ])
```

which is input to the MATLAB `eval` command

```
>> L1 = 0.1; L2 = 0.1;
>> leg = eval( dh.command('leg') )
>> leg
leg =
leg:: 3 axis, RRR, stdDH, slowRNE
+---+-----------+-----------+-----------+-----------+-----------+
| j |   theta   |     d     |     a     |   alpha   |  offset   |
+---+-----------+-----------+-----------+-----------+-----------+
| 1 |       q1| |        0| |        0| |    1.5708| |    1.5708|
| 2 |       q2 |        0| |      0.1| |        0| |        0|
| 3 |       q3 |        0| |     -0.1| |        0| |   -1.5708|
+---+-----------+-----------+-----------+-----------+-----------+
tool:    t = (0, 0, 0), RPY/zyx = (0, -90, 0) deg
```

after first setting the length of each leg segment to 100 mm in the MATLAB workspace.

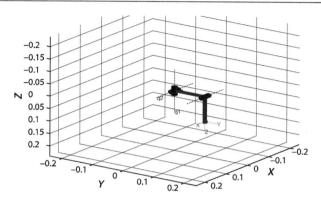

Fig. 7.18.
Robot leg in its zero angle pose. Note that the z-axis points downward

We perform a quick sanity check of our robot. For zero joint angles the foot is at

```
>> transl( leg.fkine([0,0,0]) )
ans =
        0    0.1000    0.1000
```

as we designed it. We can visualize the zero-angle pose

```
>> leg.plot([0,0,0], 'nobase', 'noshadow', 'notiles')
>> set(gca, 'Zdir', 'reverse'); view(137,48);
```

which is shown in Fig. 7.18. Now we should test that the other joints result in the expected motion. Increasing q_1

```
>> transl( leg.fkine([0.2,0,0]) )
ans =
    -0.0199    0.0980    0.1000
```

results in motion in the xy-plane, and increasing q_2

```
>> transl( leg.fkine([0,0.2,0]) )
ans =
    -0.0000    0.0781    0.1179
```

results in motion in the yz-plane, as does increasing q_3

```
>> transl( leg.fkine([0,0,0.2]) )
ans =
    -0.0000    0.0801    0.0980
```

We have now created and verified a simple robot leg.

Motion of One Leg

The next step is to define the path that the end-effector of the leg, its foot, will follow. The first consideration is that the end-effector of all feet move backwards at the same speed in the ground plane – propelling the robot's body forward without its feet slipping. Each leg has a limited range of movement so it cannot move backward for very long. At some point we must reset the leg – lift the foot, move it forward and place it on the ground again. The second consideration comes from static stability – the robot must have at least three feet on the ground at all times so each leg must take its turn to reset. This requires that any leg is in contact with the ground for ¾ of the cycle and is resetting for ¼ of the cycle. A consequence of this is that the leg has to move much faster during reset since it has a longer path and less time to do it in.

The required trajectory is defined by the via points

```
>> xf = 50; xb = -xf;   y = 50; zu = 20; zd = 50;
>> path = [xf y zd; xb y zd; xb y zu; xf y zu; xf y zd] * 1e-3;
```

where xf and xb are the forward and backward limits of leg motion in the x-direction (in units of mm), y is the distance of the foot from the body in the y-direction, and zu and zd

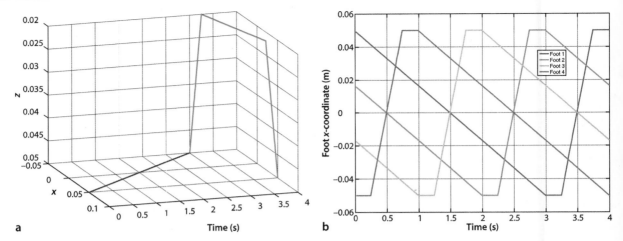

a   Time (s)   **b**   Time (s)

are respectively the height of the foot motion in the z-direction for foot up and foot down. In this case the foot moves from 50 mm forward of the hip to 50 mm behind. When the foot is down, it is 50 mm below the hip and it is raised to 20 mm below the hip during reset. The points in `path` define a complete cycle: the start of the stance phase, the end of stance, top of the leg lift, top of the leg return and the start of stance. This is shown in Fig. 7.19a.

Next we sample the multi-segment path at 100 Hz

```
>> p = mstraj(path, [], [0, 3, 0.25, 0.5, 0.25]', path(1,:), 0.01, 0);
```

In this case we have specified a vector of desired segment times rather than maxi-mum joint velocities.► The final three arguments are the initial leg configuration, the sample interval and the acceleration time. This trajectory has a total time of 4 s and therefore comprises 400 points.

We apply inverse kinematics to determine the joint angle trajectories required for the foot to follow the computed Cartesian trajectory. This robot is under-actuated so we use the generalized inverse kinematics `ikine` and set the mask so as to solve only for end-effector translation

```
>> qcycle = leg.ikine( SE3(p), 'mask', [1 1 1 0 0 0] );
```

We can view the motion of the leg in animation

```
>> leg.plot(qcycle, 'loop')
```

to verify that it does what we expect: slow motion along the ground, then a rapid lift, forward motion and foot placement. The `'loop'` option displays the trajectory in an endless loop and you need to type control-C to stop it.

Fig. 7.19. a Trajectory taken by a single foot. Recall from Fig. 7.17 that the z-axis is downward. The red segments are the leg reset. **b** The x-direction motion of each leg (offset vertically) to show the gait. The leg reset is the period of high x-direction velocity

This way we can ensure that the reset takes exactly one quarter of the cycle.

Motion of Four Legs

Our robot has width and length

```
>> W = 0.1; L = 0.2;
```

We create multiple instances of the leg by cloning the `leg` object we created earlier, and providing different base transforms so as to attach the legs to different points on the body

```
>> legs(1) = SerialLink(leg, 'name', 'leg1');
>> legs(2) = SerialLink(leg, 'name', 'leg2', 'base', SE3(-L, 0, 0));
>> legs(3) = SerialLink(leg, 'name', 'leg3', 'base', SE3(-L, -W, 0) ↵
   *SE3.Rz(pi));
>> legs(4) = SerialLink(leg, 'name', 'leg4', 'base', SE3(0, -W, 0) ↵
   *SE3.Rz(pi));
```

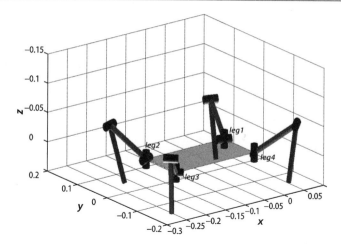

Fig. 7.20.
The walking robot

The result is a vector of `SerialLink` objects. Note that legs 3 and 4, on the left-hand side of the body have been rotated about the z-axis so that they point away from the body.

As mentioned earlier each leg must take its turn to reset. Since the trajectory is a cycle, we achieve this by having each leg run the trajectory with a phase shift equal to one quarter of the total cycle time. Since the total cycle has 400 points, each leg's trajectory is offset by 100, and we use modulo arithmetic to index into the cyclic gait for each leg. The result is the gait pattern shown in Fig. 7.19b.

The core of the walking program is

```
clf; k = 1;
while 1
    legs(1).plot( gait(qcycle, k, 0,   false) );
    if k == 1, hold on; end
    legs(2).plot( gait(qcycle, k, 100, false) );
    legs(3).plot( gait(qcycle, k, 200, true) );
    legs(4).plot( gait(qcycle, k, 300, true) );
    drawnow
    k = k+1;
end
```

where the function

```
gait(q, k, ph, flip)
```

returns the $k+ph^{th}$ element of q with modulo arithmetic that considers q as a cycle. The argument `flip` reverses the sign of the joint 1 motion for the legs on the left-hand side of the robot. A snapshot from the simulation is shown in Fig. 7.20. The entire implementation, with some additional refinement, is in the file `examples/walking.m` and detailed explanation is provided by the comments.

7.6 Wrapping Up

In this chapter we have learned how to determine the forward and inverse kinematics of a serial-link manipulator arm. Forward kinematics involves compounding the relative poses due to each joint and link, giving the pose of the robot's end-effector relative to its base. Commonly the joint and link structure is expressed in terms of Denavit-Hartenberg parameters for each link. Inverse kinematics is the problem of determining the joint coordinates given the end-effector pose. For simple robots, or those with six joints and a spherical wrist we can compute the inverse kinematics using an analytic solution. This inverse is not unique and the robot may have several joint configurations that result in the same end-effector pose.

For robots which do not have six joints and a spherical wrist we can use an iterative numerical approach to solving the inverse kinematics. We showed how this could be applied to an under-actuated 4-joint SCARA robot and a redundant 7-link robot. We also touched briefly on the topic of singularities which are due to the alignment of joint axes.

We also learned about creating paths to move the end-effector smoothly between poses. Joint-space paths are simple to compute but in general do not result in straight line paths in Cartesian space which may be problematic for some applications. Straight line paths in Cartesian space can be generated but singularities in the workspace may lead to very high joint rates.

Further Reading

Serial-link manipulator kinematics are covered in all the standard robotics textbooks such as the Robotics Handbook (Siciliano and Khatib 2016), Siciliano et al. (2009), Spong et al. (2006) and Paul (1981). Craig's text (2005) is also an excellent introduction to robot kinematics and uses the modified Denavit-Hartenberg notation, and the examples in the third edition are based on an older version of the Robotics Toolbox. Lynch and Park (2017) and Murray et al. (1994) cover the product of exponential approach. An emerging alternative to Denavit-Hartenberg notation is URDF (unified robot description format) which is described at http://wiki. ros.org/urdf.

Siciliano et al. (2009) provide a very clear description of the process of assigning Denavit-Hartenberg parameters to an arbitrary robot. The alternative approach described here based on symbolic factorization was described in detail by Corke (2007). The definitive values for the parameters of the Puma 560 robot are described in the paper by Corke and Armstrong-Hélouvry (1995).

Robotic walking is a huge field in its own right and the example given here is very simplistic. Machines have been demonstrated with complex gaits such as running and galloping that rely on dynamic rather than static balance. A good introduction to legged robots is given in the Robotics Handbook (Siciliano and Khatib 2016, § 17). Robotic hands, grasping and manipulation is another large topic which we have not covered – there is a good introduction in the Robotics Handbook (Siciliano and Khatib 2016, §37, 38).

Parallel-link manipulators were mentioned only briefly on page 192 and have advantages such as increased actuation force and stiffness (since the actuators form a truss-like structure). For this class of mechanism the inverse kinematics is usually closed-form and it is the forward kinematics that requires numerical solution. Useful starting points for this class of robots are the handbook (Siciliano and Khatib 2016, §18), a brief section in Siciliano et al. (2009) and Merlet (2006).

Closed-form inverse kinematic solutions can be derived symbolically by writing down a number of kinematic relationships and solving for the joint angles, as described in Paul (1981). Software packages to automatically generate the forward and inverse kinematics for a given robot have been developed and these include Robotica (Nethery and Spong 1994) which is now obsolete, and SYMORO (Khalil and Creusot 1997) which is now available as open-source.

Historical. The original work by Denavit and Hartenberg was their 1955 paper (Denavit and Hartenberg 1955) and their textbook (Hartenberg and Denavit 1964). The book has an introduction to the field of kinematics and its history but is currently out of print, although a version can be found online. The first full description of the kinematics of a six-link arm with a spherical wrist was by Paul and Zhang (1986).

MATLAB and Toolbox Notes

The workhorse of the Toolbox is the `SerialLink` class which has considerable functionality and very many methods – we will use it extensively for the remainder of Part III. The classes `ETS2` and `ETS3` used in the early parts of this chapter were designed to illustrate principles as concisely as possible and have limited functionality, but the names of their methods are the same as their equivalents in the `SerialLink` class.

The `plot` method draws a *stick figure* robot and needs only Denavit-Hartenberg parameters. However the joints depicted are associated with the link frames and don't necessarily correspond to physical joints on the robot, but that is a limitation of the Denavit-Hartenberg parameters. A small number of robots have more realistic 3-dimensional models defined by STL files and these can be displayed using the `plot3d`. The models shipped with the Toolbox were created by Arturo Gil and are also shipped with his ARTE Toolbox.

The numerical inverse kinematics method `ikine` can handle over- and under-actuated robot arms, but does not handle joint coordinate limits which can be set in the `SerialLink` object. The alternative inverse kinematic method `ikcon` respects joint limits but requires the MATLAB Optimization Toolbox™.

The MATLAB Robotics System Toolbox™ provides a `RigidBodyTree` class to represent a serial-link manipulator, and this also supports branched mechanisms such as a humanoid robot. It also provides a general `InverseKinematics` class to solve inverse kinematic problems and can handle joint limits.

Exercises

1. Forward kinematics for planar robot from Sect. 7.1.1.
   a) For the 2-joint robot use the `teach` method to determine the two sets of joint angles that will position the end-effector at (0.5, 0.5).
   b) Experiment with the three different models in Fig. 7.2 using the `fkine` and `teach` methods.
   c) Vary the models: adjust the link lengths, create links with a translation in the y-direction, or create links with a translation in the x- and y-direction.
2. Experiment with the `teach` method for the Puma 560 robot.
3. Inverse kinematics for the 2-link robot on page 206.
   a) Compute forward and inverse kinematics with a_1 and a_2 as symbolic rather than numeric values.
   b) What happens to the solution when a point is out of reach?
   c) Most end-effector positions can be reached by two different sets of joint angles. What points can be reached by only one set?
4. Compare the solutions generated by `ikine6s` and `ikine` for the Puma 560 robot at different poses. Is there any difference in accuracy? How much slower is `ikine`?
5. For the Puma 560 at configuration qn demonstrate a configuration change from elbow up to elbow down.
6. For a Puma 560 robot investigate the errors in end-effector pose due to manufacturing errors.
   a) Make link 2 longer by 0.5 mm. For 100 random joint configurations what is the mean and maximum error in the components of end-effector pose?
   b) Introduce an error of 0.1 degrees in the joint 2 angle and repeat the analysis above.
7. Investigate the redundant robot models `mdl_hyper2d` and `mdl_hyper3d`. Manually control them using the `teach` method, compute forward kinematics and numerical inverse kinematics.

8. If you have the MATLAB Optimization Toolbox™ experiment with the `ikcon` method which solves inverse kinematics for the case where the joint coordinates have limits (modeling mechanical end stops). Joint limits are set with the `qlim` property of the `Link` class.
9. Drawing a 'B' (page 220)
    a) Change the size of the letter.
    b) Write a word or sentence.
    c) Write on a vertical plane.
    d) Write on an inclined plane.
    e) Change the robot from a Puma 560 to the Fanuc 10L.
    f) Write on a sphere. Hint: Write on a tangent plane, then project points onto the sphere's surface.
    g) This writing task does not require 6DOF since the rotation of the pen about its axis is not important. Remove the final link from the Puma 560 robot model and repeat the exercise.
10. Walking robot (page 221)
    a) Shorten the reset trajectory by reducing the leg lift during reset.
    b) Increase the stride of the legs.
    c) Figure out how to steer the robot by changing the stride length on one side of the body.
    d) Change the gait so the robot moves sideways like a crab.
    e) Add another pair of legs. Change the gait to reset two legs or three legs at a time.
    f) Currently in the simulation the legs move but the body does not move forward. Modify the simulation so the body moves.
    g) A robot hand comprises a number of fingers, each of which is a small serial-link manipulator. Create a model of a hand with 2, 3 or 5 fingers and animate the finger motion.
11. Create a simulation with two robot arms next to each other, whose end-effectors are holding a basketball at diametrically opposite points in the horizontal plane. Write code to move the robots so as to rotate the ball about the vertical axis.
12. Create STL files to represent your own robot and integrate them into the Toolbox. Check out the code in `SerialLink.plot3d`.

8

Manipulator Velocity

A robot's end-effector moves in Cartesian space with a translational and rotational velocity – a spatial velocity. However that velocity is a consequence of the velocities of the individual robot joints. In this chapter we introduce the relationship between the velocity of the joints and the spatial velocity of the end-effector.

The 3-dimensional end-effector pose $\xi_E \in SE(3)$ has a rate of change which is represented by a 6-vector known as a spatial velocity that was introduced in Sect. 3.1. The joint rate of change and the end-effector velocity are related by the manipulator Jacobian matrix which is a function of manipulator configuration.

Section 8.1 uses a simple planar robot to introduce the manipulator Jacobian and then extends this concept to more general robots. Section 8.2 discusses the numerical properties of the Jacobian matrix which are shown to provide insight into the dexterity of the manipulator – the directions in which it can move easily and those in which it cannot. In Sect. 8.3 we use the inverse Jacobian to generate Cartesian paths without requiring inverse kinematics, and this can be applied to over- and under-actuated robots which are discussed in Sect. 8.4. Section 8.5 demonstrates how the Jacobian transpose is used to transform forces from the end-effector to the joints and between coordinate frames. Finally, in Sect. 8.6 the numeric inverse kinematic solution, used in the previous chapter, is introduced and its dependence on the Jacobian matrix is fully described.

8.1 Manipulator Jacobian

In the last chapter we discussed the relationship between joint coordinates and end-effector pose – the manipulator kinematics. Now we investigate the relationship between the rate of change of these quantities – between joint *velocity* and *velocity* of the end-effector. This is called the velocity or differential kinematics of the manipulator.

8.1.1 Jacobian in the World Coordinate Frame

We illustrate the basics with our now familiar 2-dimensional example, see Fig. 8.1, this time defined using Denavit-Hartenberg notation

```
>> mdl_planar2_sym
>> p2
two link:: 2 axis, RR, stdDH
+---+------------+-----------+-----------+-----------+-----------+
| j |     theta  |        d  |        a  |     alpha |    offset |
+---+------------+-----------+-----------+-----------+-----------+
|  1|         q1 |        0  |       a1  |        0  |        0  |
|  2|         q2 |        0  |       a2  |        0  |        0  |
+---+------------+-----------+-----------+-----------+-----------+
```

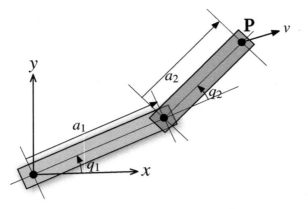

Fig. 8.1.
Two-link robot showing the end-effector position $p = (x, y)$ and the Cartesian velocity vector $\nu = \mathrm{d}p / \mathrm{d}t$

and define two real-valued symbolic variables to represent the joint angles

```
>> syms q1 q2 real
```

and then compute the forward kinematics using these

```
>> TE = p2.fkine( [q1 q2] );
```

The position of the end-effector $p = (x, y) \in \mathbb{R}^2$ is▶

```
>> p = TE.t;   p = p(1:2)
p =
  a2*cos(q1 + q2) + a1*cos(q1)
  a2*sin(q1 + q2) + a1*sin(q1)
```

The Toolbox considers robot pose in 3-dimensions using SE(3). This robot operates in a plane, a subset of SE(3), so we select $p = (x, y)$.

and we compute the derivative of p with respect to the joints variables q. Since p and q are both vectors the derivative

$$\frac{\mathrm{d}p}{\mathrm{d}q} = J(q) \tag{8.1}$$

will be a matrix – a Jacobian matrix

```
>> J = jacobian(p, [q1 q2])
J =
[ - a2*sin(q1 + q2) - a1*sin(q1),  -a2*sin(q1 + q2)]
[   a2*cos(q1 + q2) + a1*cos(q1),   a2*cos(q1 + q2)]
```

which is typically denoted by the symbol J and in this case is 2×2.

To determine the relationship between joint *velocity* and end-effector *velocity* we rearrange Eq. 8.1 as

$$\mathrm{d}p = J(q)\mathrm{d}q$$

and divide through by $\mathrm{d}t$ to obtain

$$\frac{\mathrm{d}p}{\mathrm{d}t} = J(q)\frac{\mathrm{d}q}{\mathrm{d}t}$$
$$\dot{p} = J(q)\dot{q}$$

The Jacobian matrix maps velocity from the joint coordinate or configuration space to the end-effector's Cartesian coordinate space and is itself a function of the joint coordinates.

More generally we write the forward kinematics in functional form, Eq. 7.4, as

$$^0\xi = \mathcal{K}(q)$$

A Jacobian is the matrix equivalent of the derivative – the derivative of a vector-valued function of a vector with respect to a vector. If $y = f(x)$ and $x \in \mathbb{R}^n$ and $y \in \mathbb{R}^m$ then the Jacobian is the $m \times n$ matrix

$$J = \frac{\partial f}{\partial x} = \begin{pmatrix} \frac{\partial y_1}{\partial x_1} & \cdots & \frac{\partial y_1}{\partial x_n} \\ \vdots & \ddots & \vdots \\ \frac{\partial y_m}{\partial x_1} & \cdots & \frac{\partial y_m}{\partial x_n} \end{pmatrix}$$

The Jacobian is named after Carl Jacobi, and more details are given in Appendix E.

and taking the derivative we write

$$^0\nu = {}^0J(q)\dot{q} \tag{8.2}$$

where $^0\nu = (v_x, v_y, v_z, \omega_x, \omega_y, \omega_z) \in \mathbb{R}^6$ is the spatial velocity, as discussed in Sect. 3.1.1, of the end-effector in the world frame and comprises translational and rotational velocity components. The matrix $^0J(q) \in \mathbb{R}^{6 \times N}$ is the manipulator Jacobian or the geometric Jacobian. This relationship is sometimes referred to as the instantaneous forward kinematics.

For a realistic 3-dimensional robot this Jacobian matrix can be numerically computed by the `jacob0` method of the `SerialLink` object, based on its Denavit-Hartenberg parameters. For the Puma robot in the pose shown in Fig. 8.2 the Jacobian is

```
>> J = p560.jacob0(qn)
J =
    0.1501    0.0144    0.3197         0         0         0
    0.5963    0.0000    0.0000         0         0         0
         0    0.5963    0.2910         0         0         0
         0   -0.0000   -0.0000    0.7071   -0.0000   -0.0000
         0   -1.0000   -1.0000   -0.0000   -1.0000   -0.0000
    1.0000    0.0000    0.0000   -0.7071    0.0000   -1.0000
```

τ is the task space of the robot, typically $\tau \subset \mathbf{SE}(3)$, and $\mathcal{C} \subset \mathbb{R}^N$ is the configuration or joint space of the robot where N is the number of robot joints.

and is a matrix with dimensions dim $\mathcal{T} \times$ dim $\mathcal{C}$ – in this case 6×6◄. Each row corresponds to a Cartesian degree of freedom. Each column corresponds to a joint – it is the end-effector spatial velocity created by unit velocity of the corresponding joint. In this configuration, motion of joint 1, the first column, causes motion in the world x- and y-directions and rotation about the z-axis. Motion of joints 2 and 3 cause motion in the world x- and z-directions and negative rotation about the y-axis.

Physical insight comes from Fig. 8.2 which shows the joint axes in space. Alternatively you could use the `teach` method

```
>> p560.teach(qn)
```

and jog the various joint angles and observe the change in end-effector pose.

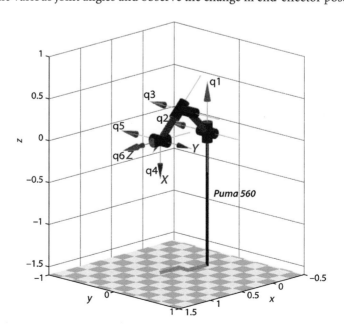

Fig. 8.2.
Puma robot in its nominal pose qn. The end-effector z-axis points in the world x-direction, and the x-axis points downward

Carl Gustav Jacob Jacobi (1804–1851) was a Prussian mathematician. He obtained a Doctor of Philosophy degree from Berlin University in 1825. In 1827 he was appointed professor of mathematics at Königsberg University and held this position until 1842 when he suffered a breakdown from overwork.

Jacobi wrote a classic treatise on elliptic functions in 1829 and also described the derivative of m functions of n variables which bears his name. He was elected a foreign member of the Royal Swedish Academy of Sciences in 1836. He is buried in Cementary I of the Trinity Church (Dreifaltigkeitskirche) in Berlin.

The 3×3 block of zeros in the top right indicates that motion of the wrist joints have no effect on the end-effector translational motion – this is a consequence of the spherical wrist and the default zero-length tool.

8.1.2   Jacobian in the End-Effector Coordinate Frame

The Jacobian computed by the method `jacob0` maps joint velocity to the end-effector spatial velocity expressed in the *world coordinate* frame. To obtain the spatial velocity in the end-effector coordinate frame we introduce the velocity transformation Eq. 3.4 from the world frame to the end-effector frame which is a function of the end-effector pose

$$
{}^E v = {}^E J_0\left({}^E \xi_0\right){}^0 J(q)\dot{q} = \underbrace{\begin{pmatrix} {}^E R_0 & 0_{3\times3} \\ 0_{3\times3} & {}^E R_0 \end{pmatrix} {}^0 J(q)}\ \dot{q} = {}^E J(q)\dot{q}
$$

which results in a new Jacobian for end-effector velocity.▶

In the Toolbox this Jacobian is computed by the method `jacobe` and for the Puma robot at the pose used above is

For historical reasons the Toolbox implementation computes the end-effector Jacobian directly and applies a velocity transform for the world frame Jacobian.

```
>> p560.jacobe(qn)
ans =
  -0.0000   -0.5963   -0.2910        0        0        0
   0.5963    0.0000    0.0000        0        0        0
   0.1500    0.0144    0.3197        0        0        0
  -1.0000         0         0   0.7071        0        0
  -0.0000   -1.0000   -1.0000  -0.0000  -1.0000        0
  -0.0000    0.0000    0.0000   0.7071   0.0000   1.0000
```

8.1.3   Analytical Jacobian

In Eq. 8.2 the spatial velocity was expressed in terms of translational and angular velocity vectors, however angular velocity is not a very intuitive concept. For some applications it can be more intuitive to consider the rotational velocity in terms of rates of change of roll-pitch-yaw angles or Euler angles. Analytical Jacobians are those where the rotational velocity is expressed in a representation other than angular velocity, commonly in terms of triple-angle rates.

Consider the case of XYZ roll-pitch-yaw angles $\Gamma = (\theta_r, \theta_p, \theta_y)$ for which the rotation matrix is

$$
R = R_x\left(\theta_y\right)R_y\left(\theta_p\right)R_z\left(\theta_r\right)
$$

$$
= \begin{pmatrix}
c\theta_p c\theta_r & -c\theta_p s\theta_r & s\theta_p \\
c\theta_y s\theta_r + c\theta_r s\theta_p s\theta_y & -s\theta_p s\theta_y s\theta_r + c\theta_y c\theta_r & -c\theta_p s\theta_y \\
s\theta_y s\theta_r - c\theta_y c\theta_r s\theta_p & c\theta_y s\theta_p s\theta_r + c\theta_r s\theta_y & c\theta_p c\theta_y
\end{pmatrix}
$$

where we use the shorthand $c\theta$ and $s\theta$ to mean $\cos\theta$ and $\sin\theta$ respectively. With some effort we can write the derivative $\dot{R}$ and recalling Eq. 3.1

$$\dot{R} = [\omega]_{\times} R$$

we can solve for ω in terms of roll-pitch-yaw angles and their rates to obtain

$$\begin{pmatrix} \omega_x \\ \omega_y \\ \omega_z \end{pmatrix} = \begin{pmatrix} s\theta_p\dot{\theta}_r + \dot{\theta}_y \\ -c\theta_p s\theta_y\dot{\theta}_r + c\theta_y\dot{\theta}_p \\ c\theta_p c\theta_y\dot{\theta}_r + s\theta_y\dot{\theta}_p \end{pmatrix}$$

which can be factored as

$$\omega = \begin{pmatrix} s\theta_p & 0 & 1 \\ -c\theta_p s\theta_y & c\theta_y & 0 \\ c\theta_p c\theta_y & s\theta_y & 0 \end{pmatrix}\begin{pmatrix} \dot{\theta}_r \\ \dot{\theta}_p \\ \dot{\theta}_y \end{pmatrix}$$

and written concisely as

$$\omega = A(\boldsymbol{\Gamma})\dot{\boldsymbol{\Gamma}}$$

This matrix A is itself a Jacobian that maps XYZ roll-pitch-yaw angle rates to angular velocity. It can be computed by the Toolbox function

```
>> rpy2jac(0.1, 0.2, 0.3)
ans =
    0.1987         0    1.0000
   -0.2896    0.9553         0
    0.9363    0.2955         0
```

where the arguments are the roll, pitch and yaw angles. The analytical Jacobian is

$$J_a(q) = \begin{pmatrix} I_{3\times3} & 0_{3\times3} \\ 0_{3\times3} & A^{-1}(\boldsymbol{\Gamma}) \end{pmatrix} J(q)$$

provided that A is not singular. A is singular when $\cos\phi = 0$ or pitch angle $\phi = \pm\frac{\pi}{2}$ and is referred to as a representational singularity. A similar approach can be taken for Euler angles using the corresponding function eul2jac.

The analytical Jacobian can be computed by passing an extra argument to the Jacobian function jacob0, for example

```
>> p560.jacob0(qn, 'eul');
```

to specify the Euler angle analytical form.

Another useful analytical Jacobian expresses angular rates as the rate of change of exponential coordinates $s = \hat{v}\theta \in \mathbf{so}(3)$

$$\omega = A(s)\dot{s}$$

where

$$A(s) = I - \frac{1-\cos\theta}{\theta}[\hat{v}]_{\times} + \frac{\theta - \sin\theta}{\theta}[\hat{v}]_{\times}^2$$

Implemented by the Toolbox functions trlog and tr2rotvec or the SE3 method torotvec.

and $\hat{v}$ and θ can be determined from the end-effector rotation matrix via the matrix logarithm. ◄

8.2   Jacobian Condition and Manipulability

We have discussed how the Jacobian matrix maps joint rates to end-effector Cartesian velocity but the inverse problem has strong practical use – what joint velocities are needed to achieve a required end-effector Cartesian velocity? We can invert Eq. 8.2 and write

$$\dot{q} = J(q)^{-1}\nu \qquad (8.3)$$

provided that J is square and nonsingular. The Jacobian is a dim $\mathcal{T} \times$ dim $\mathcal{C}$ matrix so in order to achieve a square Jacobian matrix a robot operating in the task space $\mathcal{T} \subset \mathbf{SE}(3)$, which has 6 spatial degrees-of-freedom, requires a robot with 6 joints.

8.2.1   Jacobian Singularities

A robot configuration q at which $\det(J(q)) = 0$ is described as singular or degenerate. Singularities occur when the robot is at maximum reach or when one or more axes become aligned resulting in the loss of degrees of freedom – the gimbal lock problem again.

For example at the Puma's *ready* pose the Jacobian

```
>> J = p560.jacob0(qr)
J =
    0.1500   -0.8636   -0.4318        0        0        0
    0.0203    0.0000    0.0000        0        0        0
         0    0.0203    0.0203        0        0        0
         0         0         0        0        0        0
         0   -1.0000   -1.0000        0  -1.0000        0
    1.0000    0.0000    0.0000   1.0000   0.0000   1.0000
```

is singular

```
>> det(J)
ans =
     0
```

Digging a little deeper we see that the Jacobian rank is only

```
>> rank(J)
ans =
     5
```

compared to a maximum of six for a 6×6 matrix. The rank deficiency of one means that one column is equal to a linear combination of other columns. Looking at the Jacobian it is clear that columns 4 and 6 are identical meaning that two of the wrist joints (joints 4 and 6) are aligned. This leads to the loss of one degree of freedom since motion of these joints results in the same Cartesian velocity, leaving motion in one Cartesian direction unaccounted for.▶ The function `jsingu` performs this analysis automatically, for example

For the Puma 560 robot arm joints 4 and 6 are the only ones that can become aligned and lead to singularity. The offset distances, d_j and a_j, between links prevents other axes becoming aligned.

```
>> jsingu(J)
1 linearly dependent joints:
  q6 depends on: q4
```

indicating velocity of q_6 can be expressed completely in terms of the velocity of q_4.

However if the robot is close to, but not actually at, a singularity we encounter problems where some Cartesian end-effector velocities require very high joint rates▶ – at the singularity those rates will go to infinity. We can illustrate this by choosing a configuration slightly away from `qr` which we just showed was singular. We set q_5 to a small but nonzero value of 5 deg

We observed this effect in Fig. 7.12.

```
>> qns = qr; qns(5) = 5 * pi/180
qns =
         0    1.5708   -1.5708        0    0.0873        0
```

and the Jacobian is now

```
>> J=p560.jacob0(qns);
```

To achieve relatively slow end-effector motion of 0.1 m s^{-1} in the z-direction requires

```
>> qd = inv(J)*[0 0 0.1 0 0 0]';
>> qd'
ans =    -0.0000    -4.9261     9.8522     0.0000    -4.9261          0
```

very high-speed motion of the shoulder and elbow – the elbow would have to move at 9.85 rad s^{-1} or nearly 600 deg s^{-1}. The reason is that although the robot is no longer at a singularity, the determinant of the Jacobian is still very small

```
>> det(J)
ans =
    -1.5509e-05
```

Alternatively we can say that its condition number is very high

```
>> cond(J)
ans =
    235.2498
```

and the Jacobian is *poorly conditioned*.

However for some motions, such as rotation in this case, the poor condition of the Jacobian is not problematic. If we wished to rotate the tool about the y-axis then

```
>> qd = inv(J)*[0 0 0 0 0.2 0]';
>> qd'
ans =     0.0000    -0.0000          0     0.0000    -0.2000          0
```

the required joint rates are very modest.

This particular joint configuration is therefore good for certain motions but poor for others.

8.2.2 Manipulability

Consider the set of generalized joint velocities with a unit norm

$$\dot{q}^T \dot{q} = 1$$

which lie on the surface of a hypersphere in the N-dimensional joint velocity space. Substituting Eq. 8.3 we can write

$$\nu^T \left(J(q)J(q)^T \right)^{-1} \nu = 1 \tag{8.4}$$

which is the equation of points on the surface of an ellipsoid within the dim $\mathcal{T}$-dimensional end-effector velocity space. If this ellipsoid is close to spherical, that is, its radii are of the same order of magnitude then all is well – the end-effector can achieve arbitrary Cartesian velocity. However if one or more radii are very small this indicates that the end-effector cannot achieve velocity in the directions corresponding to those small radii.

We will load the numerical, rather than symbolic model, for the planar robot arm of Fig. 8.1

```
>> mdl_planar2
```

which allows us to plot the velocity ellipse for an arbitrary pose

```
>> p2.vellipse([30 40], 'deg')
```

We can also interactively explore how its shape changes with configuration by

```
>> p2.teach('callback', @(r,q) r.vellipse(q), 'view', 'top')
```

which is shown in Fig. 8.3.

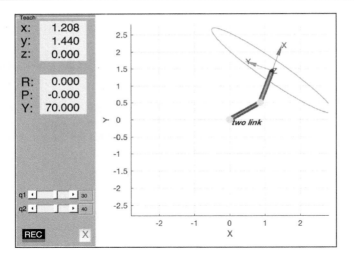

Fig. 8.3.
Two-link robot with overlaid velocity ellipse

For a robot with a task space $\mathcal{T} \subset \mathbf{SE}(3)$ Eq. 8.4 describes a 6-dimensional ellipsoid which is impossible to visualize. However we can extract that part of the Jacobian relating to *translational* velocity▸ in the world frame

Since we can only plot three dimensions.

```
>> J = p560.jacob0(qns);
>> J = J(1:3, :);
```

and plot the corresponding velocity ellipsoid

```
>> plot_ellipse(J*J')
```

which is shown in Fig. 8.4a. The Toolbox provides a shorthand for this

```
>> p560.vellipse(qns, 'trans');
```

We see that the end-effector can achieve higher velocity in the *y*- and *z*-directions than in the *x*-direction. Ellipses and ellipsoids are discussed in more detail in Sect. C.1.4.

The *rotational* velocity ellipsoid for the near singular case

```
>> p560.vellipse(qns, 'rot')
```

is shown in Fig. 8.4b and is an elliptical plate with almost zero thickness.▸ This indicates an inability to rotate about the direction corresponding to the small radius, which in this case is rotation about the *x*-axis. This is the degree of freedom that was lost – both joints 4 and 6 provide rotation about the world *z*-axis, joint 5 provides provides rotation about the world *y*-axis, but none allow rotation about the world *x*-axis.

This is much easier to see if you change the viewpoint interactively.

The shape of the ellipsoid describes how *well-conditioned* the manipulator is for making certain motions. Manipulability is a succinct scalar measure that describes how spherical the ellipsoid is, for instance the ratio of the smallest to the largest radius.▸ The Toolbox method `maniplty` computes Yoshikawa's manipulability measure

The radii are the square roots of the eigenvalues of the $J(q)J(q)^T$ as discussed in Sect. C.1.4.

$$m = \sqrt{\det\left(\boldsymbol{JJ}^T\right)}$$

which is proportional to the volume of the ellipsoid. For example

```
>> m = p560.maniplty(qr)
m =
     0
```

indicates a singularity. If the method is called with no output arguments it displays the volume of the translational and rotational velocity ellipsoids

```
>> p560.maniplty(qr)
Manipulability: translation 0.00017794, rotation 0
```

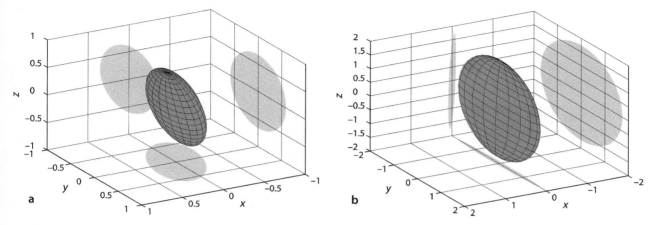

Fig. 8.4. End-effector velocity ellipsoids. **a** Translational velocity ellipsoid for the nominal pose (m s^{-1}); **b** rotational velocity ellipsoid for a near singular pose (rad s^{-1}), the ellipsoid is an elliptical plate

The manipulability measure combines translational and rotational velocity information which have different units. The options `'trans'` and `'rot'` can be used to compute manipulability on just the translational or rotational velocity respectively.

which indicates very poor manipulability for translation and zero for rotation. At the nominal pose the manipulability is higher◄

```
>> p560.maniplty(qn)
Manipulability: translation 0.111181, rotation 2.44949
```

In practice we find that the seemingly large workspace of a robot is greatly reduced by joint limits, self collision, singularities and regions of reduced manipulability. The manipulability measure discussed here is based only on the kinematics of the mechanism. The fact that it is easier to move a small wrist joint than the larger waist joint suggests that mass and inertia should be taken into account and such manipulability measures are discussed in Sect. 9.2.7.

8.3 Resolved-Rate Motion Control

Resolved-rate motion control is a simple and elegant algorithm to generate straight line motion by exploiting Eq. 8.3

$$\dot{q} = J(q)^{-1}\nu$$

to map or *resolve* desired Cartesian velocity to joint velocity without explicitly requiring inverse kinematics as we used earlier. For now we will assume that the Jacobian is square (6 × 6) and nonsingular but we will relax these constraints later.

The motion control scheme is typically implemented in discrete-time form as

$$\dot{q}^*\langle k\rangle = J(q\langle k\rangle)^{-1}\nu^* \qquad (8.5)$$
$$q^*\langle k+1\rangle \leftarrow q\langle k\rangle + \delta_t\dot{q}^*\langle k\rangle$$

where δ_t is the sample interval. The first equation computes the required joint velocity as a function of the current joint configuration and the desired end-effector velocity ν^*. The second performs forward rectangular integration to give the desired joint angles for the next time step, $q^*\langle k+1\rangle$.

An example of the algorithm is implemented by the Simulink® model

```
>> sl_rrmc
```

In this model we assume that the robot is perfect, that is, the actual joint angles are equal to the desired joint angles q^*. The issue of tracking error is discussed in Sect. 9.1.7.

shown in Fig. 8.5. The Cartesian velocity is a constant 0.05 m s^{-1} in the y-direction. The `Jacobian` block has as its input the current manipulator joint angles and outputs a 6 × 6 Jacobian matrix. This is inverted and multiplied by the desired velocity to form the desired joint rates. The robot is modeled by a discrete-time integrator – an ideal velocity controller.◄

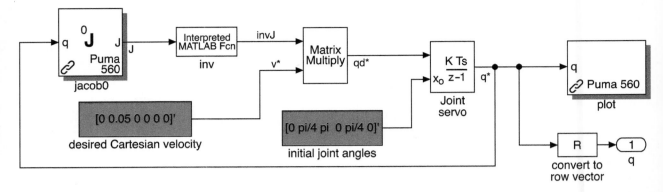

Running the simulation

```
>> r = sim('sl_rrmc');
```

Fig. 8.5. The Simulink® model sl_rrmc for resolved-rate motion control for constant end-effector velocity

we see an animation of the manipulator end-effector moving at constant velocity in Cartesian space. Simulation results are returned in the simulation object r from which we extract time and joint coordinates

```
>> t = r.find('tout');
>> q = r.find('yout');
```

We apply forward kinematics to determine the end-effector position

```
>> T = p560.fkine(q);
>> xyz = transl(T);
```

which we then plot▸ as a function of time

```
>> mplot(t, xyz(:,1:3))
```

The function mplot is a Toolbox utility that plots columns of a matrix in separate subgraphs.

which is shown in Fig. 8.6a. The Cartesian motion is 0.05 m s⁻¹ in the y-direction as demanded but we observe some small and unwanted motion in the x- and z-directions.
  The motion of the first three joints

```
>> mplot(t, q(:,1:3))
```

is shown in Fig. 8.6b and are not linear with time – reflecting the changing kinematic configuration of the arm.
  The approach just described, based purely on integration, suffers from an accumulation of error which we observed as the unwanted x- and z-direction motion in Fig. 8.6a. We can eliminate this by changing the algorithm to a *closed-loop* form based on the difference between the desired and actual pose

$$\dot{q}^*\langle k\rangle \leftarrow K_p J\big(q\langle k\rangle\big)^{-1} \Delta\Big(\mathcal{K}\big(q\langle k\rangle\big), \xi^*\langle k\rangle\Big) \tag{8.6}$$

where K_p is a proportional gain, $\Delta(\cdot) \in \mathbb{R}^6$ is a spatial displacement▸ and the desired pose $\xi^*\langle k\rangle$ is a function of time.
  A Simulink example to demonstrate this for a circular path is

See Sect. 3.1.4 for definition.

```
>> sl_rrmc2
```

shown in Fig. 8.7 and the tool of a Puma 560 robot traces out a circle of radius 50 mm. The x-, y- and z-coordinates as a function of time are computed and converted to a homogeneous transformation by the blocks in the grey area. The *difference* between the desired pose and the current pose from forward kinematics using the $\Delta(\cdot)$ operator is computed by the tr2delta block. The result is a spatial displacement, a translation and a rotation described by a 6-vector which is used as

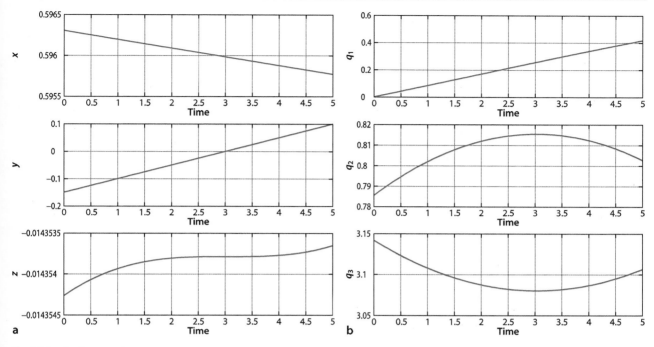

Fig. 8.6. Resolved-rate motion control, Cartesian and joint coordinates versus time. **a** Cartesian end-effector motion. Note the small, but unwanted motion in the x- and z-directions; **b** joint motion

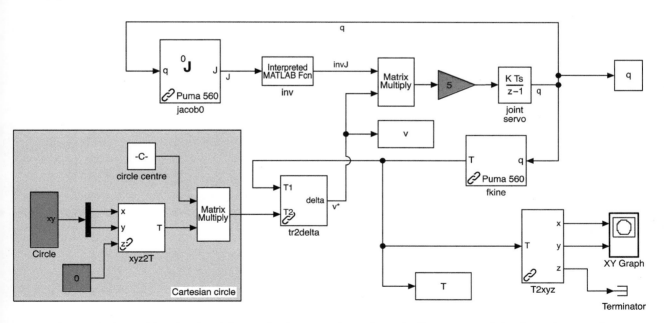

Fig. 8.7. The Simulink® model sl_rrmc2 for closed-loop resolved-rate motion control with circular end-effector motion

the desired spatial velocity to drive the end-effector toward the desired pose. The Jacobian matrix is computed from the current manipulator joint angles and is inverted so as to transform the desired spatial velocity to joint angle rates. These are scaled by a proportional gain, to become the desired joint-space velocity that will correct any Cartesian error.

8.3.1   Jacobian Singularity

For the case of a square Jacobian where $\det(J(q)) = 0$ we cannot solve Eq. 8.3 direct-ly. One strategy to deal with singularity is to replace the inverse with the damped inverse

$$\dot{q} = (J(q) + \lambda I)^{-1} \nu$$

where λ is a small constant added to the diagonal which places a *floor* under the de-terminant. This will introduces some error in $\dot{q}$, which integrated over time could lead to a significant discrepancy in tool position but the closed-loop resolved-rate motion scheme of Eq. 8.6 would minimize this.

An alternative is to use the pseudo-inverse of the Jacobian J^+ which has the property

$$J^+ J = I$$

just as the inverse does. It is defined as

$$J^+ = (J^T J)^{-1} J^T$$

and is readily computed using the MATLAB® builtin function `pinv`.▶ The solution

This is the left generalized- or pseudoin-verse, see Sect. F.1.1 for more details.

$$\dot{q} = J(q)^+ \nu$$

provides a least-squares solution for which $\|J\dot{q} - \nu\|$ is smallest.▶

Yet another approach is to delete from the Jacobian all those columns that are lin-early dependent on other columns. This is effectively locking the joints correspond-ing to the deleted columns and we now have an under-actuated system which we treat as per the next section.

A matrix expression like $v = J\dot{q}$ is a sys-tem of scalar equations which we can solve for $\dot{q}$. At singularity some of the equations are the same, leading to more unknowns than equations, and therefore an infinite number of solutions. The pseudo-inverse computes a solution that satisfies the equation and has the minumum norm.

8.4   Under- and Over-Actuated Manipulators

So far we have assumed that the Jacobian is square. For the nonsquare cases it is help-ful to consider the velocity relationship

$$\nu = J(q)\dot{q}$$

in the diagrammatic form shown in Fig. 8.8. The Jacobian is a $6 \times N$ matrix, the joint velocity is an N-vector, and ν is a 6-vector.

The case of $N < 6$ is referred to as an under-actuated robot, and $N > 6$ is over-ac-tuated or redundant. The under-actuated case cannot be solved because the system of equations is under-constrained but the system can be *squared up* by deleting some rows of ν and J – accepting that some Cartesian degrees of freedom are not controllable given the low number of joints. For the over-actuated case the system of equations is under-constrained and the best we can do is find a least-squares solution as described in the previous section. Alternatively we can *square up* the Jacobian to make it invert-ible by deleting some columns – effectively *locking* the corresponding joints.

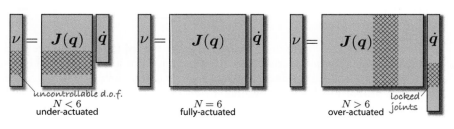

Fig. 8.8.
Schematic of Jacobian, ν and $\dot{q}$ for different cases of N. The *hatched* areas represent matrix regions that could be deleted in order to create a square sub-system capable of solution

8.4.1   Jacobian for Under-Actuated Robot

An under-actuated robot has $N < 6$, and a Jacobian that is taller than it is wide. For example a 2-joint manipulator at a nominal pose

```
>> mdl_planar2
>> qn = [1 1];
```

has the Jacobian

```
>> J = p2.jacob0(qn)
J =
   -1.7508   -0.9093
    0.1242   -0.4161
         0         0
         0         0
         0         0
    1.0000    1.0000
```

We cannot solve the inverse problem Eq. 8.3 using the pseudo-inverse since it will attempt to satisfy motion constraints that the manipulator cannot meet. For example the desired motion of 0.1 m s^{-1} in the x-direction gives the required joint velocity

```
>> qd = pinv(J) * [0.1 0 0 0 0 0]'
qd =
   -0.0698
    0.0431
```

which results in end-effector velocity

```
>> xd = J*qd;
>> xd'
ans =
    0.0829   -0.0266        0        0        0   -0.0266
```

This has the desired motion in the x-direction but undesired motion in y-axis translation and z-axis rotation. The end-effector rotation cannot be independently controlled (since it is a function of q_1 and q_2) yet this solution has taken it into account in the least-squares solution.

We have to confront the reality that we have *only* two degrees of freedom which we will use to control just v_x and v_y. We rewrite Eq. 8.2 in partitioned form as

$$
\begin{pmatrix} v_x \\ v_y \\ v_z \\ \omega_x \\ \omega_y \\ \omega_z \end{pmatrix} = \left(\frac{\mathbf{J}_{xy}}{\mathbf{J}_0} \right) \begin{pmatrix} \dot{q}_1 \\ \dot{q}_2 \end{pmatrix}
$$

and taking the top partition, the first two rows, we write

$$
\begin{pmatrix} v_x \\ v_y \end{pmatrix} = \mathbf{J}_{xy} \begin{pmatrix} \dot{q}_1 \\ \dot{q}_2 \end{pmatrix}
$$

where $\mathbf{J}_{xy}$ is a 2×2 matrix. We invert this

$$
\begin{pmatrix} \dot{q}_1 \\ \dot{q}_2 \end{pmatrix} = \mathbf{J}_{xy}^{-1} \begin{pmatrix} v_x \\ v_y \end{pmatrix}
$$

which we can solve if $\det(\mathbf{J}_{xy}) \neq 0$.

```
>> Jxy = J(1:2,:);
>> qd = inv(Jxy) * [0.1 0]'
qd =
   -0.0495
   -0.0148
```

which results in end-effector velocity

```
>> xd = J*qd;
>> xd'
ans =
     0.1000     0.0000          0          0          0    -0.0642
```

We have achieved the desired x-direction motion with no unwanted motion apart from the z-axis rotation which is unavoidable – we have used the two degrees of freedom to control x- and y-translation, not z-rotation.

8.4.2   Jacobian for Over-Actuated Robot

An over-actuated or redundant robot has $N > 6$, and a Jacobian that is wider than it is tall. In this case we rewrite Eq. 8.3 to use the left pseudo-inverse

$$\dot{q} = J(q)^+ \nu \tag{8.7}$$

which, of the infinite number of solutions possible, will yield the one for which $\|\dot{q}\|$ is smallest – the minimum-norm solution.

We will demonstrate this for the left arm of the Baxter robot from Sect. 7.2.2.4 at a nominal pose

```
>> mdl_baxter
>> TE = SE3(0.8, 0.2, -0.2) * SE3.Ry(pi);
>> q = left.ikine(TE)
```

and its Jacobian

```
>> J = jacob0(left, q);
>> about J
J [double] : 6x7 (336 bytes)
```

is a 6×7 matrix. Now consider that we want the end-effector to move at 0.2 m s^{-1} in the x-, y- and z-directions. Using Eq. 8.7 we compute the required joint rates

```
>> xd = [0.2 0.2 0.2 0 0 0]';
>> qd = pinv(J) * xd;
>> qd'
ans =
     0.0895   -0.0464   -0.4259    0.6980   -0.4248    1.0179    0.2998
```

We see that all joints have nonzero velocity and contribute to the desired end-effector motion.▶

This Jacobian has seven columns and a rank of six

```
>> rank(J)
ans =
     6
```

and therefore a null space▶ whose basis has just one vector

```
>> N = null(J)
N =
   -0.2244
   -0.1306
    0.6018
    0.0371
   -0.7243
    0.0653
    0.2005
```

In the case of a Jacobian matrix any joint velocity that is a linear combination of its null-space vectors will result in *no* end-effector motion. For this robot there is only one vector and we can show that this *null-space joint motion* causes no end-effector motion

```
>> norm( J * N(:,1))
ans =
   2.6004e-16
```

If the robot end-effector follows a repetitive path using RRMC the joint angles may *drift* over time and *not* follow a repetitive path, potentially moving toward joint limits. We can use null-space control to provide additional constraints to prevent this.

See Appendix B.

This is remarkably useful because it allows Eq. 8.7 to be written as

$$\dot{q} = \underbrace{J(q)^+ \nu}_{\text{end-effector motion}} + \underbrace{NN^+ \dot{q}_{\text{null}}}_{\text{null-space motion}}$$ (8.8)

where the matrix $NN^+ \in \mathbb{R}^{N \times N}$ *projects* the desired joint motion into the null space so that it will not affect the end-effector Cartesian motion, allowing the two motions to be superimposed.

Null-space motion can be used for highly-redundant robots to avoid collisions between the links and obstacles (including other links), or to keep joint coordinates away from their mechanical limit stops. Consider that in addition to the desired Cartesian velocity xd we wish to simultaneously increase joint 5 in order to move the arm away from some obstacle. We set a desired joint velocity

```
>> qd_null = [0 0 0 0 1 0 0]';
```

and project it into the null space

```
>> qp = N * pinv(N) * qd_null;
>> qp'
   0.1625    0.0946   -0.4359   -0.0269    0.5246   -0.0473   -0.1452
```

A scaling has been introduced but this joint velocity, or a scaled-up version of, this will increase the joint 5 angle without changing the end-effector pose. Other joints move as well – they provide the required compensating motion in order that the end-effector pose is not disturbed as shown by

```
>> norm( J * qp)
ans =
   1.9541e-16
```

A highly redundant snake robot like that shown in Fig. 8.9 would have a null space with 14 dimensions (20-6). This can be used to control the shape of the arm which is critical when moving within confined spaces.

Fig. 8.9.
20-DOF snake-robot arm: 2.5 m reach, 90 mm diameter and payload capacity of 25 kg (image courtesy of OC Robotics)

8.5   Force Relationships

In Sect. 3.2.2 we introduced wrenches $W = (f_x, f_y, f_z, m_x, m_y, m_z) \in \mathbb{R}^6$ which are a vector of forces and moments.

8.5.1   Transforming Wrenches to Joint Space

The manipulator Jacobian transforms joint velocity to an end-effector spatial velocity according to Eq. 8.2 and the Jacobian transpose transforms a wrench applied at the end-effector to torques and forces experienced at the joints►

$$Q = {}^0J(q)^{T}\,{}^0W \tag{8.9}$$

Derived through the principle of virtual work, see for instance Spong et al. (2006, sect. 4.10).

where W is a wrench in the world coordinate frame and Q is the generalized joint force vector. The elements of Q are joint torque or force for revolute or prismatic joints respectively.

The mapping for velocity, from end-effector to joints, involves the inverse Jacobian which can potentially be singular. The mapping of forces and torques, from end-effector to joints, is different – it involves the transpose of the Jacobian which can never be singular. We exploit this property in the next section to solve the inverse-kinematic problem numerically.

If the wrench is defined in the end-effector coordinate frame then we use instead

$$Q = {}^EJ(q)^{T}\,{}^EW \tag{8.10}$$

For the Puma 560 robot in its nominal pose, see Fig. 8.2, a force of 20 N in the world y-direction results in joint torques of

```
>> tau = p560.jacob0(qn)' * [0 20 0 0 0 0]';
>> tau'
ans =
   11.9261    0.0000    0.0000         0         0         0
```

The force pushes the arm *sideways* and only the waist joint will rotate in response – experiencing a torque of 11.93 N m due to a lever arm effect. A force of 20 N applied in the world x-direction results in joint torques of

```
>> tau = p560.jacob0(qn)' * [20 0 0 0 0 0]';
>> tau'
ans =
    3.0010    0.2871    6.3937         0         0         0
```

which is pulling the end-effector away from the base which results in torques being applied to the first three joints.

8.5.2   Force Ellipsoids

In Sect. 8.2.2 we introduced the velocity ellipse and ellipsoid which describe the directions in which the end-effector is best able to move. We can perform a similar analysis for the forces and torques at the end-effector – the end-effector wrench. We start with a set of generalized joint forces with a unit norm

$$Q^T Q = 1$$

and substituting Eq. 8.9 we can write

$$W^T\left(J(q)J(q)^T\right)W = 1$$

which is the equation of points on the surface of a 6-dimensional ellipsoid in the end-effector wrench space. For the planar robot arm of Fig. 8.1 we can plot this ellipse

```
>> p2.fellipse([30 40], 'deg')
```

or we can interactively explore how its shape changes with configuration by

```
>> p2.teach(qn, 'callback', @(r,q) r.fellipse(q), 'view', 'top')
```

If this ellipsoid is close to spherical, that is, its radii are of the same order of magnitude then the end-effector can achieve an arbitrary wrench. However if one or more radii are very small this indicates that the end-effector cannot exert a force along, or a moment about, the axes corresponding to those small radii.

The force and velocity ellipsoids provide complementary information about how well suited the configuration of the arm is to a particular task. We know from personal experience that to throw an object quickly we have our arm outstretched and orthogonal to the throwing direction, whereas to lift something heavy we hold our arms close in to our body.

8.6 Inverse Kinematics: a General Numerical Approach

In Sect. 7.2.2.1 we solved the inverse kinematic problem using an explicit solution that required the robot to have 6 joints and a spherical wrist. For the case of robots which do not meet this specification, for example those with more or less than 6 joints, we need to consider a numerical solution. Here we will develop an approach based on the forward kinematics and the Jacobian transpose which we can compute for any manipulator configuration since these functions have no singularities.

8.6.1 Numerical Inverse Kinematics

The principle is shown in Fig. 8.10 where the robot in its current configuration is drawn solidly and the desired configuration is faint. From the overlaid pose graph the error between actual ξ_E and desired pose ξ_E^* is ξ_Δ which can be described by a spatial displacement as discussed in Sect. 3.1.4

$$^E\boldsymbol{\Delta} = \Delta\left(\xi_E, \xi_E^*\right) = \left(\boldsymbol{t}, \hat{\boldsymbol{v}}\theta\right) \in \mathbb{R}^6$$

where the current pose is computed using forward kinematics $\xi_E = \mathcal{K}(\boldsymbol{q})$.

Imagine a *special* spring between the end-effector of the two poses which is pulling (and twisting) the robot's end-effector toward the desired pose with a wrench proportional to the spatial displacement

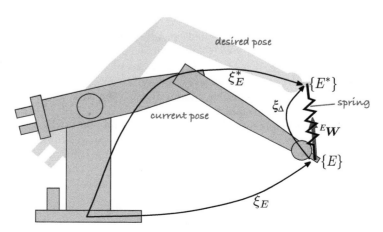

Fig. 8.10.
Schematic of the numerical inverse kinematic approach, showing the current ξ_E and the desired ξ_E^* manipulator pose

$$^{E}\boldsymbol{W} = \gamma\,^{E}\boldsymbol{\Delta} \tag{8.11}$$

which is *resolved* to generalized joint forces

$$\boldsymbol{Q} = {}^{E}\!\boldsymbol{J}(\boldsymbol{q})^{T}\,{}^{E}\boldsymbol{W}$$

using the Jacobian transpose Eq. 8.10. We assume that this virtual robot has no joint motors only viscous dampers so the joint velocity will be proportional to the applied forces

$$\dot{\boldsymbol{q}} = \boldsymbol{Q}/B$$

where B is the joint damping coefficient (assuming all dampers are the same). Putting all this together we can write

$$\dot{\boldsymbol{q}} = \frac{1}{B}\boldsymbol{J}(\boldsymbol{q})^{T}\,\Delta\!\left(\mathcal{K}(\boldsymbol{q}),\xi_{E}^{*}\right)$$

which gives the joint velocities that will drive the forward kinematic solution toward the desired end-effector pose. This can be solved iteratively by

$$\delta_{q}\langle k\rangle = \alpha\,\boldsymbol{J}(\boldsymbol{q}\langle k\rangle)^{T}\,\Delta\!\left(\mathcal{K}(\boldsymbol{q}\langle k\rangle),\xi_{E}^{*}\right) \tag{8.12}$$
$$\boldsymbol{q}\langle k+1\rangle \leftarrow \boldsymbol{q}\langle k\rangle + \delta_{q}\langle k\rangle$$

until the norm of the update $\|\delta_{q}\langle k\rangle\|$ is sufficiently small and where $\alpha > 0$ is a well-chosen constant. Since the solution is based on the Jacobian transpose rather than inverse the algorithm works when the Jacobian is nonsquare or singular. In practice however this algorithm is slow to converge and very sensitive to the choice of α.

More practically we can formulate this as a least-squares problem in the world coordinate frame and minimize the scalar cost

$$E = \boldsymbol{\Delta}^{T}\boldsymbol{M}\boldsymbol{\Delta}$$

where $\boldsymbol{M} = \mathrm{diag}(\boldsymbol{m}) \in \mathbb{R}^{6\times6}$ and $\boldsymbol{m}$ is the mask vector introduced in Sect. 7.2.2.3. The update becomes

$$\delta_{q}\langle k\rangle = \left(\boldsymbol{J}(\boldsymbol{q}\langle k\rangle)^{T}\,\boldsymbol{M}\,\boldsymbol{J}(\boldsymbol{q}\langle k\rangle)\right)^{-1}\boldsymbol{J}(\boldsymbol{q}\langle k\rangle)^{T}\,\boldsymbol{M}\,\Delta\!\left(\mathcal{K}(\boldsymbol{q}\langle k\rangle),\xi_{E}^{*}\right)$$

which is much faster to converge but can behave poorly near singularities. We remedy this by introducing a damping constant λ

$$\delta_{q}\langle k\rangle = \left(\boldsymbol{J}(\boldsymbol{q}\langle k\rangle)^{T}\,\boldsymbol{M}\,\boldsymbol{J}(\boldsymbol{q}\langle k\rangle) + \lambda\boldsymbol{I}_{N\times N}\right)^{-1}\boldsymbol{J}(\boldsymbol{q}\langle k\rangle)^{T}\,\boldsymbol{M}\,\Delta\!\left(\mathcal{K}(\boldsymbol{q}\langle k\rangle),\xi_{E}^{*}\right)$$

which ensures that the term being inverted can never be singular.

An effective way to choose λ is to test whether or not an iteration reduces the error, that is if $\|\delta_{q}\langle k\rangle\| < \|\delta_{q}\langle k-1\rangle\|$. If the error is reduced we can decrease λ in order to speed convergence. If the error has increased we revert to our previous estimate of $\boldsymbol{q}\langle k\rangle$ and increase λ. This adaptive damping factor scheme is the basis of the well-known Levenberg-Marquardt optimization algorithm.

This algorithm is implemented by the `ikine` method and works well in practice. As with all optimization algorithms it requires a reasonable initial estimate of $\boldsymbol{q}$ and this can be explicitly given using the option `'q0'`. A brute-force search for an initial value can be requested by the option `'search'`. The simple Jacobian-transpose approach of Eq. 8.12 can be invoked using the option `'transpose'` along with the value of α.

8.7   Advanced Topics

8.7.1   Computing the Manipulator Jacobian Using Twists

In Sect. 7.1.2.2 we computed the forward kinematics as a product of exponentials based on the screws representing the joint axes in a zero-joint angle configuration. It is easy to differentiate the product of exponentials with respect to motion about each screw axis which leads to the Jacobian matrix

$$^0J^\nu = \left(S_1 \quad \mathrm{Ad}\!\left(e^{[S_1]q_1}\right)S_2 \quad \cdots \quad \mathrm{Ad}\!\left(e^{[S_1]q_1}\cdots e^{[S_{N-1}]q_{N-1}}\right)S_N \right)$$

for velocity in the world coordinate frame. The Jacobian is very elegantly expressed and can be easily built up column by column. Velocity in the end-effector coordinate frame is related to joint velocity by the Jacobian matrix

$$^EJ^\nu = \mathrm{Ad}\!\left(^E\xi_0\right){}^0J^\nu$$

where Ad $(\cdot)$ is the adjoint matrix introduced in Sect. 3.1.2.

> However, compared to the Jacobian of Sect. 8.1, these Jacobians give the velocity of the end-effector as a *velocity twist*, not a spatial velocity as defined on page 65.

To obtain the Jacobian that gives spatial velocity as described in Sect. 8.1 we must apply a velocity transformation

$$^0J = \begin{pmatrix} I_{3\times3} & -\left[{}^0t_E\right]_\times \\ 0_{3\times3} & I_{3\times3} \end{pmatrix} {}^0J^\nu$$

8.8   Wrapping Up

Jacobians are an important concept in robotics, relating changes in one space to changes in another. We previously encountered Jacobians for estimation in Chap. 6 and will use them later for computer vision and control.

In this chapter we have learned about the manipulator Jacobian which describes the relationship between the rate of change of joint coordinates and the spatial velocity of the end-effector expressed in either the world frame or the end-effector frame. We showed how the inverse Jacobian can be used to resolve desired Cartesian velocity into joint velocity as an alternative means of generating Cartesian paths for under- and over-actuated robots. For over-actuated robots we showed how null-space motions can be used to move the robot's joints without affecting the end-effector pose. The numerical properties of the Jacobian tell us about manipulability, that is how well the manipulator is able to move, or exert force, in different directions. At a singularity, indicated by linear dependence between columns of the Jacobian, the robot is unable to move in certain directions. We visualized this by means of the velocity and force ellipsoids.

We also created Jacobians to map angular velocity to roll-pitch-yaw or Euler angle rates, and these were used to form the analytic Jacobian matrix. The Jacobian transpose is used to map wrenches applied at the end-effector to joint torques, and also to map wrenches between coordinate frames. It is also the basis of numerical inverse kinematics for arbitrary robots and singular poses.

Further Reading

The manipulator Jacobian is covered by almost all standard robotics texts such as the robotics handbook (Siciliano and Khatib 2016), Lynch and Park (2017), Siciliano et al. (2008), Spong et al. (2006), Craig (2005), and Paul (1981). An excellent discussion of manipulability and velocity ellipsoids is provided by Siciliano et al. (2009), and the most common manipulability measure is that proposed by Yoshikawa (1984). Computing the manipulator Jacobian based on Denavit-Hartenberg parameters, as used in this Toolbox, was first described by Paul and Shimano (1978).

The resolved-rate motion control scheme was proposed by Whitney (1969). Extensions such as pseudo-inverse Jacobian-based control are reviewed by Klein and Huang (1983) and damped least-squares methods are reviewed by Deo and Walker (1995).

MATLAB and Toolbox Notes

The MATLAB Robotics System Toolbox™ describes a serial-link manipulator using an instance of the `RigidBodyTree` class. Jacobians can be computed using the class method `GeometricJacobian`.

Exercises

1. For the simple 2-link example (page 230) compute the determinant symbolically and determine when it is equal to zero. What does this mean physically?
2. For the Puma 560 robot can you devise a configuration in which three joint axes are parallel?
3. Derive the analytical Jacobian for Euler angles.
4. Velocity and force ellipsoids for the two link manipulator (page 236, 245). Perhaps using the interactive `teach` method with the `'callback'` option:
   a) What configuration gives the best manipulability?
   b) What configuration is best for throwing a ball in the positive x-direction?
   c) What configuration is best for carrying a heavy weight if gravity applies a force in the negative y-direction?
   d) Plot the velocity ellipse (x- and y-velocity) for the two-link manipulator at a grid of end-effector positions in its workspace. Each ellipsoid should be centered on the end-effector position.
5. Velocity and force ellipsoids for the Puma manipulator (page 237)
   a) For the Puma 560 manipulator find a configuration where manipulability is greater than at `qn`.
   b) Use the `teach` method with the `'callback'` option to interactively animate the ellipsoids. You may need to use the `'workspace'` option to `teach` to prevent the ellipsoid being truncated.
6. Resolved-rate motion control (page 237)
   a) Experiment with different Cartesian translational and rotational velocity demands, and combinations.
   b) Extend the Simulink system of Fig. 8.6 to also record the determinant of the Jacobian matrix to the workspace.
   c) In Fig. 8.6 the robot's motion is simulated for 5 s. Extend the simulation time to 10 s and explain what happens.
   d) Set the initial pose and direction of motion to mimic that of Sect. 7.3.4. What happens when the robot reaches the singularity?
   e) Replace the Jacobian inverse block in Fig. 8.5 with the MATLAB function `pinv`.
   f) Replace the Jacobian inverse block in Fig. 8.5 with a damped least squares function, and investigate the effect of different values of the damping factor.

g) Replace the Jacobian inverse block in Fig. 8.5 with a block based on the MATLAB function `lscov`.

7. The model `mdl_p8` describes an 8-joint robot (PPRRRRRR) comprising an xy-base (PP) carrying a Puma arm (RRRRRR).
   a) Compute a Cartesian end-effector path and use numerical inverse kinematics to solve for the joint coordinates. Analyze how the motion is split between the base and the robot arm.
   b) With the end-effector at a constant pose explore null-space control. Set a velocity for the mobile base and see how the arm configuration accomodates that.
   c) Develop a null-space controller that keeps the last six joints in the middle of their working range by using the first two joints to position the base of the Puma. Modify this so as to maximize the manipulability of the P8 robot.
   d) Consider now that the Puma robot is mounted on a nonholonomic robot, create a controller that generates appropriate steering and velocity inputs to the mobile robot (challenging).
   e) For an arbitrary pose and end-point spatial velocity we will move six joints and lock two joints. Write an algorithm to determine which two joints should be locked.

8. The model `mdl_hyper3d(20)` is a 20-joint robot that moves in 3-dimensional space.
   a) Explore the capabilities of this robot.
   b) Compute a Cartesian end-effector trajectory that traces a circle on the ground, and use numerical inverse kinematics to solve for the joint coordinates.
   c) Add a null-space control strategy that keeps all joint angles close to zero while it is moving.
   d) Define an end-effector pose on the ground that the robot must reach after passing through two holes in vertical planes. Can you determine the joint configuration that allows this?

9. Write code to compute the Jacobian of a robot represented by a `SerialLink` object using twists as described in Sect. 8.7.1.

10. Consider the Puma 560 robot moving in the xz-plane. Divide the plane into 2-cm grid cells and for each cell determine if it is reachable, and if it is then determine the manipulability for the first three joints of the robot arm and place that value in the corresponding grid cell. Display a heat map of the robot's manipulability in the plane.

9

Dynamics and Control

In this chapter we consider the dynamics and control of a serial-link manipulator arm. The motion of the end-effector is the composition of the motion of each link, and the links are ultimately moved by *forces* and *torques* exerted by the joints. Section 9.1 describes the key elements of a robot joint control system that enables a single joint to follow a desired trajectory; and the challenges involved such as friction, gravity load and varying inertia.

Each link in the serial-link manipulator is supported by a reaction force and torque from the preceding link, and is subject to its own weight as well as the reaction forces and torques from the links that it supports. Section 9.2 introduces the *rigid-body* equations of motion, a set of coupled dynamic equations, that describe the joint torques necessary to achieve a particular manipulator state. These equations can be factored into terms describing inertia, gravity load and gyroscopic coupling which provide insight into how the motion of one joint exerts a disturbance force on other joints, and how inertia and gravity load varies with configuration and payload. Section 9.3 introduces the forward dynamics which describe how the manipulator moves, that is, how its configuration evolves with time in response to forces and torques applied by the joints and by external forces such as gravity. Section 9.4 introduces control systems that compute the required joint forces based on the desired trajectory as well as the rigid-body dynamic forces. This enables improved control of the end-effector trajectory, despite changing robot configuration, as well as compliant motion. Section 9.5 covers an important application of what we have learned about joint control – series-elastic actuators for human-safe robots.

9.1 Independent Joint Control

A robot drive train comprises an actuator or motor, and a transmission to connect it to the link. A common approach to robot joint control is to consider each joint or axis as an independent control system that attempts to accurately follow its joint angle trajectory. However as we shall see, this is complicated by various *disturbance* torques due to gravity, velocity and acceleration coupling, and friction that act on the joint. A very common control structure is the nested control loop. The outer loop is responsible for maintaining position and determines the velocity of the joint that will minimize position error. The inner loop is responsible for maintaining the velocity of the joint as demanded by the outer loop.

9.1.1 Actuators

The vast majority of robots today are driven by rotary electric motors (Fig. 9.1). Large industrial robots typically use brushless servo motors while small laboratory or hobby robots use brushed DC motors or stepper motors. Manipulators for very large payloads as used in mining, forestry or construction are typically hydraulically driven using electrically operated hydraulic valves – electro-hydraulic actuation.

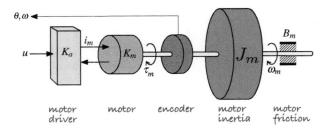

Fig. 9.1.
Key components of a robot-joint actuator. A demand voltage u controls the current i_m flowing into the motor which generates a torque τ_m that accelerates the rotational inertia J_m and is opposed by friction $B_m \omega_m$. The encoder measures rotational speed and angle

Electric motors can be either current or voltage controlled.► Here we assume current control where a motor driver or amplifier provides current

$$i_m = K_a u$$

that is linearly related to the applied control voltage u and where K_a is the transconductance of the amplifier with units of $A\,V^{-1}$. The torque generated by the motor is proportional to current

$$\tau_m = K_m i_m$$

where K_m is the motor torque constant with units of $N\,m\,A^{-1}$. The torque accelerates the rotational inertia J_m, due to the rotating part of the motor itself, which has a rotational velocity of ω. Frictional effects are modeled by B_m.

Current control is implemented by an electronic constant current source, or a variable voltage source with feedback of actual motor current. A variable voltage source is most commonly implemented by a pulse-width modulated (PWM) switching circuit. Voltage control requires that the electrical dynamics of the motor due to its resistance and inductance, as well as back EMF, must be taken into account when designing the control system.

9.1.2   Friction

Any rotating machinery, motor or gearbox, will be affected by friction – a force or torque that *opposes* motion. The net torque from the motor is

$$\tau' = \tau_m - \tau_f$$

where τ_f is the friction torque which is function of velocity

$$\tau_f = B\omega + \tau_C \tag{9.1}$$

where the slope $B > 0$ is the viscous friction coefficient and the offset is Coulomb friction. The latter is frequently modeled by the nonlinear function

$$\tau_C = \begin{cases} \tau_C^+ & \omega > 0 \\ 0 & \omega = 0 \\ \tau_C^- & \omega < 0 \end{cases} \tag{9.2}$$

In general the friction coefficients depend on the direction of rotation and this asymmetry is more pronounced for Coulomb than for viscous friction.

The total friction torque as a function of rotational velocity is shown in Fig. 9.2. At very low speeds, highlighted in grey, an effect known as stiction becomes evident. The applied torque must exceed the stiction torque before rotation can occur – a process known as *breaking stiction*. Once the machine is moving the stiction force rapidly decreases and viscous friction dominates.

There are several sources of friction *experienced* by the motor. The first component is due to the motor itself: its bearings and, for a brushed motor, the brushes rubbing on the commutator. The friction parameters are often provided in the motor manufacturer's data sheet. Other sources of friction are the gearbox and the bearings that support the link.

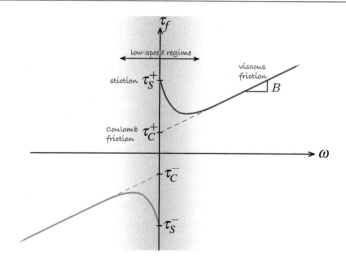

Fig. 9.2.
Typical friction versus speed characteristic. The *dashed lines* depict a simple piecewise-linear friction model characterized by slope (viscous friction) and intercept (Coulomb friction). The low-speed regime is *shaded* and shown in exaggerated fashion

Charles-Augustin de Coulomb (1736–1806) was a French physicist. He was born in Angoulême to a wealthy family and studied mathematics at the Collége des Quatre-Nations under Pierre Charles Monnier, and later at the military school in Méziéres. He spent eight years in Martinique involved in the construction of Fort Bourbon and there he contracted tropical fever.

Later he worked at the shipyards in Rochefort which he used as laboratories for his experiments in static and dynamic friction of sliding surfaces. His paper *Théorie des machines simples* won the Grand Prix from the Académie des Sciences in 1781. His later research was on electromagnetism and electrostatics and he is best known for the formula on electrostatic forces, named in his honor, as is the SI unit of charge. After the revolution he was involved in determining the new system of weights and measures.

9.1.3   Effect of the Link Mass

A motor in a robot arm does not exist in isolation, it is connected to a link as shown schematically in Fig. 9.3. The link has two obvious significant effects on the motor – it adds extra inertia and it adds a torque due to the weight of the arm and both vary with the configuration of the joint.

With reference to the simple 2-joint robot shown in Fig. 9.4 consider the first joint which is directly attached to the first link which is colored red. If we assume the mass of the red link is concentrated at its center of mass (CoM) the extra inertia of the link will be $m_1 r_1^2$. The motor will also experience the inertia of the blue link and this will depend on the value of q_2 – the inertia of the arm when it is straight is greater than the inertia when it is folded.

We also see that gravity acting on the center of mass of the red link will create a torque on the joint 1 motor which will be proportional to $\cos q_1$. Gravity acting on the center of mass of the blue link also creates a torque on the joint 1 motor, and this is more pronounced since it is acting at a greater distance from the motor – the *lever arm effect* is greater.

These effects are clear from even a cursory examination of Fig. 9.4 but the reality is even more complex. Jumping ahead to material we will cover in the next section, we can use the Toolbox◄ to determine the torque acting on each of the joints as a function of the position, velocity and acceleration of the joints

This requires the MATLAB Symbolic Math Toolbox™.

```
>> mdl_twolink_sym
>> syms q1 q2 q1d q2d q1dd q2dd real
>> tau = twolink.rne([q1 q2], [q1d q2d], [q1dd q2dd]);
```

and the result is a symbolic 2-vector, one per joint, with surprisingly many terms which we can summarize as:

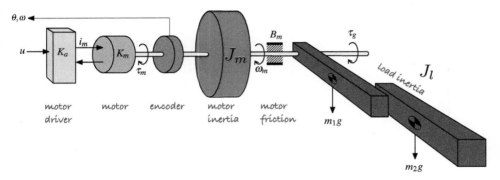

Fig. 9.3.
Robot joint actuator with attached links. The center of mass of each link is indicated by ◐

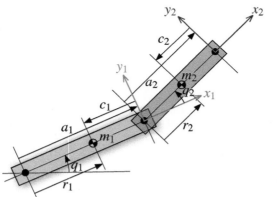

Fig. 9.4. Notation for rigid-body dynamics of two-link arm showing link frames and relevant dimensions. The center of mass (CoM) of each link is indicated by ◐. The CoM is a distance of r_i from the axis of joint i, and c_i from the origin of frame $\{i\}$ as defined in Fig. 7.5 – therefore $r_i = a_i + c_i$

$$\tau_1 = M_{11}(q_2)\ddot{q}_1 + \underbrace{M_{12}(q_2)\ddot{q}_2 + C_1(q_2)\dot{q}_1\dot{q}_2 + C_2(q_2)\dot{q}_2^2 + g(q_1, q_2)}_{\text{disturbance}}$$

$$M_{11} = m_1\left(a_1^2 + 2a_1c_1 + c_1^2\right) + m_2\left(a_1^2 + (a_2 + c_2)^2 + (2a_1a_2 + 2a_1c_2)\cos q_2\right)$$

$$M_{12} = m_2(a_2 + c_2)(a_2 + c_2 + a_1\cos q_2) \tag{9.3}$$

$$C_1 = -2a_1m_2(a_2 + c_2)\sin q_2$$

$$C_2 = -a_1m_2(a_2 + c_2)\sin q_2$$

$$g = (a_1m_1 + a_1m_2 + c_1m_1)\cos q_1 + (a_2m_2 + c_2m_2)\cos(q_1 + q_2)$$

We have already discussed the first and last terms in a qualitative way – the inertia is dependent on q_2 and the gravity torque g is dependent on q_1 and q_2. What is perhaps most surprising is that the torque applied to joint 1 depends on the velocity and the acceleration of q_2 and this will covered in more detail in Sect. 9.2.

In summary, the effect of joint motion in a series of mechanical links is nontrivial. The motion of any joint is affected by the motion of *all* the other joints and for a robot with many joints this becomes quite complex.

9.1.4   Gearbox

Electric motors are compact and efficient and can rotate at very high speed, but produce very low torque. Therefore it is common to use a reduction gearbox to tradeoff speed for increased torque. For a prismatic joint the gearbox might convert rotary motion to linear. The disadvantage of a gearbox is increased cost, weight, friction, backlash, mechanical noise and, for harmonic gears, torque ripple. Very high-performance robots, such as those used in high-speed electronic assembly, use expensive high-torque motors with a direct drive or a very low gear ratio achieved using cables or thin metal bands rather than gears.

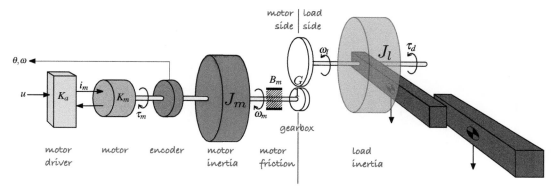

Fig. 9.5. Schematic of complete robot joint including gearbox. The effective inertia of the links is shown as J_l and the disturbance torque due to the link motion is τ_d

For example if you turned the motor shaft by hand you would *feel* the inertia of the load through the gearbox but it would be reduced by G^2.

Table 9.1. Relationship between load and motor referenced quantities for reduction gear ratio G

${}^l J = G^2 \, {}^m J$
${}^l B = G^2 \, {}^m B$
${}^l \tau_C = G \, {}^m \tau_C$
${}^l \tau = G \, {}^m \tau$
${}^l \omega = {}^m \omega / G$
${}^l \dot{\omega} = {}^m \dot{\omega} / G$

Figure 9.5 shows the complete drive train of a typical robot joint. For a $G:1$ reduction drive the torque at the link is G times the torque at the motor. For rotary joints the quantities measured at the link, reference frame l, are related to the motor referenced quantities, reference frame m, as shown in Table 9.1. The inertia of the load is reduced by a factor of G^2 and the disturbance torque by a factor of G.

There are two components of inertia *seen* by the motor. The first is due to the rotating part of the motor itself, its rotor. It is denoted J_m and is a constant intrinsic characteristic of the motor and the value is provided in the motor manufacturer's data sheet. The second component is the variable load inertia J_l which is the inertia of the driven link and all the other links that are attached to it. For joint j this is element M_{jj} of the configuration dependent inertia matrix of Eq. 9.3.

9.1.5 Modeling the Robot Joint

The complete motor drive comprises the motor to generate torque, the gearbox to amplify the torque and reduce the effects of the load, and an encoder to provide feedback of position and velocity. A schematic of such a device is shown in Fig. 9.6.

Collecting the various equations above we can write the torque balance on the motor shaft as

$$K_m K_a u - B' \omega - \tau'_C(\omega) - \frac{\tau_d(\boldsymbol{q})}{G} = J' \dot{\omega} \tag{9.4}$$

where B', τ'_C and J' are the effective total viscous friction, Coulomb friction and inertia due to the motor, gearbox, bearings and the load

$$B' = B_m + \frac{B_l}{G^2}, \quad \tau'_C = \tau_{C,m} + \frac{\tau'_{C,l}}{G}, \quad J' = J_m + \frac{J_l}{G^2} \tag{9.5}$$

In order to analyze the dynamics of Eq. 9.4 we must first linearize it, and this can be done simply by setting all additive constants to zero

$$J' \dot{\omega} + B' \omega = K_m K_a u$$

and then applying the Laplace transformation

$$s J' \Omega(s) + B' \Omega(s) = K_m K_a U(s)$$

where $\Omega(s)$ and $U(s)$ are the Laplace transform of the time domain signals $\omega(t)$ and $u(t)$ respectively. This can be rearranged as a linear transfer function

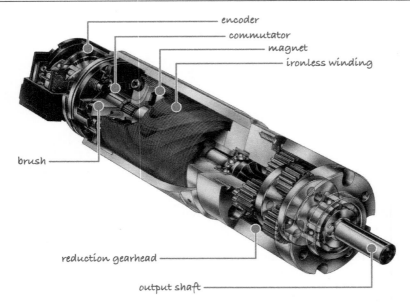

Fig. 9.6.
Schematic of an integrated motor-encoder-gearbox assembly (courtesy of maxon precision motors, inc.)

$$\frac{\Omega(s)}{U(s)} = \frac{K_m K_a}{J's + B'}$$

relating motor speed to control input, and has a single pole▸ at $s = -B'/J'$.

The mechanical pole.

We will use data for joint 2 – the shoulder – of the Puma 560 robot since its parameters are well known and are listed in Table 9.2. In the absence of other information we will take $B' = B_m$. The link inertia M_{22} experienced by the joint 2 motor as a function of configuration is shown in Fig. 9.16c and we see that it varies significantly – from 3.66 to 5.21 kg m². Using the mean value of the extreme inertia values, which is 4.43 kg m², the effective inertia is

$$J' = J_m + \frac{1}{G^2} M_{22}$$

$$= 200 \times 10^{-6} + \frac{4.43}{(107.815)^2}$$

$$= 200 \times 10^{-6} + 380 \times 10^{-6} = 580 \times 10^{-6} \text{ kg m}^2$$

and we see that the inertia of the link referred to the motor side of the gearbox is comparable to the inertia of the motor itself.

The Toolbox can automatically generate▸ a dynamic model suitable for use with the MATLAB control design tools

This requires the Control Systems Toolbox™.

```
>> tf = p560.jointdynamics(qn);
```

is a vector of continuous-time linear-time-invariant (LTI) models, one per joint, computed for the particular pose qn. For the shoulder joint we are considering here that transfer function is

```
>> tf(2)
ans =
              1
    ---------------------
    0.0005797 s + 0.000817
Continuous-time transfer function.
```

which is similar to that above except that it does not account for K_m and K_a since these are not parameters of the Link object. Once we have a model of this form we can plot the step response and use a range of standard control system design tools.

Table 9.2.
Motor and drive parameters for Puma 560 shoulder joint with respect to the motor side of the gearbox (Corke 1996b)

Parameter	Symbol	Value	Unit
Motor torque constant	K_m	0.228	N m A^{-1}
Motor inertia	J_m	200×10^{-6}	kg m^2
Drive viscous friction	B_m	817×10^{-6}	N m s rad^{-1}
Drive Coulomb friction	τ_C^+	0.126	N m
	τ_C^-	−0.709	N m
Gear ratio	G	107.815	
Maximum torque	τ_{max}	0.900	N m
Maximum speed	$\dot{q}_{max}$	165	rad s^{-1}

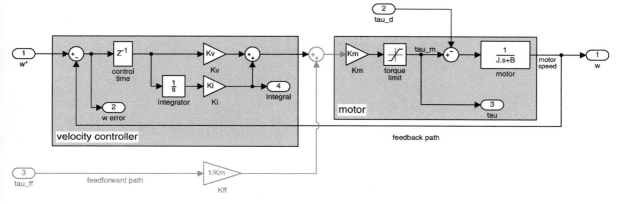

Fig. 9.7. Velocity control loop, Simulink model `vloop`

9.1.6 Velocity Control Loop

A very common approach to controlling the position output of a motor is the nested control loop. The outer loop is responsible for maintaining position and determines the velocity of the joint that will minimize position error. The inner loop – the velocity loop – is responsible for maintaining the velocity of the joint as demanded by the outer loop. Motor speed control is important for all types of robots, not just arms. For example it is used to control the speed of the wheels for car-like vehicles and the rotors of a quadrotor as discussed in Chap. 4.

The Simulink® model is shown in Fig. 9.7. The input to the motor driver is based on the error between the demanded and actual velocity.◄ A delay of 1 ms is included to model the computational time of the velocity loop control algorithm and a saturator models the finite maximum torque that the motor that can deliver.

We first consider the case of proportional control where $K_i = 0$ and

> The motor velocity is typically computed by taking the difference in motor position at each sample time, and the position is measured by a shaft encoder. This can be problematic at very low speeds where the encoder tick rate is lower than the sample rate. In this case a better strategy is to measure the time between encoder ticks.

$$u^* = K_v\left(\dot{q}^* - \dot{q}\right) \tag{9.6}$$

To test this velocity controller we create a test harness

```
>> vloop_test
```

with a trapezoidal velocity demand which is shown in Fig. 9.8. Running the simulator

```
>> sim('vloop_test');
```

and with a little experimentation we find that a gain of $K_v = 0.6$ gives satisfactory performance as shown in Fig. 9.9. There is some minor overshoot at the discontinuity but less gain leads to increased velocity error and more gain leads to oscillation – as always control engineering is all about tradeoffs.

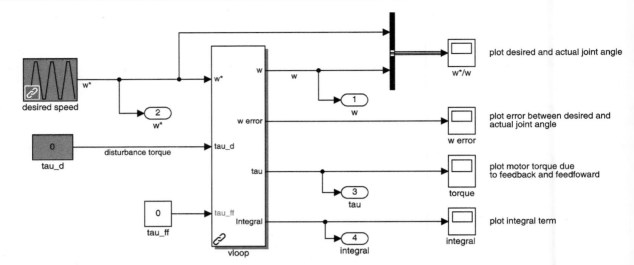

Fig. 9.8. Test harness for the velocity control loop, Simulink model `vloop_test`. The input `tau_d` is used to simulate a disturbance torque acting on the joint

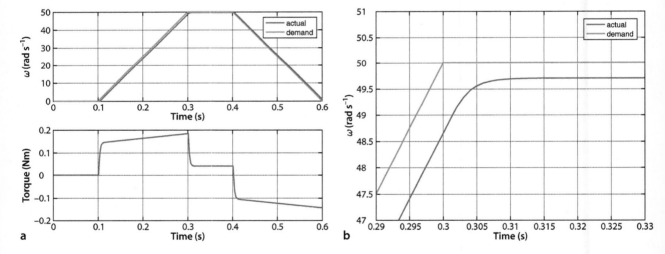

Fig. 9.9. Velocity loop with a trapezoidal demand. a Response; b closeup of response

We also observe a very slight steady-state error – the actual velocity is less than the demand at all times. From a classical control system perspective the velocity loop contains no integrator block and is classified as a Type 0 system – a characteristic of Type 0 systems is they exhibit a finite error for a constant input. More intuitively we can argue that in order to move at constant speed the motor must generate a finite torque to overcome friction, and since motor torque is proportional to velocity error there must be a finite velocity error.

Now we will investigate the effect of inertia variation on the closed-loop response. Using Eq. 9.5 and the data from Fig. 9.16c we find that the minimum and maximum joint inertia at the motor are 515×10^{-6} and 648×10^{-6} kg m^2 respectively. Figure 9.10 shows the velocity tracking error using the control gains chosen above for various values of link inertia. We can see that the tracking error decays more slowly for larger inertia, and is showing signs of instability for the case of zero link inertia. For a case where the inertia variation is more extreme the gain should be chosen to achieve satisfactory closed-loop performance at both extremes.

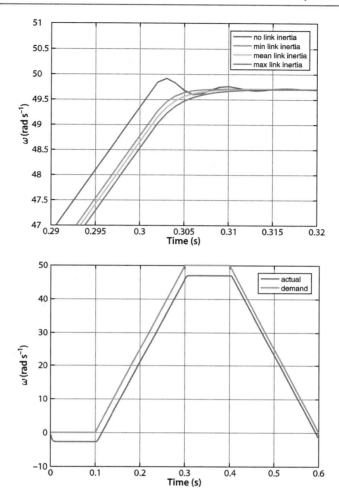

Fig. 9.10.
Velocity loop response with a trapezoidal demand for varying inertia M_{22}

Fig. 9.11.
Velocity loop response to a trapezoidal demand with a gravity disturbance of 20 N m

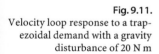

Motor limits. Electric motors are limited in both torque and speed. The maximum torque is defined by the maximum current the drive electronics can provide. A motor also has a maximum rated current beyond which the motor can be damaged by overheating or demagnetization of its permanent magnets which irreversibly reduces its torque constant. As speed increases so does friction and the maximum speed is $\omega_{max} = \tau_{max} / B$.

The product of motor torque and speed is the mechanical output power and also has an upper bound. Motors can tolerate some overloading, peak power and peak torque, for short periods of time but the sustained rating is significantly lower than the peak.

Figure 9.15a shows that the gravity torque on this joint varies from approximately -40 to 40 N m. We now add a disturbance torque equal to just half that maximum amount, 20 N m applied on the load side of the gearbox. We do this by setting a non-zero value in the `tau_d` block and rerunning the simulation. The results shown in Fig. 9.11 indicate that the control performance has been badly degraded – the tracking error has increased to more than 2 rad s^{-1}. This has the same root cause as the very small error we saw in Fig. 9.9 – a Type 0 system exhibits a finite error for a constant input or a constant disturbance.

There are three common approaches to counter this error. The first, and simplest, is to increase the gain. This will reduce the tracking error but push the system toward instability and increase the overshoot.

The second approach, commonly used in industrial motor drives, is to add integral action – adding an integrator changes the system to Type 1 which has zero error for a constant input or constant disturbance. We change Eq. 9.6 to a proportional-integral controller

$$u^* = \left(K_v + \frac{K_i}{s}\right)(\dot{q}^* - \dot{q}), \quad K_i > 0$$

In the Simulink model of Fig. 9.7 this is achieved by setting `Ki` to a nonzero value. With some experimentation we find the gains $K_v = 1$ and $K_i = 10$ work well and the performance is shown in Fig. 9.12. The integrator state evolves over time to cancel out the disturbance term and we can see the error decaying to zero. In practice the disturbance varies over time and the integrator's ability to track it depends on the value of the integral gain K_i. In reality other disturbances affect the joint, for instance Coulomb friction and torques due to velocity and acceleration coupling. The controller needs to be well tuned so that these have minimal effect on the tracking performance.

As always in engineering there are some tradeoffs. The integral term can lead to increased overshoot so increasing K_i usually requires some compensating reduction of K_v. If the joint actuator is pushed to its performance limit, for instance the torque limit is reached, then the tracking error will grow with time since the motor acceleration will be lower than required. The integral of this increasing error will grow leading to a condition known as integral windup. When the joint finally reaches its destination the large accumulated integral keeps driving the motor forward until the integral decays – leading to large overshoot. Various strategies are employed to combat this, such as limiting the maximum value of the integrator, or only allowing integral action when the motor is close to its setpoint.

These two approaches are collectively referred to as disturbance rejection and are concerned with reducing the effect of an unknown disturbance. However if we think about the problem in its robotics context the gravity disturbance is not unknown. In Sect. 9.1.3 we showed how to compute the torque due to gravity that acts on each joint. If we know this torque, and the motor torque constant, we can *add* it to the output of the PI controller.▶

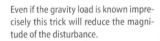

Even if the gravity load is known imprecisely this trick will reduce the magnitude of the disturbance.

The third approach is therefore to predict the disturbance and cancel it out – a strategy known as torque feedforward control. This is shown by the red wiring in Fig. 9.7 and can be demonstrated by setting the `tau_ff` block of Fig. 9.8 to the same, or approximately the same, value as the disturbance.

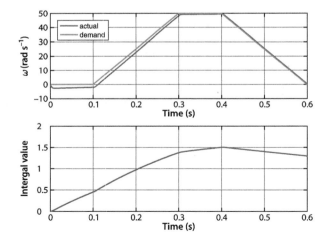

Fig. 9.12.
Velocity loop response to a trapezoidal demand with a gravity disturbance of 20 N m and proportional-integral control

Back EMF. A spinning motor acts like a generator and produces a voltage V_b called the back EMF which opposes the current flowing into the motor. Back EMF is proportional to motor speed $V_b = K_m\omega$ where K_m is the motor torque constant whose units can also be interpreted as V s rad^{-1}. When this voltage equals the maximum possible voltage from the drive electronics then no more current can flow into the motor and torque falls to zero. This provides a practical upper bound on motor speed, and torque at high speeds.

9.1.7   Position Control Loop

The outer loop is responsible for maintaining position and we use a proportional controller◀ based on the error between actual and demanded position to compute the desired speed of the motor

Another common approach is to use a proportional-integral-derivative (PID) controller for position but it can be shown that the D gain of this controller is related to the P gain of the inner velocity loop.

$$\dot{q}^* = K_p\left(q^*(t) - q\right) \tag{9.7}$$

A Simulink model is shown in Fig. 9.13 and the position demand $q^*(t)$ comes from an LSPB trajectory generator that moves from 0 to 0.5 rad in 1 s with a sample rate of 1 000 Hz. Joint position is obtained by integrating joint velocity, obtained from the motor velocity loop via the gearbox. The error between the motor and desired position provides the velocity demand for the inner loop.

We load this control loop model

```
>> ploop_test
```

Fig. 9.13. Position control loop, Simulink model `ploop_test`. **a** Test harness for following an LSPB angle trajectory. **b** The position loop `ploop` which is a proportional controller around the inner velocity loop of Fig. 9.7

and its performance is tuned by adjusting the three gains: K_p, K_v, K_i in order to achieve good tracking performance along the trajectory. For $K_p = 40$ the tracking and error responses are shown in Fig. 9.14a. We see that the final error is zero but there is some tracking error along the path where the motor position lags behind the demand. The error between the demand and actual curves is due to the cumulative velocity error of the inner loop which has units of angle.

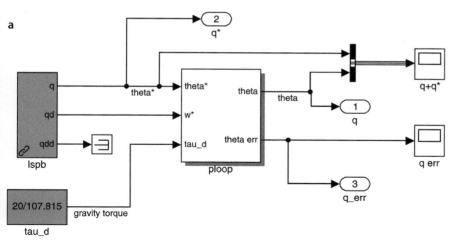

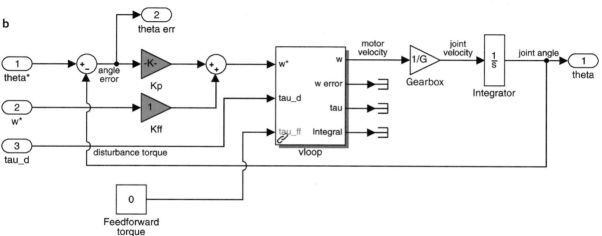

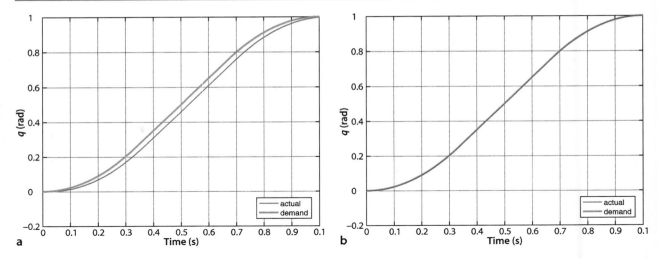

Fig. 9.14. Position loop following an LSPB trajectory. **a** Proportional control only **b** proportional control plus velocity demand feedforward

The position loop, like the velocity loop is based on classical negative feedback. Having zero position error while tracking a ramp would mean zero demanded velocity to the inner loop which is actually contradictory. More formally, we know that a Type 1 system⁴ exhibits a constant error to a ramp input. If we care about reducing this tracking error there are two common remedies. We can add an integrator to the position loop – making it a proportional-integral controller but this gives us yet another parameter to tune. A simple and effective alternative is velocity feedforward control – we add the desired velocity to the output of the proportional control loop, which is the input to the velocity loop. The LSPB trajectory function computes velocity as a function of time as well as position. The time response with velocity feedforward is shown in Fig. 9.14b and we see that tracking error is greatly reduced.

Since the model contains an integrator after the velocity loop.

9.1.8   Independent Joint Control Summary

A common structure for robot joint control is the nested control loop. The inner loop uses a proportional or proportional-integral control law to generate a torque so that the actual velocity closely follows the velocity demand. The outer loop uses a proportional control law to generate the velocity demand so that the actual position closely follows the position demand. Disturbance torques due to gravity and other dynamic coupling effects impact the performance of the velocity loop as do variation in the parameters of the plant being controlled, and this in turn leads to errors in position tracking. Gearing reduces the magnitude of disturbance torques by $1/G$ and the variation in inertia and friction by $1/G^2$ but at the expense of cost, weight, increased friction and mechanical noise.

The velocity loop performance can be improved by adding an integral control term, or by feedforward of the disturbance torque which is largely predictable. The position loop performance can also be improved by feedforward of the desired joint velocity. In practice control systems use both feedforward and feedback control. Feedforward is used to inject signals that we can compute, in this case the joint velocity, and in the earlier case the gravity torque. Feedback control compensates for all remaining sources of error including variation in inertia due to manipulator configuration and payload, changes in friction with time and temperature, and all the disturbance torques due to velocity and acceleration coupling. In general the use of feedforward allows the feedback gain to be reduced since a large part of the demand signal now comes from the feedforward.

9.2 Rigid-Body Equations of Motion

Consider the motor which actuates the jth revolute joint of a serial-link manipulator. From Fig. 7.5 we recall that joint j connects link $j-1$ to link j. The motor exerts a torque that causes the outward link, j, to rotationally accelerate but it also exerts a reaction torque on the inward link $j-1$. Gravity acting on the outward links j to N exert a weight force, and rotating links also exert gyroscopic forces on each other. The inertia that the motor *experiences* is a function of the configuration of the outward links.

The situation at the individual link is quite complex but for the *series* of links the result can be written elegantly and concisely as a set of coupled differential equations in matrix form

$$Q = M(q)\ddot{q} + C(q, \dot{q})\dot{q} + F(\dot{q}) + G(q) + J(q)^T W \qquad (9.8)$$

where q, $\dot{q}$ and $\ddot{q}$ are respectively the vector of generalized joint coordinates, velocities and accelerations, M is the joint-space inertia matrix, C is the Coriolis and centripetal coupling matrix, F is the friction force, G is the gravity loading, and Q is the vector of generalized actuator forces associated with the generalized coordinates q. The last term gives the joint forces due to a wrench W applied at the end-effector and J is the manipulator Jacobian. This equation describes the manipulator rigid-body dynamics and is known as the inverse dynamics – given the pose, velocity and acceleration it computes the required joint forces or torques.

The recursive form of the inverse dynamics does not explicitly calculate the matrices M, C and G of Eq. 9.8. However we can use the recursive Newton-Euler algorithm to calculate these matrices and the Toolbox functions `inertia` and `coriolis` use Walker and Orin's (1982) 'Method 1'. While the recursive forms are computationally efficient for the inverse dynamics, to compute the coefficients of the individual dynamic terms (M, C and G) in Eq. 9.8 is quite costly – $O(N^3)$ for an N-axis manipulator.

These equations can be derived using any classical dynamics method such as Newton's second law and Euler's equation of motion, as discussed in Sect. 3.2.1, or a Lagrangian energy-based approach. A very efficient way for computing Eq. 9.8 is the recursive Newton-Euler algorithm which starts at the base and working outward adds the velocity and acceleration of each joint in order to determine the velocity and acceleration of each link. Then working from the tool back to the base, it computes the forces and moments acting on each link and thus the joint torques. The recursive Newton-Euler algorithm has $O(N)$ complexity and can be written in functional form as

$$Q = \mathcal{D}^{-1}(q, \dot{q}, \ddot{q}) \qquad (9.9)$$

Not all robot arm models in the Toolbox have dynamic parameters, see the "dynamics" tag in the output of the models() command, or use models('dyn') to list models with dynamic parameters. The Puma 560 robot is used for the examples in this chapter since its dynamic parameters are reliably known.

In the Toolbox it is implemented by the `rne` method of the `SerialLink` object. Consider the Puma 560 robot

```
>> mdl_puma560
```

at the nominal pose, and with zero joint velocity and acceleration. To achieve this state, the required generalized joint forces, or joint torques in this case, are

```
>> Q = p560.rne(qn, qz, qz)
Q =
   -0.0000   31.6399    6.0351    0.0000    0.0283         0
```

Since the robot is not moving (we specified $\dot{q} = \ddot{q} = 0$) these torques must be those required to *hold the robot up* against gravity. We can confirm this by computing the torques required in the absence of gravity

```
>> Q = p560.rne(qn, qz, qz, 'gravity', [0 0 0])
ans =
     0     0     0     0     0     0
```

by overriding the object's default gravity vector.

Like most Toolbox methods `rne` can operate on a trajectory

```
>> q = jtraj(qz, qr, 10)
>> Q = p560.rne(q, 0*q, 0*q)
```

which has returned

```
>> about(Q)
Q [double] : 10x6 (480 bytes)
```

a 10 × 6 matrix with each row representing the generalized force required for the corresponding row of q. The joint torques corresponding to the fifth time step are

```
>> Q(5,:)
ans =
     0.0000   29.8883    0.2489         0         0         0
```

Consider now a case where the robot is moving. It is *instantaneously* at the nominal pose but joint 1 is moving at 1 rad s^{-1} and the acceleration of all joints is zero. Then in the absence of gravity, the required joint torques

```
>> p560.rne(qn, [1 0 0 0 0 0], qz, 'gravity', [0 0 0])
   30.5332    0.6280   -0.3607   -0.0003   -0.0000         0
```

are nonzero. The torque on joint 1 is that needed to overcome friction which always opposes the motion. More interesting is that torques need to be exerted on joints 2, 3 and 4. This is to oppose the gyroscopic effects (centripetal and Coriolis forces) – referred to as velocity coupling torques since the rotational velocity of one joint has induced a torque on several other joints.

The elements of the matrices M, C, F and G are complex functions of the link's kinematic parameters $(\theta_j, d_j, a_j, \alpha_j)$ and inertial parameters. Each link has ten independent inertial parameters: the link mass m_j; the center of mass (COM) r_j with respect to the link coordinate frame; and six second moments which represent the inertia of the link about the COM but with respect to axes aligned with the link frame {j}, see Fig. 7.5. We can view the dynamic parameters of a robot's link by

```
>> p560.links(1).dyn
Revolute(std): theta=q, d=0, a=0, alpha=1.5708, offset=0
  m    = 0
  r    = 0              0              0
  I    = | 0              0              0              |
         | 0              0.35           0              |
         | 0              0              0              |
  Jm   = 0.0002
  Bm   = 0.00148
  Tc   = 0.395        (+)  -0.435       (-)
  G    = -62.61
  qlim = -2.792527 to 2.792527
```

which in order are: the kinematic parameters, link mass, COM position, link inertia matrix, motor inertia, motor friction, Coulomb friction, reduction gear ratio and joint angle limits.

The remainder of this section examines the various matrix components of Eq. 9.8.

9.2.1   Gravity Term

$$Q = M(q)\ddot{q} + C(q,\dot{q})\dot{q} + F(\dot{q}) + \boxed{G(q)} + J(q)^T W$$

We start our detailed discussion with the gravity term because it is generally the dominant term in Eq. 9.8 and is present even when the robot is stationary or moving slowly. Some robots use counterbalance weights▸ or even springs to reduce the *gravity* torque that needs to be provided by the motors – this allows the motors to be smaller and thus lower in cost.

In the previous section we used the rne method to compute the gravity load by setting the joint velocity and acceleration to zero. A more convenient approach is to use the gravload method

Counterbalancing will however increase the inertia associated with a joint since it adds additional mass at the end of a lever arm, and increase the overall mass of the robot.

```
>> gravload = p560.gravload(qn)
gravload =
   -0.0000   31.6399    6.0351    0.0000    0.0283         0
```

The `SerialLink` object contains a default gravitational acceleration vector which is initialized to the nominal value for Earth◄

The `'gravity'` option for the `SerialLink` constructor can change this.

```
>> p560.gravity'
ans =
        0         0    9.8100
```

We could change gravity to the lunar value

```
>> p560.gravity = p560.gravity/6;
```

resulting in reduced joint torques

```
>> p560.gravload(qn)
ans =
    0.0000    5.2733    1.0059    0.0000    0.0047         0
```

or we could turn our lunar robot upside down

```
>> p560.base = SE3.Rx(pi);
>> p560.gravload(qn)
ans =
    0.0000   -5.2733   -1.0059   -0.0000   -0.0047         0
```

and see that the torques have changed sign. Before proceeding we bring our robot back to Earth and right-side up

```
>> mdl_puma560
```

The torque exerted on a joint due to gravity acting on the robot depends very strongly on the robot's pose. Intuitively the torque on the shoulder joint is much greater when the arm is stretched out horizontally

```
>> Q = p560.gravload(qs)
Q =
   -0.0000   46.0069    8.7722    0.0000    0.0283         0
```

than when the arm is pointing straight up

```
>> Q = p560.gravload(qr)
Q =
        0   -0.7752    0.2489         0         0         0
```

The gravity torque on the elbow is also very high in the first pose since it has to support the lower arm and the wrist. We can investigate how the gravity load on joints 2 and 3 varies with joint configuration by

```
1   [Q2,Q3] = meshgrid(-pi:0.1:pi, -pi:0.1:pi);
2   for i=1:numcols(Q2),
3       for j=1:numcols(Q3);
4           g = p560.gravload([0 Q2(i,j) Q3(i,j) 0 0 0]);
5           g2(i,j) = g(2);
6           g3(i,j) = g(3);
7       end
8   end
9   surfl(Q2, Q3, g2); surfl(Q2, Q3, g3);
```

Joseph-Louis Lagrange (1736–1813) was an Italian-born (Giuseppe Lodovico Lagrangia) French mathematician and astronomer. He made significant contributions to the fields of analysis, number theory, classical and celestial mechanics. In 1766 he succeeded Euler as the director of mathematics at the Prussian Academy of Sciences in Berlin, where he stayed for over twenty years, producing a large body of work and winning several prizes of the French Academy of Sciences. His treatise on analytical mechanics "Mécanique Analytique" first published in 1788, offered the most comprehensive treatment of classical mechanics since Newton and formed a basis for the development of mathematical physics in the nineteenth century. In 1787 he became a member of the French Academy, was the first professor of analysis at the École Polytechnique, helped drive the decimalization of France, was a member of the Legion of Honour and a Count of the Empire in 1808. He is buried in the Panthéon in Paris.

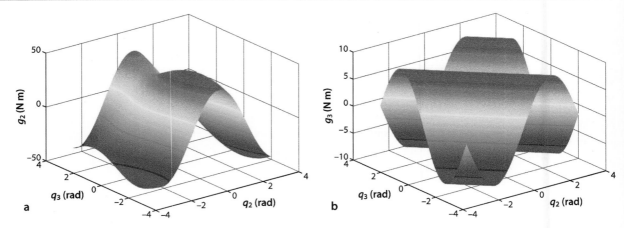

and the results are shown in Fig. 9.15. The gravity torque on joint 2 varies between ±40 N m and for joint 3 varies between ±10 N m. This type of analysis is very important in robot design to determine the required torque capacity for the motors.

Fig. 9.15. Gravity load variation with manipulator pose. **a** Shoulder gravity load, $g_2(q_2, q_3)$; **b** elbow gravity load $g_3(q_2, q_3)$

9.2.2   Inertia Matrix

$$Q = \boxed{M(q)\ddot{q}} + C(q, \dot{q})\dot{q} + F(\dot{q}) + G(q) + J(q)^T W$$

The joint-space inertia is a positive definite, and therefore symmetric, matrix ▶

The diagonal elements of this inertia matrix includes the motor armature inertias, multiplied by G^2.

```
>> M = p560.inertia(qn)
M =
    3.6594   -0.4044    0.1006   -0.0025    0.0000   -0.0000
   -0.4044    4.4137    0.3509    0.0000    0.0024    0.0000
    0.1006    0.3509    0.9378    0.0000    0.0015    0.0000
   -0.0025    0.0000    0.0000    0.1925    0.0000    0.0000
    0.0000    0.0024    0.0015    0.0000    0.1713    0.0000
   -0.0000    0.0000    0.0000    0.0000    0.0000    0.1941
```

which is a function of the manipulator configuration. The diagonal elements M_{jj} describe the inertia *experienced* by joint j, that is, $Q_j = M_{jj}\ddot{q}_j$. Note that the first two diagonal elements, corresponding to the robot's waist and shoulder joints, are large since motion of these joints involves rotation of the heavy upper- and lower-arm links. The off-diagonal terms $M_{ij} = M_{ji}$, $i \neq j$ are the products of inertia and represent coupling of acceleration from joint j to the generalized force on joint i.

We can investigate some of the elements of the inertia matrix and how they vary with robot configuration using the simple (but slow ▶) commands

```
1  [Q2,Q3] = meshgrid(-pi:0.1:pi, -pi:0.1:pi);
2  for i=1:numcols(Q2)
3      for j=1:numcols(Q3)
4          M = p560.inertia([0 Q2(i,j) Q3(i,j) 0 0 0]);
5          M11(i,j) = M(1,1);
6          M12(i,j) = M(1,2);
7      end
8  end
9  surfl(Q2, Q3, M11); surfl(Q2, Q3, M12);
```

Displaying the value of the robot object
>> p560 displays a tag slowRNE or fastRNE. The former indicates all calculations are done in MATLAB code. Build the MEX version, provided in the mex folder, to enable the fastRNE mode which is around 100 times faster.

The results are shown in Fig. 9.16 and we see significant variation in the value of M_{11} which changes by a factor of

```
>> max(M11(:)) / min(M11(:))
ans =
    2.1558
```

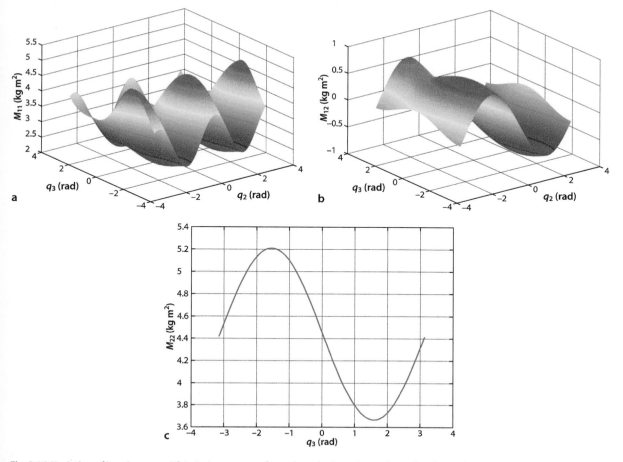

Fig. 9.16. Variation of inertia matrix elements as a function of manipulator pose. **a** Joint 1 inertia as a function of joint 2 and 3 angles $M_{11}(q_2, q_3)$; **b** product of inertia $M_{12}(q_2, q_3)$; **c** joint 2 inertia as a function of joint 3 angle $M_{22}(q_3)$

This is important for robot design since, for a fixed maximum motor torque, inertia sets the upper bound on acceleration which in turn effects path following accuracy.

The off-diagonal term M_{12} represents coupling between the angular acceleration of joint 2 and the torque on joint 1. That is, if joint 2 accelerates then a torque will be exerted on joint 1 and vice versa.

9.2.3   Coriolis Matrix

$$Q = M(q)\ddot{q} + \boxed{C(q, \dot{q})\dot{q}} + F(\dot{q}) + G(q) + J(q)^T W$$

The Coriolis matrix C is a function of joint coordinates and joint velocity. The centripetal torques are proportional to $\dot{q}_j^2$, while the Coriolis torques are proportional to $\dot{q}_i \dot{q}_j$. For example, at the nominal pose with the elbow joint moving at 1 rad s^{-1}

```
>> qd = [0 0 1 0 0 0];
```

the Coriolis matrix is

```
>> C = p560.coriolis(qn, qd)
C =
      0.8992   -0.2380   -0.2380    0.0005   -0.0375    0.0000
     -0.0000    0.9106    0.9106         0   -0.0036         0
      0.0000    0.0000   -0.0000         0   -0.0799         0
     -0.0559    0.0000    0.0000   -0.0000    0.0000   -0.0000
     -0.0000    0.0799    0.0799   -0.0000         0         0
      0.0000         0         0    0.0000         0         0
```

The off-diagonal terms $C_{i,j}$ represent coupling of joint j velocity to the generalized force acting on joint i. $C_{2,3} = 0.9106$ represents significant coupling from joint 3 velocity to torque on joint 2 – rotation of the elbow exerting a torque on the shoulder. Since the elements of this matrix represents a coupling from velocity to joint force they have the same dimensions as viscous friction or damping, however the sign can be positive or negative. The joint torques due to the motion of just this one joint are

```
>> C*qd'
ans =
    -0.2380
     0.9106
    -0.0000
     0.0000
     0.0799
          0
```

9.2.4   Friction

$$Q = M(q)\ddot{q} + C(q,\dot{q})\dot{q} + \boxed{F(\dot{q})} + G(q) + J(q)^T W$$

For most electric drive robots friction is the next most dominant joint force after gravity.▶

The Toolbox models friction within the `Link` object. The friction values are lumped and motor referenced, that is, they apply to the motor side of the gearbox. Viscous friction is a scalar that applies for positive and negative velocity.▶ Coulomb friction is a 2-vector comprising (Q_C^+, Q_C^-). The dynamic parameters of the Puma robot's first link are shown on page 264 as link parameters `Bm` and `Tc`. The online documentation for the `Link` class describes how to set these parameters.

For the Puma robot joint friction varied from 10 to 47% of the maximum motor torque for the first three joints (Corke 1996b).

In practice some mechanisms have a velocity dependent friction characteristic.

9.2.5   Effect of Payload

Any real robot has a specified maximum payload which is dictated by two dynamic effects. The first is that a mass at the end of the robot will increase the inertia *experienced* by the joint motors and which reduces acceleration and dynamic performance. The second is that mass generates a weight force which all the joints need to support. In the worst case the increased gravity torque component might exceed the rating of one or more motors. However even if the rating is not exceeded there is less torque available for acceleration which again reduces dynamic performance.

As an example we will add a 2.5 kg point mass to the Puma 560 which is its rated maximum payload. The center of mass of the payload cannot be at the center of the wrist coordinate frame, that is inside the wrist, so we will offset it 100 mm in the z-direction of the wrist frame. We achieve this by modifying the inertial parameters of the robot's last link▶

This assumes that the last link itself has no mass which is a reasonable approximation.

```
>> p560.payload(2.5, [0 0 0.1]);
```

The inertia at the nominal pose is now

```
>> M_loaded = p560.inertia(qn);
```

and the *ratio* with respect to the unloaded case, computed earlier, is

```
>> M_loaded ./ M
ans =
    1.3363    0.9872    2.1490   49.3960   80.1821    1.0000
    0.9872    1.2667    2.9191    5.9299   74.0092    1.0000
    2.1490    2.9191    1.6601   -2.1092   66.4071    1.0000
   49.3960    5.9299   -2.1092    1.0647   18.0253    1.0000
   83.4369   74.0092   66.4071   18.0253    1.1454    1.0000
    1.0000    1.0000    1.0000    1.0000    1.0000    1.0000
```

We see that the diagonal elements have increased significantly, for instance the elbow joint inertia has increased by 66% which reduces the maximum acceleration by nearly 40%. Reduced acceleration impairs the robot's ability to accurately follow a high speed path. The inertia of joint 6 is unaffected since this added mass lies on the axis of this joint's rotation. The off-diagonal terms have increased significantly, particularly in rows and columns four and five. This indicates that motion of joints 4 and 5, the wrist joints, which are swinging the offset mass give rise to large reaction forces that are *felt* by all the other robot joints.

The gravity load has also increased by some significant factors

```
>> p560.gravload(qn) ./ gravload
ans =
    0.3737      1.5222      2.5416      18.7826     86.8056          NaN
```

at the elbow and wrist. Note that the values for joints 1, 4 and 6 are invalid since they are each the quotient of numbers that are almost zero. We set the payload of the robot back to zero before proceeding

```
>> p560.payload(0)
```

9.2.6   Base Force

A moving robot exerts a wrench on its base – its weight as well as reaction forces and torques as the arm moves around. This wrench is returned as an optional output argument of the `rne` method, for example

```
>> [Q,Wb] = p560.rne(qn, qz, qz);
```

The wrench

```
>> Wb'
ans =
        0   -0.0000   230.0445   -48.4024   -31.6399   -0.0000
```

needs to be applied to the base to keep it in equilibrium. The vertical force of 230 N is the total weight of the robot which has a mass of

```
>> sum([p560.links.m])
ans =
   23.4500
```

There is also a moment about the *x*- and *y*-axes since the center of mass of the robot in this configuration is not over the origin of the base coordinate frame.

The base forces are important in situations where the robot does not have a rigid base such as on a satellite in space, on a boat, an underwater vehicle or even on a vehicle with soft suspension.

9.2.7   Dynamic Manipulability

In Sect. 8.2.2 we discussed a kinematic measure of manipulability, that is, how well configured the robot is to achieve velocity in any Cartesian direction. The force ellipsoid of Sect. 8.5.2 describes how well the manipulator is able to accelerate in different Cartesian directions but is based on the kinematic, not dynamic, parameters of the robot arm. Following a similar approach, we consider the set of generalized joint forces with unit norm

$$Q^T Q = 1$$

From Eq. 9.8 and ignoring gravity and assuming $\dot{q} = 0$ we write

$$Q = M\ddot{q}$$

Differentiating Eq. 8.2 and still assuming $\dot{q} = 0$ we write

$$\dot{\nu} = J(q)\ddot{q}$$

Combining these we write

$$\dot{\nu}^T\left(JM^{-1}M^{-T}J^T\right)^{-1}\dot{\nu} = 1$$

or more compactly

$$\dot{\nu}^T M_x^{-1}\dot{\nu} = 1$$

which is the equation of a hyperellipsoid in Cartesian acceleration space. For example, at the nominal pose

```
>> J = p560.jacob0(qn);
>> M = p560.inertia(qn);
>> Mx = (J * inv(M) * inv(M)' * J');
```

If we consider just the translational acceleration, that is the top left 3×3 submatrix of M_x

```
>> Mx = Mx(1:3, 1:3);
```

this is a 3-dimensional ellipsoid

```
>> plot_ellipse( Mx )
```

which is plotted in Fig. 9.17. The major axis of this ellipsoid is the direction in which the manipulator has maximum acceleration at this configuration. The radii of the ellipse are the square roots of the eigenvalues

```
>> sqrt(eig(Mx))
ans =
    0.4412
    0.1039
    0.1677
```

and the direction of maximum acceleration is given by the first eigenvector. The ratio of the minimum to maximum radius

```
>> min(ans)/max(ans)
ans =
    0.2355
```

is a measure of the nonuniformity of end-effector acceleration.► It would be unity for isotropic acceleration capability. In this case acceleration capability is good in the x- and z-directions, but poor in the y-direction.

The 6-dimensional ellipsoid has dimensions with different units: m s^{-2} and rad s^{-2}. This makes comparison of all 6 radii problematic.

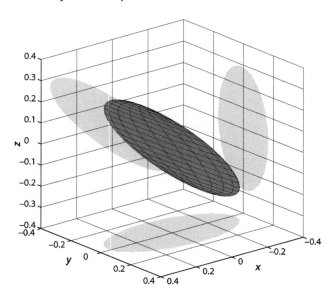

Fig. 9.17.
Spatial acceleration ellipsoid for Puma 560 robot in nominal pose

The scalar dynamic manipulability measure proposed by Asada is similar but considers the ratios of the eigenvalues of

$$\ddot{x}^T J^{-T} M J^{-1} \ddot{x} = 1$$

and returns a uniformity measure $m \in [0, 1]$ where 1 indicates uniformity of acceleration in all directions. For this example

```
>> p560.maniplty(qn, 'asada')
ans =
    0.2094
```

9.3   Forward Dynamics

To determine the motion of the manipulator in response to the forces and torques applied to its joints we require the forward dynamics or integral dynamics. Rearranging the equations of motion Eq. 9.8 we obtain the joint acceleration

$$\ddot{q} = M^{-1}(q)\Big(Q - C(q, \dot{q})\dot{q} - F(\dot{q}) - G(q) - J(q)^T W\Big) \qquad (9.10)$$

and M is always invertible. This function is computed by the `accel` method of the `SerialLink` class

```
qdd = p560.accel(q, qd, Q)
```

given the joint coordinates, joint velocity and applied joint torques. This functionality is also encapsulated in the Simulink block `Robot` and an example of its use is

```
>> sl_ztorque
```

which is shown in Fig. 9.18. The torque applied to the robot is zero and the initial joint angles is set as a parameter of the `Robot` block, in this case to the *zero-angle pose*. The simulation is run

```
>> r = sim('sl_ztorque');
```

and the joint angles as a function of time are returned in the object `r`

```
>> t = r.find('tout');
>> q = r.find('yout');
```

We can show the robot's motion in animation

```
>> p560.plot(q)
```

and see it collapsing under gravity since there are no torques to counter gravity and hold in upright. The shoulder falls and swings back and forth as does the elbow, while the waist joint rotates because of Coriolis coupling. The motion will slowly decay as the energy is dissipated by viscous friction.

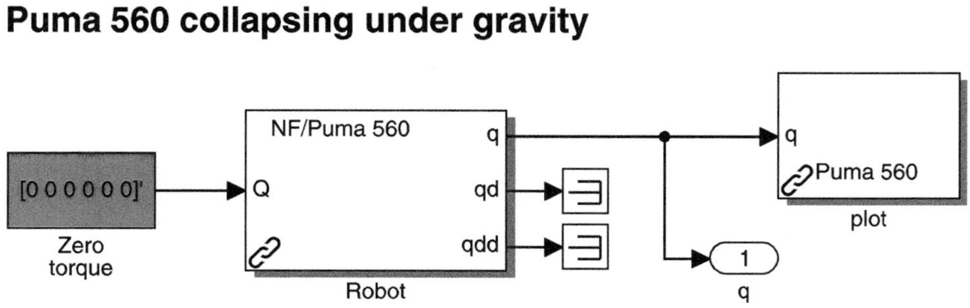

Puma 560 collapsing under gravity

Fig. 9.18.
Simulink model `sl_ztorque` for the Puma 560 manipulator with zero joint torques. This model removes Coulomb friction in order to simplify the numerical integration

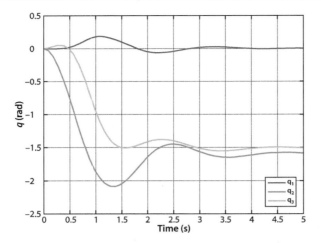

Fig. 9.19.
Joint angle trajectory for
Puma 560 robot with zero
Coulomb friction collapsing
under gravity from initial joint
configuration `qz`

Alternatively we can plot the joint angles as a function of time

```
>> plot(t, q(:,1:3))
```

and this is shown in Fig. 9.19. The method `fdyn` can be used as a nongraphical alternative to Simulink and is described in the online documentation.

This example is rather unrealistic and in reality the joint torques would be computed by some control law as a function of the actual and desired robot joint angles. This is the topic of the next section.

> Coulomb friction is a strong nonlinearity and can cause difficulty when using numerical integration routines to solve the forward dynamics. This is usually manifested by very long integration times. Fixed-step solvers tend to be more tolerant, and these can be selected through the Simulink `Simulation+Model Configuration Parameters+Solver` menu item.
>
> The default Puma 560 model, defined using `mdl_puma560`, has nonzero viscous and Coulomb friction parameters for each joint. Sometimes it is useful to zero the friction parameters for a robot and this can be achieved by
>
> ```
> >> p560_nf = p560.nofriction();
> ```
>
> which returns a copy of the robot object that is similar in all respects except that the Coulomb friction is zero. Alternatively we can set Coulomb *and* viscous friction coefficients to zero
>
> ```
> >> p560_nf = p560.nofriction('all');
> ```

9.4   Rigid-Body Dynamics Compensation

In Sect. 9.1 we discussed some of the challenges for independent joint control and introduced the concept of feedforward to compensate for the gravity disturbance torque. Inertia variation and other dynamic coupling forces were not explicitly dealt with and were left for the feedback controller to handle. However inertia and coupling torques can be computed according to Eq. 9.8 given knowledge of joint angles, joint velocities and accelerations, and the inertial parameters of the links. We can incorporate these torques into the control law using one of two *model-based* approaches: feedforward control, and computed torque control. The structural differences are contrasted in Fig. 9.20 and Fig. 9.21.

9.4.1   Feedforward Control

The torque feedforward controller shown in Fig. 9.20 is given by

$$Q^* = \underbrace{M\big(q^*\big)\ddot{q}^* + C\big(q^*,\dot{q}^*\big)\dot{q}^* + F\big(\dot{q}^*\big) + G\big(q^*\big)}_{\text{feedforward}} + \underbrace{\Big\{K_v\big(\dot{q}^* - \dot{q}^{\#}\big) + K_p\big(q^* - q^{\#}\big)\Big\}}_{\text{feedback}}$$

$$= \mathcal{D}^{-1}\big(q^*,\dot{q}^*,\ddot{q}^*\big) + \Big\{K_v\big(\dot{q}^* - \dot{q}^{\#}\big) + K_p\big(q^* - q^{\#}\big)\Big\} \qquad (9.11)$$

where K_p and K_v are the position and velocity gain (or damping) matrices respectively, and $\mathcal{D}^{-1}(\cdot)$ is the inverse dynamics function. The gain matrices are typically diagonal. The feedforward term provides the joint forces required for the desired manipulator state $(q^*, \dot{q}^*, \ddot{q}^*)$ and the feedback term compensates for any errors due to uncertainty in the inertial parameters, unmodeled forces or external disturbances.

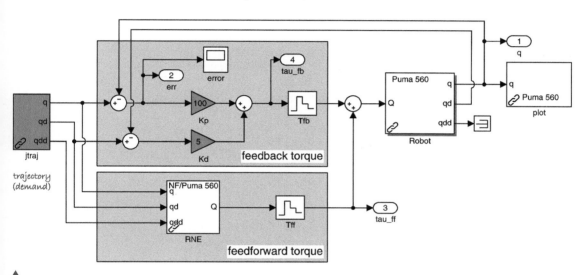

Fig. 9.20. The Simulink model `sl_fforward` for Puma 560 with torque feedforward control. The blocks with the staircase icons are zero-order holds

Fig. 9.21. Robotics Toolbox example `sl_ctorque`, computed torque control

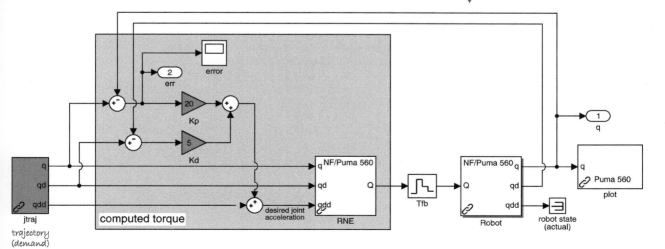

We can also consider that the feedforward term linearizes the nonlinear dynamics about the operating point $(q^*, \dot{q}^*, \ddot{q}^*)$. If the linearization is ideal then the dynamics of the error $e = q^* - q^\#$ can be obtained by combining Eq. 9.8 and 9.11

$$M\left(q^*\right)\ddot{e} + K_v\dot{e} + K_p e = 0 \tag{9.12}$$

For well chosen K_p and K_v the error will decay to zero but the joint errors are coupled▶ and their dynamics are dependent on the manipulator configuration.   Due to the nondiagonal matrix M.

To test this controller using Simulink we first create a `SerialLink` object

```
>> mdl_puma560
```

and then load the torque feedforward controller model

```
>> sl_fforward
```

The feedforward torque is computed using the RNE block and added to the feedback torque computed from position and velocity error. The desired joint angles and velocity are generated using a `jtraj` block. Since the robot configuration changes relatively slowly the feedforward torque can be evaluated at a greater interval, T_{ff}, than the error feedback loops, T_{fb}. In this example we use a zero-order hold block sampling at the relatively low sample rate of 20 Hz.

We run the simulation by pushing the Simulink play button or

```
>> r = sim('sl_fforward');
```

9.4.2   Computed Torque Control

The computed torque controller is shown in Fig. 9.21. It belongs to a class of controllers known as inverse dynamic control. The principle is that the nonlinear system is cascaded with its inverse so that the overall system has a constant unity gain. In practice the inverse is not perfect so a feedback loop is required to deal with errors.

The computed torque control is given by

$$Q = M(q)\left\{\ddot{q}^* + K_v\left(\dot{q}^* - \dot{q}^\#\right) + K_p\left(q^* - q^\#\right)\right\} + C(q^*, \dot{q}^*)\dot{q}^* + F(\dot{q}^*) + G(q^*)$$
$$= \mathcal{D}^{-1}\left(q^*, \dot{q}^*, \left(\ddot{q}^* + K_v\left(\dot{q}^* - \dot{q}^\#\right) + K_p\left(q^* - q^\#\right)\right)\right) \tag{9.13}$$

where K_p and K_v are the position and velocity gain (or damping) matrices respectively, and $\mathcal{D}^{-1}(\cdot)$ is the inverse dynamics function.

In this case the inverse dynamics must be evaluated at each servo interval, although the coefficient matrices M, C, and G could be evaluated at a lower rate since the robot configuration changes relatively slowly. Assuming ideal modeling and parameterization the error dynamics of the system are obtained by combining Eq. 9.8 and 9.13

$$\ddot{e} + K_v\dot{e} + K_p e = 0 \tag{9.14}$$

where $e = q^* - q^\#$. Unlike Eq. 9.12 the joint errors are uncoupled and their dynamics are therefore independent of manipulator configuration. In the case of model error there will be some coupling between axes, and the right-hand side of Eq. 9.14 will be a nonzero forcing function.

Using Simulink we first create a `SerialLink` object and then load the computed torque controller

```
>> mdl_puma560
>> sl_ctorque
```

The desired joint angles and velocity are generated using a `jtraj` block whose parameters are the initial and final joint angles. We run the simulation by pushing the Simulink play button or

```
>> r = sim('sl_ctorque');
```

9.4.3   Operational Space Control

The control strategies so far have been posed in terms of the robot's joint coordinates – its configuration space. Equation 9.8 describes the relationship between joint position, velocity, acceleration and applied forces or torques. However we can also express the dynamics of the end-effector in the Cartesian operational space where we consider the end-effector as a rigid body with inertia that actuator and disturbance forces and torques act on. We can reformulate Eq. 9.8 in operational space as

$$\Lambda(x)\ddot{x} + \mu(x, \dot{x})\dot{x} + p(x) = W \tag{9.15}$$

where $x \in \mathbb{R}^6$ is the manipulator Cartesian pose and Λ is the end-effector inertia which is subject to a gyroscopic and Coriolis force μ and gravity load p and an applied control wrench W. These operational space terms are related to those we have already discussed by

$$\Lambda(x) = J(q)^{-T} J(q) J(q)^{-1}$$
$$\mu(x, \dot{x}) = J(q)^{-T} C(q, \dot{q}) - \Lambda(q)\dot{J}(q)\dot{q}$$
$$p(x) = J(q)^{-T} g(q)$$

Imagine the task of wiping a table when the table's height is unknown and its surface is only approximately horizontal. The robot's z-axis is vertical so to achieve the task we need to move the end-effector along a path in the xy-plane to achieve coverage and hold the wiper at a constant orientation about the z-axis. Simultaneously we maintain a constant force in the z-direction to hold the wiper against the table and a constant torque about the x- and y-axes in order to conform to the orientation of the table top. The first group of axes are position controlled, and the second group are force controlled. Each Cartesian degree of freedom can be either position or force controlled. The operational space control allows independent control of position and forces along and about the axes of the operational space coordinate frame.

A Simulink model of the controller and a simplified version of this scenario can be loaded by

```
>> sl_opspace
```

and is shown in Fig. 9.22. It comprises a position-control loop and a force-control loop whose results are summed together and used to drive the operational space robot model – details can be found by opening that block in the Simulink diagram. In this simulation the operational space coordinate frame is parallel to the end-effector coordinate frame. Motion is position controlled in the x- and y-directions and about the x-, y- and z-axes of this frame – the robot moves from its initial pose to a nearby pose using 5 out of the 6 Cartesian DOF. ◀

The robot model and the compliance specification are set by the model's `InitFcn` callback function. The setpoints are the red user adjustable boxes in the top-level diagram.

Motion is force controlled in the z-direction with a setpoint of –5 N. To achieve this the controller moves the end-effector downward in order to decrease the force. It moves in free space until it touches the surface at $z = -0.2$ which is modeled as a stiffness of 100 N m^{-1}. Results in Fig. 9.23 show the x- and y-position moving toward the goal and the z-position decreasing and the simulated sensed force decreasing after contact. The controller is able to simultaneously satisfy position and force constraints.

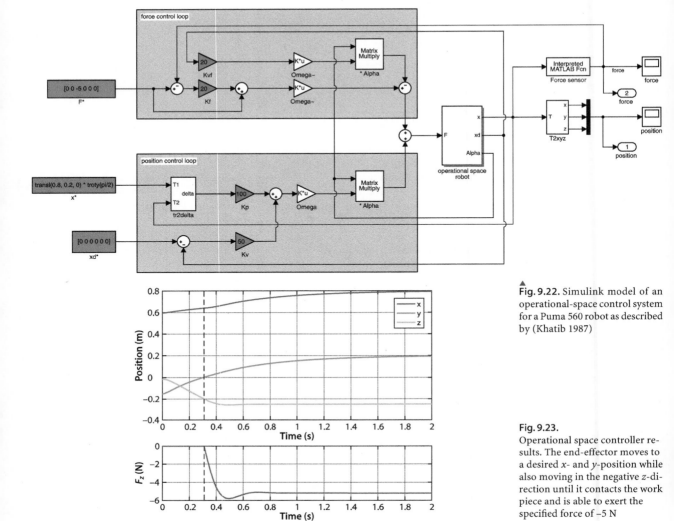

Fig. 9.22. Simulink model of an operational-space control system for a Puma 560 robot as described by (Khatib 1987)

Fig. 9.23.
Operational space controller results. The end-effector moves to a desired x- and y-position while also moving in the negative z-direction until it contacts the work piece and is able to exert the specified force of –5 N

9.5   Applications

9.5.1   Series-Elastic Actuator (SEA)

For high-speed robots the elasticity of the links and the joints becomes a significant dynamic effect which will affect path following accuracy. Joint elasticity is typically caused by elements of the transmission such as: longitudinal elasticity of a toothed belt or cable drive, a harmonic gearbox which is inherently elastic, or torsional elasticity of a motor shaft. In dynamic terms, as shown schematically in Fig. 9.24, the problem arises because the force is applied to one side of an elastic element and we wish to control the position of the other side – the actuator and sensor are not colocated. More complex still, and harder to analyze, is the case where the elasticity of the links must be taken into account.

However there are advantages in having some flexibility between the motor and the load. Imagine a robot performing a task that involves the gripper picking an object off a table whose height is uncertain.▶ A simple strategy to achieve this is to move down until the gripper touches the table, close the gripper and then lift up. However at the instant of contact a large and discontinuous force will be exerted on the robot which has the potential to damage the object or the robot. This is particularly problematic for robots with large

Or the robot is not very accurate.

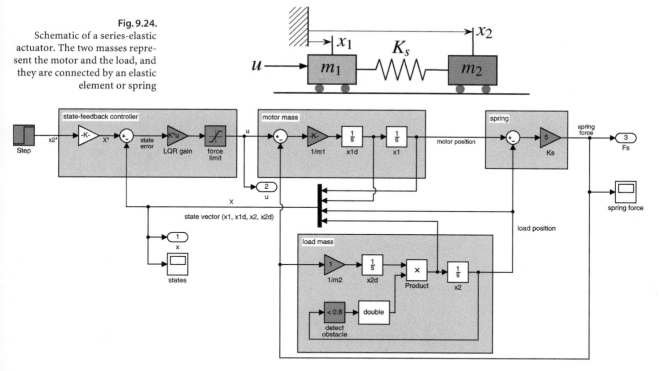

Fig. 9.24.
Schematic of a series-elastic actuator. The two masses represent the motor and the load, and they are connected by an elastic element or spring

Fig. 9.25. Simulink model `sl_sea` of a series-elastic actuator colliding with an obstacle

inertia that are moving quickly – the kinetic energy must be instantaneously dissipated. An elastic element – a spring – between the motor and the joint would help here. At the moment of contact the spring would start to compress and the kinetic energy is transferred to potential energy in the spring – the robot control system has time to react and stop or reverse the motors. We have changed the problem from a damaging hard impact to a soft impact. In addition to shock absorption, the deformation of the spring provides a means of determining the force that the robot is exerting. This capability is particularly useful for robots that interact closely with people since it makes the robot less dangerous in case of collision, and a spring is simple technology that cannot fail. For robots that must exert a force as part of their task, this is a simpler approach than the operational space controller introduced in Sect. 9.4.3. However position control is now more challenging because there is an elastic element between the motor and the load.

Consider the 1-dimensional case shown in Fig. 9.24 where the motor is represented by a mass m_1 to which a controllable force u is applied. ◀ It is connected via a linear elastic element or spring to the load mass m_2. If we apply a positive force to m_1 it will move to the right and compress the spring, and this will exert a positive force on m_2 which will also move to the right. Controlling the position of m_2 is not trivial since this system has no friction and is marginally stable. It can be stabilized by feedback of position and velocity of the motor *and* of the load – all of which are potentially measurable.

In robotics such a system, built into a robot joint, is known as a series-elastic actuator or SEA. The Baxter robot of Fig. 7.1b includes SEAs in some of its joints.

A Simulink model of an SEA system can be loaded by

```
>> sl_sea
```

and is shown in Fig. 9.25. A state-feedback LQR controller has been designed using MATLAB and requires input of motor and load position and velocity which form a vector x in the Simulink model. Fig. 9.26 shows a simulation of the model moving the load m_2 to a position $x_2^* = 1$. In the first case there is no obstacle and it achieves the goal with minimal overshoot, but note the complex force profile applied to m_1. In the second case the load mass is stopped at $x_2 = 0.8$ and the elastic force changes to accomodate this.

In a real robot this is a rotary system with a torsional spring.

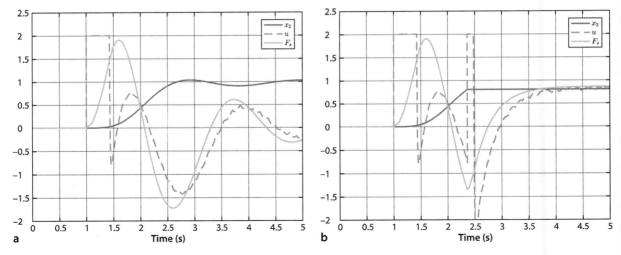

Fig. 9.26. Response of the series-elastic actuator to a unit-step demand at $t = 1$ s, showing load position (m), motor force (N) and spring force (N). **a** Moving to $x_2^* = 1$ with no collision; **b** moving to $x_2^* = 1$ with an obstacle at $x_2 = 0.8$ which is reached at $t \approx 2.3$

9.6   Wrapping Up

In this Chapter we discussed approaches to robot manipulator control. We started with the simplest case of independent joint control, and explored the effect of disturbance torques and variation in inertia, and showed how feedforward of disturbances such as gravity could provide significant improvement in performance. We then learned how to model the forces and torques acting on the individual links of a serial-link manipulator. The equations of motion or inverse dynamics compute the joint forces required to achieve particular joint velocity and acceleration. The equations have terms corresponding to inertia, gravity, velocity coupling, friction and externally applied forces. We looked at the significance of these terms and how they vary with manipulator configuration and payload. The equations of motion provide insight into important issues such as how the velocity or acceleration of one joint exerts a disturbance force on other joints which is important for control design. We then discussed the forward dynamics which describe how the configuration evolves with time in response to forces and torques applied at the joints by the actuators and by external forces such as gravity. We extended the feedforward notion to full model-based control using torque feedforward, computed torque and operational-space controllers. Finally we discussed series-elastic actuators where a compliant element between the robot motor and the link enables force control and people-safe operation.

Further Reading

The engineering design of motor control systems is covered in mechatronics textbooks such as Bolton (2015). The dynamics of serial-link manipulators is well covered by all the standard robotics textbooks such as Paul (1981), Spong et al. (2006), Siciliano et al. (2009) and the Robotics Handbook (Siciliano and Khatib 2016). The efficient recursive Newton-Euler method we use today is the culmination of much research in the early 1980s and described in Hollerbach (1982). The equations of motion can be derived via a number of techniques, including Lagrangian (energy based), Newton-Euler, d'Alembert (Fu et al. 1987; Lee et al. 1983) or Kane's method (Kane and Levinson 1983). However the computational cost of Lagrangian methods (Uicker 1965; Kahn 1969) is enormous, $O(N^4)$, which made it infeasible for real-time use on computers of that era and many simplifications and approximation had to be made. Orin et al. (1979) proposed an alternative approach based on the Newton-Euler (NE) equations of rigid-body motion applied to each link. Armstrong (1979) then showed how recursion could be applied resulting in

$O(N)$ complexity. Luh et al. (1980) provided a recursive formulation of the Newton-Euler equations with linear and angular velocities referred to link coordinate frames which resulted in a thousand-fold improvement in execution time making it practical to implement in real-time. Hollerbach (1980) showed how recursion could be applied to the Lagrangian form, and reduced the computation to within a factor of 3 of the recursive NE form, and Silver (1982) showed the equivalence of the recursive Lagrangian and Newton-Euler forms, and that the difference in efficiency was due to the representation of angular velocity.

The forward dynamics, Sect. 9.3, is computationally more expensive. An $O(N^3)$ method was proposed by Walker and Orin (1982) and is used in the Toolbox. Featherstone's (1987) articulated-body method has $O(N)$ complexity but for $N < 9$ is more expensive than Walker's method.

Critical to any consideration of robot dynamics is knowledge of the inertial parameters, ten per link, as well as the motor's parameters. Corke and Armstrong-Hélouvry (1994, 1995) published a meta-study of Puma parameters and provide a consensus estimate of inertial and motor parameters for the Puma 560 robot. Some of this data was obtained by painstaking disassembly of the robot and determining the mass and dimensions of the components. Inertia of components can be estimated from mass and dimensions by assuming mass distribution, or it can be measured using a bifilar pendulum as discussed in Armstrong et al. (1986).

Alternatively the parameters can be estimated by measuring the joint torques or the base reaction force and moment as the robot moves. A number of early works in this area include Mayeda et al. (1990), Izaguirre and Paul (1985), Khalil and Dombre (2002) and a more recent summary is Siciliano and Khatib (2016, § 6). Key to successful identification is that the robot moves in a way that is sufficiently exciting (Gautier and Khalil 1992; Armstrong 1989). Friction is an important dynamic characteristic and is well described in Armstrong's (1988) thesis. The survey by Armstrong-Hélouvry et al. (1994) is a very readable and thorough treatment of friction modeling and control. Motor parameters can be obtained directly from the manufacturer's data sheet or determined experimentally, without having to remove the motor from the robot, as described by Corke (1996a). The parameters used in the Toolbox Puma model are the best estimates from Corke and Armstrong-Hélouvry (1995) and Corke (1996a).

The discussion on control has been quite brief and has strongly emphasized the advantages of feedforward control. Robot joint control techniques are well covered by Spong et al. (2006), Craig (2005) and Siciliano et al. (2009) and summarized in Siciliano and Khatib (2016, § 8). Siciliano et al. have a good discussion of actuators and sensors as does the, now quite old, book by Klafter et al. (1989). The control of flexible joint robots is discussed in Spong et al. (2006). Adaptive control can be used to accomodate the time-varying inertial parameters and there is a large literature on this topic but some good early references include the book by Craig (1987) and key papers include Craig et al. (1987), Spong (1989), Middleton and Goodwin (1988) and Ortega and Spong (1989). The operational-space control structure was proposed in Khatib (1987). There has been considerable recent interest in series-elastic as well as variable stiffness actuators (VSA) whose position and stiffness can be independently controlled much like our own muscles – a good collection of articles on this technology can be found in the special issue by Vanderborght et al. (2008).

Dynamic manipulability is discussed in Spong et al. (2006) and Siciliano et al. (2009). The Asada measure used in the Toolbox is described in Asada (1983).

Historical and general. Newton's second law is described in his master work *Principia Nautralis* (mathematical principles of natural philosophy), written in Latin but an English translation is available on line at http://www.archive.org/details/newton-spmathema00newtrich. His writing on other subjects, including transcripts of his notebooks, can be found online at http://www.newtonproject.sussex.ac.uk.

Exercises

1. Independent joint control (page 258ff)
   a) Investigate different values of `Kv` and `Ki` as well as demand signal shape and amplitude.
   b) Perform a root-locus analysis of `vloop` to determine the maximum permissible gain for the proportional case. Repeat this for the PI case.
   c) Consider that the motor is controlled by a voltage source instead of a current source, and that the motor's impedance is 1 mH and 1.6 Ω. Modify `vloop` accordingly. Extend the model to include the effect of back EMF.
   d) Increase the required speed of motion so that the motor torque becomes saturated. With integral action you will observe a phenomena known as integral windup – examine what happens to the state of the integrator during the motion. Various strategies are employed to combat this, such as limiting the maximum value of the integrator, or only allowing integral action when the motor is close to its setpoint. Experiment with some of these.
   e) Create a Simulink model of the Puma robot with each joint controlled by `vloop` and `ploop`. Parameters for the different motors in the Puma are described in Corke and Armstrong-Hélouvry (1995).
2. The motor torque constant has units of N m A^{-1} and is equal to the back EMF constant which has units of V s rad^{-1}. Show that these units are equivalent.
3. Simple two-link robot arm of Fig. 9.4
   a) Plot the gravity load as a function of both joint angles. Assume $m_1 = 0.45$ kg, $m_2 = 0.35$ kg, $r_1 = 8$ cm and $r_2 = 8$ cm.
   b) Plot the inertia for joint 1 as a function of q_2. To compute link inertia assume that we can model the link as a point mass located at the center of mass.
4. Run the code on page 265 to compute gravity loading on joints 2 and 3 as a function of configuration. Add a payload and repeat.
5. Run the code on page 266 to show how the inertia of joints 1 and 2 vary with payload?
6. Generate the curve of Fig. 9.16c. Add a payload and compare the results.
7. By what factor does this inertia vary over the joint angle range?
8. Why is the manipulator inertia matrix symmetric?
9. The robot exerts a wrench on the base as it moves (page 269). Consider that the robot is sitting on a frictionless horizontal table (say on a large air puck). Create a simulation model that includes the robot arm dynamics and the sliding dynamics on the table. Show that moving the arm causes the robot to translate and spin. Can you devise an arm motion that moves the robot base from one position to another and stops?
10. Overlay the dynamic manipulability ellipsoid on the display of the robot. Compare this with the force ellipsoid from Sect. 8.5.2.
11. Model-based control (page 273ff)
   a) Compute and display the joint tracking error for the torque feedforward and computed torque cases. Experiment with different motions, control parameters and sample rate T_{fb}.
   b) Reduce the rate at which the feedforward torque is computed and observe its effect on tracking error.
   c) In practice the dynamic model of the robot is not exactly known, we can only invert our best estimate of the rigid-body dynamics. In simulation we can model this by using the `perturb` method, see the online documentation, which returns a robot object with inertial parameters varied by plus and minus the specified percentage. Modify the Simulink models so that the RNE block is using a robot model with parameters perturbed by 10%. This means that the inverse dynamics are computed for a slightly different dynamic model to the robot under control and shows the effect of model error on control performance. Investigate the effects on error for both the torque feedforward and computed torque cases.

d) Expand the operational-space control example to include a sensor that measures all the forces and torques exerted by the robot.on an inclined table surface. Move the robot end-effector along a circular path in the xy-plane while exerting a constant downward force – the end-effector should move up and down as it traces out the circle. Show how the controller allows the robot tool to conform to a surface with unknown height and surface orientation.

12. Series-elastic actuator (page 276)

    a) Experiment with different values of stiffness for the elastic element and control parameters. Try to reduce the settling time.

    b) Modify the simulation so that the robot arm moves to touch an object at unknown distance and applies a force of 5 N to it.

    c) Plot the frequency response function $X_2(s)/X_1(s)$ for different values of K_s, m_1 and m_2.

    d) Simulate the effect of a collision between the load and an obstacle by adding a step to the spring force.

Part IV Computer Vision

IV Computer Vision

*Vision is the process of discovering from images
what is present in the world and where it is.*
David Marr

Almost all animal species use eyes – in fact evolution has *invented* the eye many times over. Figure IV.1 shows a variety of eyes from nature: the compound eye of a fly, the main and secondary eyes of a spider, the reflector-based eyes of a scallop, and the lens-based eye of a human. Vertebrates have two eyes, but spiders and scallops have many eyes.

Even very simple animals, bees for example, with brains comprising just 10^6 neurons (compared to our 10^{11}) are able to perform complex and life critical tasks such as finding food and returning it to the hive using vision (Srinivasan and Venkatesh 1997). This is despite the very high biological cost of owning an eye: the complex eye itself, muscles to move it, eyelids and tear ducts to protect it, and a large visual cortex (relative to body size) to process its data.

Our own experience is that eyes are very effective sensors for recognition, navigation, obstacle avoidance and manipulation. Cameras mimic the function of an eye and we wish to use cameras to create vision-based competencies for robots – to use digital images to recognize objects and navigate within the world. Figure IV.2 shows a robot with a number of different types of cameras.

Technological development has made it feasible for robots to use cameras as eyes. For much of the history of computer vision, dating back to the 1960s, electronic cameras were cumbersome and expensive and computer power was inadequate. Today CMOS cameras for cell phones cost just a few dollars each, and our mobile and personal computers come standard with massive parallel computing power. New algorithms, cheap sensors and plentiful computing power make vision a practical sensor today.

In Chap. 1 we defined a robot as

a goal oriented machine that can sense, plan and act

and this part of the book is concerned with sensing using vision, or visual perception. Whether a robot works in a factory or a field it needs to sense its world in order to plan its actions.

In this part of the book we will discuss the process of vision from start to finish: from the light falling on a scene, being reflected, gathered by a lens, turned into a digital image and processed by various algorithms to extract the information required to support the robot competencies listed above. These steps are depicted graphically in Fig. IV.3.

Fig. IV.1. a Robber fly, *Holocephala fusca*; **b** jumping spider, *Phidippus putnami* (**a** and **b** courtesy Thomas Shahan, thomasshanan.com). **c** Scallop (courtesy Sönke Johnsen), each of the small blue spheres is an eye. **d** Human eye

Development of the eye. It is believed that all animal eyes share a common ancestor in a proto-eye that evolved 540 million years ago. However major evolutionary advances seem to have occurred in just the last few million years. The very earliest eyes, called eyespots, were simple patches of photoreceptor protein in single-celled animals. Multi-celled animals evolved multi-cellular eyespots which could sense the brightness of light but not its direction. Gradually the eyespot evolved into a shallow cup shape which gave a limited ability to discriminate directional brightness according to which cells were illuminated. The pit deepened, the opening became smaller, and the number of photoreceptor cells increased, forming a pin-hole camera that was capable of distinguishing shapes. Next came an overgrowth of transparent cells to protect the eyespot which led to a filled eye chamber and eventually the eye as we know it today. The lensed eye has evolved independently seven different times across species. Nature has evolved ten quite distinct eye designs including those shown above.

Fig. IV.2.
A cluster of cameras on an outdoor mobile robot: forward looking stereo pair, side looking wide angle camera, overhead panoramic camera mirror (CSIRO mobile robot)

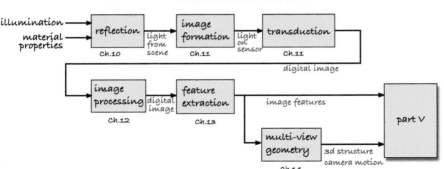

Fig. IV.3.
Steps involved in image processing

In Chap. 10 we start by discussing light, and in particular color because it is such an important characteristic of the world that we perceive. Although we learn about color at kindergarten it is a complex topic that is often not well understood. Next, in Chap. 11, we discuss how an image of the world is formed on a sensor and converted to a digital image that can be processed by a computer. Fundamental image processing algorithms are covered in Chap. 12 and provide the foundation for the feature extraction algorithms discussed in Chap. 13. Feature extraction is a problem in data reduction, in extracting the *essence* of the scene from the massive amount of pixel data. For example, how do we determine the coordinate of the round red object in the scene, which can be described with perhaps just 4 bytes, given the millions of bytes that comprise an image. To solve this we must address many important subproblems such as "what is red?", "how do we distinguish red pixels from nonred pixels?", "how do we describe the shape of the red pixels?", "what if there are more than one red object?" and so on.

As we progress through these chapters we will encounter the limitations of using just a single camera to view the world. Once again biology shows the way – multiple eyes are common and have great utility. This leads us to consider using multiple views of the world, from a single moving camera or multiple cameras observing the scene from different viewpoints. This is discussed in Chap. 14 and is particularly important for understanding the 3-dimensional structure of the world. All of this sets the scene for describing how vision can be used for closed-loop control of arm-type and mobile robots which is the subject of the next and final part of the book.

10

Light and Color

*I cannot pretend to feel impartial about colours.
I rejoice with the brilliant ones
and am genuinely sorry for the poor browns.*
Winston Churchill

In ancient times it was believed that the eye radiated a cone of visual flux which mixed with visible objects in the world to create a sensation in the observer – like the sense of touch, but at a distance – this is the extromission theory. Today we consider that light from an illuminant falls on the scene, some of which is reflected into the eye of the observer to create a perception about that scene. The light that reaches the eye, or the camera, is a function of the illumination impinging on the scene and the material property known as reflectivity.

This chapter is about light itself and our perception of light in terms of brightness and color. Section 10.1 describes light in terms of electro-magnetic radiation and mixtures of light as continuous spectra. Section 10.2 provides a brief introduction to colorimetry, the science of color perception, human trichromatic color perception and how colors can be represented in various color spaces. Section 10.3 covers a number of advanced topics such as color constancy, gamma correction and white balancing. Section 10.4 has two worked application examples concerned with distinguishing different colored objects in an image and the removal of shadows in an image.

10.1 Spectral Representation of Light

Around 1670, Sir Isaac Newton discovered that white light was a mixture of different colors. We now know that each of these colors is a single frequency or wavelength of electro-magnetic radiation. We perceive the wavelengths between 400 and 700 nm as different colors as shown in Fig. 10.1.

In general the light that we observe is a mixture of many wavelengths and can be represented as a function $E(\lambda)$ that describes intensity as a function of wavelength λ. Monochromatic light, such as emitted by a laser comprises a single wavelength in which case E is an impulse.

The most common source of light is incandescence which is the emission of light from a hot body such as the Sun or the filament of a traditional light bulb. In physics

Fig. 10.1.
The spectrum of visible colors as a function of wavelength in nanometers. The visible range depends on viewing conditions and the individual but is generally accepted as being 400–700 nm. Wavelengths greater than 700 nm are termed infra-red and those below 400 nm are ultra-violet

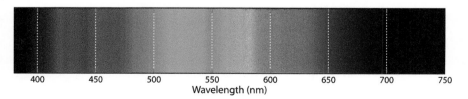

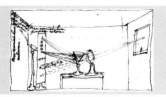

Spectrum of light. During the plague years of 1665–1666 Isaac Newton developed his theory of light and color. He demonstrated that a prism could decompose white light into a spectrum of colors, and that a lens and a second prism could recompose the multi-colored spectrum into white light. Importantly he showed that the color of the light did not change when it was reflected from different objects, from which he concluded that color is an intrinsic property of light not the object. (Newton's sketch to the left)

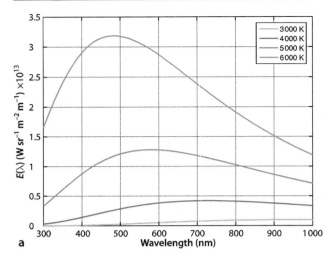

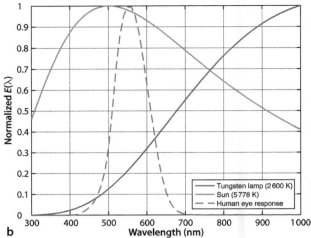

a   Wavelength (nm)

b   Wavelength (nm)

this is modeled as a blackbody radiator or Planckian source. The emitted power as a function of wavelength λ is given by Planck's radiation formula

$$E(\lambda) = \frac{2hc^2}{\lambda^5\left(e^{hc/k\lambda T} - 1\right)} \text{ W sr}^{-1} \text{ m}^{-2} \text{ m}^{-1} \tag{10.1}$$

where T is the absolute temperature (K) of the source, h is Planck's constant, k is Boltzmann's constant, and c the speed of light.◢ This is the power emitted per steradian◢ per unit area per unit wavelength.

We can plot the emission spectra for a blackbody at different temperatures. First we define a range of wavelengths

```
>> lambda = [300:10:1000]*1e-9;
```

in this case from 300 to 1 000 nm, and then compute the blackbody spectra

```
>> for T=3000:1000:6000
>>   plot( lambda, blackbody(lambda, T)); hold all
>> end
```

as shown in Fig. 10.2a. We can see that as temperature increases the maximum amount of power increases and the wavelength at which the peak occurs decreases. The total amount of power radiated (per unit area) is the area under the blackbody curve and is given by the Stefan-Boltzman law

$$P(\lambda) = \frac{2\pi^5 k^4}{15c^2 h^3} T^4 \text{ W m}^{-2}$$

and the wavelength corresponding to the peak of the blackbody curve is given by Wien's displacement law

$$\lambda_{max} = \frac{2.8978 \times 10^{-3}}{T} \text{ m}$$

The wavelength of the peak decreases as temperature increases and in familiar terms this is what we observe when we heat an object. It starts to glow faintly red at around 800 K and moves through orange and yellow toward white as temperature increases.◢

Fig. 10.2. Blackbody spectra. **a** Blackbody emission spectra for temperatures from 3 000–6 000 K. **b** Blackbody emissions for the Sun (5 778 K), a tungsten lamp (2 600 K) and the response of the human eye – all normalized to unity for readability

$c = 2.998 \times 10^8 \text{ m s}^{-1}$
$h = 6.626 \times 10^{-34} \text{ Js}$
$k = 1.381 \times 10^{-23} \text{ J K}^{-1}$

Solid angle is measured in steradians, a full sphere is 4π sr.

Incipient red heat	770 – 820 K
dark red heat	920 – 1 020 K
bright red heat	1 120 – 1 220 K
yellowish red heat	1 320 – 1 420 K
incipient white heat	1 520 – 1 620 K
white heat	1 720 – 1 820 K

Infra-red radiation was discovered in 1800 by William Herschel (1738–1822) the German-born British astronomer. He was Court Astronomer to George III; built a series of large telescopes; with his sister Caroline performed the first sky survey discovering double stars, nebulae and the planet Uranus; and studied the spectra of stars. Using a prism and thermometers to measure the amount of heat in the various colors of sunlight he observed that temperature increased from blue to red, and increased even more beyond red where there was no visible light. (Image from Herschel 1800)

Sir Humphry Davy demonstrated the first electrical incandescent lamp using a platinum filament in 1802. Sir Joseph Swan demonstrated his first light bulbs in 1850 using carbonized paper filaments. However it was not until advances in vacuum pumps in 1865 that such lamps could achieve a useful lifetime. Swan patented a carbonized cotton filament in 1878 and a carbonized cellulose filament in 1881. His lamps came into use after 1880 and the Savoy Theatre in London was completely lit by electricity in 1881. In the USA Thomas Edison did not start research into incandescent lamps until 1878 but he patented a long-lasting carbonized bamboo filament the next year and was able to mass produce them. The Swan and Edison companies merged in 1883.

The light bulb subsequently became the dominant source of light on the planet but is now being phased out due to its poor energy efficiency. (Photo by Douglas Brackett, Inv., Edisonian.com)

The filament of a tungsten lamp has a temperature of 2 600 K and glows *white hot*. The Sun has a surface temperature of 5 778 K. The spectra of these sources

```
>> lamp = blackbody(lambda, 2600);
>> sun = blackbody(lambda, 5778);
>> plot(lambda, [lamp/max(lamp) sun/max(sun)])
```

are compared in Fig. 10.2b. The tungsten lamp curve is much lower in magnitude, but has been scaled up (by 56) for readability. The peak of the Sun's emission is around 500 nm and it emits a significant amount of power in the visible part of the spectrum. The peak for the tungsten lamp is at a much longer wavelength and perversely most of its power falls in the infra-red band which we perceive as heat not light.

10.1.1   Absorption

The Sun's spectrum at ground level on the Earth has been measured and tabulated

```
>> sun_ground = loadspectrum(lambda, 'solar');
>> plot(lambda, sun_ground)
```

Fig. 10.3. a Modified solar spectrum at ground level (*blue*). The dips in the solar spectrum correspond to various water absorption bands. CO_2 absorbs radiation in the infra-red region, and ozone O_3 absorbs strongly in the ultra-violet region. The Sun's blackbody spectrum is shown in *dashed blue* and the response of the human eye is shown in *red*. **b** Transmission through 5 m of water. The longer wavelengths, reds, have been strongly attenuated

and is shown in Fig. 10.3a. It differs markedly from that of a blackbody since some wavelengths have been absorbed more than others by the atmosphere. Our eye's peak sensitivity has evolved to be closely aligned to the peak of the spectrum of atmospherically filtered sunlight.

Transmittance T is the inverse of absorptance, and is the fraction of light passed as a function of wavelength and distance traveled. It is described by Beer's law

$$T = 10^{-Ad} \tag{10.2}$$

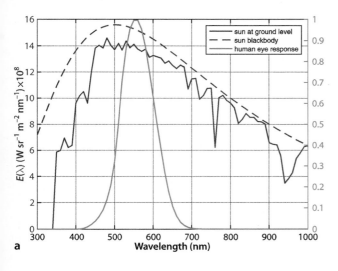

a

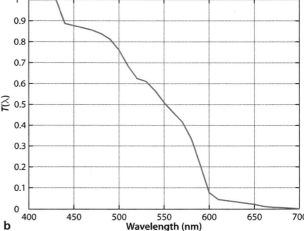

b

where A is the absorption coefficient in units of m^{-1} which is a function of wavelength, and d is the optical path length. The absorption spectrum $A(\lambda)$ for water is loaded from tabulated data

```
>> [A, lambda] = loadspectrum([400:10:700]*1e-9, 'water');
```

and the transmission through 5 m of water is

```
>> d = 5;
>> T = 10.^(-A*d);
>> plot(lambda, T);
```

which is plotted in Fig. 10.3b. We see that the red light is strongly attenuated which makes the object appear more blue. Differential absorption of wavelengths is a significant concern when imaging underwater and we revisit this topic in Sect. 10.3.4.

10.1.2   Reflectance

Surfaces reflect incoming light. The reflection might be specular (as from a mirror-like surface, see page 337), or Lambertian (diffuse reflection from a matte surface, see page 309). The fraction of light that is reflected $R \in [0, 1]$ is the reflectivity, reflectance or albedo of the surface and is a function of wavelength. White paper for example has a reflectance of around 70%. The reflectance spectra of many materials have been measured and tabulated.► Consider for example the reflectivity of a red house brick

From http://speclib.jpl.nasa.gov/ weathered red brick (0412UUUBRK).

```
>> [R, lambda] = loadspectrum([100:10:10000]*1e-9, 'redbrick');
>> plot(lambda, R);
```

which is plotted in Fig. 10.4 and shows that it reflects red light more than blue.

10.1.3   Luminance

The light reflected from a surface, its luminance, has a spectrum given by

$$L(\lambda) = E(\lambda)R(\lambda) \ W \ m^{-2} \tag{10.3}$$

where E is the incident illumination and R is the reflectance. The illuminance of the Sun in the visible region is

```
>> lambda = [400:700]*1e-9;
>> E = loadspectrum(lambda, 'solar');
```

Fig. 10.4. Reflectance of a weathered red house brick (data from ASTER, Baldridge et al. 2009). **a** Full range measured from 300 nm visible to 10 000 nm (infra-red); **b** closeup of visible region

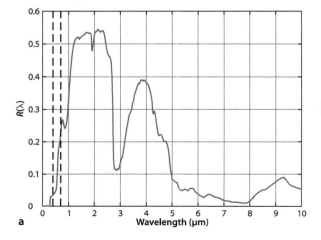

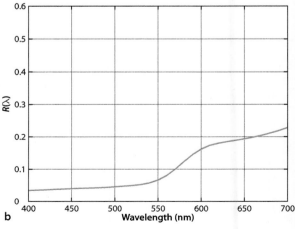

a   Wavelength (μm)

b   Wavelength (nm)

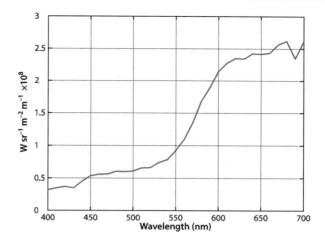

Fig. 10.5.
Luminance of the weathered red house brick under illumination from the Sun at ground level, based on data from Fig. 10.3a and 10.4b

at ground level. The reflectivity of the brick is

```
>> R = loadspectrum(lambda, 'redbrick');
```

and the light reflected from the brick is

```
>> L = E .* R;
>> plot(lambda, L);
```

which is shown in Fig. 10.5. It is this spectrum that is interpreted by our eyes as the color red.

10.2 Color

Color is the general name for all sensations arising from the activity of the retina of the eye and its attached nervous mechanisms, this activity being, in nearly every case in the normal individual, a specific response to radiant energy of certain wavelengths and intensities.
T. L. Troland, Report of Optical Society of America Committee on Colorimetry 1920–1921

We have described the spectra of light in terms of power as a function of wavelength, but our own perception of light is in terms of subjective quantities such as brightness and color. Light that is visible to humans lies in the range of wavelengths from 400 nm (violet) to 700 nm (red) with the colors blue, green, yellow and orange in between, as shown in Fig. 10.1.

The brightness we associate with a particular wavelengths is given by the luminosity function with units of lumens per watt. For our daylight (photopic) vision the luminosity as a function of wavelength has been experimentally determined, tabulated and forms the basis of the 1931 CIE standard that represents the average human observer.◄ The photopic luminosity function is provided by the Toolbox

This is the photopic response for a light-adapted eye using the cone photoreceptor cells. The dark adapted, or scotopic response, using the eye's monochromatic rod photoreceptor cells is different, and peaks at around 510 nm.

```
>> human = luminos(lambda);
>> plot(lambda,  human)
```

Radiometric and photometric quantities. Two quite different sets of units are used when discussing light: radiometric and photometric. Radiometric units are used in Sect. 10.1 and are based on quantities like power which are expressed in familiar SI units such as Watts.

Photometric units are analogs of radiometric units but take into account the *visual sensation* in the observer. Luminous power or luminous flux is the *perceived* power of a light source and is measured in *lumens* (abbreviated to lm) rather than *Watts*.

A 1 W monochromatic light source at 555 nm, the peak response, *by definition* emits a luminous flux of 683 lm. By contrast a 1 W light source at 800 nm emits a luminous flux of 0 lm – it causes no visual sensation at all.

A 1 W incandescent lightbulb however produces a perceived visual sensation of less than 15 lm or a luminous efficiency of 15 lm W^{-1}. Fluorescent lamps achieve efficiencies up to 100 lm W^{-1} and white LEDs up to 150 lm W^{-1}.

and is shown in Fig. 10.7a. Consider two light sources emitting the same power (in watts) but one has a wavelength of 550 nm (green) and the other has a wavelength of 450 nm (blue). The perceived brightness of these two lights is quite different, in fact the blue light appears only

```
>> luminos(450e-9) / luminos(550e-9)
ans =
    0.0382
```

or 3.8% as bright as the green one. The silicon sensors used in digital cameras have strong sensitivity in the red and infra-red part of the spectrum.▶

The LED on an infra-red remote control can be seen as a bright light in most digital cameras – try this with your mobile phone camera and TV remote. Some security cameras provide infra-red scene illumination for covert night time monitoring. Note that some cameras are fitted with infra-red filters to prevent the sensor becoming saturated by ambient infra-red radiation.

10.2.1   The Human Eye

Our eyes contain two types of light-sensitive cells as shown in Fig. 10.6. Rod cells are much more sensitive than cone cells but respond to intensity only and are used at night. In normal daylight conditions our cone photoreceptors are active and these are color sensitive. Humans are trichromats and have three types of cones that respond to different parts of the spectrum. They are referred to as long (L), medium (M) and short (S) according to the wavelength of their peak response, or more commonly as red, green and blue. The spectral response of rods and cones has been extensively studied and the response of human cone cells can be loaded

```
>> cones = loadspectrum(lambda, 'cones');
>> plot(lambda, cones)
```

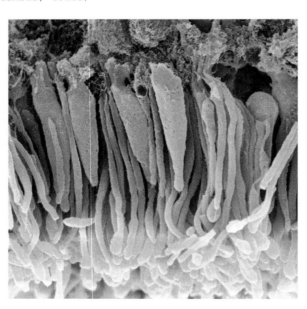

Fig. 10.6.
A colored scanning electron micrograph of rod cells (*white*) and cone cells (*yellow*) in the human eye. The cells diameters are in the range 0.5–4 µm. The cells contain different types of light-sensitive opsin proteins. Surprisingly the rods and cones are not on the surface of the retina, they are behind that surface which is a network of nerves and blood vessels

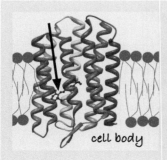

cell body

Opsins are the photoreceptor molecules used in the visual systems of all animals. They belong to the class of G protein-coupled receptors (GPCRs) and comprise seven helices that pass through the cell's membrane. They change shape in response to particular molecules outside the cell and initiate a cascade of chemical signaling events inside the cell that results in a change in cell function. Opsins contain a chromophore, a light-sensitive molecule called retinal derived from vitamin A, that stretches across the opsin. When retinal absorbs a photon its changes its shape which deforms the opsin and activates the cell's signalling pathway. The basis of all vision is a fortuitous genetic mutation 700 million years ago that made a chemical sensing receptor light sensitive. There are many opsin variants across the animal kingdom – our rod cells contain rhodopsin and our cone cells contain photopsins. The American biochemist George Wald (1906–1997) received the 1967 Nobel Prize in Medicine for his discovery of retinal and characterizing the spectral absorbance of photopsins. (Image by Dpyran from Wikipedia, the chromophore is indicated by the arrow)

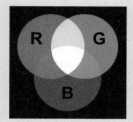

The **trichromatic theory** of color vision suggests that our eyes have three discrete types of receptors that when stimulated produce the sensations of red, green and blue, and that all color sensations are "psychological mixes" of these fundamental colors. It was first proposed by the English scientist Thomas Young (1773–1829) in 1802 but made little impact. It was later championed by Hermann von Helmholtz and James Clerk Maxwell. The figure on left shows how beams of red, green and blue light mix. Helmholtz (1821–1894) was a prolific German physician and physicist. He invented the opthalmascope for examing the retina in 1851, and in 1856 he published the "Handbuch der physiologischen Optik" (Handbook of Physiological Optics) which contained theories and experimental data relating to depth perception, color vision, and motion perception. Maxwell (1831–1879) was a Scottish scientist best known for his electro-magnetic equations, but who also extensively studied color perception, color-blindness, and color theory. His 1860 paper "On the Theory of Colour Vision" won a Rumford medal, and in 1861 he demonstrated color photography in a Royal Institution lecture.

The **opponent color theory** holds that colors are perceived with respect to two axes: red-green and blue-yellow. One clue comes from color after-images – staring at a red square and then a white surface gives rise to a green after-image. Another clue comes from language – we combine color words to describe mixtures, for example redish-blue, but we never describe a reddish-green or a blueish-yellow. The theory was first mooted by the German writer Johann Wolfgang von Goethe (1749–1832) in his 1810 "Theory of Colours" but later had a strong advocate in Karl Ewald Hering (1834–1918), a German physiologist who also studied binocular perception and eye movements. He advocated opponent color theory over trichromatic theory and had acrimonious debates with Helmholtz on the topic.

In fact both theories hold. Our eyes have three types of color or sensing cells but the early processing in the retinal ganglion layer appears to convert these signals into an opponent color representation.

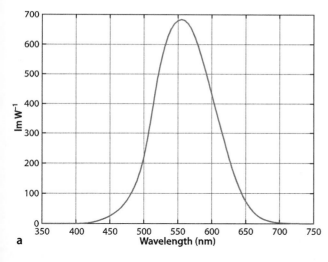

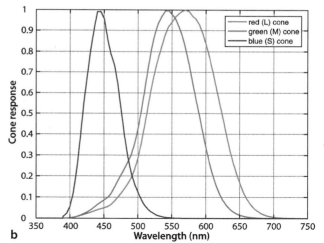

Fig. 10.7. a Luminosity curve for the standard human observer. The peak response is 683 lm W^{-1} at 555 nm (*green*). **b** Spectral response of human cones (normalized)

The different spectral characteristics are due to the different photopsins in the cone cell.

More correctly the output is proportional to the total number of photons captured by the photosite since the last time it was read. See page 364.

where `cones` has three columns corresponding to the L, M and S cone responses and each row corresponds to the wavelength in `lambda`. The spectral response of the cones $L(\lambda)$, $M(\lambda)$ and $S(\lambda)$ are shown in Fig. 10.7b.

The retina of the human eye has a central or foveal region which is only 0.6 mm in diameter, has a 5 degree field of view and contains most of the 6 million cone cells: 65% sense red, 33% sense green and only 2% sense blue. We unconsciously scan our high-resolution fovea over the world to build a large-scale mental image of our surrounds. In addition there are 120 million rod cells, which are also motion sensitive, distributed over the retina.

The sensor in a digital camera is analogous to the retina, but instead of rod and cone cells there is a regular array of light-sensitive photosites (or pixels) on a silicon chip. Each photosite is of the order 1–10 μm square and outputs a signal proportional to the intensity of the light falling over its area. For a color camera the photosites are covered by color filters which pass either red, green or blue light to the photosites. The spectral response of the filters is the functional equivalent of the cones' response $M(\lambda)$ shown in Fig. 10.7b. A very common arrangement of color filters is the Bayer pattern shown

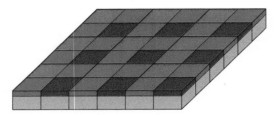

Fig. 10.8.
Bayer filtering. The *grey blocks* represent the array of light-sensitive silicon photosites over which is an array of red, green and blue filters. Invented by Bryce E. Bayer of Eastman Kodak, U.S. Patent 3,971,065.

in Fig. 10.8. It uses a regular 2×2 photosite pattern comprising two green filters, one red and one blue.▶

Each pixel therefore cannot provide independent measurements of red, green and blue but it can be estimated. For example, the amount of red at a blue sensitive pixel is obtained by interpolation from its red filtered neighbors. More expensive "3 CCD" cameras can make independent measurements at each pixel since the light is split by a set of prisms, filtered and presented to one CCD array for each primary color. Digital camera raw image files contain the actual outputs of the Bayer-filtered photosites.

3×3 or 4×4 arrays of filters allow many interesting camera designs. Using more than 3 different color filters leads to a multispectral camera with better color resolution, a range of neutral density (grey) filters leads to high dynamic range camera, or these various filters can be mixed to give a camera with better dynamic range and color resolution.

10.2.2   Measuring Color

The path taken by the light entering the eye shown in Fig. 10.9a. The spectrum of the luminance $L(\lambda)$ is a function of the light source and the reflectance of the object as given by Eq. 10.3. The response from each of the three cones is

$$
\begin{aligned}
\rho &= \int_\lambda L(\lambda) M_r(\lambda) \mathrm{d}\lambda \\
\gamma &= \int_\lambda L(\lambda) M_g(\lambda) \mathrm{d}\lambda \\
\beta &= \int_\lambda L(\lambda) M_b(\lambda) \mathrm{d}\lambda
\end{aligned}
\tag{10.4}
$$

where $M_r(\lambda)$, $M_g(\lambda)$ and $M_b(\lambda)$ are the spectral response of the red, green and blue cones respectively as shown in Fig. 10.7b. The response is a 3-vector (ρ, γ, β) which is known as a tristimulus.

For the case of the red brick the integrals correspond to the areas of the solid color regions in Fig. 10.9b. We can compute the tristimulus by approximating the integrals of Eq. 10.4 as a summation with $d\lambda = 1$ nm

```
>> sum( (L*ones(1,3)) .* cones * 1e-9)
ans =
    16.3571    10.0665     2.8225
```

The dominant response is from the L cone, which is unsurprising since we know that the brick is red.

An arbitrary continuous spectrum is an infinite-dimensional vector and cannot be uniquely represented by just 3 parameters but it is clearly *sufficient* for our species and allowed us to thrive in a variety of natural environments. A consequence of this choice of representation is that many *different* spectra will produce the *same* visual stimulus and these are referred to as metamers. More important is the corollary – an arbitrary visual stimulus can be generated by a mixture of just three monochromatic stimuli. These are the three primary colors we learned about as children.◀ There is no unique set of primaries – any three will do so long as none of them can be matched by a combination of the others. The CIE has defined a set of monochromatic primaries and their wavelengths are given in Table 10.1.

Primary colors are not a fundamental property of light – they are a fundamental property of the observer. There are three primary colors only because we, as trichromats, have three types of cones. Birds would have four primary colors and dogs would have two.

Table 10.1. The CIE 1976 primaries (Commission Internationale de L'Éclairage 1987) are spectral colors corresponding to the emission lines in a mercury vapor lamp

	red	green	blue
λ (nm)	700.0	546.1	435.8

Lightmeters, illuminance and luminance. A photographic lightmeter measures luminous flux which has units of lm m^{-2} or lux (lx). The luminous intensity I of a point light source is the luminous flux per unit solid angle measured in lm sr^{-1} or candelas (cd). The illuminance E falling normally onto a surface is

$$
E = \frac{I}{d^2} \text{ lx}
$$

where d is the distance between source and the surface. Outdoor illuminance on a bright sunny day is approximately 10 000 lx. Office lighting levels are typically around 1 000 lx and moonlight is 0.1 lx.

The luminance or *brightness* of a surface is

$$
L_s = E_i \cos\theta \text{ nt}
$$

which has units of cd m^{-2} or nit (nt), and where E_i is the incident illuminance at an angle θ to the surface normal.

Color blindness, or color deficiency, is the inability to perceive differences between some of the colors that others can distinguish. Protanopia, deuteranopia, tritanopia refer to the absence of the L, M and S cones respectively. More common conditions are protanomaly, deuteranomaly and tritanomaly where the cone pigments are mutated and the peak response frequency changed. It is most commonly a genetic condition since the red and green photopsins are coded in the X chromosome. The most common form (occurring in 6% of males including the author) is deuteranomaly where the M-cone's response is shifted toward the red end of the spectrum resulting in reduced sensitivity to greens and poor discrimination of hues in the red, orange, yellow and green region of the spectrum.

The English scientist **John Dalton (1766–1844)** confused scarlet with green and pink with blue. He hypothesized that the vitreous humor in his eyes was tinted blue and instructed that his eyes be examined after his death. This revealed that the humors were perfectly clear but DNA recently extracted from his preserved eye showed that he was a deuteranope. Color blindness was once referred to as Daltonism.

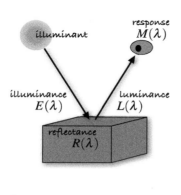

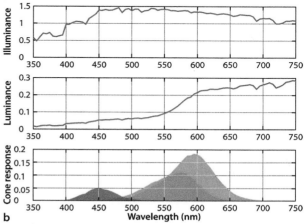

Fig. 10.9.
The tristimulus pathway. **a** Path of light from illuminant to the eye. **b** Within the eye three filters are applied and the total output of these filters, the areas shown in *solid color*, are the tristimulus value

10.2.3   Reproducing Colors

A computer or television display is able to produce a variable amount of each of three primaries at every pixel. The primaries for a cathode ray tube (CRT) are created by exciting phosphors on the back of the screen with a controlled electron beam. For a liquid crystal display (LCD) the colors are obtained by color filtering and attenuating white light emitted by the backlight, and an OLED display comprises a stack of red, green and blue LEDs at each pixel. The important problem is to determine how much of each primary is required to match a given tristimulus.

We start by considering a monochromatic stimulus of wavelength λ_S which is defined as

$$L(\lambda) = \begin{cases} L_\lambda & \text{if } \lambda = \lambda_S \\ 0 & \text{otherwise} \end{cases}$$

The response of the cones to this stimulus is given by Eq. 10.4 but because $L(\cdot)$ is an impulse we can drop the integral to obtain the tristimulus

$$\rho = L_\lambda M_r(\lambda_S)$$
$$\gamma = L_\lambda M_g(\lambda_S) \qquad\qquad (10.5)$$
$$\beta = L_\lambda M_b(\lambda_S)$$

The units are chosen such that equal quantities of the primaries appear to be white.

Consider next three monochromatic primary light sources denoted **R, G** and **B** with wavelengths λ_r, λ_g and λ_b and intensities R, G and B respectively.◄ The tristimulus from these light sources is

The notion of primary colors is very old, but their number (anything from two to six) and their color was the subject of much debate. Much of the confusion was due to there being additive primaries (red, green and blue) that are used when mixing lights, and subtractive primaries (cyan, magenta, yellow) used when mixing paints or inks. Whether or not black and white were primary colors was also debated.

$$\rho = RM_r(\lambda_r) + GM_r(\lambda_g) + BM_r(\lambda_b)$$
$$\gamma = RM_g(\lambda_r) + GM_g(\lambda_g) + BM_g(\lambda_b) \tag{10.6}$$
$$\beta = RM_b(\lambda_r) + GM_b(\lambda_g) + BM_b(\lambda_b)$$

For the perceived color of these three light sources combined to match that of the monochromatic stimulus the two tristimuli must be equal. We equate Eq. 10.5 and Eq. 10.6 and write compactly in matrix form as

$$L_\lambda \begin{pmatrix} M_r(\lambda_S) \\ M_g(\lambda_S) \\ M_b(\lambda_S) \end{pmatrix} = \begin{pmatrix} M_r(\lambda_r) & M_r(\lambda_g) & M_r(\lambda_b) \\ M_g(\lambda_r) & M_g(\lambda_g) & M_g(\lambda_b) \\ M_b(\lambda_r) & M_b(\lambda_g) & M_b(\lambda_b) \end{pmatrix} \begin{pmatrix} R \\ G \\ B \end{pmatrix}$$

and then solve for the required amounts of primary colors

$$\begin{pmatrix} R \\ G \\ B \end{pmatrix} = L_\lambda \begin{pmatrix} M_r(\lambda_r) & M_r(\lambda_g) & M_r(\lambda_b) \\ M_g(\lambda_r) & M_g(\lambda_g) & M_g(\lambda_b) \\ M_b(\lambda_r) & M_b(\lambda_g) & M_b(\lambda_b) \end{pmatrix}^{-1} \begin{pmatrix} M_r(\lambda_S) \\ M_g(\lambda_S) \\ M_b(\lambda_S) \end{pmatrix} \tag{10.7}$$

This tristimulus has a spectrum comprising three impulses (one per primary), yet has the same visual appearance as the original continuous spectrum – this is the basis of trichromatic matching. The 3 × 3 matrix is constant, but depends upon the spectral response of the cones to the chosen primaries (λ_r, λ_g, λ_b).

The right-hand side of Eq. 10.7 is simply a function of λ_S which we can write in an even more compact form

$$\begin{pmatrix} R \\ G \\ B \end{pmatrix} = \begin{pmatrix} \bar{r}(\lambda_S) \\ \bar{g}(\lambda_S) \\ \bar{b}(\lambda_S) \end{pmatrix} \tag{10.8}$$

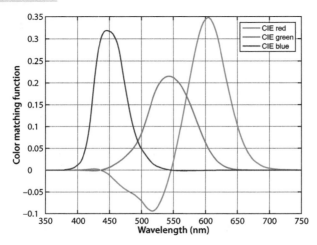

Fig. 10.10.
The 1931 color matching functions for the standard observer, based on the CIE 1976 standard primaries

Color matching experiments are performed using a light source comprising three adjustable lamps that correspond to the primary colors and whose intensity can be individually adjusted. The lights are mixed and diffused and compared to some test color. In color matching notation the primaries, the lamps, are denoted by **R**, **G** and **B**, and their intensities are R, G and B respectively. The three lamp intensities are adjusted by a human subject until they appear to match the test color. This is denoted

$$\mathbf{C} \equiv R\mathbf{R} + G\mathbf{G} + B\mathbf{B}$$

which is read as the visual stimulus **C** (the test color) is matched by, or looks the same as, a mixture of the three primaries with brightness R, G and B. The notation $R\mathbf{R}$ can be considered as the lamp **R** at intensity R.

Experiments show that color matching obeys the algebraic rules of additivity and linearity which is known as Grassmann's laws. For example two light stimuli $\mathbf{C}_1$ and $\mathbf{C}_2$

$$\mathbf{C}_1 \equiv R_1\mathbf{R} + G_1\mathbf{G} + B_1\mathbf{B}$$
$$\mathbf{C}_2 \equiv R_2\mathbf{R} + G_2\mathbf{G} + B_2\mathbf{B}$$

when mixed will match

$$\mathbf{C}_1 + \mathbf{C}_2 \equiv (R_1 + R_2)\mathbf{R} + (G_1 + G_2)\mathbf{G} + (B_1 + B_2)\mathbf{B}$$

where $\bar{r}(\lambda), \bar{g}(\lambda), \bar{b}(\lambda)$ are known as color matching functions. These functions have been empirically determined from human test subjects and tabulated for the standard CIE primaries listed in Table 10.1. They can be loaded using the function cmfrgb

```
>> lambda = [400:700]*1e-9;
>> cmf = cmfrgb(lambda);
>> plot(lambda, cmf);
```

and are shown graphically in Fig. 10.10. Each curve indicates how much of the corresponding primary is required to match the monochromatic light of wavelength λ.

For example to create the sensation of light at 500 nm (green) we would need

```
>> green = cmfrgb(500e-9)
green =
   -0.0714    0.0854    0.0478
```

Surprisingly this requires a significant *negative* amount of the red primary and this is problematic since a light source cannot have a negative luminance.

We reconcile this by adding some white light ($R = G = B = w$, see Sect. 10.2.8) so that the tristimulus values are all positive. For instance

```
>> white = -min(green) * [1 1 1]
white =
    0.0714    0.0714    0.0714
>> feasible_green = green + white
feasible_green =
         0    0.1567    0.1191
```

If we looked at this color side-by-side with the desired 500 nm green we would say that the generated color had the correct hue but was not as *saturated*.

Saturation refers to the purity of the color. Spectral colors are *fully saturated* but become less saturated (more pastel) as increasing amounts of white is added. In this case we have mixed in a stimulus of light (7%) grey.

This leads to a very important point about color reproduction – it is *not* possible to reproduce every possible color using just three primaries. This makes intuitive sense since a color is properly represented as an infinite-dimensional spectral function and a 3-vector can only approximate it. To understand this more fully we need to consider chromaticity spaces.

The Toolbox function cmfrgb can also compute the CIE tristimulus for an arbitrary spectrum. The luminance spectrum of the redbrick illuminated by sunlight at ground level was computed on page 291 and its tristimulus is

```
>> RGB_brick = cmfrgb(lambda, L)
RGB_brick =
    0.0155    0.0066    0.0031
```

These are the respective amounts of the three CIE primaries that are perceived – by the average human – as having the same color as the original brick under those lighting conditions.

10.2.4   Chromaticity Space

The tristimulus values describe color as well as brightness. Relative tristimulus values are obtained by normalizing the tristimulus values

$$r = \frac{R}{R+G+B}, \; g = \frac{G}{R+G+B}, \; b = \frac{B}{R+G+B} \tag{10.9}$$

which results in chromaticity coordinates r, g and b that are invariant to overall brightness. By definition $r + g + b = 1$ so one coordinate is redundant and typically only r and g are considered. Since the effect of intensity has been eliminated the 2-dimensional quantity (r, g) represents *color*.

We can plot the locus of spectral colors, the colors of the rainbow, on the chromaticity diagram using a variant of the color matching functions

```
>> [r,g] = lambda2rg( [400:700]*1e-9 );
>> plot(r, g)
>> rg_addticks
```

which results in the horseshoe-shaped curve shown in Fig. 10.11. The Toolbox function lambda2rg computes the color matching function Fig. 10.10 for the specified wavelength and then converts the tristimulus value to chromaticity coordinates using Eq. 10.9.

The CIE primaries listed in Table 10.1 can be plotted as well

```
>> primaries = lambda2rg( cie_primaries() );
>> plot(primaries(:,1), primaries(:,2), 'o')
```

and are shown as circles in Fig. 10.11.

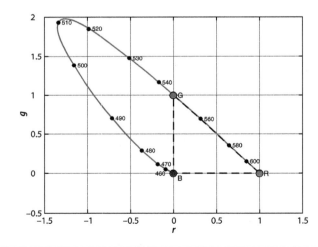

Fig. 10.11.
The spectral locus on the r-g chromaticity plane. Monochromatic stimuli lie on the locus and the wavelengths (in nm) are marked. The *straight line* joining the extremities is the purple boundary and is the locus of saturated purples. All possible colors lie on, or within, this locus. The CIE standard primary colors are marked and the *dashed line* indicates the gamut of colors that can be represented by these primaries

Colorimetric standards. Colorimetry is a complex topic and standards are very important. Two organizations, CIE and ITU, play a leading role in this area.

The Commission Internationale de l'Eclairage (CIE) or International Commission on Illumination was founded in 1913 and is an independent nonprofit organization that is devoted to worldwide cooperation and the exchange of information on all matters relating to the science and art of light and lighting, color and vision, and image technology. The CIE's eighth session was held at Cambridge, UK, in 1931 and established international agreement on colorimetric specifications and formalized the *XYZ* color space. The CIE is recognized by ISO as an international standardization body. See http://www.cie.co.at for more information and CIE datasets.

The International Telecommunication Union (ITU) is an agency of the United Nations and was established to standardize and regulate international radio and telecommunications. It was founded as the International Telegraph Union in Paris on 17 May 1865. The International Radio Consultative Committee or CCIR (Comité Consultatif International des Radiocommunications) became, in 1992, the Radiocommunication Bureau of ITU or ITU-R. It publishes standards and recommendations relevant to colorimetry in its BT series (broadcasting service television). See http://www.itu.int for more detail.

Grassmann's center of gravity law states that a mixture of two colors lies along a line between those two colors on the chromaticity plane. A mixture of N colors lies within a region bounded by those colors. Considered with respect to Fig. 10.11 this has significant implications. Firstly, since all color stimuli are combinations of spectral stimuli all real color stimuli must lie on or inside the spectral locus. Secondly, any colors we create from mixing the primaries can only lie *within* the triangle bounded by the primaries – the color gamut. It is clear from Fig. 10.11 that the CIE primaries define only a small subset of all possible colors – within the dashed triangle. Very many real colors *cannot* be created using these primaries, in particular the colors of the rainbow which lie on the spectral locus from 460–545 nm. In fact no matter where the primaries are located, not all possible colors can be produced.◄ In geometric terms there are no three points within the gamut that form a triangle that includes the entire gamut. Thirdly, we observe that much of the locus requires a negative amount of the red primary and cannot be represented.

We could increase the gamut by choosing different primaries, perhaps using a different green primary would make the gamut larger, but there is the practical constraint of finding a light source (LED or phosphor) that can efficiently produce that color.

We revisit the problem from page 297 concerned with displaying 500 nm green and Figure 10.12 shows the chromaticity of the spectral green color

```
>> green_cc = lambda2rg(500e-9)
green_cc =
   -1.1558    1.3823
>> plot2(green_cc, 's')
```

as a star-shaped marker. White is by definition $R = G = B = 1$ and its chromaticity

```
>> white_cc = tristim2cc([1 1 1])
white_cc =
    0.3333    0.3333
>> plot2(white_cc, 'o')
```

is plotted as a hollow circle. According to Grassmann's law the mixture of our desired green and white must lie along the indicated green line. The chromaticity of the feasible green computed earlier is indicated by a square, but is outside the *displayable* gamut of the nonstandard primaries used in this example. The least saturated displayable green lies at the intersection of the green line and the gamut boundary and is indicated by the triangular marker.

Luminance here has different meaning to that defined in Sect. 10.1.3 and can be considered synonymous to brightness here.

The units are chosen such that equal quantities of the primaries are required to match the equal-energy white stimulus.

Earlier we said that there are no three points within the gamut that form a triangle that includes the entire gamut. The CIE therefore proposed, in 1931, a system of *imaginary nonphysical primaries* known as **X**, **Y** and **Z** that totally enclose the spectral locus of Fig. 10.11. **X** and **Z** have zero luminance – the luminance is contributed entirely by **Y**▾. All real colors can thus be matched by positive amounts of these three primaries.◄ The corresponding tristimulus values are denoted (X, Y, Z).

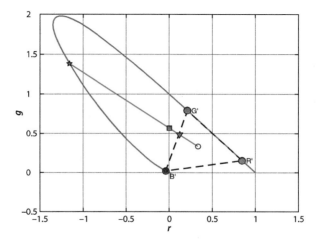

Fig. 10.12.
Chromaticity diagram showing the color gamut for nonstandard primaries at 600, 555 and 450 nm. 500 nm green (*star*), equal-energy white (*circle*), a feasible green (*square*) and a displayable green (*triangle*). The locus of different saturated greens in shown as a *green line*

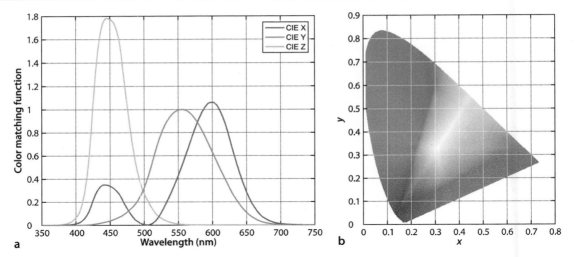

Fig. 10.13. a The color matching functions for the standard observer, based on the imaginary primaries **X**, **Y** (intensity) and **Z** are tabulated by the CIE. **b** Colors on the xy-chromaticity plane

The XYZ color matching functions defined by the CIE

```
>> cmf = cmfxyz(lambda);
>> plot(lambda, cmf);
```

are shown graphically in Fig. 10.13a. This shows the amount of each CIE XYZ primary that is required to match a spectral color and we note that these curves are never negative. The corresponding chromaticity coordinates are

$$x = \frac{X}{X+Y+Z}, \; y = \frac{Y}{X+Y+Z}, \; z = \frac{Z}{X+Y+Z} \qquad (10.10)$$

and once again $x + y + z = 1$ so only two parameters are required – by convention y is plotted against x in a chromaticity diagram. The spectral locus can be plotted in a similar way as before

```
>> [x,y] = lambda2xy(lambda);
>> plot(x, y);
```

A more sophisticated plot, showing the colors within the spectral locus, can be created

```
>> showcolorspace('xy')
```

and is shown▶ in Fig. 10.13b. These coordinates are a *standard* way to represent color for graphics, printing and other purposes. For example the chromaticity coordinates of peak green (550 nm) is

```
>> lambda2xy(550e-9)
ans =
    0.3016    0.6923
```

and the chromaticity coordinates of a standard tungsten illuminant at 2 600 K is

```
>> lamp = blackbody(lambda, 2600);
>> lambda2xy(lambda, lamp)
ans =
    0.4677    0.4127
```

The colors depicted in figures such as Fig. 10.1 and 10.13b can only approximate the true color due to the gamut limitation of the technology you use to view the book: the inks used to print the page or your computer's display. No display technology has a gamut large enough to present an accurate representation of the chromaticity at every point.

10.2.5 Color Names

Chromaticity coordinates provide a quantitative way to describe and compare colors, however humans refer to colors by name. Many computer operating systems contain a database or file▶ that maps human understood names of colors to their correspond-

The file is named `/etc/rgb.txt` on most Unix-based systems.

Colors are important to human beings and there are over 4 000 color-related words in the English language. The ancient Greeks only had words for black, white, red and yellowish-green. All languages have words for black and white, and red is the next most likely color word to appear in a language followed by yellow, green, blue and so on. We also associate colors with emotions, for example red is angry and blue is sad but this varies across cultures. In Asia orange is generally a positive color whereas in the west it is the color of road hazards and bulldozers. Chemistry and technology has made a huge number of colors available to us in the last 700 years yet with this choice comes confusion about color naming – people may not necessarily agree on the linguistic tag to assign to a particular color. (Word cloud by tagxedo.com using data from Steinvall 2002)

ing (R, G, B) tristimulus values. The Toolbox provides a copy of a such a file and an interface function `color-name`. For example, we can query a color name that includes a particular substring

```
>> colorname('?burnt')
ans =
    'burntsienna'    'burntumber'
```

The RGB tristimulus value of burnt Sienna is

```
>> colorname('burntsienna')
ans =
    0.5412    0.2118    0.0588
```

with the values normalized to the interval [0, 1]. We could also request xy-chromaticity coordinates

```
>> bs = colorname('burntsienna', 'xy')
bs =
    0.5568    0.3783
```

With reference to Fig. 10.13, we see that this point is in the red-brown part of the colorspace and not too far from the color of chocolate

```
>> colorname('chocolate', 'xy')
ans =
    0.5318    0.3988
```

We can also solve the inverse problem. Given a tristimulus value

```
>> colorname([0.2 0.3 0.4])
ans =
darkslateblue
```

we obtain the name of the closest, in Euclidean terms, color.

10.2.6   Other Color and Chromaticity Spaces

A color space is a 3-dimensional space that contains all possible tristimulus values – all colors and all levels of brightness. If we think of this in terms of coordinate frames as discussed in Sect. 2.2 then there are an infinite number of choices of Cartesian frame with which to define colors. We have already discussed two different Cartesian color spaces: RGB and XYZ. However we could also use polar, spherical or hybrid coordinate systems.

The 2-dimensional chromaticity spaces r-g or x-y do not account for brightness – we normalized it out in Eq. 10.9 and Eq. 10.10. Brightness, frequently referred to as luminance in this context, is denoted by Y and the definition from ITU Recommendation 709

$$Y^{709} = 0.2126R + 0.7152G + 0.0722B \tag{10.11}$$

is a weighted sum of the RGB-tristimulus values and reflects the eye's high sensitivity to green and low sensitivity to blue. Chromaticity plus luminance leads to 3-dimensional color spaces such as rgY or xyY.

Humans seem to more naturally consider chromaticity in terms of two characteristics: hue and saturation. Hue is the dominant color, the closest spectral color, and saturation refers to the purity, or absence of mixed white. Stimuli on the spectral locus are completely saturated while those closer to its centroid are less saturated. The concepts of hue and saturation are illustrated in geometric terms in Fig. 10.14.

The color spaces that we have discussed lack easy interpretation in terms of hue and saturation so alternative color spaces have been proposed. The two most commonly known are HSV and CIE L*C*h. In color-space notation H is hue, S is saturation which is also known as C or chroma. The intensity dimension is named either V for value or L for lightness but they are computed quite differently. ◄

L^* is a nonlinear function of relative luminance and approximates the nonlinear response of the human eye. Value is given by $V = \frac{1}{2}(\min R, G, B + \max R, G, B)$.

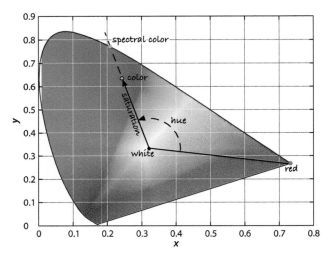

Fig. 10.14.
Hue and saturation. A line is
extended from the white point
through the chromaticity in
question to the spectral locus.
The angle of this line is hue, and
saturation is the length of the
vector normalized with respect
to distance to the locus

The function `colorspace` can be used to convert between different color spaces. For example the hue, saturation and intensity for each of pure red, green and blue RGB tristimulus value▸ is

```
>> colorspace('RGB->HSV', [1, 0, 0])
ans =
        0    1    1
>> colorspace('RGB->HSV', [0, 1, 0])
ans =
      120    1    1
>> colorspace('RGB->HSV', [0, 0, 1])
ans =
      240    1    1
```

This function assumes that *RGB* values are gamma encoded ($\gamma = 0.45$), see Sect. 10.3.6. The particular numerical values chosen here are invariant under gamma encoding. The builtin MATLAB function `rgb2hsv` does not assume gamma encoded values and represents hue in different units.

In each case the saturation is 1, the colors are pure, and the intensity is 1. As shown in Fig. 10.14 hue is represented as an angle in the range $[0, 360)°$ with red at 0° increasing through the spectral colors associated with decreasing wavelength (orange, yellow, green, blue, violet). If we reduce the amount of the green primary

```
>> colorspace('RGB->HSV', [0, 0.5, 0])
ans =
  120.0000    1.0000    0.5000
```

we see that intensity drops but hue and saturation are unchanged.▸ For a medium grey

For very dark colors numerical problems lead to imprecise hue and saturation coordinates.

```
>> colorspace('RGB->HSV', [0.4, 0.4, 0.4])
ans =
  240.0000        0    0.4000
```

the saturation is zero, it is only a mixture of white, and the hue has no meaning since there is no color. If we add the green to the grey

```
>> colorspace('RGB->HSV', [0, 0.5, 0] + [0.4, 0.4, 0.4])
ans =
  120.0000    0.5556    0.9000
```

we have the green hue and a medium saturation value.

The `colorspace` function can also be applied to a color image

```
>> flowers = iread('flowers4.png', 'double');
>> about flowers
flowers [double] : 426x640x3 (6.5 MB)
```

which is shown in Fig. 10.15a and comprises several different colored flowers and background greenery. The image `flowers` has 3 dimensions and the third is the color plane that selects the red, green or blue pixels.

Fig. 10.15. Flower scene. **a** Original color image; **b** hue image; **c** saturation image. Note that the white flowers have low saturation (they appear dark); **d** intensity or monochrome image; **e** a* image (green to red); **f** b* image (blue to yellow)

To convert the image to hue, saturation and value is simply

```
>> hsv = colorspace('RGB->HSV', flowers);
>> about hsv
hsv [double] : 426x640x3 (6.5 MB)
```

and the result is another 3-dimensional matrix but this time the color planes represent hue, saturation and value. We can display these planes

```
>> idisp( hsv(:,:,1) )
>> idisp( hsv(:,:,2) )
>> idisp( hsv(:,:,3) )
```

as images which are shown in Fig. 10.15b, c and d respectively. In the hue image dark represents red and bright white represents violet. The red flowers appear as both a very small hue angle (dark) and a very large angle close to 360°. The yellow flowers and the green background can be seen as distinct hue values. The saturation image shows that the red and yellow flowers are highly saturated, while the green leaves and stems are less saturated. The white flowers have very low saturation, since by definition the color white contains a lot of white.

A limitation of many color spaces is that the *perceived* color difference between two points is not directly related to their Euclidean distance. In some parts of the chromaticity space two distant points might appear quite similar, whereas in another region two close points might appear quite different. This has led to the development of perceptually uniform color spaces such as the CIE L*u*v* (CIELUV) and L*a*b* spaces.

The colorspace function can convert between thirteen different color spaces including L*a*b*, L*u*v*, YUV and YC_BC_R. To convert this image to L*a*b* color space follows the same pattern

```
>> Lab = colorspace('RGB->Lab', flowers);
>> about Lab
Lab [double] : 426x640x3 (6.5 MB)
```

Relative to a white illuminant, which this function assumes as CIE D_{65} with Y = 1. a*b* are not invariant to overall luminance.

which again results in an image with 3 dimensions. The chromaticity◄ is encoded in the a* and b* planes.

```
>> idisp( Lab(:,:,2) )
>> idisp( Lab(:,:,3) )
```

and these are shown in Fig. 10.15e and f respectively. L*a*b* is an opponent color space where a* spans colors from green (black) to red (white) while b* spans blue (black) to yellow (white), with white at the origin where a* = b* = 0.

10.2.7   Transforming between Different Primaries

The CIE standards were defined in 1931 which was well before the introduction of color television in the 1950s. The CIE primaries in Table 10.1 are based on the emission lines of a mercury lamp which are highly repeatable and suitable for laboratory use. Early television receivers used CRT monitors where the primary colors were generated by phosphors that emit light when bombarded by electrons. The phosphors used, and their colors has varied over the years in pursuit of brighter displays. An international agreement, ITU recommendation 709, defines the primaries for high definition television (HDTV) and these are listed in Table 10.2.

This raises the problem of converting tristimulus values from one sets of primaries to another. Consider for example that we wish to display an image, where the tristimulus values are with respect to CIE primaries, on a screen that uses ITU Rec. 709 primaries. Using the notation we introduced earlier we define two sets of primaries: $\mathbf{P}_1, \mathbf{P}_2, \mathbf{P}_3$ with tristimulus values (S_1, S_2, S_3), and $\mathbf{P}_1', \mathbf{P}_2', \mathbf{P}_3'$ with tristimulus values (S_1', S_2', S_3'). We can always express one set of primaries as a linear combination▸ of the other

The coefficients can be negative so the new primaries do not have to lie within the gamut of the old primaries.

$$
\begin{pmatrix} \mathbf{P}_1 \\ \mathbf{P}_2 \\ \mathbf{P}_3 \end{pmatrix} = \begin{pmatrix} a_{11} & a_{21} & a_{31} \\ a_{12} & a_{22} & a_{32} \\ a_{13} & a_{23} & a_{33} \end{pmatrix} \begin{pmatrix} \mathbf{P}_1' \\ \mathbf{P}_2' \\ \mathbf{P}_3' \end{pmatrix}
\tag{10.12}
$$

and since the two tristimuli match then

$$
\begin{pmatrix} S_1' & S_2' & S_3' \end{pmatrix} \begin{pmatrix} \mathbf{P}_1' \\ \mathbf{P}_2' \\ \mathbf{P}_3' \end{pmatrix} \equiv \begin{pmatrix} S_1 & S_2 & S_3 \end{pmatrix} \begin{pmatrix} \mathbf{P}_1 \\ \mathbf{P}_2 \\ \mathbf{P}_3 \end{pmatrix}
\tag{10.13}
$$

Substituting Eq. 10.12, equating tristimulus values and then transposing we obtain

$$
\begin{pmatrix} S_1' \\ S_2' \\ S_3' \end{pmatrix} = \begin{pmatrix} a_{11} & a_{21} & a_{31} \\ a_{12} & a_{22} & a_{32} \\ a_{13} & a_{23} & a_{33} \end{pmatrix}^T \begin{pmatrix} S_1 \\ S_2 \\ S_3 \end{pmatrix} = \mathbf{C} \begin{pmatrix} S_1 \\ S_2 \\ S_3 \end{pmatrix}
\tag{10.14}
$$

which is simply a linear transformation of tristimulus values.

Consider the concrete problem of transforming from CIE primaries to XYZ tristimulus values. We know from Table 10.2 the CIE primaries in terms of XYZ primaries

```
>> C = [ 0.7347, 0.2653, 0; 0.2738, 0.7174, 0.0088; 0.1666,
0.0089, 0.8245]'
C =
    0.7347    0.2738    0.1666
    0.2653    0.7174    0.0089
         0    0.0088    0.8245
```

which is exactly the first three columns of Table 10.2. The transform is therefore

$$
\begin{pmatrix} X \\ Y \\ Z \end{pmatrix} = \mathbf{C} \begin{pmatrix} R \\ G \\ B \end{pmatrix}
$$

Recall from page 299 that luminance is contributed entirely by the $\mathbf{Y}$ primary. It is common to apply the constraint that unity R, G, B values result in unity luminance Y and a white with a specified chromaticity. We will choose D_{65} white whose

Table 10.2.
xyz-chromaticity of standard primaries and whites. The CIE primaries of Table 10.1 and the more recent ITU recommendation 709 primaries defined for HDTV. D_{65} is the white of a blackbody radiator at 6500 K, and E is equal-energy white

	R_{CIE}	G_{CIE}	B_{CIE}	R_{709}	G_{709}	B_{709}	D_{65}	E
x	0.7347	0.2738	0.1666	0.640	0.300	0.150	0.3127	0.3333
y	0.2653	0.7174	0.0089	0.330	0.600	0.060	0.3290	0.3333
z	0.0000	0.0088	0.8245	0.030	0.100	0.790	0.3582	0.3333

chromaticity is given in Table 10.2 and which we will denote (x_w, y_w, z_w). We can now write

$$\frac{1}{y_w}\begin{pmatrix} x_w \\ y_w \\ z_w \end{pmatrix} = \mathbf{C}\begin{pmatrix} J_R & 0 & 0 \\ 0 & J_G & 0 \\ 0 & 0 & J_B \end{pmatrix}\begin{pmatrix} 1 \\ 1 \\ 1 \end{pmatrix}$$

where the left-hand side has $Y = 1$ and we have introduced a diagonal matrix $\mathbf{J}$ which scales the luminance of the primaries. We can solve for the elements of $\mathbf{J}$

$$\begin{pmatrix} J_R \\ J_G \\ J_B \end{pmatrix} = \mathbf{C}^{-1}\begin{pmatrix} x_w \\ y_w \\ z_w \end{pmatrix}\frac{1}{y_w}$$

Substituting real values we obtain

```
>> J = inv(C) * [0.3127 0.3290 0.3582]' * (1/0.3290)
J =
    0.5609
    1.1703
    1.3080
>> C * diag(J)
ans =
    0.4121    0.3204    0.2179
    0.1488    0.8395    0.0116
         0    0.0103    1.0785
```

The middle row of this matrix leads to the luminance relationship

$$Y = 0.1488R + 0.8395G + 0.0116B$$

which is similar to Eq. 10.11. The small variation is due to the different primaries used – CIE in this case versus Rec. 709 for Eq. 10.11.

The *RGB* tristimulus value of the redbrick was computed earlier and we can determine its *XYZ* tristimulus

```
>> XYZ_brick = C * diag(J) * RGB_brick';
ans =
    0.0092
    0.0079
    0.0034
```

which we convert to chromaticity coordinates by Eq. 10.10

```
>> tristim2cc(XYZ_brick')
xybrick =
    0.4483    0.3859
```

Referring to Fig. 10.13b we see that this *xy*-chromaticity lies in the red region and is named

```
>> colorname(ans, 'xy')
ans =
sandybrown
```

which is plausible for a "weathered red brick".

10.2.8   What Is White?

In the previous section we touched on the subject of white. White is both the absence of color and also the sum of all colors. One definition of white is *standard daylight* which is taken as the mid-day Sun in Western/Northern Europe which has been tabulated by the CIE as illuminant D_{65}. It can be closely approximated by a blackbody radiator at 6 500 K

```
>> d65 = blackbody(lambda, 6500);
>> lambda2xy(lambda, d65)
ans =
    0.3136    0.3243
```

which we see is close to the D_{65} chromaticity given in Table 10.2.

Another definition is based on white light being an equal mixture of all spectral colors. This is represented by a uniform spectrum

```
>> ee = ones(size(lambda));
```

which is also known as the equal-energy stimulus and has chromaticity

```
>> lambda2xy(lambda, ee)
ans =
    0.3334    0.3340
```

which is close to the defined value of (⅓, ⅓).

10.3   Advanced Topics

Color is a large and complex subject, and in this section we will briefly introduce a few important remaining topics. Color temperature is a common way to describe the spectrum of an illuminant, and the effect of illumination color on the apparent color of an object is the color constancy problem which is very real for a robot using color cues in an environment with natural lighting. White balancing is one way to overcome this. Another source of color change, in media such as water, is the absorption of certain wavelengths. Most cameras actually implement a nonlinear relationship, called gamma correction, between actual scene luminance and the output tristimulus values. Finally we look at a more realistic model of surface reflection which has both specular and diffuse components, each with different spectral characteristics.

10.3.1   Color Temperature

Photographers often refer to the color temperature of a light source – the temperature of a black body whose spectrum according to Eq. 10.1 is most similar to that of the light source. The color temperature of a number of common lighting conditions are listed in Table 10.3. We describe low-color-temperature illumination as warm – it appears reddy orange to us. High-color-temperature is more harsh – it appears as brilliant white perhaps with a tinge of blue.

Light source	Color temperature (K)
Candle light	1 900
Dawn/dusk sky	2 000
40 W tungsten lamp	2 600
100 W tungsten lamp	2 850
Tungsten halogen lamp	3 200
Direct sunlight	5 800
Overcast sky	6 000 – 7 000
Standard daylight (sun + blue sky)	6 500
Hazy sky	8 000
Clear blue sky	10 000 – 30 000

Table 10.3.
Color temperatures of some common light sources

Scene luminance is the product of illuminance and reflectance but reflectance is key to scene understanding since it can be used as a proxy for the type of material. Illuminance can vary in intensity and color
across the scene and this complicates image understanding. Unfortunately separating luminance into
illuminance and reflectance is an ill-posed problem yet humans are able to do this very well as the illusion to the right illustrates – the squares labeled A and B have the same grey level.

The American inventor and founder of Polaroid Corporation Edward Land (1909–1991) proposed the retinex theory (retinex = retina + cortex) to explain how the human visual system factorizes reflectance from
luminance. (Checker shadow illusion courtesy of Edward H. Adelson, http://persci.mit.edu/gallery)

10.3.2   Color Constancy

Studies show that human perception of what is white is adaptive and has a remarkable
ability to *tune out* the effect of scene illumination so that white objects always appear to
be white.◀ For example at night under a yellowish tungsten lamp the pages of a book still
appear white to us, but a photograph of that scene viewed later under different lighting
conditions will look yellow. All of this poses real problems for a robot that is using color to
understand the scene because the observed chromaticity varies with lighting. Outdoors a
robot has to contend with an illumination spectrum that depends on the time of day and
cloud cover as well as colored reflections from buildings and trees. This affects the luminance and apparent color of the object. To illustrate this problem we revisit the red brick

We adapt our perception of color so that
the integral, or average, over the entire
scene is grey. This works well over a color
temperature range 5 000–6 500 K.

```
>> lambda = [400:10:700]*1e-9;
>> R = loadspectrum(lambda, 'redbrick');
```

under two different illumination conditions, the Sun at ground level

```
>> sun = loadspectrum(lambda, 'solar');
```

and a tungsten lamp

```
>> lamp = blackbody(lambda, 2600);
```

and compute the *xy*-chromaticity for each case

```
>> xy_sun = lambda2xy(lambda, sun .* R)
xy_sun =
    0.4760    0.3784
>> xy_lamp = lambda2xy(lambda, lamp .* R)
xy_lamp =
    0.5724    0.3877
```

and we can see that the chromaticity, or apparent color, has changed significantly.
These values are plotted on the chromaticity diagram in Fig. 10.16.

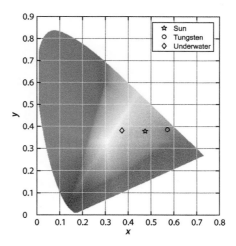

Fig. 10.16.
Chromaticity of the red-brick
under different illumination
conditions

10.3.3   White Balancing

Photographers need to be aware of the illumination color temperature. An incandescent lamp appears more yellow than daylight so a photographer would place a blue filter on the camera to attenuate the red part of the spectrum to compensate. We can achieve a similar function by choosing the matrix **J**

$$\begin{pmatrix} R' \\ G' \\ B' \end{pmatrix} = \begin{pmatrix} J_R & 0 & 0 \\ 0 & J_G & 0 \\ 0 & 0 & J_B \end{pmatrix} \begin{pmatrix} R \\ G \\ B \end{pmatrix}$$

to adjust the gains of the color channels.► For example, boosting J_B would compensate for the lack of blue under tungsten illumination. This is the process of white balancing – ensuring the appropriate chromaticity of objects that we know are white (or grey).

Typically $J_G = 1$ and J_R and J_B are adjusted.

Some cameras allow the user to set the color temperature of the illumination through a menu, typically with options for tungsten, fluorescent, daylight and flash which select different preset values of **J**. In manual white balancing the camera is pointed at a grey or white object and a button is pressed. The camera adjusts its channel gains **J** so that equal tristimulus values are produced $R' = G' = B'$ which as we recall results in the desired white chromaticity. For colors other than white these corrections introduces some color error but this nevertheless has a satisfactory appearance to the eye. Automatic white balancing is commonly used and involves heuristics to estimate the color temperature of the light source but it can be fooled by scenes with a predominance of a particular color.

The most practical solution is to use the tristimulus values of three objects with known chromaticity in the scene. This allows the matrix C in Eq. 10.14 to be estimated directly, mapping the tristimulus values from the sensor to XYZ coordinates which are an absolute lighting-independent representation of surface reflectance. From this the chromaticity of the illumination can also be estimated. This approach is used for the panoramic camera on the Mars Rover where the calibration target shown in Fig. 10.17 can be imaged periodically to update the white balance under changing Martian illumination.

10.3.4   Color Change Due to Absorption

A final and extreme example of problems with color occurs underwater. For example consider a robot trying to find a docking station identified by colored targets. As discussed earlier in Sect. 10.1.1 water acts as a filter that absorbs more red light than blue light. For an object underwater this filtering affects both the illumination falling on

Fig. 10.17.
The calibration target used for the Mars Rover's PanCam. Regions of known reflectance and chromaticity (red, yellow, green, blue and shades of grey) are used to set the white balance of the camera. The central stalk has a very low reflectance and also serves as a sundial. In the best traditions of sundials it bears a motto (photo courtesy NASA/JPL/Cornell/Jim Bell)

the object and the reflected light, the luminance, on its way to the camera. Consider again the red brick

```
>> [R,lambda] = loadspectrum([400:5:700]*1e-9, 'redbrick');
```

which is now 1 m underwater and with a camera a further 1 m from the brick. The illumination on the water's surface is that of sunlight at ground level

```
>> sun = loadspectrum(lambda, 'solar');
```

The absorption spectrum of water is

```
>> A = loadspectrum(lambda, 'water');
```

and the total optical path length through the water is

```
>> d = 2;
```

The transmission T is given by Beer's law Eq. 10.2.

```
>> T = 10 .^ (-d*A);
```

and the resulting luminance of the brick is

```
>> L = sun .* R .* T;
```

which is shown in Fig. 10.18. We see that the longer wavelengths, the reds, have been strongly attenuated. The apparent color of the brick is

```
>> xy_water = lambda2xy(lambda, L)
xy_water =
    0.3738     0.3814
```

which is also plotted in the chromaticity diagram of Fig. 10.16. The brick appears much more blue than it did before. In reality underwater imaging is more complex than this due to the scattering of light by tiny suspended particles which reflect ambient light into the camera that has not been reflected from the target.

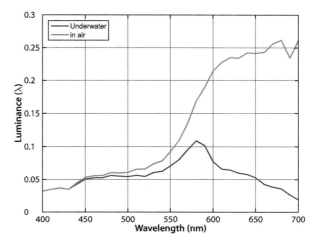

Fig. 10.18.
Spectrum of the red brick luminance when viewed underwater. The spectrum without the water absorption is shown in *red*

Lambertian reflection. A non-mirror-like or matte surface is a diffuse reflector and the amount of light reflected at a particular angle from the surface normal is proportional to the cosine of the reflection angle θ_r. This is known as Lambertian reflection after the Swiss mathematician and physicist Johann Heinrich Lambert (1728–1777). A consequence is that the object has the same apparent brightness at all viewing angles. A powerful example of this is the moon which appears as a disc of uniform brightness despite it being a sphere with its surface curved away from us. See also specular reflection on page 337. (Moon image courtesy of NASA)

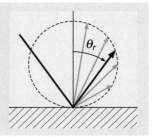

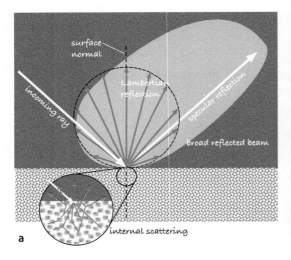

Fig. 10.19. Dichromatic reflection. **a** Some incoming light undergoes specular reflection from the surface, while some penetrates the surface is scattered, filtered and re-emitted in all directions according to the Lambertian reflection model. **b** Specular surface reflection can be seen clearly in the *nonred highlight areas* on the two tomatoes, these are reflections of the ceiling lights (courtesy of Distributed Robot Garden project, MIT)

10.3.5   Dichromatic Reflectance

The simple reflectance model introduced in Sect. 10.1.3 is suitable for objects with matte surfaces (e.g. paper, unfinished wood) but if the surface is somewhat shiny the light reflected from the object will have two components – the dichromatic reflection model – as shown in Fig. 10.19a. One component is the illuminant specularly reflected from the surface without spectral change – the interface or Fresnel reflection. The other is light that interacts with the surface: penetrating, scattering, undergoing selective spectral absorbance and being re-emitted in all directions as modeled by Lambertian reflection. The relative amounts of these two components depends on the material and the geometry of the light source, observer and surface normal.

A good example of this can be seen in Fig. 10.19b. Both tomatoes appear red which is due to the scattering lightpath where the light has interacted with the surface of the fruit. However each fruit has an area of specular reflection that appears to be white, the color of the light source, not the surface of the fruit.

The real world is more complex still due to inter-reflections. For example green light reflected from the leaves will fall on the red fruit and be scattered. Some of that light will be reflected off the green leaves again, and so on – nearby objects influence each other's color in complex ways. To achieve photorealistic results in computer graphics all these effects need to be modeled based on detailed knowledge of surface reflection properties and the geometry of all surfaces. In robotics we rarely have this information so we need to develop algorithms that are robust to these effects.

10.3.6   Gamma

CRT monitors were once ubiquitous and the luminance produced at the face of the display was nonlinearly related to the control voltage V according to

$$L = V^\gamma \tag{10.15}$$

where $\gamma \approx 2.2$. To correct for this early video cameras applied the inverse nonlinearity $V = L^{1/\gamma}$ to their output signal which resulted in a system that was linear from end to end.▶ Both transformations are commonly referred to as gamma correction though

Some cameras have an option to choose gamma as either 1 or 0.45 ($= 1 / 2.2$).

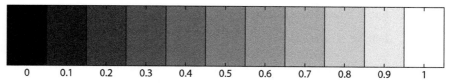

Fig. 10.20.
The linear intensity wedge

Gamma encoding and decoding are often referred to as gamma compression and gamma decompression respectively, since the encoding operation compresses the range of the signal, while decoding decompresses it.

Macintosh computers are an exception and prior to MacOS 10.6 used $\gamma = 1.8$ which made colors appear brighter and more vivid.

more properly the camera-end operation is gamma encoding and the display-end operation is gamma decoding.◄

LCD displays have a stronger nonlinearity than CRTs but correction tables are applied within the display to make it follow the *standard* $\gamma = 2.2$ behavior of the obsolete CRT.►

To show the effect of display gamma we create a simple test pattern

```
>> wedge = [0:0.1:1];
>> idisp(wedge)
```

that is shown in Fig. 10.20 and is like a photographer's *greyscale step wedge*. If we display this on our computer screen it will appear differently to the one printed in the book. We will most likely observe a large change in brightness between the second and third block – the effect of the gamma decoding nonlinearity Eq. 10.15 in the display of your computer.

If we apply gamma encoding

```
>> idisp( wedge .^ (1/2.2) )
```

we observe that the intensity changes appear to be more linear and closer to the one printed in the book.

> The chromaticity coordinates of Eq. 10.9 and Eq. 10.10 are computed as ratios of tristimulus values which are linearly related to luminance in the scene. The nonlinearity applied to the camera output must be corrected, gamma decoded, *before* any colometric operations. The Toolbox function `igamm` performs this operation. Gamma decoding can also be performed when an image is loaded using the `'gamma'` option to the function `iread`.

The JPEG file header (JFIF file format) has a tag `Color Space` which is set to either `sRGB` or `Uncalibrated` if the gamma or color model is not known. See page 363.

Today most digital cameras◄ encode images in sRGB format (IEC 61966-2-1 standard) which uses the ITU Rec. 709 primaries and a gamma encoding function of

$$E' = \begin{cases} 12.92L, & L \leq 0.0031308 \\ 1.055L^{1/2.4} - 0.055, & L > 0.0031308 \end{cases}$$

which comprise a linear function for small values and a power law for larger values. The overall gamma is approximately 2.2.

> The important property of colorspaces such as *HSV* or *xyY* is that the chromaticity coordinates are invariant to changes in intensity. Many digital video cameras provide output in *YUV* or *YC$_B$C$_R$* format which has a luminance component *Y* and two other components which are often mistaken for chromaticity coordinates – they are not. They are in fact color difference signals such that *U*, $C_B \propto B' - Y'$ and *V*, $C_R \propto R' - Y'$ where *R'*, *B'* are gamma *encoded* tristimulus values, and *Y'* is gamma *encoded* intensity. The gamma nonlinearity means that *UV* or $C_B C_R$ will not be a constant as overall lighting level changes.
>
> The tristimulus values from the camera must be first converted to linear tristimulus values, by applying the appropriate gamma decoding, and then computing chromaticity. There is no shortcut.

10.4   Application: Color Image

10.4.1   Comparing Color Spaces [examples/colorspaces]

In this section we bring together many of the concepts and tools introduced in this chapter. We will compare the chromaticity coordinates of the colored squares (squares 1–18) of the Color Checker chart shown in Fig. 10.21 using the *xy*- and $L^*a^*b^*$-color spaces. We compute chromaticity from first principles using the spectral reflectance information for each square which is provided with the Toolbox

```
>> lambda = [400:5:700]*1e-9;
>> macbeth = loadspectrum(lambda, 'macbeth');
```

which has 24 columns, one per square of the test chart. We load the relative power spectrum of the D_{65} standard white illuminant

```
>> d65 = loadspectrum(lambda, 'D65') * 3e9;
```

and scale it to a brightness comparable to sunlight as shown in Fig. 10.3a. Then for each nongrey square

```
1    >> for i=1:18
2          L = macbeth(:,i) .* d65;
3          tristim = max(cmfrgb(lambda, L), 0);
4          RGB = igamm(tristim, 0.45);
5
6          XYZ(i,:) = colorspace('XYZ<-RGB', RGB);
7          Lab(i,:) = colorspace('Lab<-RGB', RGB);
8    end
```

we compute the luminance spectrum (line 2), use the CIE color matching functions to determine the eye's tristimulus response and impose the gamut limits (line 3) and then apply a gamma encoding (line 4) since the `colorspace` function expects gamma encoded RGB data. This is converted to the *XYZ* color space (line 6), and the $L^*a^*b^*$ color space (line 7). Next we convert *XYZ* to *xy* by dividing *X* and *Y* each by $X + Y + Z$, and extract the a^*b^* columns

```
>> xy = XYZ(:,1:2) ./ (sum(XYZ,2)*[1 1]);
>> ab = Lab(:,2:3);
```

giving two matrices, each 18 × 2, with one row per colored square. Finally we plot these points on their respective color planes

```
>> showcolorspace(xy', 'xy');
>> showcolorspace(ab', 'Lab');
```

and the results are displayed in Fig. 10.22. We see, for example, that square 15 is closer to 9 and further from 7 in the a^*b^* plane. The $L^*a^*b^*$ color space was designed so that the Euclidean distance between points is proportional to the color difference perceived by humans. If we are using algorithms to distinguish objects by color then $L^*a^*b^*$ would be preferred over *RGB* or *XYZ*.

Fig. 10.21.
The Gretag Macbeth Color Checker is an array of 24 printed color squares (numbered left to right, top to bottom), which includes different greys and colors as well as spectral simulations of skin, sky, foliage etc. Spectral data for the squares is provided with the toolbox

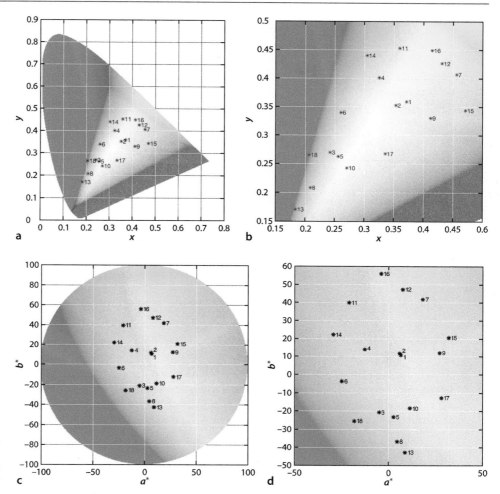

Fig. 10.22.
Color Checker chromaticities.
a *xy*-space; **b** *xy*-space zoomed;
c *a*b*-space; **d** *a*b*-space zoomed

10.4.2   Shadow Removal [examples/shadow]

For a robot vision system that operates outdoors shadows are a significant problem as we can see in Fig. 10.23a. Shadows cause surfaces of the same type to appear quite different and this is problematic for a robot trying to use vision to understand the scene and plan where to drive. Even more problematic is that this effect is not constant, it varies with the time of day and cloud condition. The image in Fig. 10.23b has had the effects of shadowing removed, and we can now see very clearly the different types of terrain – grass and gravel.

The key to removing shadows comes from the observation that the bright parts of the scene are illuminated directly by the sun while the darker shadowed regions are illuminated by the sky. Both the sun and the sky can be modeled as blackbody radiators with color temperatures as listed in Table 10.3. Shadows therefore have two defining characteristics: they are dark and they have a slight blue tint.

We model the camera using Eq. 10.4 but model the spectral response of the camera's color sensors as Dirac functions $M_x(\lambda) = \delta(\lambda - \lambda_x)$ which allows us to eliminate the integrals

$$R = E(\lambda_R)R(\lambda_R)M_R(\lambda_R)$$
$$G = E(\lambda_G)R(\lambda_G)M_G(\lambda_G)$$
$$B = E(\lambda_B)R(\lambda_B)M_B(\lambda_B)$$

For each pixel we compute chromaticity coordinates $r = R\,/\,G$ and $b = B\,/\,G$ which are invariant to change in illumination magnitude.

$$r = \frac{E(\lambda_R)R(\lambda_R)M_R(\lambda_R)}{E(\lambda_G)R(\lambda_G)M_R(\lambda_G)} = \frac{\frac{2hc^2}{\lambda^5\left(e^{hc/k\lambda_R T}-1\right)}R(\lambda_R)M_R(\lambda_R)}{\frac{2hc^2}{\lambda^5\left(e^{hc/k\lambda_G T}-1\right)}R(\lambda_G)M_R(\lambda_G)}$$

To simplify further we apply the Wien approximation, eliminating the -1 term, which is a reasonable approximation for color temperatures in the range under consideration, and now we can write

$$r \approx \frac{e^{hc/k\lambda_G T}R(\lambda_R)M_R(\lambda_R)}{e^{hc/k\lambda_R T}R(\lambda_G)M_G(\lambda_G)} = e^{hc(1/\lambda_G - 1/\lambda_R)/kT}\frac{M_R(\lambda_R)}{M_G(\lambda_G)}\frac{R(\lambda_R)}{R(\lambda_G)}$$

which is a function of color temperature T and various constants: physical constants c, h and k; sensor response wavelength λ_x and magnitude $M_x(\lambda_x)$, and material properties $R(\lambda_x)$. Taking the logarithm we obtain the very simple form

$$\log r = c_1 - \frac{c_2}{T} \tag{10.16}$$

and repeating the process for blue chromaticity we can write

$$\log b = c_1' - \frac{c_2'}{T} \tag{10.17}$$

Every color pixel $(R, G, B) \in \mathbb{R}^3$ can be mapped to a point $(\log r, \log b) \in \mathbb{R}^2$ and as the color temperature changes the points will all move along lines with a slope of c_2'/c_2. Therefore a projection onto the orthogonal direction, a line with slope c_2/c_2', results in a 1-dimensional quantity

$$s = -c_2 \log r + c_2' \log b$$

that is invariant to the color temperature of the illuminant. We can compute this for every pixel in an image

```
>> im = iread('parks.jpg', 'gamma', 'sRGB');
>> gs = invariant(im, 0.7, 'noexp');
>> idisp(gs)
```

and the result is shown in Fig. 10.23b. The pixels have a greyscale value that is a complex function of material reflectance and camera sensor properties. The arguments to the function are the color image, the slope of the line in radians and a flag to return the logarithm s rather than its exponent.

Fig. 10.23. Shadows create confounding effects in images. **a** View of a park with strong shadows; **b** the shadow invariant image in which the variation lighting has been almost entirely removed (Corke et al. 2013)

To achieve this result we have made some approximations and a number of rather strong assumptions: the camera has a linear response from scene luminance to *RGB* tristimulus values, the color channels of the camera have nonoverlapping spectral response, and the scene is illuminated by blackbody light sources. The first assumption means that we need to use a camera with $\gamma = 1$ or apply gamma decoding to the image before we proceed. The second is far from true, especially for the red and green channels of a color camera, yet the method works well in practice. The biggest effect is that the points move along a line with a slope different to c_2'/c_2 but we can estimate the slope empirically by looking at a set of shadowed and nonshadowed pixels corresponding to the same material in the scene

```
>> theta = esttheta(im)
```

which will prompt you to select a region and returns an angle which can be passed to invariant. The final assumption means that the technique will not work for nonincandescent light sources, or where the scene is partly illuminated by reflections from colored surfaces. More details are provided in the MATLAB function source code.

10.5   Wrapping Up

We have learned that the light we see is electro-magnetic radiation with a mixture of wavelengths, a continuous spectrum, which is modified by reflectance and absorption. The spectrum elicits a response from the eye which we interpret as color – for humans the response is a tristimulus, a 3-vector that represents the outputs of the three different types of cones in our eye. A digital color camera is functionally equivalent. The tristimulus can be considered as a 1-dimensional brightness coordinate and a 2-dimensional chromaticity coordinate which allows colors to be plotted on a plane. The spectral colors form a locus on this plane and all real colors lie within this locus. Any three primary colors form a triangle on this plane which is the gamut of those primaries. Any color within the triangle can be matched by an appropriate mixture of those primaries. No set of primaries can define a gamut that contains all colors. An alternative set of imaginary primaries, the CIE *XYZ* system, does contain all real colors and is the standard way to describe colors. Tristimulus values can be transformed using linear transformations to account for different sets of primaries. Nonlinear transformations can be used to describe tristimulus values in terms of human-centric qualities such as hue and saturation. We also discussed the definition of white, color temperature, color constancy, the problem of white balancing, the nonlinear response of display devices and how this effects the common representation of images and video.

We learned that the colors and brightness we perceive is a function of the light source and the surface properties of the object. While humans are quite able to "factor out" illumination change this remains a significant challenge for robotic vision systems. We finished up by showing how to remove shadows in an outdoor color image.

Infra-red cameras. Consumer cameras are functionally equivalent to the human eye and are sensitive to the visible spectrum. Cameras are also available that are sensitive to infra-red and a number of infra-red bands are defined by CIE: IR-A (700−1 400 nm), IR-B (1 400−3 000 nm), and IR-C (3 000 nm−1 000 μm). In common usage IR-A and IR-B are known as near infra-red (NIR) and short-wavelength infra-red (SWIR) respectively, and the IR-C subbands are medium-wavelength (MWIR, 3 000−8 000 nm) and long-wavelength (LWIR, 8 000−15 000 nm). LWIR cameras are also called thermal or thermographic cameras.

Ultraviolet cameras typically work in the near ultra-violet region (NUV, 200−380 nm) and are used in industrial applications such as detecting corona discharge from high-voltage electrical systems.

Hyperspectral cameras have more more than three classes of photoreceptor, they sample the incoming spectrum at many points typically from infra-red to ultra-violet and with tens or even hundreds of spectral bands. Hyperspectral cameras are used for applications including aerial survey classification of land-use and identification of the mineral composition of rocks.

Further Reading

At face value color is a simple concept that we learn in kindergarten but as we delve in we find it is a fascinating and complex topic with a massive literature. In this chapter we have only begun to scrape the surface of photometry and colorimetry. Photometry is the part of the science of radiometry concerned with measurement of visible light. It is challenging for engineers and computer scientists since it makes use of uncommon units such as lumen, steradian, nit, candela and lux. One source of complexity is that words like intensity and brightness are synonyms in everyday speech but have very specific meanings in photometry. Colorimetry is the science of color perception and is also a large and complex area since human perception of color depends on the individual observer, ambient illumination and even the field of view. Colorimetry is however critically important in the design of cameras, computer displays, video equipment and printers. Comprehensive online information about computer vision is available through CVonline at http://homepages.inf.ed.ac.uk/rbf/CVonline, and the material in this chapter is covered by the section *Image Physics*.

The computer vision textbooks by Gonzalez and Woods (2008) and Forsyth and Ponce (2011) each have a discussion on color and color spaces. The latter also has a discussion on the effects of shading and inter-reflections. The book by Gevers et al. (2012) is solid introduction to color vision theory and covers the dichromatic reflectance model in detail. It also covers computer vision algorithms that deal with the challenges of color constancy. The Retinex theory is described in Land and McCann (1971) and MATLAB implementations can be found at http://www.cs.sfu.ca/~colour/code. Other resources related to color constancy can be found at http://colorconstancy.com.

Readable and comprehensive books on color science include Koenderink (2010), Hunt (1987) and from a television or engineering perspective Benson (1986). A more conversational approach is given by Hunter and Harold (1987), which also covers other aspects of appearance such as gloss and luster. The CIE standard (Commission Internationale de l'Éclairage 1987) is definitive but hard reading. The work of the CIE is ongoing and its standards are periodically updated at www.cie.co.at. The color matching functions were first tabulated in 1931 and revised in 1964.

Charles Poynton has for a long time maintained excellent online tutorials about color spaces and gamma at http://www.poynton.com. His book (Poynton 2012) is an excellent and readable introduction to these topics while also discussing digital video systems in great depth.

General interest. Crone (1999) covers the history of theories of human vision and color. How the human visual system works, from the eye to perception, is described in two very readable books Stone (2012) and Gregory (1997). Land and Nilsson (2002) describes the design principles behind animal eyes and how characteristics such as acuity, field of view and low light capability are optimized for different species.

Data Sources

The Toolbox contains a number of data files describing various spectra which are summarized in Table 10.4. Each file has as its first column the wavelength in meters. The files have different wavelength ranges and intervals but the helper function `loadspectrum` interpolates the data to the user specified range and sample interval.

Several internet sites contain spectral data in tabular format and this is linked from the book's web site. This includes reflectivity data for over 2 000 materials provided by NASA's online ASTER spectral library 2.0 (Baldridge et al. 2009) at http://speclib.jpl.nasa.gov and the Spectral Database from the University of Eastern Finland Color Research Laboratory at http://uef.fi/en/spectral. Data on cone responses and CIE color matching functions is available from the Colour & Vision Research Laboratory at University College London at http://cvrl.org. CIE data is also available online at http://cie.co.at.

Table 10.4.
Various spectra provided with the Toolbox. Relative luminosity values lie in the interval [0, 1], and relative spectral power distribution (SPD) are normalized to a value of 1.0 at 550 nm. These files can be loaded using the Toolbox `loadspectrum` function

Filename	Units	Description
cones	Rel. luminosity	Spectral response of human cones
bb2	Rel. luminosity	Spectral response of Sony ICX 204AK sensor used in Point Grey BumbleBee2 camera
photopic	Rel. luminosity	CIE 1924 photopic response
scotopic	Rel. luminosity	CIE 1951 scoptic response
redbrick	Reflectivity	Reflectivity spectrum of a weathered red brick
macbeth	Reflectivity	Reflectivity of the Gretag-Macbeth Color Checker array (24 squares), see Fig. 10.21
solar	$W\,m^{-2}\,m^{-1}$	Solar spectrum at ground level
water	$1\,m^{-1}$	Light absorption spectrum of water
D65	Rel. SPD	CIE standard D_{65} illuminant

Exercises

1. You are a blackbody radiator! Plot your own blackbody emission spectrum. What is your peak emission frequency? What part of the EM spectrum is this? What sort of sensor would you use to detect this?
2. Consider a sensor that measures the amount of radiated power P_1 and P_2 at wavelengths λ_1 and λ_2 respectively. Write an equation to give the temperature T of the blackbody in terms of these quantities.
3. Using the Stefan-Boltzman law compute the power emitted per square meter of the Sun's surface. Compute the total power output of the Sun.
4. Use numerical integration to compute the power emitted in the visible band 400–700 nm per square meter of the Sun's surface.
5. Why is the peak luminosity defined as 683 lm W^{-1}?
6. Given typical outdoor illuminance as per page 294 determine the luminous intensity of the Sun.
7. Sunlight at ground level. Of the incoming radiant power determine, in percentage terms, the fraction of infra-red, visible and ultra-violet light.
8. Use numerical integration to compute the power emitted in the visible band 400–700 nm per square meter for a tungsten lamp at 2 600 K. What fraction is this of the total power emitted?
9. Plot and compare the human photopic and scotopic spectral response.
    a) Compare the response curves of human cones and the RGB channels of a color camera. Use `cones.dat` and `bb2.dat`.
10. Can you create a metamer for the red brick?
11. Prove Grassmann's center of gravity law mentioned on page 297.
12. On the xy-chromaticity plane plot the locus of a blackbody radiator with temperatures in the range 1 000–10 000 K.
13. Plot the XYZ primaries on the rg-plane.
14. For Fig. 10.12 determine the chromaticity of the feasible green.
15. Determine the tristimulus values for the red brick using the Rec. 709 primaries.
16. Take a picture of a white object using incandescent illumination. Determine the average RGB tristimulus value and compute the xy-chromaticity. How far off white is it? Determine the color balance matrix J to correct the chromaticity. What is the chromaticity of the illumination?
17. What is the name of the color of the red brick when viewed underwater (page 308).
18. Image a target like Fig. 10.17 that has three colored patches of known chromaticity. From their observed chromaticity determine the transform from observed tristimulus values to Rec. 709 primaries. What is the chromaticity of the illumination?

19. Consider an underwater application where a target d meters below the surface is observed through m meters of water, and the water surface is illuminated by sunlight. From the observed chromaticity can you determine the true chromaticity of the target? How sensitive is this estimate to incorrect estimates of m and d? If you knew the true chromaticity of the target could you determine its distance?

20. Is it possible that two different colors look the same under a particular lighting condition? Create an example of colors and lighting that would cause this?

21. Use one of your own pictures and the approach of Sect. 10.4.1. Can you distinguish different objects in the picture?

22. Show analytically or numerically that scaling a tristimulus value has no effect on the chromaticity. What happens if the chromaticity is computed on gamma encoded tristimulus values?

23. Create an interactive tool with sliders for R, G and B that vary the color of a displayed patch. Now modify this for sliders X, Y and Z or x, y and Y.

24. Take a color image and determine how it would appear through 1, 5 and 10 m of water.

25. Determine the names of the colors in the Gretag-Macbeth color checker chart.

26. Plot the color-matching function components shown in Fig. 10.10 as a 3D curve. Rotate it to see the locus as shown in Fig. 10.11.

11

Image Formation

In this chapter we discuss how images are formed and captured, the first step in robot and human perception of the world. From images we can deduce the size, shape and position of objects in the world as well as other characteristics such as color and texture which ultimately lead to recognition.

It has long been known that a simple pin-hole is able to create a perfect inverted image on the wall of a darkened room. Some marine mollusks, for example the Nautilus, have pin-hole camera eyes. All vertebrates have a lens that forms an inverted image on the retina where the light-sensitive cells rod and cone cells, shown previously in Fig. 10.6, are arranged. A digital camera is similar in principle – a glass or plastic lens forms an image on the surface of a semiconductor chip with an array of light-sensitive devices to convert light to a digital image.

The process of image formation, in an eye or in a camera, involves a *projection* of the 3-dimensional world onto a 2-dimensional surface. The depth information is lost and we can no longer tell from the image whether it is of a large object in the distance or a smaller closer object. This transformation from 3 to 2 dimensions is known as perspective projection and is discussed in Sect. 11.1. Section 11.2 introduces the topic of camera calibration, the estimation of the parameters of the perspective transformation. Sections 11.3 to 11.5 introduce alternative types of cameras capable of wide-angle, panoramic or light-field imaging. Section 11.6 introduces some advanced concepts such as projecting lines and conics, and non-perspective cameras.

11.1 Perspective Camera

11.1.1 Perspective Projection

A small hole in the wall of a darkened room will cast a dim inverted image of the outside world on the opposite wall – a so-called pin-hole camera. The pin-hole camera produces a very dim image since its radiant power is the scene luminance in units of W m^{-2} multiplied by the area of the pin hole. Figure 11.1a shows that only a small fraction of the light leaving the object finds its way to the image. A pin-hole camera has no focus adjustments – all objects are in focus irrespective of distance.

In the 5th century BCE, the philosopher Mozi in ancient China mentioned the effect of an inverted image forming through a pinhole. A camera obscura is a darkened room where a dim inverted image of the world is cast on the wall by light entering through a small hole. They were popular tourist attractions in Victorian times, particularly in Britain, and many are still operating today. (Image on the right from the Drawing with Optical Instruments collection at http://vision.mpiwg-berlin.mpg.de/elib)

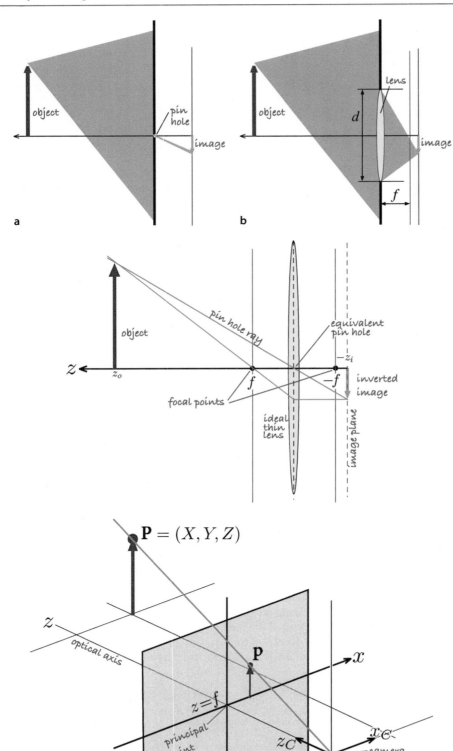

Fig. 11.1.
Light gathering ability of
a pin-hole camera and **b** a lens

Fig. 11.2.
Image formation geometry for
a thin convex lens shown in
2-dimensional cross section.
A lens has two focal points at a
distance of f on each side of the
lens. By convention the camera's
optical axis is the z-axis

Fig. 11.3.
The central-projection model.
The image plane is at a distance
f in front of the camera's origin,
and on which a noninverted image is formed. The camera's coordinate frame is right-handed
with the z-axis defining the center of the field of view

The key to brighter images is to use an objective lens, as shown in Fig. 11.1b, which collects light from the object over a larger area and directs it to the image. A convex lens can form an image just like a pinhole and the fundamental geometry of image formation for a thin lens◄ is shown in Fig. 11.2. The positive z-axis is the camera's optical axis.

The z-coordinate of the object and its image, with respect to the lens center, are related by the thin lens equation

Real camera lenses comprise multiple lens elements but still have focal points on each side of the compound lens assembly.

$$\frac{1}{z_o} + \frac{1}{z_i} = \frac{1}{f} \tag{11.1}$$

where z_o is the distance to the object, z_i the distance to the image, and f is the focal length of the lens.◄ For $z_o > f$ an inverted image is formed on the image plane at $z_i < -f$.

The inverse of focal length is known as diopter. For thin lenses placed close together their combined diopter is close to the sum of their individual diopters.

In a camera the image plane is fixed at the surface of the sensor chip so the focus ring of the camera moves the lens along the optical axis so that it is a distance z_i from the image plane – for an object at infinity $z_i = f$. The downside of using a lens is the need to focus. Our own eye has a single convex lens made from transparent crystallin proteins, and focus is achieved by muscles which change its shape – a process known as accomodation. A high-quality camera lens is a compound lens comprising multiple glass or plastic lenses.

In computer vision it is common to use the central perspective imaging model shown in Fig. 11.3. The rays converge on the origin of the camera frame {C} and a noninverted image is *projected* onto the image plane located at $z = f$. The z-axis intersects the image plane at the principal point which is the origin of the 2D image coordinate frame. Using similar triangles we can show that a point at the world coordinates $\mathbf{P} = (X, Y, Z)$ is projected to the image point $\mathbf{p} = (x, y)$ by

$$x = f\frac{X}{Z}, \; y = f\frac{Y}{Z} \tag{11.2}$$

which is a projective transformation, or more specifically a perspective projection. This mapping from the 3-dimensional world to a 2-dimensional image has consequences that we can see in Fig. 11.4 – parallel lines converge and circles become ellipses.

More formally we can say that the transformation, from the world to the image plane has the following characteristics:

Lens aperture. The *f-number* of a lens, typically marked on the rim, is a dimensionless quantity $F = f/d$ where d is the diameter of the lens (often denoted ϕ on the lens rim). The *f*-number is *inversely* related to the light gathering ability of the lens. To reduce the

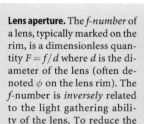

amount of light falling on the image plane the effective diameter is reduced by a mechanical aperture, or iris, which *increases* the *f*-number. Illuminance on the image plane is inversely proportional to F^2 since it depends on light gathering area. To reduce illuminance by a factor of 2, the *f*-number must be increased by a factor of $\sqrt{2}$ or "one stop". The *f*-number graduations increase by $\sqrt{2}$ at each stop. An *f*-number is conventionally written in the form $f/1.4$ for $F = 1.4$.

1. It performs a mapping from 3-dimensional space to the 2-dimensional image plane: $\mathcal{P} : \mathbb{R}^3 \mapsto \mathbb{R}^2$.
2. Straight lines in the world are projected to straight lines on the image plane.
3. Parallel lines in the world are projected to lines that intersect at a vanishing point as shown in Fig. 11.4a. In drawing, this effect is known as foreshortening. The exception are fronto-parallel lines – lines lying in a plane parallel to the image plane – which always remain parallel.

Focus and depth of field. Ideally a group of light rays from a point in the scene meet at a point in the image. With imperfect focus the rays instead form a finite sized spot called the circle of confusion which is the point spread function of the optical system. By convention, if the size of the circle is around that of a pixel then the image is *acceptably* focused.

A pin-hole camera has no focus control and always creates a focused image of objects irrespective of their distance. A lens does not have this property – the focus ring changes the distance between the lens and the image plane and must be adjusted so that the object of interest is *acceptably* focused. Photographers refer to depth of field which is the range of object distances for which *acceptably* focused images are formed. Depth of field is high for small aperture settings where the lens is more like a pin-hole, but this means less light and noisier images or longer exposure time and motion blur. This is the photographer's dilemma!

4. Conics◄ in the world are projected to conics on the image plane. For example, a circle is projected as a circle or an ellipse as shown in Fig. 11.4b.
5. The size (area) of a shape is not preserved and depends on distance.
6. The mapping is not one-to-one and no unique inverse exists. That is, given (x, y) we cannot uniquely determine (X, Y, Z). All that can be said is that the world point lies somewhere along the red projecting ray shown in Fig. 11.3. This is an important topic that we will return to in Chap. 14.
7. The transformation is not conformal – it does not preserve shape since internal angles are not preserved. Translation, rotation and scaling are examples of conformal transformations.

Fig. 11.4. The effect of perspective transformation. **a** Parallel lines converge, **b** circles become ellipses

Conic sections, or conics, are a family of curves obtained by the intersection of a plane with a cone. They include circles, ellipses, parabolas and hyperbolas.

11.1.2   Modeling a Perspective Camera

We can write the image-plane point coordinates in homogeneous form $\tilde{p} = (\tilde{x}, \tilde{y}, \tilde{z})$ where

$$\tilde{x} = fX, \; \tilde{y} = fY, \; \tilde{z} = Z$$

or in compact matrix form as

$$\tilde{p} = \begin{pmatrix} f & 0 & 0 \\ 0 & f & 0 \\ 0 & 0 & 1 \end{pmatrix} \begin{pmatrix} X \\ Y \\ Z \end{pmatrix} \tag{11.3}$$

where the nonhomogeneous image-plane coordinates are

$$x = \frac{\tilde{x}}{\tilde{z}}, \; y = \frac{\tilde{y}}{\tilde{z}}$$

These are often referred to as the retinal image-plane coordinates. For the case where $f = 1$ the coordinates are referred to as the normalized, retinal or canonical image-plane coordinates.

If we write the world coordinate in homogeneous form as well $^C\tilde{P} = (X, Y, Z, 1)^T$ then the perspective projection can be written in *linear* form as

$$\tilde{p} = \begin{pmatrix} f & 0 & 0 & 0 \\ 0 & f & 0 & 0 \\ 0 & 0 & 1 & 0 \end{pmatrix} {}^C\tilde{P} \tag{11.4}$$

or

$$\tilde{p} = C\,{}^C\tilde{P} \tag{11.5}$$

where C is a 3×4 matrix known as the camera matrix. Note that we have written ${}^C\tilde{P}$ to highlight the fact that this is the coordinate of the point with respect to the camera frame $\{C\}$. The tilde indicates homogeneous quantities and Sect. C.2 provides a refresher on homogeneous coordinates. The camera matrix can be factored

$$\tilde{p} = \begin{pmatrix} f & 0 & 0 \\ 0 & f & 0 \\ 0 & 0 & 1 \end{pmatrix} \begin{pmatrix} 1 & 0 & 0 & 0 \\ 0 & 1 & 0 & 0 \\ 0 & 0 & 1 & 0 \end{pmatrix} {}^C\tilde{P}$$

where the second matrix is the projection matrix.

The Toolbox allows us to create a model of a central-perspective camera. For example

```
>> cam = CentralCamera('focal', 0.015);
```

returns an instance of a CentralCamera object with a 15 mm lens. By default the camera is at the origin of the world frame with its optical axis pointing in the world z-direction as shown in Fig. 11.3. We define a world point

```
>> P = [0.3, 0.4, 3.0]';
```

in units of meters and the corresponding image-plane coordinates are

```
>> cam.project(P)
ans =
    0.0015
    0.0020
```

The point on the image plane is at (1.5, 2.0) mm with respect to the principal point. This is a very small displacement but it is commensurate with the size of a typical image sensor.

In general the camera will have an arbitrary pose ξ_C with respect to the world coordinate frame as shown in Fig. 11.5. The position of the point with respect to the camera is

$$^C\!P = (\ominus\xi_C)\cdot{}^0\!P \tag{11.6}$$

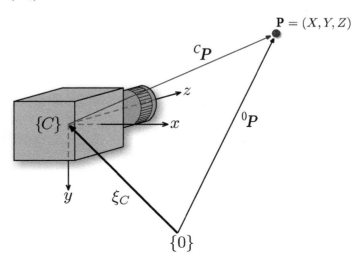

Fig. 11.5.
Camera coordinate frames

or using homogeneous coordinates

$${}^{C}\boldsymbol{P} = \boldsymbol{T}_{C}^{-1}\,{}^{0}\boldsymbol{P}$$

We can easily demonstrate this by moving our camera 0.5 m to the *left*

```
>> cam.project(P, 'pose', SE3(-0.5, 0, 0) )
ans =
    0.0040
    0.0020
```

where the third argument is the pose of the camera ξ_C as a homogeneous transformation. We see that the x-coordinate has increased from 1.5 mm to 4.0 mm, that is, the image point has moved to the *right*.

11.1.3   Discrete Image Plane

In a digital camera the image plane is a $W \times H$ grid of light-sensitive elements called photosites that correspond directly to the picture elements (or pixels) of the image as shown in Fig. 11.6. The pixel coordinates are a 2-vector (u, v) of nonnegative integers and by convention the origin is at the top-left hand corner of the image plane. In

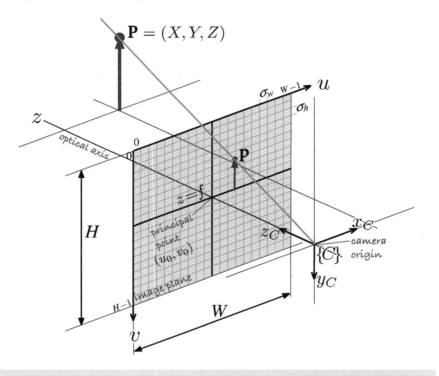

Fig. 11.6.
Central projection model showing image plane and discrete pixels

Image sensor. The light-sensitive cells in a camera chip, the photosites (see page 364), are commonly square with a side length in the range 1–10 μm. Professional cameras have large photosites for increased light sensitivity whereas cellphone cameras have small sensors and therefore small less-sensitive photosites. The ratio of the number of horizontal to vertical pixels is the aspect ratio and is commonly 4:3 or 16:9 (see page 366). The dimension of the sensor is measured diagonally across the array and is commonly expressed in inches, e.g. ⅓, ¼ or ½ inch. However the active sensing area of the chip has a diagonal that is typically around ⅔ of the given dimension.

MATLAB® the top-left pixel is (1, 1). The pixels are uniform in size and centered on a regular grid so the pixel coordinate is related to the image-plane coordinate by

$$u = \frac{x}{\rho_w} + u_0, \; v = \frac{y}{\rho_h} + v_0$$

where ρ_w and ρ_h are the width and height of each pixel respectively, and (u_0, v_0) is the principal point – the pixel coordinate of the point where the optical axis intersects the image plane with respect to the new origin. We can write Eq. 11.4 in terms of pixel coordinates by prepending a camera parameter matrix K

$$\tilde{p} = \underbrace{\begin{pmatrix} 1/\rho_w & 0 & u_0 \\ 0 & 1/\rho_h & v_0 \\ 0 & 0 & 1 \end{pmatrix}}_{K} \begin{pmatrix} f & 0 & 0 & 0 \\ 0 & f & 0 & 0 \\ 0 & 0 & 1 & 0 \end{pmatrix} {}^C\tilde{P} \tag{11.7}$$

The matrix K is often written with a finite value at $K_{1,2}$ to represent skew. This accounts for the fact that the u- and v-axes are not orthogonal, which with precise semiconductor fabrication processes is quite unlikely.

where $\tilde{p} = (\tilde{u}, \tilde{v}, \tilde{w})$ is the homogeneous coordinate of the world point P in pixel coordinates.◂ The nonhomogeneous image-plane pixel coordinates are

$$u = \frac{\tilde{u}}{\tilde{w}}, \; v = \frac{\tilde{v}}{\tilde{w}} \tag{11.8}$$

For example if the pixels are 10 µm square and the pixel array is 1 280 × 1 024 pixels with its principal point at image-plane coordinate (640, 512) then

```
>> cam = CentralCamera('focal', 0.015, 'pixel', 10e-6, ↵
   'resolution', [1280 1024], 'centre', [640 512], 'name', 'mycamera')
cam =
name: mycamera [central-perspective]
  focal length:   0.015
  pixel size:     (1e-05, 1e-05)
  principal pt:   (640, 512)
  number pixels:  1280 x 1024
  pose:           t = (0,0,0), RPY/yxz = (0,0,0) deg
```

which displays the parameters of the camera model including the camera pose T. The corresponding nonhomogeneous image-plane coordinates of the previously defined world point are

```
>> cam.project(P)
ans =
   790
   712
```

11.1.4   Camera Matrix

Combining Eq. 11.6 and Eq. 11.7 we can write the camera projection in general form as

$$\tilde{p} = \underbrace{\begin{pmatrix} f/\rho_w & 0 & u_0 \\ 0 & f/\rho_h & v_0 \\ 0 & 0 & 1 \end{pmatrix} \begin{pmatrix} 1 & 0 & 0 & 0 \\ 0 & 1 & 0 & 0 \\ 0 & 0 & 1 & 0 \end{pmatrix}}_{\text{intrinsic}} \underbrace{\left({}^0T_C \right)^{-1} \tilde{P}}_{\text{extrinsic}} \tag{11.9}$$

$$= K P_0 \, {}^0T_C^{-1} \tilde{P}$$

$$= C \tilde{P}$$

The terms f/ρ_w and f/ρ_h are the focal length expressed in units of pixels.

where all the terms are rolled up into the camera matrix C◂. This is a 3 × 4 homogeneous transformation which performs scaling, translation and perspective projection. It is often also referred to as the projection matrix or the camera calibration matrix.

We have already mentioned the fundamental ambiguity with perspective projection, that we cannot distinguish between a large distant object and a smaller closer object. We can rewrite Eq. 11.9 as

$$\tilde{p} = (CH^{-1})(H\tilde{P}) = C'\tilde{P}'$$

where H is an arbitrary nonsingular 3×3 matrix. This implies that an infinite number of camera C' and world point coordinate $\tilde{P}'$ combinations will result in the same image-plane projection $\tilde{p}$.

This illustrates the essential difficulty in determining 3-dimensional world coordinates from 2-dimensional projected coordinates. It can only be solved if we have information about the camera or the 3-dimensional object.

The projection can also be written in functional form as

$$p = \mathcal{P}(P, K, \xi_C) \tag{11.10}$$

where P is the point coordinate vector in the world frame. K is the camera parameter matrix and comprises the intrinsic parameters which are the innate characteristics of the camera and sensor such as f, ρ_w, ρ_h, u_0 and v_0. ξ_C is the pose of the camera and comprises a minimum of six parameters – the extrinsic parameters – that describe camera translation and orientation in $\mathbf{SE}(3)$.

There are 5 intrinsic and 6 extrinsic parameters – a total of 11 independent parameters to describe a camera. The camera matrix has 12 elements so one degree of freedom, the overall scale factor, is unconstrained and can be arbitrarily chosen. In practice the camera parameters are not known and must be estimated using a camera calibration procedure which we will discuss in Sect. 11.2.

The camera intrinsic parameter matrix $\mathbb{K}$ for this camera is

```
>> cam.K
ans =
   1.0e+03 *
   1.5000         0    0.6400
        0    1.5000    0.5120
        0         0    0.0010
```

The camera matrix is implicitly created when the Toolbox camera object is constructed and for this example is

```
>> cam.C
ans =
   1.0e+03 *
   1.5000         0    0.6400         0
        0    1.5000    0.5120         0
        0         0    0.0010         0
```

The **field of view** of a lens is an open rectangular pyramid, a frustum, that subtends angles θ_h and θ_v in the horizontal and vertical planes respectively. A *normal lens* is one with a field of view around 50°, while a wide angle lens has a field of view >60°. Beyond 110° it is difficult to create a lens that maintains perspective projection, so nonperspective fisheye lenses are required.

For very wide-angle lenses it is more common to describe the field of view as a solid angle which is measured in units of steradians (or sr). This is the area of the field of view projected onto the surface of a unit sphere. A hemispherical field of view is 2π sr and a full spherical view is 4π sr. If we approximate the camera's field of view by a cone with apex angle θ the corresponding solid angle is $2\pi(1 - \cos\theta/2)$ sr. A camera with a field of view greater than a full hemisphere is termed omnidirectional or panoramic.

The camera matrix $C \subset \mathbb{R}^{3 \times 4}$ has some important structure and properties:

- It can be partitioned $C = (M \mid c_4)$ into a nonsingular matrix $M \subset \mathbb{R}^{3 \times 3}$ and a vector, where $c_4 = -Mc$ and c is the world origin in the camera frame. We can recover this by $c = -M^{-1}c_4$.
- The null space of C is $\tilde{c}$.
- A pixel at coordinate p corresponds to a ray in space parallel to the vector $M^{-1}\tilde{p}$.
- The matrix $M = KR$ is the product of the camera intrinsics and the camera inverse orientation. We can perform an RQ-decomposition of $M = RQ$ where R is an upper-triangular matrix (which is K) and an orthogonal matrix Q (which is R). ◄
- The bottom row of C defines the principal plane, which is parallel to the image plane and contains the camera origin.
- If the rows of M are vectors m_i then
  - m_3^T is a vector normal to the principal plane and parallel to the optical axis and Mm_3^T is the principal point in homogeneous form.
  - if the camera has zero skew, that is $K_{1,2} = 0$, then
    $(m_1 \times m_3) \cdot (m_2 \times m_3) = 0$
  - and, if the camera has square pixels, that is $\rho_u = \rho_v$, then
    $\|m_1 \times m_3\| = \|m_2 \times m_3\| = f/\rho$

Yes, R has two different meanings here. MATLAB does not provide an RQ-decomposition but it can be determined by transforming the inputs to, and results of, the builtin MATLAB QR-decomposition function qr. There are many subtleties in doing this though: negative scale factors in the K matrix or $\det R = -1$, see Hartley and Zisserman (2003), or the implementation of $invC$ for details.

The field of view of a camera is a function of its focal length f. A wide-angle lens has a small focal length, a telephoto lens has a large focal length, and a zoom lens has an adjustable focal length. The field of view can be determined from the geometry of Fig. 11.6. In the horizontal direction the half-angle of view is

$$\frac{\theta_h}{2} = \tan^{-1}\frac{W/2\rho_w}{f}$$

where W is the number of pixels in the horizontal direction. We can then write

$$\theta_h = 2\tan^{-1}\frac{W\rho_w}{2f}, \; \theta_v = 2\tan^{-1}\frac{H\rho_h}{2f} \tag{11.11}$$

We note that the field of view is also a function of the dimensions of the camera chip which is $W\rho_w \times H\rho_h$. The field of view is computed by the fov method of the camera object

```
>> cam.fov() * 180/pi
ans =
    46.2127    37.6930
```

in degrees in the horizontal and vertical directions respectively.

11.1.5 Projecting Points

The $CentralCamera$ class is a subclass of the $Camera$ class and inherits the ability to project multiple points or lines. Using the Toolbox we create a 3×3 grid of points in the xy-plane with overall side length 0.2 m and centered at $(0, 0, 1)$

```
>> P = mkgrid(3, 0.2, 'pose', SE3(0, 0, 1.0));
```

which returns a 3×9 matrix with one column per grid point where each column comprises the coordinates in X, Y, Z order. The first four columns are

```
>> P(:,1:4)
ans =
    -0.1000    -0.1000    -0.1000         0
    -0.1000         0     0.1000    -0.1000
     1.0000     1.0000     1.0000     1.0000
```

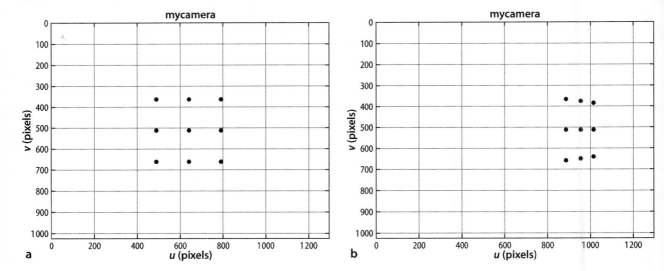

Fig. 11.7. Two views of a planar grid of points. **a** Frontal view, **b** oblique view

By default `mkgrid` generates a grid in the xy-plane that is centered at the origin. The optional last argument is a homogeneous transformation that is applied to the default points and allows the plane to be arbitrarily positioned and oriented.

The image-plane coordinates of the vertices are

```
>> cam.project(P)
ans =
    490    490    490    640    640    640    790    790    790
    362    512    662    362    512    662    362    512    662
```

which can also be plotted

```
>> cam.plot(P)
```

giving the virtual camera view shown in Fig. 11.7a. The camera pose

```
>> Tcam = SE3(-1,0,0.5)*SE3.Ry(0.9);
```

results in an oblique view of the plane

```
>> cam.plot(P, 'pose', Tcam)
```

shown in Fig. 11.7b. We can clearly see the effect of perspective projection which has distorted the shape of the square – the top and bottom edges, which are parallel lines, have been projected to lines that converge at a vanishing point.

The vanishing point for a line can be determined from the projection of its ideal line. The top and bottom lines of the grid are parallel to the world x-axis or the vector $(1, 0, 0)$. The corresponding ideal line has homogeneous coordinates $(1, 0, 0, 0)$ and exists at infinity due to the final zero element. The vanishing point is the projection of this vector

```
>> cam.project([1 0 0 0]', 'pose', Tcam)
ans =
    1.0e+03 *
    1.8303
    0.5120
```

which is $(1\,803, 512)$ and just to the right of the visible image plane.

The `plot` method can optionally return the image-plane coordinates

```
>> p = cam.plot(P, 'pose', Tcam)
```

just like the `project` method. For the oblique viewing case the image-plane coordinates

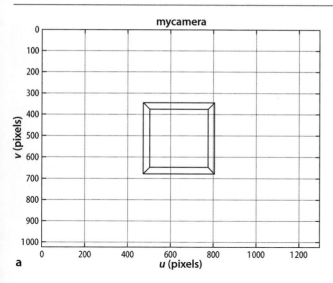
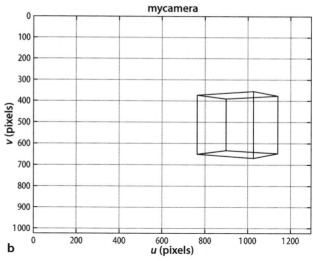

Fig. 11.8. Line segment representation of a cube. **a** Frontal view, **b** oblique view

```
>> p(:,1:4)
ans =
    887.7638   887.7638   887.7638   955.2451
    364.3330   512.0000   659.6670   374.9050
```

This is not strictly true for CMOS sensors where transistors reduce the light-sensitive area by the fill factor – the fraction of each photosite's area that is light sensitive.

have a fractional component which means that the point is not projected to the center of the pixel. However a pixel responds to light equally◄ over its surface area so the discrete pixel coordinate can be obtained by rounding.

A 3-dimensional object, a cube, can be defined and projected in a similar fashion. The vertices of a cube with side length 0.2 m and centered at (0, 0, 1) can be defined by

```
>> cube = mkcube(0.2, 'pose', SE3(0, 0, 1) );
```

which returns a 3×8 matrix with one column per vertex. The image-plane points can be plotted as before by

```
>> cam.plot(cube);
```

Alternatively we can create an *edge* representation of the cube by

```
>> [X,Y,Z] = mkcube(0.2, 'pose', SE3(0, 0, 1), 'edge');
```

and display it

```
>> cam.mesh(X, Y, Z)
```

as shown in Fig. 11.8 along with an oblique view generated by

```
>> Tcam = SE3(-1,0,0.5)*SE3.Ry(0.8);
>> cam.mesh(X, Y, Z, 'pose', Tcam);
```

The elements of the mesh (i, j) have coordinates $(X_{i,j}, Y_{i,j}, Z_{i,j})$.

The edges are in the same 3-dimensional mesh format◄ as generated by MATLAB built-in functions such as `sphere`, `ellipsoid` and `cylinder`.

Successive calls to `plot` will redraw the points or line segments and provides a simple method of animation. The short piece of code

```
1   theta = [0:500]/100*2*pi;
2   [X,Y,Z] = mkcube(0.2, [], 'edges');
3   for th=theta
4       T_cube = SE3(0, 0, 1.5)*SE3.rpy(th*[1.1 1.2 1.3])
5       cam.mesh( X, Y, Z, 'objpose', T_cube ); drawnow
6   end
```

shows a cube tumbling in space. The cube is defined with its center at the origin and its vertices are transformed at each time step.

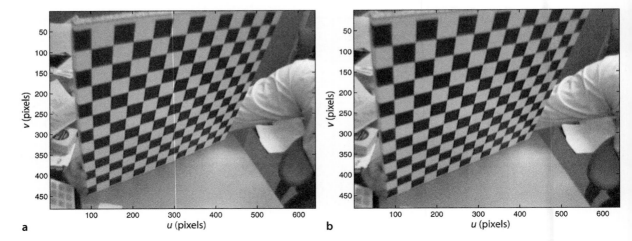

Fig. 11.9. Lens distortion. **a** Distorted image, the curvature of the top row of the squares is quite pronounced, **b** undistorted image. This is calibration image #19 from Bouguet's Camera Calibration Toolbox (Bouguet 2010)

11.1.6   Lens Distortion

No lenses are perfect and the low-cost lenses used in many webcams are far from perfect. Lens imperfections result in a variety of distortions including chromatic aberration (color fringing), spherical aberration or astigmatism (variation in focus across the scene), and geometric distortions where points on the image plane are displaced from where they should be according to Eq. 11.3.

Geometric distortion is generally the most problematic effect that we encounter for robotic applications, and comprises two components: radial and tangential. Radial distortion causes image points to be translated along radial lines from the principal point. The radial error is well approximated by a polynomial

$$\delta r = k_1 r^3 + k_2 r^5 + k_3 r^7 + \cdots \tag{11.12}$$

where r is the distance of the image point from the principal point. Barrel distortion occurs when magnification decreases with distance from the principal point which causes straight lines near the edge of the image to curve outward. Pincushion distortion occurs when magnification increases with distance from the principal point and causes straight lines near the edge of the image to curve inward. Tangential distortion, or decentering distortion, occurs at right angles to the radii but is generally less significant than radial distortion. Examples of a distorted and undistorted image are shown in Fig. 11.9.

The coordinate of the point (u, v) after distortion is given by

$$u^d = u + \delta_u, \; v^d = v + \delta_v \tag{11.13}$$

where the displacement is

$$\begin{pmatrix} \delta_u \\ \delta_v \end{pmatrix} = \underbrace{\begin{pmatrix} u\left(k_1 r^2 + k_2 r^4 + k_3 r^6 + \cdots\right) \\ v\left(k_1 r^2 + k_2 r^4 + k_3 r^6 + \cdots\right) \end{pmatrix}}_{\text{radial}} + \underbrace{\begin{pmatrix} 2p_1 uv + p_2\left(r^2 + 2u^2\right) \\ p_1\left(r^2 + 2v^2\right) + 2p_2 uv \end{pmatrix}}_{\text{tangential}} \tag{11.14}$$

This displacement vector can be plotted for different values of (u, v) as shown in Fig. 11.13b. The vectors indicate the displacement required to *correct* the distortion at different points in the image, in fact $(-\delta_u, -\delta_v)$, and shows dominant radial distortion.

In practice three coefficients are sufficient to describe the radial distortion and the distortion model is parameterized by $(k_1, k_2, k_3, p_1, p_2)$ which are considered as addi-

It has taken humankind a long time to understand light, color and human vision. The Ancient greeks had two schools of thought. The emission theory, supported by Euclid and Ptolemy, held that sight worked by the eye emitting rays of light that interacted with the world somewhat like the sense of touch. The intromission theory, supported by Aristotle and his followers, had physical forms entering the eye from the object.

Euclid of Alexandria (325–265) arguably got the geometry of image formation correct, but his rays emananted from the eye, not the object. Claudius Ptolemy (100–170) wrote Optics and discussed reflection, refraction, and color but today there remains only a poor Arabic translation of his work.

The Arab philosopher Hasan Ibn al-Haytham (aka Alhazen, 965–1040) wrote a seven-volume treatise Kitab al-Manazir (Book of Optics) around 1020. He combined the mathematical rays of Euclid, the medical knowledge

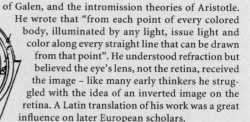

of Galen, and the intromission theories of Aristotle. He wrote that "from each point of every colored body, illuminated by any light, issue light and color along every straight line that can be drawn from that point". He understood refraction but believed the eye's lens, not the retina, received the image – like many early thinkers he struggled with the idea of an inverted image on the retina. A Latin translation of his work was a great influence on later European scholars.

It was not until 1604 that geometric optics and human vision came together when the German astronomer and mathematician Johannes Kepler (1571–1630) published Astronomiae Pars Optica (The Optical Part of Astronomy). He was the first to recognize that images are projected inverted and reversed by the eye's lens onto the retina – the image being corrected later "in the hollows of the brain". (Image from Astronomiae Pars Optica, Johannes Kepler, 1604)

tional intrinsic parameters. Distortion can be modeled by the `CentralCamera` class using the `'distortion'` option, for example

```
>> cam = CentralCamera('focal', 0.015, 'pixel', 10e-6, ...
       'resolution', [1280 1024], 'centre', [512 512], ...
       'distortion', [k1 k2 k3 p1 p2] )
```

11.2   Camera Calibration

The camera projection model Eq. 11.9 has a number of parameters that in practice are unknown. In general the principal point is *not* at the center of the photosite array. The focal length of a lens is only accurate◄ to 4% of what it purports to be, and is only correct if the lens is focused at infinity. It is also common experience that the intrinsic parameters change if a lens is detached and reattached, or adjusted for focus or aperture.◄ The only intrinsic parameters that it may be possible to obtain are the photosite dimensions ρ_w and ρ_h from the sensor manufacturer's data sheet. The extrinsic parameters, the camera's pose, raises the question of where exactly is the center point of the camera.

According to ANSI Standard PH3.13-1958 "Focal Length Marking of Lenses".

Changing focus shifts the lens along the optical axis. In some designs, changing focus rotates the lens so if it is not perfectly symmetric this will move the distortions with respect to the image plane. Changing the aperture alters the parts of the lens that light rays pass through and hence the distortion that they incur.

Camera calibration is the process of determining the camera's intrinsic parameters and the extrinsic parameters with respect to the world coordinate system. Calibration techniques rely on sets of world points whose relative coordinates are known and whose corresponding image-plane coordinates are also known. State-of-the-art techniques such as Bouguet's Calibration Toolbox for MATLAB (Bouguet 2010) simply require a number of images of a planar chessboard target such as shown in Fig. 11.12. From this, as discussed in Sect. 11.2.4, the intrinsic parameters (including distortion parameters) can be estimated as well as the relative pose of the chessboard in each image. Classical calibration techniques require a single view of a 3-dimensional calibration target but are unable to estimate the distortion model. These methods are however easy to understand and they start our discussion in the next section.

11.2.1   Homogeneous Transformation Approach

The homogeneous transform method allows direct estimation of the camera matrix C in Eq. 11.9. The elements of this matrix are functions of the intrinsic and extrinsic parameters. Setting $\tilde{p} = (u, v, 1)$, expanding equation Eq. 11.9 and substituting into Eq. 11.8 we can write

$$C_{11}X + C_{12}Y + C_{13}Z + C_{14} - C_{31}uX - C_{32}uY - C_{33}uZ - C_{34}u = 0$$
$$C_{21}X + C_{22}Y + C_{23}Z + C_{24} - C_{31}vX - C_{32}vY - C_{33}vZ - C_{34}v = 0$$
(11.15)

where (u, v) are the pixel coordinates corresponding to the world point (X, Y, Z) and C_{ij} are elements of the unknown camera matrix.

Calibration requires a 3-dimensional target such as shown in Fig. 11.10. The position of the center of each marker (X_i, Y_i, Z_i), $i \in [1, N]$ with respect to the target frame $\{T\}$ must be known, but $\{T\}$ itself is not known. An image is captured and the *corresponding* image-plane coordinates (u_i, v_i) are determined. Assuming that $C_{34} = 1$ we stack the two equations of Eq. 11.15 for each of the N markers to form the matrix equation

$$\begin{pmatrix} X_1 & Y_1 & Z_1 & 1 & 0 & 0 & 0 & 0 & -u_1X_1 & -u_1Y_1 & -u_1Z_1 \\ 0 & 0 & 0 & 0 & X_1 & Y_1 & Z_1 & 1 & -v_1X_1 & -v_1Y_1 & -v_1Z_1 \\ & & & & & \vdots & & & & & \\ X_N & Y_N & Z_N & 1 & 0 & 0 & 0 & 0 & -u_NX_N & -u_NY_N & -u_NZ_N \\ 0 & 0 & 0 & 0 & X_N & Y_N & Z_N & 1 & -v_NX_N & -v_NY_N & -v_NZ_N \end{pmatrix} \begin{pmatrix} C_{11} \\ C_{12} \\ \vdots \\ C_{33} \end{pmatrix} = \begin{pmatrix} u_1 \\ v_1 \\ \vdots \\ u_N \\ v_N \end{pmatrix}$$
(11.16)

which can be solved for the camera matrix elements $C_{11} \cdots C_{33}$. Equation 11.16 has 11 unknowns and for solution requires that $N \geq 6$. Often more than 6 points will be used leading to an over-determined set of equations which is solved using least squares.

If the points are coplanar then the left-hand matrix of Eq. 11.16 becomes rank deficient. This is why the calibration target must be 3-dimensional, typically an array of dots or squares on two or three planes as shown in Fig. 11.10.

We will illustrate this with an example. The calibration target is a cube, the markers are its vertices and its coordinate frame $\{T\}$ is parallel to the cube faces with its origin at the center of the cube. The coordinates of the markers with respect to $\{T\}$ are

```
>> P = mkcube(0.2);
```

Now the calibration target is at some "unknown pose" $^C\xi_T$ with respect to the camera which we choose to be

```
>> T_unknown = SE3(0.1, 0.2, 1.5) * SE3.rpy(0.1, 0.2, 0.3);
```

Next we create a perspective camera whose parameters we will attempt to estimate

```
>> cam = CentralCamera('focal', 0.015, ...
      'pixel', 10e-6, 'resolution', [1280 1024], 'noise', 0.05);
```

We have also specified that zero-mean Gaussian noise with $\sigma = 0.05$ is added to the (u, v) coordinates to model camera noise and errors in the computer vision algorithms. The image-plane coordinates of the calibration target at its "unknown" pose are

```
>> p = cam.project(P, 'objpose', T_unknown);
```

Now using just the object model P and the observed image features p we estimate the camera matrix

```
>> C = camcald(P, p)
maxm residual 0.066733 pixels.
C =
    853.0895  -236.9378   634.2785   740.0438
    222.6439   986.6900   295.7327   712.0152
     -0.1304     0.0610     0.6495     1.0000
```

The maximum residual in this case is less than 0.1 pixel, that is, the worst error between the projection of a world point using the camera matrix C and the actual image-plane location is very small.

Where is the camera's center? A compound lens has many cardinal points including focal points, nodal points, principal points and planes, entry and exit pupils. The entrance pupil is a point on the optical axis of a compound lens system that is its center of perspective or its *no-parallax point*. We could consider it to be the *virtual pinhole*. Rotating the camera and lens about this point will not change the relative geometry of targets at different distances in the perspective image.

Rotating about the entrance pupil is important in panoramic photography to avoid parallax errors in the final, stitched panorama. A number of web pages are devoted to discussion of techniques for determining the position of this point. Some sites even tabulate the position of the entrance pupil for popular lenses. Much of this online literature refers to this point incorrectly as the *nodal point* even though the techniques given do identify the *entrance pupil*.

Depending on the lens design, the entrance pupil may be behind, within or in front of the lens system.

Fig. 11.10.
A 3D calibration target showing its coordinate frame {T}. The centroids of the circles are taken as the calibration points. Note that the calibration circles are situated on three planes (photo courtesy of Fabien Spindler)

Linear techniques such as this cannot estimate lens distortion parameters. The distortion will introduce errors into the camera matrix elements but for many situations this might be acceptably low. Distortion parameters are often estimated using a nonlinear optimization over all parameters, typically 16 or more, with the linear solution used as the initial parameter estimate.

11.2.2   Decomposing the Camera Calibration Matrix

The elements of the camera matrix are functions of the intrinsic and extrinsic parameters. However given a camera matrix most of the parameter values can be recovered.

The null space of C is the world origin in the camera frame. Using data from the example above this is

```
>> null(C)'
ans =
    0.0809    -0.1709    -0.8138    0.5495
```

which is expressed in homogeneous coordinates that we can convert to Cartesian form

```
>> h2e(ans)'
ans =
    0.1472    -0.3110    -1.4809
```

which is close to the true value

```
>> T_unknown.inv.t'
ans =
    0.1464    -0.3105    -1.4772
```

To recover orientation as well as the intrinsic parameters we can *decompose* the previously estimated camera matrix

```
>> est = invcamcal(C)
est =
name: invcamcal [central-perspective]
  focal length:    1504
  pixel size:      (1, 0.9985)
  principal pt:    (646.8, 504.4)
  pose:            t = (0.147, -0.311, -1.48), RPY/zyx↵
    = (-1.87, -12.4, -16.4) deg
```

which returns a `CentralCamera` object with its parameters set to values that result in the same camera matrix. We note immediately that the focal length is very large compared to the true focal length of our lens which was 0.015 m, and that the pixel sizes are very large. From Eq. 11.9 we see that focal length and pixel dimensions always appear together as factors f/ρ_w and f/ρ_h.▶ The function `invcamcal` has set $\rho_w = 1$ but the ratios of the estimated parameters

```
>> est.f/est.rho(1)
ans =
    1.5044e+03
```

are very close to the ratio for the true parameters of the camera

```
>> cam.f/cam.rho(2)
ans =
    1.500e+03
```

The small error in the estimated parameter values is due to the noisy image-plane coordinate values that we used in the calibration process.

The pose of the estimated camera is with respect to the calibration target $\{T\}$ and is therefore $^T\hat{\xi}_C$. The true pose of the target with respect to the camera is $^C\xi_T$. If our estimation is accurate then $^C\xi_T \oplus {}^T\hat{\xi}_C$ will be 0. We earlier set the variable `T_unknown` equal to $^C\xi_T$ and for our example we find that

```
>> trprint(T_unknown*est.T)
t = (4.13e-05, -4.4e-05, -0.00386),↵
   RPY/zyx = (0.296, 0.253, -0.00557) deg
```

which is the relative pose between the true and estimated camera pose. The camera pose is estimated to better than 5 mm in position and a fraction of a degree in orientation.

We can plot the calibration markers as small red spheres

```
>> hold on; plot_sphere(P, 0.03, 'r')
>> trplot(eye(4,4), 'frame', 'T', 'color', 'b', 'length', 0.3)
```

as well as $\{T\}$ which we have set at the world origin. The estimated pose of the camera can be superimposed

```
>> est.plot_camera()
```

and the result is shown in Fig. 11.11.◥ The problem of determining the pose of a camera with respect to a calibration object is an important problem in photogrammetry known as the camera location determination problem.

11.2.3 Pose Estimation

The pose estimation problem is to determine the pose $^C\xi_T$ of a target's coordinate frame $\{T\}$ with respect to the camera. The geometry of the target is known, that is, we know the position of a number of points (X_i, Y_i, Z_i), $i \in [1, N]$ on the target with respect to $\{T\}$. The camera's intrinsic parameters are also known. An image is captured and the *corresponding* image-plane coordinates (u_i, v_i) are determined using computer vision algorithms.

Estimating the pose using (u_i, v_i), (X_i, Y_i, Z_i) and camera intrinsic parameters is known as the Perspective-n-Point problem or PnP for short. It is a simpler problem than camera calibration and decomposition because there are fewer parameters to estimate. To illustrate pose estimation we will create a calibrated camera with known parameters

```
>> cam = CentralCamera('focal', 0.015, 'pixel', 10e-6, ...
      'resolution', [1280 1024], 'centre', [640 512]);
```

The object whose pose we wish to determine is a cube with side lengths of 0.2 m and the coordinates of the markers with respect to $\{T\}$ are

```
>> P = mkcube(0.2);
```

These quantities have units of pixels since ρ has units of m pixel^{-1}. It is quite common in the literature to consider $\rho = 1$ and the focal length is given in pixels. If the pixels are not square then different focal lengths f_u and f_v must be used for the horizontal and vertical directions respectively.

The option `'frustum'` shows the camera as a rectangular pyramid, such as shown in Fig. 11.13a, rather than a camera icon.

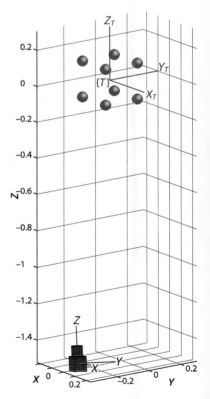

Fig. 11.11. Calibration target points and estimated camera pose with respect to the target frame $\{T\}$ which is assumed to be at the origin

which we can consider a simple geometric model of the object. The object is at some arbitrary but unknown pose $^C\xi_T$ pose with respect to the camera

```
>> T_unknown = SE3(0.1, 0.2, 1.5) * SE3.rpy(0.1, 0.2, 0.3);
>> T_unknown.print
t = (0.1, 0.2, 1.5), RPY/zyx = (5.73, 11.5, 17.2) deg
```

The image-plane coordinates of the object's points at its unknown pose are

```
>> p = cam.project(P, 'objpose', T_unknown);
```

Now using just the object model P, the observed image features p and the calibrated camera cam we estimate the relative pose $^C\xi_T$ of the object

```
>> T_est = cam.estpose(P, p).print
t = (0.1, 0.2, 1.5), RPY/zyx = (5.73, 11.5, 17.2) deg
```

which is the same (to four decimal places) as the unknown pose T_unknown of the object.

In reality the image features coordinates will be imperfectly estimated by the vision system and we would model this by adding zero-mean Gaussian noise to the image feature coordinates as we did in the camera calibration example.

Fig. 11.12. Example frames from Bouguet's Calibration Toolbox showing the calibration target in many different orientations. These are images 2, 5, 9, 18 from the Calibration Toolbox example

11.2.4 Camera Calibration Toolbox

A popular and practical tool for calibrating cameras using a planar chessboard target is the Camera Calibration Toolbox. A number of images, typically twenty, are taken of the target at different distances and orientations as shown in Fig. 11.12.

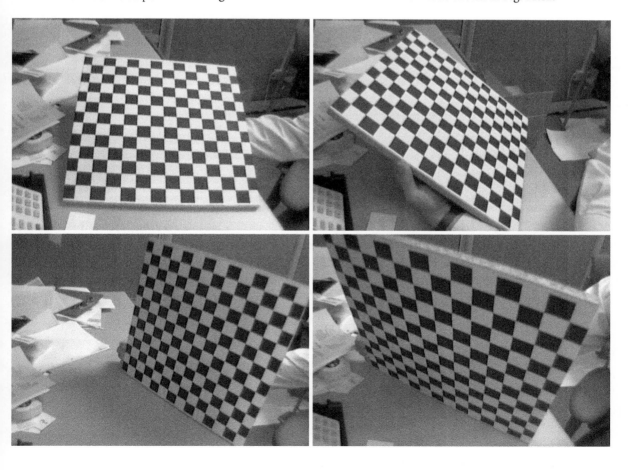

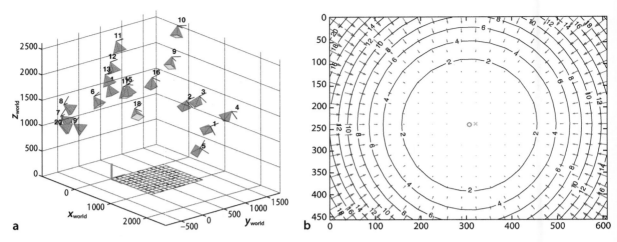

Fig. 11.13. Calibration results from the example in Bouguet's Calibration Toolbox. **a** The estimated camera pose relative to the target for each calibration image, **b** the distortion map with vectors showing how points in the image will move due to distortion

The calibration tool is launched by

```
>> calib_gui
```

and a graphical user interface (GUI) is displayed. ◢ The first step is to load the images using the (Image Names) button. The second step is the (Extract Grid Corners) button which prompts you to pick the corners of the calibration target in each of the images. This is a little tedious but needs to be done carefully. The final step, the (Calibration) button, uses the calibration target information to estimate the camera parameter values

The GUI is optional, and the Toolbox functions can be called from inside your own programs. The function `calib_gui_normal` shows the mapping from GUI button names to Calibration Toolbox function names. Note that most of the functions are actually scripts and program state variables are kept in the workspace.

```
Focal Length:          fc = [ 657.39071    657.74678 ]↵
  ± [ 0.37195    0.39793 ]
Principal point:       cc = [ 303.22367    242.74729 ]↵
  ± [ 0.75632    0.69189 ]
Skew:             alpha_c = [ 0.00000 ] ± [ 0.00000  ]↵
     => angle of pixel axes = 90.00000 ± 0.00000 degrees
Distortion:            kc = [ -0.25541    0.12617    -0.00015    0.00006↵
  0.00000 ] ± [ 0.00290    0.01154    0.00016    0.00015    0.00000 ]
Pixel error:          err = [ 0.13355    0.13727 ]
```

For each parameter the uncertainty (3σ bounds) is estimated and displayed.

The camera pose relative to the target is estimated for each calibration image and can be displayed using the (Show Extrinsic) button. This target-centric view is shown in Fig. 11.13a indicates the estimated relative pose of the camera for each input image.

The distortion vector `kc` contains the parameters in the order $(k_1, k_2, p_1, p_2, k_3)$ – note that k_3 is out of sequence. The distortion map can be displayed by

```
>> visualize_distortions
```

and is shown in Fig. 11.13b. This indicates the displacement from true to distorted image-plane coordinates which in this case is predominately in the radial direction. This is consistent with k_1 and k_2 being orders of magnitude greater than p_1 and p_2 which is typical for most lenses. The (Undistort Image) button can be used to undistort a set of images and a distorted and undistorted image are compared in Fig. 11.9b. The details of this transformation using image warping will be discussed in Sect. 12.7.4.

11.3   Wide Field-of-View Imaging

We have discussed perspective imaging in quite some detail since it is the model of our own eyes and almost all cameras that we encounter. However perspective imaging constrains us to a fundamentally limited field of view. The thin lens equation (11.1) is singular for points with $Z = f$ which limits the field of view to at most one hemisphere – real lenses

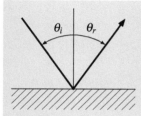

Specular reflection occurs with a mirror-like surface. Incoming rays are reflected such that the angle of incidence equals the angle of reflection or $\theta_r = \theta_i$. Speculum is Latin for mirror and speculum metal (⅔ copper, ⅓ tin) is an alloy that can be highly polished. It was used by Newton and Herschel for the curved mirrors in their reflecting telescopes. See also *Lambertian reflection* on page 309. (The 48 inch speculum mirror from Herschel's 40 foot telescope, completed in 1789, is now in the British Science Museum, photo by Mike Peel (mikepeel.net) licensed under CC-BY-SA)

Fig. 11.14.
Images formation by reflection from a curved surface (*Cloud Gate*, Chicago, Anish Kapoor, 2006). Note that straight lines have become curves

achieve far less. As the focal length decreases radial distortion is increasingly difficult to eliminate and eventually a limit is reached beyond which lenses cannot practically be built. The only way forward is to drop the constraint of perspective imaging. In Sect. 11.3.1 we describe the geometry of image formation with wide-angle lens systems.

An alternative to refractive optics is to use a reflective surface to form the image such as shown in Fig. 11.14. Newtonian telescopes are based on reflection from concave mirrors rather than refraction by lenses. Mirrors are free of color fringing and are easier to scale to larger sizes than a lens. Nature has also evolved reflective optics – the spookfish and some scallops (see page 285) have eyes based on reflectors formed from guanine crystals. In Sect. 11.3.2 we describe the geometry of image formation with a combination of lenses and mirrors.

The cost of cameras is decreasing so an alternative approach is to combine the output of multiple cameras into a single image, and this is briefly described in Sect. 11.5.1.

11.3.1   Fisheye Lens Camera

A fisheye lens image in shown in Fig. 11.17, and we create a model using the notation shown in Fig. 11.15 where the camera is positioned at the origin **O** and its optical axis is the z-axis. The world point **P** is represented in spherical coordinates (R, θ, ϕ), where θ is the angle outward from the optical axis and ϕ is the angle of rotation around the optical axis. We can write

$$R = \sqrt{X^2 + Y^2 + Z^2}, \; \theta = \cos^{-1}\frac{R}{Z}, \; \phi = \tan^{-1}\frac{Y}{X}$$

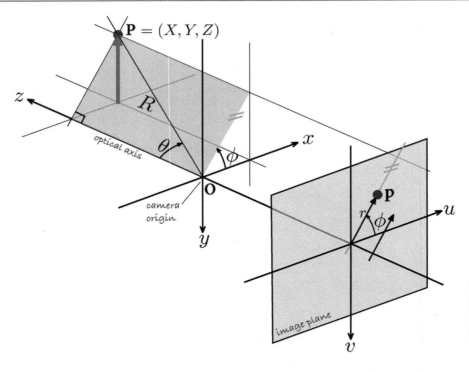

Fig. 11.15.
Image formation for a fisheye lens camera. The world point **P** is represented in spherical coordinates (R, θ, ϕ) with respect to the camera's origin

Mapping		Equation
Equiangular		$r = k\theta$
Stereographic		$r = k\tan(\theta/2)$
Equisolid		$r = k\sin(\theta/2)$
Polynomial		$r = k_1\theta + k_2\theta^2 + \cdots$

Table 11.1.
Fisheye lens projection models

On the image plane of the camera we represent the projection **p** in polar coordinates (r, ϕ) with respect to the principal point, where $r = r(\theta)$. The Cartesian image-plane coordinates are

$$u = r(\theta)\cos\phi, \quad v = r(\theta)\sin\phi$$

and the exact nature of the function $r(\theta)$ depends on the type of fisheye lens. Some common projection models are listed in Table 11.1 and all have a scaling parameter k.

Using the Toolbox we can create a fisheye camera

```
>> cam = FishEyeCamera('name', 'fisheye', ...
        'projection', 'equiangular', ...
        'pixel', 10e-6, ...
        'resolution', [1280 1024])
```

which returns an instance of a `FishEyeCamera` object which is a subclass of the Toolbox's `Camera` object and polymorphic with the `CentralCamera` class discussed earlier. If k is not specified, as in this example, then it is computed such that a hemispheric field of view is projected into the maximal circle on the image plane. As is the case for perspective cameras the parameters such as principal point and pixel dimensions are generally not known and must be estimated using a calibration procedure.

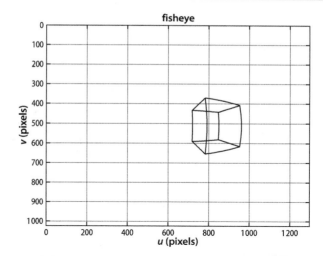

Fig. 11.16.
A cube projected using the
`FishEyeCamera` class. The
straight edges of the cube are
curves on the image plane

Fig. 11.17.
Fisheye lens image. Note that
straight lines in the world are
no longer projected as straight
lines. Note also that the field of
view is mapped to a circular region on the image plane

We create an edge-based model of a cube with side length 0.2 m

```
>> [X,Y,Z] = mkcube(0.2, 'centre', [0.2, 0, 0.3], 'edge');
```

and project it to the fisheye camera's image plane

```
>> cam.mesh(X, Y, Z)
```

and the result is shown in Fig. 11.16. We see that straight lines in the world are no longer straight lines in the image.

Wide angle lenses are available with 180° and even 190° field of view, however they
have some practical drawbacks. Firstly, the spatial resolution is lower since the camera's pixels are spread over a wider field of view. We also note from Fig. 11.17 that the
field of view is a circular region which means that nearly 25% of the rectangular image plane is effectively wasted. Secondly, outdoors images are more likely to include
bright sky so the camera will automatically reduce its exposure which can result in
some nonsky parts of the scene being very underexposed.

11.3.2   Catadioptric Camera

A catadioptric imaging system comprises both reflective and refractive elements▸, a mirror and a lens, as shown in Fig. 11.18a. An example catadioptric image is shown in Fig. 11.18b.

From the Greek for curved mirrors (catoptrics) and lenses (dioptrics).

The geometry shown in Fig. 11.19 is fairly complex. A ray is constructed from the point **P** to the focal point of the mirror at **O** which is the origin of the camera system. The ray has an elevation angle of

$$\theta = \tan^{-1}\frac{Z}{X^2+Y^2} + \frac{\pi}{2}$$

upward from the optical axis and intersects the mirror at the point **M**. The reflected ray makes an angle ψ with respect to the optical axis which is a function of the incoming ray angle, that is $\psi(\theta)$. The relationship between θ and ψ is determined by the tangent to the mirror at the point **M** and is a function of the shape of the mirror. Many different mirror shapes are used for catadioptric imaging including spherical, parabolic, elliptical and hyberbolic. In general the function $\psi(\theta)$ is nonlinear but an interesting class of mirror is the equiangular mirror for which

$$\theta = \alpha\psi$$

The reflected ray enters the camera lens at angle ψ from the optical axis, and from the lens geometry we can write

$$r = \lambda\tan\psi$$

which is the distance from the principal point. The polar coordinate of the image-plane point is $p = (r, \phi)$ and the corresponding Cartesian coordinate is

$$u = r\cos\phi, \quad v = r\sin\phi$$

where ϕ is the azimuth angle

$$\phi = \tan^{-1}\frac{Y}{X}$$

In Fig. 11.19 we have assumed that all rays pass through a single focal point or viewpoint – **O** in this case. This is referred to as central imaging and the resulting image

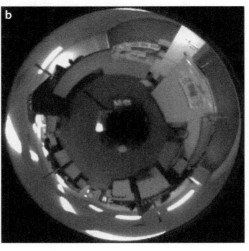

Fig. 11.18.
Catadioptric imaging. **a** A catadioptric imaging system comprising a conventional perspective camera is looking upward at the mirror; **b** Catadioptric image. Note the dark spot in the center which is the support that holds the mirror above the lens. The floor is in the center of the image and the ceiling is at the edge (photos by Michael Milford)

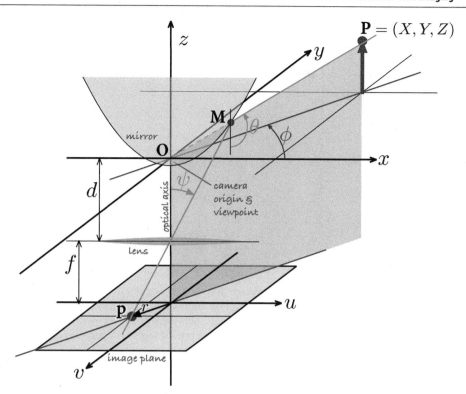

Fig. 11.19.
Catadioptric image formation. A ray from point **P** at elevation angle θ and azimuth ϕ toward **O** is reflected from the mirror surface at **M** and is projected by the lens on to the image plane at **p**

can be correctly transformed to a perspective image. The equiangular mirror does not meet this constraint and is therefore a noncentral imaging system – the focal point varies with the angle of the incoming ray and lies along a short locus within the mirror known as the caustic. Conical, spherical and equiangular mirrors are all noncentral. In practice the variation in the viewpoint is very small compared to the world scale and many such mirrors are well approximated by the central model.

The Toolbox provides a model for catadioptric cameras. For example we can create an equiangular catadioptric camera

```
>> cam = CatadioptricCamera('name', 'panocam', ...
           'projection', 'equiangular', ...
           'maxangle', pi/4, ...
           'pixel', 10e-6, ...
           'resolution', [1280 1024])
```

which returns an instance of a `CatadioptricCamera` object which is a subclass of the Toolbox's `Camera` object and polymorphic with the `CentralCamera` class discussed earlier. The option `maxangle` specifies the maximum elevation angle θ from which the parameters α and f are determined such that the maximum elevation angle corresponds to a circle that maximally fits the image plane. The parameters can be individually specified using the options `'alpha'` and `'focal'`. Other supported projection models include parabolic and spherical and each camera type has different options as described in the online documentation.

We create an edge-based cube model

```
>> [X,Y,Z] = mkcube(1, 'centre', [1, 1, 0.8], 'edge');
```

which we project onto the image plane

```
>> cam.mesh(X, Y, Z)
```

and the result is shown in Fig. 11.20.

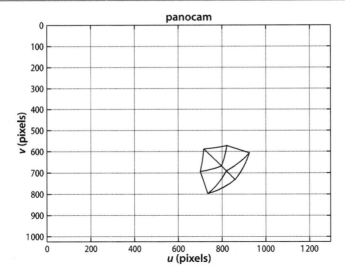

Fig. 11.20.
A cube projected with an equi-angular catadioptric camera

Catadioptric cameras have the advantage that they can view 360° in azimuth but they also have some practical drawbacks. They share many of the problems of fisheye lenses such as reduced spatial resolution, wasted image-plane pixels and exposure control. In some designs there is also a blind spot due to the mirror support which is commonly a central stalk or a number of side supports.

11.3.3   Spherical Camera

The fisheye lens and catadioptric systems just discussed guide the light rays from a large field of view onto an image plane. Ultimately the 2-dimensional image plane is a limiting factor and it is advantageous to consider instead an image *sphere* as shown in Fig. 11.21.

The world point **P** is projected by a ray to the origin of a unit sphere. The projection is the point **p** where the ray intersects the surface of the sphere. If we write $p = (x, y, z)$ then

$$x = \frac{X}{R}, \; y = \frac{Y}{R}, \text{ and } z = \frac{Z}{R} \tag{11.17}$$

where $R = \sqrt{X^2 + Y^2 + Z^2}$ is the radial distance to the world point. The surface of the sphere is defined by $x^2 + y^2 + z^2 = 1$ so one of the three Cartesian coordinates is redundant. A minimal two-parameter representation for a point on the surface of a sphere (ϕ, θ) comprises the angle of colatitude measured down from the North pole

$$\theta = \sin^{-1} r, \;\; \theta \in [0, \pi] \tag{11.18}$$

where $r = \sqrt{x^2 + y^2}$, and the azimuth angle (or longitude)

$$\phi = \tan^{-1} \frac{y}{x}, \;\; \phi \in [-\pi, \pi) \tag{11.19}$$

Conversely, the Cartesian coordinates for the point $\mathbf{p} = (\phi, \theta)$ are given by

$$x = \sin\theta \cos\phi, \; y = \sin\theta \sin\phi, \; z = \cos\theta \tag{11.20}$$

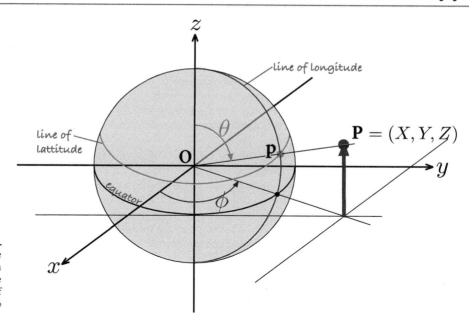

Fig. 11.21.
Spherical image formation. The world point **P** is mapped to **p** on the surface of the unit sphere and represented by the angles of colatitude θ and longitude ϕ

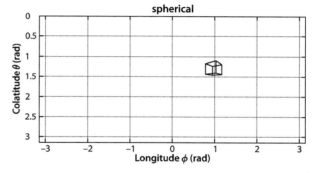

Fig. 11.22.
Cube projected by a spherical camera. The spherical image plane is represented in Cartesian coordinates

Using the Toolbox we can create a spherical camera

```
>> cam = SphericalCamera('name', 'spherical')
```

which returns an instance of a `SphericalCamera` object which is a subclass of the Tool-box's `Camera` object and polymorphic with the `CentralCamera` class discussed earlier.

As previously we can create an edge-based cube model

```
>> [X,Y,Z] = mkcube(1, 'centre', [2, 3, 1], 'edge');
```

and project it onto the sphere

```
>> cam.mesh(X, Y, Z)
```

and this is shown in Fig. 11.22. To aid visualization the spherical image plane has been unwrapped into a rectangle – lines of longitude and latitude are displayed as vertical and horizontal lines respectively. The top and bottom edges correspond to the north and south poles respectively.

It is not yet possible to buy a spherical camera although prototypes have been demonstrated in several laboratories. The spherical camera is more useful as a conceptual construct to simplify the discussion of wide-angle imaging. As we show in the next section we can transform images from perspective, fisheye or catadioptric camera onto the sphere where we can treat them in a uniform manner.

11.4   Unified Imaging

We have introduced a number of different imaging models in this chapter. Now we will discuss how to transform an image captured with one type of camera to the image that would have been captured with a different type of camera. For example, given a fisheye lens projection we will generate the corresponding projection for a spherical camera or a perspective camera. The unified imaging model provides a powerful framework to consider very different types of cameras such as standard perspective, catadioptric and many types of fisheye lens.

The unified imaging model is a two-step process and the notation is shown in Fig. 11.23. The first step is spherical projection of the world point **P** to the surface of the unit sphere **p′** as discussed in the previous section and described by Eq. 11.17 to Eq. 11.18. The view point **O** is the center of the sphere which is a distance m from the image plane along its normal z-axis. The single view point implies a *central* camera.

In the second step the point $\mathbf{p'} = (\theta, \phi)$ is reprojected to the image plane **p** using the view point **F** which is at a distance ℓ along the z-axis above **O**. The polar coordinates of the image-plane point are $\mathbf{p} = (r, \phi)$ where

$$r = \frac{(\ell + m)\sin\theta}{\ell - \cos\theta} \tag{11.21}$$

The unified imaging model has only two parameters m and ℓ and these are a function of the type of camera as listed in Table 11.2. For a perspective camera the two view points **O** and **F** are coincident and the geometry becomes the same as the central perspective model shown in Fig. 11.3.

For catadioptric cameras with mirrors that are conics the focal point **F** lies between the center of the sphere and the north pole, that is, $0 < \ell < 1$. This projection model is somewhat simpler than the catadioptric camera geometry shown in Fig. 11.19. The imaging parameters are written in terms of the conic parameters eccentricity ε and latus rectum $4p$.▶

The length of a chord parallel to the directrix and passing through the focal point.

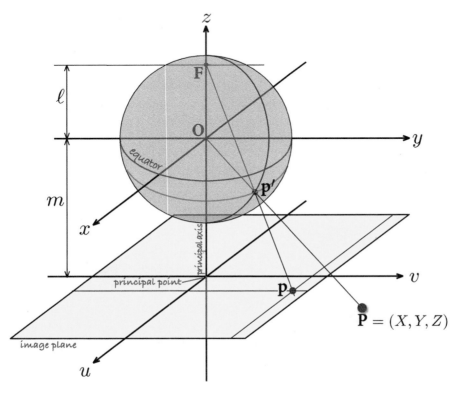

Fig. 11.23.
Unified imaging model of Geyer and Daniilidis (2000)

Table 11.2.
Unified imaging model parameters ℓ and m according to camera type. ε is the eccentricity of the conic and $4p$ is the latus rectum

Imaging	ℓ	m
Perspective	0	f
Stereographic	1	f
Fisheye	>1	f
Catadioptric (elliptical, $0 < \varepsilon < 1$)	$\dfrac{2\varepsilon}{1+\varepsilon^2}$	$\dfrac{2\varepsilon(2p-1)}{1+\varepsilon^2}$
Catadioptric (parabolic, $\varepsilon = 1$)	1	$2p-1$
Catadioptric (hyperbolic, $\varepsilon > 1$)	$\dfrac{2\varepsilon}{1+\varepsilon^2}$	$\dfrac{2\varepsilon(2p-1)}{1+\varepsilon^2}$

The projection with $\mathbf{F}$ at the north pole is known as stereographic projection and is used in many fields to project the surface of a sphere onto a plane. Many fisheye lenses are extremely well approximated by $\mathbf{F}$ above the north pole.

11.4.1   Mapping Wide-Angle Images to the Sphere

We can use the unified imaging model in reverse. Consider an image captured by a wide field of view camera such as the fisheye image shown in Fig. 11.24a. If we know the location of $\mathbf{F}$ then we can project each point from the image onto the sphere to create a spherical image, even though we do not have a spherical camera.

In order to achieve this inverse mapping we need to know some parameters of the camera that captured the image. A common feature of images captured with a fisheye lens or catadioptric camera is that the outer bound of the image is a circle. This circle can be found and its center estimated quite precisely – this is the principal point. A variation of the camera calibration procedure of Sect. 11.2.4 is applied which uses corresponding world and image-plane points from the planar calibration target shown in Fig. 11.24a. This particular camera has a field of view of 190 degrees and its calibration parameters have been estimated to be: principal point (528.1214, 384.0784), $\ell = 2.7899$ and $m = 996.4617$.

We will illustrate this using the image shown in Fig. 11.24a

```
>> fisheye = iread('fisheye_target.png', 'double', 'grey');
```

and we also define the domain of the input image

```
>> [Ui,Vi] = imeshgrid(fisheye);
```

We will use image warping to achieve this mapping. Warping is discussed in detail in Sect. 12.7.4 but we will preview the approach here. The output domain covers the entire sphere with longitude from $-\pi$ to $+\pi$ radians and colatitude from 0 to π radians with 500 steps in each direction

```
>> n = 500;
>> theta_range = linspace(0, pi, n);
>> phi_range = linspace(-pi, pi, n);
>> [Phi,Theta] = meshgrid(phi_range, theta_range);
```

For warping we require a function that returns the coordinates of a point in the input image given the coordinates of a point in the output spherical image. This function is the second step of the unified imaging model Eq. 11.21 which we implement as

```
>> r = (l+m)*sin(Theta) ./ (l-cos(Theta));
```

from which the corresponding Cartesian coordinates in the input image are

```
>> U = r.*cos(Phi) + u0;
>> V = r.*sin(Phi) + v0;
```

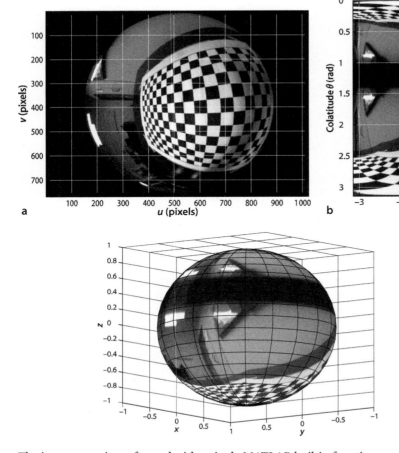

a

b

▲
Fig. 11.24. Fisheye image of a planar calibration target. **a** Fisheye image (image courtesy of Peter Hansen); **b** Image warped to (ϕ, θ) coordinates

Fig. 11.25.
Fisheye image mapped to the unit sphere. We can clearly see the planar grid lying on a table, the ceiling light, a door and a whiteboard

The image warp is performed with a single MATLAB builtin function

```
>> spherical = interp2(Ui, Vi, fisheye, U, V);
```

where the first three arguments define the input domain, and the last two arguments are the coordinates for which grey-scale values will be interpolated from the input image and returned. We display the result

```
>> idisp(spherical)
```

which is shown in Fig. 11.24b. The image appears reflected about the equator and this is because the mapping from a point on the image plane to the sphere is double valued – since **F** is above the north pole the ray intersects the sphere twice. The top and bottom row of this image corresponds to the principal point, while the dark band above the equator corresponds to the circular outer edge of the input image.

The image is extremely distorted but this coordinate system is very convenient to texture map onto a sphere

```
>> sphere
>> h = findobj('Type', 'surface');
>> set(h, 'CData', flipud(spherical), 'FaceColor', 'texture');
>> colormap(gray)
```

and this is shown in Fig. 11.25. Using the MATLAB figure toolbar we can rotate the sphere and look at the image from different view points.

Any wide-angle image that can be expressed in terms of central imaging parameters can be similarly projected onto a sphere. So too can multiple perspective images obtained from a camera array such as shown in Fig. 11.27.

11.4.2   Mapping from the Sphere to a Perspective Image

Given a spherical image we now want to reconstruct a perspective view in a particular direction. We can think of this as looking out, from inside the sphere, at a small surface area which is close to flat and approximates a perspective camera view. This is the second step of the unified imaging model where $\mathbf{F}$ is at the center of the sphere in which case Fig. 11.23 becomes similar to Fig. 11.3. The perspective camera's optical axis is the negative z-axis of the sphere.

For this example we will use the spherical image created in the previous section. We wish to create a perspective image of $1\,000 \times 1\,000$ pixels and with a field-of-view of 45°. The field of view can be written in terms of the image width W and the unified imaging parameter m as

$$\theta_{\mathrm{FOV}} = 2\tan^{-1}\frac{W}{2m}$$

For a 45° field-of-view we require

```
>> W = 1000;
>> m = W / 2 / tan(45/2*pi/180)
m =
    1.2071e+03
```

and for perspective projection we require

```
>> l = 0;
```

We also require the principal point to be in the center of the image

```
>> u0 = W/2; v0 = W/2;
```

The domain of the output image will be

```
>> [Uo,Vo] = meshgrid(0:W-1, 0:W-1);
```

The polar coordinate (r, ϕ) of each point in the output image is

```
>> [phi,r ]= cart2pol(Uo-u0, Vo-v0);
```

and the corresponding spherical coordinates (ϕ, θ) are

```
>> Phi_o = phi;
>> Theta_o = pi - atan(r/m);
```

We now warp from spherical coordinates to the perspective image plane

```
>> perspective = interp2(Phi, Theta, spherical, Phi_o, Theta_o);
```

and the result

```
>> idisp(perspective)
```

is shown in Fig. 11.26a. This is the view from a perspective camera at the center of the sphere looking down through the south pole. We see that the lines on the chessboard calibration target are now straight as we would expect from a perspective image.

Of course we are not limited to just looking along the negative z-axis of the sphere. In Fig. 11.25 we can see some other features of the room such as a door, a whiteboard and some ceiling lights. We can point our virtual perspective camera in their direction by first rotating the spherical image

From a single wide-angle image we can create a perspective view in any direction without having any mechanical pan/tilt mechanism – it's just computation. In fact multiple users could look in different directions simultaneously.

```
>> spherical = sphere_rotate(spherical, SE3.Ry(0.9)*SE3.Rz(-1.5) );
```

so that the negative z-axis now points toward the distant wall. Repeating the warp process we obtain the result shown in Fig. 11.26b in which we can clearly see a door and a whiteboard. ◄

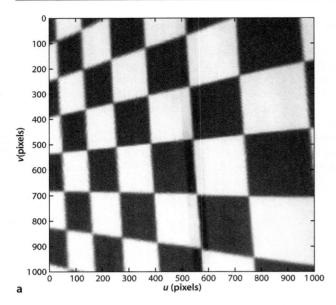

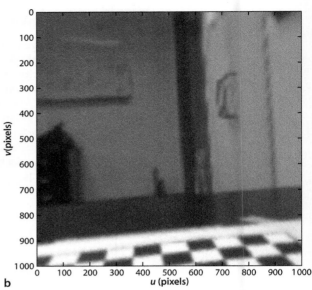

a   b

The original wide-angle image contains a lot of detail though it can be hard to see because of the distortion. After mapping the image to the sphere we can create a virtual perspective camera view (where straight lines in the world are straight) along any line of sight. This is only possible if the original image was taken with a central camera that has a single viewpoint. In theory we cannot create a perspective image from a noncentral wide-angle image but in practice if the caustic is small the parallax errors introduced into the perspective image will be negligible.

Fig. 11.26. Perspective projection of spherical image Fig. 11.25 with a field of view of 45°. **a** Note that the lines on the chessboard are now straight. **b** This view is looking toward the door and whiteboard

11.5   Novel Cameras

11.5.1   Multi-Camera Arrays

The cost of cameras and computation continues to fall making it feasible to do away with the unusual and expensive lenses and mirrors discussed so far, and instead use computation to stitch together the images from a number of cameras onto a cylindrical or spherical image plane. One such camera is shown in Fig. 11.27a and uses five cameras to capture a 360° panoramic view as shown in Fig. 11.27c. The camera in Fig. 11.27b uses six cameras to achieve an almost spherical field of view.

These camera arrays are not central cameras since light rays converge on the focal points of the individual cameras, not the center of the camera assembly. This can be problematic when imaging objects at short range but in typical use the distance between camera focal points, the caustic, is small compared to distances in the scene. The different viewpoints do have a real advantage however when it comes to capturing the light field.

11.5.2   Light-Field Cameras

As we discussed in the early part of this chapter a traditional perspective camera – analog or digital – captures a representation of a scene using the two dimensions of the film or sensor. We can think of the captured image as a 2-dimensional function $\mathcal{L}(X, Y)$ that describes the light emitted by the 3D scene. The function is scalar $\mathcal{L}(\cdot) \in \mathbb{R}$ for the monochrome case and vector-valued $\mathcal{L}(\cdot) \in \mathbb{R}^3$ for a tristimulus color representation.▶

We could add extra dimensions to represent polarization of the light.

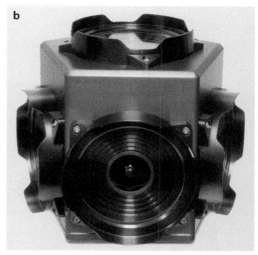

Fig. 11.27. Omnidirectional camera array. **a** Five perspective cameras provide a 360° panorama with a 72° vertical field of view (camera by Occam Vision Group). **b** Panoramic camera array uses six perspective cameras to provide 90% of a spherical field of view. **c** A seamless panoramic image (3 760 × 480 pixels) as output by the camera **a** (photographs **a** and **c** by Edward Pepperell; image **b** courtesy of Point Grey Research)

Fig. 11.28. An 8×12 camera array as described in Wilburn et al. (2005) (photo courtesy of Marc Levoy, Stanford University)

The word plenoptic comes from the Latin word plenus meaning full or complete.

The pin-hole camera of Fig. 11.1 allows only a very small number of light rays to pass through the aperture, yet space is filled with innumerable light rays that provide a richer and more complete description of the world. This detailed geometric distribution of light is called the plenoptic function.◄ Luminance is really a function of position and direction in 3-dimensional space, for example $\mathcal{L}(X, Y, Z, \theta, \phi)$.

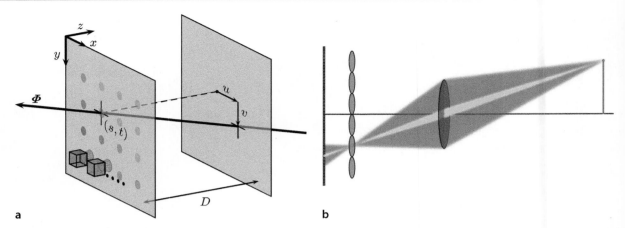

a   b

Unintuitively lines in 3D space have only four parameters, see Sect. C.1.2.2, so the plenoptic function can be written as $\mathcal{L}(s, t, u, v)$ using the 2-plane parameterization shown in Fig. 11.29a.◄ The traditional camera image is just a 2-dimensional *slice* of the full plenoptic function.

Although the concepts behind the light field have been around for decades, it is only in recent years that the technology to capture light fields has become widely available. Early light-field cameras were arrays of regular cameras arranged in a plane, such as shown in Fig. 11.28, or on a sphere surrounding the scene, but these tended to be physically large, complex and expensive to construct. More recently low-cost and compact light-field cameras based on microlens arrays have come on to the market. One selling point for consumer light-field cameras has been the ability to refocus the image *after* taking the picture but the light-field image has many other virtues including synthesizing novel views, 3D reconstruction, low-light imaging and seeing through particulate obscurants.

The microlens array is a regular grid of tiny lenses, typically comprising hundreds of thousands of lenses, which is placed a fraction of a millimeter above the surface of the camera's photosensor array. The main objective lens focuses an image onto the surface of the microlens array as shown in Fig. 11.29b. The microlens directs incoming light to one of a small, perhaps 8 × 8, patch of pixels according to its direction. The resulting image captures information about both the origin of the ray (the lenslet) and its direction (the particular pixel beneath the lenslet). By contrast, in a standard perspective camera all the rays, irrespective of direction, contribute to the value of the pixel. The light-field camera pixels are sometimes referred to as *raxels* and the resolution of these cameras is typically expressed in megarays.

The raw image from the sensor array looks like Fig. 11.30a but can be *decoded* into a 4-dimensional light field, as shown in Fig. 11.30b, and used to render novel views.

Fig. 11.29. a The light ray Φ passes through the image plane at point (u, v) and the center of the camera at (s, t). This is similar to the central projection model shown in Fig. 11.3. Any ray can be described by two points, in this case (u, v) and (s, t). **b** Path of light rays from object through main objective lens and lenslet array to the pixel array (figures courtesy Donald G. Dansereau)

If the scene contains obstructions then the rays become finite-length line segments, this increases the dimensionality of the light field from 4D to 5D. However to record such a light field the camera would have to be simultaneously at every position in the scene without obscuring anything, which is impossible.

11.6   Advanced Topics

11.6.1   Projecting 3D Lines and Quadrics

In Sect. 11.1 we projected 3D points to the image plane, and we projected 3D line segments by simply projecting their endpoints and joining them on the image plane. How would we project an arbitrary line in 3-dimensional space?

The first issue we confront is how to represent such a line and there are many possibilities which are discussed in Sect. C.1.2.2. One useful parameterization is Plücker coordinates – a 6-vector with many similarities to twists.

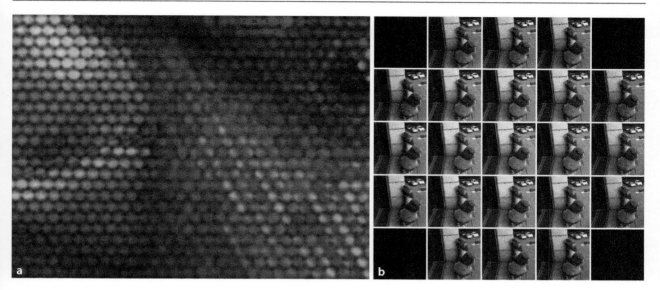

Fig. 11.30. a Closeup of image formed on the sensor array by the lenslet array; **b** array of images rendered from the light field for different camera view points (figures courtesy Donald G. Dansereau)

We can easily create a Plücker line using the Toolbox. A line that passes through $(0, 0, 1)$ and $(1, 1, 1)$ would be

```
>> L = Plucker([0 0 1], [1 1 1])
L =
  { -1   1   0; -1  -1   0 }
```

which returns a `Plucker` object that is represented as a 6-vector with two components: a moment vector and a direction vector. Options can be used to specify a line using a point and a direction or the intersection of two planes. The direction of the Plücker line is the vector

```
>> L.w'
ans =
     -1    -1     0
```

The `Plucker` object also has methods for plotting as well as determining the intersection with planes or other Plücker lines. There are many representations of a Plücker line including the 6-vector used above, a minimal 4-vector, and a skew-symmetric 4×4 matrix computed using the `L` method. The latter is used to project the line by

$$\ell = \vee \left(\boldsymbol{C}[\boldsymbol{L}]_\times \boldsymbol{C}^T \right)$$

where $\boldsymbol{C}$ is the camera matrix, and results in a 2-dimensional line expressed in homogeneous coordinates. Observing this line with the default camera

```
>> cam = CentralCamera('default');
>> l = cam.project(L)'
l =
       1    -1     0
```

results in a diagonal line across the image plane. We could plot this using `plot_homline` or on the camera's virtual image plane by

```
>> cam.plot(l)
```

Quadrics, short for quadratic surfaces, are a rich family of 3-dimensional surfaces. There are 17 standard types including spheres, ellipsoids, hyperboloids, paraboloids, cylinders and cones all described by

$$\tilde{\boldsymbol{x}}^T \boldsymbol{Q} \tilde{\boldsymbol{x}} = 0$$

where $Q \in \mathbb{R}^{4 \times 4}$ is symmetric. The *outline* of the quadric is projected to the image plane by

$$c^* = CQ^*C^T$$

where $(\cdot)^*$ represents the adjugate operation, see Appendix B, and c is a matrix representing a conic section on the image plane

$$\tilde{p}^T c \tilde{p} = 0$$

which can be written as

$$Au^2 + Buv + Cv^2 + Du + Ev + F = 0$$

where

$$c = \begin{pmatrix} A & B/2 & D/2 \\ B/2 & C & E/2 \\ D/2 & E/2 & F \end{pmatrix}$$

The determinant of the top-left submatrix indicates the type of conic: negative for a hyperbola, 0 for a parabola and positive for an ellipse.

To demonstrate this we will define a camera looking toward the origin

```
>> cam = CentralCamera('default', 'pose', SE3(0.2,0.1, -5)*SE3.Rx(0.2));
```

and define a unit sphere at the origin

```
>> Q = diag([1 1 1 -1]);
```

then compute its projection to the image plane

```
>> Qs = inv(Q)*det(Q); % adjugate
>> cs = cam.C * Qs * cam.C';
>> c = inv(cs)*det(cs);  % adjugate
```

which is a 3×3 matrix describing a conic. The determinant

```
>> det(c(1:2,1:2))
ans =
   2.2862e+14
```

is positive indicating an ellipse, and a simple way to plot this is using the Symbolic Math Toolbox™

```
>> syms x y real
>> ezplot([x y 1]*c*[ x y 1]', [0 1024 0 1024])
>> set(gca, 'Ydir', 'reverse')
```

11.6.2   Nonperspective Cameras

The camera matrix Eq. 11.9 represents a special subset of all possible camera matrices – finite projective or Euclidean cameras – where the left-hand 3×3 matrix is nonsingular. The camera projection matrix C from Eq. 11.9 can be written generally as

$$C = \begin{pmatrix} c_{11} & c_{12} & c_{13} & c_{14} \\ c_{21} & c_{22} & c_{23} & c_{24} \\ c_{31} & c_{32} & c_{33} & 1 \end{pmatrix}$$

which has arbitrary scale so one element, typically $C_{3,4}$ is set to one – this matrix has 11 unique elements or 11 DOF. We could think of every possible matrix as corresponding to some type of camera, but most of them would produce wildly distorted images.

Orthographic or parallel projection is a simple perspective-free projection of 3D points onto a plane, like a "plan view". For small objects close to the camera this projection can be achieved using a telecentric lens. The apparent size of an object is independent of its distance.

For the case of an aerial robot flying high over relatively flat terrain the variation of depth, the depth relief, Δ_Z is small compared to the average depth of the scene $\bar{Z}$, that is $\Delta_Z \ll \bar{Z}$. We can use a scaled-orthographic projection which is an orthographic projection followed by uniform scaling $m = f/\bar{Z}$.

These two nonperspective cameras are special cases of the more general affine camera model which is described by a matrix of the form

$$C = \begin{pmatrix} c_{11} & c_{12} & c_{13} & c_{14} \\ c_{21} & c_{22} & c_{23} & c_{24} \\ 0 & 0 & 0 & 1 \end{pmatrix}$$

that can be factorized as

$$C = \underbrace{\begin{pmatrix} m_x & s & 0 \\ 0 & m_y & 0 \\ 0 & 0 & 1 \end{pmatrix}}_{\text{intrinsic}} \underbrace{\begin{pmatrix} 1 & 0 & 0 & 0 \\ 0 & 1 & 0 & 0 \\ 0 & 0 & 0 & 1 \end{pmatrix}}_{\text{projection}} \underbrace{\begin{pmatrix} R & t \\ 0 & 1 \end{pmatrix}}_{\text{extrinsic}}$$

It can be shown that the principal point is undefined for such a camera model, simplifying the intrinsic matrix, but we have introduced a skew parameter to handle the case of nonorthogonal sensor axes. The projection matrix is different compared to the perspective case in Eq. 11.9 – the column of zeros has moved from column 4 to column 3. This zero column effectively deletes the third row and column of the extrinsic matrix resulting in

$$C = \underbrace{\begin{pmatrix} m_x & s & 0 \\ 0 & m_y & 0 \\ 0 & 0 & 1 \end{pmatrix}}_{\text{3 DOF}} \underbrace{\begin{pmatrix} [R]_{2\times3} & \begin{matrix} t_x \\ t_y \end{matrix} \\ 000 & 1 \end{pmatrix}}_{\text{5 DOF}}$$

The 2×3 submatrix of the rotation matrix has 6 elements but 3 constraints – the two rows have unit norms and are orthogonal – and therefore has 3 DOF.

and has at most 8 DOF◄. The independence from depth is very clear since t_z does not appear. The case where skew $s = 0$ and $m_x = m_y = 1$ is orthographic projection and has only 5 DOF, while the scaled-orthographic case when $s = 0$ and $m_x = m_y$ has 6 DOF. The case where $m_x \neq m_y$ is known as weak perspective projection, although this term is sometimes also used to describe scaled-orthographic projection.

11.7 Wrapping Up

We have discussed the first steps in the computer vision process – the formation of an image of the world and its conversion to an array of pixels which comprise a digital image. The images with which we are familiar are perspective projections of the world in which 3 dimensions are compressed into 2 dimensions. This leads to ambiguity about object size – a large object in the distance looks the same as a small object that is close. Straight lines and conics are unchanged by this projection but shape distortion occurs – parallel lines can appear to converge and circles can appear as ellipses. We have modeled the perspective projection process and described it in terms of eleven parameters – intrinsic and extrinsic. Geometric lens distortion adds additional lens parameters. Camera calibration is the process of estimating these parameters and two approaches have been introduced. We also discussed pose estimation where the pose of an object of known geometry can be estimated from a perspective projection obtained using a calibrated camera.

Perspective images are limited in their field of view and we discussed several wide-angle imaging systems based on the fisheye lens, catadioptrics and multiple cameras. We also discussed the ideal wide-angle camera, the spherical camera, which is currently still a theoretical construct. However it can be used as an intermediate representation in the unified imaging model which provides one model for almost all camera geometries. We used the unified imaging model to convert a fisheye camera image to a spherical image and then to a perspective image along a specified view axis. Finally we covered some more recent camera developments such as panoramic camera arrays and light-field cameras.

In this chapter we treated imaging as a problem of pure geometry with a small number of world points or line segments. In the next chapter we will discuss the acquisition and processing of images sourced from files, cameras and the web.

Further Reading and Resources

Computer vision textbooks such as Szeliski (2011), Hartley and Zisserman (2003), Forsyth and Ponce (2011) and Gonzalez and Woods (2008) provide deeper coverage of the topics introduced in this chapter. Many topics in geometric computer vision have also been studied by the photogrammetric community, but different language is used. For example camera calibration is known as camera resectioning, and pose estimation is known as space resectioning. The revised classic textbook by DeWitt and Wolf (2000) is a thorough and readable introduction to photogrammetry.

Camera calibration. The homogeneous transformation calibration (Sutherland 1974) approach of Sect. 11.2.1 is also known as the direct linear transform (DLT) in the photogrammetric literature. The Toolbox implementation `camcald` requires that the centroids of the calibration markers have already been determined which is a nontrivial problem (Corke 1996b, § 4.2). It also cannot estimate lens distortion. Wolf (1974) describes extensions to the linear camera calibration with models that include up to 18 parameters and suitable nonlinear optimization estimation techniques. A more concise description of nonlinear calibration is provided by Forsyth and Ponce (2011). Hartley and Zisserman (2003) describe how the linear calibration model can be obtained using features such as lines within the scene.

There are a number of good camera calibration toolboxes available on the web. The MATLAB Toolbox, discussed in Sect. 11.2.4, is by Jean-Yves Bouguet and available from http://www.vision.caltech.edu/bouguetj/calib_doc. It has extensive online documentation and includes example calibration images which were used in Sect. 11.2.4. Several tools build on this and automatically find the chessboard target which is otherwise tedious to locate in every image, for example the AMCC and RADOCC Toolboxes and the MATLAB Camera Calibrator App included with the Computer Vision System Toolbox™. The MATLAB Toolbox by Janne Heikkilä is available at http://www.ee.oulu.fi/~jth/calibr/ and works for planar or 3D targets with circular dot features and estimates lens distortion.

Photogrammetry is the science of understanding the geometry of the world from images. The techniques were developed by the French engineer Aimé Laussedat (1819–1907) working for the Army Corps of Engineers in the 1850s. He produced the first measuring camera and developed a mathematical analysis of photographs as perspective projections. He pioneered the use of aerial photography as a surveying tool to map Paris – using rooftops as well as unmanned balloons and kites.

Photogrammetry is normally concerned with making maps from images acquired at great distance but the subfield of close-range or terrestrial photogrammetry is concerned with camera to object distances less than 100 m which is directly relevant to robotics. (Image from La Métrophotographie, Aimé Laussedat, 1899)

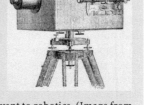

Pose estimation is a classic and hard problem in computer vision and for which there exists a very large literature. The approaches can be broadly divided into analytic and iterative solutions. Assuming that lens distortion has been corrected the analytic solutions for three and four noncollinear points are given by Fischler and Bolles (1981), DeMenthon and Davis (1992) and Horaud et al. (1989). Typically multiple solutions exist but for four coplanar points there is a unique solution. Six or more points always yield unique solutions, as well as the intrinsic camera calibration parameters. Iterative solutions were described by Rosenfeld (1959) and Lowe (1991). A more recent discussion based around the concept of bundle adjustment is provided by Triggs et al. (2000). The pose estimation in the Toolbox is a wrapper around an efficient noniterative perspective n-point pose estimator described by Lepetit et al. (2009) and available at http://cvlab.epfl.ch/EPnP. Pose estimation requires a geometric model of the object and such computer vision approaches are known as *model-based vision*. An interesting historical perspective on model-based vision is the 1987 video by the late Joe Mundy which is available at http://www.archive.org/details/JosephMu1987.

Wide field-of-view cameras. There is recent and growing interest in this type of camera and today good quality lightweight fisheye lenses and catadioptric camera systems are available. Nayar (1997) provides an excellent motivation for, and introduction to, wide-angle imaging. A very useful online resource is the catadioptric sensor design page at http://www.math.drexel.edu/~ahicks/design and a page of links to research groups, companies and workshops at http://www.cis.upenn.edu/~kostas/omni.html. Equiangular mirror systems were described by Chahl and Srinivasan (1997) and Ollis et al. (1999). Nature's solution, the reflector-based scallop eye, is described in Colicchia et al. (2009). A number of workshops on Omnidirectional Vision have been held, starting in 2000, and their proceedings are a useful introduction to the field. The book of Daniilidis and Klette (2006) is a collection of papers on nonperspective imaging and Benosman and Kang (2001) is another, earlier, published collection of papers. Some information is available through CVonline at http://homepages.inf.ed.ac.uk/rbf/CVonline in the section *Image Physics*.

A MATLAB Toolbox for calibrating wide-angle cameras by Davide Scaramuzza is available at https://sites.google.com/site/scarabotix/ocamcalib-toolbox. It is inspired by, and similar in usage, to Bouguet's Toolbox for perspective cameras. Another MATLAB Toolbox, by Juho Kannala, handles wide angle central cameras and is available at http://www.ee.oulu.fi/~jkannala/calibration.

The unified imaging model was introduced by Geyer and Daniilidis (2000) in the context of catadioptric cameras. Later it was shown (Ying and Hu 2004) that many fisheye cameras can also be described by this model. The fisheye calibration of Sect. 11.4.1 was described by Hansen et al. (2010) who estimates ℓ and m rather than a polynomial function $r(\theta)$ as does Scaramuzza's Toolbox.

There is a huge and growing literature on light-field imaging but as yet no textbook. A great introduction to light fields and its application to robotics is the thesis by Dansereau (2014). The same author has a MATLAB Toolbox available at http://mathworks.com/matlabcentral/fileexchange/49683. An interesting description of an early camera array is given by Wilburn et al. (2005) and the associated video demonstrates many capabilities. Light-field imaging is a subset of the larger, and growing, field of computational photography.

Toolbox Notes

The Toolbox camera classes `CentralCamera`, `FishEyeCamera` and `SphericalCamera` are all derived from the abstract superclass `Camera`. Common methods of all classes are shown in Table 11.3.

Method	Description
`p = cam.project(P)`	Project the world point P to the image plane p
`cam.plot(P)`	Plot the world points defined by the columns of P
`cam.mesh(X,Y,Z)`	Plot the mesh defined by X, Y and Z
`cam.showpose(T)`	Show the camera at specified pose
`cam.clf`	Clear the current camera view
`cam.hold`	Hold the current camera view, future calls to `plot` add to the camera view
`cam.name`	Property: name of camera
`cam.T`	Property: default camera transform (read and write)

Table 11.3.
Common methods for all Toolbox camera classes

Option	Description
`'name', name`	The name of the camera which is displayed in the window's title bar
`'resolution', npix`	The dimensions of the image. `npix` is a scalar for a square image or a 2-vector
`'centre', pp`	The coordinate of the principal point
`'pixel', rho`	The dimensions of the pixel. `rho` is a scalar for a square pixel or a 2-vector
`'noise', sigma`	The standard-deviation of Gaussian noise added to the image-plane coordinates
`'pose', T`	The default pose of the camera. Default is as shown in Fig. 11.2
`'image', im`	Set camera image-plane dimensions according to the image dimensions and display the image

Table 11.4.
Common options for camera class constructors

The virtual camera view has similar behavior to a MATLAB figure. By default `plot` and `mesh` will redraw the camera's view. If no camera view exists one will be created. The methods `clf` and `hold` are analogous to the MATLAB commands `clf` and `hold`.

The constructor of all camera classes accepts a number of option arguments which are listed in Table 11.4. Specific camera subclasses have unique options which are described in the online documentation. With no arguments the default `CentralCamera` parameters are for a 1 024 × 1 024 image, 8 mm focal length lens and 10 μm square pixels. If the principal point is not set explicitly it is assumed to be in the middle of the image plane.

Exercises

1. Create a central camera and a cube target and visualize it for different camera and cube poses. Create and visualize different 3D mesh shapes such as created by the MATLAB functions `cylinder` and `sphere`.
2. Write a script to fly the camera in an orbit around the cube, always facing toward the center of the cube.
3. Write a script to fly the camera through the cube.
4. Create a central camera with lens distortion and which is viewing a 10 × 10 planar grid of points. Vary the distortion parameters and see the effect this has on the shape of the projected grid. Create pincushion and barrel distortion.
5. Repeat the homogeneous camera calibration exercise of Sect. 11.2.1 and the decomposition of Sect. 11.2.2. Investigate the effect of the number of calibration points, noise and camera distortion on the calibration residual and estimated target pose.

6. Determine the solid angle for a rectangular pyramidal field of view that subtends angles θ_h and θ_v.

7. Do example 1 from Bouguet's Camera Calibration Toolbox.

8. Calibrate the camera on your computer.

9. Derive Eq. 11.14.

10. For the camera calibration matrix decomposition example (Sect. 11.2.2) determine the roll-pitch-yaw orientation error between the true and estimated camera pose.

11. Pose estimation (Sect. 11.2.3)
    a) Repeat the pose estimation exercise for different object poses (closer, further away).
    b) Repeat for different levels of camera noise.
    c) What happens as the number of points is reduced?
    d) Does increasing the number of points counter the effects of increased noise?
    e) Change the intrinsic parameters of the camera `cam` before invoking the `est-pose` method. What is the effect of changing the focal length and the principal point by say 5%.

12. Repeat exercises 2 and 3 for the fisheye camera and the spherical camera.

13. With reference to Fig. 11.19 derive the function $\psi(\theta)$ for a parabolic mirror.

14. With reference to Fig. 11.19 derive the equation of the equiangular mirror $z(x)$ in the xz-plane.

15. Quadrics
    a) Write a routine to plot a quadric given a 4×4 matrix. Hint use `meshgrid` and `isosurface`.
    b) Write code to compute the quadric matrix for a sphere at arbitrary location and of arbitrary radius.
    c) Write code to compute the quadric matrix for an arbitrary circular cylinder.
    d) Write numeric MATLAB code to plot the planar conic section described by a 3×3 matrix.

16. Project an ellipsoidal or spherical quadric to the image plane. The result will be the implicit equation for a conic – write code to plot the implicit equation.

12

Images and Image Processing

Image processing is a computational process that transforms one or more input images into an output image. Image processing is frequently used to enhance an image for *human* viewing or interpretation, for example to improve contrast. Alternatively, and of more interest to robotics, it is the foundation for the process of feature extraction which will be discussed in much more detail in the next chapter.

An image is a rectangular array of picture elements (pixels) so we will use a MATLAB® matrix to represent an image in the workspace. This allows us to use MATLAB's powerful and efficient armory of matrix operators and functions.

We start in Sect. 12.1 by describing how to load images into MATLAB from sources such as files (images and movies), cameras and the internet. Next, in Sect. 12.2, we introduce image histograms which provide useful information about the distribution of pixel values. We then discuss various classes of image processing algorithms. These algorithms operate pixel-wise on a single image, a pair of images, or on local groups of pixels within an image and we refer to these as monadic, diadic, and spatial operations respectively. Monadic and diadic operations are covered in Sect. 12.3 and 12.4. Spatial operators are described in Sect. 12.5 and include operations such as smoothing, edge detection, and template matching. A closely related technique is shape-specific filtering or mathematical morphology and this is described in Sect. 12.6. Finally in Sect. 12.7 we discuss shape changing operations such as cropping, shrinking, expanding, as well as more complex operations such as rotation and generalized image warping.

Robots will always gather imperfect images of the world due to noise, shadows, reflections and uneven illumination. In this chapter we discuss some fundamental tools and "tricks of the trade" that can be applied to real-world images.

12.1 Obtaining an Image

Today digital images are ubiquitous since cameras are built into our digital devices and images cost almost nothing to create and share. We each have ever growing personal collections and access to massive online collections of digital images such as Google Images, Picasa or Flickr. We also have access to live image streams from other people's cameras – there are tens of thousands of webcams around the world capturing images and broadcasting them on the internet, as well images of Earth from space, the Moon and Mars.

12.1.1 Images from Files

We start with images stored in files since it is very likely that you already have lots of images stored on your computer. In this chapter we will work with some images provided with the Toolbox, but you can easily substitute your own images. We import an image into the MATLAB workspace using the Toolbox function `iread`

```
>> street = iread('street.png');
```

which returns a matrix

```
>> about(street)
street [uint8] : 851x1280 (1.1 MB)
```

that belongs to the class `uint8` – the elements of the matrix are unsigned 8-bit integers in the interval [0,255]. The elements are referred to as pixel values or grey values and are the gamma-encoded▸ luminance of that point in the original scene. For this 8-bit image the pixel values vary from 0 (darkest) to 255 (brightest). The image is shown in Fig. 12.1a.

The matrix has 851 rows and 1 280 columns. We normally describe the dimensions of an image in terms of its width × height, so this would be a 1 280 × 851 pixel image.

> Gamma encoding and decoding is discussed in Sect. 10.3.6. Use the 'gamma' option for `iread` to perform gamma decoding and obtain pixel values proportional to scene luminance.

> In Chap. 11 we wrote the coordinates of a pixel as (u, v) which are the horizontal and vertical coordinates respectively. In MATLAB this is the matrix element (v, u) – note the reversal of coordinates. Note also that the top-left pixel is (1, 1) in MATLAB not (0, 0).

For example the pixel at image coordinate (300, 200) is

```
>> street(200,300)
ans =
   42
```

which is quite dark – the pixel corresponds to a point in the closest doorway.

There are some subtleties when working with `uint8` values in MATLAB which we can demonstrate by defining two `uint8` values

```
>> a = uint8(100)
a =
  100
>> b = uint8(200)
b =
  200
```

Arithmetic on `uint8` values obeys the rules of `uint8` arithmetic

```
>> a+b
ans =
  255
>> a-b
ans =
    0
```

and values are clipped to the interval 0 to 255. For division

```
>> a/b
ans =
    1
```

the result has been rounded up to an integer value. For some image processing operations that we will consider later it is useful to consider the pixel values as floating-point

A very large number of **image file formats** have been developed and are comprehensively catalogued at http://en.wikipedia.org/wiki/Image_file_formats. The most popular is JPEG which is used for digital cameras and webcams. TIFF is common in many computer systems and often used for scanners. PNG and GIF are widely used on the web. The internal format of these files are complex but a large amount of good quality open-source software exists in a variety of languages to read and write such files. MATLAB is able to read many of these image file formats.

A much simpler set of formats, widely used on Linux systems, are PBM, PGM and PPM (generically PNM) which represent images without compression, and optionally as readable ASCII text. A host of open-source tools such as ImageMagick provide format conversions and image manipulation under Linux, MacOS X and Windows. (word map by tagxedo.com)

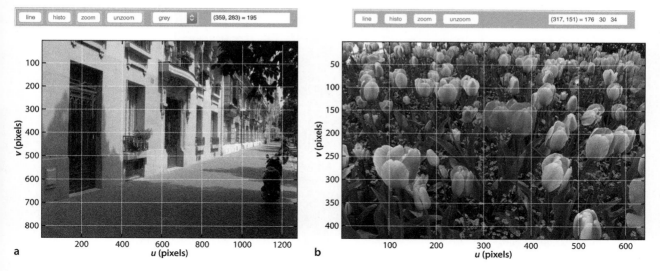

Fig. 12.1. The `idisp` image browsing window. The top right shows the coordinate and value of the last pixel clicked on the image. The *buttons* at the top left allow the pixel values along a line to be plotted, a histogram to be displayed, or the image to be zoomed. **a** Greyscale image; **b** color image

Lossless means that the compressed image, when uncompressed, will be exactly the same as the original image.

The example images are kept within the `images` folder of the Machine Vision Toolbox distribution which is automatically searched by the `iread` function.

The colormap controls the mapping of pixel values to the displayed intensity or color.

numbers for which more familiar arithmetic rules apply. In this case each pixel is an 8-byte MATLAB double precision number in the range [0, 1] and we can specify this as an option when we load the image

```
>> streetd = iread('street.png', 'double');
>> about streetd
streetd [double] : 851x1280 (8.7 MB)
```

or by applying the function `idouble` to the integer image.

The image was read from a file called `street.png` which is in portable network graphics (PNG) format – a lossless compression format◄ widely used on the internet. The function `iread` searches for the image in the current folder, and then in each folder along your MATLAB path.➤ This particular image has no color, it is a greyscale or monochromatic image.

A tool that we will use a lot in this part of the book is `idisp`

```
>> idisp(street)
```

which displays the matrix as an image and allows interactive inspection of pixel values as shown in Fig. 12.1. Clicking on a pixel will display the pixel coordinate and its grey value – integer or floating point – in the top right of the window. The image can be zoomed (and unzoomed), we can display a histogram or the intensity profile along a line between any two selected points, or change the color map.◄ It has many options and these are described in the online documentation.

We can just as easily load a color image

```
>> flowers = iread('flowers8.png');
>> about(flowers)
flowers [uint8] : 426x640x3 (817920 bytes)
```

which is a $426 \times 640 \times 3$ matrix of `uint8` values as shown in Fig. 12.1b. We can think of this as a 426×640 matrix of RGB tristimulus values, each of which is a 3-vector. For example the pixel at (318, 276)

```
>> pix = flowers(276,318,:)
ans(:,:,1) =
    57
ans(:,:,2) =
    91
ans(:,:,3) =
   198
```

has a tristimulus value (57, 91, 198) but has been displayed by MATLAB in an unusual and noncompact manner. This is because the pixel value is

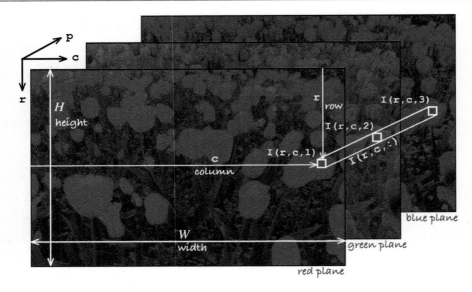

Fig. 12.2.
Color image shown as a 3-dimensional structure with dimensions: row, column, and color plane

```
>> about(pix)
pix [uint8] : 1x1x3 (3 bytes)
```

a $1 \times 1 \times 3$ matrix. The first two dimensions are called singleton dimensions and we can *squeeze* them out

```
>> squeeze(pix)'
ans =
    57    91   198
```

which results in a more familiar 3-vector. This pixel corresponds to one of the small blue flowers and has a large blue component. We can display the image and examine it interactively using idisp and clicking on a pixel will display its tristimulus value.

The tristimulus values are of type uint8 in the range [0, 255] but the image can be converted to double precision values in the range [0, 1] using the 'double' option to iread or by applying the function idouble to the integer color image, just as for a greyscale image.

The image is a matrix with three dimensions and the third dimension as shown in Fig. 12.2 is known as the color plane index. For example

```
>> idisp( flowers(:,:,1) )
```

would display the red color plane as a greyscale image that shows the red stimulus at each pixel. The index 2 or 3 would select the green or blue plane respectively.

The option 'grey' ensures that a greyscale image is returned irrespective of whether or not the file contains a color image.► The option 'gamma' performs gamma decoding and returns a linear image where the greylevels, or tristimulus values, are proportional to the luminance of the original scene.

The iread function can also accept a wildcard filename allowing it to load a sequence of files. For example

```
>> seq = iread('seq/*.png');
>> about(seq)
seq [uint8] : 512x512x9 (2.4 MB)
```

loads nine images in PNG format from the folder seq. The result is a $H \times W \times N$ matrix and the last index represents the image number within the sequence. That is seq(:,:,k) is the kth image in the sequence and is a 512×512 greyscale image. In terms of Fig. 12.2 the images in the sequence extend in the p direction. If the images were color then the result would be a $H \times W \times 3 \times N$ matrix where the last index represents the image number within the sequence, and the third index represents the color plane.

Using ITU Rec. 709 by default. See also the Toolbox monadic operator imono.

JPEG employs *lossy* compression to reduce the size of the file. Unlike normal file compression (eg. zip, rar, etc.) and decompression, the decompressed image isn't the same as the original image and this allows much greater levels of compression. JPEG compression exploits limitations of the human eye and discards information that won't be noticed such as very small color changes (which are perceived less accurately than small changes in brightness) and fine texture. It is very important to remember that JPEG is intended for compressing images that will be *viewed by humans*. The loss of color detail and fine texture may be problematic for computer algorithms that analyze images.

JPEG was designed to work well for natural scenes but it does not do so well on lettering and line drawings with high spatial-frequency content. The degree of loss can be varied by adjusting the so-called quality factor which allows a tradeoff between image quality and file size. JPEG can be used for greyscale or color images.

What is commonly referred to as a JPEG file, often with an extension of `.jpg` or `.jpeg`, is more correctly a JPEG JFIF file. JFIF is the format of the file that holds a JPEG-compressed image as well as metadata. EXIF file format (Exchangeable Image File Format) is a standard for camera related metadata such as camera settings, time, location and so on. This metadata can be retrieved as a second output argument to `iread` as a cell array, or by using a command-line utility such as `exiftool` (http://www.sno.phy.queensu.ca/~phil/exiftool). See the Independent JPEG group web site http://www.ijg.org for more details.

If `iread` is called with no arguments, a file browsing window pops up allowing navigation through the file system to find the image. The function also accepts a URL allowing it to load an image, but not a sequence, from the web. The function can read most common image file formats including JPEG, TIFF, GIF, PNG, PGM, PPM, PNM.

Many image file formats also contain rich metadata – data about the data in the file. The JPEG files generated by most digital cameras are particularly comprehensive and the metadata can be retrieved by providing a second output argument to `iread`

```
>> [im,md]=iread('church.jpg');
>> md
md = ...
                Width: 1280
               Height: 851
                 Make: 'Panasonic'
                Model: 'DMC-FZ30'
          Orientation: 1
        DigitalCamera: [1x1 struct]
```

and the DigitalCamera substructure has additional details about the camera settings for the particular image

```
>> md.DigitalCamera
ans = ...
           ExposureTime: 0.0025
                FNumber: 8
        ISOSpeedRatings: 80
                  Flash: 'Flash did not fire...'
            FocalLength: 7.4000
```

More details and options for `iread` are described in the online documentation.

12.1.2 Images from an Attached Camera

Most laptop computers today have a builtin camera for video conferencing. For computers without a builtin camera an external camera can be easily attached via a USB or FireWire connection. The means of accessing a camera is operating system specific. ◄

The Toolbox provides a simple interface to a camera for MacOS, Linux and Windows but more general support requires the Image Acquisition Toolbox.

A list of all attached cameras and their resolution can be obtained by

```
>> VideoCamera('?')
```

We open a particular camera

```
>> cam = VideoCamera('name')
```

which returns an instance of a `VideoCamera` object that is a subclass of the `ImageSource` class. If name is not provided the first camera found is used. The constructor accepts a number of additional arguments such as `'grey'` which ensures that the returned image is greyscale irrespective of the camera type, `'gamma'`

Photons to pixel values. A lot goes on inside a camera. Over a fixed time interval the number of photons falling on a photosite follows a Poisson distribution. The mean number of photons *and* the variance are proportional to the luminance – this variance appears as *shot noise* on the pixel value. A fraction of these photons are converted to electrons – this is the quantum efficiency of the sensor – and they accumulate in a charge well at the photosite. The number of photons captured is proportional to surface area but not all of a photosite is light sensitive due to the presence of transistors and other devices – the fraction of the photosite's area that is sensitive is called the fill factor and for CMOS sensors can be less than 50%, but this can be improved by fabricating micro-lenses above each photosite.

The charge well also accumulates thermally generated electrons, the dark current, which is proportional to temperature and is a source of noise – extreme low light cameras are cooled. Another source of noise is pixel nonuniformity due to adjacent pixels having a different gain or offset – uniform illumination therefore leads to pixels with different values which appears as additive noise. The charge well has a maximum capacity and with excessive illumination surplus electrons can overflow into adjacent charge wells leading to flaring and bleeding.

At the end of the exposure interval the accumulated charge (thermal- and photo-electrons) is read. For low-cost CMOS sensors the charge wells are connected sequentially via a switching network to one or more on-chip analog to digital converters. This results in a rolling shutter and for high speed relative motion this leads to tearing or jello effect as shown to the right. More expensive CMOS and CCD sensors have a global shutter – they make a temporary snapshot copy of the charge in a buffer which is then digitized sequentially.

The exposure on the sensor is

$$H = qL\frac{T}{N^2}\ \text{lx s}$$

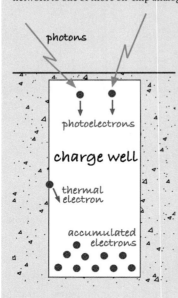

where L is scene luminance (in nit), T is exposure time, N is the f-number (inverse aperture diameter) and $q \approx 0.7$ is a function of focal length, lens quality and vignetting. Exposure time T has an upper bound equal to the inverse frame rate. To avoid motion blur a short exposure time is needed, which leads to darker and noisier images.

The integer pixel value is

$$x = kH$$

where k is a gain related to the ISO setting[a] of the camera. To obtain an sRGB (see page 311) image with an average value of 118[b] the required exposure is

$$H = \frac{10}{S_{\text{SOS}}}\ \text{lx s}$$

where S_{SOS} is the ISO rating – standard output sensitivity (SOS) – of the digital camera. Higher ISO increases image brightness by greater amplification of the measured charge but the various noise sources are also amplified leading to increased image noise which is manifested as graininess.

In photography the camera settings that control image brightness can be combined into an exposure value (EV)

$$\text{EV} = \log_2\frac{N^2}{T}$$

and all combinations of f-number and shutter speed that have the same EV value yield the same exposure. This allows a tradeoff between aperture (depth of field) and exposure time (motion blur). For most low-end cameras the aperture is fixed and the camera controls exposure using T instead of relying on an expensive, and slow, mechanical aperture. A difference of 1 EV is a factor of two change in exposure which photographers refer to as a *stop*. Increasing EV results in a darker image – most DSLR cameras allow you to manually adjust EV relative to what the camera's lightmeter has determined.

---

[a] Which is backward compatible with historical scales (ASA, DIN, ISO) devised to reflect the sensitivity of chemical films for cameras – a higher number reflected a more sensitive or "faster" film.

[b] 18% saturation, middle grey, of 8-bit pixels with gamma of 2.2.

---

to apply gamma decoding, and `'framerate'` which sets the number of frames captured per second.

The dimensions of the image returned by the camera are given by the `size` method

```
>> cam.size()
```

and an image is obtained using the `grab` method

```
>> im = cam.grab();
```

which waits until the next frame becomes available.► With no output arguments the acquired image is displayed using `idisp`.

Since the frames are generated at a rate of R per second as specified by the `'framerate'` option, then the worst case wait is uniformly distributed in the interval $[0, 1/R]$. Ungrabbed images don't accumulate, they are discarded.

The **dynamic range** of a sensor is the ratio of its largest value to its smallest value. For images it is useful to express the $\log_2$ of this ratio which makes it equivalent to the photographic concepts of stops or exposure value. Each photosite contains a charge well in which photon-generated electrons are captured during the exposure period (see page 364). The charge well has a finite capacity before the photosite saturates and this defines the maximum value. The minimum number of electrons is not zero but a finite number of thermally generated electrons.

An 8-bit image has a dynamic range of around 8 stops, a high-end 10-bit camera has a range of 10 stops, and photographic film is perhaps in the range 10–12 stops but is quite nonlinear.

At a particular state of adaptation, the human eye has a range of 10 stops, but the total adaptation range is an impressive 20 stops. This is achieved by using the iris and slower (tens of minutes) chemical adaptation of the sensitivity of rod cells. Dark adaptation to low luminance is slow, whereas adaptation from dark to bright is faster but sometimes painful.

Video file formats. Just as for image files there are a large number of different file formats for videos. The most common formats are MPEG and AVI. It is important to distinguish between the format of the file (the *container*), technically AVI is a file format, and the type of compression (the *codec*) used on the images within the file.

MPEG and AVI format files can be converted to a sequence of frames as individual files using tools such as FFmpeg and `convert` from the ImageMagick suite. The individual frames can then be loaded individually into MATLAB for processing using `iread`. The Toolbox `Movie` class provides a more convenient way to read frames directly from common movie formats without having to first convert the movies to a set of individual frames. (word map by tagxedo.com)

12.1.3 Images from a Movie File

In Sect. 12.1.1 we loaded an image sequence into memory where each image came from a separate image file. More commonly image sequences are stored in a movie file format such as MPEG4 or AVI and it may not be practical or possible to keep the whole sequence in memory.

The Toolbox supports reading frames from a movie file stored in any of the popular formats such as AVI, MPEG and MPEG4. For example we can open a movie file

```
>> cam = Movie('traffic_sequence.mpg');
traffic_sequence.mpg
720 x 576 @ 30 fps; 351 frames, 11.7 sec
cur frame 1/351 (skip=1)
```

which returns a `Movie` object that is an instance of a subclass of the `ImageSource` class and therefore polymorphic with the `VideoCamera` class just described. This movie has 350 frames and was captured at 30 frames per second.

The size of each frame within the movie is

```
>> cam.size()
ans =
   720    576
```

and the next frame is read from the movie file by

```
>> im = cam.grab();
>> about(im)
im [uint8] : 576x720x3 (1244160 bytes)
```

which is a 720×576 color image. With these few primitives we can write a very simple movie player

```
1    while 1
2        im = cam.grab;
3        if isempty(im) break; end
4        image(im); drawnow
5    end
```

where the test at line 3 is to detect the end of file, in which case `grab` returns an empty matrix.

The methods `nframes` and `framerate` provide the total number of frames and the number of frames per second. The methods `skiptotime` and `skiptoframe` provide an ability to select particular frames within the movie.

12.1.4   Images from the Web

The term web camera has come to mean *any* USB or Firewire connected local camera but here we use it to refer to an *internet* connected camera that runs a web server that can deliver images on request. There are tens of thousands of these web cameras around the world that are pointed at scenes from the mundane to the spectacular. Given the URL of a webcam from Axis Communications▸ we can acquire an image from a camera anywhere in the world and place it in a matrix in our MATLAB workspace.

For example we can connect to a camera at Dartmouth College in New Hampshire

```
>> cam = AxisWebCamera('http://wc2.dartmouth.edu');
```

which returns an `AxisWebCamera` object which is an instance of a subclass of the `ImageSource` class and therefore polymorphic with the `VideoCamera` and `Movie` classes previously described.

The image size in this case is

```
>> cam.size()
ans =
    480    640
```

and the next image is obtained by

```
>> im = cam.grab();
```

which returns a color image such as the one shown in Fig. 12.3. Webcams are configured by their owner to take pictures periodically, anything from once per second to once per minute. Repeated access will return the same image until the camera takes its next picture.

> Webcams support a variety of options that can be embedded in the URL and there is no standard for these. This Toolbox function only supports webcams from Axis Communications.

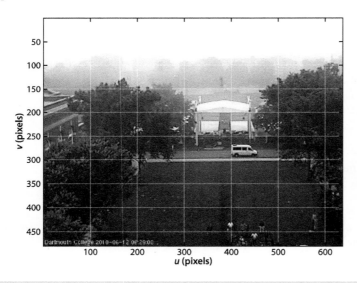

Fig. 12.3.
An image from the Dartmouth University webcam which looks out over the main college green

> **Aspect ratio** is the ratio of an image's width to its height. It varies widely across different imaging and display technologies. For 35 mm film it is 3:2 (1.5) which matches a 4 × 6" (1.5) print. Other print sizes have different aspect ratios: 5 × 7" (1.4), and 8 × 10" (1.25) which require cropping the vertical edges of the image in order to fit.
>
> TV and early computer monitors used 4:3 (1.33), for example the ubiquitous 640 × 480 VGA format. HDTV has settled on 16:9 (1.78). Modern digital SLR cameras typically use 1.81 which is close to the ratio for HDTV. In movie theatres very-wide images are preferred with aspect ratios of 1.85 or even 2.39. CinemaScope was developed by 20th Century Fox from the work of Henri Chrétien in the 1920s. An anamorphic lens on the camera compresses a wide image into a standard aspect ratio in the camera, and the process is reversed at the projector.

12.1.5   Images from Maps

You can access satellite views and road maps of anywhere on the planet from inside MATLAB. First create an instance of an `EarthView` object

```
>> ev = EarthView('key', YOUR_KEY);
```

To obtain an API key you need to register on the Google Developers Console and agree to abide by Google's terms and conditions of usage.

where YOUR_KEY is your 86-character Google API key◄. To grab a satellite image of my university is simply

```
>> ev.grab(-27.475722,153.0285, 17);
```

Zoom level is an integer and a value of zero returns a view that covers the entire Earth. Every increase by one doubles the resolution in the *x*- and *y*-directions.

which is shown in Fig. 12.4a, and the arguments are latitude, longitude and a zoom level◄. With no output arguments the result will be displayed using `idisp`. If the coordinates are unknown we can perform a lookup by name

```
>> ev.grab('QUT brisbane', 17)
```

Instead of a satellite view we can select a road map view

```
>> ev.grab(-27.475722,153.0285, 15, 'map');
```

which shows rich mapping information such as road and place names. A simpler representation is given by

```
>> ev.grab(-27.475722,153.0285, 15, 'roads');
```

which is shown in Fig. 12.4 as a binary image where white pixels correspond to roads and everything else is black – we could use this occupancy grid for robot path planning as discussed in Chap. 5.

12.1.6   Images from Code

When debugging an algorithm it can be very helpful to start with a perfect and simple image before moving on to more challenging real-world images. You could draw such an image with your favorite drawing package and import it to MATLAB, or draw it directly in MATLAB. The Toolbox function `testpattern` generates simple images with a variety of patterns including lines, grids of dots or squares, intensity ramps and intensity sinusoids. For example

```
>> im = testpattern('rampx', 256, 2);
>> im = testpattern('siny', 256, 2);
>> im = testpattern('squares', 256, 50, 25);
>> im = testpattern('dots', 256, 256, 100);
```

Fig. 12.4.
Aerial views of Brisbane, Australia. **a** Color aerial image; **b** binary image or occupancy grid where white pixels are driveable roads (images provided by Google, CNES/Astrium, Sinclair Knight Merz & Fugro)

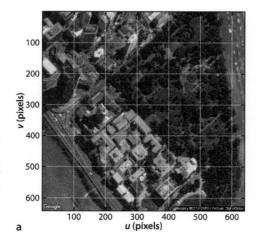
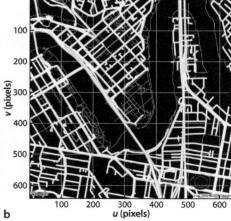

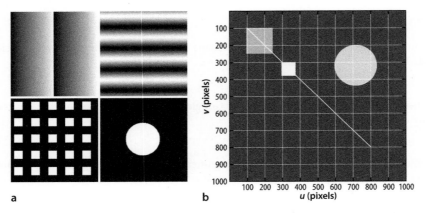

Fig. 12.5.
Images from code. **a** Some Toolbox generated test patterns; **b** Simple image created from graphical primitives

are shown in Fig. 12.5a. The second argument is the size of the of the created image, in this case they are all 256 × 256 pixels, and the remaining arguments are specific to the type of pattern requested. See the online documentation for details.

We can also construct an image from simple graphical primitives.▶ First we create a blank *canvas* containing all black pixels (pixel value of zero)

An image/matrix can be edited using the command `openvar('canvas')` which brings up a spreadsheet-like interface.

```
>> canvas = zeros(1000, 1000);
```

and then we create two squares

```
>> sq1 = 0.5 * ones(150, 150);
>> sq2 = 0.9 * ones(80, 80);
```

The first has pixel values of 0.5 (medium grey) and is 40 × 40. The second is smaller (just 20 × 20) but brighter with pixel values of 0.9. Now we can paste these onto the canvas

```
>> canvas = ipaste(canvas, sq1, [100 100]);
>> canvas = ipaste(canvas, sq2, [300 300]);
```

where the last argument specifies the canvas coordinate (u, v) where the pattern will be pasted – the top-left corner of the pattern on the canvas. We can also create a circle

```
>> circle = 0.6 * kcircle(120);
```

of radius 30 pixels with a grey value of 0.6. The Toolbox function `kcircle` returns a square matrix

```
>> size(circle)
ans =
    61    61
```

of zeros with a centered maximal disk of values set to one. We can paste that on to the canvas as well

```
>> canvas = ipaste(canvas, circle, [600, 200]);
```

Finally, we draw a line segment onto our canvas

```
>> canvas = iline(canvas, [100 100], [800 800], 0.8);
```

which extends from (100, 100) to (800, 800) and its pixels are all set to 0.8. The result

```
>> idisp(canvas)
```

is shown in Fig. 12.5b. We can clearly see that the shapes have different brightness, and we note that the line and the circle show the effects of quantization which results in a *steppy* or jagged shape.▶

In computer graphics it is common to apply anti-aliasing where edge pixels and edge-adjacent pixels are set to fractional grey values which give the impression of a smoother line.

Note that all these functions take coordinates expressed in (u, v) notation not MATLAB row column notation. The top-left pixel is (1, 1) not (0, 0).

12.2   Image Histograms

The distribution of pixel values provides useful information about the quality of the image and the composition of the scene. We obtain the distribution by computing the histogram of the image which indicates the number of times each pixel value occurs. For example the histogram of the image, shown in Fig. 12.8a, is computed and displayed by

```
>> church = iread('church.png', 'grey');
>> ihist( church )
```

and the result is shown in Fig. 12.6a. We see that the grey values (horizontal axis) span the range from 5 to 238 which is close to the full range of possible values. If the image was under-exposed the histogram area would be shifted to the left. If the image was over-exposed the histogram would be shifted to the right and many pixels would have the maximum value. A cumulative histogram is shown in Fig. 12.6b and its use will be discussed in the next section. Histograms can also be computed for color images, in which case the result is three histograms – one for each color channel.

In this case distribution of pixel values is far from uniform and we see that there are three significant peaks. However if we look more closely we see lots of very minor peaks. The concept of a peak depends on the scale at which we consider the data. We can obtain the histogram as a pair of vectors

```
>> [n,v] = ihist(church);
```

where the elements of n are the number of times pixels occur with the value of the corresponding element of v. The Toolbox function peak will automatically find the position of the peaks

```
>> [~,x] = peak(n, v)
>> about x
x [double] : 1x58 (464 bytes)
```

and in this case has found 58 peaks most of which are quite minor. Peaks that are *significant* are not only greater than their immediate neighbors they are greater than all other values *nearby* – the problem now is to specify what we mean by nearby. For example the peaks that are greater than all other values within ±25 pixel values in the horizontal direction are

```
>> [~,x] = peak(n, v, ,scale', 25)
x =
    213    147    41
```

which are the three significant peaks that we observe by eye. The critical part of finding the peaks is choosing the appropriate scale. Peak finding is a topic that we will encounter again later and is also discussed in Appendix J.

Fig. 12.6. Church scene. **a** Histogram, **b** cumulative histogram before and after normalization

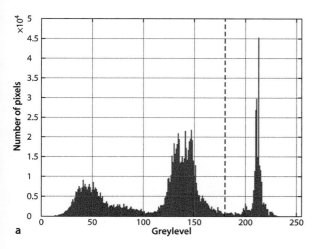

a

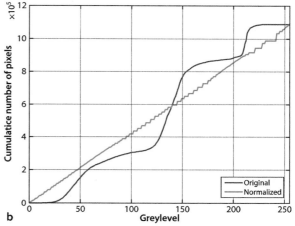

b

The peaks in the histogram correspond to particular populations of pixels in the image. The lowest peak corresponds to the dark pixels which generally belong to the ground and the roof. The middle peak generally corresponds to the sky pixels, and the highest peak generally corresponds to the white walls. However each of the scene elements has a distribution of grey values and for most real scenes we cannot simply map grey level to a scene element. For example some sky pixels are brighter than some wall pixels, and a very small number of ground pixels are brighter than some sky and wall pixels.

12.3   Monadic Operations

Monadic image-processing operations are shown schematically in Fig. 12.7. The result is an image of the same size $W \times H$ as the input image, and each output pixel is a function of the corresponding input pixel

$$O[u, v] = f(I[u, v]), \quad \forall(u, v) \in I$$

One useful class of monadic functions changes the type of the pixel data. For example to change from `uint8` (integer pixels in the range [0, 255]) to double precision values in the range [0, 1] we use the Toolbox function `idouble`

```
>> imd = idouble(church);
```

and vice versa

```
>> im = iint(imd);
```

A color image has 3-dimensions which we can also consider as a 2-dimensional image where each pixel value is a 3-vector. A monadic operation can convert a color image to a greyscale image where each output pixel value is a scalar representing the luminance of the corresponding input pixel

```
>> grey = imono(flowers);
```

The inverse operation is

```
>> color = icolor(grey);
```

which returns a 3-dimensional color image where each color plane is equal to `grey` – when displayed it still appears as a monochrome image. We can create a color image where the red plane is equal to the input image by

```
>> color = icolor(grey, [1 0 0]);
```

which is a red tinted version of the original image.

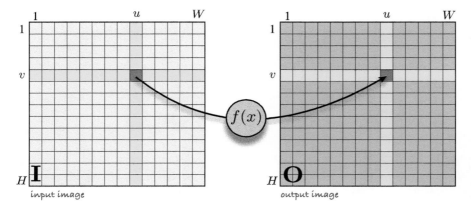

input image   *output image*

Fig. 12.7.
Monadic image processing operations. Each output pixel is a function of the corresponding input pixel (shown in *red*)

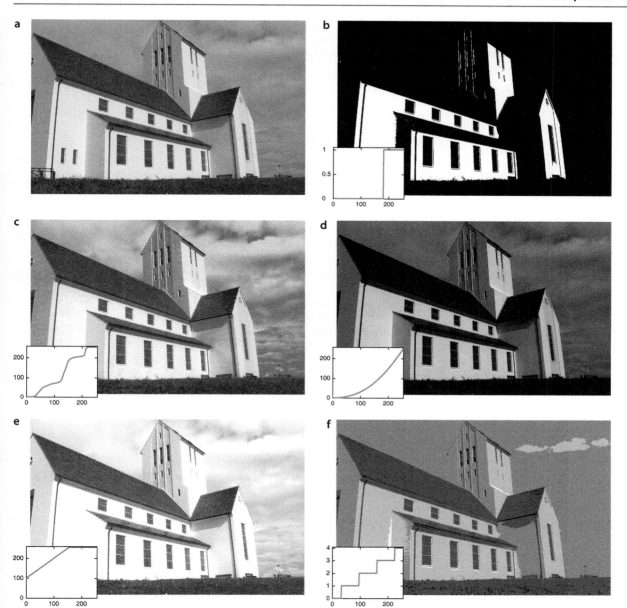

Fig. 12.8. Some monadic image operations: **a** original, **b** thresholding, **c** histogram normalized, **d** gamma correction, **e** brightness increase, **f** posterization. *Inset* in each figure is a graph showing the mapping from image grey level on the horizontal axis to the output value on the vertical axis

A very common monadic operation is thresholding. This is a logical monadic operation which separates the pixels into two classes according to their intensity

```
>> bright = (church >= 180);
>> idisp(bright)
```

and the resulting image is shown in Fig. 12.8b where all pixels that lie in the interval [180, 255] are shown as white. Such images, where the pixels have only two values are known as binary images. Looking at the image histogram in Fig. 12.6a we see that the grey value of 180 lies midway between the second and third peak which is a good approximation to the optimal strategy for separating pixels belonging to these two populations.

The variable `bright` is of type logical where the pixels have values of only true or false. MATLAB automatically converts these to one and zero respectively when used in arithmetic operations and the `idisp` function does likewise.

Many monadic operations are concerned with altering the distribution of grey levels within the image. Sometimes an image does not span the full range of available grey levels, for example the image is under- or over-exposed. We can apply a linear mapping to the grey-scale values

```
>> im = istretch(church);
```

which ensures that pixel values span the full range▶ which is either [0, 1] or [0, 255] depending on the class of the image.

A more sophisticated version is histogram normalization or histogram equalization

```
>> im = inormhist(church);
```

which is based on the cumulative distribution

```
>> ihist(church, 'cdf');
```

as shown in Fig. 12.6b. Mapping the original image via the normalized cumulative distribution ensures that the cumulative distribution of the resulting image is linear – all grey values occur an equal number of times. The result is shown in Fig. 12.8c and now details of wall and sky texture which had a very small grey-level variation have been accentuated.

> Operations such as `istretch` and `inormhist` can *enhance* the image from the perspective of a human observer, but it is important to remember that no new information has been added to the image. Subsequent image processing steps will not be improved.

As discussed in Sect. 10.3.6 the output of a camera is generally gamma encoded so that the pixel value is a nonlinear function L^γ of the luminance sensed at the photosite. Such images can be gamma decoded by a nonlinear monadic operation

```
>> im = igamm(church, 1/0.45);
```

that raises each pixel to the specified power as shown in Fig. 12.8d, or

```
>> im = igamm(church, 'sRGB');
```

to decode images with the sRGB standard gamma encoding.▶

Another simple nonlinear monadic operation is posterization or banding. This pop-art effect is achieved by reducing the number of grey levels

```
>> idisp( church/64 )
```

as shown in Fig. 12.8f. Since integer division is used the resulting image has pixels with values in the range [0, 3] and therefore just four different shades of grey. Finally, since an image is represented by a matrix any MATLAB element-wise matrix function or operator is a monadic operator, for example unary negation, scalar multiplication or addition, or functions such `abs` or `sqrt`.

The histogram of such an image will have gaps. If M is the maximum possible pixel value, and $N < M$ is the maximum value in the image then the stretched image will have at most N unique pixel values, meaning that $M - N$ values cannot occur.

The gamma correction has now been applied twice: once by the `igamm` function and once in the display device. This makes the resulting image appear to have unnaturally high contrast.

12.4   Diadic Operations

Diadic operations are shown schematically in Fig. 12.9. Two input images result in a single output image, and all three images are of the same size. Each output pixel is a function of the corresponding pixels in the two input images

$$O[u, v] = f(I_1[u, v], I_2[u, v]), \quad \forall (u, v) \in I_1$$

Examples of useful diadic operations include binary arithmetic operators such as addition, subtraction, element-wise multiplication, or builtin MATLAB diadic matrix functions such as `max`, `min`, `atan2`.

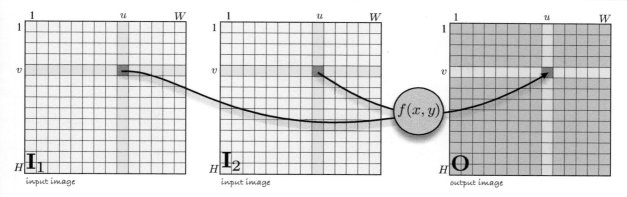

Fig. 12.9. Diadic image processing operations. Each output pixel is a function of the two corresponding input pixels (shown in *red*)

Subtracting one `uint8` image from another results in another `uint8` image even though the result is potentially negative. MATLAB quite properly clamps values to the interval [0, 255] so subtracting a larger number from a smaller number will result in zero not a negative value. With addition a result greater than 255 will be set to 255. To remedy this, the images should be first converted to signed integers using the MATLAB function `cast` or to floating-point values using the Toolbox function `idouble`.

We will illustrate diadic operations with two examples. The first example is chroma-keying – a technique commonly used in television to superimpose the image of a person over some background, for example a weather presenter superimposed over a weather map. The subject is filmed against a blue or green background which makes it quite easy, using just the pixel values, to distinguish between background and the subject. We load an image of a subject taken in front of a green screen

```
>> subject = iread('greenscreen.jpg', 'double');
```

and this is shown in Fig. 12.10a. We compute the chromaticity coordinates Eq. 10.9

```
>> linear = igamm(subject, 'sRGB');
>> [r,g] = tristim2cc(linear);
```

after first converting the gamma encoded color image to linear tristimulus values. In this case g alone is sufficient to distinguish the background pixels. A histogram of values

```
>> ihist(g)
```

shown in Fig. 12.10b indicates a large population of pixels around 0.55 which is the background and another population which belongs to the subject. We can safely say that the subject corresponds to any pixel for which $g < 0.45$ and create a *mask* image

```
>> mask = g < 0.45;
>> idisp(mask)
```

where a pixel is true (equal to one and displayed as white) if it is part of the subject as shown in Fig. 12.10c. We need to apply this mask to all three color planes so we replicate it

```
>> mask3 = icolor( idouble(mask) );
```

The image of the subject without the background is

```
>> idisp(mask3 .* subject);
```

Next we load the desired background image

```
>> bg = iread('road.png', 'double');
```

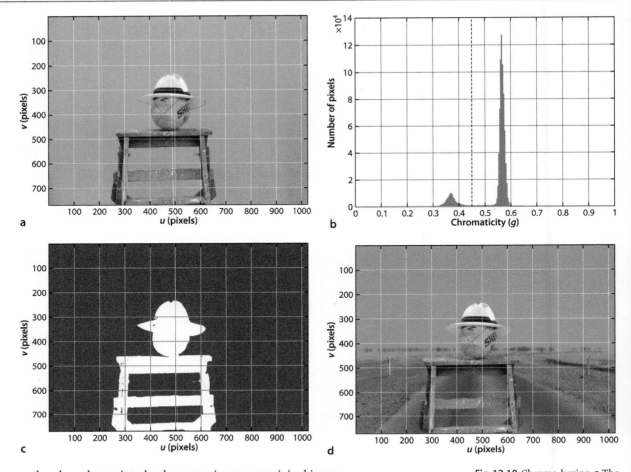

and scale and crop it to be the same size as our original image

```
>> bg = isamesize(bg, subject);
```

and display it with a *cutout* for the subject

```
>> idisp(bg .* (1-mask3))
```

Finally we add the subject with no background, to the background with no subject to obtain the subject superimposed over the background

```
>> idisp( subject.*mask3  + bg.*(1-mask3) );
```

which is shown in Fig. 12.10d. The technique will of course fail if the subject contains any colors that match the color of the background.▶ This example could be solved more compactly using the Toolbox per-pixel switching function `ipixswitch`

```
>> ipixswitch(mask, subject, bg);
```

where all arguments are images of the same width and height, and each output pixel is selected from the corresponding pixel in the second or third image according to the logical value of the corresponding pixel in the first image.

Distinguishing foreground objects from the background is an important problem in robot vision but the terms foreground and background are ill-defined and application specific. In robotics we rarely have the luxury of a special background as we did for the chroma-key example. We could instead take a picture of the scene without a foreground object present and consider this to be the background, but that requires that we have special knowledge about when the foreground object is not present. It also assumes that

Fig. 12.10. Chroma-keying. **a** The subject against a green background; **b** a histogram of green chromaticity values; **c** the computed mask image where true is white; **d** the subject *masked* into a background scene (photo courtesy of Fiona Corke)

In the early days of television a blue screen was used. Today a green background is more popular because of problems that occur with blue eyes and blue denim clothing.

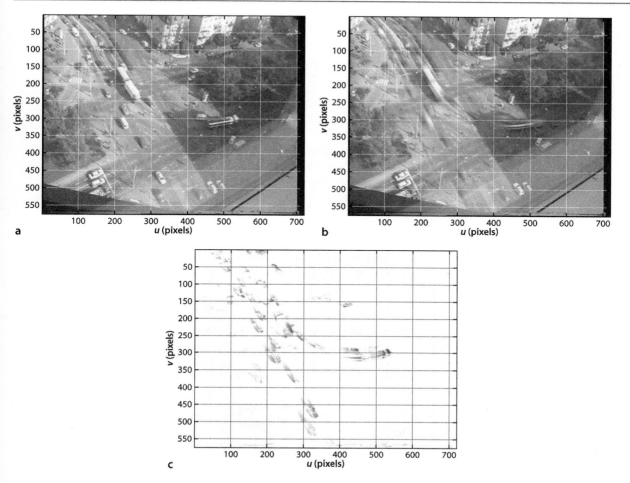

Fig. 12.11. Example of motion detection for the traffic sequence at frame 200. **a** The current image; **b** the estimated background image; **c** the difference between the current and estimated background images where *white* is zero, *red* and *blue* are negative and positive values respectively and magnitude is indicated by *color intensity*

the background does not vary over time. Variation is a significant problem in real-world scenes where ambient illumination and shadows change over quite short time intervals, and the scene may be structurally modified over very long time intervals.

In the next example we process an image sequence and *estimate* the background even though there are a number of objects moving in the scene. We will use a recursive algorithm that updates the estimated background image $\hat{\boldsymbol{B}}$ at each time step based on the previous estimate and the current image

$$\hat{\boldsymbol{B}}\langle k+1\rangle \leftarrow \hat{\boldsymbol{B}}\langle k\rangle + c\left(\boldsymbol{I}\langle k\rangle - \hat{\boldsymbol{B}}\langle k\rangle\right)$$

where k is the time step and $c(\cdot)$ is a monadic image saturation function

$$c(x) = \begin{cases} \sigma, & x > \sigma \\ x, & -\sigma \leq x \leq \sigma \\ -\sigma, & x < -\sigma \end{cases}$$

To demonstrate this we open a movie showing traffic moving through an intersection

```
>> vid = Movie('traffic_sequence.mpg', 'grey', 'double');
vid =
traffic_sequence.mpg
720 x 576 @ 30 fps; 351 frames, 11.7 sec
cur frame 1/351 (skip=1)
```

and initialize the background to the first image in the sequence

```
>> bg = vid.grab();
```

then the main loop is

```
1    sigma = 0.02;
2    while 1
3        im = vid.grab;
4        if isempty(im) break; end; % end of file?
5        d = im-bg;
6        d = max(min(d, sigma), -sigma);   % apply c(.)
7        bg = bg + d;
8        idisp(bg);
9    end
```

One frame from this sequence is shown in Fig. 12.11a. The estimated background image shown in Fig. 12.11b reveals the static elements of the scene and the moving vehicles have become a faint blur. Subtracting the scene from the estimated background creates an image where pixels are bright where they are different to the background as shown in Fig. 12.11c. Applying a threshold to the absolute value of this difference image shows the area of the image where there is motion. Of course where the cars are stationary for long enough they will become part of the background.

12.5   Spatial Operations

Spatial operations are shown schematically in Fig. 12.12. Each pixel in the output image is a function of all pixels in a *region* surrounding the corresponding pixel in the input image

$$O[u, v] = f\big(I[u + i, v + j]\big), \quad \forall (i, j) \in \mathcal{W}, \quad \forall (u, v) \in I$$

where $\mathcal{W}$ is known as the window, typically a $w \times w$ square region with odd side length $w = 2h + 1$ where $h \in \mathbb{Z}^+$ is the half-width. In Fig. 12.12 the window includes all pixels in the red shaded region. Spatial operations are powerful because of the variety of possible functions $f(\cdot)$, linear or nonlinear, that can be applied. The remainder of this section discusses linear spatial operators such as smoothing and edge detection, and some nonlinear functions such as rank filtering and template matching. The following section covers a large and important class of nonlinear spatial operators known as mathematical morphology.

12.5.1   Linear Spatial Filtering

A very important linear spatial operator is correlation

$$O[u, v] = \sum_{(i,j)\in\mathcal{W}} I[u + i, v + j]K[i, j], \, \forall (u, v) \in I \tag{12.1}$$

where $K \in \mathbb{R}^{w\times w}$ is the kernel and the elements are referred to as the filter coefficients. For every output pixel the corresponding window of pixels from the input image $\mathcal{W}$ is multiplied element-wise with the kernel K. The center of the window and kernel is considered to be coordinate $(0, 0)$ and $i, j \in [-h, h] \subset \mathbb{Z} \times \mathbb{Z}$. This can be considered as the weighted sum of pixels within the window where the weights are defined by the kernel K. Correlation is often written in operator form as

$$O = K \otimes I$$

A closely related operation is convolution

$$O[u, v] = \sum_{(i,j)\in\mathcal{W}} I[u - i, v - j]K[i, j], \quad \forall (u, v) \in I \tag{12.2}$$

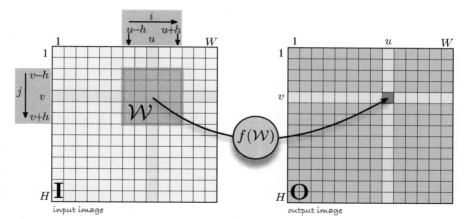

Fig. 12.12.
Spatial image processing operations. The *red shaded region* shows the window $\mathcal{W}$ that is the set of pixels used to compute the output pixel (show in *red*)

where $K \in \mathbb{R}^{w \times w}$ is the convolution kernel. Note that the sign of the i and j indices has changed in the first term. Convolution is often written in operator form as

$$\mathbf{O} = \mathbf{K} * \mathbf{I}$$

As we will see convolution is the workhorse of image processing and the kernel K can be chosen to perform functions such as smoothing, gradient calculation or edge detection.

Convolution is computationally expensive – an $N \times N$ input image with a $w \times w$ kernel requires $w^2 N^2$ multiplications and additions. In the Toolbox convolution is performed using the function `iconvolve`

```
O = iconvolve(K, I);
```

If `I` has multiple color planes then so will the output image – each output color plane is the convolution of the corresponding input plane with the kernel `K`.

12.5.1.1 Smoothing

Consider a convolution kernel which is a square 21×21 matrix containing equal elements

```
>> K = ones(21,21) / 21^2;
```

and of unit volume, that is, its values sum to one. The result of convolving an image with this kernel is an image where each output pixel is the mean of the pixels in

Correlation or convolution. These two terms are often used loosely and they have similar, albeit distinct, definitions. Convolution is the spatial domain equivalent of frequency domain multiplication and the kernel is the impulse response of a frequency domain filter. Convolution also has many useful mathematical properties outlined in the adjacent box. The difference in indexing between Eq. 12.1 and Eq. 12.2 is equivalent to reflecting the kernel – flipping it horizontally and vertically about its center point. Many kernels are symmetric in which case correlation and convolution yield the same result. However edge detection is always based on nonsymmetric kernels so we must take care to apply convolution. We will only use correlation for template matching in Sect. 12.5.2.

Properties of convolution. Convolution obeys the familiar rules of algebra, it is commutative

$$A * B = B * A$$

associative

$$A * B * C = (A * B) * C = A * (B * C)$$

distributive (superposition applies)

$$A * (B + C) = A * B + A * C$$

linear

$$A * (\alpha B) = \alpha (A * B)$$

and shift invariant – the spatial equivalent of time invariance in 1D signal processing – the result of the operation is the same everywhere in the image.

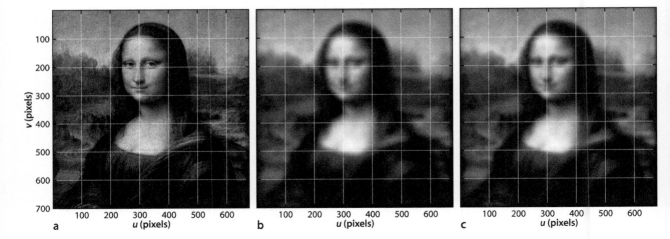

Fig. 12.13. Smoothing. a Original image; b smoothed with a 21 × 21 averaging kernel; c smoothed with a 31 × 31 Gaussian $G(\sigma = 5)$ kernel

a corresponding 21 × 21 neighborhood in the input image. As you might expect this averaging

```
>> mona = iread('monalisa.png', 'double', 'grey');
>> idisp( iconvolve(mona, K) );
```

leads to smoothing, blurring or *defocus*► which we see in Fig. 12.13b. Looking very carefully we will see some faint horizontal and vertical lines – an artifact known as ringing. A more suitable kernel for smoothing is the 2-dimensional Gaussian function

Defocus involves a kernel which is a 2-dimensional Airy pattern or sinc function. The Gaussian function is similar in shape, but is always positive whereas the Airy pattern has low amplitude negative going rings.

$$\mathbf{G}(u, v) = \frac{1}{2\pi\sigma^2} e^{-\frac{u^2 + v^2}{2\sigma^2}} \qquad (12.3)$$

which is symmetric about the origin and the volume under the curve is unity. The spread of the Gaussian is controlled by the standard deviation parameter σ. Applying this kernel to the image

```
>> K = kgauss(5);
>> idisp( iconvolve(mona, K) );
```

produces the result shown in Fig. 12.13c. Here we have specified the standard deviation of the Gaussian to be 5 pixels. The discrete approximation to the Gaussian is

```
>> about(K)
K [double] : 31x31 (7688 bytes)
```

a 31 × 31 kernel. Smoothing can be achieved conveniently using the Toolbox function `ismooth`

```
>> idisp( ismooth(mona, 5) )
```

Blurring is a counter-intuitive image processing operation since we typically go to a lot of effort to obtain a clear and crisp image. To deliberately *ruin it* seems, at face value, somewhat reckless. However as we will see later, Gaussian smoothing turns out to be extremely useful.

The kernel is itself a matrix and therefore we can display it as an image

```
>> idisp( K );
```

which is shown in Fig. 12.14a. We clearly see the large value at the center of the kernel and that it falls off smoothly in all directions. We can also display the kernel as a surface

```
>> surfl(-15:15, -15:15, K);
```

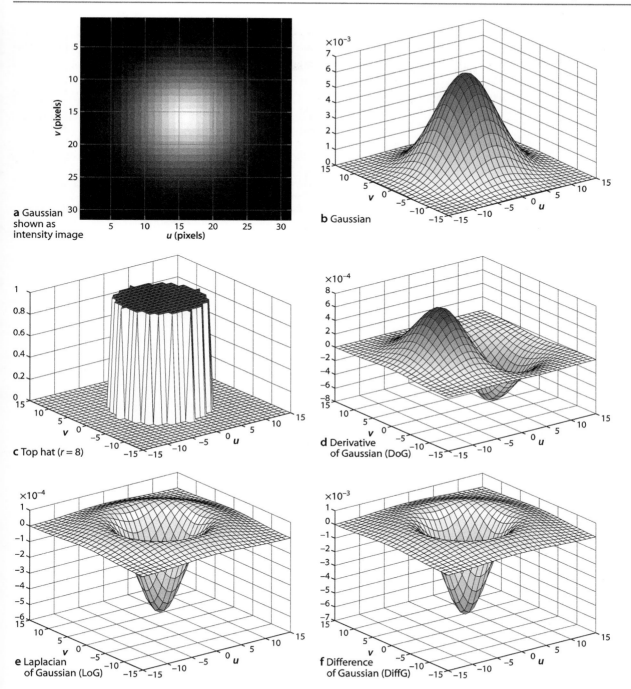

a Gaussian shown as intensity image

b Gaussian

c Top hat ($r = 8$)

d Derivative of Gaussian (DoG)

e Laplacian of Gaussian (LoG)

f Difference of Gaussian (DiffG)

Fig. 12.14. Gallery of commonly used convolution kernels. $h = 15$, $\sigma = 5$

as shown in Fig. 12.14b. A crude approximation to the Gaussian is the top hat kernel which is cylinder with vertical sides rather than a smooth and gentle fall off in amplitude. The function `kcircle` creates a kernel which can be considered a unit height cylinder of specified radius

```
>> K = kcircle(8, 15);
```

as shown in Fig. 12.14c. The arguments specify a radius of 8 pixels within a window of half width $h = 15$.

How wide is my Gaussian? When choosing a Gaussian kernel we need to consider the standard deviation, usually defined by the task, and the dimensions of the kernel $\mathcal{W} \in \mathbb{R}^{w \times w}$ that contains the discrete Gaussian function. Computation time is proportional to w^2 so ideally we want the window to be no bigger than it needs to be. The Gaussian decreases monotonically in all directions but never reaches zero. Therefore we choose the half-width h of the window such that value of the Gaussian is less than some threshold outside the $w \times w$ convolution window.

At the edge of the window, a distance h from the center, the value of the Gaussian will be $e^{-h^2/2\sigma^2}$. For $\sigma = 1$ and $h = 2$ the Gaussian will be $e^{-2} \approx 0.14$, for $h = 3$ it will be $e^{-4.5} \approx 0.01$, and for $h = 4$ it will be $e^{-8} \approx 3.4 \times 10^{-4}$. If h is not specified the Toolbox chooses $h = 3\sigma$. For $\sigma = 1$ that is a 7×7 window which contains all values of the Gaussian greater than 1% of the peak value.

Properties of the Gaussian. The Gaussian function $G(\cdot)$ has some special properties. The convolution of two Gaussians is another Gaussian

$$\mathbf{G}(\sigma_1) * \mathbf{G}(\sigma_2) = \mathbf{G}\left(\sqrt{\sigma_1^2 + \sigma_2^2}\right)$$

For the case where $\sigma_1 = \sigma_2 = \sigma$ then

$$\mathbf{G}(\sigma) * \mathbf{G}(\sigma) = \mathbf{G}\left(\sqrt{2}\sigma\right)$$

The 2-dimensional Gaussian is separable – it can be written as the product of two 1-dimensional Gaussians

$$\mathbf{G}(u, v) = \left| \frac{1}{\sigma\sqrt{2\pi}} e^{-\frac{u^2}{2\sigma^2}} \right| \left| \frac{1}{\sigma\sqrt{2\pi}} e^{-\frac{v^2}{2\sigma^2}} \right|$$

This implies that convolution with a 2-dimensional Gaussian can be computed by convolving each row with a 1-dimensional Gaussian, and then each column. The total number of operations is reduced to $2wN^2$, better by a factor of w. A Gaussian also has the same shape in the spatial and frequency domains.

12.5.1.2   Boundary Effects

A difficulty with all spatial operations occurs when the window is close to the edge of the input image as shown in Fig. 12.15. In this case the output pixel is a function of a window that contains pixels *beyond the edge* of the input image – these pixels have no defined value. There are several common remedies to this problem. Firstly, we can assume the pixels beyond the image have a particular value. A common choice is zero and this is the default behavior implemented by the Toolbox function `iconvolve`. We can see the effect of this in Fig. 12.13 where the borders of the smoothed image are dark due to the influence of these zeros.

Another option is to consider that the result is invalid when the window crosses the boundary of the image. Invalid output pixels are shown hatched out in Fig. 12.15. The result is an output image of size $(W - 2h) \times (H - 2h)$ which is slightly smaller than the input image. This option can be selected by passing the option `'valid'` to `iconvolve`.

12.5.1.3   Edge Detection

Frequently we are interested in finding the edges of objects in a scene. Consider the image

```
>> castle = iread('castle.png', 'double', 'grey');
```

shown in Fig. 12.16a. It is informative to look at the pixel values along a 1-dimensional profile through the image. A horizontal profile of the image at $v = 360$ is

```
>> p = castle(360,:);
```

which is a vector that we can plot

```
>> plot(p);
```

against the horizontal coordinate u in Fig. 12.16b. The clearly visible tall spikes correspond to the white letters and other markings on the sign. Looking at one of the spikes more closely, Fig. 12.16c, we see the intensity profile across the vertical stem of the letter T. The background intensity ≈ 0.3 and the bright intensity ≈ 0.9 but will depend on lighting levels. However the very rapid increase over the space of just a few pixels is distinctive and a more reliable indication of an edge than any decision based on the actual grey levels.

The first-order derivative along this cross-section is

$$p'[v] = p[v] - p[v - 1]$$

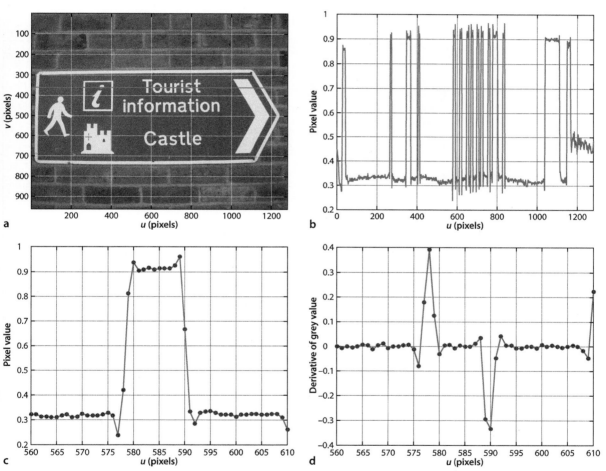

Fig. 12.15.
For the case where the window $\mathcal{W}$ *falls off the edge* of the input image the output pixel at (u, v) is not defined. The *hatched pixels* in the output image are all those for which the output value is not defined

Fig. 12.16. Edge intensity profile.
a Original image; **b** greylevel profile along horizontal line $v = 360$; **c** closeup view of the spike at $u \approx 580$; **d** derivative of **c** (image from the ICDAR 2005 OCR dataset; Lucas 2005)

which can be computed using the MATLAB function `diff`

```
>> plot(diff(p))
```

and is shown in Fig. 12.16d. The signal is nominally zero with clear nonzero responses at the edges of an object, in this case the edges of the stem of the letter T.

The derivative at point v can also be written as a *symmetrical* first-order difference

$$p'[v] = \frac{1}{2}\big(p[v+1] - p[v-1]\big)$$

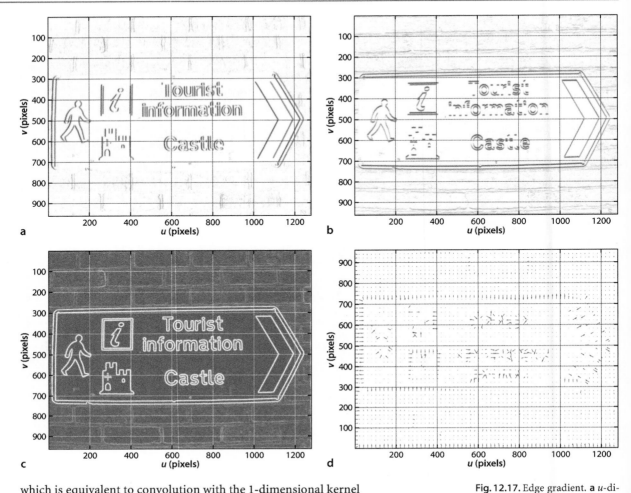

which is equivalent to convolution with the 1-dimensional kernel

$$K = \begin{pmatrix} \frac{1}{2} & 0 & -\frac{1}{2} \end{pmatrix}$$

Convolving the image with this kernel

```
>> K = [0.5 0 -0.5];
>> idisp( iconvolve(castle, K), 'invsigned')
```

produces a result very similar to that shown in Fig. 12.17a in which vertical edges, high horizontal gradients, are clearly seen.

Since this kernel has signed values the result of the convolution will also be signed, that is, the gradient at a pixel can be positive or negative as shown in Fig. 12.17a,b. `idisp` always displays the minimum, most negative, value as black and the maximum, most positive, value as white. Zero would therefore appear as middle grey. The `'signed'` option to `idisp` uses red and blue shading to clearly indicate sign – zero is black, negative pixels are red, positive pixels are blue and the intensity of the color is proportional to pixel magnitude. The `'invsigned'` option is similar except that zero is indicated by white.

Many convolution kernels have been proposed for computing horizontal gradient. A popular choice is the Sobel kernel▶

Fig. 12.17. Edge gradient. **a** u-direction gradient; **b** v-direction gradient; **c** gradient magnitude; **d** gradient direction. Gradients shown with *blue* as positive, *red* as negative and *white* as zero

This kernel is commonly written with the signs reversed which is correct for correlation. For convolution the kernel must be written as shown here.

```
>> Du = ksobel
Du =
    0.1250         0   -0.1250
    0.2500         0   -0.2500
    0.1250         0   -0.1250
```

and we see that each row is a scaled version of the 1-dimensional kernel $\mathtt{K}$ defined above. The overall result is a weighted sum of the horizontal gradient for the current row, and the rows above and below. Convolving our image with this kernel

```
>> idisp( iconvolve(castle, Du), 'invsigned')
```

generates the horizontal gradient image shown in Fig. 12.17a which highlights vertical edges. Vertical gradient is computed using the transpose of the kernel

```
>> idisp( iconvolve(castle, Du'), 'invsigned')
```

Filters can be designed to respond to edges at any arbitrary angle. The Sobel kernel itself can be considered as an image and rotated using $\mathtt{irotate}$. To obtain angular precision generally requires a larger kernel is required such as that generated by $\mathtt{kdgauss}$.

and highlights horizontal edges◄ as shown in Fig. 12.17b. The notation used for gradients varies considerably in the literature. Most commonly the horizontal and vertical gradient are denoted respectively as $\partial \mathbf{I}/\partial u$, $\partial \mathbf{I}/\partial v$; $\nabla_u \mathbf{I}$, $\nabla_v \mathbf{I}$ or $\mathbf{I}_u$, $\mathbf{I}_v$. In operator form this is written

$$\mathbf{I}_u = \mathbf{D} * \mathbf{I}$$
$$\mathbf{I}_v = \mathbf{D}^T * \mathbf{I}$$

where $\mathbf{D}$ is a derivative kernel such as Sobel.

Taking the derivative of a signal accentuates high-frequency noise, and all images have noise as discussed on page 364. At the pixel level noise is a stationary random process – the values are not correlated between pixels. However the features that we are interested in such as edges have correlated changes in pixel value over a larger spatial scale as shown in Fig. 12.16c. We can reduce the effect of noise by smoothing the image before taking the derivative

$$\mathbf{I}_u = \mathbf{D} * \big(\mathbf{G}(\sigma) * \mathbf{I}\big)$$

Instead of convolving the image with the Gaussian and *then* the derivative, we exploit the associative property of convolution to write

$$\mathbf{I}_u = \mathbf{D} * \big(\mathbf{G}(\sigma) * I\big) = \underbrace{\big(\mathbf{D} * \mathbf{G}(\sigma)\big)}_{\text{DoG}} * \mathbf{I}$$

Carl Friedrich Gauss (1777–1855) was a German mathematician who made major contributions to fields such as number theory, differential geometry, magnetism, astronomy and optics. He was a child prodigy, born in Brunswick, Germany, the only son of uneducated parents. At the age of three he corrected, in his head, a financial error his father had made, and made his first mathematical discoveries while in his teens. Gauss was a perfectionist and a hard worker but not a prolific writer. He refused to publish anything he did not consider complete and above criticism. It has been suggested that mathematics could have been advanced by fifty years if he had published all of his discoveries. According to legend Gauss was interrupted in the middle of a problem and told that his wife was dying – he responded "Tell her to wait a moment until I am through".

The normal distribution, or Gaussian function, was not one of his achievements. It was first discovered by de Moivre in 1733 and again by Laplace in 1778. The SI unit for magnetic flux density is named in his honor.

We convolve the image with the *derivative of the Gaussian* (DoG) which can be obtained numerically by

```
Gu = iconvolve( Du, kgauss(sigma) , 'full');
```

or analytically by taking the derivative, in the *u*-direction, of the Gaussian Eq. 12.3 yielding

$$\mathbf{G}_u(u, v) = -\frac{u}{2\pi\sigma^4} e^{-\frac{u^2+v^2}{2\sigma^2}} \qquad (12.4)$$

which is computed by the Toolbox function `kdgauss` and is shown in Fig. 12.14d.

The standard deviation σ controls the *scale* of the edges that are detected. For large σ, which implies increased smoothing, edges due to fine texture will be attenuated leaving only the edges of large features. This ability to find edges at different spatial scale is important and underpins the concept of scale space that we will discuss in Sect. 13.3.2. Another interpretation of this operator is as a spatial *bandpass filter* since it is a cascade of a low-pass filter (smoothing) with a high-pass filter (differentiation).

Computing the horizontal and vertical components of gradient at each pixel

```
>> Iu = iconvolve( castle, kdgauss(2) );
>> Iv = iconvolve( castle, kdgauss(2)' );
```

allows us to compute the magnitude of the gradient at each pixel

```
>> m = sqrt( Iu.^2 + Iv.^2 );
```

This *edge-strength* image shown in Fig. 12.17c reveals the edges very distinctly. The direction of the gradient at each pixel is

```
>> th = atan2( Iv, Iu);
```

and is best viewed as a sparse quiver plot

```
>> quiver(1:20:numcols(th), 1:20:numrows(th), ...
       Iu(1:20:end,1:20:end), Iv(1:20:end,1:20:end))
```

as shown in Fig. 12.17d. The edge direction plot is much noisier than the magnitude plot. Where the edge gradient is strong, on the border of the sign or the edges of letters, the direction is normal to the edge, but the fine-scale brick texture appears as almost random edge direction. The gradient images can be computed conveniently using the Toolbox function

```
>> [du,dv] = isobel( castle, kdgauss(2) );
```

where the last argument overrides the default Sobel kernel.

A well known and very effective edge detector is the Canny edge operator. It uses the edge magnitude and direction that we have just computed and performs two additional steps. The first is nonlocal maxima suppression. Consider the gradient magnitude image of Fig. 12.17c as a 3-dimensional surface where height is proportional to brightness as shown in Fig. 12.18. We see a series of hills and ridges and we wish to find the

Pierre-Simon Laplace (1749–1827) was a French mathematician and astronomer who consolidated the theories of mathematical astronomy in his five volume *Mécanique Céleste* (Celestial Mechanics). While a teenager his mathematical ability impressed d'Alembert who helped to procure him a professorship. When asked by Napoleon why he hadn't mentioned God in his book on astronomy he is reported to have said "Je n'avais pas besoin de cette hypothèse-là" ("I have no need of that hypothesis"). He became a count of the Empire in 1806 and later a marquis.

The Laplacian operator, a second-order differential operator, and the Laplace transform are named after him.

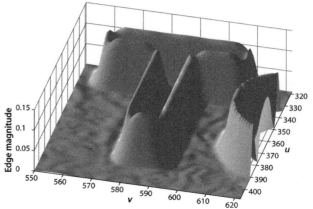

Fig. 12.18.
Closeup of gradient magnitude around the letter T shown as a 3-dimensional surface

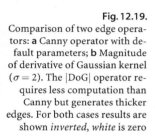

Fig. 12.19.
Comparison of two edge operators: **a** Canny operator with default parameters; **b** Magnitude of derivative of Gaussian kernel ($\sigma = 2$). The |DoG| operator requires less computation than Canny but generates thicker edges. For both cases results are shown *inverted*, *white* is zero

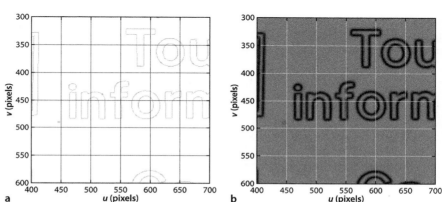

pixels that lie along the *ridge lines*. By examining pixel values in a local neighborhood *normal* to the edge direction, that is in the direction of the edge gradient, we can find the maximum value and set all other pixels to zero. The result is a set of nonzero pixels corresponding to peaks and ridge lines. The second step is hysteresis thresholding. For each nonzero pixel that exceeds the upper threshold a chain is created of adjacent pixels that exceed the lower threshold. Any other pixels are set to zero.

To apply the Canny operator to our example image is straightforward

```
>> edges = icanny(castle, 2);
```

and returns an image where the edges are marked by nonzero intensity values corresponding to gradient magnitude at that pixel as shown in Fig. 12.19a. We observe that the edges are much thinner than those for the magnitude of derivative of Gaussian operator which is shown in Fig. 12.19b. In this example $\sigma = 2$ for the derivative of Gaussian operation. The hysteresis threshold parameters can be set with optional arguments.

Difference of Gaussians. The Laplacian of Gaussian can be approximated by the difference of two Gaussian functions

$$\mathbf{DoG}(u, v; \sigma_1, \sigma_1) = \mathbf{G}(\sigma_1) - \mathbf{G}(\sigma_2) = \frac{1}{2\pi\sigma_1^2\sigma_2^2}\left(\sigma_2^2 e^{-\frac{u^2+v^2}{2\sigma_1^2}} - \sigma_1^2 e^{-\frac{u^2+v^2}{2\sigma_2^2}}\right)$$

where $\sigma_1 > \sigma_2$ and commonly $\sigma_1 = 1.6\sigma_2$. This is computed by the Toolbox function `kdog`. Figure 12.13e and f shows the LoG and DiffG kernels respectively.

This approximation is useful in scale-space sequences which will be discussed in Sect. 13.3.2. Consider an image sequence $\mathbf{I}\langle k\rangle$ where $\mathbf{I}\langle k+1\rangle = \mathbf{G}(\sigma) \otimes \mathbf{I}\langle k\rangle$, that is, the images are increasingly smoothed. The difference between any two images in the sequence is therefore equivalent to DiffG$(\sqrt{2}\sigma, \sigma)$ applied to the original image.

So far we have considered an edge as a point of high gradient, and nonlocal maxima suppression has been used to *search* for the maximum value in local neighborhoods. An alternative means to find the point of maximum gradient is to compute the second derivative and determine where this is zero. The Laplacian operator

$$\nabla^2 \mathbf{I} = \frac{\partial^2 \mathbf{I}}{\partial u^2} + \frac{\partial^2 \mathbf{I}}{\partial v^2} = \mathbf{I}_{uu} + \mathbf{I}_{vv} \tag{12.5}$$

is the sum of the second spatial derivative in the horizontal and vertical directions. For a discrete image this can be computed by convolution with the Laplacian kernel

```
>> L = klaplace()
L =
     0     1     0
     1    -4     1
     0     1     0
```

which is isotropic – it responds equally to edges in any direction. The second derivative is even more sensitive to noise than the first derivative and is again commonly used in conjunction with a Gaussian smoothed image

$$\nabla^2 \mathbf{I} = \mathbf{L} * \big(\mathbf{G}(\sigma) * \mathbf{I}\big) = \underbrace{\big(\mathbf{L} * \mathbf{G}(\sigma)\big)}_{\text{LoG}} * \mathbf{I} \tag{12.6}$$

which we combine into the Laplacian of Gaussian kernel (LoG), and **L** is the Laplacian kernel given above. This can be written analytically as

Fig. 12.20. Laplacian of Gaussian. **a** Laplacian of Gaussian; **b** closeup of **a** around the letter T where *blue* and *red* colors indicate positive and negative values respectively; **c** a horizontal cross-section of the LoG through the stem of the T; **d** closeup of the zero-crossing detector output at the letter T

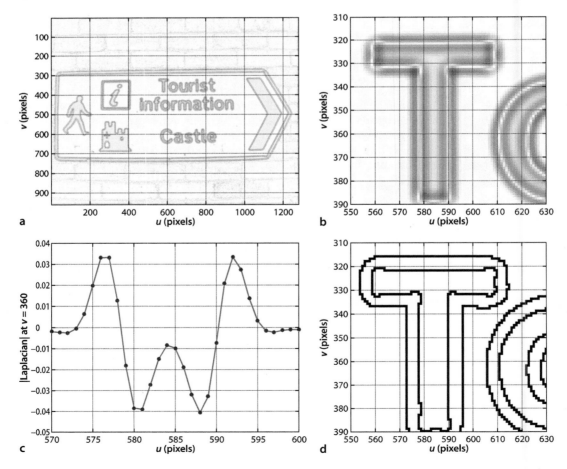

$$\mathbf{LoG}(u, v) = \frac{\partial^2 \mathbf{G}}{\partial u^2} + \frac{\partial^2 \mathbf{G}}{\partial v^2} \tag{12.7}$$

$$= \frac{1}{\pi \sigma^4} \left(\frac{u^2 + v^2}{2\sigma^2} - 1 \right) e^{-\frac{u^2 + v^2}{2\sigma^2}} \tag{12.8}$$

which is known as the Marr-Hildreth operator or the *Mexican hat kernel* and is shown in Fig. 12.14e.

We apply this kernel to our image by

```
>> lap = iconvolve( castle, klog(2) );
```

and the result is shown in Fig. 12.20a and b. The maximum gradient occurs where the second derivative is zero but a significant edge is a zero crossing from a strong positive value (blue) to a strong negative value (red). Consider the closeup view of the Laplacian of the letter T shown in Fig. 12.20b. We generate a horizontal cross-section of the stem of the letter T at $v = 360$

```
>> p = lap(360,570:600);
>> plot(570:600, p, '-o');
```

which is shown in Fig. 12.20c. We see that the zero values of the second derivative lies *between* the pixels. A zero crossing detector selects pixels adjacent to the zero crossing points

```
>> zc = zcross(lap);
```

and this is shown in Fig. 12.20d. We see that the edges appear twice. Referring again to Fig. 12.20c we observe a weak zero crossing in the interval $u \in [573, 574]$ and a much more definitive zero crossing in the interval $u \in [578, 579]$.

A fundamental limitation of all edge detection approaches is that intensity edges do not necessarily delineate the boundaries of objects. The object may have poor contrast with the background which results in weak boundary edges. Conversely the object may have a stripe on it which is not its edge. Shadows frequently have very sharp edges but are not real objects. Object texture will result in a strong output from an edge detector at points not just on its boundary, as for example with the bricks in Fig. 12.16b.

12.5.2   Template Matching

In our discussion so far we have used kernels that represent mathematical functions such as the Gaussian and its derivative and its Laplacian. We have also considered the convolution kernel as a matrix, as an image and as a 3-dimensional surface as shown in Fig. 12.14. In this section we will consider that the kernel *is an image* or a part of an image which we refer to as a template. In template matching we wish to find which parts of the input image are most similar to the template.

Template matching is shown schematically in Fig. 12.21. Each pixel in the output image is given by

$$O[u, v] = s(\mathbf{T}, \mathcal{W}), \quad \forall (u, v) \in \mathbf{I}$$

where $\mathbf{T}$ is the $w \times w$ template, the pattern of pixels we are looking for, with odd side length $w = 2h + 1$, and $\mathcal{W}$ is the $w \times w$ window centered at (u, v) in the input image. The function $s(\mathbf{I}_1, \mathbf{I}_2)$ is a scalar measure that describes the *similarity* of two equally sized images $\mathbf{I}_1$ and $\mathbf{I}_2$.

These measures can be augmented with a Gaussian weighting to deemphasize the differences that occur at the edges of the two windows.

A number of common similarity measures◄ are given in Table 12.1. The most intuitive are computed simply by computing the pixel-wise difference $T - \mathcal{W}$ and taking

David Marr (1945–1980) was a British neuroscientist and psychologist who synthesized results from psychology, artificial intelligence, and neurophysiology to create the discipline of Computational Neuroscience. He studied mathematics at Trinity College, Cambridge and his Ph.D. in physiology was concerned with modeling the function of the cerebellum. His key results were published in three journal papers between 1969 and 1971 and formed a theory of the function of the mammalian brain much of which remains relevant today. In 1973 he was a visiting scientist in the Artificial Intelligence Laboratory at MIT and later became a professor in the Department of Psychology. His attention shifted to the study of vision and in particular the so-called early visual system.

He died of leukemia at age 35 and his book *Vision: A computational investigation into the human representation and processing of visual information* (Marr 2010) was published after his death.

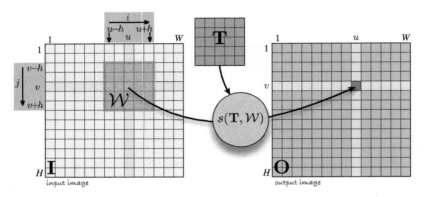

Fig. 12.21.
Spatial image processing operations. The *red shaded region* shows the window $\mathcal{W}$ that is the set of pixels used to compute the output pixel (shown in *red*)

the sum of the absolute differences (SAD) or the sum of the squared differences (SSD). These metrics are zero if the images are identical and increase with dissimilarity. It is not easy to say what value of the measure constitutes a poor match but a ranking of similarity measures can be used to determine the *best* match.

More complex measures such as normalized cross-correlation yield a score in the interval $[-1, +1]$ with $+1$ for identical regions. In practice a value greater than 0.8 is considered to be a good match. Normalized cross correlation is computationally more expensive – requiring multiplication, division and square root operations. Note that it is possible for the result to be undefined if the denominator is zero, which occurs if the elements of either I_1 or I_2 are identical.

If $I_2 \equiv I_1$ then it is easily shown that $\text{SAD} = \text{SSD} = 0$ and $\text{NCC} = 1$ indicating a perfect match. To illustrate we will use the Mona Lisa's eye as a 51×51 template

```
>> mona = iread('monalisa.png', 'double', 'grey');
>> T = mona(170:220, 245:295);
```

and evaluate the three common measures

```
>> sad(T, T)
ans =
     0
>> ssd(T, T)
ans =
     0
>> ncc(T, T)
ans =
     1
```

Now consider the case where the two images are of the same scene but one image is darker than the other – the illumination or the camera exposure has changed. In this case $I_2 = \alpha I_1$ and now

```
>> sad(T, T*0.9)
ans =
   111.1376
>> ssd(T, T*0.9)
ans =
     5.6492
```

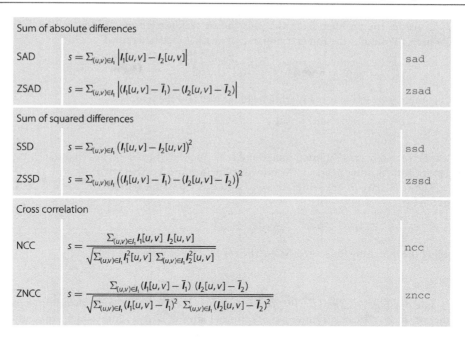

Table 12.1.
Similarity measures for two equal sized image regions I_1 and I_2. The Z-prefix indicates that the measure accounts for the zero-offset or the difference in mean of the two images (Banks and Corke 2001). $\bar{I}_1$ and $\bar{I}_2$ are the mean of image regions I_1 and I_2 respectively. Toolbox functions are indicated in the last column

Sum of absolute differences		
SAD	$s = \Sigma_{(u,v)\in I_1} \left| I_1[u,v] - I_2[u,v] \right|$	sad
ZSAD	$s = \Sigma_{(u,v)\in I_1} \left| (I_1[u,v] - \bar{I}_1) - (I_2[u,v] - \bar{I}_2) \right|$	zsad
Sum of squared differences		
SSD	$s = \Sigma_{(u,v)\in I_1} \left(I_1[u,v] - I_2[u,v] \right)^2$	ssd
ZSSD	$s = \Sigma_{(u,v)\in I_1} \left((I_1[u,v] - \bar{I}_1) - (I_2[u,v] - \bar{I}_2) \right)^2$	zssd
Cross correlation		
NCC	$s = \dfrac{\Sigma_{(u,v)\in I_1} I_1[u,v]\ I_2[u,v]}{\sqrt{\Sigma_{(u,v)\in I_1} I_1^2[u,v]\ \Sigma_{(u,v)\in I_1} I_2^2[u,v]}}$	ncc
ZNCC	$s = \dfrac{\Sigma_{(u,v)\in I_1} (I_1[u,v] - \bar{I}_1)\ (I_2[u,v] - \bar{I}_2)}{\sqrt{\Sigma_{(u,v)\in I_1} (I_1[u,v] - \bar{I}_1)^2\ \Sigma_{(u,v)\in I_1} (I_2[u,v] - \bar{I}_2)^2}}$	zncc

these measure indicate a degree of dissimilarity. However the normalized cross-correlation

```
>> ncc(T, T*0.9)
ans =
     1
```

is invariant to the change in intensity.

Next consider that the pixel values have an offset◄ so that $I_2 = I_1 + \beta$ and we find that

This could be due to an incorrect black level setting. A camera's black level is the value of a pixel corresponding to no light and is often >0.

```
>> sad(T, T+0.1)
ans =
   260.1000
>> ssd(T, T+0.1)
ans =
    26.0100
>> ncc(T, T+0.1)
ans =
     0.9974
```

all measures now indicate a degree of dissimilarity. The problematic offset can be dealt with by first subtracting from each of T and W their mean value

```
>> zsad(T, T+0.1)
ans =
   3.5670e-12
>> zssd(T, T+0.1)
ans =
   4.8935e-27
>> zncc(T, T+0.1)
ans =
     1.0000
```

and these measures now all indicate a perfect match. The z-prefix denotes variants of the similarity measures described above that are invariant to intensity offset. Only the ZNCC measure

```
>> zncc(T, T*0.9+0.1)
ans =
     1.0000
```

is invariant to both gain and offset variation. All these methods will fail if the images have even a small change in relative rotation or scale.

Consider the problem from the well known children's book "Where's Wally" or "Where's Waldo" – the fun is trying to find Wally's face in a crowd

```
>> crowd = iread('wheres-wally.png', 'double');
>> idisp(crowd)
```

Fortunately we know roughly what he looks like and the template

```
>> wally = iread('wally.png', 'double');
>> idisp(wally)
```

was extracted from a different image and scaled so that the head is approximately the same width as other heads in the crowd scene (around 21 pixel wide).

The similarity of our template `wally` to every possible window location is computed by

```
>> S = isimilarity(wally, crowd, @zncc);
```

using the matching measure ZNCC. The result

```
>> idisp(S, 'colormap', 'jet', 'bar')
```

is shown in Fig. 12.22 and the pixel color indicates the ZNCC similarity as indicated by the color bar. We can see a number of spots of high similarity (white) which are candidate positions for Wally. The peak values, with respect to a local 3×3 window, are

```
>> [mx,p] = peak2(S, 1, 'npeaks', 5);
>> mx
mx =
    0.5258    0.5230    0.5222    0.5032    0.5023
```

in descending order. The second argument specifies the window half-width $h = 1$ and the third argument specifies the number of peaks to return. The largest value 0.5258 is the similarity of the strongest match found. These matches occur at the coordinates (u, v) given by the second return value `p` and we can highlight these points on the scene

```
>> idisp(crowd);
>> plot_circle(p, 30, 'edgecolor', 'g')
>> plot_point(p, 'sequence', 'bold', 'textsize', 24, 'textcolor', 'y')
```

using green circles that are numbered sequentially. The best match at (261, 377) is in fact the correct answer – we found Wally! It is interesting to look at the other highly ranked candidates. Numbers two and three at the bottom of the image are people also wearing baseball caps who look quite similar.

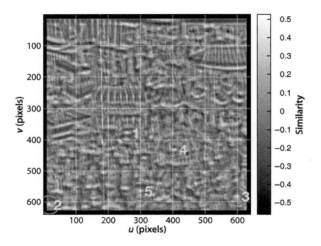

Fig. 12.22.
Similarity image S with top five Wally candidates marked. The *color bar* indicate the similarity scale. Note the border of indeterminate values where the template window falls off the edge of the input image

There are some important points to note from this example. The images have quite low resolution and the template is only 21 × 25 – it is a very crude likeness to Wally. The match is not a strong one – only 0.5258 compared to the maximum possible value of 1.0 and there are several contributing factors. The matching measure is not invariant to scale, that is, as the relative scale (zoom) changes the similarity score falls quite quickly. In practice perhaps a 10–20% change in scale between **T** and $\mathcal{W}$ can be tolerated. For this example the template was only approximately scaled. Secondly, not all Wallys are the same. Wally in the template is facing forward but the Wally we found in the image is looking to our left. Another problem is that the square template typically includes pixels from the background as well as the object of interest. As the object moves the background pixels may change, leading to a lower similarity score. This is known as the mixed pixel problem and is discussed in the next section. Ideally the template should bound the object of interest as tightly as possible. In practice another problem arises due to perspective distortion. A square pattern of pixels in the center of the image will appear keystone shaped at the edge of the image and thus will match less well with the square template.

A common problem with template matching is that false matches can occur. In the example above the second candidate had a similarity score only 0.5% lower than the first, the fifth candidate was only than 5% lower. In practice a number of rules are applied before a match is accepted: the similarity must exceed some threshold and the first candidate must exceed the second candidate by some factor to ensure there is no ambiguity.

Another approach is to bring more information to bear on the problem such as known motion of the camera or object. For example if we were tracking Wally from frame to frame in an image sequence then we would pick the best Wally closest to the previous location he was found. Alternatively we could create a motion model, typically a constant velocity model which assume he moves approximately the same distance and direction from frame to frame. In this way we could predict his future position and pick the Wally closest to that predicted position, or only search in the vicinity of the predicted position in order to reduce computation. We would also have to deal with practical difficulties such as Wally stopping, changing direction or being temporarily obscured.

12.5.2.1 Nonparameteric Local Transforms

Nonparametric similarity measures are more robust to the mixed pixel problem and we can apply a local transform to the image and template before matching. Two common transforms from this class are the census transform and the rank transform.

The census transform maps pixel values from a local region to an integer considered as a bit string – each bit corresponds to one pixel in the region as shown in Fig. 12.23. If a pixel is greater than the center pixel its corresponding bit is set to one, else it is zero. For a $w \times w$ window the string will be $w^2 - 1$ bits long.◄ The two bit strings are compared using a Hamming distance which is the number of bits that are different.

For a 32-bit integer `uint32` this limits the window to 5 × 5 unless a sparse mapping is adopted (Humenberger et al. 2009). A 64-bit integer `uint64` supports a 7 × 7 window.

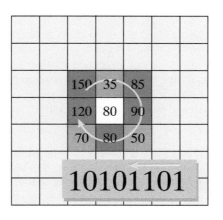

Fig. 12.23.
Example of census and rank transform for a 3 × 3 window. Pixels are marked *red* or *blue* if they are less than or greater than or equal to the center pixel respectively. These boolean values are then packed into a binary word, in the direction shown, from least significant bit upwards. The census value is 10101101_2 or decimal 173. The rank transform value is the total number of one bits and is 5

This can be computed by counting the number of set bits in the exclusive-or of the two bit strings. Thus very few arithmetic operations are required compared to the more conventional methods – no square roots or division – and such algorithms are amenable to implementation in special purpose hardware or FPGAs. Another advantage is that intensities are considered relative to the center pixel of the window making it invariant to overall changes in intensity or gradual intensity gradients.

The rank transform maps the pixel values in a local region to a scalar which is the number of elements in the region that are greater than the center pixel. This measure captures the *essence* of the region surrounding the center pixel, and like the census transform it is invariant to overall changes in intensity since it is based on local relative grey-scale values.

These transforms are typically used as a pre-processing step applied to each of the images before using a simple classical similarity measure such as SAD. The Toolbox function isimilarity supports these metrics using the 'census' and 'rank' options.

12.5.3   Nonlinear Operations

Another class of spatial operations is based on nonlinear functions of pixels within the window. For example

```
>> out = iwindow(mona, ones(7,7), 'var');
```

computes the variance of the pixels in *every* 7×7 window. The arguments specify the window size and the builtin MATLAB function var. The function is called with a 49×1 vector argument comprising the pixels in the window arranged as a column vector and the function's return value becomes the corresponding output pixel value. This operation acts as an edge detector since it has a low value for homogeneous regions irrespective of their brightness. It is however computationally expensive because the var function is called over 470 000 times. Any MATLAB function, builtin or your own M-file, that accepts a vector input and returns a scalar can be used in this way.

Rank filters sort the pixels within the window by value and return the specified element from the sorted list. The maximum value over a 5×5 window about each pixel is the first ranked pixel in the window

```
>> mx = irank(mona, 1, 2);
```

where the arguments are the rank and the window half-width $h = 2$. The median over a 5×5 window is the twelfth in rank

```
>> med = irank(mona, 12, 2);
```

and is useful as a filter to remove impulse-type noise and for the Mona Lisa image this significantly reduces the fine surface cracking. A more powerful demonstration is to add significant impulse noise to a copy of the Lena image

```
>> lena = iread('lena.pgm', 'double'); spotty = lena;
>> npix = prod(size(lena));
>> spotty(round(rand(5000,1)*(npix-1)+1)) = 0;
>> spotty(round(rand(5000,1)*(npix-1)+1)) = 1.0;
>> idisp(spotty)
```

and this is shown in Fig. 12.24a. We have set 5 000 random pixels to be zero, and another 5 000 random pixels to the maximum value. This type of noise is often referred to as impulse noise or salt and pepper noise. We apply a 3×3 median filter

```
>> idisp( irank(spotty, 5, 1) )
```

and the result shown in Fig. 12.24b is considerably improved. A similar effect could have been obtained by smoothing but that would tend to blur the image, median filtering has the advantage of preserving edges in the scene.

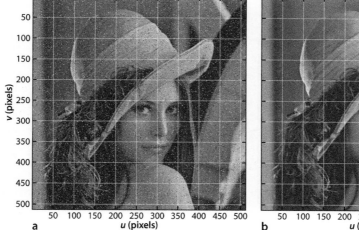

Fig. 12.24.
Median filter cleanup of impulse
noise. **a** Noise corrupted image;
b median filtered result

The third argument to `irank` can be a matrix instead of a scalar and this allows for some very powerful operations. For example

```
>> M = ones(3,3);
>> M(2,2) = 0
M =
     1    1    1
     1    0    1
     1    1    1
>> mxn = irank(lena, 1, M);
```

specifies the first in rank (maximum) over a *subset* of pixels from the window corresponding to the nonzero elements of M. In this case M specifies the eight neighboring pixels but not the center pixel. The result `mxn` is the maximum of the eight neighbors of each corresponding pixel in the input image. We can use this

```
>> idisp(lena > mxn)
```

to display all those points where the pixel value is greater than its local neighbors which performs nonlocal maxima suppression. These correspond to local maxima, or peaks if the image is considered as a surface. This mask matrix is very similar to a structuring element which we will meet in the next section.

12.6   Mathematical Morphology

Mathematical morphology is a class of nonlinear spatial operators shown schematically in Fig. 12.25. Each pixel in the output matrix is a function of a *subset* of pixels in a region surrounding the corresponding pixel in the input image

$$\mathbf{O}[u,v] = f\big(\mathbf{I}[u+i,v+j]\big), \ \ \forall(i,j) \in \mathcal{S}, \ \ \forall(u,v) \in \mathbf{I} \tag{12.9}$$

where $\mathcal{S}$ is the structuring element, an arbitrary small binary image. For implementation purposes this is embedded in a rectangular window with odd side lengths. The structuring element is similar to the convolution kernel discussed previously except that now it controls *which pixels* in the neighborhood the function $f(\cdot)$ is applied to – it specifies a subset of pixels within the window. The selected pixels are those for which the corresponding values of the structuring element are nonzero – these are shown in red in Fig. 12.25. Mathematical morphology, as its name implies, is concerned with the form or *shape* of objects in the image.

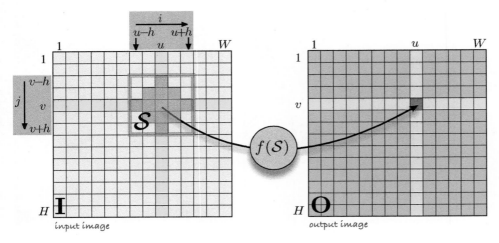

Fig. 12.25.
Morphological image processing operations. The operation is
defined only for the selected elements (*red*) within the structuring element (*red outlined square*)

The easiest way to explain the concept is with a simple example, in this case a synthetic binary image created by the script

```
>> eg_morph1
>> idisp(im)
```

which is shown, repeated, down the first column of Fig. 12.26. The structuring element
is shown in red at the end of each row. If we consider the top most row, the structuring element is a square

```
>> S = ones(5,5);
```

and is applied to the original image using the minimum operation

```
>> mn = imorph(im, S, 'min');
```

and the result is shown in the second column. For each pixel in the input image we
take the *minimum* of all pixels in the 5 × 5 window. If *any* of those pixels are zero the
resulting pixel will be zero. We can see this in animation by

```
>> morphdemo(im, S, 'min')
```

The result is dramatic – two objects have disappeared entirely and the two squares have
become separated and smaller. The two objects that disappeared were not *consistent*
with the shape of the structuring element. This is where the connection to morphology or shape comes in – only shapes that could *contain* the structuring element will
be present in the output image.

The structuring element could define any shape: a circle, an annulus, a 5-point-
ed star, a line segment 20 pixels long at 30° to the horizontal, or the silhouette of a
duck. Mathematical morphology allows very powerful shape-based filters to be created. The second row shows the results for a larger 7 × 7 structuring element which
has resulted in the complete elimination of the small square and the further reduction of the large square. The third row shows the results for a structuring element
which is a horizontal line segment 14 pixel wide, and the only remaining shapes are
long horizontal lines.

The operation we just performed is often known as erosion since large objects are
eroded and become smaller – in this case the 5 × 5 structuring element has caused
two pixels▶ to be *shaved off* all the way around the perimeter of each shape. The small
square, originally 5 × 5, is now only 1 × 1. If we repeated the operation the small square
would disappear entirely, and the large square would be reduced even further.

The inverse operation is dilation which makes objects larger. In Fig. 12.26 we apply dilation to the second column results

▶ The half width of the structuring element.

image   after erosion   after erosion then dilation

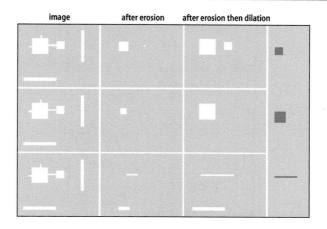

Fig. 12.26.
Mathematical morphology example. Pixels are either 0 (*grey*) or 1 (*white*). Each column corresponds to processing using the structuring element, shown at the end in *red*. The first column is the original image, the second column is after erosion by the structuring element, and the third column is after the second column is dilated by the structuring element

```
>> mx = imorph(mn, S, 'max');
```

and the results are shown in the third column. For each pixel in the input image we take the *maximum* of all pixels in the 5 × 5 window. If *any* of those neighbors is one the resulting pixel will be one. In this case we see that the two squares have returned to their original size, but the large square has lost its protrusions.

Morphological operations are often written in operator form. Erosion is

$$O = I \ominus S$$

where in Eq. 12.9 $f(\cdot) = \min(\cdot)$, and dilation is

$$O = I \oplus S$$

where in Eq. 12.9 $f(\cdot) = \max(\cdot)$. These operations are also known as Minkowski subtraction and addition respectively.

Erosion and dilation are related by

$$A \oplus B = \overline{\overline{A} \ominus B'}$$

where the bar denotes the logical complement of the pixel values, and the prime denotes reflection about the center pixel. Essentially this states that eroding the white pixels is the same as dilating the dark pixels and vice versa. For morphological operations

$$\left(A \oplus S_1\right) \oplus S_2 = A \oplus \left(S_1 \oplus S_2\right)$$
$$\left(A \ominus S_1\right) \ominus S_2 = A \ominus \left(S_1 \oplus S_2\right)$$

which means that successive erosion or dilation with a structuring element is equivalent to the application of a single larger structuring element, but the former is computationally cheaper.◄ The shorthand functions

For example a 3 × 3 square structuring element applied twice is equivalent to 5 × 5 square structuring element. The former involves $2 \times (3 \times 3 \times N^2) = 18N^2$ operations whereas the later involves $5 \times 5 \times N^2 = 25N^2$ operations.

```
>> out = ierode(im, S);
>> out = idilate(im, S);
```

can be used instead of the low-level function `imorph`.

The sequence of operations, erosion then dilation, is known as opening since it opens up gaps. In operator form it is written as

$$I \circ S = (I \ominus S) \oplus S$$

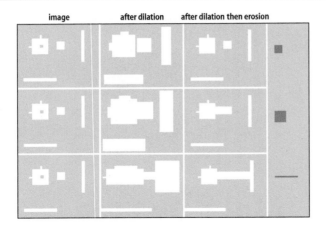

image   after dilation   after dilation then erosion

Fig. 12.27.
Mathematical morphology example. Pixels are either 0 (*grey*) or 1 (*white*). Each row corresponds to processing using the structuring element, shown at the end in *red*. The first column is the original image, the second column is after dilation by the structuring element, and the third column is after the second column is eroded by the structuring element

Not only has the opening selected particular shapes but it has also *cleaned up* the image: the squares have been separated and the protrusions on the large square have been removed since they are not consistent with the shape of the structuring element.

In Fig. 12.27 we perform the operations in the opposite order, dilation then erosion. In the first row no shapes have been lost, they grew then shrank, and the large square still has its protrusions. The hole has been filled since it is not consistent with the shape of the structuring element. In the second row, the larger structuring element has caused the two squares to join together. This sequence of operations is referred to as closing since it closes gaps and is written in operator form as

$$I \bullet S = (I \oplus S) \ominus S$$

Note that in the bottom row the two line segments have remained attached to the edge, this is due to the default behavior in handling edge pixels.

Opening and closing▸ are implemented by the Toolbox functions `iopen` and `iclose` respectively. Unlike erosion and dilation repeated application of opening or closing is futile since those operations are idempotent

$$(I \circ S) \circ S = I \circ S$$
$$(I \bullet S) \bullet S = I \bullet S$$

These names make sense when considering what happens to white objects against a black background. For black objects the operations perform the inverse function.

These operations can also be applied to greyscale images to emphasize particular shaped objects in the scene prior to an operation like thresholding. A circular structuring element of radius R can be considered as a ball of radius R rolling on the intensity surface. Dilation, or the maximum operation, is the surface defined by the center of the ball rolling over the top of the input image intensity surface. Erosion, or the minimum operation, is the surface defined by the center of the ball rolling on the underside of the input image intensity surface.

12.6.1   Noise Removal

A common use of morphological opening is to remove noise in an image. The image

```
>> objects = iread('segmentation.png');
```

shown in Fig. 12.28a is a noisy binary image from the output of a rather poor object segmentation operation.▸ We wish to remove the dark pixels that do not belong to the objects and we wish to fill in the holes in the four dark rectangular objects.

Image segmentation and binarization is discussed in Sect. 13.1.1.

Dealing with edge pixels. The problem of a convolution window near the edge of an input image was discussed on page 381. Similar problems exist for morphological spatial operations, and the Toolbox functions `imorph`, `irank` and `iwindow` support the option `'valid'` as does `iconvolve`. Other options cause the returned image to be the same size as the input image:

- `'replicate'` (default) the border pixel is replicated, that is, the value of the closest border pixel is used.
- `'none'` pixels beyond the border are not included in the set of pixels specified by the structuring element.
- `'wrap'` the image is assumed to wrap around, left to right, top to bottom.

We choose a symmetric *circular* structuring element of radius 3

```
>> S = kcircle(3)
S =
     0     0     0     1     0     0     0
     0     1     1     1     1     1     0
     0     1     1     1     1     1     0
     1     1     1     1     1     1     1
     0     1     1     1     1     1     0
     0     1     1     1     1     1     0
     0     0     0     1     0     0     0
```

and apply a closing operation to fill the holes in the objects

```
>> closed = iclose(objects, S);
```

and the result is shown in Fig. 12.28b. The holes have been filled, but the noise pixels have grown to be small circles and some have agglomerated. We eliminate these by an opening operation

```
>> clean = iopen(closed, S);
```

and the result shown in Fig. 12.28c is a considerably cleaned up image. If we apply the operations in the inverse order, opening then closing

```
>> opened = iopen(objects, S);
>> closed = iclose(opened, S);
```

the results shown in Fig. 12.28d are much poorer. Although the opening has removed the isolated noise pixels it has removed large chunks of the targets which cannot be restored.

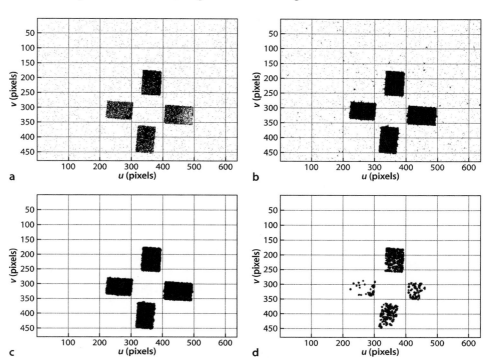

Fig. 12.28.
Morphological cleanup. **a** Original image, **b** original after opening, **c** opening then closing, **d** closing then opening. Structuring element is a *circle* of radius 3. Color map is inverted, set pixels are shown as *black*

12.6.2   Boundary Detection

The top-hat transform uses morphological operations to detect the edges of objects.
Continuing the example from above, and using the image `clean` shown in Fig. 12.28c
we compute its erosion using a circular structuring element

```
>> eroded = imorph(clean, kcircle(1), 'min');
```

The objects in this image are slightly smaller since the structuring element has caused
one pixel to be *shaved off* the outside of each object. Subtracting the eroded image
from the original

```
>> idisp(clean-eroded)
```

results in a layer of pixels around the edge of each object as shown in Fig. 12.29.

12.6.3   Hit or Miss Transform

The hit or miss transform uses a variation on the morphological structuring element.
Its values are zero, one or *don't care* as shown in Fig. 12.30a. The zero and one pixels
must exactly match the underlying image pixels in order for the result to be a one, as
shown in Fig. 12.30b. If there is any mismatch of a one or zero as shown in Fig. 12.30c
then the result will be zero. The Toolbox implementation is very similar to the morphological function, for example

```
out = hitormiss(image, S);
```

where the don't care elements of the structuring element are set to the special MATLAB
value `NaN`.

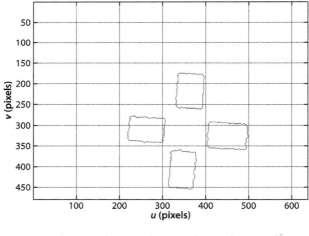

Fig. 12.29.
Boundary detection by morphological processing. Results are
shown inverted, *white* is zero

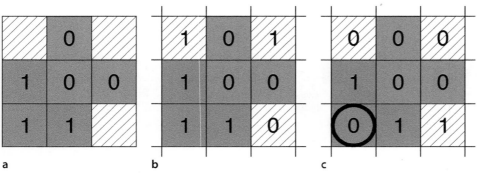

Fig. 12.30.
Hit or miss transform. **a** The
structuring element has values of
zero (*red*), one (*blue*), or don't
care (*hatched*); **b** an example of
a hit; **c** an example of a miss, the
pixel circled is inconsistent with
the structuring element

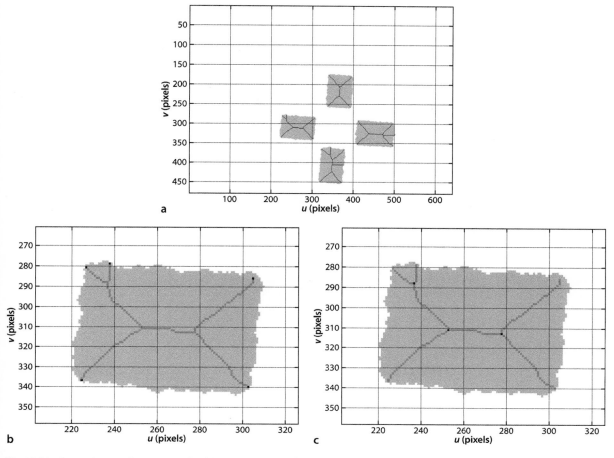

Fig. 12.31. Hit or miss transform operations. **a** Skeletonization; **b** end-point detection; **c** triple-point join detection. The images are shown inverted with the original binary image superimposed in *grey*. The end- and triplepoints are shown as *black pixels*

The hit or miss transform can be used iteratively with a sequence of structuring elements to perform complex operations such as skeletonization and linear feature detection. The skeleton of the objects is computed by

```
>> skeleton = ithin(clean);
```

and is shown in Fig. 12.31a. The lines are a single pixel wide and are the edges of a generalized Voronoi diagram – they delineate sets of pixels according to the shape boundary they are closest to. We can then find the endpoints of the skeleton

```
>> ends = iendpoint(skeleton);
```

and also the triplepoints

```
>> joins = itriplepoint(skeleton);
```

which are points at which three lines join. These are shown in Fig. 12.31b and c respectively.

For path planning in Sect. 5.2.1 we used a slow iterative wavefront approach to compute the distance transform. For this case a two pass algorithm can be used and if you have the MATLAB Image Processing Toolbox or VLFeat installed the faster functions `bwdist` or `vl_imdisttf` respectively will be used.

12.6.4 Distance Transform [examples/chamfer_match.m]

We discussed the distance transform in Sect. 5.2.1 for robot path planning. Given an occupancy grid it computed the distance of every free cell from the goal location. The distance transform we discuss here◄ operates on a binary image and the output value, corresponding to every zero pixel in the input image, is the Euclidean distance to the nearest nonzero pixel.

Consider the problem of fitting a model to a shape in an image. We create the outline of a rotated square

```
>> im = testpattern('squares', 256, 256, 128);
>> im = irotate(im, -0.3);
>> edges = icanny(im) > 0;
```

which is shown in Fig. 12.32a and then compute the distance transform

```
>> dx = distancexform(edges, 'euclidean');
```

which is shown in Fig. 12.32b.

An initial estimate of the square is shown with a red line. The value of the distance transform at any point on this red square indicates how far away it is from the nearest

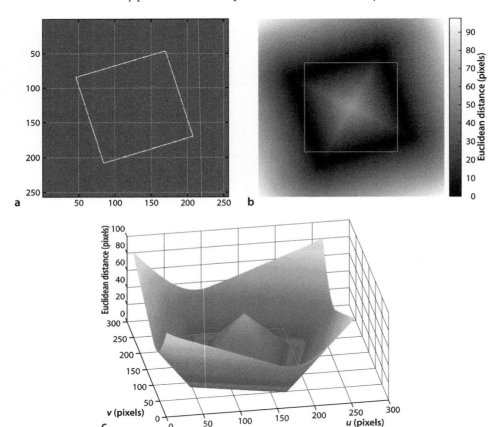

Fig. 12.32.
Distance transform. **a** Input binary image; **b** distance transformed input image with overlaid model square; **c** distance transform as a surface

The **distance transform** of a binary image has a value at each pixel equal to the distance from that pixel to the nearest nonzero pixel in the input image. The distance metric is typically either Euclidean (L_2 norm) or Manhattan distance (L_1 norm). It is zero for pixels that are nonzero in the input image. This transform is closely related to the **signed distance function** whose value at any point is the distance of that point to the nearest boundary of a shape, and is positive inside the shape and negative outside the shape. The figure shows the signed distance function for a unit circle,

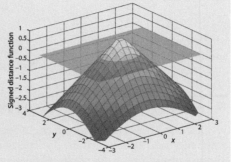

and has a value of zero, indicated by the red plane, at the object boundary. If we consider a shape to be defined by its signed distance transform then its zero contour defines the shape boundary.

point on the original square. If we summed the distance transform for every point on the red square, or even just the vertices, we obtain a total distance measure which will only be zero when our model square overlays the original square. The total distance is a cost function which we can minimize using an optimization routine that adjusts the position, orientation and size of the square. Considering the distance transform as a 3-dimensional surface in Fig. 12.32c, our problem is analogous to dropping an extensible square hoop into the valley of the distance transform. Note that the distance transform only needs to be computed once, and during model fitting the cost function is simply a lookup of the computed distance. This is an example of chamfer matching and a full example, with optimization, is given in `examples/chamfer_match.m`.

12.7 Shape Changing

The final class of image processing operations that we will discuss are those that change the shape or size of an image.

12.7.1 Cropping

The simplest shape change of all is selecting a rectangular region from an image which is the familiar *cropping* operation. Consider the image

```
>> mona = iread('monalisa.png');
```

shown in Fig. 12.33a from which we interactively specify a region of interest or ROI

```
>> [eyes,roi] = iroi(mona);
>> idisp(eyes)
```

by clicking and dragging a selection box over the image. In this case we selected the eyes, and the corners of the selected region can be optionally returned and in this case was

```
>> roi
roi =
    239    359
    237    294
```

where the columns are the (u, v) coordinates for the top-left and bottom-right corners respectively. The rows are the u- and v-span respectively. The function can be used noninteractively by specifying a ROI

```
>> smile = iroi(mona, [265 342; 264 286]);
```

which in this case selects the Mona Lisa's smile shown in Fig. 12.33b.

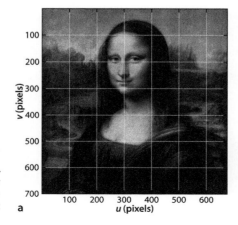

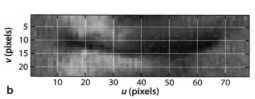

Fig. 12.33.
Example of region of interest or image cropping. **a** Original image, **b** selected region of interest

12.7.2 Image Resizing

Often we wish to reduce the dimensions of an image, perhaps because the large number of pixels results in long processing time or requires too much memory. We demonstrate this with a high-resolution image

```
>> roof = iread('roof.jpg', 'grey');
>> about(roof)
roof [uint8] : 1668x2009 (3351012 bytes)
```

which is shown in Fig. 12.34a. The simplest means to reduce image size is subsampling or decimation which selects every m^{th} pixel in the u- and v-direction, where $m \in \mathbb{Z}^+$ is the subsampling factor. For example with $m = 2$ an $N \times N$ image becomes an $N / 2 \times N / 2$ images which has one quarter the number of pixels of the original image.

For this example we will reduce the image size by a factor of seven in each direction

```
>> smaller = roof(1:7:end,1:7:end);
```

using standard MATLAB indexing syntax to select every seventh row and column. The result is shown is shown in Fig. 12.34b and we observe some pronounced curved lines on the roof which were not in the original image. These are artifacts of the sampling process. Subsampling reduces the spatial sampling rate of the image which can lead to spatial aliasing of high-frequency components due to texture or sharp edges. To ensure that the Shannon-Nyquist sampling theorem is satisfied an anti-aliasing low-

Fig. 12.34. Image scaling example. **a** Original image; **b** subsampled with $m = 7$, note the axis scaling; **c** subsampled with $m = 7$ after smoothing; **d** image **c** restored to original size by pixel replication

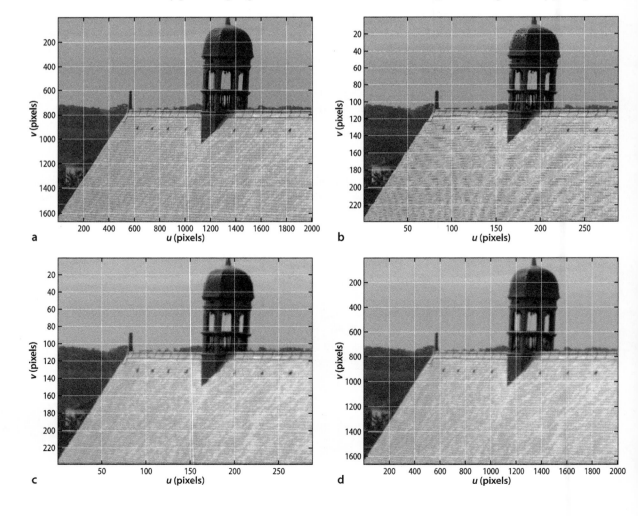

Any realizable low-pass filter has a finite response above its *cutoff* frequency. In practice the cutoff frequency is selected to be far enough below the theoretical cutoff that the filter's response at the Nyquist frequency is *sufficiently* small. As a rule of thumb for subsampling with $m = 2$ a Gaussian with $\sigma = 1$ is used.

pass spatial filter must be applied to reduce the spatial bandwidth of the image before it is subsampled.◄ This is another use for image blurring and the Gaussian kernel is a suitable low-pass filter for this purpose. The combined operation of smoothing and subsampling is implemented in the Toolbox by

```
>> smaller =  idecimate(roof, 7);
```

and the results for $m = 7$ are shown in Fig. 12.34c. We note that the curved line artifacts are no longer present.

The inverse operation is pixel replication, where each input pixel is replicated as an $m \times m$ tile in the output image

```
>> bigger = ireplicate( smaller, 7 );
```

which is shown in Fig. 12.34d and appears a little *blocky* along the edge of the roof and along the skyline. The decimation stage removed 98% of the pixels and restoring the image to its original size has not added any new information. However we could make the image easier on the eye by smoothing out the tile boundaries

```
>> smoother = ismooth( bigger, 4);
```

We can perform the same function using the Toolbox function iscale which scales an image by an arbitrary factor $m \in \mathbb{R}^+$ for example

```
>> smaller = iscale(lena, 0.1);
>> bigger = iscale(smaller, 10);
```

The second argument is the scale factor and if $m < 1$ the image will be reduced, and if $m > 1$ it will be expanded.

12.7.3   Image Pyramids

An important concept in computer vision, and one that we return to in the next chapter is scale space. The Toolbox function ipyramid returns a pyramidal decomposition of the input image

```
>> p = ipyramid( imono(mona) )
p =
  Columns 1 through 11
    [700x677 double] [350x339 double] ... [2x2 double]  [0.0302]
```

as a MATLAB cell array containing images at successively lower resolutions. Note that the last element is the 1×1 resolution version – a single dark grey pixel! These images are pasted into a composite image which is displayed in Fig. 12.35.

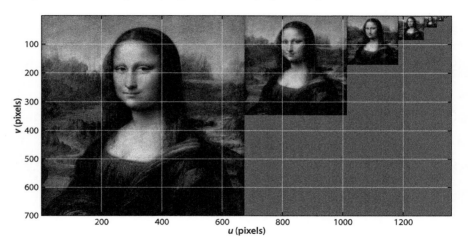

Fig. 12.35.
Image pyramid, a succession of images each half (by side length) the resolution of the one to the left

An image pyramid is the basis of many so-called coarse-to-fine strategies. Consider the problem of looking for a pattern of pixel values that represent some object of interest. The smallest image can be searched very quickly for the object since it comprises only a small number of pixels. The search is then refined using the next larger image but we now know which area of that larger image to search. The process is repeated until the object is located in the highest resolution image.

12.7.4   Image Warping

Image warping is a transformation of the pixel *coordinates* rather than the pixel values. Warping can be used to scale an image up or down in size, rotate an image or apply quite arbitrary shape changes. The coordinates of a pixel in the new view (u', v') are expressed as functions

$$u' = f_u(u, v), \quad v' = f_v(u, v) \tag{12.10}$$

of the coordinates in the original view.

Consider a simple example where the image is reduced in size by a factor of 4 in both directions and offset so that its origin, its top-left corner, is shifted to the coordinate (100, 200). We can express this concisely as

$$u' = u/4 + 100, \quad v' = v/4 + 200 \tag{12.11}$$

First we read the image and establish a pair of coordinate matrices▶ that span the domain of the input image, the set of all possible (u, v)

```
>> mona = iread('monalisa.png', 'double', 'grey');
>> [Ui,Vi] = imeshgrid(mona);
```

The coordinate matrices are such that $U(u, v) = u$ and $V(u, v) = v$ and are a common construct in MATLAB see the documentation for `meshgrid`.

and another pair that span the domain of the output image, which we choose arbitrarily to be 400 × 400, the set of all possible (u', v')

```
>> [Up,Vp] = imeshgrid(400, 400);
```

Now, for *every* pixel in the output image the corresponding coordinate in the input image is given by the inverse of the functions f_u and f_v. For our example the inverse of Eq. 12.11 is

$$u = 4(u' - 100), \quad v = 4(v' - 200) \tag{12.12}$$

which is implemented in matrix form in MATLAB as

```
>> U = 4*(Up-100); V = 4*(Vp-200);
```

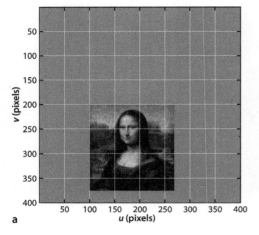

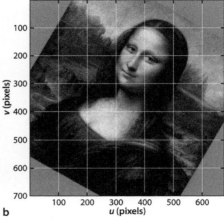

Fig. 12.36.
Warped images. **a** Scaled and shifted; **b** rotated by 30° about its center. Pixels displayed as *red* were set to a value of NaN by `interp2` – they were not interpolated from any image pixels

We can now warp the input image using the MATLAB function `interp2`

```
>> little_mona = interp2(Ui, Vi, mona, U, V);
```

and the result is shown in Fig. 12.36a. Note that `interp2` requires a floating-point image.

Some subtle things happen under the hood. Firstly, while (u', v') are integer coordinates the input image coordinates (u, v) will not necessarily be integers. The pixel values must be interpolated◄ from neighboring pixels in the input image. Secondly, not all pixels in the output image have corresponding pixels in the input image as illustrated in Fig. 12.37. Fortunately for us `interp2` handles all these issues and pixels that do not exist in the input image are set to `NaN` in the output image which we have displayed as red. In case of mappings that are extremely distorted it may be that many adjacent output pixels map to the same input pixel and this leads to pixelation or *blockyness* in the output image.

Now let's try something a bit more ambitious and rotate the image by 30° into an output image of the same size as the input image

```
>> [Up,Vp] = imeshgrid(mona);
```

We want to rotate the image about its center but since the origin of the input image is the top-left corner we must first change the origin to the center, then rotate and then move the origin back to the top-left corner.◄ The warp equation is therefore

$$\begin{pmatrix} u' \\ v' \end{pmatrix} = \underbrace{\begin{pmatrix} \cos\pi/6 & -\sin\pi/6 \\ \sin\pi/6 & \cos\pi/6 \end{pmatrix}}_{R(\pi/6)} \begin{pmatrix} u - u_c \\ v - v_c \end{pmatrix} + \begin{pmatrix} u_c \\ v_c \end{pmatrix} \tag{12.13}$$

where (u_c, v_c) is the coordinate of the image center and $R(\frac{\pi}{6})$ is a rotation matrix in $\mathbf{SE}(2)$. This can be rearranged into the *inverse form* and implemented as

```
>> R = SO2(pi/6).R; uc = 256; vc = 256;
>> U = R(1,1)*(Up-uc) + R(2,1)*(Vp-vc) + uc;
>> V = R(1,2)*(Up-uc) + R(2,2)*(Vp-vc) + vc;
>> twisted_mona = interp2(Ui, Vi, mona, U, V);
```

and the result is shown in Fig. 12.36b. Note the direction of rotation – our definition of the x- and y-axes (parallel to the u- and v-axes respectively) is such that the z-axis is defined as being into the page making a clockwise rotation a positive angle. Also note that the corners of the original image have been lost, they fall outside the bounds of the output image.

The function `iscale` uses image warping to change image scale, and the function `irotate` uses warping to perform rotation. The example above could be achieved by

```
>> twisted_mona = irotate(mona, pi/6);
```

Finally we will revisit the lens distortion example from Sect. 11.2.4. The distorted image from the camera is the input image and will be warped to remove the distortion. We are in luck since the distortion model Eq. 11.13 is already in the inverse form. Recall that

$$u' = u + \delta_u$$
$$v' = v + \delta_v$$

where the distorted coordinates are denoted with a prime and δ_u and δ_v are functions of (u, v).

Different interpolation modes can be selected by a trailing argument to `interp2` but the default option is bilinear interpolation. A pixel at coordinate $(u + \delta_u, v + \delta_v)$ where $u, v \in \mathbb{Z}^+$ and $\delta_u, \delta_v \in [0, 1)$ is a linear combination of the pixels (u, v), $(u + 1, v)$, $(u, v + 1)$ and $(u + 1, v + 1)$. The interpolation function acts as a weak anti-aliasing filter, but for very large reductions in scale the image should be smoothed first using a Gaussian kernel.

This is the application of a twist as discussed in Chap. 2.

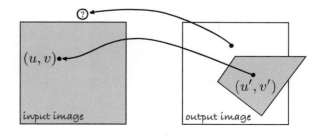

Fig. 12.37.
Coordinate notation for image warping. The pixel (u', v') in the output image is sourced from the pixel at (u, v) in the input image as indicated by the *arrow*. The warped image is not necessarily polygonal, nor entirely contained within the output image

First we load the distorted image and build the coordinate matrices for the distorted and undistorted images

```
>> distorted = iread('Image18.tif', 'double');
>> [Ui,Vi] = imeshgrid(distorted);
>> Up = Ui; Vp = Vi;
```

and then load the results of the camera calibration

```
>> load  Bouguet
```

For readability we unpack the required parameters from the Calibration Toolbox variables cc, fc and kc

```
>> k = kc([1 2 5]); p = kc([3 4]);
>> u0 = cc(1); v0 = cc(2);
>> fpix_u = fc(1); fpix_v = fc(2);
```

for radial and tangential distortion vectors, principal point and focal length in pixels. Next we convert pixel coordinates to normalized image coordinates▸

In units of meters with respect to the camera's principal point.

```
>> u = (Up-u0) / fpix_u;
>> v = (Vp-v0) / fpix_v;
```

The radial distance of the pixels from the principal point is then

```
>> r = sqrt( u.^2 + v.^2 );
```

and the pixel coordinate errors due to distortion are

```
>> delta_u = u .* (k(1)*r.^2 + k(2)*r.^4 + k(3)*r.^6) + ...
   2*p(1)*u.*v + p(2)*(r.^2 + 2*u.^2);
>> delta_v = v .* (k(1)*r.^2 + k(2)*r.^4 + k(3)*r.^6) + ...
   p(1)*(r.^2 + 2*v.^2) + 2*p(2)*u.*v;
```

The distorted pixel coordinates in metric units are

```
>> ud = u + delta_u;   vd = v + delta_v;
```

which we convert back to pixel coordinates

```
>> U = ud * fpix_u + u0;
>> V = vd * fpix_v + v0;
```

and finally apply the warp

```
>> undistorted = interp2(Ui, Vi, distorted, U, V);
```

The results are shown in Fig. 12.38. The change is quite subtle, but is most pronounced at the edges and corners of the image where r is the greatest.

Fig. 12.38. Warping to undistort an image. **a** Original distorted image; **b** corrected image. Note that the top edge of the target has become a straight line (example from Bouguet's Camera Calibration Toolbox, image number 18)

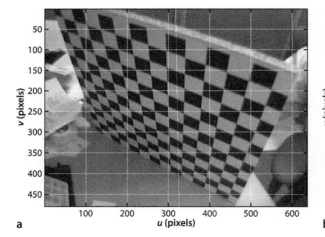

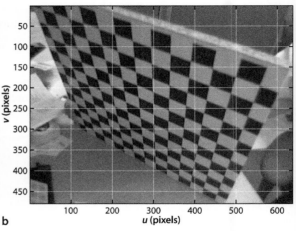

a   u (pixels)

b   u (pixels)

12.8   Wrapping Up

In this chapter we learned how to acquire images from a variety of sources such as image files, movie files, video cameras and the internet, and load them into the MATLAB workspace. Once there we can treat them as matrices, the principal MATLAB datatype, and conveniently manipulate them. The elements of the image matrices can be integer, floating-point or logical values. Next we discussed many processing operations and a taxonomy of these is shown in Table 12.2. Operations on a single image include: unary arithmetic operations, type conversion, various color transformations and grey-level stretching; nonlinear operations such as histogram normalization and gamma encoding or decoding; and logical operations such as thresholding. We also discussed operations on pairs of images such as green screening, background estimation and moving object detection.

The largest and most diverse class of operations are spatial operators. We discussed convolution which can be used to smooth an image and to detect edges. Linear operations are defined by a kernel matrix which can be chosen to perform functions such as image smoothing (to reduce the effect of image noise or as a low-pass anti-aliasing filter prior to decimation) or for edge detection. Nonlinear spatial operations were used for template matching, computing rank statistics (including the median filter which eliminates impulse noise) and mathematical morphology which filters an image based on shape and can be used to cleanup binary images. A variant form, the hit or miss transform, can be used iteratively to perform functions such as skeletonization.

Finally we discussed shape changing operations such as regions of interest, scale changing and the problems that can arise due to aliasing, and generalized image warping which can be used for scaling, translation, rotation or undistorting an image. All these image processing techniques are the foundations of feature extraction algorithms that we discuss in the next chapter.

Further Reading

Image processing is a large field and this chapter has provided an introduction to many of the most useful techniques from a robotics perspective. More comprehensive coverage of the topics introduced here and others such as greyscale morphology, image restoration, wavelet and frequency domain methods, and image compression can be found in Szeliski (2011), Nixon and Aguado (2012), Forsyth and Ponce (2011) and Gonzalez and Woods (2011). Online information about computer vision is available through CVonline at http://homepages.inf.ed.ac.uk/rbf/CVonline, and the material in this chapter is covered under the section *Image Transformations and Filters*.

Edge detection is a subset of image processing but one with huge literature of its own. Forsyth and Ponce (2011) have a comprehensive introduction to edge detection and a useful discussion on the limitations of edge detection. Nixon and Aguado (2012) also cover phase congruency approaches to edge detection and compare various edge detectors. The Sobel kernel for edge detection was described in an unpublished 1968 publication from the Stanford AI lab by Irwin Sobel and Jerome Feldman: *A 3 × 3 Isotropic Gradient Operator for Image Processing*. The Canny edge detector was originally described in Canny (1983, 1987).

Nonparametric measures for image similarity became popular in the 1990s with with a number of key papers such as Zabih and Woodfill (1994), Banks and Corke (2001), Bhat and Nayar (2002). The application to real-time image processing systems using high-speed logic such as FPGAs has been explored by several groups (Corke et al. 1999; Woodfill and Von Herzen 1997).

Operation	RVC Toolbox	MATLAB
Monadic		
▪ Type conversion	`iint, idouble`	
▪ Unary functions		**`abs, sqrt, exp`**
▪ Unary operators		`-, ~, .^2,` `imcomplement`
▪ Image and constant		`+, -, *, /, >, >=, <, <=,` `imadd, imsubtract,` `immultiply, imdivide`
▪ Greyscale mapping	`igamm, istretch, inormhist`	`imadjust, stretchlim, histeq`
▪ Color	`imono, icolor, colorspace,` `tristim2cc`	`imapplymatrix,` **`rgb2hsv`**, **`hsv2rgb`**, `rgb2lab,` `lab2rgb, rgb2xyz, xyz2rgb, lab2xyz, xyz2lab`
Diadic		
▪ Binary operators		`+, -, .*, ./, &, |,` `imadd, imsubtract,` `immultiply, imdivide, imlincomb`
▪ Binary functions	`ipixswitch`	**`atan2`**, **`min`**, **`max`** `etc`
Spatial		
▪ Linear	`iconv, ismooth`	**`conv2`**, `imgaussfilt`
▪ Edge	`isobel, icanny`	`edge`
▪ Kernels	`kgauss, ksobel, kdgauss, klaplace,` `klog, kdog, kcircle`	`fspecial`
▪ Nonlinear	`irank, ivar, iwindow`	`ordfilt2, medfilt2, stdfilt`
▪ TemplateMatching	`isimilarity, sad, ssd, ncc, zsad,` `zssd, zncc`	`normxcorr2`
▪ Morphological	`imorph, ierode, idilate, iopen,` `iclose, hitormiss, ithin, iendpoint,` `itriplepoint`	`bwmorph, imerode, imdilate, imopen, imclose,` `bwhitmiss`
▪ Distance transform	`distancexform`[a]	`bwdist`
Shape changing		
▪ Scale	`idecimate, ireplicate, iscale`	`imresize`
▪ Rotate	`iscale`	`imrotate`
▪ Cropping	`iroi, isamesize, itrim, ipad`	`imcrop`
▪ Warping	`homtrans`	**`interp2,`** `imwarp, imtransform`
Creating images		
▪ from image file	`iread`	**`imread`**
▪ from camera	`VideoCamera`[a]	`videoinput`
▪ from web camera	`AxisWebCamera`	
▪ from movie file	`Movie`	**`VideoReader`**
▪ from Google maps	`EarthView`	
▪ from code	`testpattern, ipaste, kcircle, iline`	
General		
Display	`idisp`	**`imshow`**, `ImageViewer app`
Analysis	`ihist, iline`	
Writing image		**`imwrite`**

Notes: Bold means available in base MATLAB, others require additional MATLAB toolboxes. RVC Toolbox functions sometimes wrap base MATLAB functions.
[a] Function has extra functionality if MATLAB toolboxes are installed.

◄ **Table 12.2.**
Summary of image processing
algorithms discussed in this
chapter

Mathematical morphology is another very large topic and we have only scraped the surface and important techniques such as greyscale morphology and watersheds have not been covered at all. The general image processing books mentioned above have useful discussion on this topic. Most of the specialist books in this field are now quite old but Shih (2009) is a good introduction and the book by Dougherty and Latufo (2003) has a more hands on tutorial approach.

The approach to computer vision covered in this book is often referred to as bottom-up processing. This chapter has been about *low-level* vision techniques which are operations on pixels. The next chapter is about *high-level* vision techniques where sets of pixels are grouped and then described so as to represent objects in the scene.

Sources of Image Data

All the images used in this part of the book are provided with the Toolbox in the images folder of the Machine Vision Toolbox.

There are thousands of online webcams as well as a number of sites that aggregate them and provide lists categorized by location, for example Opentopia, EarthCam and WatchThisCam. Most of these sites do not connect you directly to the web camera so the URL of the camera has to be dug out of the HTML page source. The root part of the URL (before the first single slash) is required for the AxisWebCamera class. Some of the content on these list pages can be rather dubious – so beware.

MATLAB Notes

Table 12.2 shows the image processing functions that have been discussed in this chapter and the equivalent functions from several toolboxes available from MathWorks: Image Processing Toolbox™, Image Acquisition Toolbox™ and Computer Vision System Toolbox™. There are many additional functions from these toolboxes that are not listed here. The RVC toolbox is open source and free, but its development is limited and the code is written for understanding rather than performance. In contrast the MathWorks' toolboxes are supported products and many have GPU support, can be used in Simulink or be used for automatic code generation. The companion to Gonzalez and Woods (2008) is their MATLAB based book from 2009 (Gonzalez et al. 2009) which provides a detailed coverage of image processing using MATLAB and includes functions that extend the IPT. These are provided for free as P-code format (no source or help available) or as M-files for purchase but are now quite dated.

The image processing search term at MATLAB CENTRAL http://www.mathworks.com/matlabcentral/fileexchange lists thousands of files.

General Software Tools

There are many high quality software tools for image and video manipulation outside the MATLAB environment. OpenCV at http://opencv.org is a mature open-source computer vision software project with over 2 500 algorithms, interfaces for C++, C, Python and Java and runs on Windows, Linux, Mac OS, iOS and Android. There are now several books about OpenCV and Kaehler and Bradski (2016) is the second edition of a popular book that provides a good introduction to the software and to computer vision in general.

ImageMagick http://www.imagemagick.org is a cross-platform collection of libraries and command-line tools for image format conversion (over 100 formats) and is useful for batch operations on large sets of images. For video manipulation FFmpeg http://www.ffmpeg.org is an excellent and comprehensive cross-platform tool. It supports conversion between video formats as well as videos to still images and vice versa.

Exercises

1. Become familiar with `idisp` for greyscale and color images. Explore pixel values in the image as well as the zoom, line and histogram buttons. Use `iroi` to extract the Mona Lisa's smile.

2. Look at the histogram of greyscale images that are under, well and over exposed. For a color image look at the histograms of the RGB color channels for scenes with different dominant colors. Combine real-time image capture with computation and display of the histogram.

3. Create two copies of a greyscale image into workspace variables A and B. Write code to time how long it takes to compute the difference of A and B using the MATLAB shorthand A−B or using two nested `for` loops. Use the functions `tic` and `toc` to perform the timing.

4. Grab some frames from the camera on your computer or from a movie file and display them.

5. Write a loop that grabs a frame from your camera and displays it. Add some effects to the image before display such as "negative image", thresholding, posterization, false color, edge filtering etc.

6. Given a scene with luminance of 800 nit and a camera with ISO of 1 000, $q = 0.7$ and f-number of 2.2 what exposure time is needed so that the average grey level of the 8-bit image is 150?

7. Images from space, page 367
   a) Obtain a map of the roads in your neighborhood. Use this to find a path between two locations, using the robot motion planners discussed in Chap. 5.
   b) For the images returned by the `EarthView` function write a function to convert pixel coordinate to latitude and longitude.
   c) Upload GPS track data from your phone and overlay it on a satellite image.

8. Motion detection
   a) Modify the Traffic example on page 375 and highlight the moving vehicles.
   b) Write a loop that performs background estimation using frames from your camera. What happens as you move objects in the scene, or let them sit there for a while? Explore the effect of changing the parameter σ.
   c) Combine concepts from motion detection and chroma-keying to put pixels from the camera where there is motion into the desert scene.

9. Convolution
   a) Compare the results of smoothing using a 21×21 uniform kernel and a Gaussian kernel. Can you observe the ringing artifact in the former?
   b) Why do we choose a smoothing kernel that sums to one?
   c) Compare the performance of the simple horizontal gradient kernel $K = (-0.5\ 0\ 0.5)$ with the Sobel kernel.
   d) Investigate filtering with the Gaussian kernel for different values of σ and kernel size.
   e) Create a 31×31 kernel to detect lines at 60 deg.
   f) Derive analytically the derivative of the Gaussian in the x-direction Eq. 12.4.
   g) Derive analytically the Laplacian of Gaussian Eq. 12.8.
   h) Derive analytically the difference of Gaussian from page 385.
   i) Show the difference between difference of Gaussian and derivative of Gaussian.

10. Show analytically the effect of an intensity scale error on the SSD and NCC similarity measures.

11. Template matching using the Mona Lisa image; convert it first to greyscale.
    a) Use `iroi` to select one of Mona Lisa's eyes as a template. The template should have odd dimensions.
    b) Use `isimilarity` to compute the similarity image. What is the best match and where does it occur? What is the similarity to the other eye? Where does the second best match occur and what is its similarity score?

c) Scale the intensity of the Mona Lisa image and investigate the effect on the peak similarity.

d) Add an offset to the intensity of the Mona Lisa image and investigate the effect on the peak similarity.

e) Repeat steps *(c)* and *(d)* for different similarity measures such as SAD, SSD, rank and census.

f) Scale the template size by different factors (use `iscale`) in the range 0.5 to 2.0 in steps of 0.05 and investigate the effect on the peak similarity. Plot peak similarity vs scale.

g) Repeat *(f)* for rotation of the template in the range −0.2 to 0.2 rad in steps of 0.05.

12. Perform the sub-sampling example on page 402 and examine aliasing artifacts around sharp edges and the regular texture of the roof tiles. What is the appropriate smoothing kernel width for a decimation by M?

13. Write a function to create Fig. 12.35 from the output of `ipyramid`.

14. Create a warp function that mimics your favorite funhouse mirror.

15. Warp the image to polar coordinates (r, θ) with respect to the center of the image, where the horizontal axis is r and the vertical axis is θ.

13 Image Feature Extraction

In the last chapter we discussed the acquisition and processing of images. We learned that images are simply large arrays of pixel values but for robotic applications images have too much *data* and not enough *information*. We need to be able to answer pithy questions such as what is the pose of the object? what type of object is it? how fast is it moving? how fast am I moving? and so on. The answers to such questions are *measurements* obtained from the image and which we call image features. Features are the *gist* of the scene and the raw material that we need for robot control.

The image processing operations from the last chapter operated on one or more input images and returned another image. In contrast *feature extraction* operates on an image and returns one or more *image features*. Features are typically scalars (for example area or aspect ratio) or short vectors (for example the coordinate of an object or the parameters of a line). Image feature extraction is a necessary first step in using image data to control a robot. It is an *information concentration* step that reduces the data rate from 10^6-10^8 bytes s^{-1} at the output of a camera to something of the order of tens of features per frame that can be used as input to a robot's control system.

In this chapter we discuss features and how to extract them from images. Drawing on image processing techniques from the last chapter we will discuss several classes of feature: regions, lines and interest points. Section 13.1 discusses region features which are contiguous groups of pixels that are homogeneous with respect to some pixel property. For example the set of pixels that represent a red object against a nonred background. Section 13.2 discusses line features which describe straight lines in the world. Straight lines are distinct and very common in man-made environments – for example the edges of doorways, buildings or roads. The final class of features are interest points which are discussed in Sect. 13.3. These are particularly distinctive *points* in a scene which can be reliably detected in different views of the same scene.

It is important to always keep in mind that image features are a summary of the information present in the pixels that comprise the image, and that the mapping from the world to pixels involves significant information loss – the perspective projection discussed in Chap. 11. We typically counter this information loss by making assumptions based on our knowledge of the environment, but our system will only ever be as good as the validity of our assumptions. For example, we might use image features to describe the position and shape of a group of red pixels that correspond to a red object. However the size feature, typically the number of pixels, does not say anything about the size of the red object in the world – we need extra information such as the distance between the camera and the object, and the camera's intrinsic parameters. We also need to assume that the object is not partially occluded – that would make the observed size less than the true size. Further we need to assume that the illumination is such that the chromaticity of the light reflected from the object is considered to be red. We might also find features in an image that do not correspond to a physical object – decorative markings, the strong edges of a shadow, or reflections in a window.

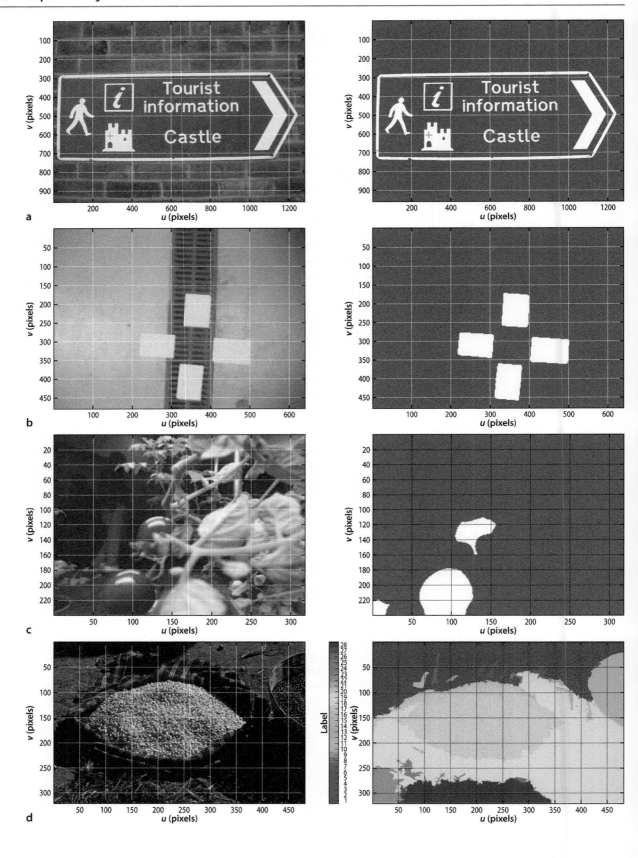

13.1   Region Features

Image segmentation is the process of partitioning an image into *application meaningful* regions as illustrated in Fig. 13.1. The aim is to segment or separate those pixels that represent objects of interest from all other pixels in the scene. This is one of the oldest approaches to scene understanding and while conceptually straightforward it is very challenging in practice. A key requirement is *robustness* which is how gracefully the method degrades as the underlying assumptions are violated, for example changing scene illumination or viewpoint.

Image segmentation is considered as three subproblems. The first is *classification* which is a decision process applied to each pixel that assigns the pixel to one of C classes $c \in \{0 \cdots C - 1\}$. Commonly we use $C = 2$ which is known as binary classification or binarization and some examples are shown in Fig. 13.1a–c. The pixels have been classified as object ($c = 1$) or not-object ($c = 0$) which are displayed as white or black pixels respectively. The classification is *always application specific* – for example the object corresponds to pixels that are bright or yellow or red or moving. Figure 13.1d is a multi-level classification where $C = 28$ and the pixel's class is reflected in its displayed color.

The underlying *assumption* in the examples of Fig. 13.1 is that regions are *homogeneous* with respect to some characteristic such as brightness, color or texture. In practice we accept that this stage is imperfect and that pixels may be misclassified – subsequent processing steps will have to deal with this.

The second step in the segmentation process is *representation* where adjacent pixels of the same class are *connected* to form spatial sets $S_1 \dots S_m$. The sets can be represented by assigning a set label to each pixel or by a list of pixel coordinates that defines the boundary of the connected set. In the third and final step, the sets S_i are *described* in terms of compact scalar or vector-valued *features* such as size, position, and shape.

13.1.1   Classification

The pixel class is represented by an integer $c \in \{0 \cdots C - 1\}$ where C is the number of classes. In this section we discuss the problem of assigning each pixel to a class. In many of the examples we will use binary classification with just two classes corresponding to not-object and object, or background and foreground.

13.1.1.1   Grey-Level Classification

A common approach to binary classification of pixels is the monadic operator

$$c[u, v] = \begin{cases} 0, & \text{if } I[u, v] < t \\ 1, & \text{if } I[u, v] \geq t \end{cases} \quad \forall (u, v) \in I$$

where the decision is based simply on the value of the pixel I. This approach is called thresholding and t is referred to as the *threshold*.

Thresholding is very simple to implement. Consider the image

```
>> castle = iread('castle.png', 'double');
```

which is shown in Fig. 13.2a. The thresholded image

```
>> idisp(castle >= 0.7)
```

is shown in Fig. 13.2c. The pixels have been quite accurately classified as corresponding to white paint or not. This classification is based on the seemingly reasonable *assumption* that the white paint objects are brighter than everything else in the image.

◀ **Fig. 13.1.**
Examples of pixel classification. The left-hand column is the input image and the right-hand column is the classification. The classification is application specific and the pixels have been classified as either object (*white*) or not-object (*black*). The objects of interest are **a** the individual letters on the sign; **b** the yellow targets; **c** the red tomatoes. **d** is a multi-level segmentation where pixels have been assigned to 28 classes that represent locally homogeneous groups of pixels in the scene

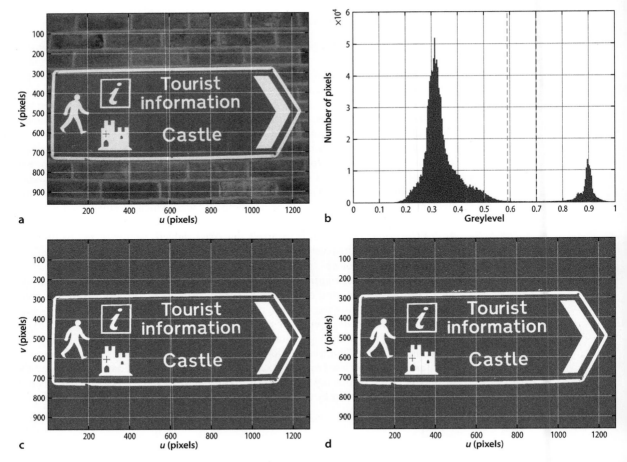

Fig. 13.2. Binary classification. **a** Original image (image sourced from the ICDAR collection; Lucas 2005); **b** histogram of greyscale pixel values, threshold values indicated, Otsu in *red*; **c** binary classification with threshold of 0.7; **d** binary classification with Otsu threshold

In the early days of computer vision, when computer power was limited, this approach was widely used – it was easier to contrive a world of white objects and dark backgrounds than to implement more sophisticated classification. Many modern industrial vision inspection systems use this simple approach since it allows the use of modest embedded computers – it works very well if the objects are on a conveyor belt of a suitable contrasting color or in silhouette at an inspection station. In a real world robot environment we generally have to work a little harder in order to achieve useful grey-level classification. An important question, and a hard one, is where did the threshold value of 0.7 come from? The most common approach is trial and error! The Toolbox function `ithresh`

```
>> ithresh(castle)
```

displays the image and a threshold slider that can be adjusted until a satisfactory result is obtained. However on a day with different lighting condition the intensity profile of the image would change

```
>> ithresh(castle*0.8)
```

and a different threshold would be required.

A more principled approach than trial and error is to analyze the histogram of the image

```
>> ihist(castle);
```

which is shown in Fig. 13.2b. The histogram has two clearly defined peaks, a bimodal distribution, which correspond to two *populations* of pixels. The smaller peak around 0.9 corresponds to the pixels that are bright and it has quite a small range of variation in

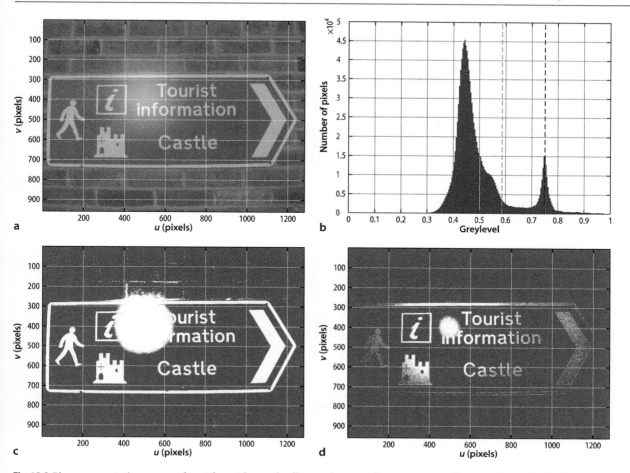

Fig. 13.3. Binary segmentation example. **a** Grey-scale image with intensity highlight, **b** histogram, **c** thresholded with Otsu's threshold at 0.59, **d** thresholded at 0.75

value. The wider and taller peak around 0.3 corresponds to pixels in the darker background of the sign and the bricks, and has a much larger variation in brightness.

To separate the two classes of pixels we choose the decision boundary, the threshold, to lie in the valley between the peaks. In this regard the choice of $t = 0.7$ is a good one. Since the valley is very wide we actually have quite a range of choice for the threshold, for example $t = 0.75$ would also work well. The optimal threshold can be computed using Otsu's method

```
>> t = otsu(castle)
t =
    0.5898
```

which separates an image into two classes of pixels in a way that minimizes the variance of values within each class and maximizes the variance of values between the classes – assuming that the histogram has just two peaks. Sadly, as we shall see, the real world is rarely this facilitating.

Consider a different image of the same scene which has a highlight

```
>> castle = iread('castle2.png', 'double');
```

and is shown in Fig. 13.3a. The histogram shown in Fig. 13.3b is similar – it is still bimodal – but we see that the peaks are wider and the valley is less deep. The pixel greylevel populations are now overlapping and unfortunately for us no single threshold can separate them. Otsu's method computes a threshold of

```
>> t = otsu(castle)
t =
    0.5859
```

and the result of applying this threshold is shown in Fig. 13.3c. The pixel classification is poor and the highlight overlaps several of the characters. The result of using a higher threshold of 0.75 is shown in Fig. 13.3d – the highlight is reduced, but not completely, but some other characters are starting to break up.

> Thresholding-based techniques are notoriously brittle – a slight change in illumination of the scene means that the thresholds we chose would no longer be appropriate. In most real scenes there is no simple mapping from pixel values to particular objects – we cannot for example choose a threshold that would select a motorbike or a duck. Distinguishing an object from the background remains a hard computer vision problem.

One alternative is to choose a local rather than a global threshold. The Niblack algorithm is widely used in optical character recognition systems and computes a local threshold

$$t[u, v] = \mu(\mathcal{W}) - k\sigma(\mathcal{W})$$

where $\mathcal{W}$ is a region about the point (u, v) and $\mu(\cdot)$ and $\sigma(\cdot)$ are the mean and standard deviation respectively. The size of the window $\mathcal{W}$ is a critical parameter and should be of a similar size to the objects we are looking for. For this example we make an assumption about the scene, that the characters are approximately 50–70 pixels tall, to choose a window half-width of 30 pixels

```
>> t = niblack(castle, -0.1, 30);
>> idisp(t)
```

where $k = -0.1$. The resulting local threshold t is shown in Fig. 13.4a. We apply the threshold pixel-wise to the original image

```
>> idisp(castle >= t)
```

resulting in the classification shown in Fig. 13.4b. All the pixels belonging to the letters have been correctly classified but compared to Fig. 13.3c there are many false positives – nonobject pixels classified as objects. Later in this section we will discuss techniques to eliminate these false positives. Note that the classification process is no longer a function of just the input pixel, it is now a complex function of the pixel and its neighbors. While we no longer need to choose t we now need to choose the parameters k and window size, and again this is usually a trial and error process that can be made to work well for a particular type of scene.

Fig. 13.4. Niblack thresholding. **a** The local threshold displayed as an image; **b** the binary segmentation

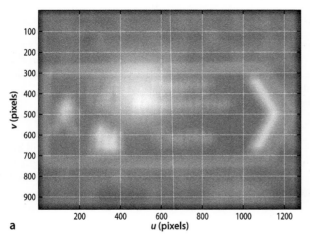

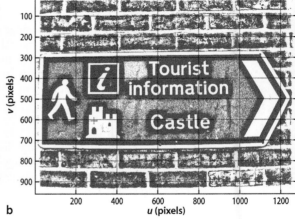

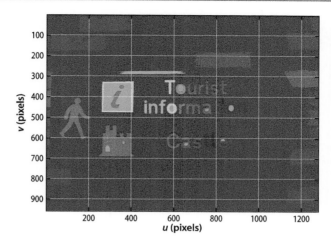

Fig. 13.5.
Segmentation using maximally
stable extremal regions (MSER).
The identified regions are
uniquely color coded

The results shown in Fig. 13.3c and d are disappointing at first glance, but we see that every character object is correctly classified at some, but not all, thresholds. In fact each object is correctly segmented for some *range* of thresholds and what we would like is the union of regions classified over the range of all thresholds. The maximally stable extremal region or MSER algorithm does exactly this. It is implemented by the Toolbox function `imser`

```
>> [mser,nsets] = imser(castle, 'area', [100 20000]);
```

and for this image

```
>> nsets
nsets =
    95
```

Although no explicit threshold has been
given `imser` has a number of parameters and in this case their default values
have given satisfactory results. See the
online documentation for details of the
parameters.

stable sets were found.◄ The other return value is an image

```
>> idisp(mser, 'colormap', 'jet')
```

which is shown in Fig. 13.5 as a false color image. Each nonzero pixel corresponds to a stable set and the value is the *label* assigned to that stable set which is displayed as a unique color. All the character objects were correctly classified. The boundary has been partly misclassified as background, and part of it has been joined to the brick texture on the right hand side of the image.

13.1.1.2 Color Classification

Color is a powerful cue for segmentation but roboticists tend to shy away from using it because of the problems with color constancy discussed in Sect. 10.3.2. In this section we consider two examples that use color images. The first is a rather primitive navigation target for an indoor UAV landing experiment

```
>> im_targets = iread('yellowtargets.png');
```

shown in Fig. 13.6a and the second from the MIT Robotic Garden project

```
>> im_garden = iread('tomato_124.jpg');
```

is shown in Fig. 13.7a. Our objective is to determine the centroids of the yellow targets and the red tomatoes respectively. The initial stages of processing are the same for each image but we will illustrate the process in detail for the image of the yellow targets as shown in Fig. 13.6.

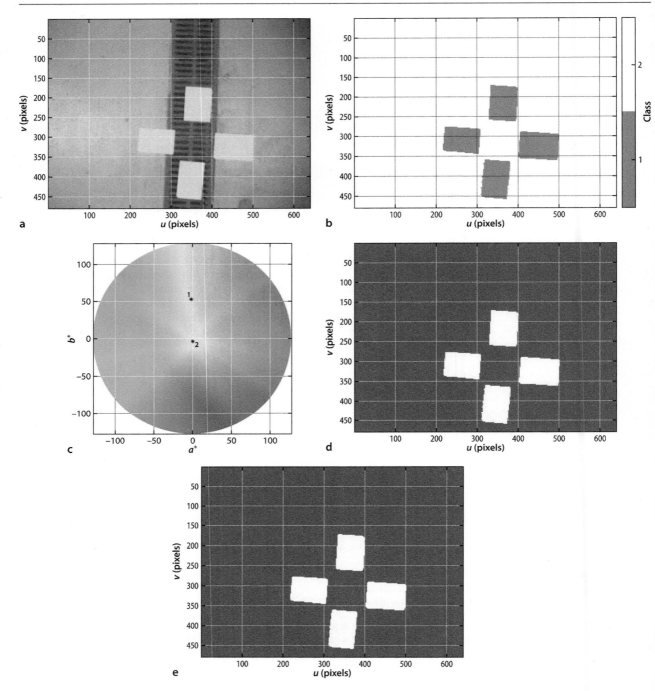

a

b

c

d

e

The Toolbox function `colorkmeans` first maps each color pixel to a point on the *xy*- or *a**b*-chromaticity plane. Then the *k*-means algorithm is used to find clusters of points on the plane and each cluster corresponds to a group of pixels with a distinguishable color. A limitation of the *k*-means algorithm is that we must specify the number of clusters to find. We will use our knowledge that this particular scene has essentially two differently colored elements: yellow targets and grey floor, metal drain cover and shadows. The pixels are clustered into *two* chromaticity classes ($C = 2$) by

Fig. 13.6. Target image example. **a** Original image; **b** pixel classification ($C = 2$) shown in false color; **c** cluster centers in the *a**b*-chromaticity space; **d** all pixels of class $c = 1$; **e** after morphological opening with a circular structuring element (radius 2)

```
>> [cls, cab,resid] = colorkmeans(im_targets, 2, 'ab');
```

We have specified a^*b^*-chromaticity since Euclidean distance in this space, used by k-means to determine the clusters, matches the human perception of difference between colors. The function returns the a^*b^*-chromaticity of the cluster centers

```
>> cab
cab =
   -0.8190     0.4783
   57.6140    -4.1910
```

as one column per cluster. We can plot these cluster centers on the a^*b^*-plane

```
>> showcolorspace(cab, 'ab');
```

which is shown in Fig. 13.6c. We see that cluster 1 is the closest to yellow

```
>> colorname(cab(:,1), 'ab')
ans =
    'gold4'
```

The residual

```
>> resid
resid =
   2.8897e+03
```

is the sum of the distance of every point from its assigned cluster centroid. Since the algorithm uses a random initialization we will obtain different clusters and classification on every run, and therefore different residuals.◂

One option is to run k-means a number of times, and take the cluster centers for which the residual is lowest.

The function colorkmeans also returns the pixel classification which we can display as an image

```
>> idisp(cls, 'colormap', flag(2), 'bar')
```

in false color◂ as shown in Fig. 13.6b. The pixels in this image have values $c = \{1, 2\}$ indicating which class the corresponding input pixels has been assigned to. We see that the yellow targets have been assigned to class $c = 1$ which is displayed as red.

We have specified a color map of length 2 since we know there are only 2 possible pixel values.

k-means clustering is computationally expensive and therefore not very well suited to real-time applications. However we can divide the process into a training phase and a classification phase. In the training phase a number of example images would be concatenated and passed to colorkmeans which would identify the centers of the clusters for each class. Subsequently we can assign pixels to their closest cluster relatively cheaply

```
>> cls = colorkmeans(im_targets, cab, 'ab');
```

The pixels belonging to class 1 can be selected

```
>> cls1 = (cls == 1);
```

which is a *logical image* that can be displayed

```
>> idisp(cls1)
```

as shown in Fig. 13.6d. All pixels of class 1 are displayed as white and correspond to the yellow targets in the original image. This binary image is a good classification but there are a few minor imperfections: some rough edges, and some tiny holes.

A morphological opening operation as discussed in Sect. 12.6 will eliminate these. We apply a symmetric structuring element of radius 2

```
>> targets_binary = iopen(cls1, kcircle(2));
```

and the result is shown in Fig. 13.6e. It shows a clean binary segmentation of the pixels into the two classes: target and not-target.

For the garden image we follow a very similar procedure. We classify the pixels into three clusters ($C = 3$) based on our knowledge that the scene contains: red tomatoes, green leaves, and dark background

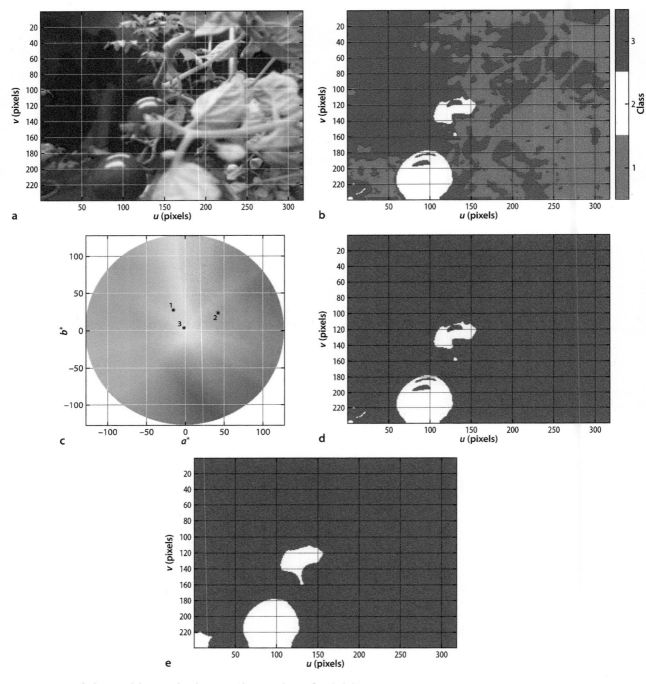

```
>> [cls, cab] = colorkmeans(im_garden, 3, 'ab');
>> cab
cab =
   -16.3326    44.0622    -1.5073
    28.8824    26.7948     3.3873
```

Fig. 13.7. Garden image example. a Original image (courtesy of Distributed Robot Garden project, MIT); b pixel classification ($C = 3$) shown in false color; c cluster centers in the a^*b^*-chromaticity space d all pixels of class $c = 2$; e after morphological closing with a circular structuring element (radius 15)

The pixel classes are shown in false color in Fig. 13.7b. Pixels corresponding to the tomato have been assigned to class $c = 2$ which are displayed as white. The cluster centers are marked on the a^*b^*-chromaticity plane in Fig. 13.7c. The name of the color closest to cluster 2 is

***k*-means clustering** is an iterative algorithm for grouping n-dimensional points into k spatial clusters. Each cluster is defined by a center point which is an n-vector c_i, $i \in [1, k]$. At each iteration all points are assigned to the *closest* cluster center, and then each cluster center is updated to be the mean of all the points assigned to the cluster.

The algorithm is implemented by the Toolbox function `kmeans`. The distance metric used is Euclidean distance. The k-means algorithm requires an initial estimate of the center of each cluster and this can be provided in various ways, see the documentation. By default `kmeans` randomly selects k of the provided points, and this means the algorithm will return different results at each invocation.

To demonstrate we choose 500 random 2-dimensional points

```
>> a = rand(2,500);
```

where `a` is a 2 × 500 matrix with one point per column. We will cluster this data into three sets

```
>> [cls,centre,r] = kmeans(a, 3);
```

where `cls` is a 500-vector whose elements specify the class of the corresponding column of `a`. `center` is a 2 × 3 matrix whose columns specify the center of each 2-dimensional cluster and `r` is the residual – the norm of the distance of every point from its assigned cluster centroid.

We plot the points in each cluster with different colors

```
>> hold on
>> for i=1:3
     plot( a(1,cls==i), a(2,cls==i), '.' );
   end
```

and it is clear that the points have been sensibly partitioned. The centroids center have been superimposed as black dots.

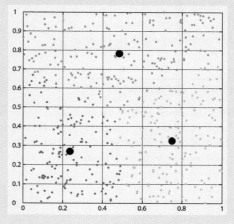

```
>> colorname(cab(:,2)', 'ab')
ans =
    'brown4'
```

The red pixels can be selected

```
>> cls2 = (cls == 2);
```

and the resulting logical image is shown in Fig. 13.7d.

This segmentation is far from perfect. Both tomatoes have *holes* due to specular reflection as discussed in Sect. 10.3.5. A few pixels at the bottom left have been erroneously classified as a tomato. We can improve the result by applying a morphological closing operation with a large circular kernel which is consistent with the shape of the tomato

```
>> tomatoes_binary = iclose(cls2, kcircle(15));
```

and the result is shown in Fig. 13.7e. The closing operation has somewhat restored the shape of the fruit but with the unwanted consequence that the group of misclassified pixels in the bottom-left corner have been enlarged. Nevertheless this image contains a workable classification of pixels into two classes: tomato and not-tomato.

The garden image illustrates two common real-world imaging artifacts: specular reflection and occlusion. The surface of the tomato is sufficiently shiny and oriented in such a way that the camera sees a reflection of the room light – these pixels are white rather than red.◄ The top tomato is also partly obscured by leaves and branches. Depending on how the application works this may or may not be a problem. Since the tomato cannot be reached from the direction the picture was taken, because of the occluding material, it might in fact be appropriate to not classify this as a tomato.

These examples have achieved a workable classification of the image pixels into object and not-object. The resulting groups of white pixels are commonly known as blobs. It is interesting to note that we have not specified any threshold or any definition of the object color, but we did have to specify the number of classes and determine which of those classes corresponded to the objects of interest.◄ We have also had to choose the sequence of image processing steps and the parameters for each of those steps, for example, the radius of the structuring element. Pixel classification is a difficult problem but we can get quite good results by exploiting knowledge of the problem, having a good collection of image processing tricks, and experience.

Observe that they have the same chromaticity as the black background, class 3 pixels, which are situated close to the white point on the a^*b^*-plane.

This is a relatively easy problem. The color of the object of interest is known (we could use the `colorname` function to find it) so we could compute the distance between each cluster center and the color of interest by name and choose the cluster that is closest.

Specular highlights in images are reflections of bright light sources and can complicate segmentation as shown in Fig. 13.3.

As discussed in Sect. 10.3.5 the light reflected by most real objects has two components: the specular surface reflectance which does not change the spectrum of the light; and the diffuse body reflectance which filters the reflected light.

There are several ways to reduce the problem of specular highlights. Firstly, move or remove the problematic light source, or move the camera. Secondly, use a diffuse light source near the camera, for instance a ring illuminator that fits around the lens of the camera. Thirdly, attenuate the specular reflection using a polarizing filter since light that is specularly reflected from a dielectric surface will be polarized.

13.1.2   Representation

In the previous section we took greyscale or color images and processed them to produce binary or blob images. Representation is the subproblem of *connecting* adjacent pixels of the same class to form spatial sets $S_1 \ldots S_m$.

Consider the binary image

```
>> im = iread('multiblobs.png');
```

which is shown

```
>> idisp(im)
```

in Fig. 13.8a. We quickly identify a number of white and black blobs in this scene but what defines a blob? It is a set of pixels of the same class that are *connected* to each other. More formally we could say a blob is a spatially contiguous region of pixels of the same class. Blobs are also known as regions or connected components.

The Toolbox can perform connected component or connectivity analysis on this binary image

```
>> [label, m] = ilabel(im);
```

The number of sets, or components, in this image is

```
>> m
m =
      11
```

comprising five white blobs and six black blobs (the background and the *holes*). These blobs are labeled from 1 to 11. The returned label matrix has the same size as the original image and each element contains the label $s \in \{1 \cdots m\}$ of the set to which the corresponding input pixel belongs. The label matrix can be displayed as an image▸ in false color

```
>> idisp(label, 'colormap', jet, 'bar')
```

We have seen a label image previously. The output of the MSER function in Fig. 13.5 is a label image.

as shown in Fig. 13.8b. Each connected region has a unique label and hence unique color. Looking at the label values in this image, or by interactively probing the displayed label matrix using idisp, we see that the background has been labeled as 1, the leftmost blob is labeled 3, and its holes are labeled 5 and 7.

To obtain an image containing just a particular blob is now very easy. To select all pixels belonging to region 3 we create a logical image

```
>> reg3 = (label==3);
>> idisp(reg3)
```

which is shown in Fig. 13.8c. The total number of pixels in this blob is given by the total number of true-valued pixels in this logical image

```
>> sum(reg3(:))
ans =
      171060
```

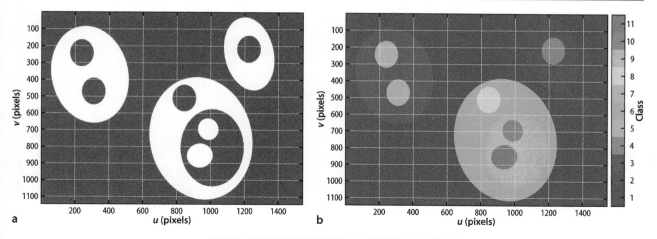

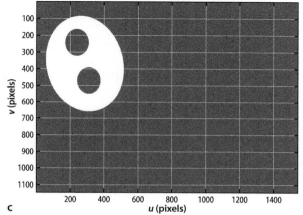

Fig. 13.8. Image labelling example. **a** Binary image; **b** labeled image; **c** all pixels with the label 3

Connectivity analysis can return additional output values

```
>> [label, m, parents, cls] = ilabel(im);
```

where the vector

```
>> parents'
ans =
    0    1    1    2    3    1    3    6    6    9    9
```

describes the topology or hierarchy of the regions. It indicates, for example, that the parent of region 4 is region 2◄ since region 4 is completely enclosed by region 2. The parent of regions 2, 3 and 5 is region 1 which is the background. Region 1 has a parent of 0 indicating that it touches the edge of the image and is not enclosed by any region. Each connected region contains pixel values of a single class and the pixel class for each region is given by

Element 4 of this array is equal to 2.

```
>> cls'
ans =
    0    1    1    0    0    1    0    0    0    1    1
```

which indicates that regions 2, 3, 6, 10 and 1 comprise pixels of class 1 (white) and regions 1, 4, 5, 7, 8 and 9 comprise pixels of class 0 (black).◄

We use the variable name `cls` rather than `class` since the latter is the name of a useful function in MATLAB.

In this example we have assumed 4-way connectivity, that is, pixels are connected within a region only through their north, south, east and west neighbors of the same class. The 8-way connectivity option allows connection via any of a pixel's eight neighbors of the same class.◄

8-way connectivity can lead to surprising results. For example a black and white chequerboard would have just two regions; all white squares are one region and all the black squares another.

Returning now to the examples from the previous section. For the colored targets

```
>> targets_label = ilabel(targets_binary);
>> idisp(targets_label, 'colormap', 'jet');
```

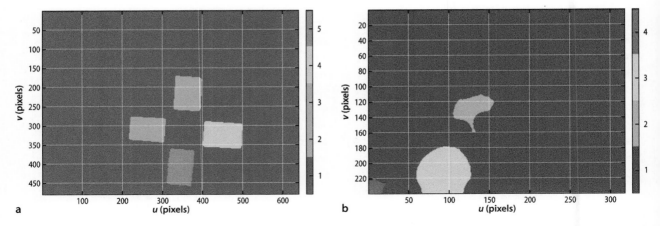

Fig. 13.9. Label images for the targets and garden examples in false color. The value of each pixel is the label of the spatially contiguous set to which the corresponding input pixel belongs

and the garden image

```
>> tomatoes_label = ilabel(tomatoes_binary);
>> idisp(tomatoes_label, 'colormap', 'jet');
```

the connected regions are shown in false color in Fig. 13.9. We are now starting to know something quantitative about these scenes: there are four yellow objects and three red objects respectively.

13.1.2.1 Graph-Based Segmentation

So far we have classified pixels based on some homogeneous characteristic of the object such as intensity or color. Consider now the complex scene

```
>> im = iread('58060.jpg');
```

shown in Fig. 13.10a. The Gestalt principle of emergence says that we identify objects as a whole rather than as a collection of parts – we see a bowl of grain rather than deducing a bowl of grain by recognizing its individual components. However when it comes to a detailed pixel by pixel segmentation things become quite subjective – different people would perform the segmentation differently based on judgment calls about what is *important*.▸ For example, should the colored stripes on the cloth be segmented? If segments represent real world objects, then the Gestalt view would be that the cloth should be just one segment. However the stripes are real, some effort was made to create them, so perhaps they should be segmented. This is why segmentation is a *hard* problem – humans cannot agree on what is correct. No computer algorithm could, or could be expected to, make this type of judgment.

The Berkeley segmentation site http://www.eecs.berkeley.edu/Research/Projects/CS/vision/bsds hosts these images plus a number of different human-made segmentations.

Nevertheless more sophisticated algorithms can do a very impressive job on complex real world scenes. The image can be represented as a graph (see Appendix I) where each pixel is a vertex and has 8 edges connecting it to its neighboring pixels. The weight of each edge is a nonnegative measure of the dissimilarity between the two pixels – the absolute value of the difference in color. The algorithm starts with every vertex assigned to its own set. At each iteration the edge weights are examined and if the vertices are in different sets but the edge weight is below a threshold the two vertex sets are merged. The threshold is a function of the size of the set and a global parameter k which sets the scale of the segmentation – a larger value of k leads to a preference for larger connected components.

For the image discussed the graph-based segmentation is given by

```
>> [label, m] = igraphseg(im, 1500, 100, 0.5);
>> m
m =
    28
>> idisp(label, 'colormap', 'jet')
```

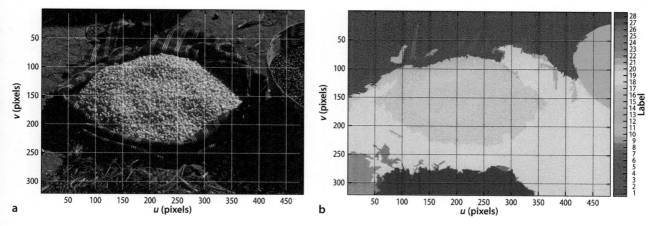

Fig. 13.10. Complex segmentation example. **a** Original color image (image from the Berkeley Segmentation Dataset; Martin et al. 2001); **b** graph-based segmentation

where `label` is a matrix, shown in Fig. 13.10b, whose elements are the region label for the corresponding input pixels. The pixel classification step has been integrated into the representation step. The arguments are a scale parameter $k = 1\,500$, the minimum component size of 100 pixels, and the standard deviation for an initial Gaussian smoothing applied to the image.

13.1.3 Description

In the previous section we learned how to find connected components in the image and how to isolate particular components such as shown in Fig. 13.8c. However this representation of the component is still just an image with logical pixel values rather than a concise *description* of its size, position and shape.

13.1.3.1 Bounding Boxes

The simplest representation of size and shape is the bounding box – the smallest rectangle with sides parallel to the u- and v-axes that encloses the region. We will illustrate this with a simple binary image

```
>> sharks = iread('sharks.png');
```

which is shown in Fig. 13.11a. As described above we will label the pixels and select all those belonging to region 2

```
>> [label, m] = ilabel(sharks);
>> blob = (label == 2);
```

and the resulting logical image is shown in Fig. 13.11b. The number of pixels in this region is simply the sum

```
>> sum(blob(:))
ans =
    7728
```

The coordinates of all the nonzero (object) pixels are the corresponding elements of

```
>> [v,u] = find(blob);
```

where u and v are each vectors of size

```
>> about(u)
u [double] : 7728x1 (61.8 kB)
```

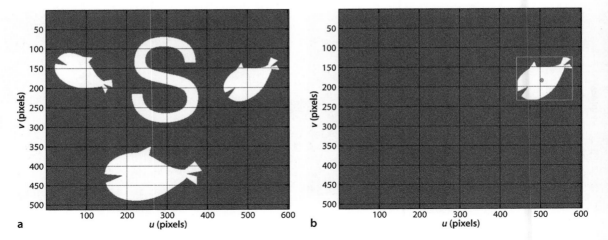

The bounds of the region◄ are

Fig. 13.11.a Sharks image. **b** Region 2 with bounding box (*green*)

```
>> umin = min(u)
umin =
    443
>> umax = max(u)
umax =
    581
>> vmin = min(v)
vmin =
    125
>> vmax = max(v)
vmax =
    235
```

This can be obtained more simply using the Toolbox function ibbox.

These bounds define a rectangle which we can superimpose on the image

```
>> plot_box(umin, vmin, umax, vmax, 'g')
```

as shown in Fig. 13.11b. The bounding box fits snugly around the blob and its center could be considered as the center of the blob. However the bounding box is not aligned with the blob, that is, its sides are not parallel with the sides of the blob. This means that as the blob rotates the size and shape of the bounding box would change even though the size and shape of the blob does not.

13.1.3.2 Moments

Moments are a rich and computationally cheap class of image features which can describe region size and location as well as shape. The moment of an image **I** is a scalar

$$m_{pq} = \sum_{(u,v) \in \mathbf{I}} u^p v^q \mathbf{I}[u,v] \tag{13.1}$$

where $(p + q)$ is the *order* of the moment. The zeroth moment $p = q = 0$ is

$$m_{00} = \sum_{(u,v) \in \mathbf{I}} \mathbf{I}[u,v] \tag{13.2}$$

and for a binary image where the background pixels are zero this is simply the number of nonzero (white) pixels – the area of the region.

Moments are calculated using the Toolbox function mpq and for the single shark the zeroth moment is

```
>> m00 = mpq(blob, 0, 0)
m00 =
        7728
```

which is the area of the region in units of pixels.

Moments can be given a physical interpretation by regarding the image function as a mass distribution. Consider the region as being made out of thin plate where each pixel has one unit of area and one unit of mass. The total mass of the region is m_{00} and the center of mass or centroid of the region is

$$u_c = \frac{m_{10}}{m_{00}}, \quad v_c = \frac{m_{01}}{m_{00}} \tag{13.3}$$

where m_{10} and m_{01} are the first-order moments. For our example the centroid of the target region is

```
>> uc = mpq(blob, 1, 0) / m00
uc =
   503.4981
>> vc = mpq(blob, 0, 1) / m00
vc =
   184.7285
```

which we can display

```
>> hold on; plot(uc, vc, 'gx', uc, vc, 'go');
```

as shown in Fig. 13.11b.

The central moments μ_{pq} are computed with respect to the centroid

$$\mu_{pq} = \sum_{(u,v) \in I} (u - u_c)^p (v - v_c)^q \, I[u,v] \tag{13.4}$$

and are invariant to the position of the region. They are related to the moments m_{pq} by

$$\mu_{10} = 0, \qquad \mu_{01} = 0$$
$$\mu_{20} = m_{20} - \frac{m_{10}^2}{m_{00}}, \quad \mu_{02} = m_{02} - \frac{m_{01}^2}{m_{00}}, \quad \mu_{11} = m_{11} - \frac{m_{10}m_{01}}{m_{00}} \tag{13.5}$$

and are computed by the Toolbox function upq.

Using the thin plate analogy again, the inertia of the region about axes parallel to the u- and v-axes and intersecting at the centroid of the region is given by the symmetric matrix

$$J = \begin{pmatrix} \mu_{20} & \mu_{11} \\ \mu_{11} & \mu_{02} \end{pmatrix} \tag{13.6}$$

The central second moments μ_{20}, μ_{02} are the moments of inertia and μ_{11} is the product of inertia. The product of inertia is nonzero if the shape is asymmetric with respect to the region's axes.

The *equivalent ellipse* is the ellipse that has the same inertia matrix as the region. For our example

```
>> u20 = upq(blob, 2, 0); u02 = upq(blob, 0, 2); u11 = upq(blob, 1, 1);
>> J = [ u20 u11; u11 u02]
J =
   1.0e+06 *
     7.8299    -2.9169
    -2.9169     4.7328
```

and we can superimpose the equivalent ellipse over the region

```
>> plot_ellipse(4*J/m00, [uc, vc], 'b');
```

and the result is shown in Fig. 13.12.

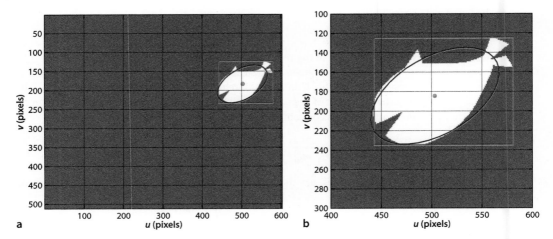

Fig. 13.12. Sharks image. a Equivalent ellipse (*blue*), centroid and bounding box (*green*) for region 2 of the targets image; b zoomed view

The eigenvalues and eigenvectors of *J* are related to the radii of the ellipse and the orientation of its major and minor axes (see Sect. C.1.4). For this example the eigenvalues

```
>> lambda = eig(J)
lambda =
   1.0e+06 *
   2.9788
   9.5838
```

are the principal moments of the region. The maximum and minimum radii of the equivalent ellipse are

$$a = 2\sqrt{\frac{\lambda_2}{m_{00}}}, \quad b = 2\sqrt{\frac{\lambda_1}{m_{00}}} \tag{13.7}$$

respectively where $\lambda_2 \geq \lambda_1$. In MATLAB this is►

MATLAB returns eigenvalues in increasing order: λ_1 then λ_2.

```
>> a = 2 * sqrt(lambda(2) / m00)
a =
   70.4313
>> b = 2 * sqrt(lambda(1) / m00)
b =
   39.2663
```

in units of pixels. These lengths are characteristic of this particular shape and are invariant to rotation. The aspect ratio of the region

```
>> b/a
ans =
   0.5575
```

is a scalar that crudely characterizes the shape and is invariant to scale and rotation.

The eigenvectors of *J* are the principal axes of the ellipse – the directions of its major and minor axes. The major, or principal, axis is the eigenvector *v* corresponding to the maximum eigenvalue. For our example this is

```
>> [x,lambda] = eig(J);
>> x
x =
   -0.5153   -0.8570
   -0.8570    0.5153
```

and since MATLAB returns eigenvalues in increasing order *v* is always the *last* column of the returned eigenvector matrix

```
>> v = x(:,end);
```

Table 13.1.
Region features and their invariance to camera motion: translation, rotation about the object's centroid and scale factor

	Translation	Rotation	Scale
Area	✓	✓	✗
Centroid	✗	✓	✓
Orientation θ	✓	✗	✓
Aspect ratio	✓	✓	✓
Circularity	✓	✓	✓
Hu moments	✓	✓	✓

Table 13.1.
Region features and their invariance to camera motion: translation, rotation about the object's centroid and scale factor

The angle of this vector with respect to the horizontal axis is

$$\theta = \tan^{-1}\frac{v_y}{v_x}$$

and for our example this is

```
>> atand( v(2)/v(1) )
ans =
    -31.0185
```

With reference to Fig. C.2b the angle increases clockwise from the horizontal since the y-axis of the image is downward so the z-axis is into the page.

degrees which indicates that the major axis of the equivalent ellipse is approximately 30 degrees above horizontal. ◀

To summarize, we have created an image containing a spatially contiguous set of pixels corresponding to one of the objects in the scene that we segmented from the original color image. We have determined its area, a box that entirely contains it, its position (the location of its centroid), its orientation and its shape (aspect ratio). The equivalent ellipse is a crude indicator of the region's shape but it is invariant to changes in position, orientation and scale. The invariance of the different blob descriptors to camera motion is summarized in Table 13.1.

13.1.3.3 Blob Features

The Toolbox provides a simpler way to perform the functions described above

```
>> f = imoments(blob)
f =
area=7728, cent=(503.5,184.7), theta=-0.54, b/a=0.558
```

which returns a `RegionFeature` object that contains many features describing this region including its area, its centroid, orientation and aspect ratio – the ratio of its minimum to maximum radius. These values are available as object properties, for example

```
>> f.uc
ans =
    503.4981
>> f.theta
ans =
    -0.5414
>> f.aspect
ans =
      0.5575
```

along with the zeroth- and first-order moments and the second-order central moments

```
>> f.moments.m00
ans =
         7728
>> f.moments.u11
ans =
    -2.9169e+06
```

The Toolbox provides a high-level function to compute features for *every* region in the image

```
>> fv = iblobs(targets_binary)
fv =
(1) area=14899, cent=(298.0,181.0), theta=1.48, b/a=0.702, ↵
 color=1, label=1, touch=0, parent=4
(2) area=7728, cent=(503.5,184.7), theta=-0.54, b/a=0.558, ↵
 color=1, label=2, touch=0, parent=4
(3) area=7746, cent=(84.2,160.7), theta=0.50, b/a=0.558, ↵
 color=1, label=3, touch=0, parent=4
(4) area=258946, cent=(306.7,252.9), theta=0.05, b/a=0.831, ↵
 color=0, label=4, touch=1, parent=0
(5) area=18814, cent=(246.8,426.9), theta=-0.02, b/a=0.559, ↵
 color=1, label=5, touch=0, parent=4
```

which returns a vector `fv` of `RegionFeature` objects. The display method shows a summary of the region's properties and the number in parentheses indicates the index within the vector. Each `RegionFeature` object contains the area, bounding box, centroid, raw and central moments, and equivalent ellipse parameters as returned by `imoments` as well as additional properties such as the class of the pixels within the region, the region label, the label of the parent region and whether or not the blob touches the edge.► We can tell that region 4 is the background since it is large, comprises zero valued pixels and touches the edge of the image.

Some examples of the properties of this class are

```
>> fv(2).umin
ans =
   443
>> fv(3).class
ans =
    1
>> fv(3).parent
ans =
        4
>> fv(3).umin
ans =
   24
>> fv(3).aspect
ans =
   0.5580
```

This calculation assumes that pixels are square and this is almost always the case with digital cameras. If not the `'aspect'` option should be provided to `iblobs` to define a nonunity pixel aspect ratio which is pixel height over pixel width.

The `RegionFeature` class also has plotting methods such as

```
>> fv(2).plot_box('g')
```

which overlays the bounding box of feature `fv(2)`, in green, on the current plot. Other plot methods include `plot_centroid` and `plot_ellipse`. All methods add to the current plot and can operate on a single object or a vector of objects, for example

```
>> fv.plot_box('r:')
```

overlays the bounding box, in dotted red, for *all* blobs in `fv`.

The `children` property is the inverse mapping of the `parent` property. It is a list of indices into the feature vector of `RegionFeature` objects which are children of this feature. For example the background blob, `fv(4)`, has as its children

```
>> fv(4).children
ans =
     1    2    3    5
```

blobs `fv(1)`, `fv(2)`, `fv(3)` and `fv(5)`.

Importantly the function `iblobs` can perform filtering. For the tomato image (page 423) we might know something about the minimum and/or maximum size of a tomato so we can set bounds on the possible area

```
>> fv = iblobs(tomatoes_binary, 'area', [1000, 5000])
fv =
(1) area=1529, cent=(132.6,132.9), theta=-0.36, b/a=0.866, ↵
  color=1, label=2, touch=0, parent=1
(2) area=3319, cent=(95.5,210.9), theta=-0.33, b/a=0.886, ↵
  color=1, label=4, touch=1, parent=0
```

which returns only blobs with an area between 1 000 and 5 000 pixels. Other filter parameters include aspect ratio and edge touching, and more details are provided in the online documentation. For the tomato image we might wish to accept only blobs that do not touch the edge

```
>> fv = iblobs(tomatoes_binary, 'touch', 0)
fv =
(1) area=1529, cent=(132.6,132.9), theta=-0.36, b/a=0.866, ↵
  color=1, label=2, touch=0, parent=1
```

The filter rules can also be cascaded, for example

```
>> fv = iblobs(tomatoes_binary, 'area', [1000, 5000], 'touch', ↵
  0, 'class', 1)
fv =
(1) area=1529, cent=(132.6,132.9), theta=-0.36, b/a=0.866, ↵
  color=1, label=2, touch=0, parent=1
```

and a blob must pass all rules in order to be accepted.

13.1.3.4   Shape from Moments

In order to recognize particular objects we need some measures of shape that are invariant to the rotation and scale of the image and provide more detail than the simple aspect ratio parameter.

For the case of planar objects, which are fronto-parallel to the camera, complex ratios of moments can be used to form a vector of invariants for recognition of planar objects irrespective of position, orientation and scale. For example the image of Fig. 13.11a has three similarly shaped regions and one that is different

```
>> [fv,L] = iblobs(sharks, 'class', 1);
>> fv =
(1) area=14899, cent=(298.0,181.0), theta=1.48, b/a=0.702, ↵
  color=1, label=1, touch=0, parent=4
(2) area=7728, cent=(503.5,184.7), theta=-0.54, b/a=0.558, ↵
  color=1, label=2, touch=0, parent=4
(3) area=7746, cent=(84.2,160.7), theta=0.50, b/a=0.558, ↵
  color=1, label=3, touch=0, parent=4
(4) area=18814, cent=(246.8,426.9), theta=-0.02, b/a=0.559, ↵
  color=1, label=5, touch=0, parent=4
```

and from the aspect ratio parameter b/a we see that blob 1 is different to blobs 2, 3 and 4. The second output argument is the label matrix and the moment invariants for the four white blobs can be computed by

```
>> for i=1:4
     H(i,:) = humoments(L == fv(i).label);
   end
>> H
H =
    0.4544    0.0238    0.0006    0.0000    0.0000   -0.0000   -0.0000
    0.2104    0.0122    0.0020    0.0006    0.0000    0.0001    0.0000
    0.2101    0.0122    0.0020    0.0006    0.0000    0.0001    0.0000
    0.2102    0.0121    0.0020    0.0006    0.0000    0.0001    0.0000
```

In practice the discrete nature of the pixel data means that the invariance will only be approximate.

which indicate the similarity of blobs 2, 3 and 4 despite their different position, orientation and scale. ◀ It also indicates the difference in shape of blob 1. This shape descriptor can be considered as a point in 7-dimensional space, and similarity to other shapes can be defined in terms of Euclidean distance in this descriptor space.

13.1.3.5   Shape from Perimeter

The shape of a region is concisely described by its boundary or perimeter pixels – sometimes called edgels. Figure 13.13 shows three common ways to represent the perimeter of a region – each will give a slightly different estimate of the perimeter length. A chain code is a list of the outermost pixels of the region whose center's are linked by short line segments. In the case of a 4-neighbor chain code the successive pixels must be adjacent and the perimeter segments have an orientation of $k \times 90°$, where $k \in \{0 \cdots 3\}$. With an 8-neighbor chain code, or Freeman chain code, the perimeter segments have an orientation of $k \times 45°$, where $k \in \{0 \cdots 7\}$. The crack code has its segments in the *cracks* between the pixels on the edge of the region and the pixels outside the region. These have orientations of $k \times 90°$, where $k \in \{0 \cdots 3\}$.

The perimeter can be encoded as a list of pixel coordinates (u_i, v_i) or very compactly as a bit string using just 2 or 3 bits to represent k for each segment. These various representations are equivalent and any representation can be transformed to another.

> Note that for chain codes the boundary follows a path that is on average half a pixel inside the true boundary and therefore underestimates the perimeter length. The error is most significant for small regions.

To enable the extra computation to trace around the boundary of the objects using 8-neighbor chain code we must give the `'boundary'` option

```
>> fv = iblobs(sharks, 'boundary', 'class', 1);
>> fv(1)
ans =
(1) area=14899, cent=(298.0,181.0), theta=1.48, b/a=0.702, ↵
  class=1, label=1, touch=0, parent=4, perim=1236.4, circ=0.136
```

and we see that two extra parameters are now displayed: `perim` and `circ` which are perimeter length and circularity respectively. The boundary is a list of edge points represented as a matrix with one column per edge point. In this case there are

```
>> about(fv(1).edge)
  [double] : 2x1085 (17.4 kB)
```

1 085 edge points and the first five points of the boundary are

```
>> fv(1).edge(:,1:5)
ans =
    285    284    283    282    281
     71     72     72     72     72
```

The displayed perimeter length of `1236.4` has had a heuristic correction applied to compensate for the underestimation due to chain coding.

The boundary can be overlaid on the current plot using the object's `plot_boundary` method

```
>> fv(1).plot_boundary('r')
```

in this case as a red line. The plotting methods can also be invoked on a feature vector

```
>> idisp(sharks)
>> fv.plot_boundary('r')
>> fv.plot_centroid()
```

which is shown in Fig. 13.14.

Moment invariants. The normalized moments

$$\eta_{pq} = \frac{\mu_{pq}}{\mu_{00}^{\gamma}}, \quad \gamma = \frac{1}{2}(p+q)+1 \quad \text{for} \quad p+q = 2, 3, \cdots \quad (13.8)$$

are invariant to translation and scale, and are computed from the central moments by the Toolbox function `npq`.

Third-order moments allow for the creation of quantities that are invariant to translation, scale and orientation within a plane. One such set of moments defined by Hu (1962) are

$$\phi_1 = \eta_{20} + \eta_{02}$$

$$\phi_2 = (\eta_{20} - \eta_{02})^2 + 4\eta_{11}^2$$

$$\phi_3 = (\eta_{30} - 3\eta_{12})^2 + (3\eta_{21} - \eta_{03})^2$$

$$\phi_4 = (\eta_{30} + \eta_{12})^2 + (\eta_{21} + \eta_{03})^2$$

$$\phi_5 = (\eta_{30} - 3\eta_{12})(\eta_{30} + \eta_{12})\left[(\eta_{30} + \eta_{12})^2 - 3(\eta_{21} + \eta_{03})^2\right]$$
$$\quad + (3\eta_{21} - \eta_{03})(\eta_{21} + \eta_{03})\left[3(\eta_{30} + \eta_{12})^2 - (\eta_{21} + \eta_{03})^2\right]$$

$$\phi_6 = (\eta_{20} - \eta_{02})\left[(\eta_{30} + \eta_{12})^2 - (\eta_{21} + \eta_{03})^2\right]$$
$$\quad + 4\eta_{11}(\eta_{30} + \eta_{12})(\eta_{21} + \eta_{03})$$

$$\phi_7 = (3\eta_{21} - \eta_{03})(\eta_{30} + \eta_{12})\left[(\eta_{30} + \eta_{12})^2 - 3(\eta_{21} + \eta_{03})^2\right]$$
$$\quad + (3\eta_{12} - \eta_{30})(\eta_{21} + \eta_{03})\left[3(\eta_{30} + \eta_{12})^2 - (\eta_{21} + \eta_{03})^2\right]$$

and computed by the Toolbox function `humoments`.

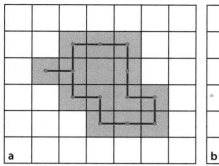
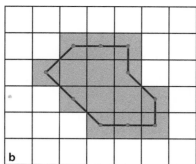
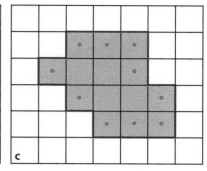

Fig. 13.13. Boundary representations with region pixels shown in *grey*, perimeter segments shown in *blue* and the center of boundary pixels marked by a *red dot*. **a** Chain code with 4 directions; **b** Freeman chain code with 8 directions; **c** crack code. The perimeter lengths for this example are respectively 14, 12.2 and 18 pixels

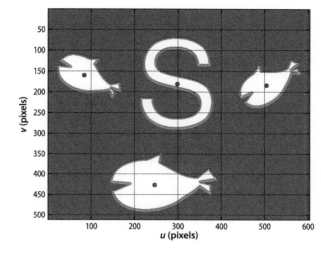

Fig. 13.14. Boundaries (*red*) and centroids of four blobs

Circularity is another commonly used and intuitive shape feature. It is defined as

$$\rho = \frac{4\pi m_{00}}{p^2} \tag{13.9}$$

where p is the region's perimeter length. Circularity has a maximum value of $\rho = 1$ for a circle, is $\rho = \frac{\pi}{4}$ for a square and zero for an infinitely long line. Circularity is also invariant to translation, rotation and scale. In the results of `iblobs` shown above we note that the circularity measure is the same for blobs 2, 3 and 4, and much lower for blob 1 due to it being effectively a long line. ◀

For small blobs, quantization effects can lead to significant errors in circularity.

Every object has one external boundary, which may include a section of the image border if the object touches the border. An object with holes has one internal boundary per hole but the Toolbox returns only the external boundary – the inner boundaries can be found as the external boundaries of the *holes* which are its child regions. Since the external boundary contains all the essential information about the shape of a region it is possible, assuming that the region has no holes, to compute the moments from the boundary using the functions `mpq_poly`, `upq_poly` and `npq_poly`.

One way to analyze the rich shape information encoded in the perimeter is to compute the distance and angle to every perimeter point with respect to the object's centroid – this is computed by the `boundary` method

```
>> [r,th] = fv(2).boundary;
>> plot([r th])
```

and is shown in Fig. 13.15a. These are computed for 400 points (default) evenly spaced along the entire perimeter of the object. Both the radius and angle signatures describe the shape of the object. The angle signature is invariant to the scale of the object while

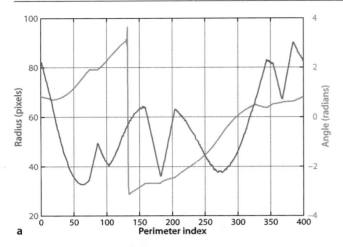

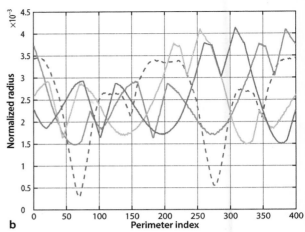

the amplitude of the radius signature scales with object size. The radius signatures of all four blobs can be compared by

Fig. 13.15. Radius signature matching. **a** Radius and angle signature for blob 2 (top left shark in Fig. 13.14), **b** normalized radius signatures for all blobs (letter S is shown *dashed*)

```
>> hold on
>> for f=fv
     [r,t] = f.boundary();
     plot(r/sum(r));
   end
```

and are shown in Fig. 13.15b. We have normalized by the sum in order to remove the effect of object scale.▶ The signatures are a function of normalized distance along the perimeter. They all start at the top left-most pixel on the object's boundary. Different objects of the same shape have identical signatures but possibly shifted horizontally and wrapped around – the first and last points in the horizontal direction are adjacent.

We could have normalized by the maximum value but normalizing by the sum is more noise tolerant.

To compare the shape profile of objects requires us to compare the signature for all possible horizontal shifts. The radius signatures of all four blobs is

```
>> b = fv.boundary
```

which is a 400 × 4 matrix with the radius signatures as columns. To compare the signature of blob 2 with all the signatures

```
>> RegionFeature.boundmatch(b(:,2), b)
ans =
     0.6494    1.0000    0.9854    0.9927
```

which indicates that the shape of blob 2 closely matches the shape of itself and blobs 3 and 4, but not the shape of blob 1. The static method `boundmatch` computes the 1-dimensional normalized cross correlation, see Table 12.1, for every possible rotation of one signature with respect to the other, and returns the highest value.

There are many variants to the approach described. The signature can be Fourier transformed and described more concisely in terms of a few Fourier coefficients. The boundary curvature can be computed which highlights corners, or the boundary can be segmented into straight lines and arcs.

13.1.3.6 Character Recognition

A particularly important class of objects are characters. Our world is filled with informative text in the form of signs and labels which provides information about the names of places and directions to travel. We process much of this unconsciously but this rich source of information is largely unavailable to robots.

If you have the Computer Vision System Toolbox™ you can access the inbuilt optical character recognition functionality. We start with the image of a sign we have used

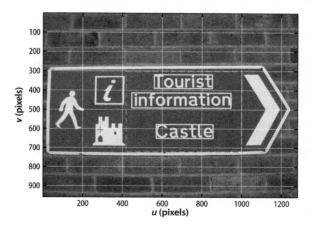

Fig. 13.16.
Optical character recognition. The *bounding boxes* of detected words are shown in *yellow*. The manually set region of interest within which to search for text is shown as a *dashed red box*

previously and apply OCR to a region of interest specified as (u, v, w, h) where (u, v) are the coordinates of the top-left of region

```
>> castle = iread('castle.png');
>> words = ocr(castle, [420 300 580 420]);
```

which returns a structure. The recognized words are

```
>> words.Text
ans =
Tourist
information
Castle
```

and its confidence about those words is

```
>> words.WordConfidences'
ans =
    0.8868    0.7884    0.8588
```

We can highlight the location of the words in the original image by

```
>> plot_box('matlab', words.WordBoundingBoxes, 'y')
```

with the result shown in Fig. 13.16. This function does require a reasonable estimate of the region in which the text is to be found.

13.1.4 Summary

We have discussed how to convert an input image, grey scale or color, into concise descriptors of regions within the scene. The criteria for what constitutes a region is *application specific*. For a tomato picking robot it would be round red regions, for landing a UAV it might be yellow targets on the ground.

The process outlined is the classical *bottom up* approach to machine vision applications and the key steps are:

1. Classifying the pixels according to the application specific criterion, for example, redness, yellowness or motion. Each pixel is assigned a class c.
2. Grouping adjacent pixels of the same class into sets, and each pixel is assigned a label S indicating the set to which it has been assigned.
3. Describing the sets in terms of features derived from their spatial extent, moments, equivalent ellipse and boundary.

These steps are a progression from *low-level* to *high-level*. The low-level operations consider pixels in isolation, whereas the high-level is concerned with more abstract concepts such as size and shape. The MSER and graphcuts algorithms are powerful because they combines steps 1 and 2 and consider regions of pixels and localized differences in order to create a segmentation.

Importantly none of these steps need be perfect. Perhaps the first step has some false positives, isolated pixels misclassified as objects, that we can eliminate by morphological operations, or reject after connectivity analysis based on their small size. The first step may also have false negatives, for example specular reflection and occlusion may cause some object pixels to be classified incorrectly as non-object. In this case we need to develop some heuristics, for instance morphological processing to fill in the gaps in the blob. Another option is to oversegment the scene – increase the number of regions and use some application-specific knowledge to merge adjacent regions. For example a specular reflection colored region might be merged with surrounding regions to create a region corresponding to the whole fruit.

For some applications it might be possible to engineer the camera position and illumination to obtain a high quality image but for a robot operating in the real world this luxury does not exist. A robot needs to glean as much useful information as it can from the image and move on.

Domain knowledge is always a powerful tool. Given that we know the scene contains tomatoes and plants, the fact that we observe a large red region that is not circular, we use our domain knowledge to infer that the fruit is occluded. We therefore might command the robot to seek the fruit that is not occluded, and then to move to another location where the fruit might not be occluded. Object segmentation remains one of the hardest aspects of machine vision and there is no silver bullet. It requires knowledge of image formation, fundamental image processing algorithms, insight, a good box of tools and patience.

13.2   Line Features

Lines are distinct visual features that are particularly common in man-made environments – for example the edges of roads, buildings and doorways. In Sect. 12.5.1.3 we discussed how image intensity gradients can be used to find edges within an image, and this section will be concerned with fitting line segments to such edges.

We will illustrate the principle using the very simple scene

```
>> im = iread('5points.png', 'double');
```

shown in Fig. 13.17a. Consider any one of these points – there are an infinite number of lines that pass through that point. If the point could vote for these lines, then each possible line passing through the point would receive one vote. Now consider

Fig. 13.17. Hough transform fundamentals. **a** Five points that define six lines; **b** the Hough accumulator array. The horizontal axis is an angle $\theta \in \mathbb{S}^1$ so we can imagine the graph wrapped around a cylinder and the left- and right-hand edges joined. The sign of ρ also changes at the *join* so the curve intersections on the left- and right-hand edges are equivalent

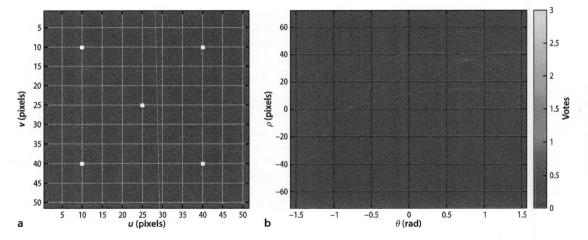

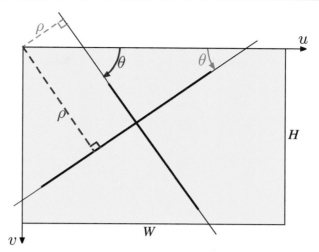

Fig. 13.18.
(θ, ρ) parameterization for two line segments. Positive quantities are shown in *blue*, negative in *red*

another point that does the same thing, casting a vote for all the possible lines that pass through it. One line (the line that both points lie on) will receive a vote from each point – a total of two votes – while all the other possible lines receive either zero or one vote.

We want to describe each line in terms of a minimum number of parameters but the standard form $v = mu + c$ is problematic for the case of vertical lines where $m = \infty$. Instead it is common to represent lines using the (ρ, θ) parameterization shown in Fig. 13.18

$$v = -u\tan\theta + \frac{\rho}{\cos\theta} \tag{13.10}$$

where $\theta \in [-\frac{\pi}{2}, \frac{\pi}{2})$ is the angle from the horizontal axis to the line, and $\rho \in [-\rho_{min}, \rho_{max}]$ is the perpendicular distance between the origin and the line. A horizontal line has $\theta = 0$ and a vertical line has $\theta = -\frac{\pi}{2}$. Any line can therefore be considered as a point (θ, ρ) in the 2-dimensional space of all possible lines.

It is not practical to vote for one out of an infinite number of lines through each point, so we consider lines drawn from a finite set. The $\theta\rho$-space is quantized and a corresponding $N_\theta \times N_\rho$ array A is used to tally the votes – the *accumulator* array. For a $W \times H$ input image

$$\rho_{max} = -\rho_{min} = \max(W, H)$$

The array A has N_ρ elements spanning the interval $\rho \in [-\rho_{max}, \rho_{max}]$ and N_θ elements spanning the interval $\theta \in [-\frac{\pi}{2}, \frac{\pi}{2})$. The indices of the array are integers $(i, j) \subset \mathbb{Z}^2$ such that

$$i \in [1, N_\theta] \mapsto \theta \in \left[-\tfrac{\pi}{2}, \tfrac{\pi}{2}\right)$$

$$j \in [1, N_\rho] \mapsto \rho \in \left[-\rho_{max}, \rho_{max}\right]$$

An edge point (u, v) votes for *all* lines that satisfy Eq. 13.10 which is all (i, j) pairs for which

$$\rho = u\sin\theta + v\cos\theta \tag{13.11}$$

and the elements $A[i, j]$ are all incremented. For every $i \in [1, N_\theta]$ the corresponding value of θ is computed, then ρ is computed according to Eq. 13.11 and mapped to a corresponding integer j. Every edge point adds a vote to N_θ elements of A that lie along a curve.

At the end of the process those elements of **A** with the largest number of votes correspond to dominant lines in the scene. For the example of Fig. 13.17a the resulting accumulator array is shown in Fig. 13.17b. Most of the array contains zero votes (dark blue) and the light curves are trails of single votes corresponding to each of the five input points. These curves intersect and those points correspond to lines with more than one vote. We see four locations where two curves intersect, resulting in cells with two votes, and these correspond to the lines joining the four outside points of Fig. 13.17a. The horizontal axis represents angle $\theta \in \mathbb{S}^1$ so the left- and right-hand ends are joined and ρ changes sign – the curve intersection points on the left- and right-hand sides of the array are equivalent. We also see two locations where three curves intersect, resulting in cells with three votes, and these correspond to the diagonal lines that include the middle point of Fig. 13.17a. This technique is known as the Hough▶ transform.

Pronounced "huff".

Consider the more complex example of a solid square rotated counter-clockwise by 0.3 rad

```
>> im = testpattern('squares', 256, 256, 128);
>> im = irotate(im, -0.3);
```

We compute the edge points

```
>> edges = icanny(im);
```

which are shown in Fig. 13.19a. The Hough transform is computed by

Fig. 13.19. Hough transform for a rotated square. **a** Edge image; **b** Hough accumulator; **c** closeup view of the Hough accumulator; **d** estimated lines overlaid on the original image

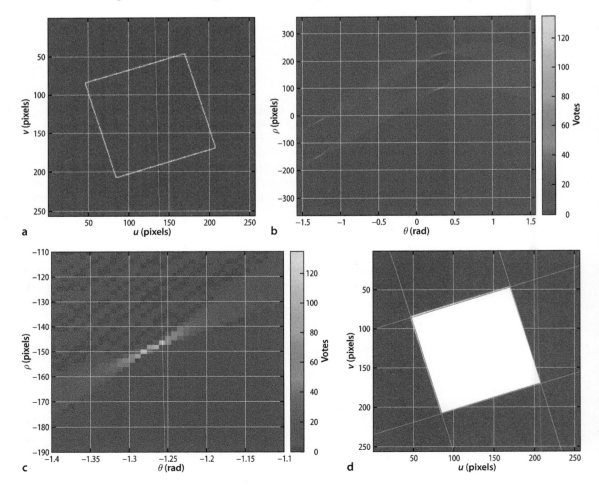

```
>> h = Hough(edges)
Hough: nd=401, ntheta=400, interp=3x3, distance=1
```

and returns an instance of the `Hough` class. Its properties include the two-dimensional vote accumulator array `A` with `nd` rows and `ntheta` columns. By default the $\theta\rho$-plane is quantized into 401×400 bins.◄ The accumulator array can be visualized as an image

```
>> h.show();
```

ρ is symmetric about zero, so including zero this is an odd number of elements. θ has a range of $[-\frac{\pi}{2}, \frac{\pi}{2})$, it is asymmetric about zero and has an even number of elements.

which is shown in Fig. 13.19b. The four bright spots correspond to dominant edges in the input image. We can see that many other possible lines have received a small number of votes as well.

The next step is to find the peaks in the accumulator array

```
>> lines = h.lines()
lines =
theta=0.298527, rho=224.412, strength=1
theta=0.306267, rho=96.2507, strength=0.962963
theta=-1.27224, rho=-20.1637, strength=0.874074
theta=-1.28026, rho=-150.256, strength=0.837037
theta=0.282667, rho=94.1344, strength=0.785185
theta=-1.25683, rho=-146.799, strength=0.77037
theta=0.318101, rho=226.064, strength=0.718519
theta=0.278998, rho=222.699, strength=0.703704
theta=-1.25286, rho=-17.337, strength=0.666667
theta=0.325784, rho=97.7635, strength=0.562963
theta=-1.29514, rho=-23.7157, strength=0.503704
```

which returns a vector of `LineFeature` objects corresponding to the lines with the most votes, as well as the number of votes associated with that line normalized with respect to the largest vote. If the function is called without output arguments the identified peaks are indicated on an image of the accumulator array.

Note that although the object has only four sides there are many more than four peaks in the accumulator array. We also note that the fourth and sixth peaks have quite similar line parameters, and this region of the accumulator is shown in more detail in Fig. 13.19c. We see several bright spots (high numbers of votes) that are close together and this is due to quantization effects. The concept of peak scale discussed on pages 369 and 390 applies here and once again we apply nonlocal maxima suppression to eliminate smaller peaks in the neighborhood of the maxima

```
>> h = Hough(edges, 'suppress', 5)
h =
Hough: nd=401, ntheta=400, interp=3x3, distance=5
```

In this case distance is five accumulator cells – the maxima suppresses smaller local maxima within a five cell radius. This leads to just four peaks

```
>> lines = h.lines()
lines =
theta=0.298527, rho=224.412, strength=1
theta=0.306267, rho=96.2507, strength=0.962963
theta=-1.27224, rho=-20.1637, strength=0.874074
theta=-1.28026, rho=-150.256, strength=0.837037
```

corresponding to the edges of the object.

Since the line parameters are quantized the `lines` method uses interpolation to refine the location of the peak (see Appendix J). By default, interpolation is performed over a 3×3 window centered on the local vote maxima. Once a peak has been found all votes within the suppression distance are zeroed so as to eliminate any close maxima, and the process is repeated for all peaks in the voting array that exceed a specified fraction of the largest peak.◄

With no argument all peaks greater than `'houghThresh'` are displayed. This defaults to 0.5 but can be set by the `'houghThresh'` option to Hough.

The detected lines can be projected onto the original image

```
>> idisp(im);
>> h.plot('b')
```

and the result is shown in Fig. 13.19d.

A real image example is

```
>> im = iread('church.png', 'grey', 'double');
>> edges = icanny(im);
>> h = Hough(edges, 'suppress', 10);
>> lines = h.lines();
```

and the strongest ten lines

```
>> idisp(im, 'dark');
>> lines(1:10).plot('g');
```

are shown in Fig. 13.20. Many strong lines in the image have been found, and lines corresponding to the roof edges and building-ground line are correct. However most of the vertical lines do not correspond to lines in the image – they are the result of disjoint sections of high gradient *voting up* a line that passes through them.

Another measure of the importance of an edge can be found by reprojecting the line onto the edge image and counting the maximum number of *contiguous* edge pixels that lie along it

```
>> lines = lines.seglength(edges);
```

which returns a vector of `LineFeature` objects similar to that returned by the `lines` method but with the property `length` set to the maximum edge segment length

```
>> lines(1)
ans =
theta=0.0237776, rho=791.008, strength=1, length=24
```

in this case 24 pixels. An edge segment is defined as an almost contiguous group of edge pixels with no gap greater than five (by default) pixels. We can then choose all those Hough peaks corresponding to segments longer than 80 pixels

```
>> k = find( lines.length > 80);
```

and then highlight those lines in blue

```
>> lines(k).plot('b--')
```

as shown in Fig. 13.20. We can see that a number of lines are converging on a perspective vanishing point to the right of the image.

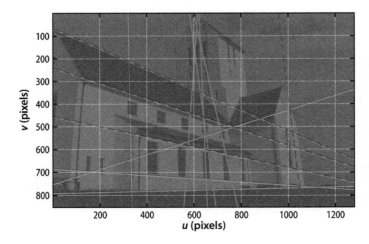

Fig. 13.20.
Hough transform of a real image. The *green lines* correspond to the ten strongest voting peaks. The overlaid *dashed blue lines* are those with an edge segment length of at least 80 pixels. Three lines meet both criteria

13.2.1   Summary

The Hough transform is elegant in principle and in practice it can either work well or infuriatingly badly. It performs poorly when the scene contains a lot of texture or the edges are indistinct. Texture causes votes to be cast widely, but not uniformly, over the accumulator array which tends to mask the true peaks. Consequently a lot of experimentation is required for the parameters of the edge detector and the Hough peak detector.

The function `Hough` has many options which are described in the online documentation. By default the strength of the vote cast by each edge point is the edge strength at that point which emphasizes stronger edges. Edge strengths less than `edgeThresh` times the maximum edge strength are considered as zero. The `Hough` object can also be constructed from an array of edge coordinates with equal votes, or from an array of edge coordinates and a vector of corresponding vote strength.

The Hough transform estimates the direction of the line by fitting lines to the edge pixels. It ignores rich information about the direction of the edge at each pixel which was discussed on page 382. The consequence of not using all the information available is poorer estimation. There is little added expense in using the direction at each pixel since we have already computed the image gradients in order to evaluate edge magnitude.

13.3   Point Features

The final class of features that we will discuss are point features. These are visually distinct points in the image that are known as interest points, salient points, keypoints or corner points. We will first introduce some classical techniques for finding interest points and then discuss more recent scale-invariant techniques.

13.3.1   Classical Corner Detectors

We recall from Sect. 12.5.1.3 that a point on a line has a strong gradient in a direction normal to the line. However gradient *along* the line is low which means that a pixel on the line will look very much like its neighbors along the line. In contrast, an *interest point* is a point that has a high image gradient in orthogonal directions. It might be single pixel that has a significantly different intensity to all of its neighbors or it might literally be a pixel on the corner of an object. Since interest points are quite distinct they have a much higher likelihood of being reliably detected in different views of the same scene. They are therefore key to multi-view techniques such as stereo and motion estimation which we will discuss in the next chapter.

The earliest corner point detector was Moravec's *interest operator*, so called because it indicated points in the scene that were *interesting* from a tracking perspective. It was based on the intuition that if a small image patch $\mathcal{W}$ is to be unambiguously located in another image it must be quite different to the same size patch at any adjacent location. Moravec defined the similarity between a region centered at (u, v) and an adjacent region, displaced by (δ_u, δ_v), as

$$s(u, v, \delta_u, \delta_v) = \sum_{(i,j) \in \mathcal{W}} \left(I[u + \delta_u + i, v + \delta_v + j] - I[u + i, v + j] \right)^2 \qquad (13.12)$$

where $\mathcal{W}$ is some local image region and typically a $W \times W$ square window. This is the SSD similarity measure from Table 12.1 that we discussed previously. Similarity is evaluated for displacements in eight cardinal◄ directions $(\delta_u, \delta_v) \in \mathcal{D}$ and the minimum value is the interest measure

N, NE, E, … W, NW or $i, j \in \{-1, 0, 1\}$.

$$C_M(u, v) = \min_{(\delta_u, \delta_v) \in \mathcal{D}} s(u, v, \delta_u, \delta_v) \qquad (13.13)$$

which has a large value only if all the displaced patches are different to the original patch. The function $C_M(\cdot)$ is evaluated for every pixel in the image and interest points are those where C_M is *high*. The main limitation of the Moravec detector is that it is nonisotropic since it examines image change, essentially gradient, in a limited number of directions. Consequently the detector can give a strong output for a point on a line, which is not desirable.

We can generalize the approach by defining the similarity as the weighted sum of squared differences between the image region and the displaced region as

$$s(u, v, \delta_u, \delta_v) = \sum_{(i,j) \in \mathcal{W}} W[i, j] \underbrace{\left| I[u + \delta_u + i, v + \delta_v + j] - I[u + i, v + j] \right|}^2$$

where W is a weighting matrix that emphasizes points closer to the center of the window $\mathcal{W}$. The indicated term can be approximated by a truncated Taylor series▸

See Appendix E.

$$I[u + \delta_u, v + \delta_v] \approx I[u, v] + I_u[u, v]\delta_u + I_v[u, v]\delta_v$$

where I_u and I_v are the horizontal and vertical image gradients respectively. We can now write

$$s(u, v, \delta_u, \delta_v) = \sum_{(i,j) \in \mathcal{W}} W[i, j] \left(I_u[u + i, v + j]\delta_u + I_v[u + i, v + j]\delta_v \right)^2$$

$$= \delta_u^2 \sum_{(i,j) \in \mathcal{W}} W[i, j] I_u^2[u + i, v + j] + \delta_v^2 \sum_{(i,j) \in \mathcal{W}} W[i, j] I_v^2[u + i, v + j]$$

$$+ \delta_u \delta_v \sum_{(i,j) \in \mathcal{W}} W[i, j] I_u[u + i, v + j] I_v[u + i, v + j]$$

which can be written compactly in quadratic form as

$$s(u, v, \delta_u, \delta_v) = \begin{pmatrix} \delta_u & \delta_v \end{pmatrix} A \begin{pmatrix} \delta_u \\ \delta_v \end{pmatrix}$$

where

$$A = \begin{pmatrix} \sum W[i, j] I_u^2[u + i, v + j] & \sum W[i, j] I_u[u + i, v + j] I_v[u + i, v + j] \\ \sum W[i, j] I_u[u + i, v + j] I_v[u + i, v + j] & \sum W[i, j] I_v^2[u + i, v + j] \end{pmatrix}$$

If the weighting matrix is a Gaussian kernel $W = G(\sigma_I)$ and we replace the summation by a convolution then

$$A = \begin{pmatrix} G(\sigma_I) * I_u^2 & G(\sigma_I) * I_u I_v \\ G(\sigma_I) * I_u I_v & G(\sigma_I) * I_v^2 \end{pmatrix} \tag{13.14}$$

which is a symmetric 2×2 matrix referred to variously as the structure tensor, autocorrelation matrix or second moment matrix. It captures the intensity structure of the local neighborhood and its eigenvalues provide a rotationally invariant description of the neighborhood. The elements of the A matrix are computed from the image gradients, squared or multiplied, and then smoothed using a weighting matrix. The latter reduces noise and improves the stability and reliability of the detector. The gradient images I_u and I_v are typically calculated using a derivative of Gaussian kernel method (Sect. 12.5.1.3) with a smoothing parameter σ_D.

An interest point (u, v) is one for which $s(\cdot)$ is high for *all* directions of the vector (δ_u, δ_v). That is, in whatever direction we move the window it rapidly becomes dissimilar to the original region. If we consider the original image I as a surface the eigenvalues of A are the principal curvatures of the surface at that point. If both eigenvalues are small then the surface is flat, that is the image region has approximately constant local intensity. If one eigenvalue is high and the other low, then the surface is ridge shaped which indicates an edge. If both eigenvalues are high the surface is sharply peaked which we consider to be a corner.◄

The Shi-Tomasi detector considers the strength of the corner, or *cornerness*, as the minimum eigenvalue

$$C_{ST}(u, v) = \min(\lambda_1, \lambda_2) \tag{13.15}$$

where λ_i are the eigenvalues of A. Points in the image for which this measure is high are referred to as "*good features to track*". The Harris detector◄ is based on this same insight but defines corner strength as

$$C_H(u, v) = \det(A) - k\,\mathrm{tr}(A) \tag{13.16}$$

and again a large value represents a strong, distinct, corner. Since $\det(A) = \lambda_1\lambda_2$ and $\mathrm{tr}(A) = \lambda_1 + \lambda_2$ the Harris detector responds when both eigenvalues are large and elegantly avoids computing the eigenvalues of A which has a somewhat higher computational cost.◄ A commonly used value for k is 0.04. Another variant is the Noble detector

$$C_N(u, v) = \frac{\det(A)}{\mathrm{tr}(A)} \tag{13.17}$$

which is arithmetically simple but potentially singular.

Typically the corner strength is computed for every pixel and results in a corner strength image. Then nonlocal maxima suppression is applied to only retain values that are greater than their immediate neighbors. A list of such points is created and sorted into descending corner strength. A threshold can be applied to only accept corners above a particular strength, or above a particular fraction of the strongest corner, or simply the N strongest corners.

The Toolbox provides a Harris corner detector which we will demonstrate using a real image

```
>> b1 = iread('building2-1.png', 'grey', 'double');
>> idisp(b1)
```

The Harris features are computed by

```
>> C = icorner(b1, 'nfeat', 200);
7497 corners found (0.8%),  200 corner features saved
```

which returns a vector of `PointFeature` objects. The detector found over 7 000 corners that were local maxima of the corner strength image and these comprised 0.8% of all pixels in the image. In this case we requested the 200 strongest corners. The vector contains the corners sorted by decreasing corner strength, and each `PointFeature` object contains the corner coordinate (u, v), the corner strength and a descriptor which comprises the unique elements of the structure tensor in vector form (A_{11}, A_{22}, A_{12}). The descriptor can be used as a simple signature of the corner to help match corresponding corners between different views.

Another approach to determining image curvature is to use the determinant of the Hessian (DoH). The Hessian is the matrix of second-order gradients at a point

$$H = \begin{pmatrix} I_{uu} & I_{uv} \\ I_{uv} & I_{vv} \end{pmatrix}$$

where $I_{uu} = \partial^2 I / \partial u^2$, $I_{vv} = \partial^2 I / \partial v^2$ and $I_{uv} = \partial^2 I / \partial u \partial v$. The determinant $\det(H)$ has a large magnitude when there is grey-level variation in two directions. However second derivatives accentuate image noise even more than first derivatives and the image must be smoothed first.

Recall from page 363 that lossy image compression such as JPEG removes high-frequency detail from the image, and this is exactly what defines a corner. Ideally corner detectors should be applied to images that have not been compressed and decompressed.

Sometimes referred to in the literature as the Plessey corner detector.

Evaluating eigenvalues for a 2×2 matrix involves solving a quadratic equation and therefore requires a square root operation.

The corners can be overlaid on the image as white squares

```
>> idisp(b1, 'dark');
>> C.plot('ws');
```

as shown in Fig. 13.21a. The 'dark' option to idisp reduces the brightness of the image to make the overlaid corner markers more visible. A closeup view is shown in Fig. 13.21b and we see the features are indeed often located on the corners of objects.

We also see that the corners tend to cluster unevenly, with a greater density in regions of high contrast and texture, and for some applications this can be problematic. To distribute corner points more evenly we can increase the distance used for nonlocal maxima suppression

```
>> Cs = icorner(b1, 'nfeat', 200, 'suppress', 10);
7497 corners found (0.7%),  200 corner features saved
```

by specifying a minimum distance between corners, in this case 10 pixels.

We can apply standard MATLAB operations and syntax to vectors of PointFeature objects, for example

```
>> length(C)
ans =
    200
```

and indexing

```
>> C(1:4)
ans =
(3,3), strength=2.97253e-05, ↵
  descrip=(0.00704357 0.00703274 0.00344741)
(600,662), strength=2.13105e-05, ↵
  descrip=(0.00568787   0.00448007 -0.000189851)
(24,277), strength=1.5516e-05, ↵
  descrip=(0.00577341   0.00361102 -0.00134506)
(54,407), strength=1.53644e-05, ↵
  descrip=(0.0062428 0.00301444 0.00016217)
```

where the display method shows the essential properties of the feature. We can also create expressions such as

```
>> C(1:5).strength
ans =
   1.0e-04 *
    0.2973    0.2131    0.1552    0.1536    0.1496
>> C(1).u
ans =
     3
```

To plot the coordinate of every fifth feature in the first 100 features is

```
>> C(1:5:100).plot()
```

The corner strength is computed at each pixel and can be optionally returned

```
>> [C,strength] = icorner(b1, 'nfeat', 200);
7497 corners found (0.7%),  200 corner features saved
```

and displayed as an image

```
>> idisp(strength, 'invsigned')
```

which is shown in Fig. 13.22a. We observe that the corner strength function is strongly positive (*blue*) for corner features and strongly negative (*red*) for linear features. A zoomed in view is shown in Fig. 13.22b which indicates that the detected corner is at the top of a peak of *cornerness* that is several pixels wide. The detected corner is a local

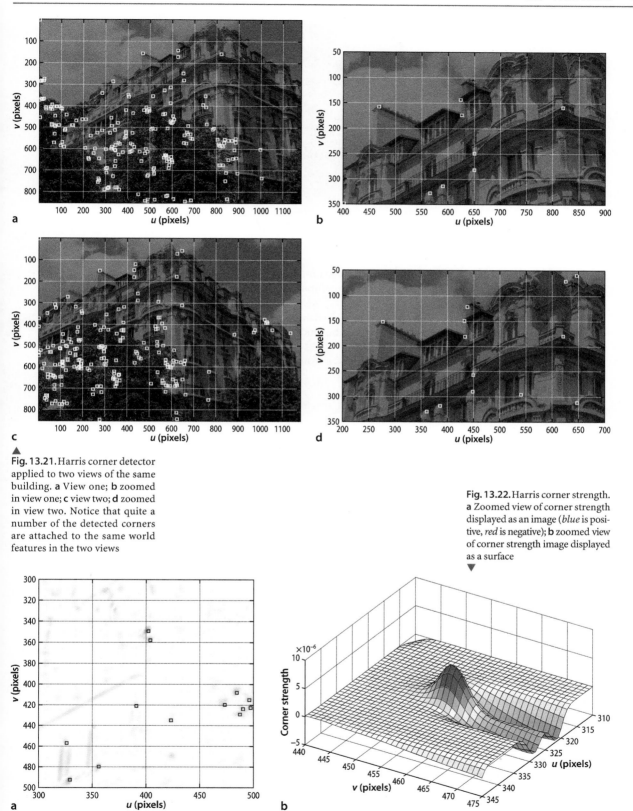

Fig. 13.21. Harris corner detector applied to two views of the same building. **a** View one; **b** zoomed in view one; **c** view two; **d** zoomed in view two. Notice that quite a number of the detected corners are attached to the same world features in the two views

Fig. 13.22. Harris corner strength. **a** Zoomed view of corner strength displayed as an image (*blue* is positive, *red* is negative); **b** zoomed view of corner strength image displayed as a surface

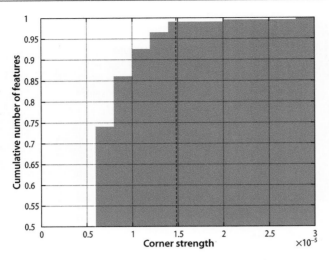

Fig. 13.23.
Cumulative histogram of corner strengths

maxima but we could use the surrounding values to estimate its location to subpixel accuracy (see Appendix J). This involves additional computation but can be enabled using the option `'interp'`.

A cumulative histogram of the strength of the 200 detected corners is shown in Fig. 13.23. The strongest corner has $C_H \approx 3 \times 10^{-5}$ but most are much weaker than this, only 2% of corners exceed half this value.

Consider another image of the same building taken from a different location

```
>> b2 = iread('building2-2.png', 'grey', 'double');
```

and the detected corners

```
>> C2 = icorner(b2, 'nfeat', 200);
7712 corners found (0.8%),  200 corner features saved
>> idisp(b2,'dark')
>> C2.plot('ws');
```

are shown in Fig. 13.21c and d. For many useful applications in robotic vision – such as tracking, mosaicing and stereo vision that we will discuss in the next chapter – it is important that corner features are detected at the same world points irrespective of variation in illumination or changes in rotation and scale between the two views. From Fig. 13.21 we see that many, but not all, of the features are indeed *attached* to the same world feature in both views.

The Harris detector is computed from image gradients and is therefore robust to offsets in illumination, and the eigenvalues of the structure tensor **A** are invariant to rotation. However the detector is not invariant to changes in scale. As we *zoom in* the gradients around the corner points become lower – the same change in intensity is spread over a larger number of pixels. This reduces the image curvature and hence the corner strength. The next section discusses a remedy for this using *scale-invariant* corner detectors.

For a color image the structure tensor is computed using the gradient images of the individual color planes which is slightly different to first converting the image to grey scale according to Eq. 10.11. In practice the use of color defies intuition – it makes surprisingly little difference for most scenes but adds significant computational cost. The `icorner` function accepts a large number of options: k, the derivative and smoothing kernel sizes σ_D and σ_I, absolute and/or relative corner strength threshold and enforcing a minimum distance between corners. The options `'st'` and `'noble'` allow computation of the corner measures Eq. 13.15 and Eq. 13.17 respectively. Details are provided in the online documentation.

13.3.2   Scale-Space Corner Detectors

The Harris corner detector introduced in the previous section works very well in practice but responds poorly to changes in scale. Unfortunately change in scale, due to changing camera to scene distance or zoom, is common in many real applications. We also notice that the Harris detector responds strongly to fine texture, such as the leaves of the trees in Fig. 13.21 but we would like to be able to detect features that are associated with larger-scale scene structure such as windows and balconies.

Figure 13.24 illustrates the fundamental principle of scale-space feature detection. We first load a synthetic image

```
>> im = iread('scale-space.png', 'double');
```

which is shown in Fig. 13.24a. The image contains four squares of different size: 5×5, 9×9, 17×17 and 33×33. The scale-space sequence is computed by applying a Gaussian kernel with increasing σ that results in the regions becoming increasingly blurred and smaller regions progressively disappearing from view. At each step in the sequence the Gaussian-smoothed image is convolved with the Laplacian kernel Eq. 12.5 which results in a strong negative responses for these bright blobs. ◄

With the Toolbox we compute the scale-space sequence by

```
>> [G,L,s] = iscalespace(im, 60, 2);
```

where the input arguments are the number of scale steps to compute, and the σ of the Gaussian kernel to be applied at each successive step. The function returns two 3-dimensional images, each a sequence of images where the last index corresponds to the scale. G is the image im at increasing levels of smoothing, L is the Laplacian of those smoothed images, and s is the corresponding scale. For example the fifth image in the Laplacian of Gaussian sequence (LoG) is displayed by

```
>> idisp(L(:,:,5), 'invsigned')
```

and has a scale of

```
>> s(5)
ans =
    4.0311
```

Figures 13.24b–e show the Laplacian of Gaussian at four different points in the scale-space sequence.

Figure 13.24f shows the magnitude of the Laplacian of Gaussian response as a function of scale, taken at the points corresponding to the center of each square in the input image. Each curve has a well defined peak, and the scale associated with the peak is proportional to the size of the region – the characteristic scale of the region.

If we consider the 3-dimensional image L as a volume then a scale-space feature point is any pixel that is a 3D maxima. That is, an element that is greater than its 26 neighbors in *all three* dimensions – its spatial neighbors at the current scale and at the scale above and below. Such points are detected by the function iscalemax

```
>> f = iscalemax(L, s)
f =
    (64,64), scale=2.91548, strength=1.96449
    (128,64), scale=4.06202, strength=1.72512
    (128,128), scale=18.1246, strength=1.54391
    (64,128), scale=8.97218, strength=1.54057
    (96,128), scale=15.5081, strength=0.345028
    (97,128), scale=14.7139, strength=0.34459
```

which returns an array of ScalePointFeature objects which are a subclass of PointFeature. Each object has properties for the feature's coordinate, strength and scale. The features are arranged in order of decreasing strength and we see that

We actually compute the difference of Gaussian approximation to the Laplacian, as illustrated in Fig. 13.27.

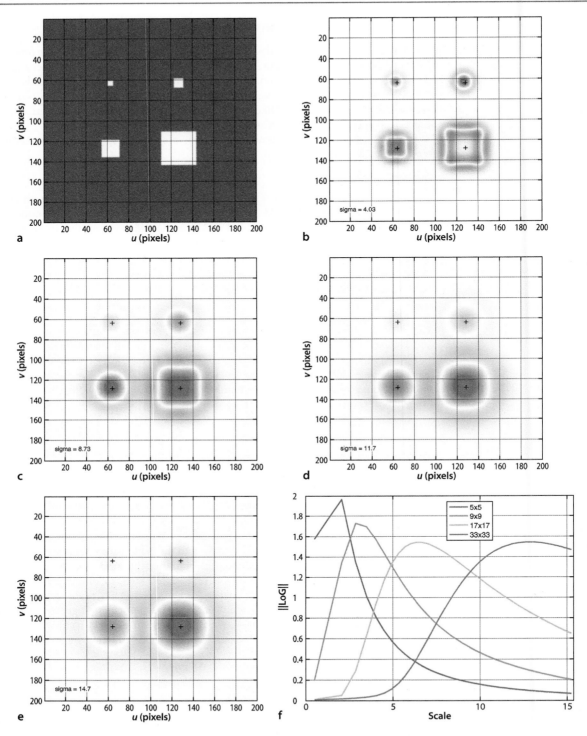

four have significant strength and correspond to the four white objects. We can superimpose the detected features on the original image

```
>> idisp(im)
>> f(1:4).plot('g+')
```

and the result is shown in Fig. 13.25.

◀ **Fig. 13.24.**
Scale-space example. **a** Synthetic image *I* with blocks of sizes 5×5, 9×9, 17×17, and 33×33; **b–e** Normalized Laplacian of Gaussian $\sigma^2 L * G(\sigma) * I$ for increasing values of scale, σ value indicated in lower left. False color is used: *red* is negative and *blue* is positive; **f** magnitude of Laplacian of Gaussian at center of each square (indicated by '+') versus σ

The scale associated with a feature can be easily visualized using circles of radius equal to the feature scale

```
>> f(1:4).plot_scale('r')
```

and the result is also shown in Fig. 13.25. We see that the identified features are located at the center of each object and that the scale of the feature is related to the size of the object. The region within the circle is known as the support region of the feature.

For a real image

```
>> im = iread('lena.pgm', 'double');
```

we compute the scale-space in eight large steps with $\sigma = 8$

```
>> [G,L] = iscalespace(im, 8, 8);
```

which we can *flatten* and display

```
>> idisp(G, 'flatten', 'wide', 'square');
>> idisp(L, 'flatten', 'wide', 'square', 'invsigned');
```

as shown in Fig. 13.26. From left to right we see the eight levels of scale. The Gaussian sequence of images becomes increasing blurry. In the Laplacian of Gaussian sequence the dark eyes are strongly positive (blue) blobs at low scale and the light colored hat becomes a strongly negative (red) blob at high scale.

Convolving the original image with a Gaussian kernel of increasing σ results in the kernel size, and therefore the amount of computation, growing at each scale step. Recalling the properties of a Gaussian from page 377, a Gaussian convolved with a

Fig. 13.25. ▶
Synthetic image with overlaid feature center and scale indicator

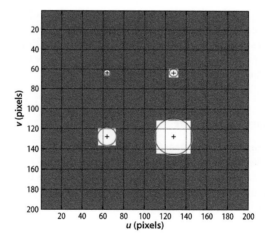

Fig. 13.26. Scale-space sequence for $\sigma = 2$, (*top*) Gaussian sequence, (*bottom*) Laplacian of Gaussian sequence
▼

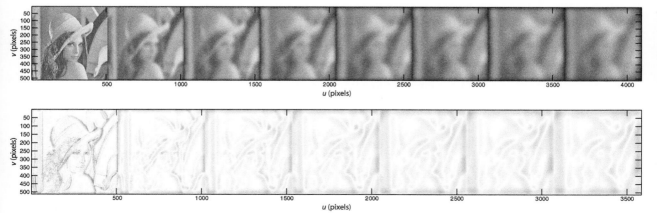

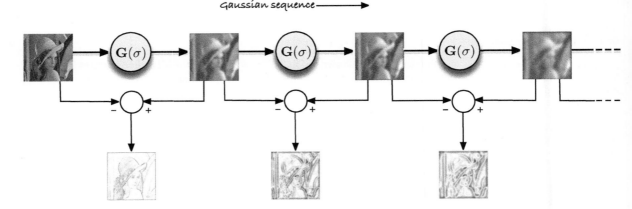

Fig. 13.27. Schematic for calculation of Gaussian and Laplacian of Gaussian scale-space sequence

Gaussian is another wider Gaussian. Instead of convolving our original image with ever wider Gaussians, we can repeatedly apply the same Gaussian to the previous result. We also recall from page 385 that the LoG kernel is approximated by the difference of two Gaussians. Using the properties of convolution we can write

$$\big(G(\sigma_1) - G(\sigma_2)\big) * I = G(\sigma_1) * I - G(\sigma_2) * I$$

where $\sigma_1 > \sigma_2$. The difference of Gaussian operator applied to the image is equivalent to the difference of the image at two different levels of smoothing. If we perform the smoothing by successive application of a Gaussian we have a sequence of images at increased levels of smoothing. The difference between successive steps in the sequence is therefore an approximation to the Laplacian of Gaussian. Figure 13.27 shows this in diagrammatic form.

13.3.2.1   Scale-Space Point Feature

The scale-space concepts just discussed underpin a number of popular feature detectors which find salient points within an image and determines their scale and also their orientation. The Scale-Invariant Feature Transform (SIFT) is based on the maxima in a difference of Gaussian sequence. The Speeded Up Robust Feature (SURF) is based on the maxima in an approximate Hessian of Gaussian sequence.

To illustrate we will compute the SURF features for the building image used previously

```
>> sf1 = isurf(b1, 'nfeat', 200)
2667 corners found (0.3%), 200 corner features saved
sf1 =
200 features (listing suppressed)
 Properties: image_id theta scale u v strength descriptor
```

which returns an array of 200 `SurfPointFeature` objects which are a subclass of `ScalePointFeature`. For example the first feature is

```
>> sf1(1)
ans = (117.587,511.978), theta=0.453513, scale=2.16257,
strength=0.0244179, descrip= ..
```

Each object includes the feature's coordinate (estimated to subpixel precision), scale, orientation, and a descriptor which is a 64-element vector. Orientation is defined by the dominant edge direction within the support region.

This image contains nearly 3 000 SURF features but, as we did earlier with the Harris corner features, we requested the 200 strongest which we plot

```
>> idisp(b1, 'dark');
>> sf1.plot_scale('g', 'clock')
```

and the result is shown in Fig. 13.28. The `plot_scale` method draws a circle around the feature's location with a radius that indicates its scale – the size of the support region. The option `'clock'` draws a radial line which indicates the orientation of the SURF feature.

Feature scale varies widely and a histogram

```
>> hist(sf1.scale, 100);
```

shown in Fig. 13.29 indicates that there are many small features associated with fine image detail and texture. The bulk of the features have a scale less than 25 pixels but some have scales over 40 pixels. The `isurf` function accepts a number of options which are described in the online documentation.

The SURF algorithm is more than just a scale-invariant feature detector, it also computes a very robust *descriptor*. The descriptor is a 64-element vector that encodes the image gradient in subregions of the support region in a way which is invariant to brightness, scale and rotation. This enables feature descriptors to be unambiguously matched to a descriptor of the same world point in another image even if their scale and orientation are quite different. The difference in position, scale and orientation of the matched features gives some indication of the relative camera motion between the two views. Matching features between scenes is crucial to the problems that we will address in the next chapter.

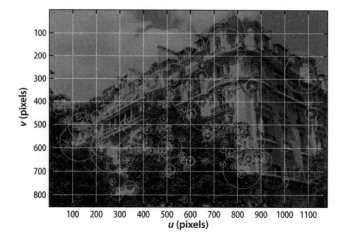

Fig. 13.28.
SURF descriptors showing the support region (scale) and orientation as a radial line

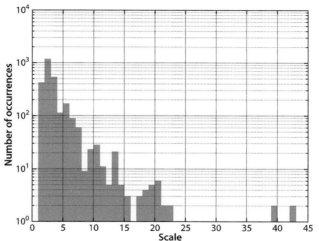

Fig. 13.29.
Histogram of feature scales shown with logarithmic vertical scale

13.4   Wrapping Up

In this chapter we have discussed the extraction of features from an image. Instead of considering the image as millions of independent pixel values we succinctly describe regions within the image that correspond to distinct objects in the world. For instance we can find regions that are homogeneous with respect to intensity or color and describe them in terms of features such as a bounding box, centroid, equivalent ellipse, aspect ratio, circularity and perimeter shape. Features have invariance properties with respect to translation, rotation about the optical axis and scale which are important for object recognition. Straight lines are common visual features in man-made environments and we showed how to find and describe distinct straight lines in an image using the Hough transform.

We also showed how to find interest points that can reliably *associate* to particular points in the world irrespective of the camera view. These are key to techniques such as camera motion estimation, stereo vision, image retrieval, tracking and mosaicing that we will discuss in the next chapter.

MATLAB Notes

The hierarchy of feature classes used in the Toolbox is shown in Fig. 13.30. A list of Toolbox functions and MATLAB equivalents is given in Table 13.2. The latter come from the Image Processing and the Computer Vision System Toolbox which have a large number of additional functions, some of which support code generation or operation with Simulink.

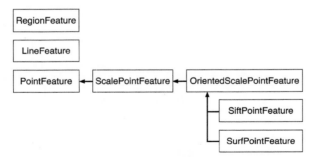

Fig. 13.30.
Feature class hierarchy used in the Toolbox

Table 13.2.
List of feature extraction functions and equivalence with MATLAB Image Processing Toolbox and Computer Vision System Toolbox functions

Operation	RVC Toolbox	MATLAB
Region features		
▪ blob analysis	`iblobs`	`regionprops, bwperim`
▪ connectivity	`ilabel`	`bwlabel`
▪ Otsu threshold	`otsu`	`graythresh`
▪ Niblack threshold	`niblack`	
Line features		
▪ Hough transform	`Hough`	`hough`
Point features		
▪ Harris	`icorner`	`detectHarrisFeatures`
▪ SURF	`isurf`	`detectSURFeatures`
▪ SIFT	`isift`	
▪ FAST		`detectFASTFeatures`
▪ MSER	`imser`[a]	`detectMSERFeatures`
▪ BRISK		`detectBRISKFeatures`

[a] Requires VLFeat to be installed.

Further Reading

This chapter has presented a classical bottom up approach for feature extraction, starting with pixels and working our way up to higher level concepts such as regions and lines. Prince (2012) and Szeliski (2011) both provide a good introduction to high-level vision using probabilistic techniques that can be applied to problems such as object recognition, for example face recognition, and image retrieval. In the last few years computer vision, particularly object recognition, has undergone a revolution using deep convolutional neural networks. These have demonstrated very high levels of accuracy in locating and recognizing objects against complex background despite changes in viewpoint and illumination and resources are available at http://deeplearning.net.

Region features. Region-based image segmentation and blob analysis are classical techniques covered in many books and papers. Gonzalez and Woods (2008) and Szeliski (2011) provide a thorough treatment of the methods introduced in this chapter, in particular thresholding and boundary descriptors. Otsu's algorithm for threshold determination was introduced in Otsu (1975), and the Niblack algorithm for adaptive thresholding was introduced in Niblack (1985). The book by Nixon and Aguado (2012) expands on material covered in this chapter and introduces techniques such as deformable templates and boundary descriptors. The Freeman chain code was first described in Freeman (1974). Flusser (2000) has shown that the seven moments proposed by Hu (1962), and described on page 434, are in fact not independent since $\phi_3 = (\phi_5^2 + \phi_7^2) / \phi_4^3$.

In addition to region homogeneity based on intensity and color it also possible to describe the texture of regions – a spatial pattern of pixel intensities whose statistics can be described (Gonzalez and Woods 2008). Regions can then be segmented according to texture, for example a smooth road versus textured grass.

Clustering of data is an important topic in machine learning (Bishop 2006). In this chapter we have used a simple implementation of k-means, which is far from state-of-the-art in clustering, and requires the number of clusters to be known in advance. More advanced clustering algorithms are hierarchical and employ data structures such as kd-trees to speed the search for neighboring points. The initialization of the cluster centers is also critical to performance. Szeliski (2011) introduces more general clustering methods as well as graph-based methods for computer vision. The graphcuts algorithm for segmentation was described by Felzenszwalb and Huttenlocher (2004) and the Toolbox graph-cuts implementation is based on code by Pedro Felzenszwalb and available at http://cs.brown.edu/~pff/segment/. The maximally stable extremal region (MSER) algorithm is described by Matas et al. (2004) and the Toolbox implementation is based on the work of Andrea Vedaldi and Brian Fulkerson which is available at http://vlfeat.org. The Berkeley Segmentation Dataset at http://www.eecs.berkeley.edu/Research/Projects/CS/vision/bsds contains numerous complex real-world images each with several human-made segmentations.

Early work on using text recognition for robotics is described by Posner et al. (2010), while Lam et al. (2015) describe the application of OCR to parsing floor plans of buildings for robot navigation. A central challenge with OCR of real-world scenes is to determine which parts of the scene contain text and should be passed to the OCR engine. A powerful text detector is the stroke width transform described by Li et al. (2014). The MATLAB `ocr` function is based on the Tesseract open-source OCR engine which is available at https://github.com/tesseract-ocr and described by Smith (2007).

Line features. The Hough transform was first first described in U.S. Patent 3,069,654 "Method and Means for Recognizing Complex Patterns" by Paul Hough, and its history is discussed in Hart (2009). The original application was automating the analysis of bubble chamber photographs and it used the problematic slope-intercept parametrization for lines. The currently known form with the (θ, ρ) parametrization was first described in Duda and Hart (1972) as a "generalized Hough transform" and is available at http://www.ai.sri.com/pubs/files/tn036-duda71.pdf. The Hough transform is covered in textbooks such as Szeliski (2011) and Gonzalez and Woods (2008). The latter has a good discussion on shape fitting in general and estimators that are robust with respect to outlier data points. The basic Hough transform has been extended in many ways and there is a large literature. A useful review of the transform and its variants is presented in Leavers (1993). The transform can be generalized to other shapes (Ballard 1981) such as circles of a fixed size where votes are cast for the coordinates of the circle's center. For circles of unknown size a three-dimensional voting array is required for the circle's center and radius.

Point features. The literature on interest operators dates back to the early work of Moravec (1980) and Förstner (Förstner and Gülch 1987; Förstner 1994). The Harris corner detector (Harris and Stephens 1988) became very popular for robotic vision application in the late 1980s since it was able to run in real-time on computers of the day and the features were quite stable (Tissainayagam and Suter 2004) from image to image. The Noble detector is described in Noble (1988). The work of Shi, Tomasi, Lucas and Kanade (Shi and Tomasi 1994; Tomasi and Kanade 1991) led to the Shi-Tomasi detector and the Kanade-Lucas-Tomasi (KLT) tracker. Good surveys of the relative performance of many corner detectors include those by Deriche and Giraudon (1993) and Mikolajczyk and Schmid (2004).

Scale-space concepts have long been known in computer vision. Koenderink (1984), Lindeberg (1993) and ter Haar Romeny (1996) are a readable introduction to the topic. Scale-space was applied to classic corner detectors creating hybrid detectors such as scale-Harris (Mikolajczyk and Schmid 2004). An important development in scale-space feature detectors was the scale-invariant feature transform (SIFT) introduced in the early 2000s by Lowe (2004) and was a significant improvement for applications such as tracking and object recognition. Unusually, and perhaps unfortunately, it was patented and could not be used in this book. Nature abhors a vacuum and an effective alternative called Speeded Up Robust Features (SURF) was developed (Bay et al. 2008). The Toolbox function `isurf` wraps a MATLAB implementation by Dirk-Jan Kroon and available at http://www.mathworks.com/matlabcentral/fileexchange/28300-opensurf-including-image-warp, which in turn is based on the OpenSurf implementations in C++ and C# by Chris Evans which is now hosted at https://github.com/amarburg. GPU-based parallel implementations have also been developed. The SIFT and SURF detectors do give different results and they are compared in Bauer et al. (2007). The Toolbox SIFT detector function `isift` returns a feature vector of class `SiftPointFeature` and is a wrapper for the MATLAB implementation from http://www.vlfeat.org which you will need to download and compile.

Many other interest point detectors and features have been, and continue to be proposed. FAST by Rosten et al. (2010) has very low computational requirements and high repeatability, and C and MATLAB software resources are available at http://www.edwardrosten.com/work/fast.html. CenSurE by Agrawal et al. (Agrawal et al. 2008) claims higher performance than SIFT, SURF and FAST at lower cost. BRIEF by Calonder et al. (2010) is not a feature detector but is a low cost and compact feature descriptor, requiring just 256 bits instead of 64 floating-point numbers per feature. Other feature descriptors include histogram of oriented gradients (HOG), oriented FAST and rotated BRIEF (ORB), binary robust invariant scaleable keypoint (BRISK), fast retina keypoint (FREAK), aggregate channel features (ACF), vector of locally aggregated descriptors (VLAD), random ferns and many many more.

Local features have many advantages and are quite stable from frame to frame, but for outdoor applications the feature locations and the descriptors vary considerably with changes in lighting conditions, see for example Valgren and Lilienthal (2010). Night and day are obvious examples but even over a period of a few hours the descriptors change considerably. Over seasons the appearance change can be drastic: trees with or without leaves; the ground covered by grass or snow, wet or dry. Enabling robots to recognize places despite their changing appearance is the research field of robust place recognition which is introduced in Lowry et al. (2015).

Exercises

1. Grey-level classification
   a) Experiment with `ithresh` on the images `castle.png` and `castle2.png`.
   b) Experiment with the Niblack algorithm and vary the value of k and window size.
   c) Apply `iblobs` to the output of the MSER segmentation. Develop an algorithm that uses the width and height of the bounding boxes to extract just those blobs that are letters.
   d) The function `imser` has many parameters: `'Delta'`, `'MinDiversity'`, `'MaxVariation'`, `'MinArea'`, `'MaxArea'`. Explore the effect of adjusting these.
   e) Apply the function `igraphcut` to the `castle2.png` image. Understand and adjust the parameters to improve performance.
   f) Load the image `adelson.png` from page 307 and attempt to segment the letters A and B.
2. Color classification
   a) Change k, the number of clusters, in the color classification examples. Is there a best value?
   b) k-means with `'random'` or `'spread'` options performs a randomized initialization. Run k-means several times and determine how different the final clusters are.
   c) Write a function that determines which of the clusters represents the targets, that is, the yellow cluster or the red cluster.
   d) Apply the function `igraphcut` to the targets and garden image. How does it perform? Understand and adjust the parameters to improve performance.
   e) Experiment with the parameters of the morphological "cleanup" used for the targets and garden images.
   f) Write code that loops over images captured from your computer's camera, applies a classification, and shows the result. The classification could be a greyscale threshold or color clustering to a pre-learned set of color clusters (see `colorkmeans`).
3. Blobs. Create an image of an object with several holes in it. You could draw it and take a picture, export it from a drawing program, or write code to generate it.
   a) Determine the outer, *inner* and total boundaries of the object.
   b) Place small objects within the holes in the objects. Write code to display the topological hierarchy of the blobs in the scene.
   c) For the same shape at different scales plot how the circularity changes as a function of scale. Explain the shape of this curve?
   d) Create a square object and plot the estimated and true perimeter as a function of the square's side length. What happens when the square is small?
   e) Create an image of a simple scene with a number of different shaped objects. Using the shape invariant features (aspect ratio, circularity) to create a simple shape classifier. How well does it perform? Repeat using the Hu moment features.
   f) Repeat the boundary matching example with some objects that you create. Modify the code to create a plot of edge-segment angle (k) versus θ and repeat the boundary matching example.
   g) Another commonly used feature, not supported by the Toolbox, is the aligned rectangle. This is the smallest rectangle whose sides are aligned with the axes of the equivalent ellipse and which entirely contains the blob. The aspect ratio of this rectangle and the ratio of the blob's area to the rectangle's area are each scale and rotation invariant features. Write code to compute this rectangle, overlay the rectangle on the image, and compute the two features.
   h) Write code to trace the perimeter of a blob.
4. Experiment with the `ocr` function.
   a) What is the effect of a larger region of interest?
   b) Capture your own image and attempt to read the text in it. How does accuracy vary with text size, contrast or orientation?

5. Hough transform
   a) Experiment with varying the size of the Hough accumulator.
   b) Experiment with using the Sobel edge operator instead of Canny.
   c) Experiment with varying the parameters `'suppress'`, `'interpSize'`, `'EdgeThresh'`, `'houghThresh'`.
   d) Apply the Hough transform to one of your own images.
   e) Write code that loops over images captured your computer's camera, finding the two dominant lines and overlaying them on the image.
6. Corner detectors
   a) Experiment with the Harris detector by changing the parameters k, σ_D and σ_I.
   b) Compare the performance of the Harris, Noble and Shi-Tomasi corner detectors.
   c) Implement the Moravec detector and compare to Harris detector.
   d) Create a smoothed second derivative I_{uu}, I_{vv} and I_{uv}.

14

Using Multiple Images

In the previous chapter we learned about corner detectors which find particularly distinctive *points* in a scene. These points can be reliably detected in different views of the same scene irrespective of viewpoint or lighting conditions. Such points are characterized by high image gradients in orthogonal directions and typically occur on the corners of objects. However the 3-dimensional coordinate of the corresponding world point was lost in the perspective projection process which we discussed in Chap. 11 – we mapped a 3-dimensional world point to a 2-dimensional image coordinate. All we know is that the world point lies along some ray in space corresponding to the pixel coordinate, as shown in Fig. 11.6. To recover the missing third dimension we need additional information. In Sect. 11.2.3 the additional information was camera calibration parameters plus a geometric object model, and this allowed us to estimate the object's 3-dimensional pose from 2-dimensional image data.

In this chapter we consider an alternative approach in which the additional information comes from *multiple* views of the same scene. As already mentioned the pixel coordinates from a single view constrain the world point to lie along some ray. If we can locate the same world point in another image, taken from a different but known pose, we can determine another ray along which that world point must lie. The world point lies at the intersection of these two rays – a process known as triangulation or 3D reconstruction. Even more powerfully, if we observe sufficient points, we can estimate the 3D motion of the camera between the views as well as the 3D structure of the world.◄

Almost! We can determine the translation of the camera only up to an unknown scale factor, that is, the translation is $\lambda t \in \mathbb{R}^3$ where the direction of t is known but λ is not.

The underlying challenge is to find the same world point in multiple images. This is the *correspondence problem*, an important but nontrivial problem that we will discuss in Sect. 14.1. In Sect. 14.2 we revisit the fundamental geometry of image formation developed in Chap. 11 for the case of a single camera. If you haven't yet read that chapter, or it's been a while since you read it, it would be helpful to (re)acquaint yourself with that material. We extend the geometry to encompass multiple image planes and show the geometric relationship between pairs of images. Stereo vision is an important technique for robotics where information from two images of a scene, taken from different viewpoints, is combined to determine the 3-dimensional structure of the world. We discuss sparse and dense approaches to stereo, and reconstruction, in some detail in Sect. 14.3. Bundle adjustment is a very general approach to combining information from many cameras and is introduced in Sect. 14.4. The 3-dimensional information that is created is typically represented as a *point cloud*, a set of 3D points, and techniques for plane fitting and alignment of such data are introduced in Sect. 14.5. For some applications we can use RGBD cameras which return depth as well as color information and the underlying principle of structured light is introduced in Sect. 14.6.

We finish this chapter, and this part of the book, with four application examples based on the concepts we have learned. Section 14.7.1 describes how we can transform an image with obvious perspective distortion into one without, effectively synthesizing the view from a virtual camera at a different location. Section 14.7.2 describes mosaicing which is the process of taking consecutive images from a moving camera and *stitching* them together to form one large virtual image. Section 14.7.3 describes image retrieval which is the problem of finding which image, from an existing set of images, is most similar to some new image. This can be used by a robot to determine whether it has visited a particular place, or seen the same object, before. Section 14.7.4 describes how we can process a sequence of images from a moving camera to locate consistent world points and to estimate the camera motion and 3-dimensional world structure.

14.1   Feature Correspondence

Correspondence is the problem of finding the pixel coordinates in two different images that correspond to the same point in the world.▶ Consider the pair of real images

```
>> im1 = iread('eiffel2-1.jpg', 'mono', 'double');
>> im2 = iread('eiffel2-2.jpg', 'mono', 'double');
```

shown in Fig. 14.1. They show the same scene viewed from two different positions using two different cameras – the pixel size, focal length and number of pixels for each image are quite different. The scenes are complex and we see immediately that determining correspondence is not trivial. More than half the pixels in each scene correspond to blue sky and it is impossible to match a blue pixel in one image to the corresponding blue pixel in the other – these pixels are insufficiently distinct. This situation is common and can occur with homogeneous image regions such as dark shadows, smooth sheets of water, snow or smooth man-made objects such as walls or the bodies of cars.

The solution is to choose only those points that are distinctive. We can use the interest point detectors that we introduced in the last chapter to find Harris corner features

```
>> hf = icorner(im1, 'nfeat', 200);
>> idisp(im1, 'dark'); hf.plot('gs');
```

or SURF features◥

```
>> sf = isurf(im1, 'nfeat', 200);
>> idisp(im1, 'dark'); sf.plot_scale('g');
```

and these are shown in Fig. 14.2. We have simplified the problem – instead of millions of pixels to deal with we have just 200 distinctive points.

Consider the general case of two sets of features points: $\{^1\boldsymbol{p}_i \in \mathbb{Z}^2, i = 1 \cdots N_1\}$ in the first image and $\{^2\boldsymbol{p}_j \in \mathbb{Z}^2, j = 1 \cdots N_2\}$ in the second image. Since these are distinctive image points we would expect a significant number of points in image one would cor-

This is another example of the data association problem.

The SURF detector cannot process a color image, it converts it to greyscale. The Harris detector computes the squared gradients for the individual color planes separately and then combines them. All detectors in the Toolbox can process an image sequence provided as a matrix with more than two dimensions. There is ambiguity between a color image and an image sequence of length three. If the image's third dimension is three it is deemed to be a color image, not a sequence. A four-dimensional image is unambiguous as a sequence of color images.

Fig. 14.1. Two views of the Eiffel tower. The images were captured approximately simultaneously using two different handheld digital cameras. **a** 7 Mpix camera with $f = 7.4$ mm; **b** 10 Mpix camera with $f = 5.2$ mm (photo by Lucy Corke). The images have quite different scale and the tower is 700 and 600 pixels tall in **a** and **b** respectively. The camera that captured image **b** is held by the person in the bottom-right corner of **a**

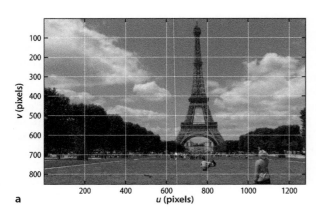

a

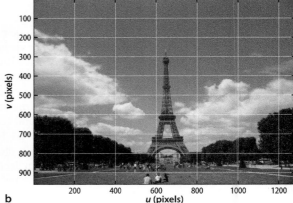

b

respond to points found in image two. The problem is to determine which $(^2u_j, {}^2v_j)$, if any, corresponds to each $(^1u_i, {}^1v_i)$.

We cannot use the feature coordinates to determine correspondence – the features will have different coordinates in each image. For example in Fig. 14.1 we see that most features are lower in the right-hand image. We cannot use the intensity or color of the pixels either. Variations in white balance, illumination and exposure setting make it highly unlikely that corresponding pixels will have the same value. Even if intensity variation was eliminated there are likely to be tens of thousands of pixels in the other image with exactly the same intensity value – it is not sufficiently unique. We need some richer way of *describing* each feature.

In practice we describe the region of pixels *around* the corner point which provides a distinctive and unique description of the corner point and its immediate surrounds – the feature descriptor. In the Toolbox the feature descriptor for a corner point is a vector – the `descriptor` property of the `PointFeature` superclass. For the Harris corner feature the descriptor

```
>> hf(1).descriptor'
ans =
     0.0805    0.0821    0.0371
```

is a 3-vector that contains the unique elements of the structure tensor Eq. 13.14. This low-dimensional descriptor is computationally cheap since the elements were already computed in order to determine corner strength. These descriptor elements are gradients which have the advantage of being robust to offsets in image intensity. The similarity of two descriptors is based on Euclidean distance and is zero for a perfect match. For example, the similarity of corner features one and two is

```
>> hf(1).distance( hf(2) )
ans =
     0.0518
```

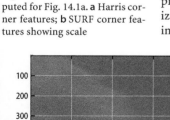

If the world point is not visible in image two then the most similar feature will be an incorrect match.

but it is difficult to know whether this value represents strong similarity or not since the units are not very intuitive. Typically we would compare feature $^1f_i \in \mathbb{R}^M$ with all features in the other image $\{^2f_j \in \mathbb{R}^M, j = 1 \cdots N_2\}$ and choose the one that is most similar. ◄ However a short descriptor vector like this is still insufficiently distinctive and prone to incorrect matching. Feature descriptors are often referred to as feature vectors.

We can create a large descriptor vector by representing the square window around the feature point as a vector. For example

```
>> hf = icorner(im1, 'nfeat', 200, 'color', 'patch', 5)
```

Fig. 14.2. Corner features computed for Fig. 14.1a. **a** Harris corner features; **b** SURF corner features showing scale

creates a 121-element descriptor vector for each corner point from the window of specified half-width around the feature point – in this case an 11×11 window. The pixel values are offset by the mean value, rearranged into a vector and then normalized to create a unit vector. We can use the ZNCC similarity measure from Table 12.1 in 1-dimensional form to compare these descriptor vectors

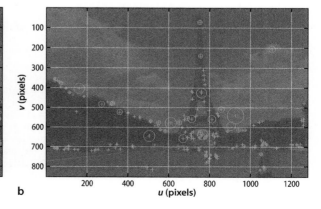

$$s = \frac{\sum_{i=1}^{N}\left(I_1[i] - \overline{I}_1\right)\left(I_2[i] - \overline{I}_2\right)}{\sqrt{\sum_{i=1}^{N}\left(I_1[i] - \overline{I}_1\right)^2 \cdot \sum_{i=1}^{N}\left(I_2[i] - \overline{I}_2\right)^2}}$$

$$= \underbrace{\frac{I_1[i] - \overline{I}_1}{\sqrt{\sum_{i=1}^{N}\left(I_1[i] - \overline{I}_1\right)^2}}}_{\hat{f}_1} \cdot \underbrace{\frac{I_2[i] - \overline{I}_2}{\sqrt{\sum_{i=1}^{N}\left(I_2[i] - \overline{I}_2\right)^2}}}_{\hat{f}_2}$$

(14.1)

which we have factored into the dot product of the descriptor unit-vectors associated with each image patch. Determining the similarity of two descriptors using normalized cross-correlation is simply the dot product of two descriptors and the resulting similarity measure $s \in [-1, 1]$ has some meaning – perfect match is $s = 1$ and $s \geq 0.8$ is typically considered a good match. For the example above

```
>> hf(1).ncc( hf(2) )
ans =
    -0.0292
```

the correlation score indicates a poor match. This descriptor is distinctive and invariant to changes in image intensity but is not invariant to scale or rotation. Other descriptors of the surrounding region that we could use include census and rank values as well as histograms of intensity or color. Histograms have the advantage of being invariant to rotation but they say nothing about the spatial relationship between the pixels, that is, the same pixel values in a completely different spatial arrangement have the same histogram.

The SURF algorithm computes a 64-element descriptor► vector to describe the feature point in a way that is scale and rotationally invariant, and based on the pixels within the feature's support region. It is created from the image in the scale-space sequence corresponding to the feature's scale and rotated according to the feature's orientation. The vector is normalized to a unit vector to increase its invariance to changes in image intensity. Similarity between descriptors is based on Euclidean distance. This descriptor is quite invariant to image intensity, scale and rotation. SURF is both a corner detector and a descriptor, whereas the Harris operator is just a corner detector which must be used with one of a number of different descriptors.►

For the remainder of this chapter we will use SURF features. They are computationally more expensive but pay for themselves in terms of the quality of matches between widely different views of the same scene. We compute SURF features for each image

A 128-element vector can be created by passing the option `'extended'` to `isurf`.

It is conceivable to use the SURF descriptor with a Harris corner point.

```
>> sf1 = isurf(im1)
sf1 =
1288 features (listing suppressed)
  Properties: theta image_id scale u v strength descriptor
>> sf2 = isurf(im2)
sf2 =
1426 features (listing suppressed)
  Properties: theta image_id scale u v strength descriptor
```

which results in two vectors of `SurfPointFeature` objects. Over a thousand corner features were found in each image.

Detectors versus descriptors. When matching world feature points, or landmarks, between different views we must first *find* points that are distinctive. This is the job of the detector and results in a coordinate (u, v) and perhaps a scale factor or orientation. The second task is to *describe* the region around the point in a way that allows it to be matched as decisively as possible with the region around the corresponding point in the other view. This is the descriptor which is typically a long vector formed from pixel values, histograms, gradients, histograms of gradient and so on. There are many detectors to choose from: Harris and variants, Shi-Tomasi, FAST, AGAST, MSER etc.; as well as many descriptors: ORB, BRISK, FREAK, CenSurE (aka STAR), HOG, ACF etc. Some algorithms such as SIFT and SURF define both a detector and a descriptor. The SIFT descriptor is a form of HOG descriptor.

Next we match the two sets of SURF features based on the distance between the SURF descriptors

```
>> m = sf1.match(sf2)
m =
644 corresponding points (listing suppressed)
```

which results in a vector of `FeatureMatch` objects that represents 644 *candidate-corresponding* points. The first five candidate correspondences◄ are

We refer to them as candidates because although they are very likely to correspond this has not yet been confirmed.

```
>> m(1:5)
ans =
(819.56, 358.557) <-> (708.008, 563.342), dist=0.002137
(1028.3, 231.748) <-> (880.14, 461.094), dist=0.004057
(1027.6, 571.118) <-> (885.147, 742.088), dist=0.004297
(927.724, 509.93) <-> (800.833, 692.564), dist=0.004371
(854.35, 401.633) <-> (737.504, 602.187), dist=0.004417
```

which shows the feature coordinate in the first and second image, as well as the Euclidean distance between the two feature vectors. The matches are ordered by decreasing similarity, and a threshold on feature similarity has been applied.

We can overlay a subset of these matches on the original image pair

```
>> idisp({im1, im2}, 'dark')
>> m.subset(100).plot('w')
```

and the result is shown in Fig. 14.3. White lines connect the matched features in each image and the lines show a consistent pattern. Most of these connections seem quite sensible, but a few are quite obviously incorrect and we will deal with these shortly. Note that we passed a cell-array of images to `idisp` which it displays horizontally tiled as a single image. The `subset` method of the `FeatureMatch` class returns a vector with the specified number of `FeatureMatch` objects sampled evenly from the original vector. If all correspondences were shown we would just see a solid white mass.

The correspondences can be obtained via an optional return value

```
>> [m,corresp] = sf1.match(sf2);
>> corresp(:,1:5)
ans =
       215         389         357        1044         853
       246         418         312        1240         765
```

which is a matrix with one column per correspondence. The first column indicates that feature 215 in image one matches feature 246 in image two and so on. In terms of workspace variables this is `sf1(215)` and `sf2(246)`.

The Euclidean distance between the matched feature descriptors is given by the `distance` property and the distribution of these, with no thresholding applied, is

```
>> m2 = sf1.match(sf2, 'all');
>> histogram(m2.distance, 'Normalization', 'cdf')
```

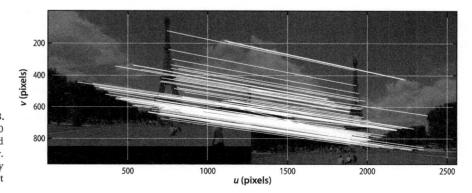

Fig. 14.3.
Feature matching. Subset (100 out of 1 664) of matches based on SURF descriptor similarity. We note that a few are clearly incorrect

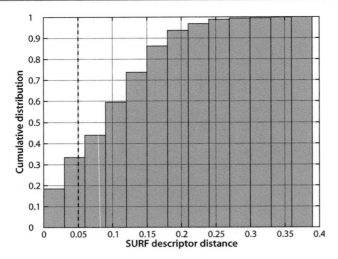

Fig. 14.4.
Cumulative distribution of
feature distance

shown in Fig. 14.4. It shows that 35% of all matches have descriptor distances below 0.05 whereas the maximum distance can be over ten times larger – such matches are less likely to be valid. We can specify a distance threshold

```
>> mm = sf1.match(sf2, 'thresh', 0.05);
```

but choosing the threshold value is always problematic. By default the method selects all matches whose distance is less than the median of all distances. Alternatively, we could choose to take the N best matches

```
>> mm = sf1.match(sf2, 'top', N);
```

Feature matching is computationally expensive – it is an $O(N^2)$ problem since every feature descriptor in one image must be compared with every feature descriptor in the other image. More sophisticated systems store the descriptors in a data structure like a kd-tree so that similar descriptors – nearest neighbors in feature space – can be easily found.

Although the quality of matching shown in Fig. 14.3 looks quite good there are a few obviously incorrect matches in this small subset. We can discern a pattern in the lines joining the corresponding points, they are slightly converging and sloping down to the right. This pattern is a function of the relative pose between the two camera views, and understanding this is key to determining which of the candidate matches are correct. That is the topic of the next section.

14.2   Geometry of Multiple Views

We start by studying the geometric relationships between images of a single point **P** observed from two different viewpoints and this is shown in Fig. 14.5. This geometry could represent the case of two cameras simultaneously viewing the same scene, or one moving camera taking a picture from two different viewpoints.► The center of each camera, the origins of {1} and {2}, plus the world point **P** defines a plane in space – the epipolar plane. The world point **P** is projected onto the image planes of the two cameras at points $^1\mathbf{p}$ and $^2\mathbf{p}$ respectively, and these points are known as conjugate points.

Consider image one. The image point $^1\mathbf{e}$ is a function of the position of camera two. The image point $^1\mathbf{p}$ is a function of the world point **P**. The camera center, $^1\mathbf{e}$ and $^1\mathbf{p}$ define the epipolar plane and hence the epipolar line $^2\ell$ in image two. By definition the conjugate point $^2\mathbf{p}$ must lie on that line. Conversely $^1\mathbf{p}$ must lie along the epipolar line in image one $^1\ell$ that is defined by $^2\mathbf{p}$ in image two.

Assuming the point does not move.

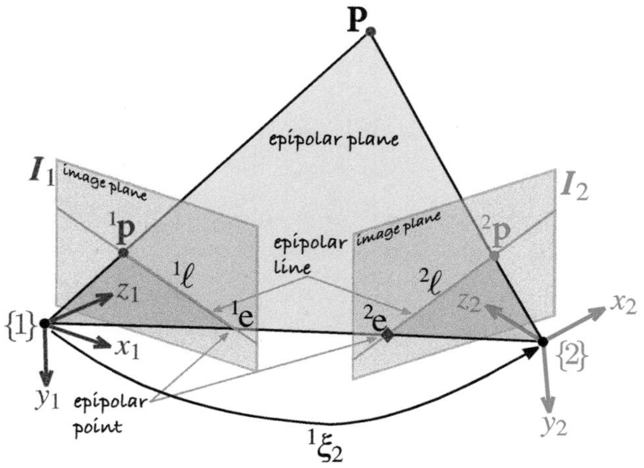

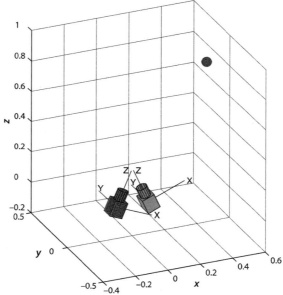

Fig. 14.5.
Epipolar geometry showing the two cameras with associated co-ordinate frames {1} and {2} and image planes. The world point **P** and the two camera centers form the epipolar plane, and the in-tersection of this plane with the image-planes form epipolar lines

Fig. 14.6.
Simulation of two cameras and a target point. The origins of the two cameras are offset along the x-axis and the cameras are *verged*, that is, their optical axes intersect

This is a very fundamental and important geometric relationship – given a point in one image we know that its conjugate is constrained to lie along a line in the other im-age. We illustrate this with a simple example that mimics the geometry of Fig. 14.5

```
>> T1 = SE3(-0.1, 0, 0)  * SE3.Ry(0.4);
>> cam1 = CentralCamera('name', 'camera 1', 'default', ...↵
   'focal', 0.002, 'pose', T1)
```

which returns an instance of the `CentralCamera` class as discussed previously in Sect. 11.1.2. Similarly for the second camera

```
>> T2 = SE3(0.1, 0,0)*SE3.Ry(-0.4);
>> cam2 = CentralCamera('name', 'camera 2', 'default', ...↵
   'focal', 0.002, 'pose', T2);
```

and the pose of the two cameras is visualized by

```
>> axis([-0.5 0.5 -0.5 0.5 0 1])
>> cam1.plot_camera('color', 'b', 'label')
>> cam2.plot_camera('color', 'r', 'label')
```

which is also shown in Fig. 14.6. We define an arbitrary world point

```
>> P=[0.5 0.1 0.8]';
```

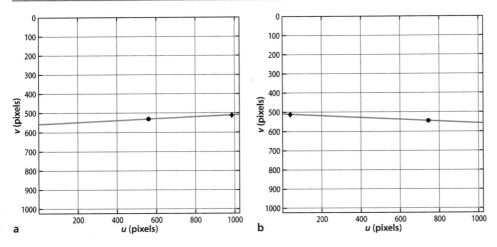

Fig. 14.7.
Epipolar geometry simulation showing the virtual image planes of two Toolbox CentralCamera objects. The perspective projection of point **P** is a *black circle*, the projection of the other camera's center is a *black ♦-marker*, and the epipolar line is shown in *red*

which we display as a small sphere

```
>> plot_sphere(P, 0.03, 'b');
```

which is shown in Fig. 14.6. We project this point to both cameras

```
>> p1 = cam1.plot(P)
p1 =
   561.6861
   532.6079
>> p2 = cam2.plot(P)
p2 =
   746.0323
   546.4186
```

and this is shown in Fig. 14.7. The epipoles are computed by projecting the center of each camera to the other camera's image plane

```
>> cam1.hold
>> e1 = cam1.plot( cam2.centre, 'Marker', 'd',↵
 'MarkerFaceColor', 'k')
e1 =
   985.0445
   512.0000
>> cam2.hold
>> e2 = cam2.plot( cam1.centre, 'Marker', 'd',↵
 'MarkerFaceColor', 'k')
e2 =
    38.9555
   512.0000
```

and these are shown in Fig. 14.7 as a black ♦-marker.

14.2.1   The Fundamental Matrix

The epipolar relationship shown graphically in Fig. 14.5 can be expressed concisely and elegantly as

$$^2\tilde{\boldsymbol{p}}^T \boldsymbol{F}\,^1\tilde{\boldsymbol{p}} = 0 \tag{14.2}$$

where $^1\tilde{\boldsymbol{p}}$ and $^2\tilde{\boldsymbol{p}}$ are the image points $^1\mathbf{p}$ and $^2\mathbf{p}$ expressed in homogeneous form and $\boldsymbol{F} \subset \mathbb{R}^{3\times3}$ is known as the fundamental matrix. We can rewrite this as

$$^2\tilde{\boldsymbol{p}}^T\,^2\tilde{\ell} = 0 \tag{14.3}$$

> **2D projective geometry in brief.** The projective plane $\mathbb{P}^2$ is the set of all points $(x_1, x_2, x_3)^T$, $x_i \in \mathbb{R}$ and x_i not all zero. Typically the 3-tuple is considered a column vector. A point $\mathbf{p} = (u, v)$ is represented in $\mathbb{P}^2$ by homogeneous coordinates $\tilde{p} = (u, v, 1)^T$. Scale is unimportant for homogeneous quantities and we express this as $\tilde{p} \simeq \lambda \tilde{p}$ where the operator $\simeq$ means equal up to a (possibly unknown) nonzero scale factor. A point in $\mathbb{P}^2$ can be represented in nonhomogeneous, or Euclidean, form $\mathbf{p} = (x_1/x_3, x_2/x_3)^T$ in $\mathbb{R}^2$. The homogeneous vector $(u, v, f)^T$, where f is the focal length in pixels, is a vector from the camera's origin that points toward the world point $\mathbf{P}$. More details are given in Sect. C.2.
>
> The Toolbox functions e2h and h2e convert between Euclidean and homogeneous coordinates for points (a column vector) or sets of points (a matrix with one column per point).

where

$$^2\tilde{\ell} \simeq F \,^1\tilde{p} \tag{14.4}$$

is the equation of a line, the epipolar line, along which conjugate point in image two must lie. This line is a function of the point coordinate $^1\tilde{p}$ in image one and Eq. 14.3 is a powerful test as to whether or not a point in image two is a possible conjugate.

Taking the transpose of both sides of Eq. 14.2 yields

$$^1\tilde{p}^T F^T \,^2\tilde{p} = 0 \tag{14.5}$$

from which we can write the epipolar line for camera one

$$^1\tilde{\ell} \simeq F^T \,^2\tilde{p} \tag{14.6}$$

in terms of a point viewed by camera two.

The fundamental matrix is a function of the camera parameters and the relative camera pose between the views

$$F \simeq K_2^{-T} [t]_\times R K_1^{-1} \tag{14.7}$$

where K_1 and K_2 are the camera intrinsic matrices defined in Eq. 11.7◄, and $^2\xi_1 \sim (R, t)$ is the relative pose of camera one with respect to camera two.➤ The fundamental matrix that relates the two views is returned by the method F of the CentralCamera class, for example

If both images were captured with the same camera then $K_1 = K_2$.

```
>> F = cam1.F( cam2 )
F =
           0    -0.0000     0.0010
     -0.0000          0     0.0019
      0.0010     0.0001    -1.0208
```

Note well that this is the inverse of what you might expect: camera two with respect to camera one, but the mathematics can be expressed more simply this way. Toolbox functions always describe camera pose with respect to the world frame.

and for the two image points computed earlier

```
>> e2h(p2)' * F * e2h(p1)
ans =
     1.1102e-16
```

we see that Eq. 14.2 holds.

The fundamental matrix has some interesting properties. It is singular with a rank of two

```
>> rank(F)
ans =
     2
```

The matrix $F \subset \mathbb{R}^{3 \times 3}$ has seven underlying parameters so its nine elements are not independent. The overall scale is not defined, and there exists a constraint that $\det(F) = 0$.

and has seven degrees of freedom.◄ The epipoles are *encoded* in the null space of the matrix. The epipole for camera one is the right null space of F

```
>> null(F)'
ans =
    -0.8873    -0.4612    -0.0009
```

in homogeneous coordinates or

```
>> e1 = h2e(ans)'
e1 =
   985.0445   512.0000
```

in Euclidean coordinates – as shown in Fig. 14.7. The epipole for camera two is the left null space► of the fundamental matrix

This is the right null space of the matrix transpose. The MATLAB function `null` returns the right null space.

```
>> null(F');
>> e2 = h2e(ans)'
e2 =
    38.9555   512.0000
```

The Toolbox can display epipolar lines using the `plot_epiline` methods of the `CentralCamera` class

```
>> cam2.plot_epiline(F, p1, 'r')
```

which is shown in Fig. 14.7 as a red line in the camera two image plane. We see, as expected, that the projection of **P** lies on this epipolar line. The epipolar line for camera one is

```
>> cam1.plot_epiline(F', p2, 'r');
```

14.2.2   The Essential Matrix

The epipolar geometric constraint can also be expressed in terms of normalized image coordinates

$$^2\tilde{\boldsymbol{x}}^T E\,^1\tilde{\boldsymbol{x}} = 0 \tag{14.8}$$

where $E \subset \mathbb{R}^{3\times 3}$ is the essential matrix and $^2\tilde{\boldsymbol{x}}$ and $^1\tilde{\boldsymbol{x}}$ are conjugate points in homogeneous normalized image coordinates.► This matrix is a simple function of the relative camera pose

For a camera with a focal length of 1 and the coordinate origin at the principal point, see page 322.

$$E \simeq [t]_\times R \tag{14.9}$$

where $^2\xi_1 \sim (R, t)$ is the relative pose of camera one with respect to camera two. The essential matrix is singular, has a rank of two, and has two equal nonzero singular values◄ and one of zero. The essential matrix has only 5 degrees of freedom and is completely defined by 3 rotational and 2 translational► parameters. For pure rotation, when $t = 0$, the essential matrix is not defined.

See Appendix B.

A 3-dimensional translation (x, y, z) with unknown scale can be considered as $(x', y', 1)$.

We recall from Eq. 11.7 that $\tilde{\boldsymbol{p}} \simeq K\tilde{\boldsymbol{x}}$ and substituting into Eq. 14.8 we write

$$^2\tilde{\boldsymbol{p}}^T \underbrace{K_2^{-T}EK_1^{-1}}_{F}\,^1\tilde{\boldsymbol{p}} = 0 \tag{14.10}$$

Equating terms with Eq. 14.2 yields a relationship between the two matrices

$$E \simeq K_2^T FK_1 \tag{14.11}$$

in terms of the intrinsic parameters of the two cameras involved.► This is implemented by the `E` method of the `CentralCamera` class

If both images were captured with the same camera then $K_1 = K_2$.

```
>> E = cam1.E(F)
E =
         0   -0.0779        0
   -0.0779        0   0.1842
         0   -0.1842   0.0000
```

where the intrinsic parameters of camera one (which is the same as camera two) are used.

Although Eq. 14.9 is written in terms of $(R, t) \sim {}^2\xi_1$ the Toolbox function returns ${}^1\xi_2$.

Like the camera matrix in Sect. 11.2.2 the essential matrix can be *decomposed* to yield the relative pose ${}^1\xi_2$ in homogeneous transformation form.◀ The inverse is not unique and in general there are two solutions

```
>> sol = cam1.invE(E)
sol(1) =
    1.0000         0         0   -0.1842
         0   -1.0000         0         0
         0         0        -1   -0.07788
         0         0         0         1

sol(2) =
    0.6967         0   -0.7174    0.1842
         0    1.0000         0         0
    0.7174         0    0.6967    0.07788
         0         0         0         1
```

The true relative pose from camera one to camera two is

```
>> inv(cam1.T) * cam2.T
ans =
    0.6967         0   -0.7174    0.1842
         0         1         0         0
    0.7174         0    0.6967    0.07788
         0         0         0         1
```

which indicates that, in this case, solution two is the correct one.

Unusually we have recovered the camera translation exactly but since $E \simeq \lambda E$ the translational part of the homogeneous transformation matrix has an unknown scale factor.◀ In this case the scale is correct because the essential matrix was determined directly from the relative pose between the cameras.

As observed by Hartley and Zisserman (2003, p 259) not even the sign of t can be determined.

In the general case we do not know the pose of the two cameras, so how do we determine the correct solution in practice? One approach is to determine whether a world point is visible. Typically we would choose a point on the optical axis in front of the first camera

```
>> Q = [0 0 10]';
```

and its projection to the first camera

```
>> cam1.project(Q)'
ans =
  429.7889   512.0000
```

is a reasonable value. We can test each of the possible relative poses in `sol` by using them to move the first camera. We can create an instance copy of the first camera with an arbitrary displacement using the `move` method

```
>> cam1.move(sol(1).T).project(Q)'
ans =
    NaN     NaN
```

and the values of `NaN` indicate that the point `Q` is not visible from this camera pose – in fact it is behind the camera. The second solution

```
>> cam1.move(sol(2).T).project(Q)'
ans =
  594.2111   512.0000
```

has a finite value and indicates that it is the valid one. We can perform this more compactly by providing a test point

```
>> sol = cam1.invE(E, Q)
sol =
    0.6967         0   -0.7174    0.1842
         0    1.0000         0         0
    0.7174         0    0.6967    0.07788
         0         0         0         1
```

in which case only the valid solution is returned.

In summary these 3×3 matrices, the fundamental and the essential matrix, encode the parameters and relative pose of the two cameras. The fundamental matrix and a point in one image defines an epipolar line in the other image along which its conjugate points must lie. The essential matrix encodes the relative pose of the two camera's centers and the pose can be extracted, with two possible values, and with translation scaled by an unknown factor. In this example the fundamental matrix was computed from known camera motion and intrinsic parameters. The real world isn't like this – camera motion is difficult to measure and the camera may not be calibrated. Instead we can estimate the fundamental matrix directly from corresponding image points.

14.2.3 Estimating the Fundamental Matrix from Real Image Data

Assume that we have N pairs of corresponding points in two views of the same scene $(^1p_i, {}^2p_i)$, $i = 1 \cdots N$. To demonstrate this we create a set of twenty random point features (within a $2 \times 2 \times 2$ m cube) whose center is located 3 m in front of the cameras ▶

> The SE3 class, a 4×4 matrix is applied to a set of 3D points expressed as a 3×20 matrix. The ∗ operator for the SE3 class does the right thing here, it first converts the second matrix to homogeneous form, performs the matrix multiplication, and then converts back to Euclidean form.

```
>> P = SE3(-1, -1, 2)*(2 *rand(3,20) );
```

and project these points onto the two camera image planes

```
>> p1 = cam1.project(P);
>> p2 = cam2.project(P);
```

If $N \geq 8$ the fundamental matrix can be estimated from these two sets of corresponding points

```
>> F = fmatrix(p1, p2)
maximum residual 2.645e-29
F =
     0.0000   -0.0000    0.0239
    -0.0000   -0.0000    0.0460
     0.0239    0.0018  -24.4896
```

where the residual is the maximum value of the left-hand side of Eq. 14.2 and is ideally zero. The value here is not zero, but it is very small, and this is due to the accumulation of errors from finite precision arithmetic. The estimated matrix has the required rank property

```
>> rank(F)
ans =
     2
```

For camera two we can plot the projected points

```
>> cam2.plot(P);
```

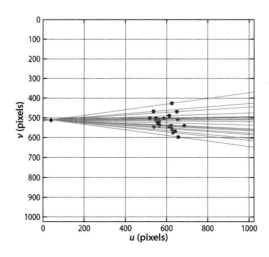

Fig. 14.8.
A pencil of epipolar lines on the camera two image plane. Note how all epipolar lines pass through the epipole which is the projection of camera one's center

and overlay the epipolar lines generated by each point in image one

```
>> cam2.plot_epiline(F, p1, 'r')
```

The example has been contrived so that the epipoles lie within the images, that is, that each camera can see the center of the other camera. A common imaging geometry is for the optical axes to be parallel, such as shown in Fig. 14.19 in which case the epipoles are at infinity (the third element of the homogeneous coordinate is zero) and all the epipolar lines are parallel.

which is shown in Fig. 14.8. We see a family or *pencil* of epipolar lines, and that every point in image two lies on an epipolar line. Note how the epipolar lines all converge on the epipole which is possible in this case◄ because the two cameras are verged as shown in Fig. 14.6.

To demonstrate the importance of correct point correspondence we will repeat the example above but introduce two *bad* data associations by swapping two elements in p2

```
>> p2(:,[8 7]) = p2(:,[7 8]);
```

The fundamental matrix estimation

```
>> fmatrix(p1, p2)
maximum residual 0.000424
ans =
     0.0000    -0.0001     0.0628
     0.0000    -0.0000     0.0098
    -0.0192     0.0511   -29.7672
```

now has a residual that is over 20 orders of magnitude larger than previously. This means that the point correspondence cannot be *explained* by the relationship Eq. 14.2.

If we knew the fundamental matrix we could test whether a pair of candidate corresponding points are in fact conjugates by measuring how far one is from the epipolar line defined by the other

```
>> epidist(F, p1(:,1), p2(:,1))
ans =
   1.5356e-13
>> epidist(F, p1(:,7), p2(:,7))
ans =
   18.8228
```

To within a defined threshold *t*. The Toolbox function epidist returns the distance between a point and an epipolar line.

which shows that point 1 is a good fit, but point 7 (which we swapped with point 8), is a poor fit. However we have to first estimate the fundamental matrix and that requires that point correspondence is known. We break this deadlock with an ingenious algorithm called Random Sampling and Consensus or RANSAC.

The underlying principle is delightfully simple. Estimating a fundamental matrix requires eight points so we randomly choose eight candidate corresponding points (the sample) and estimate *F* to create a *model*. This model is tested against all the other candidate pairs and those that fit◄ vote for this model. The process is repeated a number of times and the model that had the most supporters (the consensus) is returned. Since the sample is small the chance that it contains all valid candidate pairs is high. The point pairs that support the model are termed inliers and those that do not are outliers.

RANSAC is remarkably effective and efficient at finding the inlier set, even in the presence of large numbers of outliers (more than 50%), and is applicable to a wide range of problems. Within the Toolbox we invoke RANSAC as a *driver* of the fmatrix function

```
>> [F,in,r] = ransac(@fmatrix, [p1; p2], 1e-6, 'verbose');
15 trials
2 outliers
2.03262e-29 final residual
```

and we obtain an excellent final residual. The set of inliers is also returned

```
>> in
in =
  Columns 1 through 14
    1    2    3    4    5    6    9   10   11   12   13   14   15   16
  Columns 15 through 18
   17   18   19   20
```

and the two incorrect associations, points 7 and 8, are notably absent from this list. The third parameter to `ransac` is the threshold *t* which is used to determine whether or not a point pair supports the model. If *t* is chosen to be too small RANSAC requires many more trials than its default maximum and this requires adjustment of additional parameters. Keep in mind also that the results of RANSAC will vary from run to run due to the random subsampling performed. Using RANSAC involve some trial and error to choose the correct threshold based on the final residual and the number of outliers. There are also a number of other options that are described in the online documentation.

We return now to the pair of images of the Eiffel tower shown in Fig. 14.3. When we left off at page 464 we had found *candidate* correspondences based on descriptor similarity but there were a number of clearly incorrect matches. RANSAC is available as a method `ransac` that operates on a vector of `FeatureMatch` objects

```
>> F = m.ransac(@fmatrix, 1e-4, 'verbose')
1527 trials
312 outliers
0.000140437 final residual
F =
     0.0000    -0.0000     0.0098
     0.0000     0.0000    -0.0148
    -0.0121     0.0129     3.6393
```

A small amount of trial and error was required to settle on the tolerance of 10^{-4}. Making it smaller requires more RANSAC trials, and therefore raising the limit on the maximum number of trials allowed, but without any significant change in the result. It is also unrealistic to expect a very small residual since the real image data is subject to random error such as image sensor noise and systematic error such as lens distortion.▶

RANSAC identified 312 outliers or incorrect data associations from the SURF feature matching stage which is nearly 50% of the *candidate* matches – the preliminary matching was worse than it looked. Running RANSAC has also updated the elements of the `FeatureMatch` vector

```
>> m.show
ans =
644 corresponding points
332 inliers (51.6%)
312 outliers (48.4%)
```

which now displays the total number of inliers and outliers. Compared to page 463 the elements of the vector

```
>> m(1:5)
ans =
(819.56, 358.557) <-> (708.008, 563.342), dist=0.002137 +
(1028.3, 231.748) <-> (880.14, 461.094), dist=0.004057 -
(1027.6, 571.118) <-> (885.147, 742.088), dist=0.004297 +
(927.724, 509.93) <-> (800.833, 692.564), dist=0.004371 +
(854.35, 401.633) <-> (737.504, 602.187), dist=0.004417 +
```

Lens distortion causes points to be displaced on the image plane and this violates the epipolar geometry. Images can be corrected by warping as discussed in Sect. 12.7.4 but this is computationally expensive. A cheaper alternative is to find the coordinates of the features in the distorted image and correct those using the inverse of the distortion model Eq. 11.13.

Example of RANSAC fitting a line to data with a few erroneous, or outlier, points. The blue dashed line is the least squares best fit and is clearly biased away from the true line by the outlier data points. Despite 40% of the points not fitting the model RANSAC finds the parameters of the consensus line, the line that the largest number of points agree on

```
>> [theta,inliers] = ransac(@linefit, [x; y], 1e-3)
theta =
    3.0000   -10.0000
inliers =
    1    3    4    5    9    10
```

and the indices of the data points that support that model. [examples/linefit.m]

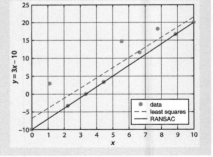

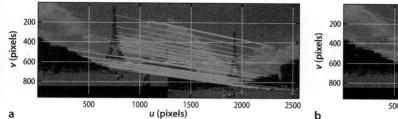
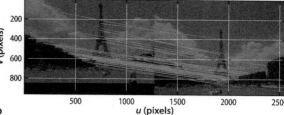

Fig. 14.9. Results of SURF feature matching after RANSAC. **a** Subset of all inlier matches; **b** subset of the outlier matches, some are quite visibly incorrect while others are more subtly wrong

Fig. 14.10. Image from Fig. 14.1a showing epipolar lines converging on the projection of the second camera's center. In this case the second camera is clearly visible in the bottom right of the image

The second match has been determined to be an outlier even though it was the second strongest candidate based on descriptor similarity. Similarity alone is not enough, the corresponding points in the two images must be *consistent* with the epipolar geometry as represented by the consensus fundamental matrix.

now have a trailing plus or minus sign to indicate whether the corresponding match is an inlier or outlier respectively.◄ We can plot some of the inliers

```
>> idisp({im1, im2});
>> m.inlier.subset(100).plot('g')
```

or some of the outliers

```
>> idisp({im1, im2});
>> m.outlier.subset(100).plot('r')
```

and these are shown in Fig. 14.9.

An alternative way to create a `CentralCamera` object is from an image

```
>> cam = CentralCamera('image', im1);
```

The size of the pixel array is inferred from the image and the intrinsic parameters are set to default values. As before, we can overlay the epipolar lines computed from the corresponding points found in the second image

```
>> cam.plot_epiline(F', m.inlier.subset(20).p2, 'g');
```

We only plot a small subset of the epipolar lines since they are too numerous and would obscure the image.

and the result is shown in Fig. 14.10. The epipolar lines intersect at the epipolar point which we can clearly see is the projection of the second camera in the first image.◄ The epipole at

```
>> h2e( null(F))
ans =
   1.0e+03 *
   1.0359
   0.6709
>> cam.plot(ans, 'bo')
```

At the focal lengths used a 20 pix displacement on the image plane corresponds to a pointing error of less than 0.5°.

is also superimposed on the plot. With two handheld cameras and a common view we have been able to pinpoint the second camera in the first image. The result is not quite perfect – there is a horizontal offset of about 20 pixels which is likely to be due to a small pointing error in one or both cameras which were handheld and only approximately synchronized.◄

14.2.4   Planar Homography

In this section we will consider a camera viewing a set of world points $\mathbf{P}_i$ that lie on a plane. They are viewed by two different cameras and the projection in the cameras are $^1\boldsymbol{p}_i$ and $^2\boldsymbol{p}_i$ respectively which are related by

$$^2\tilde{\boldsymbol{p}}_i \simeq H\,^1\tilde{\boldsymbol{p}}_i \tag{14.12}$$

where $H \subset \mathbb{R}^{3\times3}$ is a nonsingular matrix known as an homography, a planar homography, or the homography *induced* by the plane.▶

For example consider again the pair of cameras from page 465 now observing a 3×3 grid of points

```
>> Tgrid = SE3(0,0,1)*SE3.Rx(0.1)*SE3.Ry(0.2);
>> P = mkgrid(3, 1.0, 'pose', Tgrid);
```

where `Tgrid` is the pose of the grid coordinate frame $\{G\}$ and the grid points are centered in the frame's *xy*-plane. The points are projected to both cameras

```
>> p1 = cam1.plot(P, 'o');
>> p2 = cam2.plot(P, 'o');
```

and the images are shown in Fig. 14.11a and b respectively.

Just as we did for the fundamental matrix, if $N \geq 8$ we can estimate the matrix H from two sets of corresponding points

```
>> H = homography(p1, p2)
H =
    -0.4282   -0.0006   408.0894
    -0.7030    0.3674   320.1340
    -0.0014   -0.0000     1.0000
```

According to Eq. 14.12 we can predict the position of the grid points in image two from the corresponding image one coordinates

```
>> p2b = homtrans(H, p1);
```

which we can can superimpose on image two as +-symbols

```
>> cam2.hold()
>> cam2.plot(p2b, '+')
```

This is shown in Fig. 14.11b and we see that the predicted points are perfectly aligned with the actual projection of the world points. The inverse of the homography matrix

$$^1\tilde{\boldsymbol{p}}_i \simeq H^{-1}\,^2\tilde{\boldsymbol{p}}_i \tag{14.13}$$

An homography matrix has arbitrary scale and therefore 8 degrees of freedom. With respect to Eq. 14.14 the rotation, translation and normal have 3, 3 and 2 degrees of freedom respectively, for a total of 8. Homographies form a group: the product of two homographies is another homography, the identity homography is a unit matrix and an inverse operation is defined.

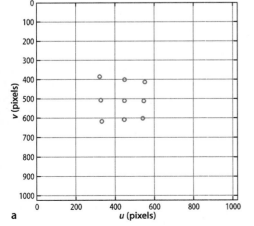
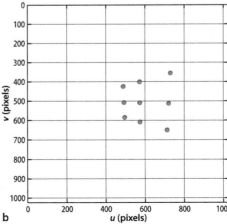

Fig. 14.11.
Views of the oblique planar grid of points from two different view points. The grid points are projected as *open circles*. *Plus signs* in **b** indicate points transformed from the camera one image plane by the homography

performs the inverse mapping, from image two coordinates to image one

```
>> p1b = homtrans(inv(H), p2);
```

The fundamental matrix constrains the conjugate point to lie along a line but the homography tells us *exactly* where the conjugate point will be in the other image – provided that the points lie on a plane.

We can use this proviso to our advantage as a test for whether or not points lie on a plane. We will add some extra world points◄ to our example

These points lie along the ray from the camera one center to an extra row of points in the grid plane. However their z-coordinates have been chosen to be 0.4, 0.5 and 0.6 m respectively.

```
>> Q = [
    -0.2302    -0.0545    0.2537
     0.3287     0.4523    0.6024
     0.4000     0.5000    0.6000  ];
```

which we plot in 3D

```
>> axis([-1 1 -1 1 0 2])
>> plot_sphere(P, 0.05, 'b')
>> plot_sphere(Q, 0.05, 'r')
>> cam1.plot_camera('color', 'b', 'label')
>> cam2.plot_camera('color', 'r', 'label')
```

and this is shown in Fig. 14.12. The new points, shown in red, are clearly not in the same plane as the original blue points. Viewed from camera one

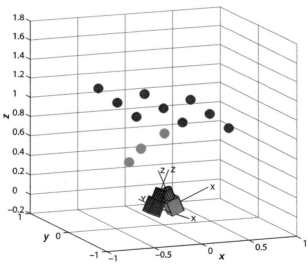

Fig. 14.12.
World view of target points and two camera poses. *Blue points* lie in a planar grid, while the *red points* appear to lie in the grid from the viewpoint of camera one

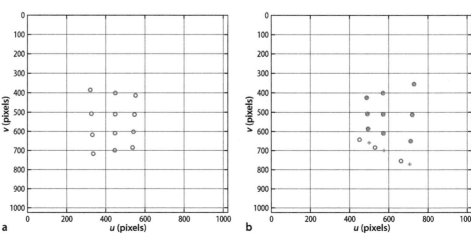

Fig. 14.13.
Views of the oblique planar grid of points from two different view points. The grid points are projected as *open circles*. *Plus signs* in **b** indicate points transformed from the camera one image plane by the homography. The bottom of row of points in each case are not coplanar with the other points

```
>> p1 = cam1.plot([P Q], 'o');
```

as shown in Fig. 14.13a, these new points *appear* as an extra row in the grid of points we used above. However in the second view

```
>> p2 = cam2.plot([P Q], 'o');
```

as shown in Fig. 14.13b these *out of plane* points no longer form a regular grid. If we apply the homography to the camera one image points

```
>> p2h = homtrans(H, p1);
```

we find where they should be in the camera two image if they belonged to the plane implicit in the homography

```
>> cam2.plot(p2h, '+')
```

We see that the original nine points overlap, but the three new points do not. We could make an automated test based on the prediction error

```
>> colnorm( homtrans(H, p1)-p2 )
ans =
    0.0000    0.0000    0.0000    0.0000    0.0000    0.0000
    0.0000    0.0000    0.0000   50.5969   46.4423   45.3836
```

which is large for these last three points – they do not belong to the plane that induced the homography.

In this example we estimated the homography based on two sets of corresponding points which were projections of known planar points. In practice we do not know in advance which points belong to the plane so we can again use RANSAC

```
>> [H,in] = ransac(@homography, [p1; p2], 0.1);
resid =
    4.0990e-13
H =
   -0.4282   -0.0006   408.0894
   -0.7030    0.3674   320.1340
   -0.0014   -0.0000     1.0000
in =
     1     2     3     4     5     6     7     8     9
```

which finds the homography that best explains the relationship between the sets of image points. It has also identified those points which support the homography and the three *out of plane points*, points 10–12, are not on the inlier list.

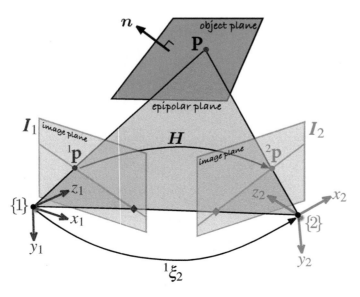

Fig. 14.14.
Geometry of homography showing two cameras with associated coordinate frames {1} and {2} and image planes. The world point **P** belongs to a plane with surface normal n. H is the homography, a 3×3 matrix that maps ${}^1\mathbf{p}$ to ${}^2\mathbf{p}$

See Sect. 11.1.2.

The geometry related to the homography is shown in Fig. 14.14. We can express the homography in normalized image coordinates◄

$$^2\tilde{x} \simeq H_E \, ^1\tilde{x}$$

where H_E is the Euclidean homography which is written

$$H_E \simeq R + \frac{t}{d}n^T \tag{14.14}$$

in terms of the camera motion $(R, t) \sim {}^2\xi_1$ and the plane $n^T P + d = 0$ with respect to frame {1}. The Euclidean and projective homographies are related by

$$H_E \simeq K^{-1}HK$$

where K is the camera intrinsic parameter matrix.

Although Eq. 14.14 is written in terms of $(R, t) \sim {}^2\xi_1$ the Toolbox function returns ${}^1\xi_2$.

As for the essential matrix the projective homography can be *decomposed* to yield the relative pose ${}^1\xi_2$ in homogeneous transformation form◄ as well as the normal to the plane. We use the `invH` method of the `CentralCamera` class

```
>> cam1.invH(H)
solution 1
    T =0.82478    -0.01907    -0.56513    -0.01966
       0.01907     0.99980    -0.00591    -0.01917
       0.56513    -0.00591     0.82498     0.19911
       0.00000     0.00000     0.00000     1.00000
    n = 0.95519     0.00998     0.29582
solution 2
    T =0.69671     0.00000    -0.71736     0.18513
       0.00000     1.00000     0.00000    -0.00000
       0.71736    -0.00000     0.69671     0.07827
       0.00000     0.00000     0.00000     1.00000
    n = -0.19676   -0.09784     0.97556
```

which returns a short structure array. Again there are multiple solutions and we need to apply additional information to determine the correct one. As usual the translational component of the transformation matrix has an unknown scale factor. We know from Fig. 14.12 that the camera motion is predominantly in the x-direction and that the plane normal is approximately parallel to the camera's optical- or z-axis and this knowledge helps us to choose solution two. The true transformation from camera one to two is

```
>> inv(T1)*T2
ans =
    0.6967         0    -0.7174    0.1842
         0    1.0000          0         0
    0.7174         0     0.6967    0.0779
         0         0          0    1.0000
```

The translation scale factor is quite close to one in this example, but in general it must be considered unknown.

and supports our choice.◄ The pose of the grid with respect to camera one is

```
>> inv(T1)*Tgrid
ans =
    0.9797    -0.0389    -0.1968    -0.2973
    0.0198     0.9950    -0.0978         0
    0.1996     0.0920     0.9756    0.9600
         0          0          0    1.0000
```

Since the points are in the *xy*-plane of the grid frame {G} the normal is the *z*-axis.

and the third column is the grid's normal◄ which matches the estimated normal associated with solution two.

We can apply this technique to a pair of real images

```
>> im1 = iread('walls-l.jpg', 'double', 'reduce', 2);
>> im2 = iread('walls-r.jpg', 'double', 'reduce', 2);
```

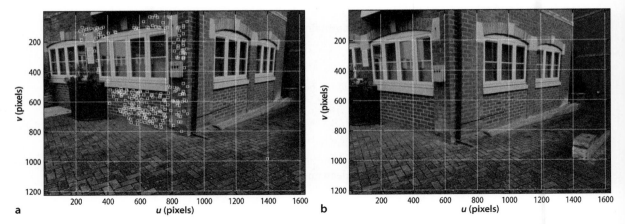

which we have downsized by a factor of 2 in each dimension and are shown in Fig. 14.15. We start by finding the SURF features

Fig. 14.15. Two pictures of a court-yard taken from different view-points. Image **b** was taken approximately 30 cm to the right of image **a**. Image **a** has superimposed features that fit a plane. The camera was handheld

```
>> sf1 = isurf(im1);
>> sf2 = isurf(im2);
```

and the candidate corresponding points

```
>> m = sf1.match(sf2, 'top', 1000)
m =
1000 corresponding points (listing suppressed)
```

then use RANSAC to find the set of corresponding points that best fits a plane in the world

```
[H,r] = m.ransac(@homography, 4)
H =
      0.8463      0.0164   -150.4748
      0.0050      1.0067     20.3413
     -0.0000     -0.0000      1.0000
r =
      1.6799
```

The number of inlier and outlier points is

```
>> m.show
ans =
1000 corresponding points
262 inliers (26.2%)
738 outliers (73.8%)
```

In this case the majority of point pairs do not fit the model, that is they do not belong to the plane that induces the homography **H**. However 262 points *do* belong to the plane and we can superimpose them on the figure

```
>> idisp(im1)
>> plot_point(m.inlier.p1, 'ys')
```

as shown in Fig. 14.15a. RANSAC has found a consensus which is the plane containing the left-hand wall. The error tolerance was set to 4 pixels to account for lens distortion and the planes being not perfectly smooth. If we remove the inlier points from the `FeatureMatch` vector, that is, we keep the outliers

```
>> m = m.outlier
```

and repeat the RANSAC homography estimation step we will find the next most dominant plane in the scene which turns out to be the right-hand wall. Planes are very common in man-made environments and we will revisit homographies and their decomposition in Sect. 14.7.1.

14.3 Stereo Vision

Stereo vision is the technique of estimating the 3-dimensional structure of the world from two images taken from different viewpoints as for example shown in Fig. 14.15. Human eyes are separated by 50–80 mm and the difference between these two viewpoints is an important, but not the only, part of how we sense distance. We will discuss two approaches known as sparse and dense stereo respectively. Sparse stereo is a natural extension of what we have learned about feature matching and recovers the world coordinate (X, Y, Z) for each corresponding point pair. Dense stereo attempts to recover the world coordinate (X, Y, Z) for *every pixel* in the image.

14.3.1 Sparse Stereo

To illustrate sparse stereo we will return to the pair of images shown in Fig. 14.15. We have already found the SURF features and established candidate correspondences between them. Now we estimate the fundamental matrix

```
>> [F,r] = m.ransac(@fmatrix,1e-4, 'verbose');
102 trials
238 outliers
0.000132333 final residual
```

which captures the relative geometry of the two views. We can display the epipolar lines for a subset of right-hand image points overlaid on the left-hand image

```
>> cam = CentralCamera('image', im1);
>> cam.plot_epiline(F', m.inlier.subset(40).p2, 'y');
```

which is shown in Fig. 14.16. In this case the epipolar lines are approximately horizontal and parallel which is expected for a camera motion that is a pure translation in the *x*-direction. Figure 14.17 shows the epipolar geometry for stereo vision. It is clear that as points move away from the camera, $\mathbf{P}$ to $\mathbf{P'}$ the conjugate points in the right-hand image moves to the right along the epipolar line.

The points $^{1}\mathbf{p}$ and $^{2}\mathbf{p}$ each define a ray in space which intersect at the world point. However to determine these rays we need to know the two poses of the camera and its intrinsic parameters. We can consider that the camera one frame {1} is the origin but the pose of camera two remains unknown. However we could estimate its pose by decomposing the essential matrix computed between the two views. We have the

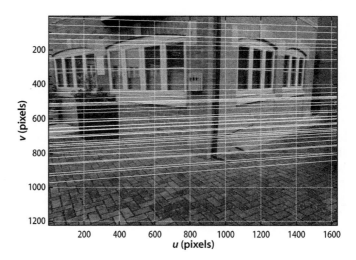

Fig. 14.16.
Image of Fig. 14.15a with epipolar lines for a subset of right image points superimposed

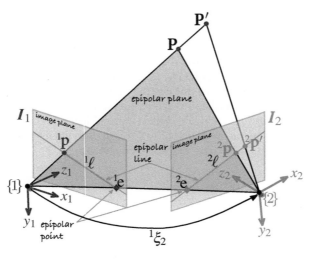

Fig. 14.17.
Epipolar geometry for stereo vision. We can see clearly that as the depth of the world point increases, from **P** to **P′**, the projection moves along the epipolar line in the second image plane

fundamental matrix, but to determine the essential matrix according to Eq. 14.11 we need the camera's intrinsic parameters. With a little sleuthing we can find them!

The focal length used when the picture was taken is stored in the metadata of the image as we discussed on page 363 and can be examined

```
>> [~,md] = iread('walls-1.jpg');
```

where md is a structure of text strings that contains various characteristics of the image – its metadata. The element DigitalCamera is a structure that describes the camera

```
>> f = md.DigitalCamera.FocalLength
f =
    4.1500
```

from which we determine the focal length is 4.15 mm.

The dimensions of the pixels $\rho_w \times \rho_h$ are not included in the image header but some web-based research on this model camera

```
>> md.Model
ans =
iPhone 5s
```

suggests that this camera has an image sensor with 1.5 μm pixels. We create a CentralCamera object based on the known focal length, pixel size and image dimension▶

```
>> cam = CentralCamera('image', im1, 'focal', f/1000, ...↵
    'pixel', 2*1.5e-6)
cam =
name: noname [central-perspective]
  focal length:    0.00415
  pixel size:      (3e-06, 3e-06)
  principal pt:    (816, 612)
  number pixels:   1632 x 1224
  pose:            t = (0,0,0), RPY/yxz = (0,0,0) deg
```

We have doubled the pixel dimensions to account for halving the image resolution when we loaded the images. A low-resolution image effectively has larger pixels.

In the absence of any other information the principal point is assumed to be in the center of the image.

The essential matrix is obtained by applying the camera intrinsic parameters to the fundamental matrix

```
>> E = cam.E(F)
E =
    0.0143    1.1448    0.2380
   -1.1286   -0.1483    6.0273
   -0.2826   -6.0536   -0.1461
```

and we then decompose it to determine the camera motion

```
>> T = cam.invE(E, [0,0,10]')
T =
      0.9999      0.0115      0.0027      6.042
     -0.0115      0.9996     -0.0255     -0.3092
     -0.0030      0.0254      0.9997      1.124
           0           0           0          1
```

We chose a test point $^1\mathbf{P} = (0, 0, 10)$, a distant point along the optical axis, to determine the correct solution for the relative camera motion. Since the camera orientation was kept fairly constant the rotational part of the transformation is expected to be close to the identity matrix as we observe, and the actual rotation

```
>> T.torpy('yxz', 'deg')
ans =
     -0.6569      1.4597      0.1561
```

We have specified a different roll-pitch-yaw rotation order *YXZ*. Given the way we have defined our axes the camera orientation with respect to the world frame is a yaw about the vertical or *y*-axis, followed by a pitch about the *x*-axis followed by a roll about the optical axis or *z*-axis.

is less than two degrees of rotation about any axis. ◄

The estimated translation t from {1} to {2} has an unknown scale factor. Once again we bring in an extra piece of information – when we took the images the camera position changed by approximately 0.3 m in the positive *x*-direction. The estimated translation has the correct direction, dominant *x*-axis motion, but the magnitude is quite wrong. We therefore scale the translation

```
>> t = T.t;
>> T.t = 0.3 * t/t(1)
T =
      0.9999      0.0115      0.0027        0.3
     -0.0115      0.9996     -0.0255    -0.01535
     -0.0030      0.0254      0.9997     0.0558
           0           0           0          1
```

and we have an estimate of $^1\xi_2$ – the relative pose of camera two with respect to camera one represented as a homogeneous transformation.

Each point **p** in an image corresponds to a ray in space ◄

$$\boldsymbol{P} = \alpha\boldsymbol{d} + \boldsymbol{P}_0, \quad \forall \alpha > 0$$

Sometimes called a raxel. Many representations exist including Plücker coordinates which are described in Sect. C.1.2.2. Determining a ray from a pixel coordinate is covered on page 327.

where $\boldsymbol{P}_0$ is the principal point of the camera and $\boldsymbol{d} \in \mathbb{R}^3, \|\boldsymbol{d}\| = 1$ is a unit-vector in the direction of the ray. Consider now the first corresponding point pair m(1). The ray from camera one is

```
>> r1 = cam.ray(m(1).p1)
r1 =
d=(0.37844, -0.0819363, 0.921992), P0=(0, 0, 0)
```

which is an instance of a Ray3D object with properties d and P0 representing d and $\boldsymbol{P}_0$ respectively. The corresponding ray from the second camera is

```
>> r2 = cam.move(T).ray(m(1).p2)
r2 =
d=(0.29936, -0.0826926, 0.95055), P0=(0.3, -0.0153494, 0.0557958)
```

where the move method returns an instance copy of the CentralCamera object cam with the relative pose we have just estimated. The two rays intersect at

```
>> [P,e] = r1.intersect(r2);
P'
ans =
      1.2134     -0.2627      2.9563
```

which is a point with a *z*-coordinate, or depth, of almost 3 m. Due to errors in the estimate of camera two's pose the two rays do not actually intersect, but their closest point is returned. At their closest point the rays are

```
>> e
e =
      0.0049
```

nearly 5 mm apart. Considering the lack of rigor in this exercise, two handheld camera shots and only approximate knowledge of the magnitude of the camera displacement, the recovered depth information is quite remarkable.▶

We draw a subset of one hundred corresponding points from the inlier set

```
>> m2 = m.inlier.subset(100);
```

and then compute the rays in world space from each camera

```
>> r1 = cam.ray( m2.p1 );
>> r2 = cam.move(T).ray( m2.p2 );
```

which are each vectors of `Ray3D` objects. Their intersection points are

```
>> [P,e] = r1.intersect(r2);
```

where P is a matrix of closest points, one per column, and the last row

```
>>   z = P(3,:);
```

is the depth coordinate. The columns of the vector e contains the distance between the rays at their closest points. We can superimpose the distance to each point on the image of the courtyard

```
>> idisp(im1)
>> plot_point(m2.p1, 'y+', 'textcolor', 'y', 'printf', {'%.1f', z});
```

which is shown in Fig. 14.18 and the feature markers are annotated with the estimated depth.

Even small errors in the estimated rotation between the camera poses will lead to large closing errors at distances of several meters. The closing error observed here would be induced by a rotational error of less than 1 deg.

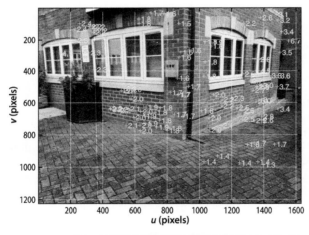

Fig. 14.18.
Image of Fig. 14.15a with depth of selected points indicated (units of meters)

Fig. 14.19.
A small stereo camera sensor mounted on a mobile robot and capable of real-time depth map generation (image courtesy of Stereolabs Inc.)

This is an example of stereopsis where we have used information from two overlapping images to infer the 3-dimensional position of points in the world. For obvious reasons the approach used here is referred to as sparse stereo because we only compute distance at a tiny subset of pixels in the image. More commonly the relative pose between the cameras would be known as would the camera intrinsic parameters.

14.3.2 Dense Stereo Matching

A stereo pair is more commonly taken simultaneously by two cameras, generally with parallel optical axes, and separated by a known distance referred to as the camera baseline. Figure 14.19 shows a typical stereo camera system which simultaneously captures images from both cameras and transfers them to a host computer for processing.

To illustrate we load the left and right images comprising a stereo pair

```
>> L = iread('rocks2-l.png', 'reduce', 2);
>> R = iread('rocks2-r.png', 'reduce', 2);
```

Fig. 14.20. The `stdisp` image browsing window. The *black cross hair* in the left-hand image has been positioned at the top right of the digit 5 on a foreground rock. Another *black cross hair* is automatically positioned at the same coordinate in the right-hand image. Clicking on the corresponding point in the right-hand image sets the *green cross-hair*, and the panel at the top indicates a horizontal shift of 70.9 pixels to the left. This stereo image pair is from the Middlebury stereo database (Scharstein and Pal 2007). The focal length f/ρ is 3 740 pixels, and the baseline is 160 mm. The images have been cropped so that the actual disparity should be offset by 274 pixels

We can interactively examine these two images together

```
>> stdisp(L, R)
```

as shown in Fig. 14.20. Clicking on a point in the left-hand image updates a pair of cross hairs that mark the *same* coordinate relative to the right-hand image. Clicking in the right-hand image sets another vertical cross hair and displays the difference between the horizontal coordinate of the two crosshairs. The cross hairs as shown are set to a point on the digit 5 written on one of the foreground rocks and we observe several things. Firstly the spot has the same vertical coordinate in both images, and this implies that the epipolar lines are horizontal. Secondly, in the right-hand image the spot has moved to the left by 70.9 pixels. If we probed more points we would see that disparity decreases for points that are further from the camera.

As shown in Fig. 14.17 the conjugate point in the right-hand image moves rightward along the epipolar line as the point depth increases. For the parallel-axis camera geometry the epipolar lines are parallel and horizontal, so conjugate points have the same v-coordinate. If the coordinates of two corresponding points are $(^Lu, ^Lv)$ and $(^Ru, ^Rv)$ then $^Rv = {}^Lv$. The displacement along the horizontal epipolar line $d = {}^Lu - {}^Ru$ where $d \geq 0$ is called disparity.

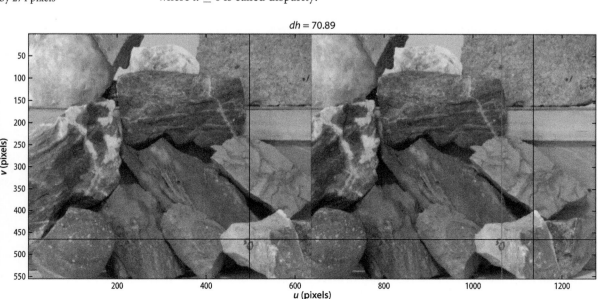

dh = 70.89

Fig. 14.21.
Stereo matching. A search window in the right image, starting at $u = {}^L u$, is moved leftward along the epipolar line $v = {}^L v$ until it matches the *template* window T from the left image

The dense stereo process is illustrated in Fig. 14.21. For the pixel at $({}^L u, {}^L v)$ in the left-hand image we know that its corresponding pixel is at some coordinate $({}^L u - d, {}^L v)$ in the right-hand image where $d \in [d_{min}, d_{max}]$. To reliably find the corresponding point for a pixel in the left-hand image we create an $N \times N$ pixel *template* region T about that pixel. As shown in Fig. 14.21 we *slide* the template window horizontally across the right-hand image. The position at which the template is most similar is considered to be the corresponding point from which disparity is calculated. Compared to the matching problem we discussed in Sect. 12.5.2, this one is much simpler because there is no change in relative scale or orientation between the two images.

The epipolar constraint means that we only need to perform a 1-dimensional search for the corresponding point. The template is moved in D steps of 1 pixel in the range $d_{min} \cdots d_{max}$. At each template position we perform a template matching operation, such as we discussed in Sect. 12.5.2, and for an $N \times N$ template these have a computational cost of $O(N^2)$. For a $W \times H$ image the total cost of dense stereo matching is $O(DWHN^2)$ which is high but feasible in real time.

To perform stereo matching for the image pair in Fig. 14.20 using the Toolbox is quite straightforward

```
>> d = istereo(L, R, [40, 90], 3);
```

The result is a matrix the same size as L and the value of each element $d[u, v]$, or `d(v,u)` in MATLAB, is the disparity at that coordinate in the left image. The corresponding pixel in the right image would be at $(u - d[u, v], v)$. We can display the disparity as an image – a disparity image

```
>> idisp(d, 'bar')
```

which is shown in Fig. 14.22. Disparity images have a distinctive ghostly appearance since all surface color and texture is absent. The third argument to `stereo` is the range of disparities to be searched, in this case from 40 to 90 pixel so the pixel values in the disparity image lie in the range [40, 90]. The disparity range was determined by examining some far and near points using `stdisp`.► The fourth argument to `istereo` is the half-width of the template, in this case we are using a 7×7 window. By default `stereo` uses the ZNCC similarity measure.

In the disparity image we can clearly see that the rocks at the bottom of the pile have a larger disparity and are closer to the camera than those at the top. There are also some errors, such as the anomalous bright values around the edges of some rocks. These pixels are indicated as being nearer than they really are. The similarity score is set to `NaN` around the edge of the image where the similarity matching template falls off the edge of the image and to `Inf` for the case where the denominator of the ZNCC similarity metric (Table 12.1) is equal to zero.► The values `NaN` and `Inf` are both displayed as red.

We could chose a range such as [0, 90] but this increases the search time: 91 disparities would have to be evaluated instead of 51. It also increases the possibility of matching errors.

This occurs if all the pixels in either template have exactly the same value.

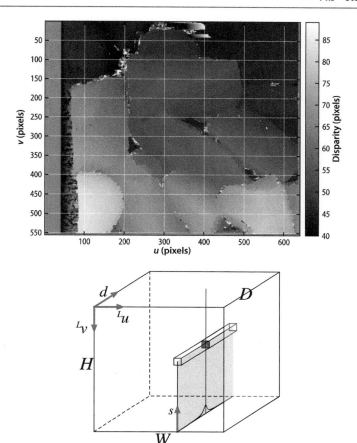

Fig. 14.22.
Disparity image for the rock pile stereo pair, where *brighter* means higher disparity or shorter range. *Red* indicates `Inf` or `NaN` values in the result where disparity could not be computed. Note the quantization in grey levels since we search for disparity in steps of one pixel

Fig. 14.23.
The disparity space image (DSI) is a 3-dimensional image where element $D(u, v, d)$ is the similarity between the support regions centered at $({}^{L}u, {}^{L}v)$ in the left image and $({}^{L}u - d, {}^{L}v)$ in the right image

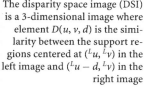

14.3.2.1 Stereo Failure Modes

The stereo function can also return the disparity space image (DSI)

```
>> [d,sim,DSI] = istereo(L, R, [40 90], 3);
```

where `sim` is an $H \times W$ matrix whose elements are the peak similarity score at the corresponding pixel and `DSI` is an $H \times W \times D$ matrix shown in Fig. 14.23

```
>> about(DSI)
DSI [double] : 555x638x51 (144468720 bytes)
```

This is a large matrix (144 Mbyte) which is why the images were reduced in size when loaded.

whose elements (u, v, d) are the similarity measure between the templates centered at (u, v) in the left image and $(u - d, v)$ in the right image.◀ The disparity image we saw earlier is simply the position of the maximum value in the d-direction evaluated at every pixel◀ and the matrix `sim` is the value of those maxima.

This is a workable but simplistic approach. A better approach is to apply regularization and estimate a function $g(u, v)$ that fits the points of maximum similarity while maintaining smoothness and continuity.

Each column in the d-direction, as shown in Fig. 14.23, holds the similarity measure versus disparity for the corresponding pixel in the left image. For the pixel at $(138, 439)$ we can plot this

```
>> plot( squeeze(DSI(439,138,:)), 'o-');
```

which is shown in Fig. 14.24a. We are using the ZNCC measure and an almost perfect match occurs at a disparity of 80 pixels, since the horizontal axis is $d - d_{min}$ and $d_{min} = 40$. Such a strong and unambiguous peak is fortunately very common. However Fig. 14.22 shows that the stereo matching process is not perfect and plots of the template similarity metric versus disparity provide insight into the causes of error.

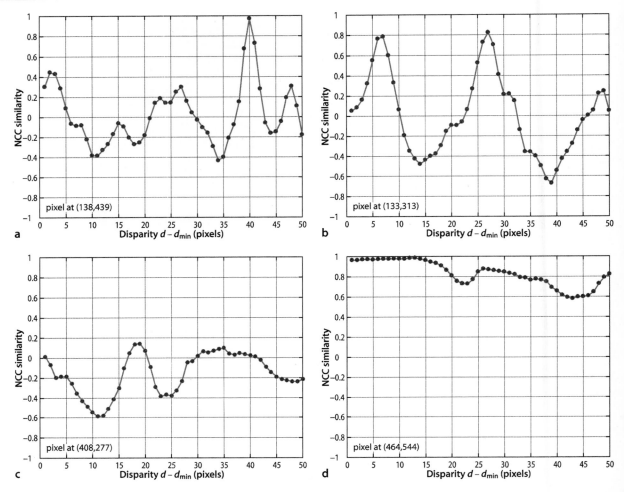

Fig. 14.24. Some typical ZNCC metric versus disparity curves. **a** Single strong peak; **b** multiple peaks; **c** weak peak; **d** broad peak

Figure 14.24b shows two peaks of almost similar amplitude and this means that the template pattern was found twice in the search region. This occurs when there are regular vertical features in the scene as is often the case in man-made scenes: brick walls, rows of windows, architectural features or a picket fence. The problem, illustrated in Fig. 14.25, is commonly known as the picket fence effect and more properly as spatial aliasing. There is no real cure for this problem▶ but we can detect its presence. The ambiguity ratio is the ratio of the height of second peak to the height of the first peak – a high-value indicate that the result is uncertain and should not be used. The chance of detecting incorrect peaks can be reduced by ensuring that the disparity range used in `istereo` is as small as possible but this requires some knowledge of the expected range of objects.

A weak match is shown in Fig. 14.24c. This typically occurs when the corresponding scene point is not visible in the right-hand view due to occlusion – also known as the missing parts problem. Occlusion is illustrated in Fig. 14.26 and it is clear that point 3 is only visible to the left camera. The stereo matching algorithm will always return the best match so if the point is occluded it will return disparity to the most similar, but wrong, template. Even though the figure is an exaggerated depiction, real images suffer this problem where the depth changes rapidly. In our example, this occurs at the edges of the rocks which is exactly where we observe the incorrect disparities in Fig. 14.22. The problem becomes more prevalent as the baseline increases. The problem also occurs when the corresponding point does not lie within the disparity search range, that is, the disparity search range is too small.

Multi-camera stereo, using more than two cameras, is a powerful method to solve this ambiguity.

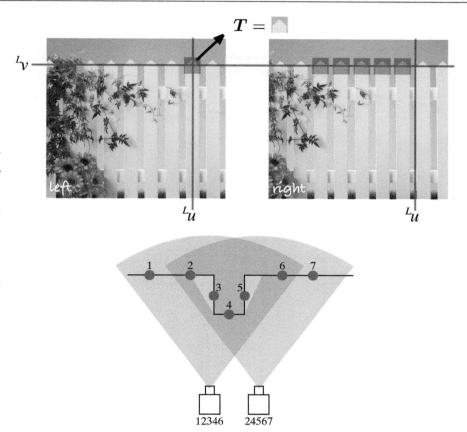

Fig. 14.25.
Picket fence effect. The template will match well at a number of different disparities. This problem occurs in any scene with repeating patterns

Fig. 14.26.
Occlusion in stereo vision. The field of view of the two cameras are shown as *colored sectors*. *Points 1* and *7* fall outside the overlapping view area and are seen by only one camera each. *Point 5* is occluded from the left camera and *point 3* is occluded from the right camera. The order of points seen by each camera is given underneath it

The problem cannot be cured but it can be detected. The simplest method is to consider the similarity score returned by the `istereo` function

```
>> idisp(sim)
```

as shown in Fig. 14.27a and we see that the erroneous disparity values correspond to low similarity scores. Disparity results where similarity is low can be discarded

```
>> ipixswitch(sim<0.7, 'yellow', d/90);
```

and this is shown in Fig. 14.27b where pixels with similarity $s < 0.7$ are displayed as yellow. The distribution of maximum similarity scores

```
>> ihist(sim(isfinite(sim)), 'normcdf');
```

is shown in Fig. 14.28. We see that only 5% of pixels have a similarity score less than 0.6, and that around 80% of pixels have a similarity score greater than 0.9.

A simple but effective way to test for occlusion is to perform the matching in two directions – left-right consistency checking. Starting with a pixel in the left-hand image the strongest match in the right-image is found. Then the strongest match to that pixel is found in the left-hand image. If this is where we started the match is considered valid. However if the corresponding point was occluded in the right image the first match will be a weak one to a different feature, and there is a high probability that the second match will be to a different pixel in the left image.

From Fig. 14.26 it is clear that pixels on the left-side of the left-hand image may not overlap at all with the right-hand image – point 1 for example is outside the field of view of the right-hand camera. This is the reason for the large number of incorrect matches on the left-hand side of the disparity image in Fig. 14.22. It is common practice to discard the d_{max} left-most columns (90 in this case) of the disparity image.

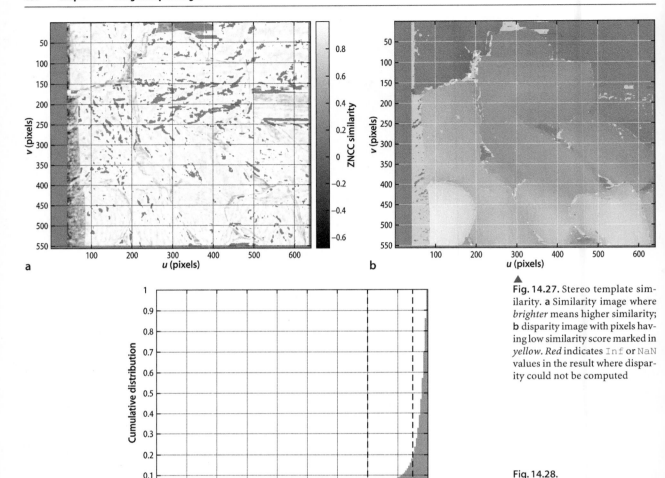

a

b

▲
Fig. 14.27. Stereo template similarity. **a** Similarity image where *brighter* means higher similarity; **b** disparity image with pixels having low similarity score marked in *yellow*. *Red* indicates `Inf` or `NaN` values in the result where disparity could not be computed

Fig. 14.28.
Cumulative probability of ZNCC scores. The probability of a score less than 0.9 is 45%

The final problem that can arise is a similarity function with a very broad peak as shown in Fig. 14.24d. The breadth makes it difficult to precisely estimate the maxima. This generally occurs when the template region has very low texture for example corresponding to the sky, dark shadows, sheets of water, snow, ice or smooth man-made objects. Simply put, in a region that is all grey, a grey template matches equally well with any number of grey candidate regions. One approach to detect this is to look at the variability of pixel values in the template using measures such as the difference between the maximum and minimum value or the variance of the pixel values. If the template has too little variance it is less likely to result in a strong peak. Measures of peak sharpness can also be used to eliminate these cases and this is discussed in the next section.

For the various problem cases just discussed disparity cannot be determined, but the problem can be detected. This is important since it allows those pixels to be marked as having no known range and this allows a robot to be prudent with respect to regions whose 3-dimensional structure cannot be reliably determined.

The design of a stereo-vision system has three degrees of freedom. The first is the baseline distance between the cameras. As it increases the disparities become larger making it possible to estimate depth to greater precision, but the occlusion problem becomes worse. Second, the disparity search range needs to be set carefully. If the maximum is too large the chance of spatial aliasing increases but if too small then

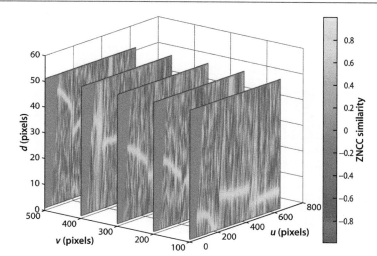

Fig. 14.29.
The disparity space image is a 3-dimensional image where element $D(u, v, d)$ is the similarity between the support regions centered at $({}^Lu, {}^Lv)$ in the left image and $({}^Lu - d, {}^Lv)$ in the right image

points close to the camera will generate incorrect and weak matches. A large disparity range also increases the computation time. Third, template size involves a tradeoff between computation time and quality of the disparity image. A small template size can pick up fine depth structure but tends to give results that are much noisier since a small template is more susceptible to ambiguous matches. A large template gives a smoother disparity image but requires greater computation. It also increases the chance that the template will contain pixels belonging to objects at different depths which is referred to as the mixed pixel problem. This tends to cause poor quality matching at the edges of objects, and the resulting disparity map appears blurred. One solution is to use a nonparametric local transform such as the rank or census transform prior to performing correlation. Since these rely on the ordering of intensity values not the values themselves they give better performance at object boundaries.

An alternative way to look at the failure modes is to use MATLAB's volume visualization functions to create horizontal slices through the disparity space image

```
>> slice(DSI, [], [100 200 300 400 500], [])
>> shading interp; colorbar
```

which is shown in Fig. 14.29. These are slices at constant v-coordinate, effectively horizontal cross sections of the scene. Within each of the ud-planes we see a bright path (high similarity values) that represents disparity $d(u)$. Note the significant discontinuities in the path for the plane at $v = 100$ which correspond to sudden changes in depth. The planes at $v = 200, 300, 400$ show that the path also fades away in places. In these regions the maximum similarity is low, there is no strong match in the right-hand image, and the most likely cause is occlusion.

14.3.3   Peak Refinement

The disparity at each pixel is an integer value $d \in [d_{\min}, d_{\max}]$ at which the greatest similarity was found. Figure 14.24a shows a single unambiguous strong peak and we can use the peak and adjacent points to refine the estimate of the peak's position. ◀

This two-dimensional peak refinement is discussed in Appendix J.

A parabola

$$s = Ad^2 + Bd + C \qquad (14.15)$$

is defined by three points and is fitted to the peak value and its two neighbors. For the ZNCC similarity measure, a maxima corresponds to the best match which means that the parabola is inverted and $A < 0$. The maximum value of the fitted parabola occurs

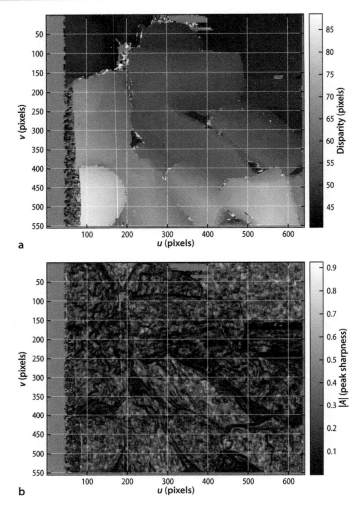

Fig. 14.30.
a Disparity image with peak refinement; **b** magnitude of the d^2 coefficient for every pixel. High values (*bright*) correspond to sharp peaks and occur where image texture is high. Broad peaks (*dark*) occur where image texture is low

when its derivative equals zero, from which we can obtain a more precise estimate of the position of the peak which is the disparity

$$\hat{d} = \frac{-B}{2A}$$

The A coefficient will have a large magnitude for a sharp peak, and a simple threshold can be used to reject broad peaks, as we will discuss in the next section.

Disparity peak refinement is enabled with the `'interp'` option

```
>> [di,sim,peak] = istereo(L, R, [40 90], 3, 'interp');
>> idisp(di)
```

and the resulting disparity image is shown in Fig. 14.30a. We see that it is much smoother than the one shown previously in Fig. 14.22. The additional optional output argument `peak` is a structure

```
>> peak
peak =
     A: [555x638 double]
     B: [555x638 double]
```

that contains the per-pixel values of the parabola coefficients. The magnitude of the A coefficient is shown as an image in Fig. 14.30b.

14.3.4   Cleaning up and Reconstruction

The result of stereo matching, such as shown in Fig. 14.22 or 14.30a, have a number of imperfections for the reasons we have just described. For robotic applications such as path planning and obstacle avoidance it is important to know the 3-dimensional structure of the world, but it is also critically important to know what we don't know. Where reliable depth information from stereo vision is missing a robot should be prudent and treat it differently to free space. We use a number of simple measures to mark elements of the disparity image as being invalid or unreliable.

We start by creating a matrix `status` the same size as `d` and initialized to one

```
>> status = ones(size(d));
```

The elements are set to different values if they correspond to specific failure conditions

```
>> [U,V] = imeshgrid(L);
>> status(isnan(d)) = 5;      % search template off the edge
>> status(U<=90) = 2;         % no overlap
>> status(sim<0.8) = 3;       % weak match
>> status(peak.A>=-0.1) = 4;  % broad peak
```

We can display this matrix as an image

```
>> idisp(status)
>> colormap( colorname({'lightgreen', 'cyan', 'blue', 'orange', 'red'}) )
```

which is shown in Fig. 14.31. The colormap is chosen to display the status values as light green for a good stereo match, cyan if the disparity search range extends beyond the left edge of the right image, blue if the peak similarity is too small, orange if the peak is too broad, and red for NaN values where the search template would fall off the edge of the image. The good news is that there are a lot of light green pixels! In fact

```
>> sum(status(:) == 1) / prod(size(status)) * 100
ans =
    57.7223
```

nearly 60% of disparity values pass our battery of quality tests. The blue pixels, indicating weak similarity, occur around the edges of rocks and are due to occlusion. The orange pixels, indicating a broad peak, occur in areas that are fairly smooth, either deep shadow between rocks or the nonrock background.

Earlier we created an interpolated disparity image `di` and now we will invalidate the disparity values that we have determined to be unreliable

```
>> di(status>1) = NaN;
```

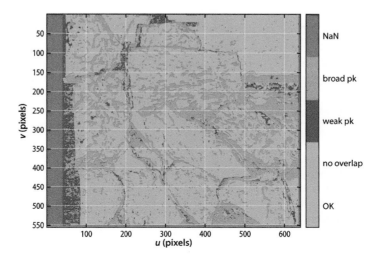

Fig. 14.31.
Stereo matching status on a per pixel basis

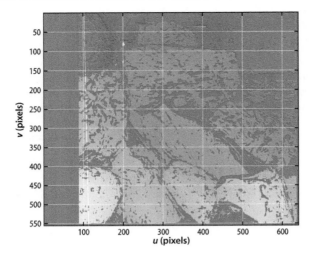

Fig. 14.32.
Interpolated disparity image
with unreliable estimates indicated in *red*

by setting them to the value NaN.► We can display this with the *unreliable* pixels
marked in red by

```
>> ipixswitch(isnan(di), 'red', di/90);
```

which is shown in Fig. 14.32.◄ This is now in useful form for a robot – it contains disparity values interpolated to better than a pixel and all unreliable values are clearly marked.

The final step is to convert the disparity values in pixels to world coordinates in
meters – a process known as 3D reconstruction. In the earlier discussion on sparse stereo we determined the world point from the intersection of two rays in 3-dimensional
space. For a parallel axis stereo camera rig as shown in Fig. 14.19 the geometry is much
simpler as illustrated in Fig. 14.33. For the red and blue triangles we can write

$$X = Z\tan\theta_1, \quad X - b = Z\tan\theta_2$$

where b is the baseline and the angles of the rays correspond to the horizontal image
coordinate ${}^i u$, $i = \{L, R\}$

$$\tan\theta_i = \frac{\rho_u\left({}^i u - u_0\right)}{f}$$

Substituting and eliminating X gives

$$Z = \frac{fb}{\rho_u\left({}^L u - {}^R u\right)} = \frac{fb}{\rho_u d}$$

which shows that depth is inversely proportional to disparity $d = {}^L u - {}^R u$ and $d > 0$.
We can also recover the X- and Y-coordinates so the 3D point coordinate is

$$X = \frac{b\left({}^L u - u_0\right)}{d}, \quad Y = \frac{b\left({}^L v - v_0\right)}{d}, \quad Z = \frac{fb}{\rho_u d} \qquad (14.16)$$

A good stereo system can estimate disparity with an accuracy of 0.2 pixels.
Distant points have a small disparity and the error in the estimated 3D coordinate
will be significant. A rule of thumb is that stereo systems typically have a maximum range of 50b.

The special floating-point value NaN (for
not a number) has the useful property
that the result of any arithmetic operation involving NaN is always NaN. Many
MATLAB functions such as max or min
ignore NaN values in the input matrix,
and plotting and graphics functions do
not display this value, leaving a hole in
the graph or surface.

The division by 90 is to convert the floating-point disparity values in the range
[40, 90] into valid greyscale values in the
range [0, 1].

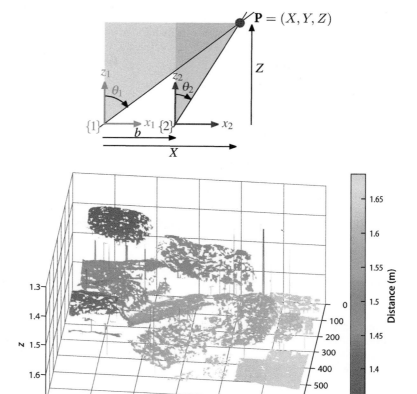

Fig. 14.33.
Stereo geometry for parallel camera axes. X and Z are measured with respect to camera one, b is the baseline

Fig. 14.34.
3-dimensional reconstruction for parallel stereo cameras. *Hotter colors* indicate parts of the surface that are further from the camera

The images shown in Fig. 14.20, from the Middlebury dataset, were taken with a very wide camera baseline. The left edge of the left-image and the right edge of the right-image have no overlap and have been cropped. Cropping N pixels from the left of the left-hand image only, reduces the disparity by N. For this stereo pair the actual disparity must be increased by 274 to account for the cropping.

The true disparity is

```
>> di = di + 274;
```

A process known as vectorizing. Using matrix and vector operations instead of `for` loops greatly increases the speed of MATLAB code execution. See http://www.mathworks.com/support/tech-notes/1100/1109.html for details.

and we compute the X-, Y- and Z-coordinate of each pixel as separate matrices to exploit MATLAB's efficient matrix operations◄

```
>> [U,V] = imeshgrid(L);
>> u0 = size(L,2)/2; v0 = size(L,1)/2;
>> b = 0.160;
>> X = b*(U-u0) ./ di; Y = b*(V-v0) ./ di; Z = 3740 * b ./ di;
```

which can be displayed as a surface

```
>> surf(Z)
>> shading interp; view(-150, 75)
>> set(gca,'ZDir', 'reverse'); set(gca,'XDir', 'reverse')
>> colormap(flipud(hot))
```

as shown in Fig. 14.34. This is somewhat unimpressive in print but by using the mouse to rotate the image using the MATLAB figure toolbar *3D rotate* option the

3-dimensionality becomes quite clear. The axis reversals are required to have z increase from our viewpoint and to maintain a right-handed coordinate frame. There are many *holes* in this surface which are the NaN values we inserted to indicate unreliable disparity values.

14.3.5   3D Texture Mapped Display

For human, rather than robot, consumption it would be nice to enhance the surface representation so that it looks less ragged. We create a median filtered image

```
>> dimf = irank(di, 41, ones(9,9));
```

where each output pixel is the median value over a 9×9 window. This has *patched* many of the smaller holes but has the undesirable side effect of blurring the underlying disparity image. Instead we will keep the original interpolated disparity image and insert the median filtered values only where a NaN exists

```
>> di = ipixswitch(isnan(di), dimf, di);
```

We perform the reconstruction again

```
>> X = b*(U-u0) ./ di;   Y = b*(V-v0) ./ di; Z = 3740 * b ./ di;
```

and plotting this as a surface we see that it looks significantly less ragged.

However we can do even better. We can *drape* the left-hand image over the 3-dimensional surface using a process called texture mapping. We reload the left-hand image, this time in color

```
>> Lcolor = iread('rocks2-l.png');
```

and render the surface with the image texture mapped

```
>> surface(X, Y, Z, Lcolor, 'FaceColor', 'texturemap', ...
   'EdgeColor', 'none', 'CDataMapping', 'direct')
>> xyzlabel
>> set(gca,'ZDir', 'reverse'); set(gca,'XDir', 'reverse')
```

which creates the image shown in Fig. 14.35. Once again it is easier to get an impression of the 3-dimensionality by using the mouse to rotate the image using the MATLAB figure toolbar *3D rotate* option.

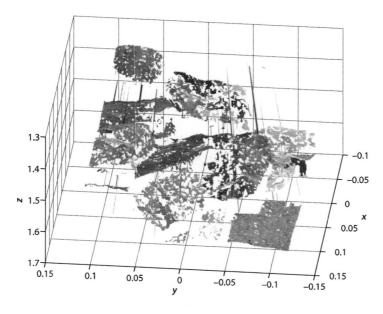

Fig. 14.35.
3-dimensional reconstruction for parallel stereo cameras with image texture mapped onto the surface

14.3.6 Anaglyphs

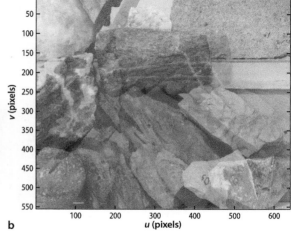

Human stereo perception of depth works because each eye views the scene from a different viewpoint. If we look at a photograph of a 3D scene we still get a perception of depth, albeit reduced, because our brain uses many visual cues besides stereo to infer depth. Since the invention of photography in the 19th century people have been fascinated by 3D photographs and movies, and the current popularity of 3D movies and availability of 3D television is further evidence of this.

The key in all 3D display technologies is to take the image from two cameras, with a similar baseline to the human eyes (approximately 8 cm) and present those images again to the corresponding eyes. Old fashioned stereograms required a binocular viewing device or could, with difficulty, be viewed by squinting at the stereo pair and crossing your eyes. More modern and convenient means of viewing stereo pairs are LCD shutter (gaming) glasses or polarized glasses which allow full-color stereo movie viewing, or head mounted displays.

An old but inexpensive method of viewing and distributing stereo information is through anaglyph images in which the left and right images are overlaid in different colors. Typically red is used for the left eye and cyan (greeny blue) for the right eye but many other color combinations are used. The red lens allows only the red part of the anaglyph image through to the left eye, while the cyan lens allows only the cyan parts of the image through to the right eye. The disadvantage is that only the scene intensity, not its color, can be portrayed. The big advantage of anaglyphs is that they can be printed on paper or imaged onto ordinary movie film and viewed with simple and cheap glasses such as those shown in Fig. 14.36a.

The rock pile stereo pair can be displayed as an anaglyph

```
>> anaglyph(L, R, 'rc')
```

which is shown in Fig. 14.36b. The argument `'rc'` indicates that left and right images are encoded in <u>r</u>ed and <u>c</u>yan respectively. Other color options include: <u>b</u>lue, <u>g</u>reen, <u>m</u>agenta and <u>o</u>range.

> **Anaglyphs.** The earliest developments occurred in France. In 1858 Joseph D'Almeida projected 3D magic lantern slide shows as red-blue anaglyphs and the audience wore red and blue goggles. Around the same time Louis Du Hauron created the first printed anaglyphs. Later, around 1890 William Friese-Green created the first 3D anaglyphic motion pictures using a camera with two lenses. Anaglyphic films called plasticons or plastigrams were a craze in the 1920s.
> Today high-resolution panoramic anaglyphs can be found on http://gigapan.com and anaglyphs of Mars can be found at http://mars.nasa.gov/mars3d.

Fig. 14.36. Anaglyphs for stereo viewing. **a** Anaglyph glasses shown with red and blue lenses, **b** anaglyph rendering of the rock scene from Fig. 14.20 with left in *red* and right in *cyan*

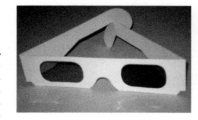

a

b

14.3.7   Image Rectification

The rock pile stereo pair of Fig. 14.20 has corresponding points on the same row in the left- and right-hand images – they are an epipolar-aligned image pair. Stereo cameras, such as shown in Fig. 14.19, are built with precision to ensure that the optical axes of the cameras are parallel and that the u- and v-axes of the two sensor chips are parallel. However there are limits to the precision of mechanical alignment and lens distortion will introduce error. Typically one or both images are warped to correct for these errors – a process known as rectification.

We will illustrate rectification using the courtyard stereo pair from Fig. 14.15

```
>> L = iread('walls-l.jpg', 'mono', 'double', 'reduce', 2);
>> R = iread('walls-r.jpg', 'mono', 'double', 'reduce', 2);
```

which we recall are far from being epipolar aligned. We first find the SURF features

```
>> sL = isurf(L);
>> sR = isurf(R);
```

and determine the candidate matches

```
>> m = sL.match(sR, 'top', 1000);
```

then determine the epipolar relationship

```
>> F = m.ransac(@fmatrix,1e-4, 'verbose');
96 trials
309 outliers
0.000305205 final residual
```

The rectification step requires the fundamental matrix as well as the set of corresponding points which is embedded in the `FeatureMatch` object m

```
>> [Lr,Rr] = irectify(F, m, L, R);
```

and returns rectified versions of the two input images. We display these using `stdisp`

```
>> stdisp(Lr, Rr)
```

which is shown in Fig. 14.37. We see that corresponding points in the scene now have the same vertical coordinate. The function `irectify` works by computing unique homographies to warp the left and right images. As we have observed previously when warping images not all of the output pixels are mapped to the input images which results in undefined pixels which are displayed here as red.

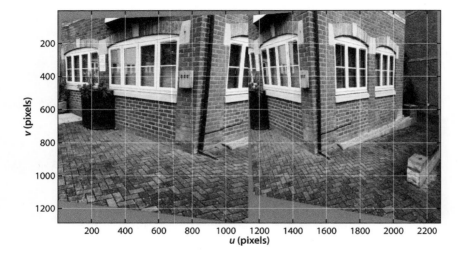

Fig. 14.37.
Rectified images of the courtyard. *Red pixels* have no corresspondance in the other image

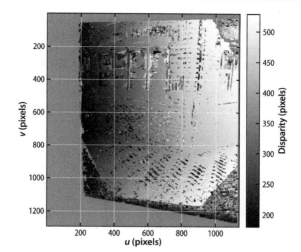

Fig. 14.38.
Dense stereo disparity image for the courtyard. The walls and ground show a clear depth gradient

We can think of these images as having come from a virtual stereo camera with parallel axes and aligned pixel rows, and they can now be used for dense stereo matching

```
>> d = istereo(Lr, Rr, [180 530], 7, 'interp');
```

and the result is shown in Fig. 14.38. The disparity range parameters were determined interactively using `stdisp(Lr, Rr)` to check the disparity at near and far points in the rectified image pair. The window half size of 7 was arrived at with a little trial and error, this value corresponding to a 15×15 window and produces a reasonably smooth result, at the expense of computation. The noisy patches at the bottom and top right are due to occlusion – world points in one image are not visible in the other. Nevertheless this is quite an impressive result – using only two images taken from a handheld camera we have been able to create a dense 3-dimensional representation of the scene.

14.4 Bundle Adjustment

In Sect. 14.3.1 we used triangulation to estimate the 3D coordinates of a number of landmark points in the world, but this was an approximation based on a guesstimate of the relative pose between the cameras. To assess the quality of our solution we can "back project" the estimated 3D landmark points onto the image planes based on the estimated camera poses◄ and the known camera model. The back-projection error is the image-plane distance between the back-projected landmark and its observed position on the image plane.

In this case we only estimated the relative pose $^1\xi_2$ but we can consider the first camera pose as the reference coordinate frame $\xi_1 = 0$ and $\xi_2 = {}^1\xi_2$.

For the previous example the back projection is

```
>> p1 = cam.project(P);
>> p2 = cam.move(T).project(P);
```

for the first and second camera respectively. The distances between the back projections and observations across both cameras is

```
>> e = colnorm( [p1-m2.p1 p2-m2.p2] );
```

with statistics

```
>> mean(e)
ans =
    0.9942
>> max(e)
ans =
    6.6765
```

which clearly indicates the approximate nature of our solution – each back-projected point is in error by up to 7 pixels. Unfortunately we do not know whether the error is in the estimated camera poses, the landmark coordinates or both. However we do know that a good estimate is one where this total back-projection error is low – ideally zero.

Bundle adjustment is an optimization process that simultaneously adjusts the camera poses and the landmark coordinates so as to minimize the total back-projection error. It uses 2D measurements from a set of images of the same scene to recover information related to the 3D geometry of the imaged scene as well as the locations and optical characteristics of the cameras. This is also called Structure from Motion (SfM) or Structure and Motion Estimation (SaM) – *structure* being the 3D landmarks in the world and *motion* being a sequence of camera poses. It is also called visual SLAM (VSLAM) since it is very similar to the pose-graph SLAM problem discussed in Sect. 6.6. That was a planar problem solved in the three dimensions of $\mathbf{SE}(2)$ whereas bundle adjustment involves camera poses in $\mathbf{SE}(3)$ and points in $\mathbb{R}^3$.

To formalize the problem consider a camera with known intrinsic parameters at N different poses $\xi_i \in \mathbf{SE}(3)$, $i = 1 \cdots N$ and a set of M landmark points $\boldsymbol{P}_j \in \mathbb{R}^3$, $j = 1 \cdots M$. At pose $\{i\}$ the camera observes $\boldsymbol{P}_j$ and the measured image-plane projection is ${}^i\boldsymbol{p}_j^{\#} \in \mathbb{R}^2$, but not all landmarks are necessarily visible from each camera pose. The notation is shown in Fig. 14.39 for two cameras.

The estimated value of the image-plane projection of the jth landmark in the ith image plane is

$$
{}^i\hat{\boldsymbol{p}}_j = \mathcal{P}\left(\hat{\xi}_i, \hat{\boldsymbol{P}}_j; \boldsymbol{K}\right)
$$

and the back-projection error is ${}^i\hat{\boldsymbol{p}}_j - {}^i\boldsymbol{p}_j^{\#}$.

Using the Toolbox we start by creating a `BundleAdjust` object

```
>> ba = BundleAdjust(cam);
```

which is passed an intrinsic model of the camera which we assume is known. Next we add estimates of the two camera poses

```
>> c1 = ba.add_camera( SE3(), 'fixed' );
>> c2 = ba.add_camera( T );
```

and indicate that the first camera pose is known and that we do not wish to optimize for it. The second camera's estimated pose is that derived earlier from the essential matrix. The method returns an integer handle to the particular camera pose which we will use below.

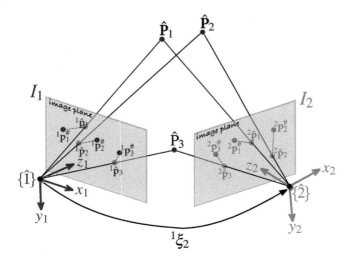

Fig. 14.39.
Bundle adjustment notation illustrated with a simple problem comprising only two cameras and three world points. The estimated camera poses and point positions are indicated, as are the estimated and measured image-plane coordinates. The reprojection errors are shown as *dashed grey lines*. The problem is solved when the variables are adjusted so that the total reprojection error is as small as possible

Next we add the estimated landmarks

```
>> for j=1:length(m2)
     lm = ba.add_landmark( P(:,j) );

     ba.add_projection(c1, lm, m2(j).p1);
     ba.add_projection(c2, lm, m2(j).p2);
   end
```

where `lm` is another integer handle, in this case to the particular landmark coordinate. Finally, we add the measurements by specifying the camera, the landmark and its projection on the image plane. The problem is now fully defined and a summary can be displayed

```
>> ba
ba =
Bundle adjustment problem:
  2 cameras
    locked cameras: 1
  100 landmarks
  200 projections
  306 dimension linear problem
  landmarks per camera: min=100.0, max=100.0, avg=100.0
  cameras per landmark: min=2.0, max=2.0, avg=2.0
```

In general only a subset of landmarks are visible from any camera, and this visibility information can be represented elegantly using a graph as shown in Fig. 14.40 where each camera pose and each landmark coordinate is a node. Edges between camera and landmark nodes represent observations, and the value of the edge is the observed image-plane coordinate. Such a graph, a Toolbox `PGraph` object, is held inside the `BundleAdjust` object and can be plotted by

```
>> ba.plot
```

and an example is shown in Fig. 14.41.

To solve this optimization problem we put all the variables we wish to adjust into a single state vector. For bundle adjustment the state vector contains camera poses and landmark coordinates

$$\boldsymbol{x} = \left\{\xi_1, \xi_2 \cdots \xi_N \middle| \boldsymbol{P}_1, \boldsymbol{P}_2 \cdots \boldsymbol{P}_M \right\} \in \mathbb{R}^{6N+3M}$$

where the **SE**(3) camera pose is represented in a vector format $\xi_i \sim (\boldsymbol{t}, \boldsymbol{r}) \in \mathbb{R}^6$ comprising translation $\boldsymbol{t} \in \mathbb{R}^3$ and rotation $\boldsymbol{r} \in \mathbb{R}^3$; and $\boldsymbol{P}_j \in \mathbb{R}^3$.

Possible representations of rotation include Euler angles, roll-pitch-yaw angles, angle-axis or exponential coordinate representations. For bundle adjustment it is common to use the vector component of a unit-quaternion which is singularity free and has only three parameters. The double cover property of unit-quaternions means that any unit-quaternion can be written with a nonnegative scalar component. By definition

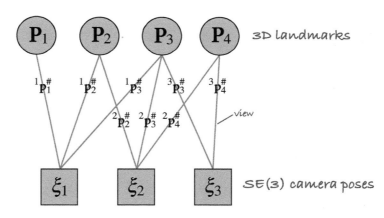

Fig. 14.40.
A simple visibility graph showing camera nodes (*red*) and landmark nodes (*blue*). Lines connecting nodes represent a view of that node on that camera, and the edge value is the observed image-plane coordinate. The landmarks here are viewed by 1, 2 or 3 cameras

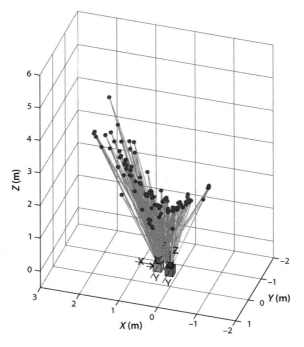

Fig. 14.41.
Bundle adjustment problem shown as an embedded graph. *Blue dots* represent landmark positions, *camera icons* represent camera pose, and *grey lines* denote observations. Camera 1 is *blue* and camera 2 is *red*

the unit-quaternion has a unit norm, so the scalar component can be easily recovered $s = \sqrt{1 - v_x^2 - v_y^2 - v_z^2}$ given the vector component.

The number of unknowns in this system is $6N + 3M$: 6 unknowns for each camera pose and 3 unknowns for the position of each landmark point. However we have up to $2NM$ equations due to the measured projections of the points on the image planes. Typically the pose of one camera is assumed to be the reference coordinate frame, and this reduces the number of unknowns to $6(N - 1) + 3M$.

In the problem we are discussing $N = 2$, but one camera is locked, and $M = 100$ so we have $6 \times (2 - 1) + 3 \times 100 = 306$ unknowns and $2 \times 2 \times 100 = 400$ equations – an overdetermined set of equations for which a solution should be possible. For our problem we can extract the state vector

```
>> x = ba.getstate;
>> about x
x [double] : 1x312 (2.5 kB)
```

which includes the pose of the fixed camera, although that will remain constant. The pose of camera two is stored in the second block of 6 elements

```
>> x(7:12)
ans =
    0.3000   -0.0153    0.0558    0.0127    0.0014   -0.0057
```

as translation followed by rotation, and the first landmark is stored in

```
>> x(13:15)
ans =
    1.2134   -0.2627    2.9563
```

Bundle adjustment is a minimization problem that finds the camera poses and landmark positions that minimize the reprojection error across all the edges

$$x^* = \arg \min_x \sum_k F_k(x)$$

where $F_k(\cdot) > 0$ is a nonnegative scalar cost associated with the graph edge k from camera i to landmark j. The reprojection error of a landmark at P_j onto the camera at pose ξ_i is

$$f_k(x) = \mathcal{P}\left(\hat{\xi}_i, \hat{P}_j; K\right) - {}^i p_j^\# \in \mathbb{R}^2$$

and the scalar cost is the squared Euclidean reprojection error

$$F_k(x) = f_k(x)^T f_k(x)$$

Although written as a function of the entire state vector $F_k(\cdot)$ only depends on two elements of that vector: ξ_i and P_j. The total error, the sum of the squared back-projection error for all edges, can be computed for any value of the state vector and for the initial conditions is

```
>> ba.errors(x)
ans =
  553.2853
```

The bundle adjustment task is to adjust the camera and landmark parameters to reduce this value. We have framed bundle adjustment as a sparse nonlinear least squares problem and this can be solved numerically if we have a sufficiently good initial estimate of x.

The first step in solving this class of problem is to linearize it. The reprojection error $f_k(x)$ can be linearized about the current state x_0 of the system

$$f_k'(\Delta) \approx f_{0,k} + J_k \Delta$$

where $f_{0,k} = f_k(x_0)$ and

$$J_k = \frac{\partial f_k(x)}{\partial x} \in \mathbb{R}^{2 \times (6N+3M)}$$

is a Jacobian matrix which depends only on the camera pose ξ_i and the landmark position P_j so is therefore mostly zeros

$$J_k = \left(0 \cdots A_i \cdots B_j \cdots 0 \right), \quad \text{where } A_i = \frac{\partial f_k(x)}{\partial \xi_i} \in \mathbb{R}^{2 \times 6}, \ B_j = \frac{\partial f_k(x)}{\partial P_j} \in \mathbb{R}^{2 \times 3}$$

The structure of the Jacobian matrix A_i is specific to the chosen representation of camera pose. The Jacobians, particularly A_i, are quite complex to derive but can be automatically generated using the MATLAB Symbolic Math Toolbox™ and the script `vision/symbolic/bundleAdjust`. Code derived from this is implemented by the `derivs` method of the `CentralCamera` class

```
[p,A,B] = cam.derivs(t, r, P);
```

Translating the camera by d and translating the point by $-d$ have an equivalent effect on the image. Therefore B_j is the negative of the first three columns of A_j.

which returns the image-plane projection and the two Jacobians in a single call, and where t and r are the camera pose and P is the landmark coordinate.◀ Linearization and Jacobians are discussed in Appendix E, and solution of sparse nonlinear equations in Appendix F.

Now that everything is in place we can solve our bundle adjustment problem

```
>> baf = ba.optimize(x);
Initial cost 553.285
  total cost 33.5955 (solved in 0.15 sec)
  total cost 33.5459 (solved in 0.051 sec)
  total cost 33.5459 (solved in 0.04 sec)
  total cost 33.5459 (solved in 0.038 sec)
  total cost 33.5459 (solved in 0.041 sec)
  total cost 33.5459 (solved in 0.037 sec)
 * 6 iterations in 0.5 seconds
 * 0.41 pixels RMS error
```

and the displayed text shows how the total cost (squared reprojection error) decreases at each iteration, reducing by over an order of magnitude. The final result has an RMS reprojection error better than half a pixel for each landmark which is impressive given that the images were captured with a phone camera and we have completely ignored lens distortion.

The result is another `BundleAdjust` object but with updated camera poses and landmark positions. We can compare the initial and final pose of camera 2

```
>> ba.getcamera(2).print('camera')
t = (0.3, -0.0153, 0.0558), RPY/yxz = (-0.657, 1.46, 0.156) deg
>> baf.getcamera(2).print('camera')
t = (0.303, -0.0158, 0.0649), RPY/yxz = (-0.685, 1.38, 0.128) deg
```

and the final coordinate of landmark 5 is

```
>> baf.getlandmark(5)'
ans =
    -0.3861    -0.0968     2.0744
```

We can also plot the result graphically

```
>> ba.plot()
```

and this is shown in Fig. 14.41. While the overall RMS error is low we can look at the final reprojection error in more detail

```
>> e = sqrt( baf.getresidual() );
>> about e
e [double] : 2x100 (1.6 kB)
```

where element (i, j) is the reprojection error in pixels for camera i and landmark j. The median error is

```
>> median( e(:) )
ans =
    0.2540
```

around a quarter of a pixel, but there are a handful of landmarks with a final reprojection error in camera one that are greater than 1 pixel

```
>> find( e(1,:) > 1 )
ans =
    90    97
```

and the worst error for camera 1

```
>> [mx,k] = max( e(1,:) )
mx =
    1.2129
k =
    90
```

of 1.2 pixels occurs for landmark 90.

Bundle adjustment finds the optimal *relative* pose and positions – not absolute. For example if all the cameras and landmarks moved 1 m in the x-direction the total reprojection error would be the same. To remedy this we can fix or *anchor* one or more cameras or landmarks – in this example we fixed the first camera. The values of the fixed poses and positions are kept in the state vector but they are not updated during the iterations – their Jacobians do not need to be computed and the Hessian matrix used to solve the update at each iteration is smaller since the rows and columns corresponding to those fixed parameters can be deleted.

The fundamental issue of scale ambiguity with monocular cameras has been mentioned a number of times and it applies here as well. A scale model of the same world with a similarly scaled camera translation is indistinguishable from the real thing. More formally, if the whole problem was scaled so that $P'_j = \lambda \, P_j$, $[\xi'_i]_t = \lambda[\xi_i]_t$ and $\lambda \neq 0$ the total reprojection error would be the same. The solution we obtained above has an arbitrary scale or value of λ – changing the initial condition for the camera poses or landmark coordinates will lead to a solution with a different scale. We can remedy this by anchoring the pose of at least two cameras, one camera and one landmark, or two landmarks.

The bundle adjustment technique, but not this implementation, allows for constraints between cameras. For example, a multi-camera rig moving through space would use constraints to ensure the fixed relative pose of the cameras at each time step. Odometry from wheels or inertial sensing could be used to constrain the distance between camera coordinate frames to enforce the correct scale, or orientation from an IMU could be used to constrain the camera attitude. In the underlying graph representation of the problem as shown in Fig. 14.40 this would involve adding additional edges between the camera nodes. Constraints could also be added between landmarks that had a known relative position, for example the corners of a window – this would involve adding additional edges between the relevant landmark nodes.

The particular problem we studied is unusual in that every camera views every landmark. In a more common situation the camera might be moving in a very large environment so any one camera will only see a small subset of landmarks. In a real-time system a limited bundle adjustment might be performed with respect to occasional frames known as key frames, and a bundle adjustment over all frames, or all keyframes, performed at a low rate in the background.

The camera matrix has an arbitrary scale factor.

In this example we have assumed the camera intrinsic parameters are known and constant. Theoretically bundle adjustment can solve for intrinsic as well as extrinsic parameters. We simply add additional parameters for each camera in the state vector and adjust the Jacobian A accordingly. However given the coupling between intrinsic and extrinsic parameters◄ this may lead to poor performance. If we chose to estimate the elements of the camera matrix C directly then the state vector would contain 11◄ rather than 6 elements for each camera. However if C_i and P_j are solutions so to is $C_i Q^{-1}$ and $Q\tilde{P}_j$ for any nonsingular matrix $Q \in \mathbb{R}^{4\times4}$. Fortunately, projection matrices for realistic cameras have well defined structure and properties as described on page 327, and these provide constraints that allow us to estimate Q. Estimating an arbitrary C_i is referred to as a projective reconstruction. This can be *upgraded* to an affine reconstruction (using an affine camera model) or a metric reconstruction (using a perspective camera model) by suitable choice of Q.

Changes in focal length and *z*-axis translation have similar image-plane effects as do change in principal point and camera *x*- and *y*-axis translation.

14.5   Point Clouds

Stereo vision results in a set of 3-dimensional world points P_i which are often referred to as a point cloud. For a robotics application we need to extract some concise meaning from the thousands or millions of points.

14.5.1   Fitting a Plane

Planes are common in our built world and for robotics useful planes include the ground (for wheeled mobile robot driving or UAV landing) and walls. Given a set of 3-dimensional coordinates, a point cloud, a simple and effective approach for finding the plane of best fit is to fit the data to an ellipsoid. The ellipsoid will have one very small radius in the direction normal to the plane – that is, it will be an elliptical plate. The inertia matrix of the points can be calculated by

$$J = \sum_{i=1}^{N} x_i x_i^T \tag{14.17}$$

where $x = P_i - \overline{P}$ are the coordinates of the points with respect to the centroid of the points $\overline{P} = \frac{1}{N}\sum_{i=1}^{N} P_i$. The ellipsoid is centered at the centroid of the point cloud. The radii of the ellipsoid are the eigenvalues of J and the eigenvector corresponding to the smallest eigenvalue is the direction of the minimum radius which is the normal to the plane.

To illustrate this we create a 10×10 grid of points in a plane

```
>> T = SE3(1,2,3) * SE3.rpy(0.3, 0.4, 0.5);
>> P = mkgrid(10, 1, 'pose', T);
>> P = P + 0.02*randn(size(P));
```

with an arbitrary orientation represented by the homogeneous transformation T, and to which some Gaussian noise has been added with $\sigma = 0.02$ m.

The mean of the point cloud is

```
>> x0 = mean(P')
ans =
     0.9967    2.0009    3.0013
```

which we subtract from all the data points

```
>> P = bsxfun(@minus, P, x0');
```

and the inertia Eq. 14.17 is simply a matrix multiplication

```
>> J = P*P'
J =
     7.8769    0.3239   -4.2585
     0.3239   10.0076    0.6153
    -4.2585    0.6153    2.4271
```

The eigenvalues are

```
>> [x,lambda] = eig(J);
>> diag(lambda)'
ans =
     0.0478   10.0553   10.2085
```

and we see two large eigenvalues corresponding to the spread of points within the plane, and one eigenvalue which is the *thickness* of the plane. The eigenvector corresponding to the first, and smallest, eigenvalue is

```
>> n = x(:,1)'
n =
     0.4789   -0.0696    0.8751
```

which is the estimated normal to the plane.

The true direction of the plane's normal is given by the third column► of the rotation matrix

```
>> T.SO3.a'

ans =
     0.4682   -0.0810    0.8799
```

Since the points lie in the frame's *xy*-plane, the normal is the frame's *z*-axis.

and we see that it is very close to the estimated normal.

The equation of a plane is the set of points x such that

$$n^T (x - x_0) = 0 \tag{14.18}$$

where n is the normal and x_0 is the centroid.

Outlier data points are problematic with this type of estimator since they significantly bias the solution. A number of approaches are commonly used but a simple one is to modify Eq. 14.17 to include a weight

$$J = \sum_{i=1}^{N} w_i x_i x_i^T$$

which is inversely related to the distance of x_i from the plane and solve iteratively. Initially all weights $w_i = 1$, and on subsequent iterations the weights► are set according to the distance of $\mathbf{P}_i$ from the plane estimated at the previous step.

Alternatively we could apply RANSAC by taking samples of three points to solve for Eq. 14.18. Section C.1.4 has more details about ellipses.

Commonly a Cauchy-Lorentz function $w = \beta^2 / (d^2 + \beta^2)$ is used where d is the distance of the point from the plane and β is the half-width. The function is smooth for $d = [0, \text{inf})$ and has a value of ½ when $d = \beta$.

14.5.2   Matching Two Sets of Points

Consider a model of some object represented by a set of points in 2- or 3-dimensions with respect to the world frame. Now consider an example of that object with a different pose and we observe a set of 2- or 3-dimensional points on the object. The task is to determine the relative pose ξ that will transform the model points to the observed data points by matching the two sets of points. ◄

The dual problem is that the camera has moved, not the object. The same technique can be applied to determine the camera motion.

More formally, given two sets of point coordinate vectors: the model $M_i \in \mathbb{R}^n$, $i \in [1, N_M]$ and some noisy observed data $D_j \in \mathbb{R}^n$, $j \in [1, N_D]$ determine the rigid-body motion from the data coordinate frame to the model frame

$$^D\xi_M^* = \arg\min_\xi \sum_{i,j} \left\| D_j - \xi \cdot M_i \right\|$$

At first glance this looks like a problem where we need to establish correspondence between the points in the two sets but we will introduce an alternative approach called iterated closest point or ICP. For each data point D_j, the corresponding model point M_i is assumed to be the closest one, that is M_i which minimizes $\| M_i - D_j \|$. Correspondence is not unique and quite commonly several points in one set can be associated with a single point in the other set, and consequently some points will be unpaired. Often the sensor returns only a subset of points in the model, for instance a laser scanner can see the front but not the back of an object. This approach to correspondence is far from perfect but it is surprisingly good in practice and *improves* the alignment of the point clouds so that in the next iteration the computed correspondences will be a little more accurate.

In robotics the problem is often considered as comprising a *model M* of a 3-dimensional object which we want to fit to the observed sensor data *D*. To illustrate we will load a version of the famous Stanford bunny ◄

This model is well known in the computer graphics community. It was created by Greg Turk and Marc Levoy in 1994 at Stanford University using a Cyberware 3030 MS scanner and a ceramic rabbit figurine. The original scan has over 30 000 points, here we use a low-resolution version.

```
>> load bunny
>> about bunny
bunny [double] : 3x453 (10.9 kB)
```

which is a cloud of 453 3-dimensional points and this will be our model

```
>> M = bunny;
```

We simulate a sensor that is observing the model with respect to a different coordinate frame by making a copy of the model and applying a transformation

```
>> T_unknown = SE3(0.2, 0.2, 0.1) * SE3.rpy(0.2, 0.3, 0.4);
>> D = T_unknown * M;
```

The first step is to compute a translation that makes the centroids of the two point clouds coincident ◄

We consider the general case where the two points clouds have different numbers of points, that is, $N_D \neq N_M$.

$$\bar{M} = \frac{1}{N_M} \sum_{i=1}^{N_M} M_i$$

$$\bar{D} = \frac{1}{N_D} \sum_{j=1}^{N_D} D_j$$

from which we compute a displacement

$$t = \bar{D} - \bar{M}$$

Next we compute correspondence. For each data point D_j we find the closest model point M_i, and for this we use the Toolbox function closest

```
>> coresp = closest(D, M);
```

where `i=corresp(j)` is the column of M that corresponds to column j of D. The next step is to compute the 3×3 moment matrix

$$W = \sum_{i,j} \left(M_i - \bar{M} \right)\left(D_j - \bar{D} \right)^T$$

which encodes the rotation between the two point sets.► The singular value decomposition is

This is the sum of a number of rank 1 matrices.

$$W = U\Sigma V^T$$

from which the rotation matrix is determined► to be

See Sect. F.1.1.

$$R = VU^T$$

The estimated relative pose between the two point clouds is $\xi_\Delta \sim (R, t)$ and the model points are transformed so that they are closer to the data points

$$M_i \leftarrow \xi \cdot M_i, \ \forall i$$
$$\xi \leftarrow \xi \oplus \xi_\Delta$$

and the process repeated until it converges. The correspondences used are unlikely to have all been correct and therefore the estimate of the relative orientation between the sets is only an approximation.

The Toolbox provides an implementation of ICP (Fig. 14.42)

```
>> [T,d] = icp(M, D, 'plot');
```

which returns the pose $^D\xi_M$

```
>> trprint(T, 'rpy', 'radian')
t = (0.2, 0.2, 0.1), RPY/zyx = (0.2, 0.3, 0.4) rad
```

which is exactly the "unknown" relative pose of the data point cloud that we chose above. The residual

```
>> d
d =
    1.7619e-15
```

is the root mean square of the errors between the transformed model points and the data. The option `'plot'` shows the model and data points at each step as well as the closest-point correspondences. ICP is a popular algorithm because it is both fast and robust.

We can demonstrate the robustness of ICP by simulating some realistic sensor errors. Firstly we will randomly remove forty points from the data

```
>> D(:,randi(numcols(D), 40,1)) = [];
```

Fig. 14.42. Iterated closest point (ICP) matching of two point clouds: model (*red*) and data (*blue*) **a** before registration, **b** after registration; observed data points have been transformed to the model coordinate frame using the inverse of the identified transformation (Stanford bunny model courtesy Stanford Graphics Laboratory)

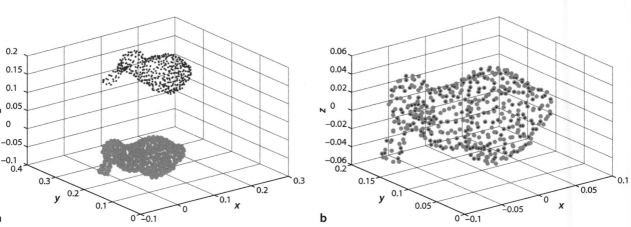

which are points in the model not observed by the sensor. Then we will add twenty spurious points that are not part of the model

```
>> D = [D 0.1*rand(3,20)+0.1];
```

and finally we will add Gaussian noise with $\sigma = 0.01$ to the data

```
>> D = D + 0.01*randn(size(D));
```

Now we fit this imperfect sensor data to the model

```
>> [T,d] = icp(M, D, 'plot', 'distthresh', 3);
```

We would expect the residual to be approximately equal to $\sqrt{N}\sigma$ where N is the number of corresponding points and σ is the standard deviation of the additive noise.

using an additional option to eliminate incorrect closest-point correspondences. The correspondences are established as described above and the median of the distances between the corresponding points is computed. In this case the correspondence is not made if the distance between the points is more than 3 times the median distance. The estimated pose $^D\xi_M$ is now

For large-scale problems the data would be kept in a *kd*-tree which reduces the time required to find the closest point.

```
>> trprint(T, 'rpy', 'radian')
t = (0.186, 0.194, 0.108), RPY/zyx = (0.125, 0.287, 0.298) rad
```

which is still close to the value computed for the ideal case but the residual

```
>> d
d =
    0.2114
```

Laser-based line projectors, so called "laser stripe"' or "line laser"', are available for just a few dollars. They comprise a low-power solid-state laser and a cylindrical lens or diffractive optical element.

is higher since an exact fit between the model and noise corrupted data is no longer possible.▼ ICP is popular, fast and robust for modest sized point clouds but the correspondence determination is an $O(N^2)$ problem which leads to computational bottlenecks for very large data sets.▼

Fig. 14.43. a Geometry of structured light showing a light projector on the left and a camera on the right; four corresponding points are marked with *dots* on the left and right images and the scene; **b** a real structured light scenario showing the light stripe falling on a simple 3D scene. The superimposed *dashed line* represents the stripe position for a plane at infinity. Disparity, left shift of the projected line relative to the dashed line, is inversely proportional to depth

14.6 Structured Light

An old, yet simple and effective, method of estimating the 3D structure of a scene is structured light. It is conceptually similar to stereo vision but we replace the left camera with a projector that emits a vertical plane of light as shown in Fig. 14.43a.▼ This is equivalent, in a stereo system, to a left-hand image that is a vertical line. The image of the line projected onto the surface viewed from the right-hand camera will be a distorted version of the line, as shown in Fig. 14.43b. The disparity between the virtual left-hand image and the actual right-hand image is a function of the depth of points along the line.

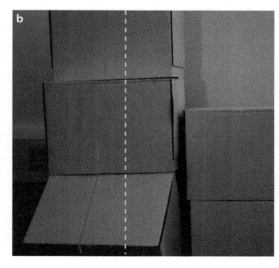

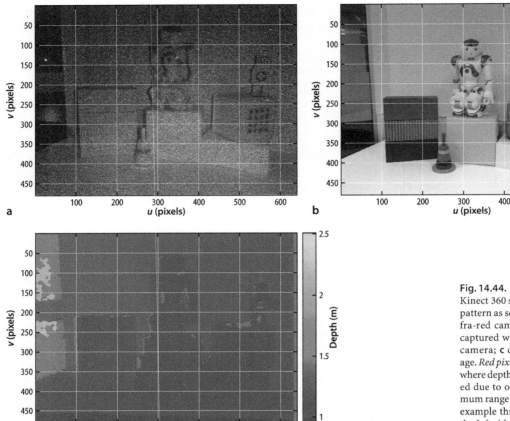

Fig. 14.44. 3D imaging with the Kinect 360 sensor. **a** Random dot pattern as seen by the Kinect's infra-red camera; **b** original scene captured with the Kinect's color camera; **c** computed depth image. *Red pixels* indicate NaN values where depth could not be computed due to occlusion or the maximum range being exceeded, as for example through the window on the left side of the scene (images courtesy William Chamberlain)

Finding the light stripe on the scene is a relatively simple vision problem. In each image row we search for the pixel corresponding to the projected stripe based on intensity or color. If the camera coordinate frames are parallel then depth is computed by Eq. 14.16.

To achieve depth estimates over the whole scene we need to move the light plane horizontally across the scene and there are many ways to achieve this: mechanically rotating the laser stripe projector, using a moving mirror to deflect the stripe or using a data projector and software to create the stripe image. However sweeping the light plane across the scene is slow and fundamentally limited by the rate at which we can acquire successive images of the scene. One way to speed up the process is to project multiple lines on the scene but then we have to solve the correspondence problem which is not simple if parts of some lines are occluded. Many solutions have been proposed but generally involve coding the lines in some way – using different colors or using a sequence of binary or grey-coded line patterns which can match 2^N lines in just N frames.

A related approach is to project a known but random pattern of *dots* onto the scene as shown in Fig. 14.44a. Each dot can be identified by the unique pattern of dots in its surrounding window. The original Kinect sensor▶ uses this approach: its left-most lens◄ projects an infra-red dot pattern using a laser with a diffractive optical element which is viewed, see Fig. 14.44a, by an infra-red sensitive camera behind the right-most lens from which the depth image shown in Fig. 14.44c is computed. The shape of the dots also varies with distance, due to imperfect focus, and this provides additional cues about the distance of a point. The middle lens is a regular color camera which provides

The Kinect for Xbox 360 and Kinect for Windows is now known as the Kinect 1, as well as sensors such as PrimeSense Carmine and Asus Xtion. The newer Kinect for Xbox One, or Kinect 2, uses per pixel time-of-flight measurement.

Looking at the front of the device.

the view shown in Fig. 14.44b. This is an example of an RGBD camera, returning an RGB color values as well as depth (D) at every pixel.

Structured light approaches work well for ranges of a few meters indoors, for textureless surfaces, and they also work in the dark. However outdoors the projected pattern is overwhelmed by ambient illumination from the sun.

Some stereo systems, such as the Intel RealSense R200, also employ a dot pattern projector, sometimes known as a speckle projector. This provides artificial texture which helps the stereo vision system when it is looking at textureless surfaces where matching is frequently weak and ambiguous as discussed in Sect. 14.3.2.1. Such a sensor has the advantage of working on textureless surfaces which are common indoors where the sun is not a problem, and outdoors using pure stereo where scene texture is usually rich.

14.7   Applications

14.7.1   Perspective Correction

Consider the image

```
>> im = iread('notre-dame.jpg', 'double');
>> idisp(im)
```

shown in Fig. 14.45. The shape of the building is significantly distorted because the camera's optical axis was not normal to the plane of the building and we see evidence of perspective foreshortening or keystone distortion. We manually pick four points, clockwise from the bottom left, that are the corners of a large rectangle on the planar face of the building

```
>> p1 = ginput(4)'
ans =
    44.1364    94.0065   537.8506   611.8247
   377.0654   152.7850   163.4019   366.4486
```

which has one column per point that contains the u- and v-coordinate. We mark these on the image of the cathedral and overlay a translucent blue keystone shape

```
>> plot_poly(p1, 'wo', 'fill', 'b', 'alpha', 0.2);
```

We use the extrema of these points to define the vertices of a rectangle in the image

```
>> mn = min(p1');
>> mx = max(p1');
>> p2 = [mn(1) mx(2); mn(1) mn(2); mx(1) mn(2); mx(1) mx(2)]';
```

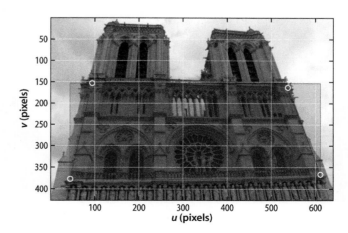

Fig. 14.45.
Photograph taken from the ground shows the effect of foreshortening which gives the building a trapezoidal appearance (also known as keystone distortion). Four points on the approximately planar face of the building have been manually picked as indicated by the *white* ○-markers (Notre Dame de Paris)

which we overlay on the image in red

```
>> plot_poly(p2, 'k', 'fill', 'r', 'alpha', 0.2)
```

The sets of points p1 and p2 are projections of world points that lie approximately in a plane so we can compute an homography

```
>> H = homography(p1, p2)
H =
    1.4003    0.3827  -136.5900
   -0.0785    1.8049   -83.1054
   -0.0003    0.0016     1.0000
```

that will transform the vertices of the blue trapezoid to the vertices of the red rectangle.▶

> An homography can also be computed from four lines in the plane, but this is not supported by the Toolbox.

$$\tilde{p}_2 \simeq H\tilde{p}_1$$

That is, the homography maps image coordinates from the distorted keystone shape to an undistorted rectangular shape.

We can apply this homography to the coordinate of every pixel in an output image in order to warp the input image. We use the Toolbox generalized image warping function

```
>> homwarp(H, im, 'full')
```

and the result shown in Fig. 14.46 is a synthetic fronto-parallel view. This is equivalent to the view that would be seen by a camera high in the air with its optical axis normal to the face of the cathedral. However points that are not in the plane, such as the left-hand side of the right bell tower have been distorted. The black pixels in the output image are due to the corresponding pixel coordinates not being present in the input image. Note that with no output argument specified the warped image is displayed using idisp.

In addition to creating this synthetic view we can decompose the homography to recover the camera motion from the actual to the virtual viewpoint and also the surface normal of the cathedral. As we saw in Sect. 14.2.4 we need to determine the camera calibration matrix so that we can convert the projective homography into a Euclidean homography. We obtain the focal length from the metadata in the EXIF-format file that holds the image

```
>> [~,md] = iread('notre-dame.jpg', 'double');
>> f = md.DigitalCamera.FocalLength
f =
    7.4000
```

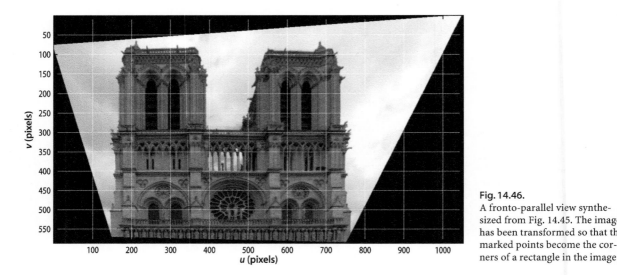

Fig. 14.46.
A fronto-parallel view synthesized from Fig. 14.45. The image has been transformed so that the marked points become the corners of a rectangle in the image

which is in units of millimeters, and the sensor is known to be 7.18×5.32 mm. We create a calibrated camera

```
>> cam = CentralCamera('image', im, 'focal', f/1000, ...
     'sensor', [7.18e-3,5.32e-3])
name: image [central-perspective]
   focal length:    0.0074
   pixel size:      (1.122e-05, 1.249e-05)
   principal pt:    (320, 213)
   number pixels:   640 x 426
   pose:            t = (0, 0, 0), RPY/yxz = (0, 0, 0) deg
```

Now we use the camera model to compute and decompose the Euclidean homography

```
>> sol = cam.invH(H, 'verbose');
solution 1
       T =  0.99958     -0.01394      0.02526     -0.07271
            0.01431      0.99979     -0.01453     -0.00041
           -0.02505      0.01488      0.99958      0.68149
            0.00000      0.00000      0.00000      1.00000
       n =  0.21602     -0.95261      0.21420
solution 2
       T =  0.98872      0.10353     -0.10820      0.10448
           -0.01647      0.79331      0.60859     -0.57151
            0.14885     -0.59994      0.78607      0.36357
            0.00000      0.00000      0.00000      1.00000
       n = -0.18131      0.32802      0.92711
```

which returns a structure array of two possible solutions for $^{1}\xi_{2}$. The coordinate frames for this example are sketched in Fig. 14.47 and shows the actual and virtual camera poses. In this case the second solution is the correct one since it represents considerable rotation about the x-axis. The camera translation vector, which is not to scale but has the correct sign,◄ is dominantly in the negative y- and positive z-direction with respect to the frame {1}. The rotation in YXZ-angle form

See Malis and Vargas (2007).

```
>> tr2rpy(sol(2).T, 'deg', 'camera')
ans =
   -1.1893   -37.4876   -7.8375
```

indicates that the camera needs to be pitched downward (pitch is rotation about the camera's x-axis) by 37 degrees to achieve the attitude of the virtual camera. The normal to the frontal plane of the church n is defined with respect to {1} and is essentially in the camera z-direction as expected.

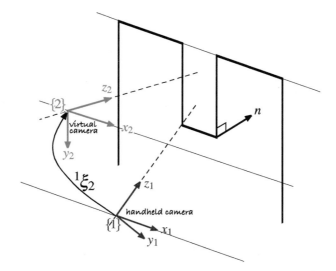

Fig. 14.47.
Notre-Dame example showing the two camera coordinate frames. The *blue frame {1}* is that of the camera that took the image, and the *red frame {2}* is the viewpoint for the synthetic fronto-parallel view

14.7.2   Mosaicing [examples/mosaic]

Mosaicing or image stitching is the process of creating a large-scale composite image from a number of overlapping images. It is commonly applied to drone and satellite images to create a seemingly continuous single picture of the Earth's surface. It can also be applied to images of the ocean floor captured from downward looking cameras on an underwater robot. The panorama generation software supplied with, or built into, digital cameras is another example of mosaicing.

The input to the mosaicing process is a sequence of overlapping images.▶ It is not necessary to know the camera calibration parameters or the pose of the camera where the images were taken – the camera can rotate arbitrarily between images and the scale can change. However for the approach that we will use the scene is assumed to be planar which is reasonable for high-altitude photography where the vertical relief▶ is small.

We will illustrate our discussion with a real example using the pair of images

```
>> im1 = iread('mosaic/aerial2-1.png', 'double', 'grey');
>> im2 = iread('mosaic/aerial2-2.png', 'double', 'grey');
```

which are each $1\,280 \times 1\,024$. We create an empty composite image that is $2\,000 \times 2\,000$

```
>> composite = zeros(2000,2000);
```

that will hold the mosaic. The essentials of the mosaicing process are shown in Fig. 14.48.

The first image is easy and we simply paste it into the top left corner

```
>> composite = ipaste(composite, im1, [1 1]);
```

of the composite image as shown in red in Fig. 14.48. The next image, shown in blue, is more complex and needs to be rotated, scaled and translated so that it correctly overlays the red image.

For this problem we assume that the scene is planar. This means that we can use an homography to relate the various camera views. The first step is to identify common feature points which are known as tie points, and we use now familiar tools

```
>> f1 = isurf(im1)
>> f2 = isurf(im2)
>> m = f1.match(f2);
```

and then RANSAC to estimate the homography

```
>> [H,in] = m.ransac(@homography, 0.2)
```

which maps $^1\mathbf{p}$ to $^2\mathbf{p}$. Now we wish to map $^2\mathbf{p}$ to its corresponding coordinate in the first image

$$^1\mathbf{p} \simeq H^{-1}\,{}^2\mathbf{p}$$

As a rule of thumb images should overlap by 60% of area in the forward direction and 30% sideways.

The ratio of the height of points above the plane to the distance of the camera from the plane.

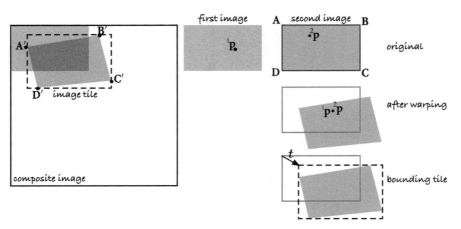

Fig. 14.48.
The first image in the sequence is shown as *red*, the second as *blue*. The second image is warped into the image tile and then blended into the composite image

The bounding box of the tile is computed by applying the homography to the image corners $A = (1, 1)$, $B = (W, 1)$, $C = (W, H)$ and $D = (1, H)$, where W and H are the width and height respectively, and finding the bounds in the u- and v-directions.

The default is NaN.

We do this for every pixel in the new image by warping

```
>> [tile,t] = homwarp(inv(H), im2, 'full', 'extrapval', 0);
```

As shown in Fig. 14.48 the warped blue image falls outside the bounds of the original blue image and the option 'full' specifies that the returned image is the minimum containing rectangle◄ of the warped image. This image is referred to as a *tile* and shown with a dashed black line. The vector t is returned by homwarp and gives the offset of the tile's coordinate frame with respect to the original image. In general not every pixel in the tile has a corresponding point in the input image and those pixels are set to zero, as specified by the fifth argument.◄

Now the tile has to be *blended* into the composite mosaic image

```
>> composite = ipaste(composite, tile, t, 'add');
```

and the result is shown in Fig. 14.49. We can clearly see several images overlaid with good alignment. The nonmapped pixels in the warped image are set to zero so adding them causes no change to the existing pixel values in the composite image.

Simply *adding* the tile into the composite image means that overlapping pixels are necessarily brighter and a number of different strategies can be used to remedy this. We could instead set pixels in the composite image from the tile only if the composite image pixels have not yet been set. Conversely we could *always* set pixels in the composite image from the nonzero pixels in the tile. Alternatively we set the composite image pixels to the mean of the tile and the composite image. This requires that we tag the tile pixels that are not mapped

```
>> [tile,t] = homwarp(inv(H), im2, 'full', 'extrapval', NaN);
```

Which ignores any pixels with the value NaN.

and then blend using the 'mean' option◄

```
>> composite = ipaste(composite, tile, t, 'mean');
```

If the images were taken with the same exposure then the edges of the tiles would not be visible. If the exposures were different the two sets of overlapping pixels have to be analyzed to determine the average intensity offset and scale factor which can be used to correct the tile before blending – a process known as tone matching.

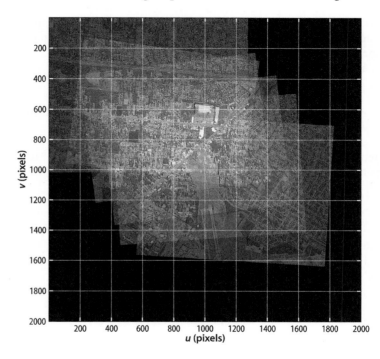

Fig. 14.49.
Example image mosaic. At the bottom of the frame we can clearly see three overlapping views of the airport runway which shows good alignment between the frames

Finally, we need to consider the effect of points in the image that are not in the ground plane such as those on a tall building. An image taken from directly overhead will show just the roof of the building, but an image taken from further away will be an oblique view that shows the side of the building. In a mosaic we want to create the illusion that we are directly above every point in the image so we should not see the sides of any building. This type of image is known as an orthophoto and unlike a perspective view, where rays converge on the camera's focal point, the rays are all parallel which implies a viewpoint at infinity.▶ At every pixel in the composite image we can choose a pixel from any of the overlapping tiles. To best approximate an orthophoto we should choose the pixel that is closest to overhead, that is, prior to warping the pixel was closest to the principal point.

Google Earth sometimes provides an imperfect orthophoto. When looking at cities we might see oblique views of buildings.

In photogrammetry this type of mosaic is referred to as an uncontrolled digital mosaic since it does not use explicit control points – manually identified corresponding features in the images. The full code is given by `mosaic` in the examples directory. The principles illustrated here can also be applied to the problem of image stabilization. The homography is used to map features in the new image to the location they had in the previous image.

14.7.3 Image Matching and Retrieval [examples/retrieval]

Given a set of images $\{\mathbf{I}_j, j = 1 \cdots N\}$ and a new image $\mathbf{I}'$ the image matching problem is to determine j such that $\mathbf{I}'$ and $\mathbf{I}_j$ are most similar. This is a difficult problem when we consider the effect of changes in viewpoint and exposure. Pixel-level similarity measures such as SSD or ZNCC that we used previously are not suitable for this problem since quite small changes in viewpoint will result in almost zero similarity.

Image matching is useful to a robot to determine if it has visited a particular place before, or seen the same object before. If those previous images have some associated semantic data such as the name of an object or the name of a place then by inference that semantic data applies to the new image. For example if a new image matches an existing image that has the semantic tag "lobby" then it implies the robot is seeing the same scene and is therefore in or close to, the lobby.

The particular technique that we will introduce is commonly referred to as "bag of words" and has been used successfully in a number of robotic applications. It builds on techniques we have previously encountered such as SURF point features and k-means clustering.

We start by loading a set of twenty images

```
>> images = iread('campus/*.jpg', 'mono');
```

as a $426 \times 640 \times 20$ array and for each of these we compute the SURF features

```
>> sf = isurf(images, 'thresh', 0);
```

which returns a MATLAB cell array whose elements are vectors of SURF features that correspond to the input images. For example

```
>> sf{1}
ans =
1407 features (listing suppressed)
  Properties: theta scale u v strength descriptor image_id
```

is a vector of 1 407 SURF feature objects corresponding to the first image in the sequence. The set of all SURF features across all images is

```
>> sf = [sf{:}]
sf =
28644 features (listing suppressed)
  Properties: theta scale u v strength descriptor image_id
```

which is a vector of nearly 30 000 SURF features objects.

Consider a particular SURF feature

```
>> sf(259)
ans =
(207.101,300.162), theta=2.31733, scale=2.1409, ↵
  strength=0.00114015, image_id=1, descrip= ..
```

and we see the `SurfPointFeature` properties discussed earlier such as centroid, scale and orientation. The property `image_id` indicates that this feature was extracted from the first image in the original image sequence. We can display that image and superimpose the feature

```
>> idisp(images(:,:,1))
>> sf(259).plot('g+')
>> sf(259).plot_scale('g', 'clock')
```

which is shown in Fig. 14.50a. The support region for this feature

```
>> sf(259).support(images)
```

is shown in Fig. 14.50b. The support region shows bricks and the edge of a window. The `support` method uses the `image_id` property to determine which of the passed images contains the feature.

The key insight behind the bag of words technique is that many of these features will describe visually similar scene elements such as leaves, corners of windows, bricks, chimneys and so on. If we consider each SURF feature descriptor as a point in a 64-dimensional space then similar descriptors will form clusters, and this is a *k*-means problem. To find 2 000 feature clusters

```
>> bag = BagOfWords(sf, 2000)
```

The `BagOfWords` class uses the MEX-file *k*-means implementation from http://www.vlfeat.org/. This uses its own random number generator and to initialize it to a known state use `vl_twister('STATE', 0.0);`.

returns a `BagOfWords` object that contains the original features, the center of each cluster, and various other information.◄ Each cluster is referred to as a visual word and is described by a 64-element SURF descriptor. The set of all visual words, 2 000 in this case, is a visual vocabulary. Just as a document comprises a set of words drawn from some vocabulary, each image comprises a collection (or *bag*) of visual words drawn from the visual vocabulary.

The clustering step assigns a visual word index to every SURF feature. For the particular feature shown above

Fig. 14.50. a Image 1 with visual word SURF feature 380 indicated by *green circle* showing scale and a radial line showing orientation direction; **b** the square support region has the same area as the circle and the horizontal axis is parallel to the orientation direction

```
>> w = bag.words(259)
w =
        1962
```

we find that the *k*-means clustering has assigned this image feature to word 1 962 in the vocabulary – it is an instance of visual word 1 962. That particular visual word appears

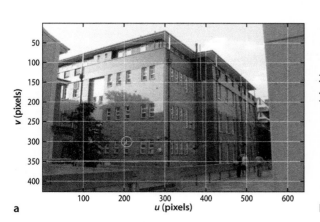

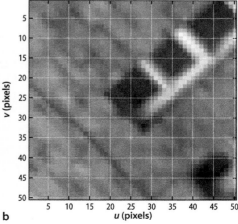

a   b

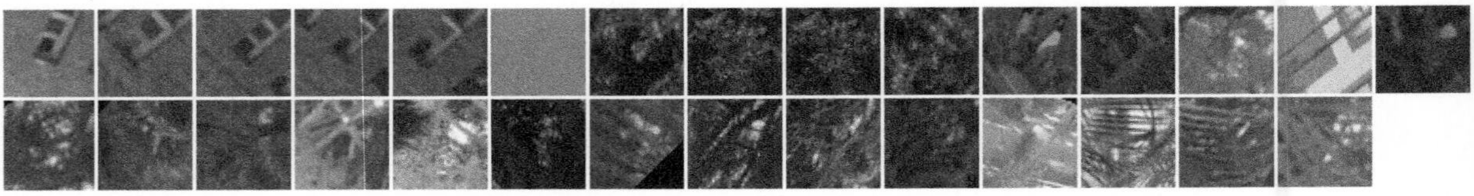

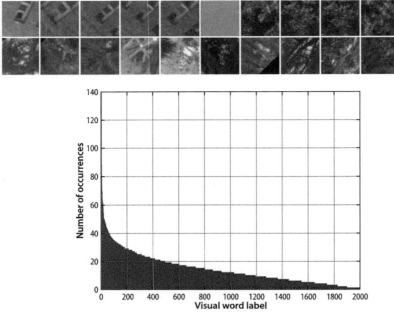

▲
Fig. 14.51. Exemplars of visual word 1 962 from the various images in which it appears. The annotation is of the form word/image

Fig. 14.52.
Histogram of the number of occurrences of each word (sorted). Note the small number of words that occur very frequently

```
>> bag.occurrence(w)
ans =
    29
```

times across the set of images, and it appears at least once in each of the images

```
>> bag.contains(w)
ans =
    1    5    7    8    9    11   12   15   16   18
```

We can display some of the different instances of word 1 962 by

```
>> bag.exemplars(w, images)
```

which is shown in Fig. 14.51. These exemplars actually look quite different, but we need to keep in mind that we are viewing them as patterns of pixels whereas the similarity is in terms of the descriptor.► The exemplars do however share some dominant horizontal and vertical structure.

Visual words occur with quite different frequencies

The descriptor comprises responses of Haar wavelet detectors computed over multiple windows within the support region.

```
>> [word,f] = bag.wordfreq();
```

where `word` is a vector containing all unique words and `f` are their corresponding frequencies. We can display these in descending order of frequency

```
>> bar( sort(f, 'descend') )
```

which is shown in Fig. 14.52. Words that occur very frequently have less meaning or power to discriminate between images. They are analogous to English words that are considered stop words in text document retrieval.► The visual stop words are removed from the bag of words

Search engines ignore words such as 'a', 'and', 'the' and so on.

```
>> bag.remove_stop(50)
Removing 2863 features associated with 50 most frequent words
>> bag
bag =
BagOfWords: 25781 features from 20 images
            1950 words, 50 stop words
```

which leaves some 26 000 SURF features behind. This method performs relabelling so that word labels are now in the interval 1 to 1 950.

Our visual vocabulary comprises K visual words and in this case $K = 1\,950$. We apply a technique from text document retrieval and describe *each* image by a word frequency vector. This is a K-element vector

$$\boldsymbol{v}_i = \left(t_1, \cdots t_j \cdots t_K\right)$$

whose elements describes the frequency of the corresponding visual words in an image.

$$t_j = \frac{n_{ij}}{n_i} \underbrace{\log \frac{N}{N_j}}_{\text{idf}} \tag{14.19}$$

where j is the visual word label, N is the total number of images in the database, N_j is the number of images which contain word j, n_i is the number of words in image i, and n_{ji} is the number of times word j appears in image i. The inverse document frequency (idf) term is a weighting that reduces the significance of words that are common across all images and which are thus less discriminatory. The weighted word frequency vectors are a property of the `BagOfWords` object and can be accessed by

```
>> M = bag.wordvector;
```

This might seem like a very large vector but it contains less than 1% of the number of elements of the original image.

which is a $1\,950 \times 20$ matrix and each column is a $1\,950$-element vector that concisely describes the corresponding image. ◄

The similarity between two images is the cosine of the angle between their corresponding word-frequency vectors

$$s\left(\boldsymbol{v}_1, \boldsymbol{v}_2\right) = \frac{\boldsymbol{v}_1 \boldsymbol{v}_2^T}{\|\boldsymbol{v}_1\|\|\boldsymbol{v}_2\|}$$

and is implemented by the `similarity` method. A value of one indicates maximum similarity. To compute the mutual similarity across this set of images (bags of words) is simply

```
>> S = bag.similarity(bag)
```

which returns a 20×20 similarity matrix where the elements `S(i,j)` indicate the similarity between the ith column and j^{th} columns of `M`, or between image i and image j. This matrix is symmetric and is best interpreted visually

```
>> idisp(S, 'bar')
```

which is shown in Fig. 14.53. The bright diagonal indicates, as a useful cross check, that image i is identical to image i. We also see that there is also some nonzero similarity between images 12 and 18, among others.

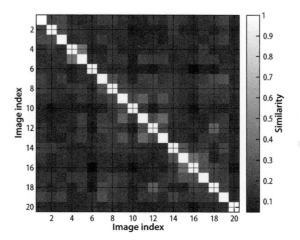

Fig. 14.53.
Similarity matrix for 20 images where *light colors* indicate strong similarity. Element (i, j) indicates the similarity between image i and image j

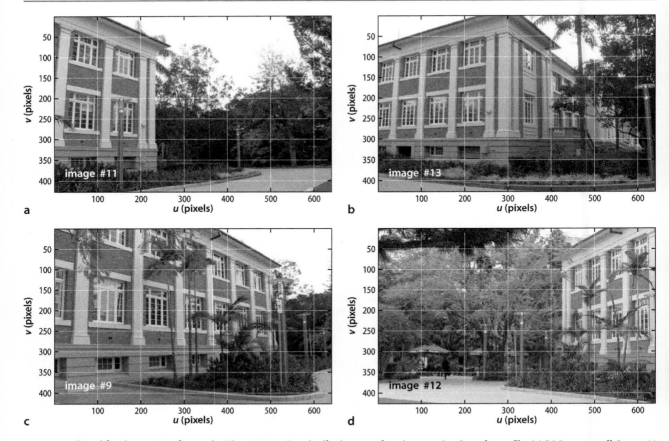

a   **b**   **c**   **d**

Consider image 11 shown in Fig. 14.54a. Its similarity to other images is given by row, or column, 11 of the similarity matrix

```
>> s = S(:,11);
```

Fig. 14.54. Image recall. Image 11 is the query, and in decreasing order of match quality we have recalled images 13, 9 and 12

which we sort into descending order of similarity

```
>> [z,k] = sort(s, 'descend');
>> [z k]
ans =
    1.0000   11.0000
    0.3722   13.0000
    0.3394    9.0000
    0.2610   12.0000
    0.2038    5.0000
       .
       .
```

where each row comprises the similarity measure and the corresponding image. Image 11 is identical to image 11 as expected, and in decreasing order of similarity we have images 13, 9, 12 and so on. These are shown in Fig. 14.54 and we see that the algorithm has recalled quite different views of the same building.

Now consider that we have some new images and we wish to determine which of the previous images is the most similar. Perhaps the robot has taken a picture and wishes to compare it to its database of existing images. The steps are broadly similar to the previous case

```
>> images2 = iread('campus/holdout/*.jpg', 'mono');
>> sf2 = isurf(images2, 'thresh', 0)
```

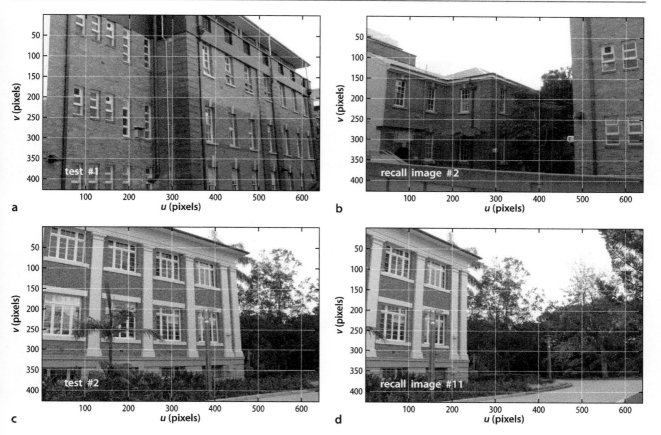

Fig. 14.55. Image recall for new images. The new query images **a** and **c** recall the database images **b** and **d** respectively

but rather than perform clustering we want to assign the features to the existing set of visual words, that is, to determine the closest visual word for each of the new feature descriptors

```
>> bag2 = BagOfWords(sf2, bag)
BagOfWords: 6530 features from 5 images
            1950 words, 50 stop words
```

This operation also removes any features words that were previously determined to be stop words, and computes the word frequency vectors◄ according to Eq. 14.19.

Which requires the image-word statistics from the existing bag of words to compute the idf weighting terms.

Finally the similarity between the images in the two bags of words is

```
>> S2 = bag.similarity(bag2);
```

which returns a 20×5 matrix where the elements $S2(i,j)$ indicates the similarity between the existing image i and new image j. The maxima in each *column* corresponds to the most similar image in the previously observed set

```
>> [z,k] = max(S2)
z =
    0.3435    0.6948    0.5427    0.5521    0.3627
k =
    2    11    16    18    20
```

New image 1 best matches image 2 in the original sequence, new image 2 matches image 11 and so on. Two of the new images and their closest existing images are shown in Fig. 14.55. The first recall has a low similarity score but is a reasonable result – the recall image includes the building from the test image at the right and another building that has many similarities.

14.7.4   Visual Odometry [examples/vodemo]

A common problem in robotics is to estimate the distance a robot has traveled, and this is a key input to all of the localization algorithms discussed in Chap. 6. For a wheeled robot we can use information from the wheel encoders but these are subject to random errors (slippage) as well as systematic errors (imprecisely known wheel radius). However for a flying or underwater robot the problem of odometry is much more difficult. Visual odometry (VO) is the process of using information from consecutive images to estimate the robot's relative motion from one camera image to the next.

We load a sequence of images taken from a car driving along a road▶

```
>> left = iread('bridge-l/*.png', 'roi', [20 750; 20 440]);
```

and the option `'roi'` selects a region of interest from each image to eliminate an irregular black border.▶ These images are unusual in having 16-bit pixels

```
>> about(left)
left [uint16] : 421x731x251 (154.5 MB)
```

and the image `im` belongs to the class `'uint16'`. Since this sequence is already nearly 200 Mbyte we do not convert it to double precision since this would quadruple the amount of memory required.

The image sequence can be displayed as an animation

```
>> ianimate(left, 'fps', 10);
```

at 10 frames per second.

For each frame we compute corner features

```
>> c = icorner(im, 'nfeat', 200, 'patch', 7);
```

and for a change we have used Harris corners since they are computationally cheaper. For this application the change in orientation and scale from frame to frame is small and Harris corner features are well suited for this purpose. The function returns a cell array with one element per input image, and each element is a vector of the 200 strongest Harris corner features per image. The image sequence can be displayed as an animation with the features overlaid

```
>> ianimate(im, c, 'fps', 10);
```

at 10 frames per second and a single frame of this sequence is shown in Fig. 14.56. The features are associated with regions of high gradient such as the edges of trees, as

This image sequence is bulky and not distributed with the main toolbox, but it can be found in the contrib2 zip file on the Toolbox website. This sequence is dataset 4 of the .enpeda.. Image Sequence Analysis Test Site (EISATS).

The black border is the result of image rectification.

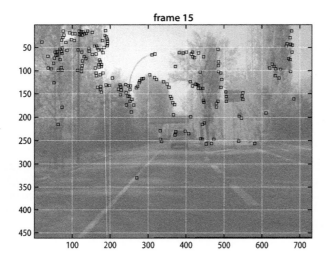

Fig. 14.56.
Frame number 15 from the `bridge-l` image sequence with overlaid features (image from .enpeda. project, Klette et al. 2011)

well as the corners of signs and cars. Watching the animation we see that the corner features *stick* reliably to world points for many frames. The motion of features in the image is known as optical flow and is a function of the camera's motion through the world and the 3-dimensional structure of the world. ◀

We will revisit optical flow in the next chapter.

The magnitude of optical flow – the speed of a world point on the image plane – is proportional to camera velocity divided by distance to the world point and therefore has a scale ambiguity – a camera moving quickly through a world with distant points yields the same flow magnitude as a slower camera moving past closer points. To resolve this we need to use additional information. For example if we knew that the points were on the road surface, that the road was flat, and the height of the camera above the road then we can resolve this unknown scale. However this assumption is quite strict and would not apply for something like a drone moving over unknown terrain. Instead we will use information from a different view of the world – the right image from a stereo camera fitted to the vehicle.

```
>> right = iread('bridge-r/*.png', 'roi', [20 750; 20 440]);
```

For each pair of left and right images we extract features, and determine correspondence by robustly matching features using descriptor similarity and the epipolar constraint implied by a fundamental matrix. Next we compute horizontal disparity between corresponding features, and assuming the cameras are fully calibrated we triangulate the image-plane coordinates to determine the world coordinates of the landmark points with respect to the left-hand camera on the vehicle. We can match the 3D point clouds at the current and previous time step using a technique like iterated closest point (ICP) in order to determine the camera pose change. This is the so-called 3D-3D approach to visual odometry and while the principle is sound it works poorly in practice. Firstly, some of the 3D points may be on other moving objects and this violates the assumption of ICP that the sensor or the object moves, but not both. Secondly, the estimated range to distant points is quite inaccurate since errors in estimated disparity become significant when disparity is small.

An alternative approach, 3D-2D matching, projects the 3D points at the current time step into the previous image and finds the camera pose that minimizes the error with respect to the observed feature coordinates – this is bundle adjustment. Typically this is done for just one image and we will choose the left image. To establish correspondence of features over time we find correspondences between left-image features that had a match with the right image and a match with features from the previous left image – again enforcing an epipolar constraint. We now know the correspondence between points in the three views of the scene as shown in Fig. 14.57.

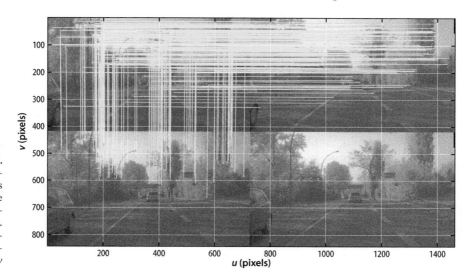

Fig. 14.57.
Feature correspondence for visual odometry. The top row is a stereo pair at the current time step, and the bottom row is a stereo pair at the previous time step. Epipolar consistent correspondences between three of the image images are shown in *yellow*

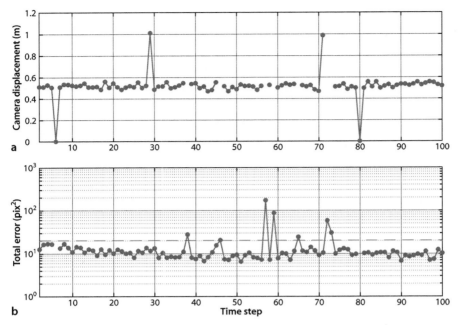

Fig. 14.58.
Visual odometry results. a Estimated displacement of the camera its z-direction (forward); b bundle adjustment final error per frame, shown with a *logarithmic vertical scale*

At each time step we set up a bundle adjustment problem that has two cameras and a number of landmarks determined from stereo triangulation. The first camera is associated with the previous time step and is fixed at the reference frame origin. The second camera is associated with the current time step and would be expected to have a translation in the positive z-axis direction. We could obtain an initial estimate of the second camera's pose by estimating and decomposing an essential matrix, but we will instead set it to the origin.

The details can be found in the example script

```
>> visodom
```

which processes 100 frames and displays graphics like Fig. 14.57 for every frame. The final results for z-axis translation are shown in Fig. 14.58a and we notice a value of around 0.5 m at each time step, but there are also some missing data points and two incorrect looking results. The bundle adjustment process returns the final squared error and this is plotted in Fig. 14.58b for each frame. For 8% of the frames that error was over 20 pix^2 (*red dashed line*) and we exclude those results. The likely source of error is incorrect point correspondences. Bundle adjustment assumes that all points in the world are fixed but in this sequence there are numerous moving objects. We used the epipolar constraint between current and previous frame to ensure that only features points consistent with a moving camera and a fixed world are in the inliner set. However when the script runs we see quite a lot of points on the car in front which are being incorrectly included in the inlier set – that car is moving but because it is a large and constant distance away those points are not inducing enough error to be considered outliers. A more sophisticated bundle adjustment algorithm would detect and reject such points. Finally there is a preponderance of points in the top part of the scene which tend to be quite distant from the cameras. A more sophisticated approach to feature detection would choose features more uniformly over the image.

The erroneous points at timesteps 29 and 71 highlight a common problem with using video data for robots. The clue is that those values are suspiciously close to exactly twice the other values. Each image in the sequence was assigned a timestamp when it was received by the computer and those timestamps can be loaded

```
>> ts = load('timestamps.dat');
```

and if we plot the difference between timestamps

```
>> plot(diff(ts))
```

we see that the average time between images is 44.6 ms but there are two spikes where the interval is twice that. The computer logging the images has skipped a frame, perhaps it was unable to write image data to memory or disk as quickly as it was arriving. So the interval between the frames was twice as long, the vehicle traveled twice as far, and the spikes on our estimated displacement are in fact correct. This is not an uncommon situation – in a robot system all data should be timestamped and timestamps should be checked to detect problems like this.

The median velocity over the valid estimates is

```
>> median(tz(ebundle<20))
ans =
    0.5201
```

in units of meters which, with the camera frame interval of 44.6 ms, indicates a vehicle speed of around 40 km h^{-1}. The variable tz is a vector of frame-to-frame displacement computed by the script, and $ebundle$ is a vector of bundle adjustment errors at each time step. The residuals from estimating the fundamental matrix between the current and previous left image are saved in the vector $efund$.

For a vehicle or robot the estimated displacements over time are not independent and are related by vehicle kino-dynamic model, and we can use this to smooth the results and discount erroneous velocity estimates. If the bundle adjuster included constraints on camera pose we could set the weighting to penalize infeasible motion in the lateral and vertical directions as well as roll and pitch motion.

14.8 Wrapping Up

This chapter has covered many topics but the aim has been to demonstrate a multiplicity of concepts that are of use in real robotic vision systems. There have been two common threads through this chapter. The first has been the use of corner features to find distinctive points in images, and matching them to the same world point in another image. The second thread has been the loss of scale in the perspective projection process and techniques based on additional sources of information to recover scale such as stereo vision, structured light or bundle adjustment.

We extended the geometry of single camera imaging to the case of two cameras and showed how corresponding points in the two images are constrained by the fundamental matrix. We showed how the fundamental matrix can be estimated from image data, the effect of incorrect data association, and how to overcome this using the RANSAC algorithm. Using camera intrinsic parameters the essential matrix can be computed and then decomposed to give the camera motion between the two view, but the translation has an unknown scale factor. With some extra information such as the magnitude of the translation, the camera motion can be estimated completely. Given the camera motion, then the 3-dimensional coordinates of points in the world can be estimated.

For the special case where world points lie on a plane they *induce* an homography that is a linear mapping of image points between images. The homography can be used to detect points that do not lie in the plane and can be decomposed to give the camera motion between the two views (translation again has an unknown scale factor) and the normal to the plane.

If the fundamental matrix is known then a pair of overlapping images can be rectified to create an epipolar-aligned stereo pair and dense stereo matching can be used to recover the world coordinates for every point. Errors due to effects such as occlusion and lack of texture were discussed as were techniques to detect these situations.

We used bundle adjustment to solve the structure and motion estimation problem – using 2D measurements from a set of images of the scene to recover information related to the 3D geometry of the scene as well as the locations of the cameras. Stereo vision is a simple case where the motion is known – fixed by the stereo baseline – and we are interested only in structure. The visual odometry problem is complementary and we are interested only in the motion of the camera, not the scene structure.

These multi-view techniques were then used in a number of application examples such as perspective correction, mosaic creation, image retrieval and visual odometry.

MATLAB and Toolbox Notes

The Toolbox uses open-source code to support SIFT (VLFeat) and SURF (OpenSURF http://www.mathworks.com/matlabcentral/fileexchange/28300) features. VLFeat (http://www.vlfeat.org) includes a number of feature detectors and other useful functions. The OpenCV library implements many feature detectors and descriptors and can be accessed in MATLAB using mexopencv (https://kyamagu.github.io/mexopencv).

The MATLAB Computer Vision System Toolbox™ (CVST) has support for stereo rectification; stereo matching; SURF, FAST and Harris feature detectors; a range of descriptors (BRISK, HOG, MSER); and point cloud processing including kd-trees, model fitting and visualization. Many CVST functions can be used inside Simulink and support automatic code generation for real-time hardware such as FPGAs.

Further Reading

3-dimensional reconstruction and camera pose estimation has been studied by the photogrammetry community since the mid nineteenth century, see page 354. 3-dimensional computer vision or *robot vision* has been studied by the computer vision and artificial intelligence communities since the 1960s. This book follows the language and nomenclature associated with the computer vision literature, but the photogrammetric literature can be comprehended with only a little extra difficulty. The similarity of a stereo camera to our own two eyes is very striking, and while we do make strong use of stereo vision it is not the only technique we use to infer distance (Cutting 1997).

Significant early work on multi-view geometry was conducted at laboratories such as Stanford, SRI International, MIT AI laboratory, CMU, JPL, INRIA, Oxford and ETL Japan in the 1980s and 1990s and led to a number of text books being published in the early 2000s. The definitive references for multiple-view geometry are Hartley and Zisserman (2003) and Ma et al. (2003). These books present quite different approaches to the same body of material. The former takes a more geometric approach while the latter is more mathematical. Unfortunately they use quite different notation, and each differs from the notation used in this book – a summary of the important notational elements is given in Table 14.1. These books all cover feature extraction (using Harris corner features, since they were published before scale invariant feature detectors such as SIFT and SURF corner detectors were developed); the geometry of one, two and N views; fundamental and essential matrices; homographies; and the recovery of 3-dimensional scene structure and camera motion through offline batch techniques. Both provide the key algorithms in pseudo-code and have some supporting MATLAB code on their associated web sites. The slightly earlier book by Faugeras et al. (2001) covers much of the same material using a fairly mathematical approach and with different notation again. The older book by Faugeras (1993) focuses on sparse stereo from line features. The recent book by Szeliski (2010) provides a very readable and deeper discussion of the topics in this chapter.

Table 14.1.
Rosetta stone. Summary of notational differences between
two other popular textbooks
and this book

Table 14.1.
Rosetta stone. Summary of notational differences between
two other popular textbooks
and this book

Object	Hartley and Zisserman 2003	Ma et al. 2003	This book
World point	$\mathbf{X}$	P	$\mathbf{P}$
Image point	$\mathbf{x}, \mathbf{x}'$	x_1, x_2	$^1\mathbf{p}, ^2\mathbf{p}$
i^{th} image point	$\mathbf{x}_i, \mathbf{x}_i'$	x_1^i, x_2^i	$^1\mathbf{p}_i, ^2\mathbf{p}_i$
Camera motion	$\mathsf{R}, \mathbf{t}$	R, T	$\boldsymbol{R, t}$
Normalized coordinates	$\mathbf{x}, \mathbf{x}'$	x_1, x_2	$(\bar{u}, \bar{v})$
Camera matrix	P	Π	$\mathbf{C}$
Homogeneous quantities	$\mathbf{x}, \mathbf{X}$	x, P	$\tilde{p}, \tilde{P}$
Homogeneous equivalence	$\mathbf{x} = P\mathbf{X}$	$\lambda x = \Pi P$ $x \sim \Pi P$	$\tilde{p} \simeq \mathbf{C}\tilde{P}$

References related to SURF and other feature detectors were previously discussed on page 456. The performance of feature detectors and their matching performance is covered in Mikolajczyk and Schmid (2005) which reviews a number of different feature descriptors including spin images and local jets.◀ Arandjelović and Zisserman (2012) discuss some important points when matching feature vectors.

A jet is a vector of higher order derivatives such as $I_{uu}, I_{vv}, I_{uv}, I_{uuu}, I_{uuv}, I_{uvv}, I_{vvv}, I_{uuuu}, I_{uuuv}, I_{uuvv}, I_{uvvv}, I_{vvvv}$ and so on (Mikolajczyk and Schmid 2005).

The RANSAC algorithm described by Fischler and Bolles (1981) is the workhorse of all the feature-based methods discussed in this chapter but fails with very small inlier ratios. A recent more robust development is vector field consensus (VFC) by Ma et al. (2014). Pilu (1997) discusses how SVD can be applied to a matrix formed from the distances between features to determine correspondence. Dellaert et al. (2000) describe a probabilistic approach to determining structure from a group of images not necessarily in order.

The term fundamental matrix was defined in the thesis of Luong (1992). The book by Xu and Zhang (1996) is a readable introduction to epipolar geometry. Epipolar geometry can also be formulated for nonperspective cameras in which case the epipolar line becomes an epipolar curve (Mičušík and Pajdla 2003; Svoboda and Pajdla 2002). For three views the geometry is described by the trifocal tensor $\mathcal{T}$ which is a $3 \times 3 \times 3$ tensor with 18 degrees of freedom that relates a point in one image to epipolar lines in two other images (Hartley and Zisserman 2003; Ma et al. 2003). An important early paper on epipolar geometry for an image sequence is Bolles et al. (1987).

The essential matrix was first described a decade earlier in a letter to Nature (Longuet-Higgins 1981) by the eminent theoretical chemist and cognitive scientist Christopher Longuet-Higgins (1923–2004). The paper describes a method of estimating the essential matrix from eight corresponding point pairs. The decomposition of the essential matrix was first described in Faugeras (1993, § 7.3.1) but is also covered in the texts Hartley and Zisserman (2003) and Ma et al. (2003). In this chapter we have estimated camera motion by first computing the essential matrix and then decomposing it. The first step requires at least eight pairs of corresponding points but algorithms such as Nistér (2003), Li and Hartley (2006) compute the motion directly from just five pairs of points. Decomposition of an homography is described by Faugeras and Lustman (1988), Hartley and Zisserman (2003), Ma et al. (2003), and the comprehensive technical report by Malis and Vargas (2007). The relationships between these matrices, camera motion, and the relevant Toolbox functions are summarized in Fig. 14.59.

Stereo cameras and stereo matching software are available today from many vendors and can provide high-resolution depth maps at more than 10 Hz on standard computers. A decade ago this was difficult and custom hardware including FPGAs was required to achieve real-time operation (Corke et al. 1999; Woodfill and Von Herzen 1997). The application of stereo vision for planetary rover navigation is discussed by Matthies (1992). More than two cameras can be used, and multi-camera stereo was introduced by Okutomi and Kanade (1993) and provides robustness to problems such as the picket fence effect.

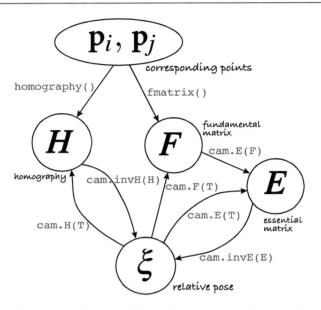

Fig. 14.59.
Toolbox functions and camera
object methods, and their inter-
relationship

Brown et al. (2003) provide a readable review of stereo vision techniques with a focus on real-time issues. An old but clearly written book on the principles of stereo vision is Shirai (1987). Scharstein and Szeliski (2002) consider the stereo process as four steps: matching, aggregation, disparity computation and refinement. The cost and performance of different algorithms for each step are compared. The example in this chapter would be described as: NCC matching, box filter aggregation, winner takes all, and subpixel refinement. The dense stereo matching algorithm presented in Sect. 14.3.2 is a very conventional correlation-based stereo algorithm. The disparity computed at each pixel is independent of other pixels but for most real scenes adjacent pixels belong to the same surface and disparity will be quite similar – this is referred to as the *smoothness constraint*. Of course disparity will be discontinuous at the edges of surfaces. Finding the shortest best-fit path through a slice of the disparity space image as shown in Fig. 14.29 will enforce the smoothness constraint in the horizontal direction. Ideally we wish to also ensure vertical smoothness as well and this can be achieved using Markov random fields (MRFs), total variation with regularizers (Pock 2008), or more efficient semi-global matching (SGM) algorithms (Hirschmüller 2008). The very popular library for efficient large-scale stereo matching (LIBELAS) by Geiger et al. (2010) uses an alternative to global optimization that provides fast and accurate results for a variety of indoor and outdoor scenes. Stereo vision involves a significant amount of computation but there is considerable scope for parallelization using multiple cores, MIMD instruction sets, GPUs, custom chips and FPGAs. The use of nonparametric local transforms is described by Zabih and Woodfill (1994) and Banks and Corke (2001).

An emerging alternative to stereo vision are cameras based on time-of-flight measurement which are dropping rapidly in cost. A pulse of infra-red light illuminates the scene and every pixel records the intensity and time delay of the reflected energy. Time-of-flight sensors include the REAL3 devices by Infineon (infineon.com) and PhotonICs from **pmd**technologies (pmdtec.com). Complete time-of-flight cameras include the Kinect for Xbox One (Kinect 2) and various from **pmd**technologies. This type of camera works well indoors and even in complete darkness, but outdoors under full sun the maximum range is limited just as it is for structured light.►

The ICP algorithm (Besl and McKay 1992) is used for a wide range of applications from robotics to medical imaging. ICP is fast but determining the correspondences via nearest neighbors is an expensive $O(N^2)$ operation. Many variations have been developed that make the approach robust to outlier data and to improve computational speed for large datasets. Salvi et al. (2007) provide a recent review and comparison of

In fact it is worse than structured light. The illumination energy is limited by eye-safety considerations and structured light concentrates that energy over a line whereas time-of-flight cameras spread it over an area.

some different algorithms. Determining the relative orientation between two sets of points is a classical problem and the SVD approach used here is described by Arun et al. (1987). Solutions based on quaternions and orthonormal rotation matrices have been described by Horn (Horn et al. 1988; Horn 1987).

Structure from motion (SfM), the simultaneous recovery of world structure and camera motion, is a classical problem in computer vision. Two useful review papers are by Huang and Netravali (1994) which provides a taxonomy of approaches, and Jebara et al. (1999). Broida et al. (1990) describe an early recursive SfM technique for a monocular camera sequence using an EKF where each world point is represented by its (X, Y, Z) coordinate. McLauchlan provides a detailed description of a variable-length state estimator for SfM (McLauchlan 1999). Azarbayejani and Pentland (1995) present a recursive approach where each world point is parameterized by a scalar, its depth with respect to the first image. A more recent algorithm with bounded estimation error is described by Chiuso et al. (2002) and also discusses the problem of scale variation. The MonoSlam system by Davison et al. (2007) is an impressive monocular SfM system that maintains a local map that includes features even when they are not currently in the field of view. A more recent extension by Newcombe et al. (2011) performs camera tracking and dense 3D reconstruction from a single moving RGB camera. The application of SfM to large-scale urban mapping is becoming increasing popular and Pollefeys et al. (2008) describe a system for offline processing of large image sets.

Bundle adjustment or structure from motion (SfM) is a big field with a large literature that cover many variants of the problem, for example robustness to outliers, and specific applications and camera types. Classical introductions include Triggs et al. (2000) and Hartley and Zisserman (2003). Recent theses by Warren (2015), Sünderhauf (2012) and Strasdat (2012) are comprehensive and readable. Unfortunately every reference uses different notation. Estimating the camera matrix for each view, computing a projective reconstruction, and then upgrading it to a Euclidean reconstruction is described by Hartley and Zisserman (2003) and Ma et al. (2003).

The SfM problem can be simplified by using stereo rather than monocular image sequences (Molton and Brady 2000; Zhang et al. 1992), or by incorporating inertial data (Strelow and Singh 2004). A readable two-part tutorial introduction to visual odometry (VO) is Scaramuzza and Fraundorfer (2011) and Fraundorfer and Scaramuzza (2012). Visual odometry is discussed by Nistér et al. (2006) using point features and monocular or stereo vision. Maimone et al. (2007) describe experience with stereo-camera VO on the Mars rover and Corke et al. (2004) describe monocular catadioptric VO for a prototype planetary rover.

Mosaicing is a process as old as photography. In the past it was highly skilled and labor intensive requiring photographs, scalpels and sandpaper. The surface of the Moon and nearby planets was mosaiced manually in the 1960s using imagery sent back by robotic spacecraft. High-quality offline mosaicing tools are available for creating panoramas, for example the Hugin open source project http://hugin.sourceforge.net and the proprietary AutoStitch.

The "bag of words" technique for image retrieval was first proposed by Sivic and Zisserman (2003) and has been used by many other researchers since. A notable extension for robotic applications is FABMAP (Cummins and Newman 2008) which explicitly accounts for the joint probability of feature occurrence and associates a probability with the image match, and is available in OpenCV. An open source version (Glover et al. 2012) is available at https://github.com/arrenglover/openfabmap. Chatfield et al. (2011) discussed some recent improvements to the bag-of-words image retrieval problem.

Image sequence analysis is the core of many real-time robotic vision systems. Real-time feature tracking across frames is described by Hager and Toyama (1998), Lucas and Kanade (1981) and is typically based on the computationally cheaper Harris detectors or the pyramidal Kanade-Lucas-Tomasi (KLT) tracker. SURF detectors are still too time consuming to use for this purpose although some C-based implementations and GPU implementations are capable of real-time performance.

Resources

The field of computer vision has progressed through the availability of standard datasets. These have enabled researchers to quantitatively compare the performance of different algorithms on the same data. One of the earliest collections of stereo image pairs was the JISCT dataset (Bolles et al. 1993). The more recent Middlebury dataset (Scharstein and Szeliski 2002) at http://vision.middlebury.edu/stereo provides an extensive collection of stereo images, at high resolution, taken at different exposure settings and including ground truth data. Stereo images from various NASA Mars rovers are available online as left+right pairs or encoded in anaglyphs. Motion datasets include classic motion sequences of indoor scenes http://vasc.ri.cmu.edu//idb/html/motion, people moving inside a building http://homepages.inf.ed.ac.uk/rbf/CAVIARDATA1, traffic scenes http://i21www.ira.uka.de/image_sequences, and from a moving vehicle http://www.mi.auckland.ac.nz/EISATS.

The popular LIBELAS library (http://www.cvlibs.net/software/libelas) for large-scale stereo matching supports parallel processing using OpenMP and has MATLAB and ROS interfaces. Various stereo vision algorithms are compared for speed and accuracy at the KITTI (www.cvlibs.net/datasets/kitti/eval_scene_flow.php) and Middlebury (vision.middlebury.edu/stereo/eval3) benchmark sites.

An implementation of the KLT feature tracker, in C, written by Stan Birchfield is available at http://www.ces.clemson.edu/~stb/klt. A GPU-based version of KLT, in C, is available at http://cs.unc.edu/~ssinha/Research/GPU_KLT. The ViSP cross-platform library includes tracking capability and can be found at https://visp.inria.fr. Pointers to SIFT and SURF implementations are given on page 456. The Epipolar Geometry Toolbox (Mariottini and Prattichizzo 2005) for MATLAB by Gian Luca Mariottini and Domenico Prattichizzo is available at http://egt.dii.unisi.it and handles perspective and catadioptric cameras. Andrew Davison's monocular visual SLAM system (MonoSLAM) for C and MATLAB is available at http://www.doc.ic.ac.uk/~ajd/software.html.

The sparse bundle adjustment software by Lourakis (users.ics.forth.gr/~lourakis/sba) is an efficient C implementation that is widely used and has a MATLAB and OpenCV wrapper. One application is Bundler (www.cs.cornell.edu/~snavely/bundler) which can perform matching of points from thousands of cameras over city scales and has enabled reconstruction of cities such as Rome (Agarwal et al. 2014), Venice and Dubrovnik. Some of these large-scale datasets are available from grail.cs.washington.edu/projects/bal and www.robots.ox.ac.uk/~vgg/data/data-mview.html. A MATLAB interface to Bundler is available at www.mathworks.com/matlabcentral/fileexchange/46341. SFMedu, a Structure from Motion System for Education (http://vision.princeton.edu/courses/SFMedu) has learning resources and MATLAB source code. Other open source solvers that can be used for sparse bundle adjustment include g^2o, SSBA and CERES, all implemented in C++. g^2o by Kümmerle et al. (2011) (github.com/RainerKuemmerle/g2o) can also be used to solve SLAM problems. SSBA by Christopher Zach is available at https://github.com/chzach/SSBA. The CERES solver from Google (ceres-solver.org) is a library for modeling and solving large complex optimization problems on desktop and mobile platforms and also supports parallel processing using OpenMP. A MATLAB interface is available at github.com/tikroeger/BA_Matlab.

Pointcloud library (PCL) (pointclouds.org) is a large-scale, open and standalone package for 2D/3D image and point cloud processing with support for feature detectors and descriptors, 3D registration, kd-trees, shape segmentation, surface meshing, visualization, camera interfaces and includes g^2o. The Point Data Abstraction Library (PDAL) (www.pdal.io) is a library and set of Unix command line tools for manipulating point cloud data.

Point clouds can be stored in a number of common open formats. Point cloud data (PCD) files are defined by Pointcloud library (PCL) (pointclouds.org) and can be imported into MATLAB using www.mathworks.com/matlabcentral/fileexchange/40382.

Polygon file format (PLY) files are designed to describe meshes but can be used to represent an unmeshed point cloud, and there are a number of great visualizers such as MeshLab and potree. PCL and PDAL can read, write and convert many point cloud file formats.

The fundamental matrix song can be found at http://danielwedge.com/fmatrix/.

Exercises

1. Corner features and matching (page 462). Examine the cumulative distribution of corner strength for Harris and SURF features. What is an appropriate way to choose strong corners for feature matching?
2. Feature matching. We could define the quality of descriptor-based feature matching in terms of the percentage of inliers after applying RANSAC.
   a) Take any image. We will match this image against various transforms of itself to explore the robustness of SURF and Harris features. The transforms are: *(a)* scale the intensity by 70%; *(b)* add Gaussian noise with standard deviation of 0.05, 0.5 and 2 grey values; *(c)* scale the size of the image by 0.9, 0.8, 0.7, 0.6 and 0.5; *(d)* rotate by 5, 10, 15, 20, 30, 40 degrees.
   b) For the Harris detector compare the performance for the structure-tensor-based feature and the patch descriptor sizes of 3×3, 7×7 and 11×11 and 15×15.
   c) Try increasing the suppression radius for SURF and Harris corners. Does the lower density of matches improve the matching performance?
   d) The Harris detector can process a color image. Does this lead to improved performance compared to the greyscale version of the same image.
   e) Is there any correlation between outlier matches and strength of the corner features involved?
3. Write the equation for the epipolar line in image two, given a point in image one.
4. Show that the epipoles are the null space of the fundamental matrix.
5. Can you determine the camera matrix C for camera two given the fundamental matrix and the camera matrix for camera one?
6. Estimating the fundamental matrix (page 470)
   a) For the synthetic data example vary the number of points and the additive Gaussian noise and observe the effect on the residual.
   b) For the Eiffel tower data observe the effect of varying the parameter to RANSAC. Repeat this with SURF features computed with a lower strength threshold (the default is 0.002).
   c) What is the probability of drawing 8 inlier points in a random sample (without replacement) from N inliers and M outliers?
7. Epipolar geometry
   a) Create two central cameras, one at the origin and the other translated in the x-direction. For a sparse fronto-parallel grid of world points display the family of epipolar lines in image two that correspond to the projected points in image one. Describe these epipolar lines? Repeat for the case where camera two is translated in the y- and z-axes and rotated about the x-, y- and z-axes. Repeat this for combinations of motion such as x- and z-translation or x-translation and y-rotation.
   b) The example of Fig. 14.16 has epipolar lines that slope slightly upward. What does this indicate about the two camera views?
8. Essential matrix (page 469)
   a) Create a set of corresponding points for a camera undergoing pure rotational motion, and compute the fundamental and essential matrix. Can you recover the rotational motion?
   b) For a case of translational and rotational motion visualize both poses that result from decomposing the essential matrix. Sketch it or use `trplot`.

9. Homography (page 477)
   a) Compute Euclidean homographies for translation in the x-, y- and z-directions and for rotation about the x-, y- and z-axes. Convert these to projective homographies and apply to a fronto-parallel grid of points. Is the resulting image motion what you would expect? Apply these homographies as a warp to a real image such as Lena.
   b) Decompose the homography of Fig. 14.15, the courtyard image, to determine the plane of the wall with respect to the camera. You will need the camera intrinsic parameters.
10. Load a reference image of this book's cover from `rvc2_cover.png`. Next, capture an image that includes the book's front cover, compute SIFT or SURF features, match them and use RANSAC to estimate an homography between the two views of the book cover. Decompose the homography to estimate rotation and translation. Put all of this into a real-time loop and continually display the pose of the book relative to the camera.
11. Sparse stereo (page 482)
    a) The ray intersection method can return the closest distance between the rays (which is ideally zero). Plot a histogram of the closing error and compute the mean and maximum error.
    b) The assumed camera translation magnitude was 30 cm. Repeat for 25 and 35 cm. Are the closing error statistics changed? Can you determine what translation magnitude minimizes this error?
12. Bundle adjustment (page 497)
    a) Vary the initial condition for the second camera, for example, set it to the identity matrix.
    b) Set the initial camera translation to 3 m in the x-direction, and scale the landmark coordinates by $10\times$. What is the final value of the back-projection error and the second camera pose.
    c) Experiment with anchoring landmarks and cameras.
    d) Derive the two Jacobians A (hard) and B.
13. Derive a relationship for depth in terms of disparity for the case of verged cameras. That is, cameras with their optical axes intersecting similar to the cameras shown in Fig. 14.6.
14. Stereo vision. Using the rock piles example (page 483)
    a) Use `idisp` to zoom in on the disparity image and examine pixel values on the boundaries of the image and around the edges of rocks.
    b) Experiment with different similarity measures and window sizes. What effects do you observe in the disparity image and computation time?
    c) Experiment with changing the disparity range. Try $[50, 90]$, $[30, 90]$, $[40, 80]$ and $[40, 100]$. What happens to the disparity image and why?
15. Using the rock piles example (page 483) obtain the disparity space image D
    a) For selected pixels (u, v) plot $D(u, v, d)$ versus d. Look for pixels that have a sharp peak, broad peak and weak peak. Repeat this for stereo computed using ZSSD similarity. For a selected row v display $D(u, v, d)$ as an image. What does this represent?
    b) For a particular pixel plot s versus d, fit a parabola around the maxima and overlay this on the plot.
    c) Use raw data from the DSI, find the second peak at each pixel and compute the ambiguity ratio
    d) Display the epipolar lines on image two for selected points in image one.
16. Download an anaglyph image and convert it into a pair of greyscale images, then compute dense stereo.
17. Variations to stereo matching
    a) Try some other stereo images, either acquired with a stereo camera or from the Middlebury dataset.

b) Perform stereo matching using the SAD rather than NCC metric. Use the `'metric'` option to `istereo`.

c) Apply the census (`icensus`) or rank transforms (`irank`) to the left and right image prior to matching using the SAD measure and investigate the matching quality. More details in Banks and Corke (2001).

18. Stereo vision. For a pair of identical cameras with a focal length of 8 mm, $1\,000 \times 1\,000$ pixels that are 10 μm square on an 80 mm baseline and with parallel optical axes:

a) Sketch the fields of views of the camera in a plan view. If the cameras are viewing a plane surface normal to the principal axes how wide is the horizontal overlapping field of view in units of pixels?

b) Assuming that disparity error is normally distributed with $2\sigma = 0.2$ pixels compute and plot the distribution of error in the z-coordinate of the reconstructed 3D points which have a mean disparity of 0.5, 1, 2, 5, 10 and 20 pixels. Draw 1 000 random values of disparity, convert these to Z and plot a histogram (distribution) of their values.

19. Mona Lisa on your wall. Acquire an image of a room in your house and display it using MATLAB. Select four points, using `ginput`, to define the corners of the virtual frame on your wall. Perhaps use the corners of an existing rectangular feature in your room such as a window, poster or picture. Estimate the appropriate homography, warp the Mona Lisa image and insert it into the original image of your room.

20. Plane fitting (page 504)

a) Test the robustness of the plane fitting algorithm to additive noise and outlier points.

b) Implement an iterative approach with weighting to minimize the effect of outliers.

c) Create a RANSAC-based plane fit algorithm that takes random samples of three points to solve for Eq. 14.18. Use the `fmatrix` and `homography` code to guide you. You will need to create a number of functions that are invoked by the `ransac_driver`.

21. ICP (page 505)

a) Run the ICP example on your computer and watch the animation.

b) Change the initial relative pose between the point clouds. Try some very large rotations.

c) Increase the noise added to the data points.

d) For the case where there are missing and/or spurious data points experiment with different values of the `'distthresh'` option.

22. Perspective correction (page 509)

a) Create a virtual view looking downward at 45° to the front of the cathedral.

b) Create a virtual view from the original camera height but with the camera rotated 20° to the left.

c) Find another real picture with perspective distortion and attempt to correct it.

23. Mosaicing (page 512)

a) Run the example file `mosaic` and watch the whole mosaic being assembled.

b) Modify the way the tile is pasted into the composite image to use pixel averaging rather than addition.

c) Modify the way the tile is pasted into the composite image so that pixels closest to the principal point are used.

d) Run the software on a set of your own overlapping images and create a panorama.

24. Image stabilization can be used to virtually stabilize an unsteady camera, perhaps one that is handheld, on a drone or on a mobile robot traversing rough terrain. Capture a short image sequence $I_1, I_2 \cdots I_N$ from an unsteady camera. For frame i, $i \geq 2$ estimate an homography with respect to frame 1, warp the image appropriately, and store it in an array. Display the stabilized image sequence using `ianimate`.

25. Bag of words (page 514)
    a) Examine the different support regions of different visual words using the `exemplars` method.
    b) Investigate the effect of changing the number of stop words.
    c) Investigate the effect of changing the size of the vocabulary. Try 1 000, 1 500, 2 500, 3 000.
    d) Build a bag of words from a set of your own images.
    e) the RootSIFT trick described by Arandjelović and Zisserman (2012).
    f) SURF rather than SIFT features.
    g) SURF corner detector with BRISK or FREAK features.
26. Visual odometry, page 520. Modify the example script to
    a) use SIFT or SURF features instead of Harris. What happens to accuracy and execution time?
    b) ensure that features are more uniformly spread over the scene, investigate the `'suppress'` option of `icorner`.
    c) plot the fundamental matrix residuals at each time step (there are two of them). Is there a pattern here? Adjust the RANSAC parameters so as to reduce the number of times bundle adjustment fails.
    d) use a robust bundle adjuster, either find one or implement one (hard).
    e) use a Kalman filter with simple vehicle dynamics to smooth the velocity estimates.
27. Learn about kd-trees. What problems in this chapter could benefit from kd-trees?

Part V Robotics, Vision and Control

V Robotics, Vision and Control

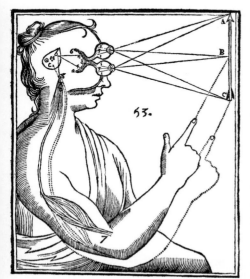

It is common to talk about a robot moving to an object, but in reality the robot is only moving to a pose at which it *expects* the object to be. This is a subtle but deep distinction. A consequence of this is that the robot will fail to grasp the object if it is not at the expected pose. It will also fail if imperfections in the robot mechanism or controller result in the end-effector not actually achieving the end-effector pose that was specified. In order for this conventional approach to work successfully we need to solve two quite difficult problems: determining the pose of the object and ensuring the robot achieves that pose.

The first problem, determining the pose of an object, is typically avoided in manufacturing applications by ensuring that the object is always precisely placed. This requires mechanical jigs and fixtures which are expensive, and have to be built and set up for every different part the robot needs to interact with – somewhat negating the flexibility of robotic automation.

The second problem, ensuring the robot can achieve a desired pose, is also far from straightforward. As we discussed in Chap. 7 a robot end-effector is moved to a pose by computing the required joint angles. This assumes that the kinematic model is accurate, which in turn necessitates high precision in the robot's manufacture: link lengths must be precise and axes must be exactly parallel or orthogonal. Further the links must be stiff so they do not deform under dynamic loading or gravity. It also assumes that the robot has accurate joint sensors and high-performance joint controllers that eliminate steady state errors due to friction or gravity loading. The nonlinear controllers we discussed in Sect. 9.4 are capable of this high performance but they require an accurate dynamic model that includes the mass, center of gravity and inertia for every link, as well as the payload.

None of these problems are insurmountable but this approach has led us along a path toward high complexity. The result is a heavy and stiff robot that in turn needs powerful actuators to move it, as well as high quality sensors and a sophisticated controller – all this contributes to a high overall cost. However we should, whenever possible, avoid solving hard problems if we do not have to. Stepping back for a moment and looking at this problem it is clear that

the root cause of the problem is that the robot cannot see what it is doing.

Consider if the robot could see the object and its end-effector, and could use that information to guide the end-effector toward the object. This is what humans call *hand-eye coordination* and what we will call vision-based control or visual servo control – the use of information from one or more cameras to guide a robot in order to achieve a task.

The pose of the target does not need to be known a priori; the robot moves toward the observed target wherever it might be in the workspace. There are numerous advantages of this approach: part position tolerance can be relaxed, the ability to deal with parts that are moving comes almost for free, and any errors in the robot's intrinsic accuracy will be compensated for.

A vision-based control system involves continuous measurement of the target and the robot using vision to create a feedback signal and moves the robot arm until the visually observed error between the robot and the target is zero. Vision-based control is quite different to taking an image, determining where the target is and then reaching for it. The advantage of continuous measurement and feedback is that it provides great robustness with respect to any errors in the system. There are of course some practical complexities. If the camera is on the end of the robot it might interfere with the task, or when the robot is close to the target the camera might be unable to focus, or the target might be obscured by the gripper.

In this part of the book we bring together much that we have learned previously: kinematics and dynamics for robot arms and mobile robots; geometric aspects of image formation; and feature extraction. The part comprises two chapters. Chapter 15 discusses the two classical approaches to visual servoing which are known as *position-based* and *image-based* visual servoing. The image coordinates of world features are used to move the robot toward a desired pose relative to the observed object. The first approach requires explicit estimation of object pose from image features, but because it is performed in a closed-loop fashion any errors in pose estimation are compensated for. The second approach involves no pose estimation and uses image-plane information directly. Both approaches are discussed in the context of a perspective camera which is free to move in $SE(3)$, and their respective advantages and disadvantages are described. The chapter also includes a discussion of the problem of determining object depth, and the use of line and ellipse image features.

Chapter 16 extends the discussion to hybrid visual-servo algorithms which overcome the limitations of the position- and image-based visual servoing by using the best features of both. The discussion is then extended to nonperspective cameras such as fisheye lenses and catadioptric optics as well as arm robots, holonomic and nonholonomic ground robots, and a flying robot.

This part of the book is pitched at a higher level than earlier parts. It assumes a good level of familiarity with the rest of the book, and the increasingly complex examples are *sketched out* rather than described in detail. The text introduces the essential mathematical and algorithmic principles of each technique, but the full details are to be found in the source code of the MATLAB® classes that implement the controllers, or in the details of the Simulink diagrams. The results are also increasingly hard to depict in a book and are best understood by running the supporting MATLAB or Simulink® code and plotting the results or watching the animations.

Chapter

15

Vision-Based Control

The task in visual servoing is to control the pose of the robot's end-effector, relative to the goal, using visual *features* extracted from an image of the goal object. As shown in Fig. 15.1 the camera may be carried by the robot or be fixed in the world. The configuration of Fig. 15.1a has the camera mounted on the robot's end-effector observing the goal, and is referred to as end-point closed-loop or eye-in-hand. The configuration of Fig. 15.1b has the camera at a fixed point in the world observing both the goal and the robot's end-effector, and is referred to as end-point open-loop. In the remainder of this book we will discuss only the eye-in-hand configuration.

The image of the goal is a function of the relative pose $^C\xi_G$. Features such as the coordinates of points, or the parameters of lines or ellipses are extracted from the image and these are also a function of the relative pose $^C\xi_G$.

There are two fundamentally different approaches to visual servo control: Position-Based Visual Servo (PBVS) and Image-Based Visual Servo (IBVS). Position-based visual servoing, shown in Fig. 15.2a, uses observed visual features, a calibrated camera and a known geometric model of the goal object to determine its pose with respect to the camera. The robot then moves toward that pose and the control is performed in task space which is commonly **SE**(3). Good algorithms exist for pose estimation but it is computationally expensive and relies critically on the accuracy of the camera calibration and the model of the object's geometry. PBVS is discussed in Sect. 15.1.

A **servo mechanism**, or servo is an automatic device that uses feedback of error between the desired and actual position of a mechanism to drive the device to the desired position. The word servo is derived from the Latin root *servus* meaning slave and the first usage was by the Frenchman J. J. L. Farcot in 1868 – "Le Servomoteur" – to describe the hydraulic and steam engines used for steering ships.

Error in position is measured by a sensor then amplified to drive a motor that generates a force to move the device to reduce the error. Servo system development was spurred by WW II with the development of electrical servo systems for fire-control applications that used electric motors and electromechanical *amplidyne* power amplifiers. Later servo amplifiers used vacuum tubes and more recently solid state power amplifiers (motor drives). Today servo mechanisms are ubiquitous and are used to position the read/write heads in optical and magnetic disk drives, the lenses in autofocus cameras, remote control toys, satellite-tracking antennas, automatic machine tools and robot joints.

"Servo" is properly a noun or adjective but has become a verb "to servo". In the context of vision-based control we use the compound verb "visual servoing".

Fig. 15.1. Visual servo configurations and relevant coordinate frames: world, end-effector {E}, camera {C} and goal {G}. **a** End-point closed-loop configuration (eye-in-hand); **b** end-point open-loop configuration

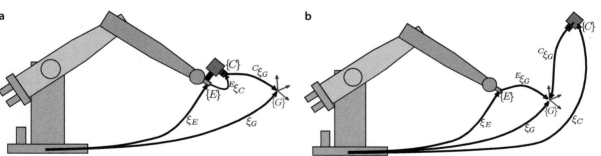

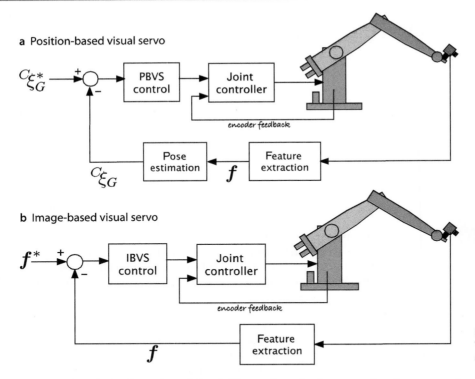

a Position-based visual servo

b Image-based visual servo

Fig. 15.2.
The two distinct classes of visual
servo system

Image-based visual servoing, shown in Fig. 15.2b, omits the pose estimation step, and uses the image features directly. The control is performed in image coordinate space $\mathbb{R}^2$. The desired camera pose with respect to the goal is defined *implicitly* by the image feature values at the goal pose. IBVS is a challenging control problem since the image features are a highly nonlinear function of camera pose. IBVS is discussed in Sect. 15.2.

15.1   Position-Based Visual Servoing

In a PBVS system the pose of the goal with respect to the camera $^C\xi_G$ is estimated. The pose estimation problem was discussed in Sect. 11.2.3 and requires knowledge of the goal object's geometry, the camera's intrinsic parameters and the observed image features. The relationships between the poses is shown as a pose graph in Fig. 15.3. We specify the desired relative pose with respect to the goal $^{C^*}\xi_G$ and wish to determine the motion ξ_Δ required to move the camera from its initial pose ξ_C to ξ_C^*. The actual pose of the goal ξ_G is not known. The indicated loop of the pose network is

$$\xi_\Delta \oplus {}^{C^*}\xi_G = {}^C\hat{\xi}_G$$

where $^C\hat{\xi}_T$ is the estimated pose of the goal relative to the camera. We rearrange this as

$$\xi_\Delta = {}^C\xi_G \ominus {}^{C^*}\hat{\xi}_G$$

which is the camera motion required to achieve the desired relative pose. The change in pose might be quite large so we do not attempt to make this movement in one step, rather we move to a point closer to $\{C^*\}$ by

$$\xi_C\langle k+1\rangle \leftarrow \xi_C\langle k\rangle \oplus \lambda\xi_\Delta\langle k\rangle$$

which is a fraction $\lambda \in (0, 1)$ of the translation and rotation required.

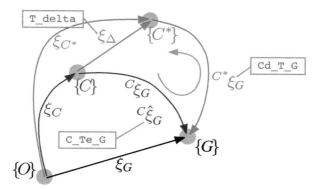

Fig. 15.3.
Relative pose network for PBVS example. Frame $\{C\}$ is the current camera pose and frame $\{C^*\}$ is the desired camera pose. The MATLAB variable names are shown in the grey boxes: an estimate (^) is indicated by 'e' and a desired value (*) by 'd'

Using the Toolbox we start by defining a camera with known parameters

```
>> cam = CentralCamera('default');
```

The goal comprises four points that form a square of side length 0.5 m that lies in the xy-plane and is centered at $(0, 0, 3)$

```
>> P = mkgrid( 2, 0.5, 'pose', SE3(0,0,3) );
```

and we assume that its pose is unknown to the control system. The camera is at some pose T_C so the image-plane projections of the world points are

```
>> p = cam.plot(P, 'pose', T_C)
```

from which the pose of the goal with respect to the camera $^C\hat{\xi}_G$ is estimated

```
>> C_Te_G = cam.estpose(P, p);
```

The required motion ξ_Δ is

```
>> T_delta = C_Te_G * inv(Cd_T_G);
```

and the fractional motion toward the goal is obtained by scaling this using its `interp` method giving the new value of the camera pose

```
>> T_C = T_C .* T_delta.interp(lambda);
```

where we ensure that the product is a proper homogeneous transformation by using the .* operator. At each time step we repeat the process, moving a fraction of the required relative pose until the motion is complete. In this way even if the goal moves, or the robot has errors and does not move as requested, the motion computed at the next time step will account for that error.

For this example we choose the initial pose of the camera in world coordinates as

```
>> T_C0 = SE3(1,1,-3)*SE3.Rz(0.6);
```

and the desired pose of the goal with respect to the camera is

```
>> Cd_T_G = SE3(0, 0, 1);
```

which has the goal 1 m in front of the camera and fronto-parallel to it. We create an instance of the PBVS class

```
pbvs = PBVS(cam, 'pose0', C_T0, 'posef', Cd_T_G, ...↵
        'axis', [-1 2 -1 2 -3 0.5])
Visual servo object: camera=default
  200 iterations, 0 history
    P= -0.25        -0.25        0.25         0.25
       -0.25        0.25         0.25        -0.25
        0            0            0            0
    C_T0:    t = ( 1,   1,  -3), R = ( 34.3775deg |  0,   0,  1)
    C*_T_G:  t = ( 0,   0,   1), R = ( 0deg |  0,   0,   0)
```

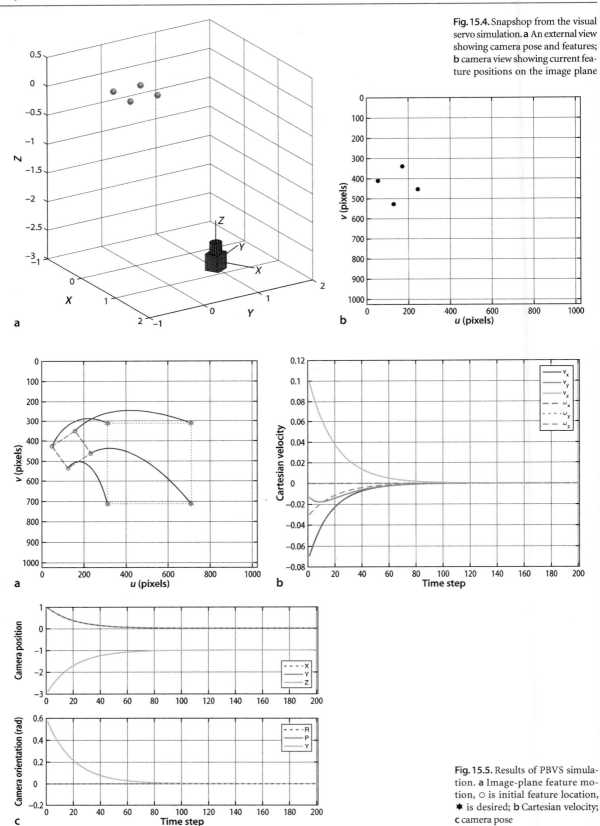

Fig. 15.4. Snapshop from the visual servo simulation. **a** An external view showing camera pose and features; **b** camera view showing current feature positions on the image plane

Fig. 15.5. Results of PBVS simulation. **a** Image-plane feature motion, ○ is initial feature location, ✱ is desired; **b** Cartesian velocity; **c** camera pose

which is a subclass of the `VisualServo` class and implements the controller outlined above. The object constructor takes a `CentralCamera` object as its argument, and contains the control algorithm required to drive this camera to achieve the desired pose relative to the goal specified by the `'posef'` option. Many additional options can be passed to this class constructor. The display method shows the coordinates of the world points, the initial camera pose, and the desired goal relative pose. The simulation is run by

```
>> pbvs.run();
```

which repeatedly calls the `step` method to simulate the motion for a single time step. The simulation animates the features moving on the image plane of the camera and the 3-dimensional visualization of the camera and the world points – as shown in Fig. 15.4. The simulation completes after a defined number of iterations or when ξ_Δ falls below some threshold.

The simulation results are stored within the object for later analysis. We can plot the path of the goal features in the image, the Cartesian velocity versus time or Cartesian position versus time

```
>> pbvs.plot_p();
>> pbvs.plot_vel();
>> pbvs.plot_camera();
```

which are shown in Fig. 15.5. We see that the feature points have followed a curved path in the image, and that the camera's translation and orientation have converged smoothly on the desired values.

15.2 Image-Based Visual Servoing

IBVS differs fundamentally from PBVS by not estimating the relative pose of the goal. The relative pose is implicit in the values of the image features. Figure 15.6 shows two views of a square goal object. The view from the initial camera pose is shown in red and it is clear that the camera is viewing the goal obliquely. The desired view is shown in blue where the camera is further from the goal and its optical axis is normal to the plane of the goal – a fronto-parallel view.

The control problem can be expressed in terms of image coordinates. The task is to move the feature points indicated by ○-markers to the points indicated by ✶-markers. The points may, but do not have to, follow the straight line paths indicated by the arrows. Moving the feature points in the image *implicitly* changes the camera pose – we have changed the problem from pose estimation to control of points in the image.

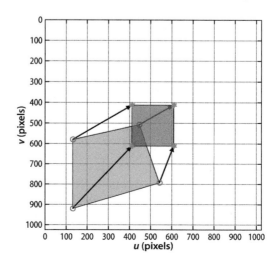

Fig. 15.6.
Two views of a square goal object. The *blue shape* is the desired view, and the *red shape* is the initial view

15.2.1   Camera and Image Motion

Consider the default camera

```
>> cam = CentralCamera('default');
```

and a world point at

```
>> P = [1 1 5]';
```

which has image coordinates

```
>> p0 = cam.project( P )
p0 =
   672
   672
```

Now if we displace the camera slightly in the x-direction the pixel coordinates will become

```
>> px = cam.project( P, 'pose', SE3(0.1,0,0) )
px =
   656
   672
```

Using the camera coordinate conventions of Fig. 11.5, the camera has moved to the right so the image point has moved to the left. The sensitivity of image motion to camera motion is

```
>> ( px - p0 ) / 0.1
ans =
   -160
      0
```

which is an approximation to the derivative $\partial p / \partial x$. It shows that 1 m of camera motion would lead to -160 pixel of feature motion in the u-direction. We can repeat this for z-axis translation

```
>> ( cam.project( P, 'pose', SE3(0, 0, 0.1) ) - p0 ) / 0.1
ans =
   32.6531
   32.6531
```

which shows equal motion in the u- and v-directions. For x-axis rotation

```
>> ( cam.project( P, 'pose', SE3.Rx(0.1) ) - p0 ) / 0.1
ans =
   40.9626
   851.8791
```

the image motion is predominantly in the v-direction. It is clear that camera motion along and about the different degrees of freedom in $\mathbf{SE}(3)$ causes quite different motion of image points. Earlier, in Eq. 11.10, we expressed perspective projection in functional form

$$p = \mathcal{P}(P, K, \xi_C)$$

and its derivative with respect to time is

$$\dot{p} = J_p(P, K, \xi_C)\nu$$

where $\nu = (v_x, v_y, v_z, \omega_x, \omega_y, \omega_z) \in \mathbb{R}^6$ is the velocity of the camera, the spatial velocity, which we introduced in Sect. 3.1. J_p is a Jacobian-like object, but because we have taken the derivative with respect to a pose $\xi \in \mathbf{SE}(3)$ rather than a vector it is technically called an *interaction matrix*. However in the visual servoing world it is more commonly called an *image Jacobian* or a *feature sensitivity matrix*.

Consider a camera moving with a body velocity $\nu = (v, \omega)$ in the world frame and observing a world point $\mathbf{P}$ with camera relative coordinates $P = (X, Y, Z)$. The velocity of the point relative to the camera frame is

$$\dot{P} = -\omega \times P - v \tag{15.1}$$

which we can write in scalar form as

$$
\begin{aligned}
\dot{X} &= Y\omega_z Z - \omega_y - v_x \\
\dot{Y} &= Z\omega_x - X\omega_z - v_y \\
\dot{Z} &= X\omega_y - Y\omega_x - v_z
\end{aligned}
\tag{15.2}
$$

The perspective projection Eq. 11.2 for normalized image-plane coordinates is

$$x = \frac{X}{Z}, \; y = \frac{Y}{Z}$$

and the temporal derivative, using the quotient rule, is

$$\dot{x} = \frac{\dot{X}Z - X\dot{Z}}{Z^2}, \; \dot{y} = \frac{\dot{Y}Z - Y\dot{Z}}{Z^2}$$

Substituting Eq. 15.2, $X = xZ$ and $Y = yZ$ we can write this in matrix form

$$
\begin{pmatrix} \dot{x} \\ \dot{y} \end{pmatrix} =
\begin{pmatrix}
-\frac{1}{Z} & 0 & \frac{x}{Z} & xy & -(1+x^2) & y \\
0 & -\frac{1}{Z} & \frac{y}{Z} & 1+y^2 & -xy & -x
\end{pmatrix}
\begin{pmatrix} v_x \\ v_y \\ v_z \\ \omega_x \\ \omega_y \\ \omega_z \end{pmatrix}
\tag{15.3}
$$

which relates camera spatial velocity to feature velocity in normalized image coordinates.
The normalized image-plane coordinates are related to the pixel coordinates by Eq. 11.7

$$u = \frac{f}{\rho_u} x + u_0, \; v = \frac{f}{\rho_v} y + v_0$$

which we rearrange as

$$x = \frac{\rho_u}{f}\bar{u}, \; y = \frac{\rho_v}{f}\bar{v} \tag{15.4}$$

where $\bar{u} = u - u_0$ and $\bar{v} = v - v_0$ are the pixel coordinates relative to the principal point. The temporal derivative is

$$\dot{x} = \frac{\rho_u}{f}\dot{\bar{u}}, \; \dot{y} = \frac{\rho_v}{f}\dot{\bar{v}} \tag{15.5}$$

and substituting Eq. 15.4 and Eq. 15.5 into Eq. 15.3 leads to

$$
\begin{pmatrix} \dot{\bar{u}} \\ \dot{\bar{v}} \end{pmatrix} =
\begin{pmatrix}
-\frac{f}{\rho_u Z} & 0 & \frac{\bar{u}}{Z} & \frac{\rho_v \bar{u}\bar{v}}{f} & -\frac{f^2 + \rho_u^2 \bar{u}^2}{\rho_u f} & \frac{\rho_v \bar{v}}{\rho_u} \\
0 & -\frac{f}{\rho_v Z} & \frac{\bar{v}}{Z} & \frac{f^2 + \rho_v^2 \bar{v}^2}{\rho_v f} & -\frac{\rho_u \bar{u}\bar{v}}{f} & -\frac{\rho_u \bar{u}}{\rho_v}
\end{pmatrix}
\begin{pmatrix} v_x \\ v_y \\ v_z \\ \omega_x \\ \omega_y \\ \omega_z \end{pmatrix}
$$

and for the typical case where $\rho_u = \rho_v = \rho$ we can express the focal length in pixels $f' = f/\rho$ and write

$$
\begin{pmatrix} \dot{\bar{u}} \\ \dot{\bar{v}} \end{pmatrix} = \underbrace{\begin{pmatrix} -\dfrac{f'}{Z} & 0 & \dfrac{\bar{u}}{Z} & \dfrac{\bar{u}\bar{v}}{f'} & -\dfrac{f'^2+\bar{u}^2}{f'} & \bar{v} \\[3mm] 0 & -\dfrac{f'}{Z} & \dfrac{\bar{v}}{Z} & \dfrac{f'^2+\bar{v}^2}{f'} & -\dfrac{\bar{u}\bar{v}}{f'} & -\bar{u} \end{pmatrix}}_{J_p(p,Z)} \begin{pmatrix} v_x \\ v_y \\ v_z \\ \omega_x \\ \omega_y \\ \omega_z \end{pmatrix}
\tag{15.6}
$$

in terms of pixel coordinates *with respect to the principal point*. We can write this in concise matrix form as

$$
\dot{p} = J_p(p, Z)v
\tag{15.7}
$$

where J_p is the 2 × 6 *image Jacobian* matrix for a point feature at coordinate $p^{\blacktriangleright}$ and camera distance Z.

The Toolbox `CentralCamera` class provides the method `visjac_p` to compute the image Jacobian and for the example above it is

```
>> J = cam.visjac_p([672; 672], 5)
J =
  -160      0     32     32   -832    160
     0   -160     32    832    -32   -160
```

This is commonly written in terms of u and v rather than $\bar{u}$ and $\bar{v}$ but we use the overbar notation to emphasize that the coordinates are with respect to the principal point, not the image origin which is typically in the top-left corner.

where the first argument is the pixel coordinate of the point of interest, and the second argument is the depth of the point. The approximate values computed on page 542 appear as columns one, three and four respectively. Image Jacobians can also be derived for line and circle features and these are discussed in Sect. 15.3.

For a given camera velocity, the velocity of the point is a function of the point's coordinate, its depth and the camera parameters. Each column of the Jacobian indicates the velocity of an image feature point caused by one unit of the corresponding component of the velocity vector. The `flowfield` method of the `CentralCamera` class shows the image-plane velocity for a grid of world points projected to the image plane for a particular camera velocity. For camera translational velocity in the x-direction the flow field is

```
>> cam.flowfield( [1 0 0 0 0 0] );
```

which is shown in Fig. 15.7a. As expected, moving the camera to the right causes all the features points to move to the left. The motion of points on the image plane is known as optical flow and can be computed from image sequences as we showed in Sect. 14.7.4. Equation 15.6 is often referred to as the optical flow equation.

For translation in the z-direction

```
>> cam.flowfield( [0 0 1 0 0 0] );
```

the points radiate outward from the principal point – the Star Trek warp effect – as shown in Fig. 15.7e. Rotation about the z-axis is

```
>> cam.flowfield( [0 0 0 0 0 1] );
```

causes the points to rotate about the principal point as shown in Fig. 15.7f.

Rotational motion about the y-axis is

```
>> cam.flowfield( [0 0 0 0 1 0] );
```

is shown in Fig. 15.7b and is very similar to the case of x-axis translation, with some small curvature for points far from the principal point. This similarity is because the first and fifth column of the image Jacobian are approximately equal in this case. For

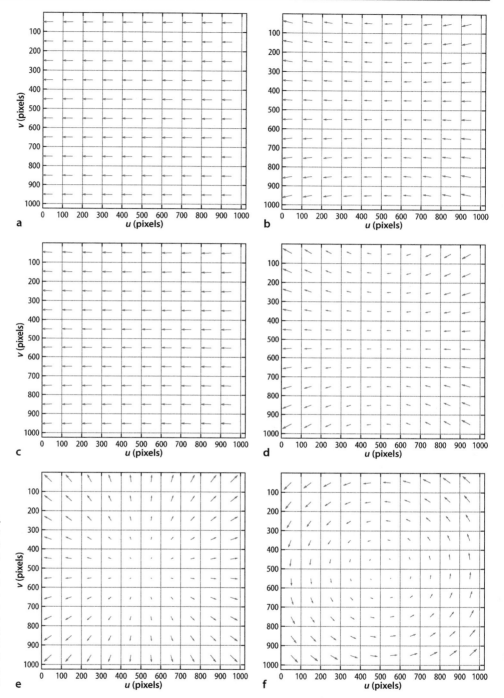

Fig. 15.7.
Image-plane velocity vectors for canonic camera velocities where all corresponding world points lie in a fronto-parallel plane. **a** x-axis translation; **b** y-axis rotation, $f = 8$ mm; **c** y-axis rotation, $f = 20$ mm; **d** y-axis rotation, $f = 4$ mm; **e** z-axis translation; **f** z-axis rotation. Note that the flow vectors are normalized – they are shown with correct relative scale within each plot, but not between plots

a point that projects to the center of the image, the principal point, and at a depth of 1 m the visual Jacobian is

```
>> cam.visjac_p(cam.pp', 1)
ans =
  -800.0000         0         0         0 -800.0000         0
         0 -800.0000         0  800.0000         0         0
```

and we see that columns one and five are exactly equal. This implies that translation in the x-direction causes the same image motion as rotation about the y-axis. You can easily demonstrate this equivalence by watching how the world moves if you translate your head to the right or rotate your head to the right – in both cases the world appears to move to the left. As the focal length increases column five

$$\lim_{f' \to \infty} \begin{pmatrix} -\dfrac{f'^2 + \bar{u}^2}{f'} \\ -\dfrac{\bar{u}\bar{v}}{f'} \end{pmatrix} = \begin{pmatrix} -f' \\ 0 \end{pmatrix}$$

approaches a scalar multiple of column one.

We can easily demonstrate this by increasing the focal length to $f = 20$ mm (the default focal length is 8 mm) and the flow field

```
>> cam.f = 20e-3;
>> cam.flowfield( [0 0 0 0 1 0] );
```

shown in Fig. 15.7c is almost identical to that of Fig. 15.7a. Conversely, for small focal lengths (wide-angle cameras) the image motion due to these camera motions will be more dissimilar

```
>> cam.f = 4e-3;
>> cam.flowfield( [0 0 0 0 1 0] );
```

and as shown in Fig. 15.7d the curvature is much more pronounced. The same applies for columns two and four except for a difference of sign – there is an equivalence between translation in the y-direction and rotation about the x-axis.►

The Jacobian matrix of Eq. 15.6 has some interesting properties. It does not depend at all on the world coordinates X or Y, only on the image-plane coordinates (u, v). However the first three columns depend on the point's depth Z and this reflects the fact that for a translating camera the image-plane velocity is inversely proportional to depth. You can easily demonstrate this to yourself – translate your head sideways and observe that near objects move more in your field of view than distant objects. However, if you rotate your head all objects, near and far, move equally in your field of view.

The matrix has a rank of two,► and therefore has a null space of dimension four. The null space comprises a set of spatial velocity vectors that individually, or in any linear combination, cause *no motion* in the image. Consider the simple case of a world point lying on the optical axis which projects to the principal point

```
>> J = cam.visjac_p(cam.pp', 1);
```

The null space of the Jacobian is

```
>> null(J)
ans =
          0          0    -0.7071          0
          0     0.7071          0          0
     1.0000          0          0          0
          0     0.7071          0          0
          0          0     0.7071          0
          0          0          0     1.0000
```

The first column indicates that motion in the z-direction, along the ray toward the point, results in no motion in the image. Nor does rotation about the z-axis, as indicated by column four. Columns two and three are more complex, combining rotation and translation. Essentially these exploit the image motion ambiguity mentioned above. Since x-axis translation causes the same image motion as y-axis rotation, column three indicates that if one is positive and the other negative the resulting image motion will be zero – that is translating left and rotating to the right.

Our visual system uses additional information from sensors to help resolve this ambiguity – proprioception from muscles in our body as well as motion estimates from the inertial sensors in our vestibular system.

The rank cannot be less than 2, even if $Z \to \infty$.

We can consider the motion of two points by stacking their Jacobians

$$
\begin{pmatrix} \dot{u}_1 \\ \dot{v}_1 \\ \dot{u}_2 \\ \dot{v}_2 \end{pmatrix} = \begin{pmatrix} J_p(\boldsymbol{p}_1, Z_1) \\ J_p(\boldsymbol{p}_2, Z_2) \end{pmatrix} \boldsymbol{\nu}
$$

to give a 4×6 matrix which will have a null space with two columns. One of these camera motions corresponds to rotation around a line joining the two points.

For three points

$$
\begin{pmatrix} \dot{u}_1 \\ \dot{v}_1 \\ \dot{u}_2 \\ \dot{v}_2 \\ \dot{u}_3 \\ \dot{v}_3 \end{pmatrix} = \begin{pmatrix} J_p(\boldsymbol{p}_1, Z_1) \\ J_p(\boldsymbol{p}_2, Z_2) \\ J_p(\boldsymbol{p}_3, Z_3) \end{pmatrix} \boldsymbol{\nu}
\tag{15.8}
$$

the matrix will be full rank, nonsingular, so long as the points are not coincident or collinear.

15.2.2 Controlling Feature Motion

So far we have shown how points move in the image plane as a consequence of camera motion. As is often the case, it is the inverse problem that is more useful – *what camera motion is needed in order to move the image features at a desired velocity?*

For the case of three points $\{(u_i, v_i),\ i = 1 \cdots 3\}$ and corresponding velocities $\{(\dot{u}_i, \dot{v}_i)\}$ we can invert Eq. 15.8

$$
\boldsymbol{\nu} = \begin{pmatrix} J_p(\boldsymbol{p}_1, Z_1) \\ J_p(\boldsymbol{p}_2, Z_2) \\ J_p(\boldsymbol{p}_3, Z_3) \end{pmatrix}^{-1} \begin{pmatrix} \dot{u}_1 \\ \dot{v}_1 \\ \dot{u}_2 \\ \dot{v}_2 \\ \dot{u}_3 \\ \dot{v}_3 \end{pmatrix}
\tag{15.9}
$$

and solve for the required camera velocity.

Given feature velocity we can compute the required camera motion, but how do we determine the feature velocity? The simplest strategy is to use a simple linear controller

$$
\dot{\boldsymbol{p}}^* = \lambda \left(\boldsymbol{p}^* - \boldsymbol{p} \right)
\tag{15.10}
$$

that *drives* the features toward their desired values $\boldsymbol{p}^*$ on the image plane. Combined with Eq. 15.9 we write

$$
\boldsymbol{\nu} = \lambda \begin{pmatrix} J_p(\boldsymbol{p}_1, Z_1) \\ J_p(\boldsymbol{p}_2, Z_2) \\ J_p(\boldsymbol{p}_3, Z_3) \end{pmatrix}^{-1} \left(\boldsymbol{p}^* - \boldsymbol{p} \right)
$$

That's it! This controller will drive the camera so that the feature points move toward the desired position in the image. It is important to note that nowhere have we required the pose of the camera or of the object◄ – everything has been computed in terms of what can be measured on the image plane.

We do require the depth Z of the points but we will come to that shortly.

For the general case where $N > 3$ points we can stack the Jacobians for all features and solve for camera motion using the pseudo-inverse ▶

Note that papers based on the task function approach such as Espiau et al. (1992) write this as actual position minus demanded position and write $-\lambda$ in Eq. 15.11 to ensure negative feedback.

$$\boldsymbol{\nu} = \lambda \begin{pmatrix} J_p(\boldsymbol{p}_1, Z_1) \\ \vdots \\ J_p(\boldsymbol{p}_N, Z_N) \end{pmatrix}^+ (\boldsymbol{p}^* - \boldsymbol{p}) \tag{15.11}$$

Note that it is possible to specify a set of feature point velocities which are inconsistent, that is, there is no possible camera motion that will result in the required image motion. In such a case the pseudo-inverse will find a solution that minimizes the norm of the feature velocity error.

The Jacobian is a first-order approximation of the relationship between camera motion and image-plane motion. Faster convergence is achieved by using a second-order approximation and it has been shown that this can be obtained very simply

$$\boldsymbol{\nu} = \frac{\lambda}{2} \left[\begin{pmatrix} J_p(\boldsymbol{p}_1, Z_1) \\ \vdots \\ J_p(\boldsymbol{p}_N, Z_N) \end{pmatrix}^+ + \begin{pmatrix} J_p\left(\boldsymbol{p}_1^*, Z_1^*\right) \\ \vdots \\ J_p\left(\boldsymbol{p}_N^*, Z_N^*\right) \end{pmatrix}^+ \right] (\boldsymbol{p}^* - \boldsymbol{p}) \tag{15.12}$$

by taking the mean of the pseudo inverse of the image Jacobians at the current and desired states.

For $N \geq 3$ the matrix can be poorly conditioned if the points are nearly coincident or collinear. In practice this means that some camera motions will cause very small image motions, that is, the motion has low perceptibility. There is strong similarity with the concept of manipulability that we discussed in Sect. 8.2.2 and we take a similar approach in formalizing it. Consider a camera spatial velocity of unit magnitude

$$\boldsymbol{\nu}^T \boldsymbol{\nu} = 1$$

and from Eq. 15.7 we can write the camera velocity in terms of the pseudo-inverse

$$\boldsymbol{\nu} = J^+ \dot{\boldsymbol{p}}$$

where $J \in \mathbb{R}^{2N \times 6}$ is the Jacobian stack and the point velocities are $\dot{\boldsymbol{p}} \in \mathbb{R}^{2N}$. Substituting yields

$$\dot{\boldsymbol{p}}^T J^{+^T} J^+ \dot{\boldsymbol{p}} = 1$$
$$\dot{\boldsymbol{p}}^T \left(JJ^T\right)^{-1} \dot{\boldsymbol{p}} = 1$$

which is the equation of an ellipsoid in the point velocity space. The eigenvectors of JJ^T define the principal axes of the ellipsoid and the singular values of J are the radii. The ratio of the maximum to minimum radius is given by the condition number of JJ^T and indicates the anisotropy of the feature motion. A high value indicates that some of the points have low velocity in response to some camera motions. An alternative to stacking all the point feature Jacobians is to select just three that, when stacked, result in the best conditioned square matrix which can then be inverted.

Using the Toolbox we start by defining a camera

```
>> cam = CentralCamera('default');
```

The goal comprises four points that form a square of side length 0.5 m that lies in the xy-plane and is centered at $(0, 0, 3)$

```
>> P = mkgrid( 2, 0.5, 'pose', SE3(0,0,3) );
```

and we assume that this goal pose is unknown to the control system. The desired position of the goal features on the image plane are a 400×400 square centered on the principal point

```
>> pd = bsxfun(@plus, 200*[-1 -1 1 1; -1 1 1 -1], cam.pp');
```

which implicitly has the square goal fronto-parallel to the camera.

The camera is at some pose $\xi_C \sim$ T_C so the image-plane projections of the world points are

```
>> p = cam.plot(P, 'pose', T_C);
```

where p and pd are each 2×4 matrices and have one column per point. We compute the image-plane error

```
>> e = pd - p;
```

and the stacked image Jacobian

```
>> J = cam.visjac_p( p, 1 );
```

is an 8×6 matrix in this case since p contains four points. The Jacobian does require the point depth which we do not know, so for now we will just choose a constant value◄. This is an important topic that we will address in Sect. 15.2.3.

The control law determines the required translational and angular velocity of the camera

```
>> v = lambda * pinv(J) * e(:);
```

where lambda is the gain, a positive number, and we take the pseudo-inverse of the nonsquare Jacobian to implement Eq. 15.11. The resulting velocity is expressed in the camera coordinate frame, and integrating it over a unit time step results in a spatial displacement of the same magnitude. The camera pose is updated by

$$\xi_C \langle k{+}1 \rangle \leftarrow \xi_C \langle k \rangle \oplus \Delta^{-1}\big(\nu\langle k \rangle\big)$$

where $\Delta^{-1}(\cdot)$ is described in Sect. 3.1.4. Using the Toolbox this is implemented as

```
>> T_C = T_C .* delta2tr(v);
```

where we ensure that the transformation remains a proper homogeneous transformation by using the .* operator.

For this example we choose the initial pose of the camera in world coordinates as

```
>> T_C0 = SE3(1,1,-3)*SE3.Rz(0.6);
```

Similar to the PBVS example we create an instance of the IBVS class

```
>> ibvs = IBVS(cam, 'pose0', T_C0, 'pstar', pd);
```

which is a subclass of the VisualServo class and implements the controller outlined above. The object constructor takes a CentralCamera object as its argument, and drives this camera to achieve the desired pose relative to the goal. The option 'pose0' specifies the initial pose of the camera and 'pstar' specifies the desired image coordinates of the features. Many additional options can be passed to this class constructor. The display method shows the coordinates of the world points, the initial absolute pose, and the desired image-plane feature coordinates. The simulation is run by

```
>> ibvs.run();
```

which repeatedly calls the step method that simulates motion for a single time step. The simulation animates the image plane of the camera as well as a 3-dimensional visualization of the camera and the world points.

Here we provide a single value which is taken as the depth of all the points. Alternatively we could provide a vector to specify the depth of each point individually.

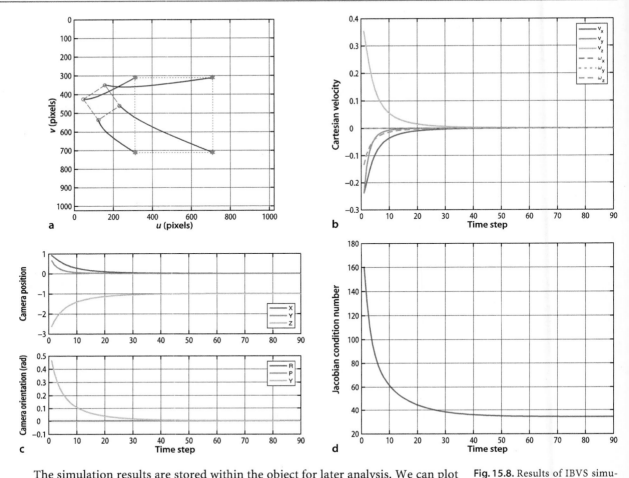

The simulation results are stored within the object for later analysis. We can plot the path of the goal features on the image plane, the Cartesian velocity versus time or Cartesian position versus time

```
>> ibvs.plot_p();
>> ibvs.plot_vel();
>> ibvs.plot_camera();
```

Fig. 15.8. Results of IBVS simulation, created by IBVS. **a** Image-plane feature motion, ○ is the initial feature position and ✳ is the desired position **b** spatial velocity components; **c** camera pose; **d** image Jacobian condition number

which are shown in Fig. 15.8. We see that the feature points have followed an almost straight-line path in the image, and the Cartesian pose has changed smoothly toward the final value. The condition number of the image Jacobian

```
>> ibvs.plot_jcond();
```

decreases over the motion indicating that the Jacobian is becoming better conditioned, and this is a consequence of the features moving further apart.

How is p^* determined? The image points can be found by demonstration, by moving the camera to the desired pose and recording the observed image coordinates. Alternatively, if the camera calibration parameters and the goal geometry are known the desired image coordinates can be computed for any specified goal pose. Note that this calculation, world point point projection, is computationally cheap and performed only once before visual servoing commences.

The IBVS system can also be expressed in terms of a Simulink model

```
>> sl_ibvs
```

which is shown in Fig. 15.9. The simulation is run by

```
>> r = sim('sl_ibvs')
```

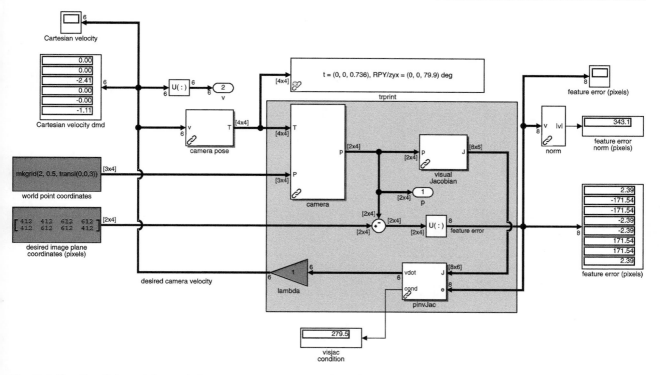

Fig. 15.9. The Simulink model `sl_ibvs` drives the feature points to the desired positions on the image plane. User adjustable parameters are in the *red blocks*

and the camera pose, image-plane feature error and camera velocity are animated. Scope blocks also plot the camera velocity and feature error against time. The initial pose of the camera is set by a parameter of the `camera pose` block, and the world points are parameters of a constant block. The `CentralCamera` object is a parameter to both the `camera` and `visual Jacobian` blocks.

The signals sent to the output ports are stored in the simulation output object `r`. For example the camera velocity on output port 2

```
>> t = r.find('tout');
>> v = r.find('yout').signals(2).values;
>> about(v)
v [double] : 501x6 (24048 bytes)
```

has one row for every simulation time step, and the columns are the camera spatial velocity components. We can plot camera velocity against time

```
>> plot(t, v)
```

The image-plane coordinates on output port 1 can also be retrieved and plotted

```
>> p = r.find('yout').signals(1).values;
>> about(p)
p [double] : 2x4x501 (32.1 kB)
>> plot2(p)
```

15.2.3   Estimating Feature Depth

Computing the image Jacobian requires knowledge of the camera intrinsics, the principal point and focal length, but in practice it is quite tolerant to errors in these. The Jacobian also requires knowledge of Z_i, the distance to, or the depth of, each point. In the simulations just discussed we have assumed that depth is known – this is easy in simulation but not so in reality. Fortunately, in practice we find that IBVS is remarkably tolerant to errors in Z.

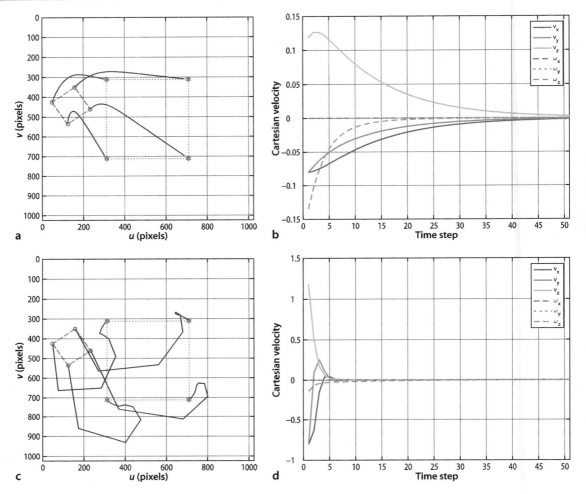

Fig. 15.10. Results of IBVS with different constant estimates of point depth: **a, b** Image and camera motion for $Z = 1$; **c, d** Image and camera motion for $Z = 10$

A number of approaches have been proposed to deal with the problem of unknown depth. The simplest is to just assume a constant value for the depth which is quite reasonable if the required camera motion is approximately in a plane parallel to the plane of the object points. To evaluate the performance of different constant estimates of point depth, we can compare the effect of choosing $Z = 1$ and $Z = 10$ for the example above where the true depth is $Z = 3$

```
>> ibvs = IBVS(cam, 'pose0', T_C0, 'pstar', pd 'depth', 1)
>> ibvs.run(50)
>> ibvs = IBVS(cam, 'pose0', T_C0, 'pstar', pd 'depth', 10)
>> ibvs.run(50)
```

and the results are plotted in Fig. 15.10. We see that the image-plane paths are no longer straight, because the Jacobian is now a poor approximation of the relationship between the camera motion and image feature motion. We also see that for $Z = 1$ the convergence is much slower than for the $Z = 10$ case. The Jacobian for $Z = 1$ overestimates the optical flow, so the inverse Jacobian underestimates the required camera velocity. Nevertheless, for quite significant errors, IBVS has converged. For the $Z = 10$ case the camera displacement at each timestep is large leading to a very jagged path.

A second approach is to use standard computer vision techniques to estimate the value for Z. If the camera intrinsic parameters were known we could use sparse stereo techniques from consecutive camera positions to estimate the depth of each feature point.

A third approach is to estimate the value of Z online using measurements of robot and image motion. We can create a simple depth estimator by rearranging Eq. 15.6 into estimation form

$$\begin{pmatrix} \dot{u} \\ \dot{v} \end{pmatrix} = \left(\begin{array}{cc|ccc} -\frac{f}{\rho_u Z} & 0 & \frac{\bar{u}}{Z} & \frac{\rho_u \bar{u}\bar{v}}{f} & -\frac{f^2 + \rho_u^2 \bar{u}^2}{\rho_u f} & \bar{v} \\ 0 & -\frac{f}{\rho_v Z} & \frac{\bar{v}}{Z} & \frac{f^2 + \rho_v^2 \bar{v}^2}{\rho_v f} & -\frac{\rho_v \bar{u}\bar{v}}{f} & -\bar{u} \end{array} \right) \begin{pmatrix} \boldsymbol{v} \\ \boldsymbol{\omega} \end{pmatrix}$$

$$= \left(\tfrac{1}{Z} J_t \mid J_\omega \right) \begin{pmatrix} \boldsymbol{v} \\ \boldsymbol{\omega} \end{pmatrix}$$

$$= \frac{1}{Z} J_t \boldsymbol{v} + J_\omega \boldsymbol{\omega}$$

which we rearrange as

$$(J_t \boldsymbol{v}) \frac{1}{Z} = \begin{pmatrix} \dot{u} \\ \dot{v} \end{pmatrix} - J_\omega \boldsymbol{\omega} \tag{15.13}$$

The right-hand side is the observed optical flow from which the expected optical flow due to rotation of the camera is subtracted – a process referred to as derotating optical flow. The remaining optical flow, after subtraction, is only due to translation. Writing Eq. 15.13 in compact form

$$\mathbf{A}\boldsymbol{\theta} = \mathbf{b} \tag{15.14}$$

we have a simple linear equation with one unknown parameter $\theta = 1 / Z$ which can be solved using least-squares.

In our example we can enable this by

```
>> ibvs = IBVS(cam, 'pose0', T_C0, 'pstar', pd 'depthest')
>> ibvs.run()
>> ibvs.plot_z()
>> ibvs.plot_p()
```

and the result is shown in Fig. 15.11. Figure 15.11b shows the estimated and true point depth versus time. The estimate depth was initially zero, a poor choice, but it has risen rapidly and then tracked the actual goal depth as the controller converges. Figure 15.11a shows the feature motion, and we see that the features initially move in the wrong direction because the error in depth has led to an image Jacobian that predicts poorly how feature points will move.

Fig. 15.11. IBVS with online depth estimator. **a** Feature paths; **b** comparison of estimated (*dashed*) and true depth (*solid*) for all four points

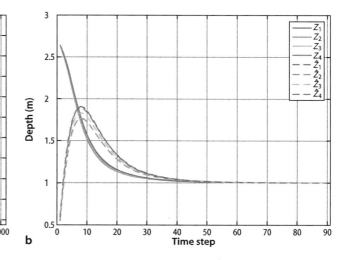

15.2.4   Performance Issues

The control law for PBVS is defined in terms of the 3-dimensional workspace so there is no mechanism by which the motion of the image features is directly regulated. For the PBVS example shown in Fig. 15.5 the feature points followed a curved path on the image plane, and therefore it is possible that they could leave the camera's field of view. For a different initial camera pose

```
>> pbvs.T0 = SE3(-2.1, 0, -3)*SE3.Rz(5*pi/4);
>> pbvs.run()
```

the result is shown in Fig. 15.12a and we see that two of the points move outside the image which would cause the PBVS control to fail.▶

By contrast the IBVS control for the same initial pose

In this simulation the image plane coordinates are still computed and used, even though they fall outside the image bounds.

```
>> ibvs = IBVS(cam, 'pose0', pbvs.T0, 'pstar', pd, 'lambda',↵
   0.002, 'niter', Inf, 'eterm', 0.5)
>> ibvs.run()
>> ibvs.plot_p();
```

gives the feature trajectories shown in Fig. 15.12b but there is no direct control over the Cartesian motion of the camera. This can sometimes result in surprising motion, particularly when the goal is rotated about the z-axis

```
>> ibvs = IBVS(cam, 'pose0', SE3(0,0, -1)*SE3.Rz(1), 'pstar', pd);
>> ibvs.run()
>> ibvs.plot_camera
```

which is shown in Fig. 15.13a,b. We see that the camera has performed an unnecessary translation along the z-axis – away from the goal and back again. This phenomenon is termed camera retreat. The resulting motion is not time optimal and can require large and possibly unachievable camera motion. An extreme example arises for a pure rotation about the optical axis by π rad

```
>> ibvs = IBVS(cam, 'pose0',  SE3(0,0, -1)*SE3.Rz(pi), ...
   'pstar', pStar, 'niter', 10);
>> ibvs.run()
>> ibvs.plot_camera
```

which is shown in Fig. 15.13c,d. The feature points are, as usual, moving in a straight line toward their desired values, but for this problem the paths all pass through the principal point which is a singularity and where IBVS will fail. The only way the goal feature points can be at the principal point is if the camera is at negative infinity, and that is where it is headed!

Fig. 15.12. Image-plane feature paths for **a** PBVS and **b** IBVS

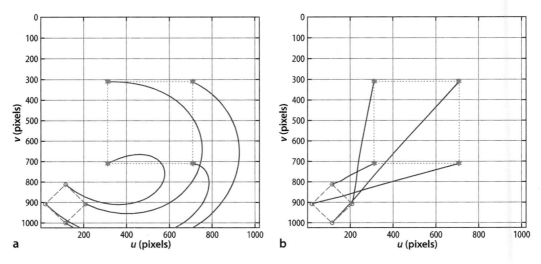

a u (pixels)   b u (pixels)

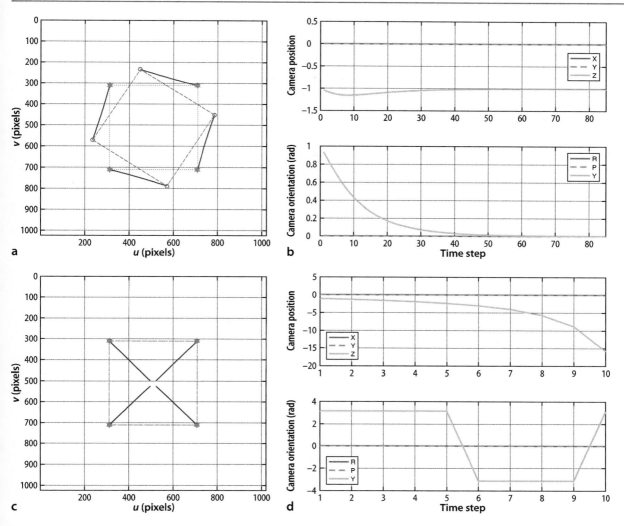

Fig. 15.13. IBVS for pure goal rotation about the optical axis. **a, b** for rotation of 1 rad; **c, d** for rotation of π rad

A final consideration is that the image Jacobian is a linearization of a highly non-linear system. If the motion at each time step is large then the linearization is not valid and the features will follow curved rather than linear paths in the image, as we saw in Fig. 15.10. This can occur if the desired feature positions are a long way from the initial positions and/or the gain λ is too high. One solution is to limit the maximum norm of the commanded velocity

$$\nu = \begin{cases} \nu_{max}\dfrac{\nu}{|\nu|} & \text{if } |\nu| > \nu_{max} \\ \nu & \text{if } |\nu| \leq \nu_{max} \end{cases}$$

The feature paths do not have to be straight lines and nor do the features have to move with asymptotic velocity – we have used these only for simplicity. Using the trajectory planning methods of Sect. 3.3 the features could be made to follow any arbitrary trajectory in the image.

In summary, IBVS is a remarkably robust approach to vision-based control. We have seen that it is quite tolerant to errors in the depth of points. We have also shown that it can produce less than optimal Cartesian paths for the case of large rotations about the optical axis. We will discuss remedies to these problems in the next chapter.

15.3   Using Other Image Features

So far we have considered only point features. In a real system we would use the feature extraction techniques discussed in Chap. 13 and the points would be the centroids of distinct regions, or Harris or SURF corner features. The points would then be used for pose estimation in a PBVS scheme, or directly in an IBVS scheme. For both PBVS or IBVS we need to solve the correspondence problem, that is, for each observed feature we must determine which desired image-plane coordinate it corresponds to. IBVS can also be formulated to work with other image features such as lines, as found by the Hough transform, or the shape of an ellipse.

15.3.1   Line Features

For a line the Jacobian is written in terms of the (ρ, θ) parameterization that we used for the Hough transform in Sect. 13.2. The rate of change of the line parameters is related to camera velocity by

$$\begin{pmatrix} \dot{\theta} \\ \dot{\rho} \end{pmatrix} = J_l \nu$$

and the Jacobian is

$$J_l = \begin{pmatrix} \lambda_\theta \sin\theta & \lambda_\theta \cos\theta & -\rho\lambda_\theta & -\rho\sin\theta & -\rho\cos\theta & -1 \\ \lambda_\rho \sin\theta & \lambda_\rho \cos\theta & -\lambda_\rho\rho & -\cos\theta(1+\rho^2) & \sin\theta(1+\rho^2) & 0 \end{pmatrix}$$

where $aX + bY + cZ + d = 0$ is the equation of the plane that contains the line and $\lambda_\theta = (a\cos\theta - b\sin\theta)/d$ and $\lambda_\rho = -(a\rho\sin\theta + b\rho\cos\theta + c)/d$. The Jacobian describes how the line parameters change as a function of camera velocity. Just as the point-feature Jacobian required some partial 3-dimensional knowledge (the point depth Z) the line-feature Jacobian requires the equation of the plane that contains the line. There are an infinite number of planes that contain the line and we choose one for which $d \neq 0$. Like a point feature, a line provides two rows of the Jacobian so we require a minimum of three lines in order to have a Jacobian of full rank.▶

We illustrate this with an example comprising three lines that all lie in the plane $Z = 3$, and we can conveniently construct three points in that plane using the `circle` function with just three boundary points

```
>> P = circle([0 0 3], 0.5, 'n', 3);
```

and use the familiar `CentralCamera` class methods to project these to the image. For each pair of points we compute the equations of the line

Interestingly a line feature provides two rows of the stacked Jacobian, yet two points which define a line segment would provide four rows.

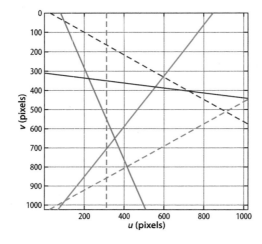

Fig. 15.14.
IBVS using line features. The image plane showing the three current lines (*solid*) and desired (*dashed*)

$$\tan\theta = \frac{v_2 - v_1}{u_1 - u_2}, \ \rho = u_1\sin\theta + v_1\cos\theta$$

The simulation is run in familiar fashion

```
>> ibvs = IBVS_l(cam, 'example');
>> ibvs.run()
```

and a snapshot of results is shown in Fig. 15.14. Note that we need to establish correspondence between the observed and desired lines.

15.3.2 Circle Features

Ellipse are described in more detail in Sect. C.1.4.

A circle in the world will be projected, in the general case, to an ellipse in the image which is described by◄

$$u^2 + E_1 v^2 - 2E_2 uv + 2E_3 u + 2E_4 v + E_5 = 0 \tag{15.15}$$

where E_i are parameters of the ellipse. The rate of change of the ellipse coefficients is related to camera velocity by

$$\begin{pmatrix} \dot{E}_1 \\ \dot{E}_2 \\ \vdots \\ \dot{E}_5 \end{pmatrix} = J_e(E, \rho)\nu$$

and the Jacobian is

$$J_e(E,\rho) = \begin{pmatrix} 2\beta E_2 - 2\alpha E_1 & 2E_1(\beta - \alpha E_2) & 2\beta E_4 - 2\alpha E_1 E_3 & 2E_4 & 2E_1 E_3 & -2E_2(E_1+1) \\ \beta - \alpha E_2 & \beta E_2 - \alpha(2E_2^2 - E_1) & \alpha(E_4 - 2E_2 E_3) + \beta E_3 & E_3 & 2E_2 E_3 - E_4 & E_1 - 2E_2^2 - 1 \\ \gamma - \alpha E_3 & \alpha(E_4 - 2E_2 E_3) + \gamma E_2 & \gamma E_3 - \alpha(2E_3^2 - E_5) & -E_2 & 1 + 2E_3^2 - E_5 & E_4 - 2E_2 E_3 \\ E_3\beta + E_2\gamma - 2\alpha E_4 & E_4\beta + E_1\gamma - 2\alpha E_2 E_4 & \beta E_5 + \gamma E_4 - 2\alpha E_3 E_4 & E_5 - E_1 & 2E_3 E_4 + E_2 & -2E_2 E_4 - E_3 \\ 2\gamma E_3 - 2\alpha E_5 & 2\gamma E_4 - 2\alpha E_2 E_5 & 2\gamma A5 - 2\alpha E_3 E_5 & -2E_4 & 2E_3 E_5 + 2E_3 & -2E_2 E_5 \end{pmatrix}$$

where $\rho = (\alpha, \beta, \gamma)$ defines a plane in world coordinates $aX + bY + cZ + d = 0$ in which the ellipse lies and $\alpha = -a/d$, $\beta = -b/d$ and $\gamma = -c/d$. Just as was the case for point and line feature Jacobians we need to provide some depth information about the goal. The Jacobian has a maximum rank of five, but this drops to three when the projection is of a circle centered in the image plane, and a rank of two if the circle has zero radius.

An advantage of the ellipse feature is that the ellipse can be computed from the set of all boundary points without needing to solve the correspondence problem. The ellipse feature can also be computed from the moments of all the points within the ellipse boundary. We illustrate this with an example of a circle comprising ten points around its circumference

```
>> P = circle([0 0 3], 0.5, 'n', 10);
```

and the `CentralCamera` class projects these to the image plane.

```
>> p = cam.project(P, 'pose', Tc);
```

where `Tc` is the current camera pose and we convert to normalized image coordinates

```
>> pn = cam.normalized( p);
```

The parameters of an ellipse are calculated using the methods of Sect. C.1.4.3

```
>> x = pn(1,:); y = pn(2,:);
>> a = [y.^2; -2*x.*y; 2*x; 2*y; ones(1,numcols(x))]';
>> b = -(x.^2)';
>> E = a\b;
```

which returns a 5-vector of ellipse parameters. The image Jacobian for an ellipse feature is computed by a method of the `CentralCamera` class

```
>> J = cam.visjac_e(E, plane);
```

where the plane containing the circle must also be specified. For this example the plane is $Z = 3$ so the plane parameters are $(0, 0, 1, -3)$.

The Jacobian is 5×6 and has a maximum rank of only 5 so we cannot uniquely solve for the camera velocity. We have at least two options. Firstly, if our final view is of a circle then we may not be concerned about rotation around the center of the circle, and in this case we can delete the sixth column of the Jacobian to make it square and set ω_z to zero. Secondly, and the approach taken in this example, is to combine the features for the ellipse and a single point▶

Here we arbitrarily choose the first point, any one will do, but we need to establish correspondence in every frame.

$$\begin{pmatrix} \dot{E}_1 \\ \dot{E}_2 \\ \vdots \\ \dot{E}_5 \\ \dot{u}_1 \\ \dot{v}_1 \end{pmatrix} = \begin{pmatrix} J_e(E) \\ J_p(p_1) \end{pmatrix} \nu$$

and the stacked Jacobian is now 7×6 and we can solve for camera velocity. As for the previous IBVS examples the desired velocity is proportional to the difference between the current and desired feature values

Fig. 15.15. IBVS using ellipse feature. **a** The image plane showing the current points (*solid*) and demanded (✳); **b** a world view showing the points and the camera

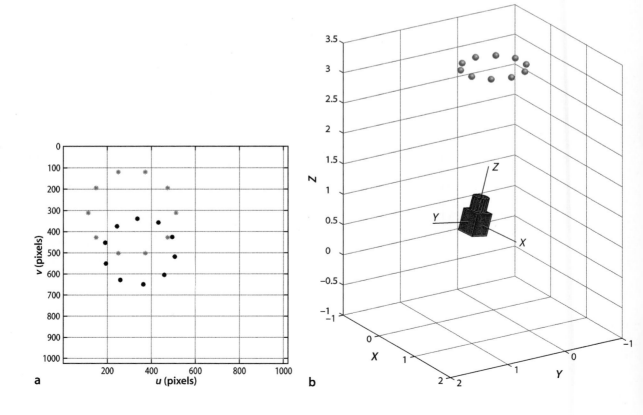

a

b

$$\begin{pmatrix} \dot{E}_1 \\ \dot{E}_2 \\ \vdots \\ \dot{E}_5 \\ \dot{u}_1 \\ \dot{v}_1 \end{pmatrix} = \lambda \begin{pmatrix} E_1^* - E_1 \\ E_2^* - E_2 \\ \vdots \\ E_5^* - E_5 \\ u_1^* - u_1 \\ v_1^* - v_1 \end{pmatrix}$$

The simulation is run in the now familiar fashion

```
>> ibvs = IBVS_e(cam, 'example');
>> ibvs.run()
```

and a snapshot of results is shown in Fig. 15.15.

15.3.3   Photometric Features

When servoing using point or line features we have to determine the error between current and desired features, and this requires determining correspondence between features in the current and the desired images. Correspondence is a complex task which can be complicated by occlusions or features leaving the camera's field of view. In photometric visual servoing we work directly with the pixel values and no correspondences are required.

The image feature is a vector that contains all the pixels in the image – current or desired – stacked into a very tall vector of height $N = W \times H$. The rate of change of the pixel values is related to the camera velocity by

$$\begin{pmatrix} \dot{I}_1 \\ \dot{I}_2 \\ \vdots \\ \dot{I}_N \end{pmatrix} = J_I(I)\nu$$

where the Jacobian is

$$J_I(I) = \begin{pmatrix} \nabla_I(p_1)J_p(p_1) \\ \nabla_I(p_2)J_p(p_2) \\ \vdots \\ \nabla_I(p_N)J_p(p_N) \end{pmatrix} \in \mathbb{R}^{N \times 6}$$

and where p_i is the image-plane coordinate of the pixel corresponding to the ith element of the feature vector, $\nabla_I(p) = (\nabla_u(p), \nabla_v(p))$ are the image gradients in the u- and v-directions at that pixel, and $J_p(\cdot)$ is the image point-feature Jacobian computed at that pixel.

The Jacobians, as always, are also a function of the point depth. If we are servoing with respect to a planar image then the depth might be approximately known and as we have remarked previously IBVS is quite robust to errors in point depth. If we are servoing with respect to a complex 3-dimensional scene then depth at each pixel will be very difficult to determine and we again rely on the inherent robustness of IBVS.

In order to converge the actual and destination images must have significant overlap. The derivation makes assumptions that the scene has Lambertian reflectance, no specular highlights, and that lighting magnitude and direction does not change over time. In practice photometric visual servoing works well even if these assumptions are not met or if the images are partially occluded during the camera motion.

15.4   Wrapping Up

In this chapter we have learned about the fundamentals of vision-based robot control, and the fundamental techniques developed over two decades up to the mid 1990s. There are two distinct configurations. The camera can be attached to the robot observing the goal (eye-in-hand) or fixed in the world observing both robot and goal. Another form of distinction is the control structure: Position-Based Visual Servo (PBVS) and Image-Based Visual Servo (IBVS). The former involves pose estimation based on a calibrated camera and a geometric model of the goal object, while the latter performs the control directly in the image plane. Each approach has certain advantages and disadvantages. PBVS performs efficient straight-line Cartesian camera motion in the world but may cause image features to leave the image plane. IBVS always keeps features in the image plane but may result in trajectories that exceed the reach of the robot, particularly if it requires a large amount of rotation about the camera's optical axis. IBVS also requires a touch of 3-dimensional information (the depth of the feature points) but is quite robust to errors in depth and it is quite feasible to estimate the depth as the robot moves. IBVS can be formulated to work not only with point features, but also with lines, ellipses and pixel values. An arbitrary number of features (which can be any mix of points, lines or ellipses) from an arbitrary number of cameras can be combined simply by stacking the relevant Jacobian matrices.

So far in our simulations we have determined the required camera velocity and moved the camera accordingly, without consideration of the mechanism to move it. In the next chapter we consider cameras attached to arm-type robots, mobile ground robots and flying robots.

Further Reading

The tutorial paper by Hutchinson et al. (1996) was the first comprehensive articulation and taxonomy of the field, and Chaumette and Hutchinson (2006) provide a more recent tutorial introduction. Chapters on visual servoing are included in Siciliano et al. (2016, § 34) and Spong et al. (2006, § 12).

It is well known that IBVS is very tolerant to errors in depth and its effect on control performance is examined in detail in Marey and Chaumette (2008). Feddema and Mitchell (1989) performed a partial 3D reconstruction to determine point depth based on observed features and known goal geometry. Papanikolopoulos and Khosla (1993) described adaptive control techniques to estimate depth, as used in this chapter. Hosoda and Asada (1994), Jägersand et al. (1996) and Piepmeier et al. (1999) have shown how the image Jacobian matrix itself can be estimated online from measurements of robot and image motion. The second-order visual servoing technique was introduced by Malis (2004).

The most common image Jacobian is based on the motion of points in the image, but it can also be derived for the parameters of lines in the image plane (Chaumette 1990; Espiau et al. 1992) and the parameters of an ellipse in the image plane (Espiau et al. 1992). Moments of binary regions have been proposed for visual servoing of planar scenes (Chaumette 2004; Tahri and Chaumette 2005). More recently the ability to servo directly from image pixel values, without segmentation or feature extraction, has been described by Collewet et al. (2008) and subsequent papers, and more recently by Bakthavatchalam et al. (2015).

The literature on PBVS is much smaller, but the paper by Westmore and Wilson (1991) is a good introduction. They use an EKF to implicitly perform pose estimation – the goal pose is the filter state and the innovation between predicted and observed feature coordinates updates the goal pose state. Hashimoto et al. (1991) present simulations to compare position-based and image-based approaches.

History and background. Visual servoing has a very long history – the earliest reference is by Shirai and Inoue (1973) who describe how a visual feedback loop can be used to correct the position of a robot to increase task accuracy. They demonstrated a system with a servo cycle time of 10 s, and this highlights a harsh reality of the field which has been the problem of real-time feature extraction. Until the late 1990s this required bulky and expensive special-purpose hardware such as that shown in Fig. 15.16. Other significant early work on industrial applications occurred at SRI International during the late 1970s (Hill and Park 1979; Makhlin 1985).

In the 1980s Weiss et al. (1987) introduced the classification of visual servo structures as either position-based or image-based. They also introduced a distinction between visual servo and dynamic look and move, the former uses only visual feedback whereas the latter uses joint feedback and visual feedback. This latter distinction is no longer in common usage and most visual servo systems today make use of joint-position and visual feedback, commonly encoder-based joint velocity loops as discussed in Chap. 9 with an outer vision-based position loop. Weiss (1984) applied adaptive control techniques for IBVS of a robot arm without joint-level feedback, but the results were limited to low degree of freedom arms due to the low-sample rate vision processing available at that time. Others have looked at incorporating the manipulator dynamics Eq. 9.8 into controllers that command motor torque directly (Kelly 1996; Kelly et al. 2002a,b) but all still require joint angles in order to evaluate the manipulator Jacobian, and the joint rates to provide damping. Feddema (Feddema and Mitchell 1989; Feddema 1989) used closed-loop joint control to overcome problems due to low visual sampling rate and demonstrated IBVS for 4-DOF. Chaumette, Rives and Espiau (Chaumette et al. 1991; Rives et al. 1989) describe a similar approach using the task function method (Samson et al. 1990) and show experimental results for robot positioning using a goal object with four features. Feddema et al. (1991) describe an algorithm to select which subset of the available features give the best conditioned square Jacobian. Hashimoto et al. (1991) have shown that there are advantages in using a larger number of features and using a pseudo-inverse to solve for velocity. Control and stability in closed-loop visual control systems was addressed by several researchers (Corke and Good 1992; Espiau et al. 1992; Papanikolopoulos et al. 1993) and feedforward predictive, rather than feedback, controllers were proposed by Corke (1994) and Corke and Good (1996).

The 1993 book edited by Hashimoto (1993) was the first collection of papers covering approaches and applications in visual servoing. The 1996 book by Corke (1996b) is now out of print but available free online and covers the fundamentals of robotics and vision for controlling the dynamics of an image-based visual servoing system. It contains an extensive, but dated, collection of references to visual servoing applications including industrial applications, camera control for tracking, high-speed planar micromanipulator, road vehicle guidance, aircraft refueling, and fruit picking. Another

Fig. 15.16.
A 19 inch VMEbus rack of hardware image processing cards, capable of 10 Mpix s^{-1} throughput or 50 Hz framerate for 512×512 images. Used by the author circa the early 1990s

important collection of papers (Kriegman et al. 1998) stems from a 1998 workshop on the synergies between control and vision: how vision can be used for control and how control can be used for vision. More recent algorithmic developments and application are covered in a collection of workshop papers by Chesi and Hashimoto (2010).

Visual servoing has been applied to a diverse range of problems that normally require human hand-eye skills such as ping-pong (Andersson 1989), juggling (Rizzi and Koditschek 1991) and inverted pendulum balancing (Dickmanns and Graefe 1988a; Andersen et al. 1993), catching (Sakaguchi et al. 1993; Buttazzo et al. 1993; Bukowski et al. 1991; Skofteland and Hirzinger 1991; Skaar et al. 1987; Lin et al. 1989), and controlling a labyrinth game (Andersen et al. 1993).

Exercises

1. Position-based visual servoing
   a) Run the PBVS example. Experiment with varying parameters such as the initial camera pose, the path fraction λ and adding pixel noise to the output of the camera.
   b) Create a Simulink model for PBVS.
   c) Use a different camera model for the pose estimation (slightly different focal length or principal point) and observe the effect on final end-effector pose.
   d) Implement an EKF based PBVS system as described in Westmore and Wilson (1991).
2. Optical flow fields
   a) Plot the optical flow fields for cameras with different focal lengths.
   b) Plot the flow field for some composite camera motions such as x- and y-translation, x- and z-translation, and x-translation and z-rotation.
3. For the case of two points the image Jacobian is 4×6 and the null space has two columns. What camera motions do they correspond to?
4. Image-based visual servoing
   a) Run the IBVS example, either command line or Simulink version. Experiment with varying the gain λ. Remember that λ can be a scalar or a diagonal matrix which allows different gain settings for each degree of freedom.
   b) Implement the function to limit the maximum norm of the commanded velocity.
   c) Experiment with adding pixel noise to the output of the camera.
   d) Experiment with different initial camera poses and desired image-plane coordinates.
   e) Experiment with different number of goal points, from three up to ten. For the cases where $N > 3$ compare the performance of the pseudo-inverse with just selecting a subset of three points (first three or random three). Can you design an algorithm that chooses a subset of points which results in the stacked Jacobian with the best condition number?
   f) Create a set of desired image-plane points that form a rectangle rather than a square. There is no perspective viewpoint from which a square appears as a rectangle (why is this?). What does the IBVS system do?
   g) Create a set of desired image-plane points that cannot be reached, for example swap two adjacent world or image points. What does the IBVS system do?
   h) Use a different camera model for the image Jacobian (slightly different focal length or principal point) and observe the effect on final end-effector pose.
   i) Implement second-order IBVS using Eq. 15.12.
   j) For IBVS we generally force points to move in straight lines but this is just a convenience. Use a trajectory generator to move the points from initial to desired position with some sideways motion, perhaps a half or full cycle of a sine wave. What is the effect on camera Cartesian motion?
   k) Implement stereo IBVS. Hint: stack the point feature Jacobians for both cameras and determine the desired feature positions on each camera's image plane.

5. Derive the image Jacobian for a pan/tilt camera head.
6. When discussing motion perceptibility we used the identity $(J_p^+)^T J_p^+ = (J_p J_p^T)^{-1}$. Prove this. Hint, use the singular value decomposition $J = U \Sigma V^T$ and remember that U and V are orthogonal matrices.
7. End-point open-loop visual servo systems have not been discussed in this book. Consider a group of goal points on the robot end-effector as well as the those on the goal object, both being observed by a single camera (challenging).
   a) Create an end-point open-loop PBVS system.
   b) Use a different camera model for the pose estimation (slightly different focal length or principal point) and observe the effect on final end-effector relative pose.
   c) Create an end-point open-loop IBVS system.
   d) Use a different camera model for the image Jacobian (slightly different focal length or principal point) and observe the effect on final end-effector relative pose.
8. Run the line-based visual servo example.
9. Ellipse-based visual servo
   a) Run the ellipse-based visual servo example.
   b) Modify to servo five degrees of camera motion using just the ellipse parameters (without the point feature).
   c) For an arbitrary shape we can compute its equivalent ellipse which is expressed in terms of an inertia matrix and a centroid. Determine the ellipse parameters of Eq. 15.15 from the inertia matrix and centroid. Create an ellipse feature visual servo to move to a desired view of the arbitrary shape (challenging).
10. Implement photometric visual servoing. Perhaps use the derivative of Gaussian kernel to compute the image gradients. Investigate performance as you servo over different translations and rotation, vary the assumed depth, and vary the parameters of the derivative kernel.

16

Advanced Visual Servoing

This chapter builds on the previous one and introduces some advanced visual servo techniques and applications. Section 16.1 introduces a hybrid visual servo method that avoids some of the limitations of the IBVS and PBVS schemes described previously.

Wide-angle cameras such as fisheye lenses and catadioptric cameras have significant advantages for visual servoing. Section 16.2 shows how IBVS can be reformulated for polar rather than Cartesian image-plane coordinates. This is directly relevant to fisheye lenses but also gives improved rotational control when using a perspective camera. The unified imaging model from Sect. 11.4 allows most cameras (perspective, fisheye and panoramic) to be represented by a spherical projection model, and Sect. 16.3 shows how IBVS can reformulated for spherical cameras.

Section 16.4 presents a number of application examples. These illustrate how visual servoing can be used with different types of cameras (perspective and spherical) and different types of robots (arm-type robots, mobile ground robots and flying robots). Examples include a 6 degree of freedom robot arm manipulating a camera; a mobile robot moving to a specific pose which could be used for navigating through a doorway or docking; and a quadrotor moving to, and hovering at, a fixed pose with respect to a goal on the ground.

16.1 XY/Z-Partitioned IBVS

In the last chapter, in Sect. 15.2.4, we encountered the problem of camera retreat in an IBVS system. This phenomenon can be explained intuitively by the fact that our IBVS control law causes feature points to move in straight lines on the image plane, but for a rotating camera the points will naturally move along circular arcs. The linear IBVS controller dynamically changes the overall image scale so that motion along an arc appears as motion along a straight line. The scale change is achieved by z-axis translation.

Partitioned methods eliminate camera retreat by using IBVS to control some degrees of freedom while using a different controller for the remaining degrees of freedom. The XY/Z hybrid schemes consider the x- and y-axes as one group, and the z-axes as another group. The approach is based on a couple of insights. Firstly, and intuitively, the camera retreat problem is a z-axis phenomenon: z-axis rotation leads to unwanted z-axis translation. Secondly, from Fig. 15.7, the image-plane motion due to x- and y-axis translational and rotation motion are quite similar, whereas the optical flow due to z-axis rotation and translation are radically different.

We partition the point feature optical flow of Eq. 15.7 so that

$$\dot{\boldsymbol{p}} = \boldsymbol{J}_{xy}\boldsymbol{\nu}_{xy} + \boldsymbol{J}_z\boldsymbol{\nu}_z \tag{16.1}$$

where $\boldsymbol{\nu}_{xy} = (v_x, v_y, \omega_x, \omega_y)$, $\boldsymbol{\nu}_z = (v_z, \omega_z)$, and $\boldsymbol{J}_{xy}$ and $\boldsymbol{J}_z$ are respectively columns {1, 2, 4, 5} and {3, 6} of $\boldsymbol{J}_p$. Since $\boldsymbol{\nu}_z$ will be computed by a different controller we can write Eq. 16.1 as

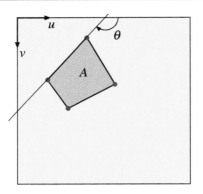

Fig. 16.1.
Image features for XY/Z partitioned IBVS control. As well as the coordinates of the four points (blue dots), we use the polygon area A and the angle of the longest line segment θ

$$\boldsymbol{\nu}_{xy} = \lambda J_{xy}^+ \left(\dot{\boldsymbol{p}}^* - J_z \nu_z \right) \tag{16.2}$$

where $\dot{\boldsymbol{p}}^*$ is the desired feature point velocity as in the traditional IBVS scheme Eq. 15.10.

The z-axis velocities v_z and ω_z are computed directly from two additional image features A and θ shown in Fig. 16.1. The first image feature $\theta \in [0, \pi)$, is the angle between the u-axis and the directed line segment joining feature points i and j. For numerical conditioning it is advantageous to select the longest line segment that can be constructed from the feature points, and allowing that this may change during the motion as the feature point configuration changes. The desired rotational rate is obtained using a simple proportional control law

$$\omega_z^* = \lambda_{\omega_z} \left(\theta^* \ominus \theta \right)$$

where the operator $\ominus$ indicates modulo-2π subtraction which is implemented by the Toolbox function `angdiff`. As always with motion on a circle there are two directions to move to achieve the goal. If the rotation is limited, for instance by a mechanical stop, then the sign of ω_z should be chosen so as to avoid motion through that stop.

The second image feature that we use is a function of the area $A \in \mathbb{R}$ of the regular polygon whose vertices are the image feature points. The advantages of this measure are: it is a scalar; it is rotation invariant[▸] thus decoupling camera rotation from z-axis translation; and it can be cheaply computed. The area of the polygon is just the zeroth-order moment, m_{00} which can be computed from the vertices using the Toolbox function `mpq_poly(p, 0, 0)`. The feature for control is the square root of area

> Rotationally invariant to rotation about the z-axis, not the x- and y-axes.

$$\sigma = \sqrt{m_{00}}$$

which has units of length, in pixels. The desired camera z-axis translation rate is obtained using a simple proportional control law

$$v_z^* = \lambda_{v_z} \left(\sigma^* - \sigma \right) \tag{16.3}$$

The features discussed above for z-axis translation and rotation control are simple and inexpensive to compute, but work best when the goal's normal is within $\pm 40°$ of the camera's optical axis. When the goal plane is not orthogonal to the optical axis its area will appear diminished, due to perspective, which causes the camera to initially approach the goal. Perspective will also change the perceived angle of the line segment which can cause small, but unnecessary, z-axis rotational motion.

The Simulink® model

```
>> sl_partitioned
```

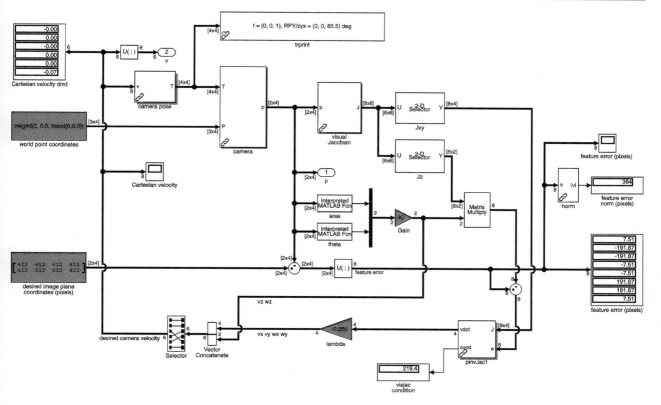

Fig. 16.2. The Simulink model `sl_partitioned` is an XY/Z-partitioned visual servo scheme, an extension of the IBVS system shown in Fig. 15.9. The initial camera pose is set in the `camera pose` block and the desired image-plane points p^* are set in the lower left red block

is shown in Fig. 16.2. The initial pose of the camera is set by a parameter of the `pose` block. The simulation is run by

```
>> sim('sl_partitioned')
```

and the camera pose, image-plane feature error and camera velocity are animated. Scope blocks also plot the camera velocity and feature error against time.

If points are moving toward the edge of the field of view the simplest way to keep them in view is to move the camera away from the scene. We define a repulsive force that acts on the camera, pushing it away as a point approaches the boundary of the image plane

$$F_z(\boldsymbol{p}) = \begin{cases} \left(\dfrac{1}{d(\boldsymbol{p})} - \dfrac{1}{d_0} \right) \dfrac{1}{d^2(\boldsymbol{p})} & \text{if } d(\boldsymbol{p}) \leq d_0 \\ 0 & \text{if } d(\boldsymbol{p}) > d_0 \end{cases}$$

where $d(\boldsymbol{p})$ is the shortest distance to the edge of the image plane from the image point coordinate $\boldsymbol{p}$, and d_0 is the width of the image zone in which the repulsive force acts. For a $W \times H$ image

$$d(\boldsymbol{p}) = \min\{u, v, W - u, H - v\} \tag{16.4}$$

Such a repulsion force could be incorporated into the z-axis translation controller

$$v_z^* = \lambda_{v_z}\left(\sigma^* - \sigma\right) - \eta\sum_{i=1}^{N} F_z\left(\boldsymbol{p}_i\right)$$

where η is a gain constant with units of damping. The repulsion force is discontinuous and may lead to chattering where the feature points oscillate in and out of the repulsive force – this can be remedied by introducing smoothing filters and velocity limiters.

16.2   IBVS Using Polar Coordinates

In Sect. 15.3 we showed image feature Jacobians for nonpoint features, but here we will show the point feature Jacobian expressed in terms of a different coordinate system. In polar coordinates the image point is written $\mathbf{p} = (r, \phi)$ where r is the distance of the point from the principal point

$$r = \sqrt{\bar{u}^2 + \bar{v}^2} \tag{16.5}$$

where we recall that $\bar{u}$ and $\bar{v}$ are the image coordinates with respect to the principal point rather than the image origin. The angle from the u-axis to a line joining the principal point to the image point is

$$\phi = \tan^{-1}\frac{\bar{v}}{\bar{u}} \tag{16.6}$$

The two coordinate representations are related by

$$\bar{u} = r\cos\phi, \ \bar{v} = r\sin\phi \tag{16.7}$$

and taking the derivatives with respect to time

$$\begin{pmatrix} \dot{\bar{u}} \\ \dot{\bar{v}} \end{pmatrix} = \begin{pmatrix} \cos\phi & -r\sin\phi \\ \sin\phi & r\cos\phi \end{pmatrix}\begin{pmatrix} \dot{r} \\ \dot{\phi} \end{pmatrix}$$

and inverting

$$\begin{pmatrix} \dot{r} \\ \dot{\phi} \end{pmatrix} = \begin{pmatrix} \cos\phi & \sin\phi \\ -\frac{1}{r}\sin\phi & \frac{1}{r}\cos\phi \end{pmatrix}\begin{pmatrix} \dot{\bar{u}} \\ \dot{\bar{v}} \end{pmatrix}$$

which we substitute into Eq. 15.6 along with Eq. 16.7 to write

$$\begin{pmatrix} \dot{r} \\ \dot{\phi} \end{pmatrix} = J_{p,p}\begin{pmatrix} v_x \\ v_y \\ v_z \\ \omega_x \\ \omega_y \\ \omega_z \end{pmatrix} \tag{16.8}$$

where the feature Jacobian is

$$J_{p,p} = \begin{pmatrix} -\frac{f}{Z}\cos\phi & -\frac{f}{Z}\sin\phi & \frac{r}{Z} & \frac{f^2+r^2}{f}\sin\phi & -\frac{f^2+r^2}{f}\cos\phi & 0 \\ \frac{f}{rZ}\sin\phi & -\frac{f}{rZ}\cos\phi & 0 & \frac{f}{r}\cos\phi & \frac{f}{r}\sin\phi & -1 \end{pmatrix} \tag{16.9}$$

This Jacobian is unusual in that it has three constant elements. In the first row the zero indicates that radius r is invariant to rotation about the z-axis. In the second row the zero indicates that polar angle is invariant to translation along the optical axis (points move along radial lines), and the negative one indicates that the angle of a feature (with respect to the u-axis) decreases with positive camera rotation. As for the Cartesian point features, the translational part of the Jacobian (the first 3 columns) are proportional to $1 / Z$. Note also that the Jacobian is undefined for $r = 0$, that is for a point on the optical axis. The interaction matrix is computed by the `visjac_p_polar` method of the `CentralCamera` class.

The desired feature velocity is a function of feature error

$$\dot{\boldsymbol{p}}^* = \lambda \begin{pmatrix} r^* - r \\ \phi^* \ominus \phi \end{pmatrix}$$

where $\ominus$ is modulo-2π subtraction for the angular component subtraction for the angular component. The choice of units (pixels and radians) means that $|r| \gg |\phi|$ and radius should be normalized

$$r = \frac{\sqrt{\bar{u}^2 + \bar{v}^2}}{\sqrt{W^2 + H^2}}$$

so that r and ϕ are of approximately the same order.

An example of IBVS using polar coordinates is implemented by the class `IBVS_polar`. We first create a canonic camera, that has normalized image coordinates

```
>> cam = CentralCamera('default')
>> T_C0 = SE3(-0.3, 0.2, -2)*SE3.Rz(pi/2);
>> vs = IBVS_polar(cam, 'T0', T_C0, 'verbose')
```

and we run run a simulation

```
>> vs.run()
```

The animation shows the feature motion in the image, and the camera and world points in a world view. The camera motion is quite different compared to the Cartesian IBVS scheme introduced in the previous chapter. For the previously problematic case of large optical-axis rotation the camera has simply moved toward the goal and rotated. The features have followed straight line paths on the $r\phi$-plane. The performance of polar IBVS is the complement of Cartesian IBVS – it generates good camera motion for the case of large rotation, but poorer motion for the case of large translation.

The methods `plot_error`, `plot_vel` and `plot_camera` can be used to show data recorded during the simulation. An additional method

Fig. 16.3. IBVS using polar coordinates. **a** Feature motion in polar ϕr-space; **b** camera motion in Cartesian space

```
>> vs.plot_features()
```

displays the path of the features in ϕr-space and this is shown in Fig. 16.3 along with the camera motion which shows no sign of camera retreat.

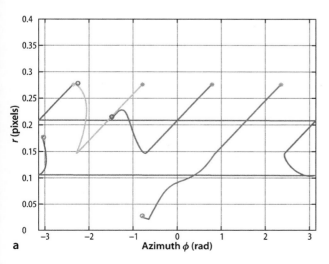

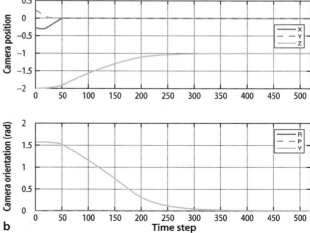

a   b

16.3   IBVS for a Spherical Camera

In Sect. 11.3 we looked at nonperspective cameras such as the fisheye lens camera and the catadioptric camera. Given the particular projection equations for any camera we can derive an image feature Jacobian from first principles. However the many different lens and mirror shapes leads to many different projection models and image Jacobians. In Sect. 11.4 we showed that feature points from any type of camera can be projected to a sphere, so we need to derive an image Jacobian for visual servo control on the sphere.

The image Jacobian for the sphere is derived in a manner similar to the perspective camera in Sect. 15.2.1. Referring to Fig. 11.21, the world point $\mathbf{P}$ is represented by the vector $\mathbf{P} = (X, Y, Z)$ in the camera frame, and is projected onto the surface of the sphere at the point $\mathbf{p} = (x, y, z)$ by a ray passing through the center of the sphere

$$x = \frac{X}{R}, \ y = \frac{Y}{R}, \ \text{and} \ z = \frac{Z}{R} \tag{16.10}$$

where $R = \sqrt{(X^2 + Y^2 + Z^2)}$ is the distance from the camera origin to the world point.

The spherical surface constraint $x^2 + y^2 + z^2 = 1$ means that one of the Cartesian coordinates is redundant so we will use a minimal spherical coordinate system comprising the angle of colatitude▶

$$\theta = \sin^{-1} r, \ \theta \in [0, \pi] \tag{16.11}$$

Colatitude is zero at the north pole and increases as we move southwards.

where $r = \sqrt{x^2 + y^2}$, and the azimuth angle (or longitude)

$$\phi = \tan^{-1} \frac{y}{x}, \ \phi \in [-\pi, \pi) \tag{16.12}$$

which yields the point feature vector $\mathbf{p} = (\theta, \phi)$.

Taking the derivatives of Eq. 16.11 and Eq. 16.12 with respect to time and substituting Eq. 15.2 as well as

$$X = R \sin\theta \cos\phi, \ Y = R \sin\theta \sin\phi, \ Z = R \cos\theta \tag{16.13}$$

we obtain, in matrix form, the spherical optical flow equation

$$\begin{pmatrix} \dot{\theta} \\ \dot{\phi} \end{pmatrix} = J_{p,s}(\theta, \phi, R) \begin{pmatrix} v_x \\ v_y \\ v_z \\ \omega_x \\ \omega_y \\ \omega_z \end{pmatrix} \tag{16.14}$$

where the image feature Jacobian is

$$J_{p,s} = \begin{pmatrix} -\dfrac{\cos\phi\cos\theta}{R} & -\dfrac{\sin\phi\cos\theta}{R} & \dfrac{\sin\theta}{R} & \sin\phi & -\cos\phi & 0 \\[2mm] \dfrac{\sin\phi}{R\sin\theta} & -\dfrac{\cos\phi}{R\sin\theta} & 0 & \dfrac{\cos\phi\cos\theta}{\sin\theta} & \dfrac{\sin\phi\cos\theta}{\sin\theta} & -1 \end{pmatrix} \tag{16.15}$$

There are similarities to the Jacobian derived for polar coordinates in the previous section. Firstly, the constant elements fall at the same place, indicating that colatitude is invariant to rotation about the optical axis, and that azimuth angle is invariant to translation along the optical axis but equal and opposite to camera rotation about the optical axis. As for all image Jacobians the translational submatrix (the first three columns) is a function of point depth $1/R$.

The Jacobian is not defined at the north and south poles where $\sin\theta = 0$ and azimuth also has no meaning at these points. This is a singularity, and as we remarked in Sect. 2.2.1.3, in the context of Euler angle representation of orientation, this is a consequence of using a minimal representation. However, in general the benefits outweigh the costs for this application.

For control purposes we follow the normal procedure of computing one 2×6 Jacobian, Eq. 16.15, for each of N feature points and stacking them to form a $2N \times 6$ matrix

$$\begin{pmatrix} \dot{\theta}_1 \\ \dot{\phi}_1 \\ \vdots \\ \dot{\theta}_N \\ \dot{\phi}_N \end{pmatrix} = \begin{pmatrix} J_1 \\ \vdots \\ J_N \end{pmatrix} \nu \qquad (16.16)$$

The control law is

$$\nu = J^+ \dot{p}^* \qquad (16.17)$$

where $\dot{p}^*$ is the desired velocity of the features in $\phi\theta$-space. Typically we choose this to be proportional to feature error

$$\dot{p}^* = \lambda\left(p^* \ominus p\right) \qquad (16.18)$$

Note that motion on this plane is in general not a great circle on the sphere – only motion along lines of longitude and the equator are great circles.

where λ is a positive gain, p is the current point in $\phi\theta$-coordinates, and p^* the desired value. This results in locally linear motion of features within the feature space. ◄ $\ominus$ denotes modulo subtraction and returns the smallest angular distance given that $\theta \in [0, \pi]$ and $\phi = [-\pi, \pi)$.

An example of IBVS using spherical coordinates (Fig. 16.4) is implemented by the class `IBVS_sph`. We first create a spherical camera

```
>> cam = SphericalCamera()
```

Fig. 16.4. IBVS using spherical camera and coordinates. **a** Feature motion in $\theta - \phi$ space; **b** four goal points projected onto the sphere in its initial pose

and then a spherical IBVS object

```
>> T_C0 = SE3(0.3, 0.3, -2)*SE3.Rz(0.4);
>> vs = IBVS_sph(cam, 'T0', T_C0, 'verbose')
```

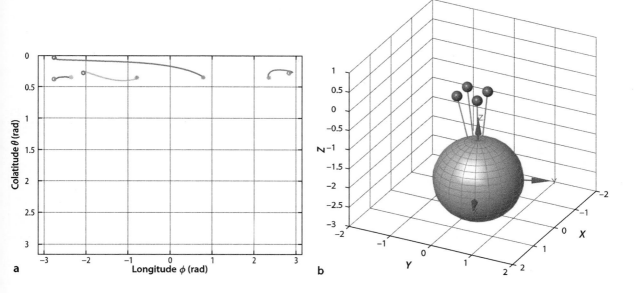

a

b

and we run run a simulation

```
>> vs.run()
```

The animation shows the feature motion on the $\phi\theta$-plane and the camera and world points in a world view. Spherical imaging has many advantages for visual servoing. Firstly, a spherical camera eliminates the need to explicitly keep features in the field of view which is a problem with both position-based visual servoing and some hybrid schemes. Secondly, we previously observed an ambiguity between the optical flow fields for R_x and $-T_y$ motion (and R_y and T_x motion) for a small field of view. For IBVS with a long focal length this can lead to slow convergence and/or sensitivity to noise in feature coordinates. For a spherical camera, with the largest possible field of view, this ambiguity is reduced.▸

Spherical cameras do not yet exist◂ but we can can project features from one or more cameras of any type onto the spherical image plane, and compute the control law in terms of spherical coordinates.

Provided that the world points are well distributed around the sphere.

The camera of Fig. 11.27b (page 349) comes close with 90% of a spherical field of view.

16.4   Applications

16.4.1   Arm-Type Robot

In this example the camera is carried by a 6-axis robot which can control all six degrees of camera motion. We will assume that the robot's joints are ideal velocity sources, that is, they move at precisely the velocity that was commanded. A modern robot is very close to this ideal, typically having high performance joint controllers using velocity and position feedback from encoders on the joints.

The *nested* control structure for a robot joint was discussed in Sect. 9.1.7. The inner velocity loop uses joint velocity feedback to ensure that the joint moves at the desired speed. The outer position loop uses joint position feedback to determine the joint speed required to follow the trajectory. In this visual servo system the position loop function is provided by the vision system. Vision sensors have a low sample rate compared to an encoder, typically 25 or 30 Hz, and often with a high latency of one or two sample times.

The Simulink model of this eye-in-hand system

```
>> sl_arm_ibvs
```

is shown in Fig. 16.5. This is a complex example that simulates not only the camera and IBVS control but also the robot, in this case the ubiquitous Puma 560 from Part III of

Fig. 16.5. The Simulink model `sl_arm_ibvs` uses IBVS to drive a Puma robot arm that is holding a camera

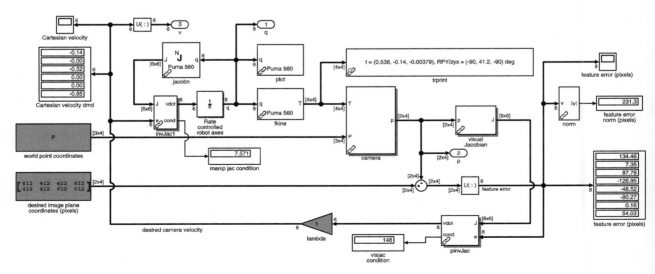

this book. The joint angles are the outputs of an integrator which represents the robot's velocity loops. These angles are input to a forward kinematics block which outputs the end-effector pose. A perspective camera with default parameters is mounted on the robot's end-effector and its axes are aligned with the end-effector coordinate frame. The camera block has one parameter which is a `CentralCamera` object, and its inputs are the camera pose world and the coordinates of the goal points which are the corners of a square in the *yz*-plane. The output image features are used to compute a Jacobian with an assumed *Z* value for every point, and also to determine the feature error in image space. The image Jacobian is inverted and a gain applied to determine the spatial velocity of the camera. The inverse manipulator Jacobian maps this to joint rates which are integrated to determine joint angles. This closed loop system drives the robot to the desired pose with respect to a square goal object.

We run this model

```
>> r = sim('sl_arm_ibvs')
```

which displays the robot moving and the image plane of a virtual camera. This model, and the others in this chapter, use the `InitFcn` callback to create variables required by the Simulation in the MATLAB® workspace◄.

The signals at the various output blocks are stored in the simulation results object `r` and the joint angles at each time step, output port one, are

```
>> q = squeeze(r.find('yout').signals(1).values)';
>> about(q)
q [double] : 60x6 (2880 bytes)
```

with one row per time step. Note that this model does not include any dynamics of the robot arm or the vision system. The joints are modeled as perfect velocity control devices, and the vision system is modeled as having no delay. This model could form the basis of more realistic system models that incorporate these real-world effects.

16.4.2 Mobile Robot

In this section we consider a camera mounted on a mobile robot moving in a planar environment. We will first consider a holonomic robot, that is one that has an omnidirectional base and can move in any direction, and then extend the solution to a non-holonomic car-like base which touches on some of the issues discussed in Chap. 4. The camera observes two or more point landmarks that have known 3-dimensional coordinates, that is, they can be placed above the plane on which the robot operates. The visual servo controller will drive the robot until its view of the landmarks matches the desired view.

16.4.2.1 Holonomic Mobile Robot

For this problem we assume a central perspective camera fixed to the robot and a number of landmarks with known locations that are continuously visible to the camera as the robot moves along the path. The vehicle's coordinate frame is such that the *x*-axis is forward and the *z*-axis is upward.

We define a perspective camera

```
>> cam = CentralCamera('default', 'focal', 0.002);
```

with a wide field of view so that it can keep the landmarks in view as it moves. The camera is mounted on the vehicle with a relative pose ${}^{B}\xi_{C}$

```
>> V_T_C = SE3(0.2, 0.1, 0.3)*SE3.Rx(-pi/4);
```

Simulink menu File+Model Properties +Callbacks+PreLoadFcn. These commands are executed once when a model is loaded.

relative to the vehicle coordinate frame. This is to the front left of the vehicle, 30 cm above ground level, with its optical axis forward but pitched upward at 45°, and its x-axis pointing to the right of the vehicle. The two landmarks are 2 m above the ground and situated at $x = 0$ and $y = \pm 1$ m

```
>> P = [0 0; 1 -1; 2 2]
```

The desired vehicle position is with the center of the rear axle at $(-2, 0, 0)$.

Since the robot operates in the xy-plane and can rotate only about the z-axis we can remove the columns from Eq. 15.6 that correspond to nonpermissible motion and write

$$\begin{pmatrix} \dot{u} \\ \dot{v} \end{pmatrix} = \begin{pmatrix} -\frac{f}{\rho_u Z} & 0 & \bar{v} \\ 0 & -\frac{f}{\rho_v Z} & -\bar{u} \end{pmatrix} \begin{pmatrix} v_x \\ v_y \\ \omega_z \end{pmatrix} \tag{16.19}$$

As for standard IBVS case we stack these Jacobians, one per landmark, and then invert the equation to solve for the vehicle velocity. Since there are only three unknown components of velocity, and each landmark contributes two equations, we need two or more feature points in order to solve for velocity.

The Simulink model

```
>> sl_omni_vs
```

is shown in Fig. 16.6 and is similar in principle to earlier models such as Fig. 16.5 and 15.9. The model is simulated by

```
>> r = sim('sl_omni_vs')
```

and displays an animation of the vehicle's path in the xy-plane and the camera view. Results are stored in the simulation results object r and can be displayed as for previous examples. The parameters and camera are defined in the properties of the model's various blocks.

Fig. 16.6. The Simulink model sl_mobile_vs drives a holonomic mobile robot to a pose using IBVS control

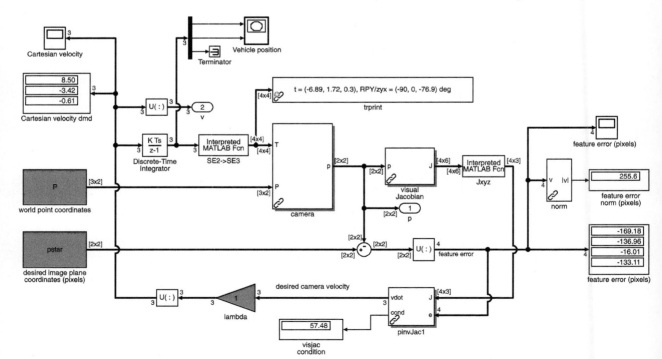

16.4.2.2 Nonholonomic Mobile Robot

The difficulties of servoing a nonholonomic mobile robot to a pose were discussed earlier and a nonlinear pose controller was introduced in Sect. 4.1.1.4. The notation for our problem is shown in Fig. 16.7 and once again we use a controller based on the polar coordinates ρ, α and β. For this control example we will use PBVS techniques to estimate the variables needed for control. We assume a central perspective camera that is fixed to the robot body frame with a relative pose ${}^B\xi_C$, a number of landmarks with known locations that are continuously visible to the camera as it moves along the path, and that the vehicle's orientation θ is also known, perhaps using a compass or some other sensor.

The Simulink model

```
>> sl_drivepose_vs
```

is shown in Fig. 16.8. The initial pose of the camera is set by a parameter of the Bicycle block. The view of the landmarks is simulated by the camera block and its output, the projected points, are input to a pose estimation block and the known locations of the landmarks are set as parameters. As discussed in Sect. 11.2.3 at least three landmarks are needed and in this example four landmarks are used. The output ${}^C\hat{\xi}_L$ is the estimated pose of the landmarks with respect to the camera. The vehicle pose in the world frame is obtained by a chain of simple transform operations $\hat{\xi}_B = \ominus {}^C\hat{\xi}_0 \ominus {}^B\xi_C$. The x- and y-components of this transform are combined with estimated heading angle◄ to yield an estimate of the vehicle's configuration $(\hat{x}, \hat{y}, \hat{\theta})$ which is input to the pose controller. The remainder of the system is essentially the same as the example from Fig. 4.11.

The simulation is run by

```
>> r = sim('sl_ibvs')
```

and the camera pose, image-plane feature error and camera velocity are animated. Scope blocks also plot the camera velocity and feature error against time. Results are stored in the simulation results object r and can be displayed as for previous examples.

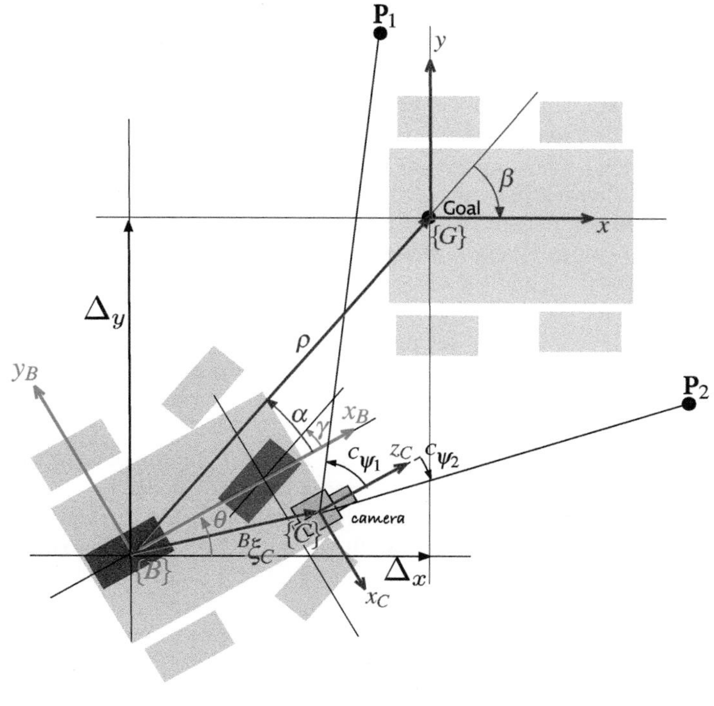

In a real system heading angle would come from a compass, in this simulation we "cheat" and simply use the true heading angle.

Fig. 16.7.
PBVS for nonholonomic vehicle (bicycle model) vehicle moving toward a goal pose: ρ is the distance to the goal, β is the angle of the goal vector with respect to the world frame, and α is the angle of the goal vector with respect to the vehicle frame. $\mathbf{P}_1$ and $\mathbf{P}_2$ are landmarks which are at bearing angles of ψ_1 and ψ_2 with respect to the camera

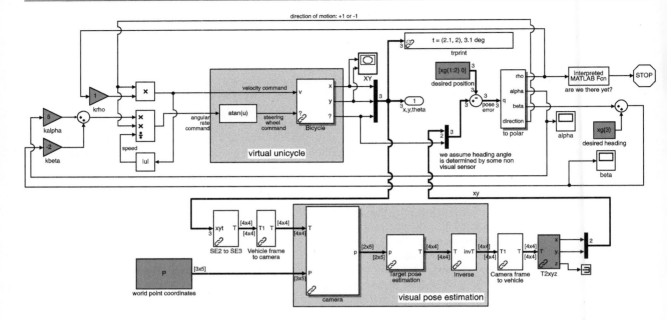

Fig. 16.8. The Simulink model `sl_drivepose_vs` drives a non-holonomic mobile robot to a pose (derived from Fig. 4.11)

16.4.3   Aerial Robot

A spherical camera is particularly suitable for platforms that move in **SE**(3) such as aerial and underwater robots. In this example we consider a spherical camera attached to a quadrotor and we will use IBVS to servo the quadrotor to a particular pose with respect to four goal points on the ground.

As we discussed in Sect. 4.3 the quadrotor is under-actuated and we cannot independently control all 6 degrees of freedom in task space. We can control position (X, Y, Z) and also yaw angle. Roll and pitch angle are manipulated to achieve translation in the horizontal plane and must be zero when the vehicle is in equilibrium. The Simulink model

```
>> sl_quadrotor_vs
```

is shown in Fig. 16.9. This controller attempts to keep the quadrotor at a constant relative pose with respect to the goal points on the ground. If the goal moves so too will the quadrotor – we could imagine a scheme like this being used to land a quadrotor on a car.

The model is a hybrid of the quadrotor controller from Fig. 4.21 and the under-actuated IBVS system of Fig. 16.6. There are however a number of key differences. Firstly, in the quadrotor control of Fig. 4.21 we used a rotation matrix to map xy-error in the world frame to the pitch and roll demand of the vehicle. This is not needed for the visual servo case since the xy-error is given in the camera, or vehicle, frame rather than the world frame. Secondly, like the mobile robot case the vehicle is under-actuated, and here the Jacobian comprises only the columns corresponding to $(v_x, v_y, v_z, \omega_z)$. Thirdly, we are using a spherical camera, so a `SphericalCamera` object is passed to the camera and visual Jacobian blocks.

Fourthly, there is coupling between the roll and pitch motion of the quadrotor and the image-plane feature coordinates. We recall how the quadrotor cannot translate without first tilting into the direction it wishes to translate, and this will cause the features to move in the image and increase the image feature error. For small amounts of roll and pitch this can be ignored but for aggressive maneuvers it must be taken into account. We can use the image Jacobian to approximate▶ the displacements in θ and ϕ as a function of displacements in camera roll and pitch angle which are rotations about the x- and y-axes respectively

This is a first-order approximation to the feature motion.

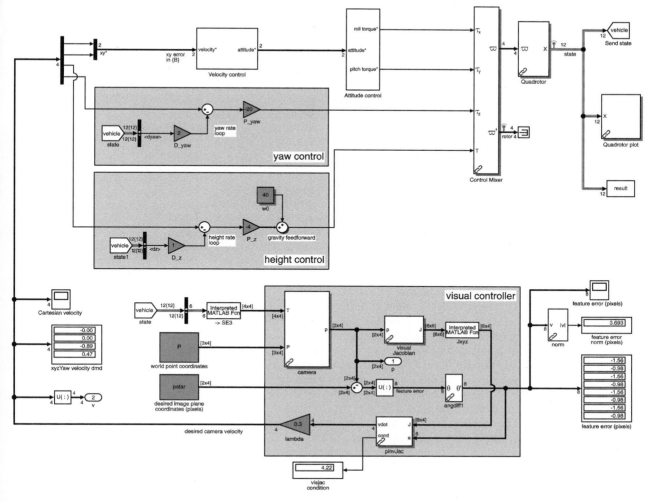

Fig. 16.9. The Simulink model `sl_quadrotor_vs`. IBVS with a spherical camera for hovering over a goal. Compared to the previous models this one has an `angdiff` block after the feature error summing junction to allow for proper handling of angles on the sphere

$$\begin{pmatrix} \Delta\theta \\ \Delta\phi \end{pmatrix} = \begin{pmatrix} \dfrac{\cos\phi\cos\theta}{\sin\theta} & \dfrac{\sin\phi\cos\theta}{\sin\theta} \\ \sin\phi & -\cos\phi \end{pmatrix}\begin{pmatrix} \Delta\theta_R \\ \Delta\theta_P \end{pmatrix}$$

and these are subtracted from the features observed by the camera to give the features that would be observed by a camera in the vehicle's body frame {B}. This scheme is sometimes referred to as feature derotation since it mimics in software the effect of a nonrotating or gimbal-stabilized camera.

Comparing Fig. 16.9 to Fig. 4.21 we see the visual controller performs the function of the outermost *position loops* for x- and y-position, altitude and yaw and generates the required velocities for the velocity loops of these degrees of freedom directly. Note that rate information is still required as input to the velocity loops and in a real robot this would be derived from an inertial measurement unit.

The simulation is run by

```
>> sim('sl_quadrotor_vs')
```

Simulink menu File+Model Properties +Callbacks+InitFcn. These commands are always executed prior to the beginning of a simulation.

and the camera pose, image-plane feature error and camera velocity are animated. Scope blocks also plot the camera velocity and feature error against time. The simulation results can be obtained from the simulation output object `out`. The initial pose of the camera is set in the model's properties◄.

16.5   Wrapping Up

Further Reading

A good introduction to advanced visual servo techniques is the tutorial article by Chaumette and Hutchinson (2007) and also the visual servoing chapter in Siciliano and Khatib (2016, § 34). Much of the interest in so-called hybrid techniques was sparked by Chaumette's paper (Chaumette 1998) which introduced the specific example that drives the camera of a point-based IBVS system to infinity for the case of goal rotation by π about the optical axis. One of the first methods to address this problem was 2.5D visual servoing, proposed by Malis et al. (1999), which augments the image-based point features with a minimal Cartesian feature. Other notable early hybrid methods were proposed by Morel et al. (2000) and Deguchi (1998) which partitioned the image Jacobian into a translational and rotational part. An homography is computed between the initial and final view (so the goal points must be planar) and then decomposed to determine a rotation and translation. Morel et al. combine this rotational information with translational control based on IBVS of the point features. Conversely, Deguchi et al. combine this translational information with rotational control based on IBVS. Since translation is only determined up to an unknown scale factor some additional means of determining scale is required.

Corke and Hutchinson (2001) presented an intuitive geometric explanation for the problem of the camera moving away from the goal during servoing, and proposed a partitioning scheme split by axes: x- and y-translation and rotation in one group, and z-translation and rotation in the other. Another approach to hybrid visual servoing is to switch rapidly between IBVS and PBVS approaches (Gans et al. 2003). The performance of several partitioned schemes is compared by Gans et al. (2003).

The polar form of the image Jacobian for point features (Iwatsuki and Okiyama 2002a; Chaumette and Hutchinson 2007) handles the IBVS failure case nicely, but results in somewhat suboptimal camera translational motion (Corke et al. 2009) – the converse of what happens for the Euclidean formulation.

The Jacobian for a spherical camera is similar to the polar form. The two angle parameterization was first described in Corke (2010) and was used for control and structure-from-motion estimation. There has been relatively little work on spherical visual servoing. Fomena and Chaumette (2007) consider the case for a single spherical object from which they extract features derived from the projection to the spherical imaging plane such as the center of the circle and its apparent radius. Tahri et al. (2009) consider spherical image features such as lines and moments. Hamel and Mahony (2002) describe kino-dynamic control of an under-actuated aerial robot using point features.

The robot manipulator dynamics Eq. 9.8 and the perspective projection Eq. 11.2 are highly nonlinear and a function of the state of the manipulator and the goal. Almost all visual servo systems consider that the robot is velocity controlled, and that the underlying dynamics are suppressed and linearized by tight control loops. As we learned in Sect. 9.1 this is the case for arm-type robots and in the quadrotor example we used a similar nested control structure. This approach is necessitated by the short time constants of the underlying mechanism and the slow sample rate and latency of any visual control loop. Modern computers and high-speed cameras make it theoretically possible to do away with axis-level velocity loops but it is far simpler to use them.

Visual servoing of nonholonomic robots is nontrivial since Brockett's theorem (1983) shows that no linear time-invariant controller can control it. The approach used in this chapter was position based which is a minor extension of the pose controller introduced in Sect. 4.1.1.4. IBVS approaches have been proposed (Tsakiris et al. 1998; Masutani et al. 1994) but require that the camera is attached to the base by a robot with a small number of degrees of freedom. Mariottini et al. (1994, 2007) describe a two-step servoing approach where the camera is rigidly attached to the base and the epipoles of the geometry defined by the current and desired camera views are explicitly servoed. Usher

(Usher et al. 2003; Usher 2005) describes a switching control law that takes the robot onto a line that passes through the desired pose, and then along the line to the pose – experimental results on an outdoor vehicle are presented. The similarity between mobile robot navigation and visual servoing problem is discussed in Corke (2001).

Resources

The controllers demonstrated in this chapter have all worked with simulated robotic systems, and have executed much slower than real time. In order to put visual control into practice we need to have fast image processing and feature extraction algorithms, as well as means of communicating with the robot hardware. Fortunately there are lots of tools and technologies to help with this: the Robot Operating System (aka, ROS www.ros.org) is a comprehensive robot software framework for creating robots, OpenCV for image processing (www.opencv.org), ViSP for creating visual trackers and controllers (www.irisa.fr/lagadic/visp). Simulink supports real-time vision through the Computer Vision System Toolbox, and the automatic synthesis of controllers that can run on your computer, can be exported to real-time hardware, or be exported as source code of a complete ROS node.

Exercises

1. XY/Z-partitioned IBVS (page 567)
   a) Investigate the generated motion for different combinations of initial camera translation and rotation, and compare to the classical IBVS scheme of the last chapter.
   b) Create a scenario where the features leave the image.
   c) Add a repulsion field to ensure that the features remain within the image.
   d) Investigate variations of Eq. 16.3. Instead of driving the difference of area to zero, try driving the ratio of current and desired area to one, or the logarithm of this ratio to zero.
2. Investigate the performance of polar and spherical IBVS for different combinations of initial camera translation and rotation, and compare to the classical IBVS scheme of the last chapter.
3. Arm-robot IBVS example (page 572)
   a) Add an offset (rotation and/or translation) between the end-effector and the camera. Your controller will need to incorporate an additional Jacobian (see Sect. 3.1.2) to account for this.
   b) Add a discrete time sampler and delay after the camera block to model the camera's frame rate and image processing time. Investigate the response as the delay is increased, and the tradeoff between gain and delay. You might like to plot a discrete-time root locus diagram for this dynamic system.
   c) Model a moving goal. Hint use the `Camera2` block from the `roblocks` library. Show the tracking error, that is, the distance between the camera and the goal.
   d) Investigate feedforward techniques to improve the control (Corke 1996b). Hint, instead of basing the control on where the goal was seen by the camera, base it on where it will be some short time into the future. How far into the future? What is a good model for this estimation? Check out the Toolbox class `AlphaBeta` for a simple to use tracking filter (challenging).
   e) An eye-in-hand camera for a docking task might have problems as the camera gets really close to the goal. How might you configure the goal points and camera to avoid this?
4. Mobile robot visual servo (page 574)
   a) For the holonomic and nonholonomic cases replace the perspective camera with a catadioptric camera.

b) For the holonomic case with a catadioptric camera, move the robot through a series of via points, each defined in terms of a set of desired feature coordinates.

c) For the nonholonomic case implement the pure pursuit and line following controllers from Chap. 4 but in this case using visual features. For pure pursuit consider the object being pursued carries one or two point features. For the line following case consider using one or two line features.

5. Display the feature flow fields, like Fig. 15.7, for the polar $r - \phi$ and spherical $\theta - \phi$ projections (Sect. 16.2 and 16.3). For the spherical case can you plot the flow vectors on the surface of a sphere?

6. Quadrotor

a) Replace the spherical camera with a perspective camera.

b) Create a controller to follow a series of point features rather than hover over a single point (challenging).

c) Create a controller to follow a series of point features rather than hover over a single point (challenging).

d) Add image feature derotation to minimize the effect of vehicle roll and pitch on the visual control.

7. Implement the 2.5D visual servo scheme by Malis (1999) (challenging).

Appendices

Installing the Toolboxes

The Toolboxes are freely available from the book's home page

http://www.petercorke.com/RVC

which also has a lot of additional information related to the book such as web links (all those printed in the book and more), code, figures, exercises and errata.

Downloading and Installing

Two toolboxes support this book: the Robotics Toolbox (RTB) and the Machine Vision Toolbox (MVTB). For the second edition of this book the relevant versions are RTB v10 and MVTB v4.

Toolboxes can be installed from .zip or .mltbx format files, with details below. Once the toolboxes are downloaded you can explore their capability using

```
>> rtbdemo
```

or

```
>> mvtbdemo
```

From .mltbx File

Since MATLAB® R2014b toolboxes can be packaged as, and installed from, files with the extension .mltbx. Download the most recent version of `robot.mltbx` or `vision.mltbx` to your computer. Using MATLAB navigate to the folder where you downloaded the file and double-click it (or right-click then select Install). The Toolbox will be installed within the local MATLAB file structure, and the paths will be appropriately configured for this, and future MATLAB sessions.

From .zip File

Download the most recent version of `robot.zip` or `vision.zip` to your computer. Use your favorite unarchiving tool to unzip the files that you downloaded.

To add the Toolboxes to your MATLAB path execute the command

```
>> addpath RVCDIR ;
>> startup_rvc
```

where `RVCDIR` is the full pathname of the directory where the folder `rvctools` was created when you unzipped the Toolbox files. The script `startup_rvc` adds various subfolders to your path and displays the version of the Toolboxes.

You will need to run the `startup_rvc` script each time you start MATLAB. Alternatively you can run `pathtool` and save the path configuration created by `startup_rvc`.

For installation from zip files, the files for both Toolboxes reside in a top-level directory called `rvctools` and beneath this are a number of subdirectories:

robot   The Robotics Toolbox.
vision   The Machine Vision Toolbox.
common   Utility functions common to the Robotics and Machine Vision
               Toolboxes.
simulink   Simulink® blocks for robotics and vision, as well as examples.
contrib   Code written by third-parties.

MEX-Files

Some functions in the Toolbox are implemented as MEX-files, that is, they are written in C for computational efficiency but are callable from MATLAB just like any other function. Source code is provided in the `mex` folder along with instructions and scripts to build the MEX-files from inside MATLAB or from the command line. You will require a C-compiler in order to build these files, but prebuilt MEX-files for a limited number of architectures are included.

Contributed Code

A number of useful functions are provided by third-parties and wrappers have been written to make them consistent with other Toolbox functions. If you attempt to access a contributed function that is not installed you will receive an error message.

The contributed code `contrib.zip` can be downloaded, expanded and then added your MATLAB path. If you installed the Toolboxes from .zip files then expand `contrib.zip` inside the folder RVCDIR.

Many of these contributed functions are part of active software projects and the downloadable file is a snapshot that has been tested and works as described in this book.

Getting Help

A Google group at http://tiny.cc/rvcforum provides answers to frequently asked questions, and has a user forum for discussing questions, issues and bugs.

License

All the non-third-party code is released under the LGPL license. This means you are free to distribute it in original or modified form provided that you keep the license and authorship information intact.

The third-party code modules are provided under various open-source licenses. The Toolbox compatibility wrappers for these modules are provided under compatible licenses.

MATLAB Versions

The Toolbox software for this book has been developed and tested using MATLAB R2015b and R2016a under Mac OS X (10.11 El Capitan). MATLAB continuously evolves so older versions of MATLAB are increasingly unlikely to work. Please do not report bugs if you are using a MATLAB version older than R2014a.

Octave

GNU Octave (www.octave.org) is an impressive piece of free software that implements a language that is close to, but not the same as, MATLAB. The Toolboxes will not work well with Octave, though with Octave 4 the incompatibilities are greatly reduced. An old version of the arm-robot functions described in Chap. 7–9 have been ported to Octave and this code is distributed in `RVCDIR/robot/octave`.

B

Linear Algebra Refresher

B.1 Vectors

A rank 1 tensor.

We will only consider real vectors◄ which are an ordered *n-tuple* of real numbers $v_1, v_2, \cdots v_n$ which is usually written as

$$v = \begin{pmatrix} v_1 \\ v_2 \\ \vdots \\ v_n \end{pmatrix} \quad \text{or} \quad v = (v_1, v_2, \cdots v_n)$$

which are a colum- and row-vector respectively. These are equivalent to an $n \times 1$ and a $1 \times n$ matrix respectively, and can be multiplied with a conforming matrix.

The numbers v_1, v_2 etc. are called the scalar components of v, and v_i is called the ith component of v. For a 3-vector we often write the elements as $v = (v_x, v_y, v_z)$.

The symbol $\mathbb{R}^n$ represents the set of ordered n-tuples of real numbers, each vector is a point in this space, that is $v \in \mathbb{R}^n$. The elements of $\mathbb{R}^2$ can be represented in a plane by a point or a directed line segment. The elements of $\mathbb{R}^3$ can be represented in a volume by a point or a directed line segment.

A vector space is an n-dimensional space whose elements are vectors *plus* the operations of addition and scalar multiplication. The addition of any two elements $a, b \in \mathbb{R}^n$ yields $(a_1 + b_1, a_2 + b_2 \cdots a_n + b_n)$ and $sa = (sa_1, sa_2 \cdots sa_n)$. Both results are element of $\mathbb{R}^n$. The negative of a vector is obtained by negating each element of the vector $-a = (-a_1, -a_2 \cdots -a_n)$.

We can use a vector to represent a point with coordinates $(x_1, x_2, \cdots x_n)$ which is called a coordinate vector. However we need to be careful because the operations of addition and scalar multiplication, while valid for vectors are meaningless for points. We can add a vector to the coordinate vector of a point to obtain the coordinate vector of another point, and we can subtract one coordinate vector from another, and the result is the is the displacement between the points.

The magnitude or length of a vector is a nonnegative scalar given by its p-norm

$$\|v\|_p = \left(\sum_{i=1}^{n} |v_i|^p \right)^{1/p}$$

The Euclidean length of a vector is given by $\|v\|_2$ which is also referred to as the L_2 norm and is generally assumed when p is omitted, for example $\|v\|$. A unit vector is one where $\|v\|_2 = 1$ and is denoted as $\hat{v}$. The L_1 norm is sum of the absolute value of the elements and is also known as the Manhattan distance, it is the distance traveled when confined to moving along the lines in a grid. The L_∞ norm is the maximum element of the vector.

The dot product of two column vectors is a scalar

$$a \cdot b = b \cdot a = a^T b = b^T a = \sum_{i=1}^{n} a_i b_i = \|a\|_2 \|b\|_2 \cos\theta$$

where θ is the angle between the vectors. $a \cdot b = 0$ when the vectors are orthogonal. For 3-vectors the cross product

$$\boldsymbol{a} \times \boldsymbol{b} = -\boldsymbol{b} \times \boldsymbol{a} = \det\begin{pmatrix} \hat{\boldsymbol{x}} & \hat{\boldsymbol{y}} & \hat{\boldsymbol{z}} \\ a_1 & a_2 & a_3 \\ b_1 & b_1 & b_3 \end{pmatrix} = [\boldsymbol{a}]_\times \boldsymbol{b} = \|\boldsymbol{a}\|_2 \|\boldsymbol{b}\|_2 \sin\theta \hat{\boldsymbol{n}}$$

where $\hat{\boldsymbol{x}}$ is a unit-vector parallel to the x-axis etc., $[\cdot]_\times$ is a skew-symmetric matrix as described in the next section, and $\hat{\boldsymbol{n}}$ is a unit-vector normal to the plane containing $\boldsymbol{a}$ and $\boldsymbol{b}$. If the vectors are parallel $\boldsymbol{a} \times \boldsymbol{b} = 0$.

B.2   Matrices

A taxonomy of matrices is shown in Fig. B.1. In this book we are concerned only with real $m \times n$ matrices▸

$$\boldsymbol{A} = \begin{pmatrix} a_{1,1} & a_{1,2} & \cdots & a_{1,n} \\ a_{2,1} & a_{2,2} & \cdots & a_{2,n} \\ \vdots & \vdots & \ddots & \\ a_{m,1} & a_{n,2} & \cdots & a_{m,n} \end{pmatrix}, \quad \boldsymbol{A} \in \mathbb{R}^{m \times n}$$

Real matrices are a subset of all matrices. For the general case of complex matrices the term Hermitian is the analog of symmetric, and unitary the analog of orthogonal. A^H denotes the Hermitian transpose, the complex conjugate transpose of the complex matrix A. Matrices are rank 2 tensors.

with m rows and n columns. If $n = m$ the matrix is square.

The transpose is

$$\boldsymbol{B} = \boldsymbol{A}^T, \quad b_{i,j} = a_{j,i} \quad \forall i, j$$

and it can be shown that

$$(\boldsymbol{A}\boldsymbol{B})^T = \boldsymbol{B}^T \boldsymbol{A}^T, \quad (\boldsymbol{A}\boldsymbol{B}\boldsymbol{C})^T = \boldsymbol{C}^T \boldsymbol{B}^T \boldsymbol{A}^T, \quad \text{etc.}$$

Fig. B.1. Taxonomy of matrices. Classes of matrices that are always singular are shown in *red*, those that are never singular are shown in *blue*

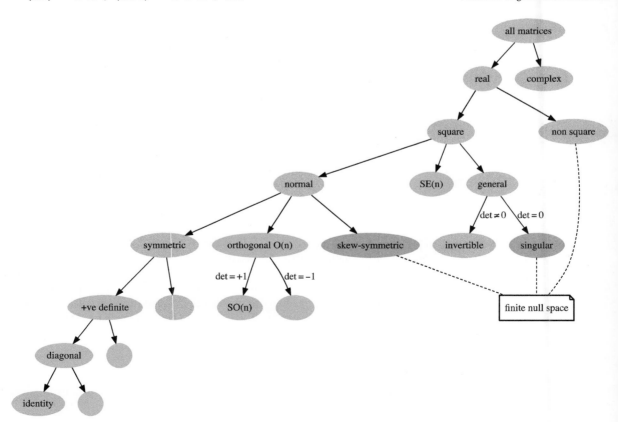

B.2.1   Square Matrices

A square matrix may have an inverse A^{-1} in which case

`Ai = inv(A)` $$AA^{-1} = A^{-1}A = I_{n \times n}$$

where

$$I_{n \times n} = \begin{pmatrix} 1 & & & \\ & 1 & & 0 \\ & & \ddots & \\ 0 & & & 1 \end{pmatrix} \in \mathbb{R}^{n \times n}$$

is the identity matrix, a unit diagonal matrix. The inverse exists provided that the matrix is nonsingular, that is, its determinant $\det(A) \neq 0$. The inverse can be computed from the matrix of cofactors. If A and B are square and nonsingular then

$$(AB)^{-1} = B^{-1}A^{-1}, \ (ABC)^{-1} = C^{-1}B^{-1}A^{-1}, \ \text{etc.}$$

and also

$$\left(A^T\right)^{-1} = \left(A^{-1}\right)^T$$

The inverse can be written as

$$A^{-1} = \frac{1}{\det(A)}\text{adj}(A)$$

where $\text{adj}(A)$ is the transpose of the matrix of cofactors and known as the adjugate or adjoint matrix and sometimes denoted by A^*. If $B = \text{adj}(A)$ then $A = \text{adj}(B)$. If A is nonsingular the adjugate can be computed by

$$\text{adj}(A) = \det(A)A^{-1}$$

For a square matrix if

$A = A^T$..............the matrix is **symmetric.** The inverse of a symmetric matrix is also symmetric. Many matrices that we encounter in robotics are symmetric, for example covariance matrices and manipulator inertia matrices.

`S = skew(v)` $A = -A^T$..........the matrix is **skew-symmetric** or **anti-symmetric.** Such a matrix has a zero diagonal, is always singular and has the property that $[av]_\times = a[v]_\times$, $[Rv]_\times = R[v]_\times R^T$ and $v^T[v]_\times = [v]_\times v = 0, \forall v$. For the 3×3 case

$$S = [v]_\times = \begin{pmatrix} 0 & -v_z & v_y \\ v_z & 0 & -v_x \\ -v_y & v_x & 0 \end{pmatrix} \tag{B.1}$$

`v = vex(S)`   and the inverse operation is

$$v = \text{vex}(S)$$

$A^{-1} = A^T$..........the matrix is **orthogonal.** The matrix is also known as orthonormal since its column vectors (and row vectors) must be of unit length and orthogonal to each other. The product of two orthogonal matrices of the same size is also an orthogonal matrix. The set of $n \times n$ orthogonal matrices forms a group $\mathbf{O}(n)$, known as the orthogonal group. The determinant of an orthogonal matrix is either $+1$ or -1. The subgroup $\mathbf{SO}(n)$ consisting of orthogonal matrices with determinant $+1$ is called the special orthogonal group. The columns (and rows) are orthogonal vectors, that is, their dot product is zero.

$A^T A = A A^T$ ....the matrix is **normal** and can be diagonalized by an orthogonal matrix U so that $U^T A U$ is a diagonal matrix. All symmetric, skew-symmetric and orthogonal matrices are normal matrices as are matrices of the form $A = B^T B = B B^T$ where B is an arbitrary matrix.

The square matrix $A \in \mathbb{R}^{n \times n}$ can be applied as a linear transformation to a vector $x \in \mathbb{R}^n$

$$x' = A x$$

which results in another vector, generally with a change in its length and direction. However there are some important special cases. If $A \in \mathbf{SO}(n)$ the transformation is isometric and the vector's *length* is unchanged $\|x'\| = \|x\|$.

In 2-dimensions if x is the set of all points lying on a circle then x' defines points that lie on an ellipse. The MATLAB® builtin demonstration

```
>> eigshow
```

shows this very clearly as you interactively drag the tip of the vector x around the unit circle.

The eigenvectors of a square matrix are those vectors x such that

`[x,e] = eig(A)`

$$A x = \lambda_i x \tag{B.2}$$

that is, their direction is unchanged when transformed by the matrix. They are simply scaled by λ_i, the corresponding eigenvalue. The matrix has n eigenvalues (the *spectrum* of the matrix) which can be real or complex. For an orthogonal matrix the eigenvalues lie on a unit circle in the complex plane, $|\lambda_i| = 1$, and the eigenvectors are all orthogonal to one another.

The eigenvalues of a real symmetric matrix are all real and we classify the matrix according to the sign of its eigenvalues

- $\lambda_i > 0, \forall i$ positive definite
- $\lambda_i \geq 0, \forall i$ positive semi-definite
- $\lambda_i < 0, \forall i$ negative definite
- otherwise   indefinite

The inverse of a positive definite matrix is also positive definite.

The matrices $A^T A$ and $A A^T$ are always symmetric and positive semidefinite. This implies than any symmetric matrix A can be written as

$$A = L L^T$$

where L is the Cholesky decomposition of A.

`L = chol(A)`

The matrix R such that

$$A = R R$$

`R = sqrtm(A)`

is the square root of A or $A^{1/2}$.

If T is any nonsingular matrix then

$$A = TBT^{-1}$$

is known as a similarity transform and A and B are said to be similar, and it can be shown that the eigenvalues are unchanged by the transformation.

If A is nonsingular then the eigenvectors of A^{-1} are the same as A and the eigenvalues of A^{-1} are the reciprocal of those of A. The eigenvalues of A^T are the same as those of A but the eigenvectors are different.

The matrix form of Eq. B.2 is

$$AX = X\Lambda$$

where $X \in \mathbb{R}^{n \times n}$ is a matrix of eigenvectors of A, arranged column-wise, and Λ is a diagonal matrix of corresponding eigenvalues. If X is not singular we can rearrange this as

$$A = X\Lambda X^{-1}$$

which is the eigenvalue or spectral decomposition of the matrix. This implies that the matrix can be diagonalized by a similarity transform

$$\Lambda = X^{-1}AX$$

If A is symmetric then X is orthogonal and we can instead write

$$A = X\Lambda X^T \tag{B.3}$$

`det(A)`   The determinant of a square matrix $A \in \mathbb{R}^{n \times n}$ is the factor by which the transformation changes changes volumes in an n-dimensional space. For 2-dimensions imagine a shape defined by points x_i with an enclosed area a. The shape formed by the points Ax_i would have an enclosed area $a\det(A)$. If A is singular the points Ax_i would lie at a single point or along a line and have zero enclosed area. In a similar way for 3-dimensions, the determinant is a scale factor applied to the volume of a set of points mapped through the transformation A.

The determinant is equal to the product of the eigenvalues

$$\det(A) = \prod_{i=1}^{n} \lambda_i$$

thus a matrix with one or more zero eigenvalues will be singular. A positive definite `trace(A)`   matrix, $\lambda_i > 0$, therefore has $\det(A) > 0$ and is not singular. The trace of a matrix is the sum of the diagonal elements

$$\mathrm{tr}(A) = \sum_{i=1}^{n} A_{ii}$$

which is also the sum of the eigenvalues

$$\mathrm{tr}(A) = \sum_{i=1}^{n} \lambda_i$$

The columns of $A = (c_1 c_2 \cdots c_n)$ can be considered as a set of vectors that define a space – the column space. Similarly, the rows of A can be considered as a set of vectors that define a space – the row space. The column rank of a matrix is the number of linearly independent columns of A. Similarly, the row rank is the number of linearly independent rows of A. The column rank and the row rank are always equal and are simply called the rank of A and the rank has an upper bound of $\min(m, n)$. The rank is the dimension of the largest nonsingular square submatrix that can be formed from A. A square matrix for which $\text{rank}(A) < n$ is said to be rank deficient or not of full rank. The rank shortfall $\min(m, n) - \text{rank}(A)$ is the nullity of A. In addition $\text{rank}(AB) \leq \min(\text{rank}(A), \text{rank}(B))$ and $\text{rank}(A + B) \leq \text{rank}(A) + \text{rank}(B)$. The matrix vv^T has rank 1 for all $v \neq 0$.

rank(A)

B.2.2   Nonsquare and Singular Matrices

For a nonsquare matrix $A \in \mathbb{R}^{m \times n}$ we can determine the left generalized inverse or pseudo inverse or Moore-Penrose pseudo inverse

$$A^+ A = I_{n \times n}$$

where $A^+ = (A^T A)^{-1} A^T$. The right generalized inverse is

$$AA^+ = I_{m \times m}$$

where $A^+ = A^T (AA^T)^{-1}$.

If the matrix A is not of full rank then it has a finite null space or kernel. A vector x lies in the null space of the matrix if

$$Ax = 0$$

More precisely this is the right-null space. A vector lies in the left-null space if

$$xA = 0$$

The left null space is equal to the right null space of A^T.

The null space is defined by a set of orthogonal basis vectors whose dimension is the nullity of A. Any linear combination of these null-space basis vectors lies in the null space.

null(A)

For a nonsquare matrix $A \in \mathbf{R}^{m \times n}$ the analog to Eq. B.2 is

$$Av_i = \sigma_i u_i$$

where $u_i \in \mathbf{R}^m$ and $v_i \in \mathbf{R}^n$ are respectively the right- and left-singular vectors of A, and σ_i its singular values. The singular values are nonnegative real numbers that are the square root of the eigenvalues of AA^T and u_i are the corresponding eigenvectors. v_i are the eigenvectors of $A^T A$.

The singular value decomposition or SVD of the matrix A is

[U,S,Vt] = svd(A)

$$A = U \Sigma V^T$$

where $U \in \mathbb{R}^{m \times m}$ and $V \in \mathbb{R}^{n \times n}$ are both orthogonal matrices comprising, as columns, the corresponding singular vectors u_i and v_i. $\Sigma \in \mathbb{R}^{m \times n}$ is a diagonal matrix of the singular values

$$\Sigma = \begin{pmatrix} \sigma_1 & & & & & \\ & \ddots & & & 0 & \\ & & \sigma_r & & & \\ & & & 0 & & \\ & 0 & & & \ddots & \\ & & & & & 0 \end{pmatrix}$$

where $r = \text{rank}(A)$ is the rank of A and $\sigma_i \geq \sigma_{i+1}$. For the case where $r < n$ the diagonal will have zero elements as shown. Columns of V^T corresponding to the zero columns

of Σ define the null space of A. The condition number of a matrix A is $\max \sigma_i / \min \sigma_i$ and a high value means the matrix is close to singular or "poorly conditioned".

The matrix quadratic form

$$s = \boldsymbol{x}^T A \boldsymbol{x} \tag{B.4}$$

is a scalar. If A is positive definite then $s = \boldsymbol{x}^T A \boldsymbol{x} > 0, \forall \boldsymbol{x} \neq 0$.

For the case that A is diagonal this can be written

$$s = \sum_{i=1}^{n} A_{ii} x_i^2$$

which is a weighted sum of squares. If A is symmetric then

$$s = \sum_{i=1}^{n} A_{ii} x_i^2 + 2 \sum_{i=1}^{n} \sum_{j=i+1}^{n} A_{ij} x_i x_j$$

the result also includes products or correlations between elements of $\boldsymbol{x}$.

The Mahalanobis distance is a weighted distance or norm

$$s = \sqrt{\boldsymbol{v}^T P^{-1} \boldsymbol{v}}$$

where $P \in \mathbb{R}^{n \times n}$ is a covariance matrix which down-weights components of $\boldsymbol{v}$ where uncertainty is high.

C

Geometry

Geometric concepts such as points, lines, ellipses and planes are critical to the fields of robotics and robotic vision. We briefly summarize key representations in both Euclidean and projective (homogeneous coordinate) space.

C.1 Euclidean Geometry

C.1.1 Points

A point in n-dimensional space is represented by an n-tuple, an ordered set of n numbers $(x_1, x_2 \cdots x_n)$ which define the coordinates of the point. The tuple can also be interpreted as a vector – a coordinate vector – from the origin to the point.

C.1.2 Lines

C.1.2.1 Lines in 2D

A line is defined by $\ell = (a, b, c)$ such that

$$ax + by + c = 0 \tag{C.1}$$

which is a generalization of the line equation we learned in school $y = mx + c$ but which can easily represent a vertical line by setting $a = 0$. $v = (a, b)$ is a vector parallel to the line, and $v = (-b, a)$ is a vector normal to the line. The line that joins two points is given by the solution to

$$\begin{pmatrix} x_1 & y_1 & 1 \\ x_2 & y_2 & 1 \end{pmatrix} \begin{pmatrix} a \\ b \\ c \end{pmatrix} = 0$$

which is found from the right-null space of the left-most term. The intersection point of two lines is

$$\begin{pmatrix} a_1 & b_1 \\ a_2 & b_2 \end{pmatrix} \begin{pmatrix} x \\ y \end{pmatrix} = - \begin{pmatrix} c_1 \\ c_2 \end{pmatrix}$$

which has no solution if the lines are parallel – the left-most term is singular.

We can also represent the line in polar form

$$\cos\theta x + \sin\theta y + \rho = 0$$

where θ is the angle from the x-axis to the line and ρ is the normal distance between the line and the origin, as shown in Fig. 13.18.

C.1.2.2   Lines in 3D and Plücker Coordinates

We can define a line by two points, **p** and **q**, as shown in Fig. C.1, which would require a total of six parameters $\ell = (q_x, q_y, q_z, p_x, p_y, p_z)$. However since these points can be arbitrarily chosen there would be an infinite set of parameters that represent the same line making it hard to determine the equivalence of two lines.

There are advantages in representing a line as

$$\ell = (\boldsymbol{\omega} \times \boldsymbol{q}, \boldsymbol{p} - \boldsymbol{q}) = (\boldsymbol{v}, \boldsymbol{\omega}) \in \mathbb{R}^6$$

where $\boldsymbol{\omega}$ is the direction of the line and $\boldsymbol{v}$ is the moment of the line – a vector from the origin to a point on the line and normal to the line. This is a Plücker coordinate vector – a six dimensional quantity subject to two constraints: the coordinates are homogeneous and thus invariant to overall scale factor; and $\boldsymbol{v} \cdot \boldsymbol{\omega} = 0$. Lines therefore have 4 degrees-of-freedom► and the Plücker coordinates lie on a 4-dimensional manifold in 6-dimensional space. Lines with $\boldsymbol{\omega} = 0$ lie at infinity and are known as ideal lines.►

This is not intuitive but consider two parallel planes and an arbitrary 3D line passing through them. The line can be described by the 2-dimensional coordinates of its intersection point on each plane – a total of four coordinates.

Ideal as in imaginery, not as in perfect.

In MATLAB® we will first define two points as column vectors

```
>> P = [2 3 4]';  Q = [3 5 7]';
```

and then create a Plücker line object

```
>> L = Plucker(P, Q)
L =
{ 1  -2  1; -1  -2  -3 }
```

which displays the v and w components. These can be accessed as properties

```
>> L.v'
ans =
     1    -2     1
>> L.w'
ans =
    -1    -2    -3
```

A Plücker line can also be represented as a skew-symmetric matrix

```
>> L.L
ans =
     0     1     2    -1
    -1     0     1    -2
    -2    -1     0    -3
     1     2     3     0
```

which can also be formed by $\tilde{\boldsymbol{p}}\tilde{\boldsymbol{q}}^T - \tilde{\boldsymbol{q}}\tilde{\boldsymbol{p}}^T$.

To plot this line we first define a region of 3D space► then plot it in blue

Since lines lines are infinite we need to specify a finite volume in which to draw it.

```
>> axis([-5 5  -5 5  -5 5]);
>> L.plot('b');
```

The line is the set of all points

$$\boldsymbol{p}(\lambda) = \frac{\boldsymbol{v} \times \boldsymbol{\omega}}{\boldsymbol{\omega} \cdot \boldsymbol{\omega}} + \lambda\boldsymbol{\omega}, \ \lambda \in \mathbb{R}$$

which can be generated parametrically in terms of a scalar parameter

```
>> L.point([0 1 2])
ans =
   -0.5714   -1.5714   -2.5714
   -0.1429   -2.1429   -4.1429
    0.2857   -2.7143   -5.7143
```

where the columns are points on the line corresponding to $\lambda = 0, 1, 2$.

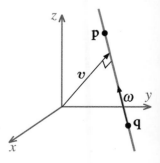

Fig. C.1. Describing a line in 3-dimensions

A point x is closest to the line when

$$\lambda = \frac{(x - q) \cdot \omega}{\omega \cdot \omega}$$

For the point $(1, 2, 3)$ the closest point on the line, and its distance, is given by

```
>> [x, d] = L.closest([1 2 3]')
x =
    3.1381
    2.5345
    1.9310
d =
    2.4495
```

The line intersects the plane $n^T x + d = 0$ at the point coordinate

$$x = \frac{v \times n - d\omega}{\omega \cdot n}$$

For the xy-plane the line intersects at

```
>> L.plane_intersect([0 0 1 0])'
ans =
    0.6667     0.3333          0
```

Two lines can be identical, coplanar or skewed. Identical lines have linearly dependent Plücker coordinates, that is, $\ell_1 = \lambda \ell_2$. If coplanar they can be parallel or intersecting and if skewed can be intersecting or not. If lines have $\omega_1 \times \omega_2 = 0$ they are parallel otherwise they are skewed.

The minimum distance between two lines is

$$d = \omega_1 \cdot v_2 + \omega_2 \cdot v_1$$

and is zero if they intersect.

The side operator is a permuted dot product

$$\text{side}\left(\ell^1, \ell^2\right) = \ell_1^1 \ell_5^2 + \ell_2^1 \ell_6^2 + \ell_3^1 \ell_4^2 + \ell_4^1 \ell_3^2 + \ell_5^1 \ell_1^2 + \ell_6^1 \ell_2^2$$

which is zero if the lines intersect or are parallel and is computed by the `side` method.

We can transform a Plücker line by the adjoint of a rigid-body motion.

$$\ell' = \text{Ad}(\xi)\ell$$

Julius Plücker (1801–1868) was a German mathematician and physicist who made contributions to the study of cathode rays and analytical geometry. He was born at Elberfeld and studied at Düsseldorf, Bonn, Heidelberg and Berlin and went to Paris in 1823 where he was influenced by the French geometry movement. In 1825 he returned to the University of Bonn, was made professor of mathematics in 1828, and professor of physics in 1836. In 1858 he proposed that the lines of the spectrum, discovered by his colleague Heinrich Geissler (of Geissler tube fame), were characteristic of the chemical substance which emitted them. In 1865, he returned to geometry and invented what was known as line geometry. He was the recipient of the Copley Medal from the Royal Society in 1866, and is buried in the Alter Friedhof (Old Cemetery) in Bonn.

C.1.3   Planes

A plane is defined by a 4-vector $\pi = (a, b, c, d)$ such that

$$ax + by + cz + d = 0$$

which can be written in point-normal form as

$$\boldsymbol{n}^T(\boldsymbol{x} - \boldsymbol{p}) = 0$$

for a plane containing a point with coordinate $\boldsymbol{p}$ and a normal $\boldsymbol{n}$, or more generally as

$$\boldsymbol{n}^T\boldsymbol{x} + d = 0$$

A plane can be defined by 3 points

$$\begin{pmatrix} x_1 & y_1 & z_1 & 1 \\ x_2 & y_2 & z_2 & 1 \\ x_3 & y_3 & z_3 & 1 \end{pmatrix} \pi = 0$$

and solved for using the right-null space of the left-most term, or by two nonparallel lines

$$\pi = \left(\boldsymbol{\omega}_1 \times \boldsymbol{\omega}_2, \boldsymbol{v}_1 \cdot \boldsymbol{\omega}_2 \right)$$

or by a line and a point with coordinate $\boldsymbol{r}$

$$\pi = \left(\boldsymbol{\omega} \times \boldsymbol{r} - \boldsymbol{v}, \boldsymbol{v} \cdot \boldsymbol{r} \right)$$

A point is defined as the intersection point of three planes

$$\begin{pmatrix} a_1 & b_1 & c_1 \\ a_2 & b_2 & c_2 \\ a_3 & b_3 & c_3 \end{pmatrix} \begin{pmatrix} x \\ y \\ z \end{pmatrix} = - \begin{pmatrix} d_1 \\ d_2 \\ d_3 \end{pmatrix}$$

The Plücker line formed by the intersection of two planes is

$$\ell = \left(\boldsymbol{n}_1 \times \boldsymbol{n}_2, d_2\boldsymbol{n}_1 - d_1\boldsymbol{n}_2 \right)$$

Fig. C.2. Ellipses. **a** Canonical ellipse centered at the origin and aligned with the x- and y-axes; **b** general form of ellipse

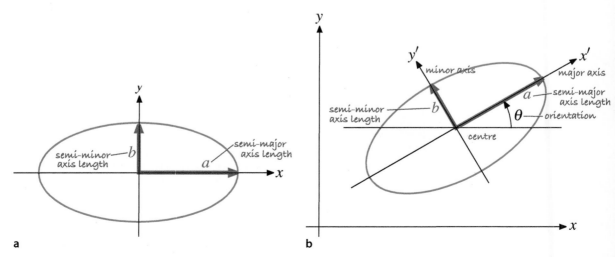

a

b

C.1.4   Ellipses and Ellipsoids

An ellipse belongs to the family of planar curves known as conics. The simplest form of an ellipse is defined implicitly

$$\frac{x^2}{a^2} + \frac{y^2}{b^2} = 1$$

and is shown in Fig. C.2a. This canonical ellipse is centered at the origin and has its major and minor axes aligned with the x- and y-axes. The radius in the x-direction is a and in the y-direction is b. The longer of the two radii is known as the semi-major axis length and the other is the semi-minor axis length.

We can write the ellipse in matrix quadratic form Eq. B.4 as

$$\begin{pmatrix} x & y \end{pmatrix} \begin{pmatrix} 1/a^2 & 0 \\ 0 & 1/b^2 \end{pmatrix} \begin{pmatrix} x \\ y \end{pmatrix} = 1$$

$$\boldsymbol{x}^T \begin{pmatrix} a^2 & 0 \\ 0 & b^2 \end{pmatrix}^{-1} \boldsymbol{x} = 1 \tag{C.2}$$

$$\boldsymbol{x}^T \boldsymbol{E}^{-1} \boldsymbol{x} = 1 \tag{C.3}$$

In the most general form E is a symmetric matrix

$$\boldsymbol{E} = \begin{pmatrix} A & C \\ C & B \end{pmatrix} \tag{C.4}$$

and its determinant $\det(E) = AB - C^2$ defines the type of conic

$$\det(\boldsymbol{E}) \begin{cases} > 0 & \text{ellipse} \\ = 0 & \text{parabola} \\ < 0 & \text{hyperbola} \end{cases}$$

An ellipse is therefore represented by a positive definite symmetric matrix E. Conversely any positive definite symmetric matrix, such as an inertia matrix or covariance matrix, can be represented by an ellipse.

Nonzero values of C change the orientation of the ellipse. The ellipse can be arbitrarily centered at $\boldsymbol{x}_c$ by writing it in the form

$$\left(\boldsymbol{x} - \boldsymbol{x}_c\right)^T \boldsymbol{E}^{-1} \left(\boldsymbol{x} - \boldsymbol{x}_c\right) = 1$$

which leads to the general ellipse shown in Fig. C.2b.

Since E is symmetric it can be diagonalized by Eq. B.3

$$\boldsymbol{E} = \boldsymbol{X} \boldsymbol{\Lambda} \boldsymbol{X}^T$$

where X is an orthogonal matrix comprising the eigenvectors of E. The inverse is

$$\boldsymbol{E}^{-1} = \boldsymbol{X} \boldsymbol{\Lambda}^{-1} \boldsymbol{X}^T$$

so the quadratic form becomes

$$\boldsymbol{x}^T \boldsymbol{X} \boldsymbol{\Lambda}^{-1} \boldsymbol{X}^T \boldsymbol{x} = 1$$

$$\left(\boldsymbol{X}^T \boldsymbol{x}\right)^T \boldsymbol{\Lambda}^{-1} \left(\boldsymbol{X}^T \boldsymbol{x}\right) = 1$$

$$\boldsymbol{x}'^T \boldsymbol{\Lambda}^{-1} \boldsymbol{x}' = 1$$

This is similar to Eq. C.3 but with the ellipse defined by the diagonal matrix Λ with respect to the rotated coordinated frame $x' = X^T x$. The major and minor ellipse axes are aligned with the eigenvectors of E. The squared radii of the ellipse are the eigenvalues of E or the diagonal elements of Λ.

For the general case of $E \in \mathbb{R}^{n \times n}$ the result is an ellipsoid in n-dimensional space. The Toolbox function `plot_ellipse` will draw an ellipse for the $n = 2$ case and an ellipsoid for the $n = 3$ case.

Alternatively the ellipse can be represented in polynomial form by writing as

$$\left(x - (x_0, y_0)\right)^T \begin{pmatrix} a & c \\ c & b \end{pmatrix} \left(x - (x_0, y_0)\right) = 1$$

and expanding to obtain

$$e_1 x^2 + e_2 y^2 + e_3 xy + e_4 x + e_5 y + e_6 = 0$$

where $e_1 = a$, $e_2 = b$, $e_3 = 2c$, $e_4 = -2(ax_0 + cy_0)$, $e_5 = -2(by_0 + cx_0)$ and $e_6 = ax_0^2 + by_0^2 + 2cx_0 y_0 - 1$. The ellipse has only five degrees of freedom, its center coordinate and the three unique elements in E. For a nondegenerate ellipse where $e_1 \neq 0$ we can rewrite the polynomial in normalized form

$$x^2 + E_1 y^2 + E_2 xy + E_3 x + E_4 y + E_5 = 0 \tag{C.5}$$

with five unique parameters.

C.1.4.1   Properties

The area of an ellipse is πab and its eccentricity is

$$\varepsilon = \frac{\sqrt{a^2 - b^2}}{a}$$

The eigenvectors of E define the principal directions of the ellipse and the square root of the eigenvalues are the corresponding radii.

Consider the ellipse

$$x \begin{pmatrix} 2 & -1 \\ -1 & 1 \end{pmatrix}^{-1} x = 1$$

which is represented in MATLAB by

```
>> E = [2 -1; -1 1];
```

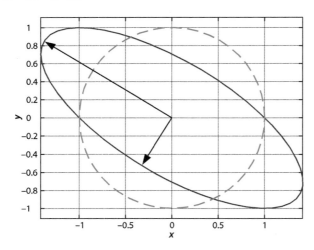

Fig. C.3.
Ellipse corresponding to a symmetric 2×2 matrix, and the unit circle shown in *red*. The arrows indicate the major and minor axes of the ellipse

We can plot this by

```
>> plot_ellipse(E)
```

which is shown in Fig. C.3.

The eigenvectors and eigenvalues of E are

```
>> [x,e] = eig(E)
x =
    -0.5257    -0.8507
    -0.8507     0.5257
e =
     0.3820          0
          0     2.6180
```

and the ellipse radii are

```
>> r = sqrt(diag(e))
r =
     0.6180
     1.6180
```

which correspond to b and a respectively. If either radius is equal to zero the ellipse is degenerate and becomes a line. If both radii are zero the ellipse is a point.

The eigenvectors are unit vectors in the minor- and major-axis directions and we will scale them by the radii to yield radius vectors which we can plot

```
>> arrow([0 0]', x(:,1)*r(1));
>> arrow([0 0]', x(:,2)*r(2));
```

The orientation of the ellipse is the angle of the major-axis with respect to the horizontal axis and is

$$\theta = \tan^{-1}\frac{x_y}{x_x}$$

For our example this is

```
>> atan2(x(2,2), x(1,2)) * 180/pi
ans =
   148.2825
```

in units of degrees.

The ellipse area is $\pi r_1 r_2$ and the ellipsoid volume is $\frac{4}{3}\pi r_1 r_2 r_3$ where the radii $r_i = \sqrt{\lambda_i}$ where λ_i are the eigenvalues of E. Since $\det(E) = \Pi\lambda_i$ the area or volume is proportional to $\sqrt{\det(E)}$.

C.1.4.2   Drawing an Ellipse

In order to draw an ellipse we first define a point coordinate $y = [x, y]^T$ on the unit circle

$$y^T y = 1$$

and rewrite Eq. C.3 as

$$x^T E^{-\frac{1}{2}} E^{-\frac{1}{2}} x = 1$$

where $E^{\frac{1}{2}}$ is the matrix square root (MATLAB function `sqrtm`). Equating these two equations we can write

$$x^T E^{-\frac{1}{2}} E^{-\frac{1}{2}} x = y^T y$$

It is clear that

$$y = E^{-\frac{1}{2}}x$$

which we can rearrange as

$$x = E^{\frac{1}{2}}y$$

which transforms a point on the unit circle to a point on an ellipse. If the ellipse is centered at x_c rather than the origin we can perform a change of coordinates

$$(x - x_c)^T E^{-\frac{1}{2}} E^{-\frac{1}{2}} (x - x_c) = 1$$

from which we write the transformation as

$$x = E^{\frac{1}{2}}y + x_c$$

Continuing the MATLAB example above

```
>> E = [2 -1; -1 1];
```

We define a set of points on the unit circle

```
>> th = linspace(0, 2*pi, 50);
>> y = [cos(th); sin(th)];
```

which we transform to points on the perimeter of the ellipse

```
>> x = (sqrtm(E) * y)';
>> plot(x(:,1), x(:,2));
```

which is encapsulated in the Toolbox function

```
>> plot_ellipse(E, [0 0])
```

An ellipsoid is described by a positive-definite symmetric 3×3 matrix. Drawing an ellipsoid is tackled in an analogous fashion and `plot_ellipse` is also able to display a 3-dimensional ellipsoid.

C.1.4.3   Fitting an Ellipse to Data

From a Set of Interior Points

We wish to find the equation of an ellipse that best fits a set of points that lie within the ellipse boundary. A common approach is to find the ellipse that has the same mass properties as the set of points. From the set of N points $x_i = (x_i, y_i)$ we can compute the moments

$$m_{00} = N$$

$$m_{10} = \sum_{i=1}^{N} x_i$$

$$m_{01} = \sum_{i=1}^{N} y_i$$

The center of the ellipse is taken to be the centroid of the set of points

$$(x_c, y_c) = \left(\frac{m_{10}}{m_{00}}, \frac{m_{01}}{m_{00}} \right)$$

which allows us to compute the central second moments

$$\mu_{20} = \sum_{i=1}^{N} (x_i - x_c)^2$$

$$\mu_{02} = \sum_{i=1}^{N} (y_i - y_c)^2$$

$$\mu_{11} = \sum_{i=1}^{N} (x_i - x_c)(y_i - y_c)$$

The inertia matrix for a general ellipse is the symmetric matrix

$$J = \begin{pmatrix} \mu_{20} & \mu_{11} \\ \mu_{11} & \mu_{02} \end{pmatrix}$$

where the diagonal terms are the moments of inertia and the off-diagonal terms are the products of inertia. Inertia can be computed more directly by

$$J = \sum_{i=1}^{N} (\boldsymbol{x} - \boldsymbol{x}_c)(\boldsymbol{x} - \boldsymbol{x}_c)^T$$

The relationship between the inertia matrix and the symmetric ellipse matrix is

$$E = \frac{4}{m_{00}} J$$

To demonstrate this we can create a set of points that lie within the ellipse used in the example above

```
1   % generate a set of points within the ellipse
2   p = [];
3   while true
4       x = (rand(2,1)-0.5)*4;
5       if norm(x'*inv(E)*x) <= 1
6           p = [p x];
7       end
8       if numcols(p) >= 500
9           break;
10      end
11  end
12  plot(p(1,:), p(2,:), '.')
13
14  % compute the moments
15  m00 = mpq_point(p, 0,0);
16  m10 = mpq_point(p, 1,0);
17  m01 = mpq_point(p, 0,1);
18  xc = m10/m00; yc = m01/m00;
19
20  % compute second moments relative to centroid
21  pp = bsxfun(@minus, p, [xc; yc]);
22
23  m20 = mpq_point(pp, 2,0);
24  m02 = mpq_point(pp, 0,2);
25  m11 = mpq_point(pp, 1,1);
26
27  % compute the moments and ellipse matrix
28  J = [m20 m11; m11 m02];
29  E_est = 4 * J / m00
```

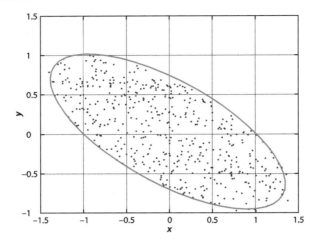

Fig. C.4.
Data points (*blue*) with a fitted ellipse (*red*).

which results in an estimate

```
>> E_est
E_est =
    1.8706   -0.9151
   -0.9151    0.9716
```

which is similar to the original value of E. The point data is shown in Fig. C.4. We can overlay the estimated ellipse on the point data

```
>> plot_ellipse(E_est, [xc yc], 'r')
```

and the result is shown in red in Fig. C.4.

From a Set of Boundary Points

We wish to find the equation of an ellipse given a set of points (x_i, y_i) that define the boundary of an ellipse. Using the polynomial form of the ellipse Eq. C.5 for each point we write this in matrix form

$$\begin{pmatrix} y_1^2 & x_1 y_1 & x_1 & y_1 & 1 \\ y_2^2 & x_2 y_2 & x_2 & y_2 & 1 \\ & & \vdots & & \\ y_N^2 & x_N y_N & x_N & y_N & 1 \end{pmatrix} \begin{pmatrix} E_1 \\ E_2 \\ E_3 \\ E_4 \\ E_5 \end{pmatrix} = \begin{pmatrix} -x_1^2 \\ -x_2^2 \\ \vdots \\ -x_N^2 \end{pmatrix}$$

and for $N \geq 5$ we can solve for the ellipse parameter vector.

C.2   Homogeneous Coordinates

A point in homogeneous coordinates, or the projective space $\mathbb{P}^n$, is represented by a coordinate vector $\tilde{x} = (\tilde{x}_1, \tilde{x}_2 \cdots \tilde{x}_{n+1})$. The Euclidean coordinates are related to the projective coordinates by

$$x_i = \frac{\tilde{x}_i}{\tilde{x}_{n+1}}, \quad i = 1 \cdots n$$

Conversely a homogeneous coordinate vector can be constructed from a Euclidean coordinate vector by

$$\tilde{x} = \left(x_1, x_2 \cdots x_n, 1 \right)$$

and the tilde is used to indicate that the quantity is homogeneous.

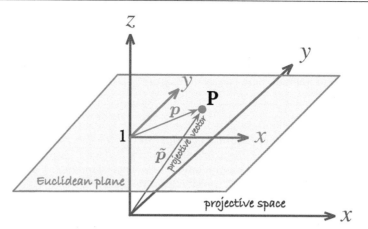

Fig. C.5.
A point P on the Euclidean plane $\mathbb{R}^2$ (*red*) is described by a coordinate vector $p \in \mathbb{R}^2$ which is equivalent to the three-dimensional vector in the projective space $\mathbb{P}^2$ (*blue*) which is the homogeneous coordinate $\tilde{p} \in \mathbb{P}^2$

The extra *degree of freedom* offered by projective coordinates has several advantages. It allows points and lines at infinity, known as ideal points and lines, to be represented using only real numbers. It also means that scale is unimportant, that is $\tilde{x}$ and $\tilde{x}' = \alpha \tilde{x}$ both represent the same Euclidean point for all $\alpha \neq 0$. We express this as $\tilde{x} \simeq \tilde{x}'$. Points in homogeneous form can also be rotated with respect to a coordinate frame and translated simply by multiplying the homogeneous coordinate by an $(n+1) \times (n+1)$ homogeneous transformation matrix.

Homogeneous vectors are important in computer vision when we consider points and lines that exist in a plane – a camera's image plane. We can also consider that the homogeneous form represents a ray in Euclidean space as shown in Fig. C.5. The relationship between points and rays is at the core of the projective transformation.

C.2.1 Two Dimensions

In two dimensions there is a duality between points and lines. In $\mathbb{P}^2$ a line is defined by a 3-tuple, $\tilde{\ell} = (\ell_1, \ell_2, \ell_3)^T$, not all zero, and the equation of the line is the set of all points

$$\tilde{\ell}^T \tilde{x} = 0$$

which expands to $\ell_1 x + \ell_2 y + \ell_3 = 0$ and can be manipulated into the more familiar representation of a line. Note that this form can represent a vertical line, parallel to the y-axis, which the familiar form $y = mx + c$ cannot. This is the point equation of a line. The nonhomogeneous vector (ℓ_1, ℓ_2) is a normal to the line, and $(-\ell_2, \ell_1)$ is parallel to the line.

A point is defined by the intersection of two lines. If we write the point equations for two lines $\tilde{\ell}_1^T \tilde{p} = 0$ and $\tilde{\ell}_2^T \tilde{p} = 0$ their intersection is the point with coordinates

$$\tilde{p} = \tilde{\ell}_1 \times \tilde{\ell}_2$$

and is known as the line equation of a point. Similarly, a line joining two points $\tilde{p}_1$ and $\tilde{p}_2$ is given by the cross-product

$$\tilde{\ell}_{12} = \tilde{p}_1 \times \tilde{p}_2$$

Consider the case of two parallel lines at 45° to the horizontal axis

```
>> l1 = [1 -1 0]';
>> l2 = [1 -1 -1]';
```

which we can plot

```
>> plot_homline(l1, 'b')
>> plot_homline(l2, 'r')
```

The intersection point of these parallel lines is

```
>> cross(l1, l2)
ans =
      1     1     0
```

This is an *ideal point* since the third coordinate is zero – the equivalent Euclidean point would be at infinity. Projective coordinates allow points and lines at infinity to be simply represented and manipulated without special logic.

The distance from a point with coordinates $\tilde{p}$ to a line $\tilde{\ell}$ is

$$d = \frac{\tilde{\ell}^T \tilde{p}}{p_3 \sqrt{\ell_1^2 + \ell_2^2}} \tag{C.6}$$

C.2.1.1   Conics

Conic sections are an important family of planar curves that includes circles, ellipses, parabolas and hyperbolas which can be described by

$$Au^2 + Buv + Cv^2 + Du + Ev + F = 0$$

or more concisely as $\tilde{p}^T c \tilde{p} = 0$ where c is a matrix

$$c = \left(\begin{array}{cc|c} A & B/2 & D/2 \\ B/2 & C & E/2 \\ \hline D/2 & E/2 & F \end{array} \right)$$

The determinant of the top-left submatrix indicates the type of conic: negative for a hyperbola, 0 for a parabola and positive for an ellipse.

C.2.2   Three Dimensions

In three dimensions there is a duality between points and planes.

C.2.2.1   Lines

Using the homogeneous representation of the two points $\tilde{p}$ and $\tilde{q}$ we can form a 4×4 skew-symmetric matrix

$$\begin{aligned} L &= \tilde{q}\tilde{p}^T - \tilde{p}\tilde{q}^T \\ &= \begin{pmatrix} 0 & v_3 & -v_2 & -\omega_1 \\ -v_3 & 0 & v_1 & -\omega_2 \\ v_2 & -v_1 & 0 & -\omega_3 \\ \omega_1 & \omega_2 & \omega_3 & 0 \end{pmatrix} \end{aligned}$$

whose 6 unique elements comprise the Plücker coordinate vector. This matrix is rank 2 and the determinant is a quadratic in the Plücker coordinates – a 4-dimensional quadric hypersurface known as the Klein quadric. All points that lie on this manifold are valid lines. Many of the relationships in Sect. C.1.2.2 (between lines and points and planes) can be expressed in terms of this matrix. This matrix is returned by the `L` method of the `Plucker` class.

For a perspective camera with a camera matrix C the 3-dimensional Plücker line represented as a 4×4 skew-symmetric matrix L is projected onto the image plane as

$$\ell = CLC^T$$

which is a homogeneous line in $\mathbb{P}^2$. This is computed automatically if a `Plucker` object is passed to the `project` method of a `CentralCamera` object.

C.2.2.2   Planes

The plane described by $\pi\tilde{x} = 0$ can be defined by a line and a point

$$\pi = L\tilde{p}$$

The join and incidence relationships are more complex than the cross products used for the 2-dimensional case. Three points define a plane and the join relationship is

$$\begin{pmatrix} \tilde{p}_1^T \\ \tilde{p}_2^T \\ \tilde{p}_3^T \end{pmatrix} \tilde{\pi} = 0$$

and the solution is found from the right-null space of the matrix. The incidence of three planes is the dual

$$\begin{pmatrix} \tilde{\pi}_1^T \\ \tilde{\pi}_2^T \\ \tilde{\pi}_3^T \end{pmatrix} \tilde{p} = 0$$

and is an ideal point, zero last component, if the planes do not intersect at a point.

C.2.2.3   Quadrics

Quadrics, short for quadratic surfaces, are a rich family of 3-dimensional surfaces. There are 17 standard types including spheres, ellipsoids, hyperboloids, paraboloids, cylinders and cones all described by

$$\tilde{x}^T Q \tilde{x} = 0$$

where $Q \in \mathbb{R}^{4\times4}$ is symmetric.

For a perspective camera with a camera matrix C the outline of the quadric is projected to the image plane by

$$c^* = CQ^*C^T$$

where c is a 3×3 matrix describing the conic, see Sect. C.2.1.1, and $(\cdot)^*$ represents the adjugate operation, see Appendix B.

C.3   Geometric Transformations

A linear transform is $y = Ax$ and an affine transform is

$$y = Ax + b \tag{C.7}$$

which comprises a linear transformation *and* a change of origin. Examples of affine transformations include translation, scaling, homothety, similarity transformation, reflection, rotation, shear mapping, and compositions of them in any combination and sequence. Every linear transformation is affine, but not every affine transformation is linear.

In homogeneous coordinates we can write Eq. C.7 as

$$\tilde{y} = H\tilde{x}, \quad \text{where } H = \begin{pmatrix} A & b \\ 0 & 1 \end{pmatrix}$$

and the transformation operates on a point with homogeneous coordinates $\tilde{x}$. If a vector is defined as the difference between two homogeneous points $\tilde{p}$ and $\tilde{q}$ then the difference $\tilde{p} - \tilde{q}$ is a 4-vector whose last element is zero, distinguishing a point from a vector.

Affine space is a generalization of Euclidean space and has no distinguished point that serves as an origin. Hence, no vector has a fixed origin and no vector can be uniquely associated to a point an affine space, there are instead displacement vectors between two points of the space. Thus it makes sense to subtract two points of the space, giving a vector, but it does not make sense to add two points of the space. Likewise, it makes sense to add a vector to a point of an affine space, resulting in a new point displaced from the starting point by that vector.

In two-dimensions the most general transformation is projective transformation projective, also known as a collineation

$$H = \begin{pmatrix} h_{11} & h_{12} & h_{13} \\ h_{21} & h_{22} & h_{23} \\ h_{31} & h_{32} & 1 \end{pmatrix}$$

which is unique up to scale and one element has been normalized to one. It has 8 degrees of freedom.

The affine transformation is a subset where the elements of the last row are fixed

$$H = \begin{pmatrix} a_{11} & a_{12} & t_x \\ a_{21} & a_{22} & t_y \\ 0 & 0 & 1 \end{pmatrix}$$

and has 6 degrees of freedom.

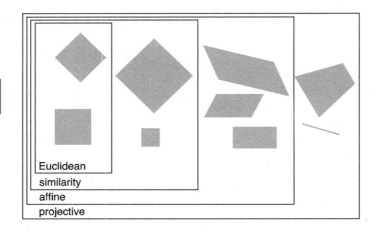

Fig. C.6.
A 2-dimensional square (*dark grey*) is operated on by various transformations from the most limited (Euclidean) to the most general (projective)

Table C.1.
For various planar transformation families the possible geometric transformations and the geometric properties which are preserved are listed

	Euclidean	Similarity	Affine	Projective
Geometric transformation				
Rotation	✓	✓	✓	✓
Translation	✓	✓	✓	✓
Reflection		✓	✓	✓
Uniform scaling		✓	✓	✓
Nonuniform scaling			✓	✓
Shear			✓	✓
Perspective projection				✓
Preserved geometric properties (invariants)				
Length	✓			
Angle	✓	✓		
Ratio of lengths	✓	✓		
Parallelism	✓	✓	✓	
Incidence	✓	✓	✓	✓
Cross ratio	✓	✓	✓	✓

The similarity transformation is further subset

$$H = \begin{pmatrix} s\boldsymbol{R} & \boldsymbol{t} \\ 0 & 1 \end{pmatrix} = \begin{pmatrix} sr_{11} & sr_{12} & t_x \\ sr_{21} & sr_{22} & t_y \\ 0 & 0 & 1 \end{pmatrix}$$

where $\boldsymbol{R} \in \mathbf{SO}(2)$ resulting in only 4 degrees of freedom. Similarity transforms, without reflection, are sometimes referred to as a Procrustes transform.

Finally the Euclidean or rigid-body transformation

$$H = \begin{pmatrix} \boldsymbol{R} & \boldsymbol{t} \\ 0 & 1 \end{pmatrix} = \begin{pmatrix} r_{11} & r_{12} & t_x \\ r_{21} & r_{22} & t_y \\ 0 & 0 & 1 \end{pmatrix} \in \mathbf{SE}(2)$$

is the most restrictive and has only 3 degrees of freedom. Some graphical examples of the effect of the various transformations on a square are shown in Fig. C.6. The possible geometric transformations for each type of transform are summarized in Table C.1 along with the geometric properties which are unchanged, or invariant, under that transformation. We see that while Euclidean is most restrictive in terms of the geometric transformations it can perform it is able to preserve important properties such as length and angle.

D Lie Groups and Algebras

We cannot go very far in the study of rotations or rigid-body motion without coming across the terms Lie groups, Lie algebras or Lie brackets – all named in honor of the Norwegian mathematician Sophus Lie. Rotations and rigid-body motion in 2- and 3-dimensions can be represented by matrices which form Lie groups and which have Lie algebras.

We will start simply by considering the set of all real 2×2 matrices $A \in \mathbb{R}^{2 \times 2}$

$$A = \begin{pmatrix} a_{11} & a_{12} \\ a_{21} & a_{22} \end{pmatrix}$$

which we could write as a linear combination of basis matrices

$$A = a_{11} \begin{pmatrix} 1 & 0 \\ 0 & 0 \end{pmatrix} + a_{12} \begin{pmatrix} 0 & 1 \\ 0 & 0 \end{pmatrix} + a_{21} \begin{pmatrix} 0 & 0 \\ 1 & 0 \end{pmatrix} + a_{22} \begin{pmatrix} 0 & 0 \\ 0 & 1 \end{pmatrix}$$

where each basis matrix represents a *direction* in a 4-dimensional space of 2×2 matrices. That is, the four axes of this space are *parallel* with each of these basis matrices. Any 2×2 matrix can be represented by a point in this space – this particular matrix is a point with the coordinates $(a_{11}, a_{12}, a_{21}, a_{22})$.

All proper rotation matrices, those belonging to **SO**(2), are a subset of points within the space of all 2×2 matrices. For this example the points lie in a 1-dimensional subset, a closed curve, in the 4-dimensional space. This is an instance of a manifold, a lower-dimensional smooth *surface* embedded within a space.

The notion of a curve in the 4-dimensional space makes sense when we consider that the **SO**(2) rotation matrix

$$A = \begin{pmatrix} \cos\theta & \sin\theta \\ -\sin\theta & \cos\theta \end{pmatrix}$$

has only one free parameter, and varying that parameter moves the point along the manifold.

Sophus Lie (1842–1899) (surname pronounced lee) was a Norwegian mathematician who obtained his Ph.D. from the University of Christiania in Oslo in 1871. He spent time in Berlin working with Felix Klein, and later contributed to Klein's Erlangen program to characterize geometries based on group theory and projective geometry. On a visit to Milan during the Franco-Prussian war he was arrested as a German spy and spent one month in prison. He is best known for his discovery that continuous transformation groups (now called Lie groups) can be understood by linearizing them and studying their generating vector spaces. He is buried in the Vår Frelsers gravlund in Oslo. (Photograph by Ludwik Szacinski)

Invoking mathematical formalism we say that rotations $\mathbf{SO}(2)$ and $\mathbf{SO}(3)$, and rigid-body motions $\mathbf{SE}(2)$ and $\mathbf{SE}(3)$ are matrix Lie groups and this has two implications. Firstly, they are an *algebraic group*, a mathematical structure comprising elements and a single operator. In simple terms, a group $\mathbf{G}$ has the following properties:

1. if g_1 and g_2 are elements of the group, that is $g_1, g_2 \in \mathbf{G}$, then the result of the group's operator $\circ$ is also an element of the group: $g_1 \circ g_2 \in \mathbf{G}$. In general, groups are not commutative so $g_1 \circ g_2 \neq g_2 \circ g_1$. For rotations and rigid-body motions the group operator $\circ$ represents composition. ◥

2. the group operator is associative, that is, $(g_1 \circ g_2) \circ g_3 = g_1 \circ (g_2 \circ g_3)$.

3. for $g \in \mathbf{G}$ there is an identity element $I \in \mathbf{G}$ such that $g \circ I = I \circ g = g$. ◥

4. for every $g \in \mathbf{G}$ there is a unique inverse $h \in \mathbf{G}$ such that $g \circ h = h \circ g = I$. ▶

> In this book's notation the $\oplus$ operator is the group operator.

> In this book's notation the identity is denoted by 0 (implying null motion) so we can say that $\xi \oplus 0 = 0 \oplus \xi = \xi$.

> In this book's notation we use the operator $\ominus \xi$ to form the inverse.

The second implication of being a Lie group is that there is a smooth (differentiable) manifold structure. At any point on the manifold we can construct tangent vectors. The set of all tangent vectors at that point form a vector space – the tangent space. This is the multidimensional equivalent to a tangent line on a curve, or a tangent plane on a solid. We can think of this as the set of all possible derivatives of the manifold at that point.

The tangent space *at the identity* is described by the Lie algebra of the group, and the basis directions of the tangent space are called the generators of the group. Points in this tangent space map to elements of the group via the exponential function. If $\mathbf{g}$ is the Lie algebra for group $\mathbf{G}$ then

$$\forall X \in \mathbf{g} \Rightarrow e^X \in \mathbf{G}$$

where the elements of $\mathbf{g}$ and $\mathbf{G}$ are matrices of the same size and which each have a specific structure.

The surface of a sphere is a manifold in 3-dimensional space and at any point on that surface we can create a tangent vector. In fact we can create an infinite number of them and they lie within a plane which is a 2-dimensional vector space – the tangent space. We can choose a set of basis directions and establish a 2-dimensional coordinate system and we can map points on the plane to points on the sphere's surface.

Now consider an arbitrary real 3×3 matrix $A \in \mathbb{R}^{3 \times 3}$

$$A = \begin{pmatrix} a_{11} & a_{12} & a_{13} \\ a_{21} & a_{22} & a_{23} \\ a_{31} & a_{32} & a_{33} \end{pmatrix}$$

which we could write as a linear combination of basis matrices

$$A = a_{11}\begin{pmatrix} 1 & 0 & 0 \\ 0 & 0 & 0 \\ 0 & 0 & 0 \end{pmatrix} + a_{12}\begin{pmatrix} 0 & 1 & 0 \\ 0 & 0 & 0 \\ 0 & 0 & 0 \end{pmatrix} \cdots + a_{33}\begin{pmatrix} 0 & 0 & 0 \\ 0 & 0 & 0 \\ 0 & 0 & 1 \end{pmatrix}$$

where each basis matrix represents a *direction* in a 9-dimensional space of 3×3 matrices. Every possible 3×3 matrix is represented by a point in this space.

Not all matrices in this space are proper rotation matrices belonging to $\mathbf{SO}(3)$, but those that do lie on a manifold since $\mathbf{SO}(3)$ is a Lie group. The null rotation, represented by the identity matrix, is one point in this space. At that point we can construct a tangent space which has only 3 dimensions. Every point in the tangent space – the derivatives of the manifold – can be expressed as a linear combination of basis matrices

$$\Omega = \omega_1 \underbrace{\begin{pmatrix} 0 & 0 & 0 \\ 0 & 0 & -1 \\ 0 & 1 & 0 \end{pmatrix}}_{G_1} + \omega_2 \underbrace{\begin{pmatrix} 0 & 0 & 1 \\ 0 & 0 & 0 \\ -1 & 0 & 0 \end{pmatrix}}_{G_2} + \omega_3 \underbrace{\begin{pmatrix} 0 & -1 & 0 \\ 1 & 0 & 0 \\ 0 & 0 & 0 \end{pmatrix}}_{G_3} \tag{D.1}$$

which is the Lie algebra of the **SO**(3) group. The bases of this space: G_1, G_2 and G_3 are called the generators of **SO**(3) and belong to **so**(3). ◄

The equivalent algebra is denoted using lower case letters and is a set of matrices.

Equation D.1 can be written as a skew-symmetric matrix parameterized by the vector $\omega = (\omega_1, \omega_2, \omega_3) \in \mathbb{R}^3$

$$\Omega = [\omega]_\times = \begin{pmatrix} 0 & -\omega_3 & \omega_2 \\ \omega_3 & 0 & -\omega_1 \\ -\omega_2 & \omega_1 & 0 \end{pmatrix} \in \textbf{so}(3)$$

and this reflects the 3 degrees of freedom of the **SO**(3) group embedded in the space of all 3×3 matrices. The 3DOF is consistent with our intuition about rotations in 3D space and also Euler's rotation theorem.

Mapping between vectors and skew-symmetric matrices is frequently required and the following shorthand notation will be used

$$[\cdot]_\times : \mathbb{R} \mapsto \textbf{so}(2),\ \mathbb{R}^3 \mapsto \textbf{so}(3)$$

$$\vee_\times(\cdot) : \textbf{so}(2) \mapsto \mathbb{R},\ \textbf{so}(3) \mapsto \mathbb{R}^3$$

The first mapping is performed by the Toolbox function `skew` and the second by `vex` (which is named after the $\vee_\times$).

The exponential of *any* matrix in **so**(3) is a valid member of **SO**(3)

$$R(\theta, \hat{\omega}) = e^{[\hat{\omega}]_\times \theta} \in \textbf{SO}(3)$$

and an efficient closed-form solution is given by Rodrigues' rotation formula

$$R(\theta, \hat{\omega}) = I + \sin\theta[\hat{\omega}]_\times + (1 - \cos\theta)[\hat{\omega}]_\times^2$$

Finally, consider an arbitrary real 4×4 matrix $A \in \mathbb{R}^{4 \times 4}$

$$A = \begin{pmatrix} a_{11} & a_{12} & a_{13} & a_{14} \\ a_{21} & a_{22} & a_{23} & a_{24} \\ a_{31} & a_{32} & a_{33} & a_{34} \\ a_{41} & a_{42} & a_{43} & a_{44} \end{pmatrix}$$

which we could write as a linear combination of basis matrices

$$A = a_{11} \begin{pmatrix} 1 & 0 & 0 & 0 \\ 0 & 0 & 0 & 0 \\ 0 & 0 & 0 & 0 \\ 0 & 0 & 0 & 0 \end{pmatrix} + a_{12} \begin{pmatrix} 0 & 1 & 0 & 0 \\ 0 & 0 & 0 & 0 \\ 0 & 0 & 0 & 0 \\ 0 & 0 & 0 & 0 \end{pmatrix} \cdots + a_{44} \begin{pmatrix} 0 & 0 & 0 & 0 \\ 0 & 0 & 0 & 0 \\ 0 & 0 & 0 & 0 \\ 0 & 0 & 0 & 1 \end{pmatrix}$$

where each basis matrix represents a *direction* in a 16-dimensional space of all possible 4×4 matrices. Every 4×4 matrix is represented by a point in this space.

Not all matrices in this space are proper homogeneous transformation matrices belonging to **SE**(3), but those that do lie on a smooth manifold. The null motion (zero rotation and translation), which is represented by the identity matrix, is one point in this space. At that point we can construct a tangent space, which has 6 dimensions in this case, and points in the tangent space can be expressed as a linear combination of basis matrices

$$\Sigma = \omega_1 \begin{pmatrix} 0 & 0 & 0 & 0 \\ 0 & 0 & -1 & 0 \\ 0 & 1 & 0 & 0 \\ 0 & 0 & 0 & 0 \end{pmatrix} + \omega_2 \begin{pmatrix} 0 & 0 & 1 & 0 \\ 0 & 0 & 0 & 0 \\ -1 & 0 & 0 & 0 \\ 0 & 0 & 0 & 0 \end{pmatrix} + \omega_3 \begin{pmatrix} 0 & -1 & 0 & 0 \\ 1 & 0 & 0 & 0 \\ 0 & 0 & 0 & 0 \\ 0 & 0 & 0 & 0 \end{pmatrix}$$

$$+ v_1 \begin{pmatrix} 0 & 0 & 0 & 1 \\ 0 & 0 & 0 & 0 \\ 0 & 0 & 0 & 0 \\ 0 & 0 & 0 & 0 \end{pmatrix} + v_2 \begin{pmatrix} 0 & 0 & 0 & 0 \\ 0 & 0 & 0 & 1 \\ 0 & 0 & 0 & 0 \\ 0 & 0 & 0 & 0 \end{pmatrix} + v_3 \begin{pmatrix} 0 & 0 & 0 & 0 \\ 0 & 0 & 0 & 0 \\ 0 & 0 & 0 & 1 \\ 0 & 0 & 0 & 0 \end{pmatrix}$$

and these generator matrices belong to the Lie algebra of the group $\mathbf{SE}(3)$ and are denoted $\mathbf{se}(3)$. This can be written in general form as

$$\Sigma = [S] = \left(\begin{array}{ccc|c} 0 & -\omega_3 & \omega_2 & v_1 \\ \omega_3 & 0 & -\omega_1 & v_2 \\ -\omega_2 & \omega_1 & 0 & v_3 \\ \hline 0 & 0 & 0 & 0 \end{array} \right) \in \mathbf{se}(3)$$

which is an augmented skew symmetric matrix parameterized by $S = (v, \omega) \in \mathbb{R}^6$ which is referred to as a twist and has physical interpretation in terms of a screw axis direction and position. The sparse matrix structure and this concise parameterization reflects the 6 degrees of freedom of the $\mathbf{SE}(3)$ group embedded in the space of all 4×4 matrices. We extend our earlier shorthand notation

$$[\cdot]\colon \mathbb{R}^3 \mapsto \mathbf{se}(2),\ \mathbb{R}^6 \mapsto \mathbf{se}(3)$$

$$\vee(\cdot)\colon \mathbf{se}(2) \mapsto \mathbb{R}^3,\ \mathbf{se}(3) \mapsto \mathbb{R}^6$$

We can use these operators to convert between a twist representation which is a 6-vector and a Lie algebra representation which is a 4×4 augmented skew-symmetric matrix. We convert the Lie algebra to the Lie group representation using

$$T(\theta, S) = e^{[S]\theta} \in \mathbf{SE}(3)$$

or the inverse using the matrix logarithm. The exponential and the logarithm each have an efficient closed form solution.

Transforming a Twist – the Adjoint Representation

We have seen that rigid-body motions can be described by a twist which represents motion in terms of a screw axis direction and position, for example in Fig. D.1 the twist S_A can be used to transform points on the body. If the screw is rigidly attached to the body which undergoes some motion in $\mathbf{SE}(3)$ the new twist is

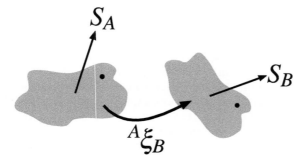

Fig. D.1.
Points in the body (*grey cloud*) can be transformed by the twist S_A. If the body and the screw axis undergo a rigid-body transformation ${}^A\xi_B$ the new twist is S_B

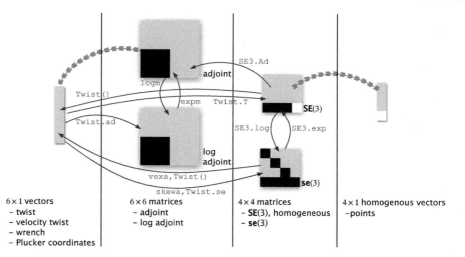

Fig. D.2.
The menagerie of **SE**(3) related quantities. Matrix values are coded as: 0 (*black*), 1 (*white*), other values (*grey*). Transformations between types are indicated by *blue arrows* with the relevant class plus method name. Operations are indicated by *red arrows*: the tail-end object operates on the head-end object and results in another object of the head-end type

6 × 1 vectors	6 × 6 matrices	4 × 4 matrices	4 × 1 homogenous vectors
– twist	– adjoint	– **SE**(3), homogeneous	–points
– velocity twist	– log adjoint	– **se**(3)	
– wrench			
– Plucker coordinates			

$$^{B}S = \mathrm{Ad}\left(^{B}\xi_{A} \right)\,^{A}S$$

where

$$\mathrm{Ad}(\xi) = \begin{pmatrix} R & [t]_{\times}R \\ 0 & R \end{pmatrix} \in \mathbb{R}^{6\times6} \tag{D.2}$$

is the adjoint representation of the rigid-body motion. Alternatively we can write

$$\mathrm{Ad}\left(e^{[S]} \right) = e^{\mathrm{ad}(S)}$$

where ad(S) is the logarithm of the adjoint and defined in terms of the twist parameters as

$$\mathrm{ad}(S) = \begin{pmatrix} [\boldsymbol{\omega}]_{\times} & [\boldsymbol{v}]_{\times} \\ 0 & [\boldsymbol{\omega}]_{\times} \end{pmatrix} \in \mathbb{R}^{6\times6}$$

The relationship between the various mathematical objects discussed are shown in Fig. D.2.

E Linearization, Jacobians and Hessians

In robotics and computer vision the equations we encounter are often nonlinear. To apply familiar and powerful analytic techniques we must work with linear or quadratic approximations to these equations. The principle is illustrated in Fig. E.1 for the 1-dimensional case, and the analytical approximations shown in red are made at $x = x_0$. The approximation equals the nonlinear function at x_0 but is increasing inaccurate as we move away from that point. This is called a local approximation since it is valid in a region local to x_0 – the size of the valid region depends on the severity of the nonlinearity. This approach can be extended to an arbitrary number of dimensions.

Scalar Function of a Scalar

The function $f\colon \mathbb{R} \mapsto \mathbb{R}$ can be expressed as a Taylor series

$$f(x_0 + \Delta) = f(x_0) + \frac{\mathrm{d}f}{\mathrm{d}x}\Delta + \tfrac{1}{2}\frac{\mathrm{d}^2 f}{\mathrm{d}x^2}\Delta^2 + \cdots$$

which we truncate to form a first-order or linear approximation

$$f'(\Delta) \approx f(x_0) + J(x_0)\Delta$$

or a second-order approximation

$$f'(\Delta) \approx f(x_0) + J(x_0)\Delta + \tfrac{1}{2}H(x_0)\Delta^2$$

where $\Delta \in \mathbb{R}$ is an infinitesimal change in x relative to the linearization point x_0, and the first and second derivatives are given by $J(x_0) = \mathrm{d}f/\mathrm{d}x|_{x_0}$ and $H(x_0) = \mathrm{d}^2 f/\mathrm{d}x^2|_{x_0}$ respectively.

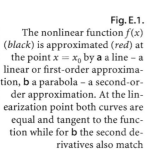

Fig. E.1.
The nonlinear function $f(x)$ (*black*) is approximated (*red*) at the point $x = x_0$ by **a** a line – a linear or first-order approximation, **b** a parabola – a second-order approximation. At the linearization point both curves are equal and tangent to the function while for **b** the second derivatives also match

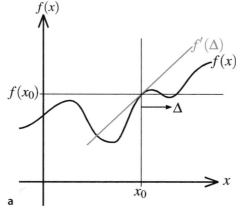

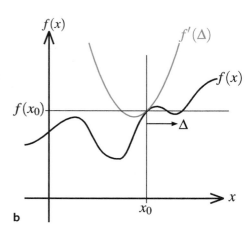

Scalar Function of a Vector

The scalar field $f(\boldsymbol{x})$: $\mathbb{R}^n \mapsto \mathbb{R}$ can be expressed as a Taylor series

$$f(\boldsymbol{x}_0 + \boldsymbol{\Delta}) = f(\boldsymbol{x}_0) + J(\boldsymbol{x}_0)\boldsymbol{\Delta} + \tfrac{1}{2}\boldsymbol{\Delta}^T H(\boldsymbol{x}_0)\boldsymbol{\Delta} + \cdots$$

which we can truncate to form a first-order or linear approximation

$$f'(\boldsymbol{\Delta}) \approx f(\boldsymbol{x}_0) + J(\boldsymbol{x}_0)\boldsymbol{\Delta}$$

or a second-order approximation

$$f'(\boldsymbol{\Delta}) \approx f(\boldsymbol{x}_0) + J(\boldsymbol{x}_0)\boldsymbol{\Delta} + \tfrac{1}{2}\boldsymbol{\Delta}^T H(\boldsymbol{x}_0)\boldsymbol{\Delta}$$

Ludwig Otto Hesse (1811–1874) was a German mathematician, born in Königsberg, Prussia, who studied under Jacobi (p. 232) and Bessel at the University of Königsberg. He taught at Königsberg, Halle, Heidelberg and finally at the newly established Polytechnic School in Munich. In 1869 he joined the Bavarian Academy of Sciences.

where $\boldsymbol{\Delta} \in \mathbb{R}^n$ is an infinitesimal change in $\boldsymbol{x} \in \mathbb{R}^n$ relative to the linearization point $\boldsymbol{x}_0$, $J \in \mathbb{R}^{1 \times n}$ is the vector version of the first derivative, and $H \in \mathbb{R}^{n \times n}$ is the Hessian – the matrix version of the second derivative.

The derivative of the function $f(\cdot)$ with respect to the vector $\boldsymbol{x}$ is

$$J(\boldsymbol{x}) = \nabla f(\boldsymbol{x}) = \begin{pmatrix} \frac{\partial f}{\partial x_1} \\ \frac{\partial f}{\partial x_2} \\ \vdots \\ \frac{\partial f}{\partial x_n} \end{pmatrix}$$

and is itself a vector that points in the direction at which the function $f(\boldsymbol{x})$ has maximal increase. It is often written as $\nabla_x f$ to make explicit that the differentiation is with respect to $\boldsymbol{x}$.

The Hessian is an $n \times n$ symmetric matrix of second derivatives

$$H = \begin{pmatrix} \frac{\partial^2 f}{\partial x_1^2} & \frac{\partial^2 f}{\partial x_1 \partial x_2} & \cdots & \frac{\partial^2 f}{\partial x_1 \partial x_n} \\ \frac{\partial^2 f}{\partial x_1 \partial x_2} & \frac{\partial^2 f}{\partial x_2^2} & \cdots & \frac{\partial^2 f}{\partial x_2 \partial x_n} \\ \vdots & \vdots & \ddots & \vdots \\ \frac{\partial^2 f}{\partial x_1 \partial x_n} & \frac{\partial^2 f}{\partial x_2 \partial x_n} & \cdots & \frac{\partial^2 f}{\partial x_n^2} \end{pmatrix}$$

The function is at a critical point when the Jacobian is not full rank. If the Hessian is positive definite then the function is at a local minimum, if negative definite then a local maximum, and if indefinite then the function is at a saddle point.

For functions which are quadratic in $\boldsymbol{x}$, as is the case for least-squares problems, it can be shown that the Hessian is

$$H(\boldsymbol{x}) = J(\boldsymbol{x})^T J(\boldsymbol{x}) + \sum_{i=1}^{m} f_i(\boldsymbol{x}) \frac{\partial^2 f_i}{\partial \boldsymbol{x}^2} \approx J(\boldsymbol{x})^T J(\boldsymbol{x})$$

which is frequently approximated by just the first term and this is key to Gauss-Newton least-squares optimization discussed in Sect. F.2.2.

Vector Function of a Vector

The vector field $\boldsymbol{f}(\boldsymbol{x})$: $\mathbb{R}^n \mapsto \mathbb{R}^m$ can be expressed as a Taylor series which can also be written as

$$f(x) = \begin{pmatrix} f_1(x) \\ f_2(x) \\ \vdots \\ f_n(x) \end{pmatrix}$$

where $f_i: \mathbb{R}^m \to \mathbb{R}$ for $i \in \{1, 2, \cdots n\}$. The derivative of f with respect to the vector x can be expressed in matrix form as a Jacobian matrix

$$J = \begin{pmatrix} \frac{\partial f_1}{\partial x_1} & \cdots & \frac{\partial f_1}{\partial x_n} \\ \vdots & \ddots & \vdots \\ \frac{\partial f_m}{\partial x_1} & \cdots & \frac{\partial f_m}{\partial x_n} \end{pmatrix}$$

which can also be written as

$$J(x) = \begin{pmatrix} \nabla f_1^T \\ \nabla f_2^T \\ \vdots \\ \nabla f_n^T \end{pmatrix}$$

This derivative is also known as the tangent map of f, denoted Tf, or the differential of f denoted Df. To make explicit that the differentiation is with respect to x this can be denoted as J_x, $T_x f$, $D_x f$ or even $\partial f / \partial x$.

The Hessian in this case is $H \in \mathbb{R}^{n \times m \times n}$ which is a 3-dimensional array called a cubix.

Deriving Jacobians

Jacobians of functions are required for many optimization algorithms as well as for the extended Kalman filter, and can be evaluated numerically or symbolically.

Consider Eq. 6.8 for the range and bearing angle of a landmark given the pose of the vehicle and the position of the landmark. We can express this as the very simple MATLAB® anonymous function

```
>> zrange = @(xi, xv, w) ...
        [ sqrt((xi(1)-xv(1))^2 + (xi(2)-xv(2))^2) + w(1);
          atan((xi(2)-xv(2))/(xi(1)-xv(1)))-xv(3) + w(2) ];
```

To estimate the Jacobian $H_{xv} = \partial h / \partial x_v$ for $x_v = (1, 2, \frac{\pi}{3})$ and $x_i = (10, 8)$ we can compute a first-order numerical difference

```
>> xv = [1, 2, pi/3]; xi = [10, 8]; w= [0,0];
>> h0 = zrange(xi, xv, w)
h0 =
    10.8167
    -0.4592
>> d = 0.001;
>> J = [ zrange(xi, xv+[1,0,0]*d, w)-h0 ...
         zrange(xi, xv+[0,1,0]*d, w)-h0, ...
         zrange(xi, xv+[0,0,1]*d,w)-h0]  /  d
J =
    -0.8320    -0.5547          0
     0.0513    -0.0769    -1.0000
```

which shares the characteristic last column with the Jacobian shown in Eq. 6.14. Note that in computing this Jacobian we have set the measurement noise w to zero. The principal difficulty with this approach is choosing d, the difference used to compute the finite-difference approximation to the derivative. Too large and the results will be quite inaccurate if the function is nonlinear, too small and numerical problems will lead to reduced accuracy.

Alternatively we can perform the differentiation symbolically. This particular function is relatively simple and the derivatives can be determined easily using differential calculus. The numerical derivative can be used as a quick check for correctness. To avoid the possibility of error, or for more complex functions we can perform the differentiation symbolically using any of a large number of computer algebra packages. Using the MATLAB Symbolic Math Toolbox™ we can declare some symbolic variables

```
>> syms xi yi xv yv thetav wr wb
```

and then evaluate the same function as above

```
>> z = zrange([xi yi], [xv yv thetav], [wr wb])
z =
        wr + ((xi - xv)/(yi - yv)^2)^(1/2)
  wb - thetav + atan((yi - yv)/(xi - xv))
```

which is simply Eq. 6.8 in MATLAB symbolic form. The Jacobian is computed by a Symbolic Math Toolbox™ function

```
>> J = jacobian(z, [xv yv thetav])
J =
[ -(2*xi - 2*xv)/(2*((xi - xv)^2 + (yi - yv)^2)^(1/2)),↵
    -(2*yi - 2*yv)/(2*((xi - xv)^2 + (yi - yv)^2)^(1/2)),   0]
[ (yi - yv)/((xi - xv)^2*((yi - yv)^2/(xi - xv)^2 + 1)),↵
    -1/((xi - xv)*((yi - yv)^2/(xi - xv)^2 + 1)), -1]
```

which has the required dimensions

```
>> about(J)
J [sym] : 2x3 (112 bytes)
```

and the characteristic last column. We could cut and paste this code into our program or automatically create a MATLAB callable function

```
>> Jf = matlabFunction(J);
```

where `Jf` is a MATLAB function handle. We can evaluate the Jacobian at the operating point given above

```
>> xv = [1, 2, pi/3]; xi = [10, 8]; w = [0,0];
>> Jf( xi(1), xv(1), xi(2), xv(2) )
ans =
   -0.8321   -0.5547         0
    0.0513   -0.0769   -1.0000
```

which is similar to the approximation above obtained numerically. The function `matlabFunction` can also write the function to an M-file. The functions `ccode` and `fcode` generate C and Fortran representations of the Jacobian.

Another interesting approach is the package ADOL-C which is an open-source tool for the automatic differentiation of C and C++ programs, that is, given a function written in C it will return a Jacobian function written in C. It is available at http://www.coin-or.org/projects/ADOL-C.xml.

F

Solving Systems of Equations

Solving systems of linear and nonlinear equations, particularly over-constrained systems, is a common problem in robotics and computer vision.

F.1 Linear Problems

F.1.1 Nonhomogeneous Systems

These are equations of the form

$$Ax = b$$

where we wish to solve for the unknown vector $x \in \mathbb{R}^n$ and $A \in \mathbb{R}^{m \times n}$ and $b \in \mathbb{R}^m$ are constants.

If $n = m$ then A is square, and if A is nonsingular then the solution is obtained using the matrix inverse

$$x = A^{-1}b$$

In practice we often encounter systems where $m > n$, that is there are more equations than unknowns. In general there will not be an exact solution but we can attempt to find the *best* solution, in a least-squares sense, which is

$$x^* = \arg\min_x \|Ax - b\|$$

That solution is given by

$$x^* = \left(A^T A\right)^{-1} A^T b = A^+ b$$

Since the inverse left multiplies b.

which is known as the pseudo inverse or more formally the left-generalized inverse. ◀
Using SVD where $A = U\Sigma V^T$ this is

$$x = V\Sigma^{-1}U^T b$$

where Σ^{-1} is simply the element-wise inverse of the diagonal elements of Σ^T.

If the matrix is singular, or the system is under constrained $n < m$, then there are infinitely many solutions. We can again use the SVD approach

$$x = V\Sigma^{-1}U^T b$$

where this time Σ^{-1} is the element-wise inverse of the *nonzero* diagonal elements of Σ, all other zeros are left in place.

In MATLAB all these problems can be solved using the backslash operator

```
>> x = A\b
```

For the problem

$$RP = Q$$

where R is an unknown rotation matrix in $\mathbf{SO}(n)$, and $P = \{p_1 \cdots p_m\} \in \mathbb{R}^{n \times m}$ and $Q = \{q_1 \cdots q_m\} \in \mathbb{R}^{n \times m}$ comprise column vectors for which $q_i = Rp_i$. We first compute the moment matrix

$$M = \sum_{i=1}^{N} q_i p_i^T$$

and take then compute the SVD $M = U\Sigma V^T$. The least squares estimate of the rotation matrix is

$$R = UV^T$$

and is guaranteed to be an orthogonal matrix.

F.1.2   Homogeneous Systems

These are equations of the form

$$Ax = 0$$

and always have the trivial solution $x = 0$. If A is square and nonsingular this is the only solution. Otherwise, if A is not of full rank, that is the matrix is nonsquare, or square and singular then there are an infinite number of solutions which are linear combinations of vectors in the right null space of A which is computed by the MATLAB function `null`.

F.2   Nonlinear Problems

Many problems in robotics and computer vision involves sets of nonlinear equations. Solution of these problems requires linearizing the equations about an estimated solution, solving for an improved solution and iterating. Linearization is discussed in Appendix E.

F.2.1   Finding Roots

Consider a set of equations expressed in the form

$$f(x) = 0$$

where $f: \mathbb{R}^n \mapsto \mathbb{R}^m$. This is a nonlinear version of the homogeneous system described above. We first linearize the equation about our best estimate of the solution x_0

$$f(x_0 + \Delta) = f(x_0) + J(x_0)\Delta + \tfrac{1}{2}\Delta^T H(x_0)\Delta + \cdots \tag{F.1}$$

where $\Delta \in \mathbb{R}^n$ is an infinitesimal change in x relative to x_0. We truncate this to form a linear approximation

$$f'(\Delta) \approx f_0 + J\Delta \tag{F.2}$$

where $f_0 = f(x_0)$ is the function value and $J = J(x_0) \in \mathbb{R}^{1 \times n}$ the Jacobian, both evaluated at the linearization point. Now we solve an approximation of our original problem $f'(\Delta) = 0$

$$f_0 + J\Delta = 0 \Rightarrow \Delta = -J^{-1}f_0$$

If $n \neq m$ then J is nonsquare and we can use the pseudo-inverse or the MATLAB backslash operator $-J \backslash f0$. The computed step Δ is based on an approximation to the original nonlinear function so $x_0 + \Delta$ will generally not be the solution but it will be closer. This leads to an iterative solution – the Newton-Raphson method:

1   **while** $\left\| f(x_0) \right\| > \varepsilon$ **do**
2   compute $f_0 = f(x_0), J(x_0)$
3       $\Delta = -J^{-1}f_0$
4       $x_0 \leftarrow x_0 + \Delta$
5   **end**

F.2.2   Nonlinear Minimization

A very common class of problems involves finding the *minimum* of a scalar function $f(x) \colon \mathbb{R}^n \mapsto \mathbb{R}$ which can be expressed as

$$x^* = \arg\min_x f(x)$$

The derivative of the linearized system Eq. F.2 is

$$\frac{\mathrm{d}f'}{\mathrm{d}\Delta} = J$$

and if we consider the function to be a multi-dimensional surface then $J(x_0)$ is vector indicating the direction and magnitude of the *slope* at $x = x_0$ so an update of

$$\Delta = -\beta J$$

will move the estimate *down hill* toward the minimum. This leads to an iterative solution called gradient descent:

1   **repeat**
2   compute $J = J(x_0)$
3       $\Delta = -\beta J$
4       $x_0 \leftarrow x_0 + \Delta$
5   **until** $\left\| \Delta \right\| < \varepsilon$

and the challenge is to choose the appropriate step size β.

If we include the second-order term from Eq. F.1 the approximation becomes

$$f'(\Delta) \approx f_0 + J\Delta + \tfrac{1}{2}\Delta^T H(x_0)\Delta$$

and to find its minima we take the derivative and set it to zero

$$\frac{\mathrm{d}f'}{\mathrm{d}\Delta} = 0 \Rightarrow J + H\Delta = 0$$

and the update is

$$\Delta = -H^{-1}J$$

This leads to another iterative solution – Newton's method. The challenge is determining the Hessian of the nonlinear system, either by numerical approximation or symbolic manipulation.

F.2.3   Nonlinear Least Squares Minimization

Very commonly the scalar function we wish to optimize is a quadratic cost function

$$F(x) = \|f(x)\|^2 = f(x)^T f(x)$$

where $f(x): \mathbb{R}^n \mapsto \mathbb{R}^m$ is some vector-valued nonlinear function which we can linearize as

$$f'(\Delta) \approx f_0 + J\Delta$$

and the scalar cost is

$$
\begin{aligned}
F(\Delta) &\approx (f_0 + J\Delta)^T (f_0 + J\Delta) \\
&\approx f_0^T f_0 + \boxed{f_0^T J\Delta + \Delta^T J^T f_0} + \Delta^T J^T J\Delta \\
&\approx f_0^T f_0 + 2 f_0^T J\Delta + \Delta^T J^T J\Delta
\end{aligned}
$$

One term is the transpose of the other, but since both result in a scalar transposition doesn't matter.

where $J^T J \in \mathbb{R}^{n \times n}$ is the *approximate* Hessian from page 618.

To minimize the error of this linearized least squares system we take the derivative with respect to Δ and set it to zero

$$\frac{dF}{d\Delta} = 0 \Rightarrow 2 f_0^T J + 2 J^T J\Delta = 0$$

which we can solve for the locally optimal update

$$\Delta = -\left(J^T J\right)^{-1} J^T f_0 \tag{F.3}$$

where we can recognize the pseudo or left generalized-inverse of J. Once again we iterate to find the solution – a Gauss-Newton iteration.

Numerical Issues

When solving Eq. F.3 we may find that the Hessian $J^T J$ is poorly conditioned or singular and this can be remedied by adding a damping term

$$\Delta = -\left(J^T J + \lambda I\right)^{-1} J^T f_0$$

which makes the system more positive definite. Since $J^T J + \lambda I$ is effectively in the denominator, increasing λ will decrease $\|\Delta\|$ and slow convergence.

How do we choose λ? We can experiment with different values but a better way is the Levenberg-Marquardt algorithm (Algorithm F.1) which adjusts λ to ensure convergence. If the error increases compared to the last step then the step is repeated with increased λ to reduce the step size. If the error decreases then λ is reduced to increase the convergence rate. The updates vary continuously between Gauss-Newton (low λ) and gradient descent (high λ).

For problems where n is large inverting the $n \times n$ approximate Hessian is expensive. Typically $m < n$ which means the Jacobian is not square and Eq. F.3 can be rewritten as

$$
\begin{array}{ll}
1 & \textbf{initialize } \lambda \\
2 & \textbf{repeat} \\
3 & \quad \text{compute } \boldsymbol{f}_0 = \boldsymbol{f}(\boldsymbol{x}_0), J = J(\boldsymbol{x}_0), H = J^T J \\
4 & \quad \Delta = -(H + \lambda I)^{-1} J^T \boldsymbol{f}_0 \\
5 & \quad \textbf{if } F(\boldsymbol{x}_0 + \Delta) < F(\boldsymbol{x}_0) \textbf{ then} \\
6 & \quad\quad - \textit{error decreased: reduce damping} \\
7 & \quad\quad \boldsymbol{x}_0 \leftarrow \boldsymbol{x}_0 + \Delta \\
8 & \quad\quad \lambda \leftarrow \lambda / c \\
9 & \quad \textbf{else} \\
10 & \quad\quad - \textit{error increased: discard and raise damping} \\
11 & \quad\quad \lambda \leftarrow c\lambda \\
12 & \quad \textbf{end} \\
13 & \textbf{until } \|\Delta\| < \varepsilon
\end{array}
$$

Algorithm F.1.
Levenberg-Marquardt algorithm, c is typically chosen in the range 2 to 10

$$
\Delta = -J^T \left(J J^T \right)^{-1} \boldsymbol{f}_0
$$

which is the right pseudo-inverse and involves inverting a smaller matrix. We can reintroduce a damping term

$$
\Delta = -J^T \left(J J^T + \lambda I \right)^{-1} \boldsymbol{f}_0
$$

and if λ is large this becomes simply

$$
\Delta \approx -\beta J^T \boldsymbol{f}_0
$$

but exhibits very slow convergence.

If $\boldsymbol{f}_k(\cdot)$ has additive noise that is zero mean, normally distributed and time invariant we have a maximum likelihood estimator of $\boldsymbol{x}$. Outlier data has a significant impact on the result since errors are squared. Robust estimators minimize the effect of outlier data and in an M-estimator

$$
F(\boldsymbol{x}) = \rho\big(\boldsymbol{f}_k(\boldsymbol{x})\big)
$$

the squared norm is replaced by a loss function $\rho(\cdot)$ which models the likelihood of its argument. Unlike the squared norm these functions flatten off for large values, and some common examples include the Huber loss function and the Tukey biweight function.

F.2.4   Sparse Nonlinear Least Squares

For a large class of problems the overall cost is the sum of quadratic costs

$$
F(\boldsymbol{x}) = \sum_k \|\boldsymbol{f}_k(\boldsymbol{x})\|^2 = \sum_k \boldsymbol{f}_k(\boldsymbol{x})^T \boldsymbol{f}_k(\boldsymbol{x}) \tag{F.4}
$$

Consider the problem of fitting a model $z = \phi(\boldsymbol{w}; \boldsymbol{x})$ where $\phi \colon \mathbb{R}^p \mapsto \mathbb{R}^m$ with parameters $\boldsymbol{x} \in \mathbb{R}^n$ to a set of data points $(\boldsymbol{w}_k, z_k)$. The error vector associated with the kth data point is

$$
\boldsymbol{f}_k(\boldsymbol{x}) = z_k - \phi\big(\boldsymbol{w}_k; \boldsymbol{x}\big) \in \mathbb{R}^m
$$

and minimizing Eq. F.4 gives the optimal model parameters $\boldsymbol{x}$.

Another example is pose-graph optimization as used for pose-graph SLAM and bundle adjustment. Edge k in the graph connects vertices i and j and has an associated cost $f_k(\cdot): \mathbb{R}^n \mapsto \mathbb{R}^m$

$$f_k(x) = \hat{e}_k(x) - e_k^{\#} \tag{F.5}$$

where $e_k^{\#}$ is the observed value of the edge parameter and $\hat{e}_k(x)$ is the estimate based on the state x of the pose graph. This is linearized

$$f_k'(\Delta) \approx f_{0,k} + J_k \Delta$$

and the squared error for the edge is

$$F_k(x) = f_k(x)^T \Omega_k f_k(x)$$

where $\Omega_k \in \mathbb{R}^{m \times m}$ is a positive-definite constant matrix▸ which we combine as

This can be used to specify the significance of the edge $\det\Omega_k$ with respect to other edges, as well as the relative significance of the elements of $f_k(\cdot)$.

$$
\begin{aligned}
F_k(\Delta) &\approx \left(f_{0,k} + J_k \Delta\right)^T \Omega_k \left(f_{0,k} + J_k \Delta\right) \\
&\approx f_{0,k}^T \Omega_k f_{0,k} + f_{0,k}^T \Omega_k J_k \Delta + \Delta^T J_k^T \Omega_k f_{0,k} + \Delta^T J_k^T \Omega_k J_k \Delta \\
&\approx c_k + 2b_k^T \Delta + \Delta^T H_k \Delta
\end{aligned}
$$

where $b_k^T = f_{0,k}^T \Omega_k J_k$ and $H_k = \Sigma_k J_k^T \Omega_k J_k$. The total cost is the sum of all edge costs

$$
\begin{aligned}
F(\Delta) &= \sum_k F_k(\Delta) \\
&\approx \sum_k \left(c_k + 2b_k^T \Delta + \Delta^T H_k \Delta\right) \\
&\approx \sum_k c_k + 2\left(\sum_k b_k^T\right)\Delta + \Delta^T \left(\sum_k H_k\right)\Delta \\
&\approx c + 2b^T \Delta + \Delta^T H \Delta
\end{aligned}
$$

where $b^T = \Sigma_k f_{0,k}^T \Omega_k J_k$ and $H = \Sigma_k J_k^T \Omega_k J_k$ are summations over the edges of the graph. Once they are computed we proceed as previously, taking the derivative with respect to Δ and setting it to zero, solving for the update Δ and iterating using Algorithm F.1.

State Vector

The state vector is a concatenation of all poses and coordinates in the optimization problem. For pose-graph SLAM it takes the form

$$x = \{\xi_1, \xi_2 \cdots \xi_N\} \in \mathbb{R}^n$$

Poses must be represented in a vector form and preferably one that is compact and singularity free. For **SE(2)** this is quite straightforward and we use $\xi \sim (x, y, \theta) \in \mathbb{R}^3$. For **SE(3)** we will use $\xi \sim (t, r) \in \mathbb{R}^6$ which comprises translation $t \in \mathbb{R}^3$ and rotation $r \in \mathbb{R}^3$. The latter can be triple angles (Euler or roll-pitch-yaw), axis-angle, exponential coordinates or the vector part of a unit-quaternion as discussed on page 499. The state vector has structure, comprising a sequence of subvectors one per pose. We denote the ith subvector of x as $x_i \in \mathbb{R}^{N_\xi}$, where $N_\xi = 3$ for **SE(2)** and $N_\xi = 6$ for **SE(3)**.

For pose-graph SLAM with landmarks, or bundle adjustment the state vector comprises poses and coordinate vectors

$$x = \{\xi_1, \xi_2 \cdots \xi_N \,|\, P_1, P_2 \cdots P_M\} \in \mathbb{R}^n$$

and the ith and j^{th} subvectors of x are denoted $x_i \in \mathbb{R}^{N_\xi}$ and $x_j \in \mathbb{R}^{N_P}$ and correspond to ξ_i and P_j respectively.

Inherent Structure

A key observation is that the error vector $f_k(x)$ for edge k depends only on the associated vertices i and j, and this means that the Jacobian

$$J_k = \frac{\partial f_k(x)}{\partial x} \in \mathbb{R}^{m \times m}$$

is mostly zeros

$$J_k = \left(0 \cdots A_i \cdots B_j \cdots 0\right), \; A_i = \frac{\partial f_k(x)}{\partial x_i}, \; B_j = \frac{\partial f_k(x)}{\partial x_j}$$

where $A_i \in \mathbb{R}^{m \times N_\xi}$ and $B_j \in \mathbb{R}^{m \times N_\xi}$ or $B_j \in \mathbb{R}^{m \times N_P}$ according to the state vector structure.

This sparse block structure means that the vector b_k and the Hessian $J_k^T \Omega_k J_k$ also have a sparse block structure as shown in Fig. F.1. The Hessian has just four small nonzero blocks so rather than compute the product $J_k^T \Omega_k J_k$, which involves many multiplications by zero, we can just compute the four nonzero blocks and add them into the Hessian for the least squares system. All blocks in a row have the same height, and in a column have the same width. For pose-graph SLAM with landmarks, or bundle adjustment the blocks are of different sizes as shown in Fig. F.1b.

If the value of an edge represents pose then Eq. F.5 must be replaced with $f_k(x) = \hat{e}_k(x) \ominus e_k^{\#}$. We generalize this with the $\boxminus$ operator to indicate that the use of $-$ or $\ominus$ as appropriate. Similarly when updating the state vector at the end of an iteration the poses must be compounded $x_0 \leftarrow x_0 \oplus \Delta$ and we generalize this to the $\boxplus$ operator. The pose-graph optimization is solved by the iteration in Algorithm F.2.

1	**repeat**
2	$\quad H \leftarrow 0, \, b \leftarrow 0$
3	$\quad$ **for each** k **do**
4	$\qquad f_{0,k}(x_0) = \hat{e}_k(x) \boxminus e_k^{\#}$
5	$\qquad (i, j) = \text{vertices}(k)$
6	$\qquad$ compute $A_i(x_i), \, B_j(x_j)$
7	$\qquad b_i \leftarrow b_i + f_{0,k}^T \Omega_k A_i$
8	$\qquad b_j \leftarrow b_j + f_{0,k}^T \Omega_k B_j$
9	$\qquad H_{i,i} \leftarrow H_{i,i} + A_i^T \Omega_k A_i$
10	$\qquad H_{i,j} \leftarrow H_{i,j} + A_i^T \Omega_k B_j$
11	$\qquad H_{j,i} \leftarrow H_{j,i} + B_j^T \Omega_k A_i$
12	$\qquad H_{j,j} \leftarrow H_{j,j} + B_j^T \Omega_k B_j$
13	$\quad$ **end**
14	$\quad \Delta = -H^{-1} b$
15	$\quad x_0 \leftarrow x_0 \boxplus \Delta$
16	**until** $\|\Delta\| < \varepsilon$

Alogorithm F.2.
Pose graph optimization. For Levenberg-Marquardt optimization replace line 14 with lines 4–12 from Algorithm F.1

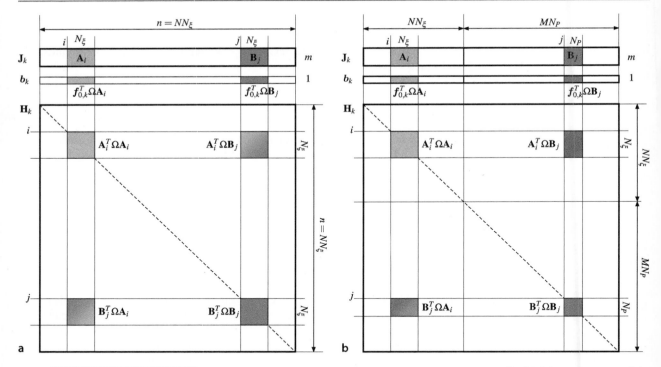

Fig. F.1. Inherent structure of the error vector, Jacobian and Hessian matrices for graph-based least-squares problems. **a** Pose-graph SLAM with N nodes representing robot pose as $\mathbb{R}^{N\xi}$; **b** bundle adjustment with N nodes representing camera pose as $\mathbb{R}^{N\xi}$ and M nodes representing landmark position as $\mathbb{R}^{N_P}$. The indices i and j denote the ith and jth block not the ith and jth row or column. *White* indicates zero values

Large Scale Problems

For pose-graph SLAM with thousands of poses or bundle adjustment with thousands of cameras and millions of landmarks the Hessian matrix will be massive leading to computation and storage challenges. The overall Hessian is the summation of many edge Hessians structured as shown in Fig. F.1 and the total Hessian for two problems we have discussed are shown in Fig. F.2. They have clear structure which we can exploit.

Firstly, in both cases the Hessian is sparse – that is, it contains mostly zeros. MATLAB has built-in support for such matrices and instead of storing all those zeros (at 8 bytes each) it simply keeps a list of the nonzero elements. All the standard matrix operations employ efficient algorithms for manipulating sparse matrices.

Secondly, for the bundle adjustment case we see that the Hessian has two block diagonal submatrices so we partition the system as

$$\begin{pmatrix} B & E \\ E^T & C \end{pmatrix}\begin{pmatrix} \Delta_\xi \\ \Delta_P \end{pmatrix} = \begin{pmatrix} b_\xi \\ b_P \end{pmatrix}$$

where B and C are block diagonal.▶ The subscripts ξ and P denote the blocks of Δ and b associated with camera poses and landmark positions respectively. We solve first for the camera pose updates Δ_ξ

A block diagonal matrix is inverted by simply inverting each of the nonzero blocks along the diagonal.

$$S\Delta_\xi = b_\xi - EC^{-1}b_P$$

where $S = B - EC^{-1}E^T$ is the Schur complement which is a symmetric positive-definite matrix that is also block diagonal. Then we solve for the update to landmark positions

$$\Delta_P = C^{-1}\left(b_P + E^T\Delta_\xi\right)$$

More sophisticated techniques exploit the fine-scale block structure to further reduce computational time, for example GTSAM (https://bitbucket.org/gtborg/gtsam) and SLAM++ (https://sourceforge.net/projects/slam-plus-plus).

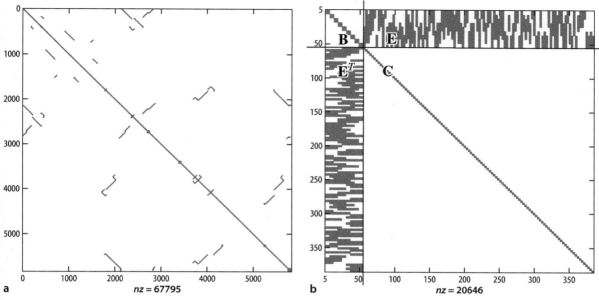

Fig. F.2. Hessian sparsity maps produced using the MATLAB spy function, the number of nonzero elements is shown beneath the plot. **a** Hessian for the posegraph SLAM problem of Fig. 6.17, the diagonal elements represent pose constraints between successive nodes due to odometry, the off-diagonal terms represent constraints due to revisiting locations (loop closures); **b** Hessian for a bundle adjustment problem with 10 cameras and 110 landmarks (vision/examples/bademo.m)

Anchoring

Optimization provides a solution where the *relative* poses and positions give the lowest overall cost, and the solution will have an arbitrary transformation with respect to a global reference frame. To obtain absolute poses and positions we must anchor or fix some nodes – assign them values with respect to the global frame and prevent the optimization from adjusting them. The appropriate way to achieve this is to remove from H and b the rows and columns corresponding to the anchored poses and positions. We then solve a lower dimensional problem for Δ' which will be shorter than x and careful book keeping is required to correctly match the subvectors of Δ' with those of x for the update.

G   Gaussian Random Variables

The 1-dimensional Gaussian function

$$g(x) = \frac{1}{\sqrt{\sigma^2 2\pi}} e^{-\frac{1}{2\sigma^2}(x-\mu)^2} \tag{G.1}$$

is described by the position of its peak μ and its width σ. The total area under the curve is unity and $g(x) > 0, \forall x$. The function can be plotted using the Toolbox function gaussfunc

```
>> x = linspace(-6, 6, 500);
>> plot(x, gaussfunc(0, 1, x), 'r' )
>> hold on
>> plot(x, gaussfunc(0, 2^2, x), '--b' )
```

and Fig. G.1 shows two Gaussians with zero mean and $\sigma = 1$ and $\sigma = 2$. Note that the second argument to gaussfunc is the variance not standard deviation.

If the Gaussian is considered to be a probability density function (PDF) then this is the well known normal distribution and the peak position μ is the mean value and the width σ is the standard deviation. A random variable drawn from a normal distribution is often written as $X \sim N(\mu, \sigma^2)$, and $N(0, 1)$ is referred to as the standard normal distribution – the MATLAB function randn draws random numbers from this distribution. To draw one hundred Gaussian random numbers with mean mu and standard deviation sigma is

```
>> g = sigma * randn(100) + mu;
```

The probability that a random value falls within an interval $x \in [x_1, x_2]$ is obtained by integration

$$P = \int_{x_1}^{x_2} g(x)dx = \Phi(x_2) - \Phi(x_1) \quad \text{where} \quad \Phi(x) = \int_{-\infty}^{x} g(x)dx$$

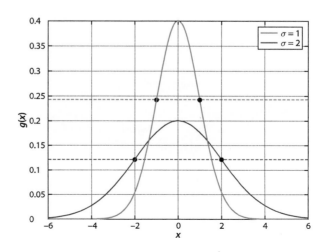

Fig. G.1.
Two Gaussian functions, both with with mean $\mu = 0$, and with standard deviation $\sigma = 1$, and $\sigma = 2$. The markers indicate the points $x = \mu \pm 1\sigma$. The *blue curve* is wider but less tall, since the total area under the curve is unity

or evaluation of the cumulative normal distribution function $\Phi(x)$. The marked points in Fig. G.1 at $\mu \pm 1\sigma$ delimit the 1σ confidence interval. The area under the curve over this interval is 0.68, so the probability of a random value being drawn from this interval is 68%.

The Gaussian can be extended to an arbitrary number of dimensions. The n-dimensional Gaussian, or multivariate normal distribution, is

$$g(\boldsymbol{x}) = \frac{1}{\sqrt{\det(\boldsymbol{P})(2\pi)^n}} e^{-\frac{1}{2}(\boldsymbol{x}-\boldsymbol{\mu})^T \boldsymbol{P}^{-1}(\boldsymbol{x}-\boldsymbol{\mu})} \tag{G.2}$$

and compared to the scalar case of Eq. G.1 $\boldsymbol{x} \in \mathbb{R}^n$ and $\boldsymbol{\mu} \in \mathbb{R}^n$ have become vectors, the squared term in the exponent has been replaced by a matrix quadratic form, and σ^2, the variance, has become a positive-definite (and hence symmetric) covariance matrix $\boldsymbol{P} \in \mathbb{R}^{n \times n}$. The diagonal elements represent the variance of x_i and the off-diagonal elements P_{ij} are the correlationss between x_i and x_j. If the variables are independent or uncorrelated the matrix $\boldsymbol{P}$ would be diagonal. The covariance matrix is symmetric and positive definite.

We can plot a 2-dimensional Gaussian

```
>> [x,y] = meshgrid(-5:0.1:5, -5:0.1:5);
>> P = diag([1 2^2]);
>> surfc(x, y, gaussfunc([0 0], P, x, y))
```

as a surface which is shown in Fig. G.2. In this case $\boldsymbol{\mu} = (0, 0)$ and $\boldsymbol{P} = \mathrm{diag}(1^2, 2^2)$ which corresponds to uncorrelated variables with standard deviation of 1 and 2 respectively. Figure G.2 also shows a number of elliptical contours – contours of constant probability density. If this 2-dimensional probability density function represents the position of a robot in the xy-plane the most likely position for the robot is at $(0, 0)$ and the size of the ellipse says something about our spatial certainty. A particular contour indicates the boundary of a region within which the robot is located with a particular probability. A large ellipse indicates we know, with that probability, that the robot is somewhere inside a large area – we have low certainty about the robot's position. Conversely, a small ellipse means that we know the robot, with the same probability, is somewhere within a much smaller area.

The contour lines are ellipses and in this example the radii in the y- and x-directions are in the ratio 2:1 as defined by the ratio of the standard deviations. For higher order Gaussians, $n > 2$, the corresponding confidence interval is the surface of an ellipsoid in n-dimensional space.

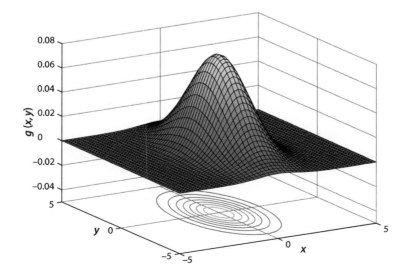

Fig. G.2.
The 2-dimensional Gaussian with covariance $\boldsymbol{P} = \mathrm{diag}(1^2, 2^2)$. Contours lines of constant probability density are shown beneath

The connection between Gaussian probability density functions and ellipses can be found in the quadratic exponent of Eq. G.2 which is the equation of an ellipse or ellipsoid◄. All the points that satisfy

It is also the definition of Mahalanobis distance, the covariance weighted distance between x and μ.

$$(x - \mu)^T P^{-1}(x - \mu) = s$$

result in a constant probability density value, that is, a contour of the 2-dimensional Gaussian. s is related to the probability by

$$s = \chi_n^2(p)$$

If we draw a vector of length n from the multivariate Gaussian each element is normally distributed. The sum of squares of independent normally distributed values is known to be distributed according to a χ^2 (chi-squared) distribution with n degrees of freedom.

which is the χ^2 distribution◄ with n degrees of freedom, 2 in this case, and p is the probability that the point x lies on the ellipse. For example the 50% confidence interval is

```
>> s = chi2inv(0.5, 2)
s =
    1.3863
```

This function requires the MATLAB Statistics and Machine Learning Toolbox™. The Robotics Toolbox provides `chi2inv_rtb` which is an approximation for the case $n = 2$.

where the first argument is the probability and the second is the number of degrees of freedom◄.

If the covariance matrix is diagonal then the ellipse is aligned with the x- and y-axes as we saw in Sect. C.1.4. This indicates that the two variables are independent and have zero correlation. Conversely a rotated ellipse indicates that the covariance is not diagonal and the two variables are correlated.

To draw a covariance ellipse we use the general approach for ellipses outlined in Sect. C.1.4 but the right-hand side of the ellipse equation is s not 1, and $E \equiv P$.

H   Kalman Filter

Consider the system shown in Fig. H.1. The physical robot is a "black box" which has a true state or pose x that evolves over time according to the applied inputs. We cannot directly measure the state, but sensors on the robot have outputs which are a function of that true state. Our challenge is: given the system inputs and sensor outputs *estimate* the unknown true state x and how certain we are of that estimate.

At face value this might seem hard, or even impossible, but there are quite a lot of things we know about system that will help us. Firstly, we know how the state evolves over time as a function of the inputs – this is the state transition◄ model $f(\cdot)$, and we know the inputs to the system u. Our model is unlikely to be perfect➤ and it is common to represent this uncertainty by an imaginary random number generator which is corrupting the system state – process noise. Secondly, we know how the sensor output depends on the state – this is the sensor model $h(\cdot)$ and its uncertainty is also modeled by an imaginary random number generator – sensor noise.

The imaginary random number sources v and w are *inside* the black box so the random numbers are also unknowable. However we can describe the characteristics of these random numbers – their *distribution* which tells us how likely it is that we will *draw* a random number with a particular value. A lot of noise in physical systems can be modeled well by the Gaussian (aka normal) distribution $\mathcal{N}(\mu, \sigma^2)$ which is characterized by a mean μ and a standard deviation σ. There are infinitely many possible distributions◄ but the Gaussian distribution has some nice mathematical properties that we will rely on. However we should never assume that noise is Gaussian – we should attempt to determine the distribution by understanding the physics of the process and the sensor, or from careful measurement and analysis.

Often called the process or motion model.

For example wheel slippage on a mobile ground robot or wind gusts for a UAV.

Which can be nonsymmetrical or have multiple peaks.

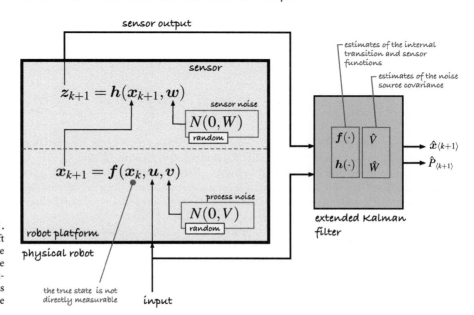

Fig. H.1.
The physical robot on the left has a true state that cannot be directly measured, however we gain a clue from the sensor output which is a function of this unknown true state

In general terms, the problem we wish to solve is:

given a model of the system $f(\cdot)$, $h(\cdot)$, $\hat{V}$ and $\hat{W}$; the known inputs applied to the system u; and some noisy sensor measurements z, find an estimate $\hat{x}$ of the system state and our uncertainty $\hat{P}$ in that estimate.

In a robotic localization context x is the unknown position or pose of the robot, u is the commands sent to the motors and z is the output of various sensors on the robot. For a ground robot x would be the pose in $\mathbf{SE}(2)$ and u would be the motor commands and z might be the measured odometry or range and bearing to landmarks. For a flying robot x would be the pose in $\mathbf{SE}(3)$ and u are the known forces applied to the airframe and z might be the measured accelerations and angular velocities.▶

The state is a vector and there are many approaches to mapping pose to a vector, especially the rotational component – Euler angles, quaternions, and exponential coordinates are commonly used.

H.1   Linear Systems – Kalman Filter

Consider the transition model described as a discrete-time linear time-invariant system

$$x\langle k+1\rangle = Fx\langle k\rangle + Gu\langle k\rangle + v\langle k\rangle \tag{H.1}$$

$$z\langle k\rangle = Hx\langle k\rangle + w\langle k\rangle \tag{H.2}$$

where k is the time step, $x \in \mathbb{R}^n$ is the state vector, and $u \in \mathbb{R}^m$ is a vector of inputs to the system at time k, for example a velocity command, or applied forces and torques. The matrix $F \in \mathbb{R}^{n\times n}$ describes the dynamics of the system, that is, how the states evolve with time. The matrix $G \in \mathbb{R}^{n\times m}$ describes how the inputs are coupled to the system states. The vector $z \in \mathbb{R}^p$ represents the outputs of the system as measured by sensors. The matrix $H \in \mathbb{R}^{p\times n}$ describes how the system states are mapped to the system outputs which we can observe.

To account for errors in the motion model (F and G) or unmodeled disturbances we introduce a Gaussian random variable $v \in \mathbb{R}^n$ termed the process noise. $v\langle k\rangle \sim N(0, V)$, that is, it has zero mean and covariance $V \in \mathbb{R}^{n\times n}$. Covariance is a matrix quantity which is the variance for a multi-dimensional distribution – it is a positive definite matrix and therefore symmetric. The sensor measurement model H is not perfect either and this is modeled by sensor measurement noise, a Gaussian random variable $w \in \mathbb{R}^p$, $w\langle k\rangle \sim N(0, W)$ and covariance $W \in \mathbb{R}^{p\times p}$.

The Kalman filter is an optimal estimator for the case where the process and measurement noise are zero-mean Gaussian noise. The filter has two steps: prediction and update. The prediction is based on the previous state and the inputs that were applied

$$\hat{x}^+\langle k+1\rangle = F\hat{x}\langle k\rangle + Gu\langle k\rangle \tag{H.3}$$

$$\hat{P}^+\langle k+1\rangle = \underbrace{F\hat{P}\langle k\rangle F^T} + \hat{V} \tag{H.4}$$

where $\hat{x}$ is the estimate of the state and $\hat{P} \in \mathbb{R}^{n\times n}$ is the estimated covariance, or uncertainty, in $\hat{x}$. The notation $^+$ makes explicit that the left-hand side is an estimate at time $k + 1$ based on information from time k. $\hat{V}$ is our best estimate of the covariance of the process noise.

The indicated term in Eq. H.4 *projects* the estimated covariance from the current time step to the next. Consider a one dimensional example where F is a scalar and the state estimate $\hat{x}\langle k\rangle$ has a PDF which is Gaussian with a mean $\bar{x}\langle k\rangle$ and a variance $\sigma^2\langle k\rangle$. The prediction equation maps the state and its Gaussian distribution to a new Gaussian distribution with a mean $F\bar{x}\langle k\rangle$ and a variance $F^2\sigma^2\langle k\rangle$. The term $FP\langle k\rangle F\langle k\rangle^T$ is the matrix form of this since

$$\mathrm{cov}(Fx) = F\,\mathrm{cov}(x)F^T \tag{H.5}$$

which scales the covariance appropriately.

The prediction of $\hat{P}$ involves the addition of two positive-definite matrices so the uncertainty will increase – this is to be expected since we have used an uncertain model to predict the future value of an already uncertain estimate. $\hat{V}$ must be a reasonable estimate of the covariance of the actual process noise. If we overestimate it, that is our estimate of process noise is larger than it really is, then we will have a large increase in uncertainty at this step, a pessimistic estimate of our certainty.

To counter this growth in uncertainty we need to introduce new information such as measurements made by the sensors since they depend on the state. The difference between what the sensors measure and what the sensors are predicted to measure is

$$\nu = z^{\#}\langle k+1 \rangle - H\hat{x}^{+}\langle k+1 \rangle \in \mathbb{R}^{p}$$

Some of this difference is due to noise in the sensor, the measurement noise, but the remainder provides valuable information related to the error between the actual and the predicted value of the state. Rather than considering this as error we refer to it more positively as *innovation* – new information.

The second step of the Kalman filter, the *update* step, maps the innovation into a correction for the predicted state, optimally tweaking the estimate based on what the sensors observed

$$\hat{x}\langle k+1 \rangle = \hat{x}^{+}\langle k+1 \rangle + K\nu \tag{H.6}$$

$$\hat{P}\langle k+1 \rangle = \hat{P}^{+}\langle k+1 \rangle - KH\hat{P}^{+}\langle k+1 \rangle \tag{H.7}$$

Uncertainty is now *decreased* or *deflated*, since new information, from the sensors, is being incorporated. The matrix

$$K = P^{+}\langle k+1 \rangle H^{T} \left[\underbrace{H P^{+}\langle k+1 \rangle H^{T} + \hat{W}} \right]^{-1} \in \mathbb{R}^{n \times p} \tag{H.8}$$

is known as the Kalman gain. The term indicated is the estimated covariance of the innovation and comprises the uncertainty in the state and the estimated measurement noise covariance. If the innovation has high uncertainty in some dimensions then the Kalman gain will be correspondingly small, that is, if the new information is uncertain then only small changes are made to the state vector. The term $H P^{+}\langle k+1 \rangle H^{T}$ in Eq. H.13 *projects* the covariance of the state estimate into the space of sensor values.

The covariance matrix must be positive-definite but after many updates the accumulated numerical errors may cause this matrix to be no longer symmetric. The positive-definite structure can be enforced by using the Joseph form of Eq. H.7

$$\hat{P}\langle k+1 \rangle = \left(I_{n \times n} - KH \right) \hat{P}^{+}\langle k+1 \rangle \left(I_{n \times n} - KH \right)^{T} + K\hat{V}K^{T}$$

but this is computationally more costly.

The equations above constitute the classical Kalman filter which is widely used in robotics, aerospace and econometric applications. The filter has a number of important characteristics. Firstly it is optimal, but only if the noise is truly Gaussian with zero mean and time invariant parameters. This is often a good assumption but not always. Secondly it is recursive, the output of one iteration is the input to the next. Thirdly, it is asynchronous. At a particular iteration if no sensor information is available we just perform the prediction step and not the update. In the case that there are different sensors, each with their own H, and different sample rates, we just apply the update with the appropriate z and H. The filter must be initialized with some reasonable value of $\hat{x}$ and $\hat{P}$, as well as good choices of the covariance matrices $\hat{V}$ and $\hat{W}$. As the filter runs the estimated covariance $\|\hat{P}\|$ decreases but never reaches zero – the minimum value can be shown to be a function of $\hat{V}$ and $\hat{W}$. The Kalman-Bucy filter is a continuous-time version of this filter.

The covariance matrix $\hat{P}$ is rich in information. The diagonal elements $\hat{P}_{ii}$ are the variance, or uncertainty, in the state x_i. The off-diagonal elements $\hat{P}_{ij}$ are the correlations between states x_i and x_j and indicate that the errors are not independent. The correlations are critical in allowing any piece of new information to *flow through* to adjust all the states that affect a particular process output.

H.2   Nonlinear Systems – Extended Kalman Filter

For the case where the system is not linear it can be described generally by two functions: the state transition (the motion model in robotics) and the sensor model

$$x\langle k{+}1\rangle = f\big(x\langle k\rangle, u\langle k\rangle, v\langle k\rangle\big) \tag{H.9}$$

$$z\langle k\rangle = h\big(x\langle k\rangle, w\langle k\rangle\big) \tag{H.10}$$

and as before we represent model uncertainty, external disturbances and sensor noise by Gaussian random variables v and w.

We linearize the state transition function about the current state estimate $\hat{x}_k$ as shown in Fig. H.2 resulting in

$$x'\langle k{+}1\rangle \approx F_x x'\langle k\rangle + F_u u\langle k\rangle + F_v v\langle k\rangle \tag{H.11}$$

$$z'\langle k\rangle \approx H_x x'\langle k\rangle + H_w w\langle k\rangle \tag{H.12}$$

where $F_x = \partial f / \partial x \in \mathbb{R}^{n \times n}, F_u = \partial f / \partial u \in \mathbb{R}^{n \times m}, F_v = \partial f / \partial v \in \mathbb{R}^{n \times n}, H_x = \partial h / \partial x \in \mathbb{R}^{p \times n}$ and $H_w = \partial h / \partial w \in \mathbb{R}^{p \times p}$ are Jacobians of the functions $f(\cdot)$ and $h(\cdot)$. Equating coefficients between Eq. H.1 and Eq. H.11 gives $F \sim F_x$, $G \sim F_u$ and $v\langle k\rangle \sim F_v v\langle k\rangle$; and between Eq. H.2 and Eq. H.12 gives $H \sim H_x$ and $w\langle k\rangle \sim H_w w\langle k\rangle$.

Taking the prediction equation Eq. H.9 with $v\langle k\rangle = 0$, and the covariance equation Eq. H.4 with the linearized terms substituted we can write the prediction step as

$$\hat{x}^+\langle k{+}1\rangle = f\big(\hat{x}\langle k\rangle, u\langle k\rangle\big)$$
$$\hat{P}^+\langle k{+}1\rangle = F_x \hat{P}\langle k\rangle F_x^T + F_v \hat{V} F_v^T$$

and the update step as

$$\hat{x}\langle k{+}1\rangle = \hat{x}^+\langle k{+}1\rangle + K\nu$$
$$\hat{P}\langle k{+}1\rangle = \hat{P}^+\langle k{+}1\rangle - KH_x \hat{P}^+\langle k{+}1\rangle$$

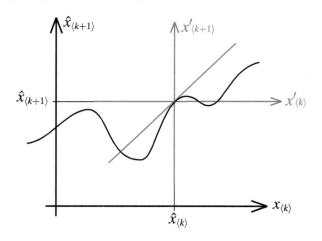

Fig. H.2.
One dimensional example illustrating how the nonlinear state transition function $f: x_k \mapsto x_{k+1}$ shown in *black* is linearized about the point $(\hat{x}\langle k\rangle, \hat{x}\langle k{+}1\rangle)$ shown in *red*

Procedure EKF

Input : $\hat{\boldsymbol{x}}\langle k\rangle \in \mathbb{R}^n$, $\hat{\boldsymbol{P}}\langle k\rangle \in \mathbb{R}^{n\times n}$, $\boldsymbol{u}\langle k\rangle \in \mathbb{R}^m$, $\boldsymbol{z}\langle k+1\rangle \in \mathbb{R}^p$, $\hat{\boldsymbol{V}} \in \mathbb{R}^{n\times n}$, $\hat{\boldsymbol{W}} \in \mathbb{R}^{p\times p}$

Output: $\hat{\boldsymbol{x}}\langle k+1\rangle \in \mathbb{R}^n$, $\hat{\boldsymbol{P}}\langle k+1\rangle \in \mathbb{R}^{n\times n}$

$-$ *linearize about* $\boldsymbol{x} = \hat{\boldsymbol{x}}\langle k\rangle$

compute Jacobians: $\boldsymbol{F}_x \in \mathbb{R}^{n\times n}$, $\boldsymbol{F}_v \in \mathbb{R}^{n\times n}$, $\boldsymbol{H}_x \in \mathbb{R}^{p\times n}$, $\boldsymbol{H}_w \in \mathbb{R}^{p\times p}$

$-$ *the prediction step*

$\qquad \hat{\boldsymbol{x}}^+\langle k+1\rangle = \boldsymbol{f}\big(\hat{\boldsymbol{x}}\langle k\rangle, \boldsymbol{u}\langle k\rangle\big)$ // *predict state at next time step*

$\qquad \hat{\boldsymbol{P}}^+\langle k+1\rangle = \boldsymbol{F}_x\hat{\boldsymbol{P}}\langle k\rangle\boldsymbol{F}_x^T + \boldsymbol{F}_v\hat{\boldsymbol{V}}\boldsymbol{F}_v^T$ // *predict cov ariance at next time step*

$-$ *the update step*

$\qquad \nu = \boldsymbol{z}\langle k+1\rangle - \boldsymbol{h}\big(\hat{\boldsymbol{x}}^+\langle k+1\rangle\big)$ // *innovation : measured* $-$ *predicted sensor value*

$\qquad \boldsymbol{K} = \boldsymbol{P}^+\langle k+1\rangle\boldsymbol{H}_x^T\Big[\boldsymbol{H}_x\boldsymbol{P}^+\langle k+1\rangle\boldsymbol{H}_x^T + \boldsymbol{H}_w\hat{\boldsymbol{W}}\boldsymbol{H}_w^T\Big]^{-1}$ // *Kalman gain*

$\qquad \hat{\boldsymbol{x}}\langle k+1\rangle = \hat{\boldsymbol{x}}^+\langle k+1\rangle + \boldsymbol{K}\nu$ // *update state estimate*

$\qquad \hat{\boldsymbol{P}}\langle k+1\rangle = \hat{\boldsymbol{P}}^+\langle k+1\rangle - \boldsymbol{K}\boldsymbol{H}_x\hat{\boldsymbol{P}}^+\langle k+1\rangle$ // *update covariance estimate*

Algorithm H.1.
Procedure EKF

where the Kalman gain is now

$$\boldsymbol{K} = \boldsymbol{P}^+\langle k+1\rangle\boldsymbol{H}_x^T\Big(\boldsymbol{H}_x\boldsymbol{P}^+\langle k+1\rangle\boldsymbol{H}_x^T + \boldsymbol{H}_w\hat{\boldsymbol{W}}\boldsymbol{H}_w^T\Big)^{-1} \tag{H.13}$$

Properly these matrices should be denoted as depending on the time step, i.e. $F_x\langle k\rangle$ but this has been dropped in the interest of readability.

These equations are only valid at the linearization point $\hat{\boldsymbol{x}}\langle k\rangle$ – the Jacobians $\boldsymbol{F}_x$, $\boldsymbol{F}_v$, $\boldsymbol{H}_x$, $\boldsymbol{H}_w$ must be computed at every iteration.◄ The full procedure is summarized in Algorithm H.1.

A fundamental problem with the extended Kalman filter is that PDFs of the random variables are no longer Gaussian after being operated on by the nonlinear functions $\boldsymbol{f}(\cdot)$ and $\boldsymbol{h}(\cdot)$. We can easily illustrate this by considering a nonlinear scalar function $y = (x + 2)^2 / 4$. We will draw a million Gaussian random numbers from the normal distribution $\mathcal{N}(5, 4)$ which has a mean of 5 and a standard deviation of 2

```
>> x = 2*randn(1000000,1) + 5;
```

and map them through our function

```
>> y = (x+2).^2 / 4;
```

and plot the probability density function of y

```
>> histogram(y, 'Normalization', 'pdf');
```

Fig. H.3.
PDF of the state x (*red*) which is Gaussian $\mathcal{N}(5, 4)$ and the PDF of the nonlinear function $y = (x + 2)^2 / 4$ (*black*). The peak and the mean of the nonlinear distribution are shown by *blue solid* and *dashed vertical lines* respectively

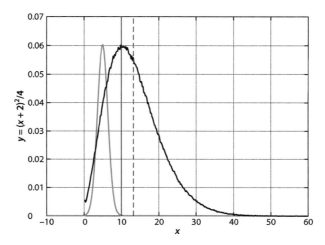

which is shown in Fig. H.3. We see that the PDF of y is substantially changed and no longer Gaussian. It has lost its symmetry so the mean value is greater than the mode. The Jacobians that appear in the EKF equations appropriately scale the covariance but the resulting non-Gaussian distribution breaks the assumptions which guarantee that the Kalman filter is an optimal estimator. Alternatives include the iterated EKF described by Jazwinski (2007) or the Unscented Kalman Filter (UKF) (Julier and Uhlmann 2004) or the sigma-point filter which uses discrete sample points (sigma points) to approximate the PDF.

Graphs

A graph is an abstract representation of a set of objects connected by links and depicted graphically as shown in Fig. I.1. Mathematically a graph is denoted $G(V, E)$ where V are the vertices or nodes, and E are the links that connect pairs of vertices and are called edges or arcs. Edges can be directed (*arrows*) or undirected as in this case. Edges can have an associated weight or cost associated with moving from one vertex to another. A sequence of edges from one vertex to another is a path, and a sequence that starts and ends at the same vertex is a cycle. An edge from a vertex to itself is a loop. Graphs can be used to represent transport, communications or social networks, and this branch of mathematics is graph theory.

The Toolbox provides a MATLAB® graph class called `PGraph` that supports embedded graphs where the vertices are associated with a point in an *n*-dimensional space.◄ To create a new graph

This class is used other Toolbox classes such as `PRM`, `Lattice`, `RRT`, `PoseGraph` and `BundleAdjust`. MATLAB 2015b introduced a built in `graph` class to represent graphs.

```
>> g = PGraph()
g =
   2 dimensions
   0 vertices
   0 edges
   0 components
```

and by default the nodes of the graph exist in a 2-dimensional space. We can add nodes to the graph

```
>> g.add_node( rand(2,1) );
>> g.add_node( rand(2,1) );
>> g.add_node( rand(2,1) );
>> g.add_node( rand(2,1) );
>> g.add_node( rand(2,1) );
```

and each has a random coordinate. The `add_node` method returns an integer identifier for the node just added. A summary of the graph is given with its display method

```
>> g
g =
   2 dimensions
   5 vertices
   0 edges
   0 components
```

and shows that the graph has 5 nodes but no edges. The nodes are numbered 1 to 5 and we add edges between pairs of nodes

```
>> g.add_edge(1, 2);
>> g.add_edge(1, 3);
>> g.add_edge(1, 4);
>> g.add_edge(2, 3);
>> g.add_edge(2, 4);
>> g.add_edge(4, 5);
>> g
g =
   2 dimensions
   5 vertices
   6 edges
   1 components
```

By default the distance between the nodes is the Euclidean distance between the vertices but this can be overridden by a third argument to `add_edge`. The methods `add_node` and `add_edge` return an integer that uniquely identifies the node or edge just created. The graph has one component, that is all the nodes are connected into one network. The graph can be plotted by

```
>> g.plot('labels')
```

as shown in Fig. I.1. The vertices are shown as blue circles, and the option `'labels'` displays the vertex index next to the circle. Edges are shown as black lines joining vertices. Many options exist to change default plotting behavior. Note that only graphs embedded in 2- and 3-dimensional space can be plotted.

The neighbors of vertex 2 are

```
>> g.neighbours(2)
ans =
     3      4      1
```

which are vertices connected to vertex 2 by edges. Each edge has a unique index and the edges connecting to vertex 2 are

```
>> e = g.edges(2)
e =
     4      5      1
```

The cost or length of these edges is

```
>> g.cost(e)
ans =
    0.9597    0.3966    0.6878
```

and clearly edge 5 has a lower cost than edges 4 and 1. Edge 5

```
>> g.vertices(5)'
ans =
     2      4
```

joins vertices 2 and 4, and vertex 4 is clearly the closest neighbor of vertex 2. Frequently we wish to obtain a node's neighboring vertices and their distances at the same time, and this can be achieved conveniently by

```
>> [n,c] = g.neighbours(2)
n =
     3      4      1
c =
    0.9597    0.3966    0.6878
```

Concise information about a node can be obtained by

```
>> g.about(1)
Node 1 #1@ (0.814724 0.905792 )
  neighbours:    >-o-> 2 3 4
  edges:    >-o-> 1 2 3
```

Arbitrary data can be attached to any node or edge by the methods `setvdata` and `setedata` respectively and retrieved by the methods `vdata` and `edata` respectively.

The vertex closest to the coordinate (0.5, 0.5) is

```
>> g.closest([0.5, 0.5])
ans =
     4
```

and the vertex closest to an interactively selected point is given by `g.pick`.

The minimum cost path between any two nodes in the graph can be computed using well known algorithms such as A* (Nilsson 1971)

```
>> g.Astar(3, 5)
ans =
     3      2      4      5
```

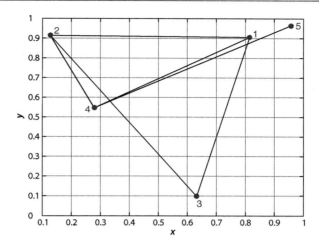

Fig. I.1.
An example graph generated by
the PGraph class

or the earlier method by Dijstrka (1959). By default the graph is treated as undirected, that is, the edges have no preferred direction. The 'directed' option causes edges to be treated as directed, and the path will only traverse edges in their specified direction which is from the first to the second argument of the method add_edge.

Methods exist to compute various other representations of the graph such as adjacency, incidence, degree and Laplacian matrices.

J Peak Finding

A commonly encountered problem is estimating the position of the peak of some discrete 1-dimensional signal $y(k)$, $k \in \mathbb{Z}$, see for example Fig. J.1a

```
>> load peakfit1
>> plot(y,    '-o')
```

Finding the peak to the nearest integer is straightforward using MATLAB's max function

```
>> [ypk,k] = max(y)
ypk =
     0.9905
  k =
       8
```

which indicates the peak occurs at the eighth element and has a value of 0.9905. In this case there is more than one peak and we can use the Toolbox function peak instead

```
>> [ypk,k] = peak(y)
ypk =
     0.9905      0.6718     -0.5799
  k =
       8      25      16
```

which has returned three maxima in descending magnitude. A common test of the quality of a peak is its magnitude and the ratio of the height of the second peak to the first peak

```
>> ypk(2)/ypk(1)
ans =
     0.6783
```

which is called the ambiguity ratio and is ideally small.

This signal is a sampled representation of a continuous underlying signal $y(x)$ and the real peak might actually lie between the samples. If we look at a zoomed version of the signal, Fig. J.1b, we can see that although the eighth point is the maximum the

Fig. J.1. Peak fitting. **a** A signal with several local maxima; **b** closeup view of the first maxima with the fit curve (*red*) and the estimated peak (*red-◊*)

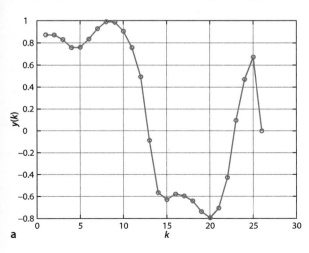

a

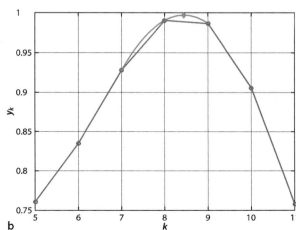

b

ninth point is only slightly lower so the peak lies somewhere between points eight and nine. A common approach is to fit a parabola

$$y = a\delta_x^2 + b\delta_x + c, \quad \delta_x \in \mathbb{R} \tag{J.1}$$

to the points surrounding the peak. For the discrete peak that occurs at (k, y_k) then $\delta_x = 0$ corresponds to k and the discrete x-coordinates on either side correspond to $\delta_x = -1$ and $\delta_x = +1$ respectively. Substituting the points $(k - 1, y_{k-1})$, (k, y_k) and $(k + 1, y_{k+1})$ into Eq. J.1 we can write three equations

$$y_{k-1} = a - b + c$$
$$y_k = c$$
$$y_{k+1} = a + b + c$$

or in compact matrix form as

$$\begin{pmatrix} y_{k-1} \\ y_k \\ y_{k+1} \end{pmatrix} = \begin{pmatrix} 1 & -1 & 1 \\ 0 & 0 & 1 \\ 1 & 1 & 1 \end{pmatrix} \begin{pmatrix} a \\ b \\ c \end{pmatrix}$$

and then solve for the parabolic coefficients

$$\begin{pmatrix} a \\ b \\ c \end{pmatrix} = \begin{pmatrix} 1 & -1 & 1 \\ 0 & 0 & 1 \\ 1 & 1 & 1 \end{pmatrix}^{-1} \begin{pmatrix} y_{k-1} \\ y_k \\ y_{k+1} \end{pmatrix} = \frac{1}{2} \begin{pmatrix} 1 & -2 & 1 \\ -1 & 0 & 1 \\ 0 & 2 & 0 \end{pmatrix} \begin{pmatrix} y_{k-1} \\ y_k \\ y_{k+1} \end{pmatrix} \tag{J.2}$$

The maxima of the parabola occurs when its derivative is zero

$$2a\delta_x + b = 0$$

and substituting the values of a and b from Eq. J.2 we find the displacement of the peak of the fitted parabola with respect to the discrete maxima

$$\delta_x = \frac{1}{2} \frac{y_{k-1} - y_{k+1}}{y_{k-1} - 2y_k + y_{k+1}}, \quad \delta_x \in (-1, 1)$$

so the refined, or interpolated, position of the maxima is at

$$\hat{x} = k + \delta_x \in \mathbb{R}$$

and the estimated value of the maxima is obtained by substituting δ_x into Eq. J.1.

The coefficient a, which is negative for a maxima, indicates the sharpness of the peak which can be useful in determining whether a peak is *sufficiently* sharp. A large magnitude of a indicates a well defined sharp peak wheras a low value indicates a very broad peak for which estimation of a refined peak may not be so accurate.

Continuing the earlier example we can use the Toolbox function `peak` to estimate the refined peak positions

```
>> [ymax,xmax] = peak(y, 'interp', 2)
ymax =
     0.9905    0.6718   -0.5799
xmax =
     8.4394   24.7299   16.2438
```

where the argument after the `'interp'` option indicates that a second-order polynomial should be fitted. The fitted parabola is shown in red in Fig. J.1b and is plotted if the option `'plot'` is given.

If the signal has superimposed noise then there are likely to be multiple peaks, many of which are quite minor, and this can be overcome by specifying the *scale* of the peak. For example the peaks that are greater than all other values within ±5 values in the horizontal direction are

```
>> peak(y, 'scale', 5)
ans =
     0.9905     0.8730     0.6718
```

In this case the result is unchanged since the signal is fairly smooth.

For a 2-dimensional signal we follow a similar procedure but instead fit a *paraboloid*

$$z = ax^2 + by^2 + cx + dy + e \tag{J.3}$$

which has five coefficients that can be calculated from the center value (the discrete maximum) and its four neighbors (north, south, east and west) using a similar procedure to above. The displacement of the estimated peak with respect to the central point is

$$\delta_x = \frac{1}{2} \frac{z_e - z_w}{2z_c - z_w - z_e}, \quad \delta_x \in (-1, 1)$$

$$\delta_y = \frac{1}{2} \frac{z_s - z_n}{2z_c - z_n - z_s} \quad \delta_y \in (-1, 1)$$

In this case the coefficients a and b represent the sharpness of the peak in the x- and y-directions, and the quality of the peak can be considered as being $\min(a, b)$.

A 2D discrete signal was loaded from `peakfit1` earlier

```
>> z
z =
   -0.0696    0.0348    0.1394    0.2436    0.3480
    0.0800    0.2000    0.3202    0.4400    0.5600
    0.0400    0.1717    0.3662    0.4117    0.5200
    0.0002    0.2062    0.8766    0.4462    0.4802
   -0.0400    0.0917    0.2862    0.3317    0.4400
   -0.0800    0.0400    0.1602    0.2800    0.4000
```

In this small example it is clear that the peak is at element (3, 4) using image coordinate convention, but programatically this is

```
>> [zmax,i] = max(z(:))
zmax =
     0.8766
i =
     16
```

Counting the elements, starting with 1 at the top-left down each column then back to the top of the next rightmost column.

and the maximum is at the sixteenth element in row-major order◄ which we convert to array subscripts

```
>> [y,x] = ind2sub(size(z), i)
y =
     4
x =
     3
```

We can find this more conveniently using the Toolbox function `peak2`

```
>> [zpk,xy]=peak2(z)
zpk =
     0.8766     0.5600
xy =
     3     5
     4     2
```

which has returned two local maxima, one per column of the returned variables. This function will return all nonlocal maxima where the size of the local region is given by the `'scale'` option. As for the 1-dimensional case we can refine the estimate of the peak

```
>> [zpk,xy]=peak2(z, 'interp')
Warning: Peak at (5,2) too close to edge of image
zpk =
    0.8839
xy =
    3.1090
    3.9637
```

that is, the peak is at element (3.1090, 3.9637). When this process is applied to image data it is referred to as subpixel interpolation.

Bibliography

Achtelik MW (2014) Advanced closed loop visual navigation for micro aerial vehicles. Ph.D. thesis, ETH Zurich

Agarwal S, Furukawa Y, Snavely N, Simon I, Curless B, Seitz SM, Szeliski R (2011) Building Rome in a day. Commun ACM 54(10):105–112

Agarwal P, Burgard W, Stachniss C (2014) Survey of geodetic mapping methods: Geodetic approaches to mapping and the relationship to graph-based SLAM. IEEE Robot Autom Mag 21(3):63–80

Agrawal M, Konolige K, Blas M (2008) CenSurE: Center surround extremas for realtime feature detection and matching. In: Forsyth D, Torr P, Zisserman A (eds) Lecture notes in computer science. Computer Vision – ECCV 2008, vol 5305. Springer-Verlag, Berlin Heidelberg, pp 102–115

Albertos P, Mareels I (2010) Feedback and control for everyone. Springer-Verlag, Berlin Heidelberg

Altmann SL (1989) Hamilton, Rodrigues, and the quaternion scandal. Math Mag 62(5):291–308

Alton K, Mitchell IM (2006) Optimal path planning under defferent norms in continuous state spaces. In: Proceedings of the IEEE International Conference on Robotics and Automation (ICRA). pp 866–872

Andersen N, Ravn O, Sørensen A (1993) Real-time vision based control of servomechanical systems. In: Chatila R, Hirzinger G (eds) Lecture notes in control and information sciences. Experimental Robotics II, vol 190. Springer-Verlag, Berlin Heidelberg, pp 388–402

Andersson RL (1989) Dynamic sensing in a ping-pong playing robot. IEEE T Robotic Autom 5(6):728–739

Antonelli G (2014) Underwater robots: Motion and force control of vehicle-manipulator systems, 3rd ed. Springer Tracts in Advanced Robotics, vol 2. Springer-Verlag, Berlin Heidelberg

Arandjelovi R, Zisserman A (2012) Three things everyone should know to improve object retrieval. In: IEEE Conference on Computer Vision and Pattern Recognition (CVPR). pp 2911–2918

Arkin RC (1999) Behavior-based robotics. MIT Press, Cambridge, Massachusetts

Armstrong WW (1979) Recursive solution to the equations of motion of an N-link manipulator. In: Proceedings of the 5th World Congress on Theory of Machines and Mechanisms, Montreal, Jul, pp 1343–1346

Armstrong BS (1988) Dynamics for robot control: Friction modelling and ensuring excitation during parameter identification. Stanford University

Armstrong B (1989) On finding exciting trajectories for identification experiments involving systems with nonlinear dynamics. Int J Robot Res 8(6):28

Armstrong B, Khatib O, Burdick J (1986) The explicit dynamic model and inertial parameters of the Puma 560 Arm. In: Proceedings of the IEEE International Conference on Robotics and Automation (ICRA), vol 3. pp 510–518

Armstrong-Hélouvry B, Dupont P, De Wit CC (1994) A survey of models, analysis tools and compensation methods for the control of machines with friction. Automatica 30(7):1083–1138

Arun KS, Huang TS, Blostein SD (1987) Least-squares fitting of 23-D point sets. IEEE T Pattern Anal 9(5):699–700

Asada H (1983) A geometrical representation of manipulator dynamics and its application to arm design. J Dyn Syst-T ASME 105:131

Astolfi A (1999) Exponential stabilization of a wheeled mobile robot via discontinuous control. J Dyn Syst-T ASME 121(1):121–126

Azarbayejani A, Pentland AP (1995) Recursive estimation of motion, structure, and focal length. IEEE T Pattern Anal 17(6):562–575

Bailey T (n.d.) Software resources. University of Sydney. http://www-personal.acfr.usyd.edu.au/tbailey

Bailey T, Durrant-Whyte H (2006) Simultaneous localization and mapping: Part II. IEEE Robot Autom Mag 13(3):108–117

Bakthavatchalam M, Chaumette F, Tahri O (2015) An improved modelling scheme for photometric moments with inclusion of spatial weights for visual servoing with partial appearance/disappearance. In: Proceedings of the IEEE International Conference on Robotics and Automation (ICRA). pp 6037–6043

Baldridge AM, Hook SJ, Grove CI, Rivera G (2009) The ASTER spectral library version 2.0. Remote Sens Environ 113(4):711–715

Ball RS (1876) The theory of screws: A study in the dynamics of a rigid body. Hodges, Foster & Co., Dublin

Ball RS (1908) A treatise on spherical astronomy. Cambridge University Press, New York

Ballard DH (1981) Generalizing the Hough transform to detect arbitrary shapes. Pattern Recogn 13(2):111–122

Banks J, Corke PI (2001) Quantitative evaluation of matching methods and validity measures for stereo vision. Int J Robot Res 20(7):512–532

Bar-Shalom Y, Fortmann T (1988) Tracking and data association. Mathematics in science and engineering, vol 182. Academic Press, London Oxford

Bar-Shalom Y, Rong Li X, Thiagalingam Kirubarajan (2001) Estimation with applications to tracking and navigation. John Wiley & Sons, Inc., Chichester

Bauer J, Sünderhauf N, Protzel P (2007) Comparing several implementations of two recently published feature detectors. In: IFAC Symposium on Intelligent Autonomous Vehicles (IAV). Toulouse

Bay H, Ess A, Tuytelaars T, Van Gool L (2008) Speeded-up robust features (SURF). Comput Vis Image Und 110(3):346–359

Benosman R, Kang SB (2001) Panoramic vision: Sensors, theory, and applications. Springer-Verlag, Berlin Heidelberg

Benson KB (ed) (1986) Television engineering handbook. McGraw-Hill, New York

Bertozzi M, Broggi A, Cardarelli E, Fedriga R, Mazzei L, Porta P (2011) VIAC expedition: Toward autonomous mobility. IEEE Robot Autom Mag 18(3):120–124

Besl PJ, McKay HD (1992) A method for registration of 3-D shapes. IEEE T Pattern Anal 14(2): 239–256

Bhat DN, Nayar SK (2002) Ordinal measures for image correspondence. IEEE T Pattern Anal 20(4): 415–423

Biber P, Straßer W (2003) The normal distributions transform: A new approach to laser scan matching. In: Proceedings of the IEEE/RSJ International Conference on intelligent robots and systems (IROS), vol 3. pp 2743–2748

Bishop CM (2006) Pattern recognition and machine learning. Information science and statistics. Springer-Verlag, New York

Blewitt M (2011) Celestial navigation for yachtsmen. Adlard Coles Nautical, London

Bolles RC, Baker HH, Marimont DH (1987) Epipolar-plane image analysis: An approach to determining structure from motion. Int J Comput Vision 1(1):7–55, Mar

Bolles RC, Baker HH, Hannah MJ (1993) The JISCT stereo evaluation. In: Image Understanding Workshop: proceedings of a workshop held in Washington, DC apr 18–21, 1993. Morgan Kaufmann, San Francisco, pp 263

Bolton W (2015) Mechatronics: Electronic control systems in mechanical and electrical engineering, 6th ed. Pearson, Harlow

Borenstein J, Everett HR, Feng L (1996) Navigating mobile robots: Systems and techniques. AK Peters, Ltd. Natick, MA, USA, Out of print and available at http://www-personal.umich.edu/~johannb/Papers/pos96rep.pdf

Borgefors G (1986) Distance transformations in digital images. Comput Vision Graph 34(3):344–371

Bostrom N (2016) Superintelligence: Paths, dangers, strategies. Oxford University Press, Oxford, 432 p

Bouguet J-Y (2010) Camera calibration toolbox for MATLAB®. http://www.vision.caltech.edu/bouguetj/calib_doc

Brady M, Hollerbach JM, Johnson TL, Lozano-Pérez T, Mason MT (eds) (1982) Robot motion: Planning and control. MIT Press, Cambridge, Massachusetts

Braitenberg V (1986) Vehicles: Experiments in synthetic psychology. MIT Press, Cambridge, Massachusetts

Bray H (2014) You are here: From the compass to GPS, the history and future of how we find ourselves. Basic Books, New York

Brockett RW (1983) Asymptotic stability and feedback stabilization. In: Brockett RW, Millmann RS, Sussmann HJ (eds) Progress in mathematics. Differential geometric control theory, vol 27. pp 181–191

Broida TJ, Chandrashekhar S, Chellappa R (1990) Recursive 3-D motion estimation from a monocular image sequence. IEEE T Aero Elec Sys 26(4):639–656

Brooks RA (1986) A robust layered control system for a mobile robot. IEEE T Robotic Autom 2(1):14–23

Brooks RA (1989) A robot that walks: Emergent behaviors from a carefully evolved network. MIT AI Lab, Memo 1091

Brown MZ, Burschka D, Hager GD (2003) Advances in computational stereo. IEEE T Pattern Anal 25(8):993–1 008

Brynjolfsson E, McAfee A (2014) The second machine age: Work, progress, and prosperity in a time of brilliant technologies. W. W. Norton & Co., New York

Buehler M, Iagnemma K, Singh S (eds) (2007) The 2005 DARPA grand challenge: The great robot race. Springer Tracts in Advanced Robotics, vol 36. Springer-Verlag, Berlin Heidelberg

Buehler M, Iagnemma K, Singh S (eds) (2010) The DARPA urban challenge. Tracts in Advanced Robotics, vol 56. Springer-Verlag, Berlin Heidelberg

Bukowski R, Haynes LS, Geng Z, Coleman N, Santucci A, Lam K, Paz A, May R, DeVito M (1991) Robot hand-eye coordination rapid prototyping environment. In: Proc ISIR, pp 16.15–16.28

Buttazzo GC, Allotta B, Fanizza FP (1993) Mousebuster: A robot system for catching fast moving objects by vision. In: Proceedings of the IEEE International Conference on Robotics and Automation (ICRA). Atlanta, pp 932–937

Calonder M, Lepetit V, Strecha C, Fua P (2010) BRIEF: Binary robust independent elementary features. In: Daniilidis K, Maragos P, Paragios N (eds) Lecture notes in computer science. Computer Vision – ECCV 2010, vol 6311. Springer-Verlag, Berlin Heidelberg, pp 778–792

Canny JF (1983) Finding edges and lines in images. MIT, Artificial Intelligence Laboratory, AI-TR-720. Cambridge, MA

Canny J (1987) A computational approach to edge detection. In: Fischler MA, Firschein O (eds) Readings in computer vision: Issues, problems, principles, and paradigms. Morgan Kaufmann, San Francisco, pp 184–203

Censi A (2008) An ICP variant using a point-to-line metric. In: Proceedings of the IEEE International Conference on Robotics and Automation (ICRA). pp 19–25

Chahl JS, Srinivasan MV (1997) Reflective surfaces for panoramic imaging. Appl Optics 31(36):8275–8285

Chatfield K, Lempitsky VS, Vedaldi A, Zisserman A (2011) The devil is in the details: An evaluation of recent feature encoding methods. In: Proceedings of the British Machine Vision Conference 2011. 12 p

Chaumette F (1990) La relation vision-commande: Théorie et application et des tâches robotiques. Ph.D. thesis, Université de Rennes 1

Chaumette F (1998) Potential problems of stability and convergence in image-based and position-based visual servoing. In: Kriegman DJ, Hager GD, Morse AS (eds) Lecture notes in control and information sciences. The confluence of vision and control, vol 237. Springer-Verlag, Berlin Heidelberg, pp 66–78

Chaumette F (2004) Image moments: A general and useful set of features for visual servoing. IEEE T Robotic Autom 20(4):713–723

Chaumette F, Hutchinson S (2006) Visual servo control 1: Basic approaches. IEEE Robot Autom Mag 13(4):82–90

Chaumette F, Hutchinson S (2007) Visual servo control 2: Advanced approaches. IEEE Robot Autom Mag 14(1):109–118

Chaumette F, Rives P, Espiau B (1991) Positioning of a robot with respect to an object, tracking it and estimating its velocity by visual servoing. In: Proceedings of the IEEE International Conference on Robotics and Automation (ICRA). Seoul, pp 2248–2253

Chesi G, Hashimoto K (eds) (2010) Visual servoing via advanced numerical methods. Lecture notes in computer science, vol 401. Springer-Verlag, Berlin Heidelberg

Chiaverini S, Sciavicco L, Siciliano B (1991) Control of robotic systems through singularities. Lecture notes in control and information sciences. Advanced Robot Control, Proceedings of the International Workshop on Nonlinear and Adaptive Control: Issues in Robotics, vol 162. Springer-Verlag, Berlin Heidelberg, pp 285–295

Chiuso A, Favaro P, Jin H, Soatto S (2002) Structure from motion causally integrated over time. IEEE T Pattern Anal 24(4):523–535

Choset HM, Lynch KM, Hutchinson S, Kantor G, Burgard W, Kavraki LE, Thrun S (2005) Principles of robot motion. MIT Press, Cambridge, Massachusetts

Colicchia G, Waltner C, Hopf M, Wiesner H (2009) The scallop's eye – A concave mirror in the context of biology. Physics Education 44(2):175–179

Collewet C, Marchand E, Chaumette F (2008) Visual servoing set free from image processing. In: Proceedings of IEEE International Conference on Robotics and Automation (ICRA). pp 81–86

Commission Internationale de L'Éclairage (1987) Colorimetry, 2nd ed. Commission Internationale de L'Eclairage, CIE No 15.2

Corke PI (1994) High-performance visual closed-loop robot control. University of Melbourne, Dept. Mechanical and Manufacturing Engineering. http://eprints.unimelb.edu.au/archive/00000547/01/thesis.pdf

Corke PI (1996a) In situ measurement of robot motor electrical constants. Robotica 14(4):433–436

Corke PI (1996b) Visual control of robots: High-performance visual servoing. Mechatronics, vol 2. Research Studies Press (John Wiley). Out of print and available at http://www.petercorke.com/bluebook

Corke PI (2001) Mobile robot navigation as a planar visual servoing problem. In: Jarvis RA, Zelinsky A (eds) Springer tracts in advanced robotics. Robotics Research: The 10th International Symposium, vol 6. IFRR, Lorne, pp 361–372

Corke PI (2007) A simple and systematic approach to assigning Denavit-Hartenberg parameters. IEEE T Robotic Autom 23(3):590–594

Corke PI (2010) Spherical image-based visual servo and structure estimation. In: Proceedings of the IEEE International Conference on Robotics and Automation (ICRA). Anchorage, pp 5550–5555

Corke PI, Armstrong-Hélouvry BS (1994) A search for consensus among model parameters reported for the PUMA 560 robot. In: Proceedings of the IEEE International Conference on Robotics and Automation (ICRA). San Diego, pp 1608–1613

Corke PI, Armstrong-Hélouvry BS (1995) A meta-study of PUMA 560 dynamics: A critical appraisal of literature data. Robotica 13(3):253–258

Corke PI, Good MC (1992) Dynamic effects in high-performance visual servoing. In: Proceedings of the IEEE International Conference on Robotics and Automation (ICRA). Nice, pp 1838–1843

Corke PI, Good MC (1996) Dynamic effects in visual closed-loop systems. IEEE T Robotic Autom 12(5):671–683

Corke PI, Hutchinson SA (2001) A new partitioned approach to image-based visual servo control. IEEE T Robotic Autom 17(4):507–515

Corke PI, Dunn PA, Banks JE (1999) Frame-rate stereopsis using non-parametric transforms and programmable logic. In: Proceedings of the IEEE International Conference on Robotics and Automation (ICRA). Detroit, pp 1928–1933

Corke PI, Strelow D, Singh S (2004) Omnidirectional visual odometry for a planetary rover. In: Proceedings of the International Conference on Intelligent Robots and Systems (IROS). Sendai, pp 4007–4012

Corke PI, Spindler F, Chaumette F (2009) Combining Cartesian and polar coordinates in IBVS. In: Proceedings of the International Conference on Intelligent Robots and Systems (IROS). St. Louis, pp 5962–5967

Corke PI, Paul R, Churchill W, Newman P (2013) Dealing with shadows: Capturing intrinsic scene appearance for image-based outdoor localisation. In: Proceedings of the IEEE/RSJ International Conference on Intelligent Robots and Systems (IROS). pp 2085–2092

Craig JJ (1987) Adaptive control of mechanical manipulators. Addison-Wesley Longman Publishing Co., Inc. Boston

Craig JJ (2005) Introduction to robotics: Mechanics and control, 3rd ed. Pearson/Prentice Hall, Upper Saddle River, New Jersey

Craig JJ, Hsu P, Sastry SS (1987) Adaptive control of mechanical manipulators. Int J Robot Res 6(2):16–28

Crombez N, Caron G, Mouaddib EM (2015) Photometric Gaussian mixtures based visual servoing. In: Proceedings of the IEEE/RSJ International Conference on Intelligent Robots and Systems (IROS). pp 5486–5491

Crone RA (1999) A history of color: The evolution of theories of light and color. Kluwer Academic, Dordrecht

Cummins M, Newman P (2008) FAB-MAP: Probabilistic localization and mapping in the space of appearance. Int J Robot Res 27(6):647

Cutting JE (1997) How the eye measures reality and virtual reality. Behav Res Meth Ins C 29(1):27–36

Daniilidis K, Klette R (eds) (2006) Imaging beyond the pinhole camera. Computational Imaging, vol 33. Springer-Verlag, Berlin Heidelberg

Dansereau DG (2014) Plenoptic signal processing for robust vision in field robotics. Ph.D. thesis, The University of Sydney

Davison AJ, Reid ID, Molton ND, Stasse O (2007) MonoSLAM: Real-time single camera SLAM. IEEE T Pattern Anal 29(6):1052–1067

Deguchi K (1998) Optimal motion control for image-based visual servoing by decoupling translation and rotation. In: Proceedings of the International Conference on Intelligent Robots and Systems (IROS). Victoria, Canada, pp 705–711

Dellaert F, Kaess M (2006) Square root SAM: Simultaneous localization and mapping via square root information smoothing. Int J Robot Res 25(12):1181–1203

Dellaert F, Seitz SM, Thorpe CE, Thrun S (2000) Structure from motion without correspondence. In: Proceedings of the IEEE Conference on Computer Vision and Pattern Recognition. Hilton Head Island, SC, pp 557–564

DeMenthon D, Davis LS (1992) Exact and approximate solutions of the perspective-three-point problem. IEEE T Pattern Anal 14(11):1100–1105

Denavit J, Hartenberg RS (1955) A kinematic notation for lower-pair mechanisms based on matrices. J Appl Mech-T ASME 22(1):215–221

Deo AS, Walker ID (1995) Overview of damped least-squares methods for inverse kinematics of robot manipulators. J Intell Robot Syst 14(1):43–68

Deriche R, Giraudon G (1993) A computational approach for corner and vertex detection. Int J Comput Vision 10(2):101–124

DeWitt BA, Wolf PR (2000) Elements of photogrammetry (with applications in GIS). McGraw-Hill, New York

Dickmanns ED (2007) Dynamic vision for perception and control of motion. Springer-Verlag, London

Dickmanns ED, Graefe V (1988a) Applications of dynamic monocular machine vision. Mach Vision Appl 1:241–261

Dickmanns ED, Graefe V (1988b) Dynamic monocular machine vision. Mach Vision Appl 1(4):223–240

Dickmanns ED, Zapp A (1987) Autonomous high speed road vehicle guidance by computer vision. In: Tenth Triennial World Congress of the International Federation of Automatic Control, vol 4. Munich, pp 221–226

Dijkstra EW (1959) A note on two problems in connexion with graphs. Numer Math 1(1):269–271

Dougherty ER, Lotufo RA (2003) Hands-on morphological image processing. Society of Photo-Optical Instrumentation Engineers (SPIE)

Duda RO, Hart PE (1972) Use of the Hough transformation to detect lines and curves in pictures. Commun ACM 15(1):11–15

Durrant-Whyte H, Bailey T (2006) Simultaneous localization and mapping: Part I. IEEE Robot Autom Mag 13(2):99–110

Espiau B, Chaumette F, Rives P (1992) A new approach to visual servoing in robotics. IEEE T Robotic Autom 8(3):313–326

Everett HR (1995) Sensors for mobile robots: Theory and application. AK Peters Ltd., Wellesley

Faugeras OD (1993) Three-dimensional computer vision: A geometric viewpoint. MIT Press, Cambridge, Massachusetts

Faugeras OD, Lustman F (1988) Motion and structure from motion in a piecewise planar environment. Int J Pattern Recogn 2(3):485–508

Faugeras O, Luong QT, Papadopoulou T (2001) The geometry of multiple images: The laws that govern the formation of images of a scene and some of their applications. MIT Press, Cambridge, Massachusetts

Featherstone R (1987) Robot dynamics algorithms. Kluwer Academic, Dordrecht

Feddema JT (1989) Real time visual feedback control for hand-eye coordinated robotic systems. Purdue University

Feddema JT, Mitchell OR (1989) Vision-guided servoing with feature-based trajectory generation. IEEE T Robotic Autom 5(5):691–700

Feddema JT, Lee CSG, Mitchell OR (1991) Weighted selection of image features for resolved rate visual feedback control. IEEE T Robotic Autom 7(1):31–47

Felzenszwalb PF, Huttenlocher DP (2004) Efficient graph-based image segmentation. Int J Comput Vision 59(2):167–181

Ferguson D, Stentz A (2006) Using interpolation to improve path planning: The Field D* algorithm. J Field Robotics 23(2):79–101

Fischler MA, Bolles RC (1981) Random sample consensus: A paradigm for model fitting with applications to image analysis and automated cartography. Commun ACM 24(6):381–395

Flusser J (2000) On the independence of rotation moment invariants. Pattern Recogn 33(9):1405–1410

Fomena R, Chaumette F (2007) Visual servoing from spheres using a spherical projection model. In: Proceedings of the IEEE International Conference on Robotics and Automation (ICRA). Rome, pp 2 080–2 085

Ford M (2015) Rise of the robots: Technology and the threat of a jobless future. Basic Books, New York

Förstner W (1994) A framework for low level feature extraction. In: Ecklundh J-O (ed) Lecture notes in computer science. Computer Vision – ECCV 1994, vol 800. Springer-Verlag, Berlin Heidelberg, pp 383–394

Förstner W, Gülch E (1987) A fast operator for detection and precise location of distinct points, corners and centres of circular features. In: ISPRS Intercommission Workshop. Interlaken, pp 149–155

Forsyth DA, Ponce J (2011) Computer vision: A modern approach, 2nd ed. Pearson, London

Fraundorfer F, Scaramuzza D (2012) Visual odometry: Part II – Matching, robustness, optimization, and applications. IEEE Robot Autom Mag 19(2):78–90

Freeman H (1974) Computer processing of line-drawing images. ACM Comput Surv 6(1):57–97

Friedman DP, Felleisen M, Bibby D (1987) The little LISPer. MIT Press, Cambridge, Massachusetts

Funda J, Taylor RH, Paul RP (1990) On homogeneous transforms, quaternions, and computational efficiency. IEEE T Robotic Autom 6(3):382–388

Gans NR, Hutchinson SA, Corke PI (2003) Performance tests for visual servo control systems, with application to partitioned approaches to visual servo control. Int J Robot Res 22(10–11):955

Gautier M, Khalil W (1992) Exciting trajectories for the identification of base inertial parameters of robots. Int J Robot Res 11(4):362

Geiger A, Roser M, Urtasun R (2010) Efficient large-scale stereo matching. In: Kimmel R, Klette R, Sugimoto A (eds) Computer vision – ACCV 2010: 10th Asian Conference on Computer Vision, Queenstown, New Zealand, November 8–12, 2010, revised selected papers, part I. Springer-Verlag, Berlin Heidelberg, pp 25–38

Geraerts R, Overmars MH (2004) A comparative study of probabilistic roadmap planners. In: Boissonnat J-D, Burdick J, Goldberg K, Hutchinson S (eds) Springer tracts in advanced robotics. Algorithmic Foundations of Robotics V, vol 7. Springer-Verlag, Berlin Heidelberg, pp 43–58

Gevers T, Gijsenij A, van de Weijer J, Geusebroek J-M (2012) Color in computer vision: Fundamentals and applications. John Wiley & Sons, Inc., Chichester

Geyer C, Daniilidis K (2000) A unifying theory for central panoramic systems and practical implications. In: Vernon D (ed) Lecture notes in computer science. Computer vision – ECCV 2000, vol 1 843. Springer-Verlag, Berlin Heidelberg, pp 445–461

Glover A, Maddern W, Warren M, Reid S, Milford M, Wyeth G (2012) OpenFABMAP: An open source toolbox for appearance-based loop closure detection. In: Proceedings of the IEEE International Conference on Robotics and Automation (ICRA). pp 4730–4735

Gonzalez R, Woods R (2008) Digital image processing, 3rd ed. Prentice Hall, Upper Saddle River, New Jersey

Gonzalez R, Woods R, Eddins S (2009) Digital image processing using MATLAB, 2nd ed. Gatesmark Publishing

Grassia FS (1998) Practical parameterization of rotations using the exponential map. Journal of Graphics Tools 3(3):29–48

Gregory RL (1997) Eye and brain: The psychology of seeing. Princeton University Press, Princeton, New Jersey

Grey CGP (2014) Humans need not apply. YouTube video, www.youtube.com/watch?v=7Pq-S557XQU

Grisetti G (n.d.) Teaching resources. Sapienza University of Rome. http://www.dis.uniroma1.it/~grisetti/teaching.html

Groves PD (2013) Principles of GNSS, inertial, and multisensor integrated navigation systems, 2nd ed. Artech House, Norwood, USA

Hager GD, Toyama K (1998) X Vision: A portable substrate for real-time vision applications. Comput Vis Image Und 69(1):23–37

Hamel T, Mahony R (2002) Visual servoing of an under-actuated dynamic rigid-body system: An image-based approach. IEEE T Robotic Autom 18(2):187–198

Hamel T, Mahony R, Lozano R, Ostrowski J (2002) Dynamic modelling and configuration stabilization for an X4-flyer. IFAC World Congress 1(2), p 3

Hansen P, Corke PI, Boles W (2010) Wide-angle visual feature matching for outdoor localization. Int J Robot Res 29(1–2):267–297

Harris CG, Stephens MJ (1988) A combined corner and edge detector. In: Proceedings of the Fourth Alvey Vision Conference. Manchester, pp 147–151

Hart PE (2009) How the Hough transform was invented [DSP history]. IEEE Signal Proc Mag 26(6):18–22

Hartenberg RS, Denavit J (1964) Kinematic synthesis of linkages. McGraw-Hill, New York, available online at http://kmoddl.library.cornell.edu/bib.php?m=23

Hartley R, Zisserman A (2003) Multiple view geometry in computer vision. Cambridge University Press, New York

Harvey P (nd) ExifTool. http://www.sno.phy.queensu.ca/~phil/exiftool

Hashimoto K (ed) (1993) Visual servoing. In: Robotics and automated systems, vol 7. World Scientific, Singapore

Hashimoto K, Kimoto T, Ebine T, Kimura H (1991) Manipulator control with image-based visual servo. In: Proceedings of the IEEE International Conference on Robotics and Automation (ICRA). Seoul, pp 2 267–2 272

Hellerstein JL, Diao Y, Parekh S, Tilbury DM (2004) Feedback control of computing systems. Wiley-IEEE Press, 456 p

Herschel W (1800) Experiments on the refrangibility of the invisible rays of the sun. Phil Trans R Soc Lond 90:284–292

Hill J, Park WT (1979) Real time control of a robot with a mobile camera. In: Proceedings of the 9th ISIR, SME. Washington, DC. Mar, pp 233–246

Hirata T (1996) A unified linear-time algorithm for computing distance maps. Inform Process Lett 58(3):129–133

Hirschmüller H (2008) Stereo processing by semiglobal matching and mutual information. IEEE Transactions on Pattern Analysis and Machine Intelligence 30(2):328–341

Hirt C, Claessens S, Fecher T, Kuhn M, Pail R, Rexer M (2013) New ultrahigh-resolution picture of Earth's gravity field. Geophys Res Lett 40:4279–4283

Hoag D (1963) Consideration of Apollo IMU gimbal lock. MIT Instrumentation Laboratory, E–1344, http://www.hq.nasa.gov/alsj/e-1344.htm

Hollerbach JM (1980) A recursive Lagrangian formulation of manipulator dynamics and a comparative study of dynamics formulation complexity. IEEE T Syst Man Cyb 10(11):730–736, Nov

Hollerbach JM (1982) Dynamics. In: Brady M, Hollerbach JM, Johnson TL, Lozano-Pérez T, Mason MT (eds) Robot motion – Planning and control. MIT Press, Cambridge, Massachusetts, pp 51–71

Horaud R, Canio B, Leboullenx O (1989) An analytic solution for the perspective 4-point problem. Comput Vision Graph 47(1):33–44

Horn BKP (1987) Closed-form solution of absolute orientation using unit quaternions. J Opt Soc Am A 4(4):629–642

Horn BKP, Hilden HM, Negahdaripour S (1988) Closed-form solution of absolute orientation using orthonormal matrices. J Opt Soc Am A 5(7):1 127–1 135

Hosoda K, Asada M (1994) Versatile visual servoing without knowledge of true Jacobian. In: Proceedings of the International Conference on Intelligent Robots and Systems (IROS). Munich, pp 186–193

Howard TM, Green CJ, Kelly A, Ferguson D (2008) State space sampling of feasible motions for high-performance mobile robot navigation in complex environments. J Field Robotics 25(6–7):325–345

Hu MK (1962) Visual pattern recognition by moment invariants. IRE T Inform Theor 8:179–187

Hua M-D, Ducard G, Hamel T, Mahony R, Rudin K (2014) Implementation of a nonlinear attitude estimator for aerial robotic vehicles. IEEE T Contr Syst T 22(1):201–213

Huang TS, Netravali AN (1994) Motion and structure from feature correspondences: A review. P IEEE 82(2):252–268

Humenberger M, Zinner C, Kubinger W (2009) Performance evaluation of a census-based stereo matching algorithm on embedded and multi-core hardware. In: Proceedings of the 19th International Symposium on Image and Signal Processing and Analysis (ISPA). pp 388–393

Hunt RWG (1987) The reproduction of colour, 4th ed. Fountain Press, Tolworth

Hunter RS, Harold RW (1987) The measurement of appearance. John Wiley & Sons, Inc., Chichester

Hutchinson S, Hager G, Corke PI (1996) A tutorial on visual servo control. IEEE T Robotic Autom 12(5):651–670

Iwatsuki M, Okiyama N (2002a) A new formulation of visual servoing based on cylindrical coordinate system with shiftable origin. In: Proceedings of the International Conference on Intelligent Robots and Systems (IROS). Lausanne, pp 354–359

Iwatsuki M, Okiyama N (2002b) Rotation-oriented visual servoing based on cylindrical coordinates. In: Proceedings of the IEEE International Conference on Robotics and Automation (ICRA). Washington, DC, May, pp 4 198–4 203

Izaguirre A, Paul RP (1985) Computation of the inertial and gravitational coefficients of the dynamics equations for a robot manipulator with a load. In: Proceedings of the IEEE International Conference on Robotics and Automation (ICRA). Mar, pp 1 024–1 032

Jägersand M, Fuentes O, Nelson R (1996) Experimental evaluation of uncalibrated visual servoing for precision manipulation. In: Proceedings of the IEEE International Conference on Robotics and Automation (ICRA). Albuquerque, NM, pp 2 874–2 880

Jarvis RA, Byrne JC (1988) An automated guided vehicle with map building and path finding capabilities. In: Robotics Research: The Fourth international symposium. MIT Press, Cambridge, Massachusetts, pp 497–504

Jazwinski AH (2007) Stochastic processes and filtering theory. Dover Publications, Mineola

Jebara T, Azarbayejani A, Pentland A (1999) 3D structure from 2D motion. IEEE Signal Proc Mag 16(3):66–84

Julier SJ, Uhlmann JK (2004) Unscented filtering and nonlinear estimation. P IEEE 92(3):401–422

Kaehler A, Bradski G (2016) Learning OpenCV: Computer vision in C++ with the OpenCV library. O'Reilly & Associates, Köln

Kaess M, Ranganathan A, Dellaert F (2007) iSAM: Fast incremental smoothing and mapping with efficient data association. In: Proceedings of the IEEE International Conference on Robotics and Automation (ICRA). pp 1670–1677

Kahn ME (1969) The near-minimum time control of open-loop articulated kinematic linkages. Stanford University, AIM-106

Kálmán RE (1960) A new approach to linear filtering and prediction problems. J Basic Eng-T Asme 82(1):35–45

Kane TR, Levinson DA (1983) The use of Kane's dynamical equations in robotics. Int J Robot Res 2(3):3–21

Karaman S, Walter MR, Perez A, Frazzoli E, Teller S (2011) Anytime motion planning using the RRT*. In: Proceedings of the IEEE International Conference on Robotics and Automation (ICRA). pp 1478–1483

Kavraki LE, Svestka P, Latombe JC, Overmars MH (1996) Probabilistic roadmaps for path planning in high-dimensional configuration spaces. IEEE T Robotic Autom 12(4):566–580

Kelly R (1996) Robust asymptotically stable visual servoing of planar robots. IEEE T Robotic Autom 12(5):759–766

Kelly A (2013) Mobile robotics: Mathematics, models, and methods. Cambridge University Press, New York

Kelly R, Carelli R, Nasisi O, Kuchen B, Reyes F (2002a) Stable visual servoing of camera-in-hand robotic systems. IEEE-ASME T Mech 5(1):39–48

Kelly R, Shirkey P, Spong MW (2002b) Fixed-camera visual servo control for planar robots. In: Proceedings of the IEEE International Conference on Robotics and Automation (ICRA). Washington, DC, pp 2 643–2 649

Khalil W, Creusot D (1997) SYMORO+: A system for the symbolic modelling of robots. Robotica 15(2):153–161

Khalil W, Dombre E (2002) Modeling, identification and control of robots. Kogan Page Science, London

Khatib O (1987) A unified approach for motion and force control of robot manipulators: The operational space formulation. IEEE T Robotic Autom 3(1):43–53

King-Hele D (2002) Erasmus Darwin's improved design for steering carriages and cars. Notes and Records of the Royal Society of London 56(1):41–62

Klafter RD, Chmielewski TA, Negin M (1989) Robotic engineering – An integrated approach. Prentice Hall, Upper Saddle River, New Jersey

Klein CA, Huang CH (1983) Review of pseudoinverse control for use with kinematically redundant manipulators. IEEE T Syst Man Cyb 13:245–250

Klein G, Murray D (2007) Parallel tracking and mapping for small AR workspaces. In: Sixth IEEE and ACM International Symposium on Mixed and Augmented Reality (ISMAR 2007). pp 225–234

Klette R, Kruger N, Vaudrey T, Pauwels K, van Hulle M, Morales S, Kandil F, Haeusler R, Pugeault N, Rabe C (2011) Performance of correspondence algorithms in vision-based driver assistance using an online image sequence database. IEEE T Veh Technol 60(5):2 012–2 026

Koenderink JJ (1984) The structure of images. Biol Cybern 50(5):363–370

Koenderink JJ (2010) Color for the sciences. MIT Press, Cambridge, Massachusetts

Koenig S, Likhachev M (2002) D* Lite. In: Proceedings of the National Conference on Artificial Intelligence, Menlo Park, CA; Cambridge, MA; London; AAAI Press; MIT Press, Cambridge, Massachusetts; 1999, pp 476–483

Koenig S, Likhachev M (2005) Fast replanning for navigation in unknown terrain. IEEE T Robotic Autom 21(3):354–363

Kriegman DJ, Hager GD, Morse AS (eds) (1998) The confluence of vision and control. Lecture notes in control and information sciences, vol 237. Springer-Verlag, Berlin Heidelberg

Kuipers JB (1999) Quaternions and rotation sequences: A primer with applications to orbits, aeroespace and virtual reality. Princeton University Press, Princeton, New Jersey

Kümmerle R, Grisetti G, Strasdat H, Konolige K, Burgard W (2011) g^2o: A general framework for graph optimization. In: Proceedings of the IEEE International Conference on Robotics and Automation (ICRA). pp 3607–3613

Lam O, Dayoub F, Schulz R, Corke P (2015) Automated topometric graph generation from floor plan analysis. In: Proceedings of the Australasian Conference on Robotics and Automation. Australasian Robotics and Automation Association (ARAA)

Lamport L (1994) LATEX: A document preparation system. User's guide and reference manual. Addison-Wesley Publishing Company, Reading

Land EH, McCann J (1971) Lightness and retinex theory. J Opt Soc Am A 61(1):1–11

Land MF, Nilsson D-E (2002) Animal eyes. Oxford University Press, Oxford

LaValle SM (1998) Rapidly-exploring random trees: A new tool for path planning. Computer Science Dept., Iowa State University, TR 98–11

LaValle SM (2006) Planning algorithms. Cambridge University Press, New York

LaValle SM (2011a) Motion planning: The essentials. IEEE Robot Autom Mag 18(1):79–89

LaValle SM (2011b) Motion planning: Wild frontiers. IEEE Robot Autom Mag 18(2):108–118

LaValle SM, Kuffner JJ (2001) Randomized kinodynamic planning. Int J Robot Res 20(5):378–400

Laussedat A (1899) La métrophotographie. Enseignement supérieur de la photographie. Gauthier-Villars, 52 p

Leavers VF (1993) Which Hough transform? Comput Vis Image Und 58(2):250–264

Lee CSG, Lee BH, Nigham R (1983) Development of the generalized D'Alembert equations of motion for mechanical manipulators. In: Proceedings of the 22nd CDC, San Antonio, Texas. pp 1205–1210

Lepetit V, Moreno-Noguer F, Fua P (2009) EPnP: An accurate $O(n)$ solution to the PnP problem. Int J Comput Vision 81(2):155–166

Li H, Hartley R (2006) Five-point motion estimation made easy. In: 18th International Conference on Pattern Recognition ICPR 2006. Hong Kong, pp 630–633

Li Y, Jia W, Shen C, van den Hengel A (2014) Characterness: An indicator of text in the wild. IEEE T Image Process 23(4):1666–1677

Li T, Bolic M, Djuric P (2015) Resampling methods for particle filtering: Classification, implementation, and strategies. IEEE Signal Proc Mag 32(3):70–86

Lin Z, Zeman V, Patel RV (1989) On-line robot trajectory planning for catching a moving object. In: Proceedings of the IEEE International Conference on Robotics and Automation (ICRA). pp 1726–1731

Lindeberg T (1993) Scale-space theory in computer vision. Springer-Verlag, Berlin Heidelberg

Lloyd J, Hayward V (1991) Real-time trajectory generation using blend functions. In: Proceedings of the IEEE International Conference on Robotics and Automation (ICRA). Seoul, pp 784–789

Longuet-Higgins H (1981) A computer algorithm for reconstruction of a scene from two projections. Nature 293:133–135

Lovell J, Kluger J (1994) Apollo 13. Coronet Books

Lowe DG (1991) Fitting parametrized three-dimensional models to images. IEEE T Pattern Anal 13(5): 441–450

Lowe DG (2004) Distinctive image features from scale-invariant keypoints. Int J Comput Vision 60(2):91–110

Lowry S, Sunderhauf N, Newman P, Leonard J, Cox D, Corke P, Milford M (2015) Visual place recognition: A survey. Robotics, IEEE Transactions on (99):1–19

Lu F, Milios E (1997) Globally consistent range scan alignment for environment mapping. Auton Robot 4:333–349

Lucas SM (2005) ICDAR 2005 text locating competition results. In: Proceedings of the Eighth International Conference on Document Analysis and Recognition, ICDAR05. pp 80–84

Lucas BD, Kanade T (1981) An iterative image registration technique with an application to stereo vision. In: International joint conference on artificial intelligence (IJCAI), Vancouver, vol 2. http://ijcai.org/Past%20Proceedings/IJCAI-81-VOL-2/PDF/017.pdf, pp 674–679

Luh JYS, Walker MW, Paul RPC (1980) On-line computational scheme for mechanical manipulators. J Dyn Syst-T ASME 102(2):69–76

Lumelsky V, Stepanov A (1986) Dynamic path planning for a mobile automaton with limited information on the environment. IEEE T Automat Contr 31(11):1058–1063

Luong QT (1992) matrice fondamentale et autocalibration en vision par ordinateur. Ph.D. thesis, Université de Paris-Sud, Orsay, France

Lynch KM, Park FC (2017) Modern robotics: Mechanics, planning, and control. Cambridge University Press, New York

Ma Y, Kosecka J, Soatto S, Sastry S (2003) An invitation to 3D. Springer-Verlag, Berlin Heidelberg

Magnusson M, Lilienthal A, Duckett T (2007) Scan registration for autonomous mining vehicles using 3D-NDT. J Field Robotics 24(10):803–827

Magnusson M, Nuchter A, Lorken C, Lilienthal AJ, Hertzberg J (2009) Evaluation of 3D registration reliability and speed – A comparison of ICP and NDT. In: Proceedings of the IEEE International Conference on Robotics and Automation (ICRA). pp 3907–3912

Mahony R, Kumar V, Corke P (2012) Multirotor aerial vehicles: Modeling, estimation, and control of quadrotor. IEEE Robot Autom Mag (19):20–32

Maimone M, Cheng Y, Matthies L (2007) Two years of visual odometry on the Mars exploration rovers. J Field Robotics 24(3):169–186

Makhlin AG (1985) Stability and sensitivity of servo vision systems. In: Proc 5th International Conference on Robot Vision and Sensory Controls – RoViSeC 5. IFS (Publications), Amsterdam, pp 79–89

Malis E (2004) Improving vision-based control using efficient second-order minimization techniques. In: Proceedings of the IEEE International Conference on Robotics and Automation (ICRA). pp 1843–1848

Malis E, Vargas M (2007) Deeper understanding of the homography decomposition for vision-based control. Research Report, RR-6303, Institut National de Recherche en Informatique et en Automatique (INRIA), 90 p, https://hal.inria.fr/inria-00174036v3/document

Malis E, Chaumette F, Boudet S (1999) 2-1/2D visual servoing. IEEE T Robotic Autom 15(2):238–250

Marey M, Chaumette F (2008) Analysis of classical and new visual servoing control laws. In: Proceedings of the IEEE International Conference on Robotics and Automation (ICRA). Pasadena, pp 3244–3249

Mariottini GL, Prattichizzo D (2005) EGT for multiple view geometry and visual servoing: Robotics vision with pinhole and panoramic cameras. IEEE T Robotic Autom 12(4):26–39

Mariottini GL, Oriolo G, Prattichizzo D (2007) Image-based visual servoing for nonholonomic mobile robots using epipolar geometry. IEEE T Robotic Autom 23(1):87–100

Marr D (2010) Vision: A computational investigation into the human representation and processing of visual information. MIT Press, Cambridge, Massachusetts

Martin D, Fowlkes C, Tal D, Malik J (2001) A database of human segmented natural images and its application to evaluating segmentation algorithms and measuring ecological statistics. Proceedings of the 8th International Conference on Computer Vision, vol 2. pp 416–423

Martins FN, Celeste WC, Carelli R, Sarcinelli-Filho M, Bastos-Filho TF (2008) An adaptive dynamic controller for autonomous mobile robot trajectory tracking. Control Eng Pract 16(11):1354–1363

Masutani Y, Mikawa M, Maru N, Miyazaki F (1994) Visual servoing for non-holonomic mobile robots. In: Proceedings of the International Conference on Intelligent Robots and Systems (IROS). Munich, pp 1 133–1 140

Mataríc MJ (2007) The robotics primer. MIT Press, Cambridge, Massachusetts

Matas J, Chum O, Urban M, Pajdla T (2004) Robust wide-baseline stereo from maximally stable extremal regions. Image Vision Comput 22(10):761–767

Matthews ND, An PE, Harris CJ (1995) Vehicle detection and recognition for autonomous intelligent cruise control. Technical Report, University of Southampton

Matthies L (1992) Stereo vision for planetary rovers: Stochastic modeling to near real-time implementation. Int J Comput Vision 8(1):71–91

Mayeda H, Yoshida K, Osuka K (1990) Base parameters of manipulator dynamic models. IEEE T Robotic Autom 6(3):312–321

McLauchlan PF (1999) The variable state dimension filter applied to surface-based structure from motion. University of Surrey, VSSP-TR-4/99

Merlet JP (2006) Parallel robots. Kluwer Academic, Dordrecht

Mettler B (2003) Identification modeling and characteristics of miniature rotorcraft. Kluwer Academic, Dordrecht

Mičušík B, Pajdla T (2003) Estimation of omnidirectional camera model from epipolar geometry. In: IEEE Conference on Computer Vision and Pattern Recognition, vol 1. Madison, pp 485–490

Middleton RH, Goodwin GC (1988) Adaptive computed torque control for rigid link manipulations. Syst Control Lett 10(1):9–16

Mikolajczyk K, Schmid C (2004) Scale and affine invariant interest point detectors. Int J Comput Vision 60(1):63–86

Mikolajczyk K, Schmid C (2005) A performance evaluation of local descriptors. IEEE T Pattern Anal 27(10):1 615–1 630

Mindell DA (2008) Digital Apollo. MIT Press, Cambridge, Massachusetts

Molton N, Brady M (2000) Practical structure and motion from stereo when motion is unconstrained. Int J Comput Vision 39(1):5–23

Montemerlo M, Thrun S (2007) FastSLAM: A scalable method for the simultaneous localization and mapping problem in robotics, vol 27. Springer-Verlag, Berlin Heidelberg

Montemerlo M, Thrun S, Koller D, Wegbreit B (2002) FastSLAM: A factored solution to the simultaneous localization and mapping problem. In: Proceedings of the AAAI National Conference on Artificial Intelligence. AAAI, Edmonton, Canada

Montemerlo M, Thrun S, Koller D, Wegbreit B (2003) FastSLAM 2.0: An improved particle filtering algorithm for simultaneous localization and mapping that provably converges. In: Proceedings of the 18th International Joint Conference on Artificial Intelligence. Morgan Kaufmann, San Francisco, pp 1151–1156

Moravec H (1980) Obstacle avoidance and navigation in the real world by a seeing robot rover. Ph.D. thesis, Stanford University

Morel G, Liebezeit T, Szewczyk J, Boudet S, Pot J (2000) Explicit incorporation of 2D constraints in vision based control of robot manipulators. In: Corke PI, Trevelyan J (eds) Lecture notes in control and information sciences. Experimental robotics VI, vol 250. Springer-Verlag, Berlin Heidelberg, pp 99–108

Muja M, Lowe DG (2009) Fast approximate nearest neighbors with automatic algorithm configuration. International Conference on Computer Vision Theory and Applications (VISAPP), Lisbon, Portugal (Feb 2009), pp 331–340

Murray RM, Sastry SS, Zexiang L (1994) A mathematical introduction to robotic manipulation. CRC Press, Inc., Boca Raton

NASA (1970) Apollo 13: Technical air-to-ground voice transcription. Test Division, Apollo Spacecraft Program Office, http://www.hq.nasa.gov/alsj/a13/AS13_TEC.PDF

Nayar SK (1997) Catadioptric omnidirectional camera. In: Proceedings of the IEEE Conference on Computer Vision and Pattern Recognition. Los Alamitos, CA, pp 482–488

Neilson S (2011) Robot nation: Surviving the greatest socio-economic upheaval of all time. Eridanus Press, New York, 124 p

Neira J, Tardós JD (2001) Data association in stochastic mapping using the joint compatibility test. IEEE T Robotic Autom 17(6):890–897

Neira J, Davison A, Leonard J (2008) Guest editorial special issue on Visual SLAM. IEEE T Robotic Autom 24(5):929–931

Nethery JF, Spong MW (1994) Robotica: A mathematica package for robot analysis. IEEE T Robotic Autom 1(1):13–20

Newcombe RA, Lovegrove SJ, Davison AJ (2011) DTAM: Dense tracking and mapping in real-time. In: Proceedings of the International Conference on Computer Vision, pp 2320–2327

Newman P (n.d.) C4B mobile robots and estimation resources. Oxford University. http://www.robots.ox.ac.uk/~pnewman/Teaching/C4CourseResources/C4BResources.html

Ng J, Bräunl T (2007) Performance comparison of bug navigation algorithms. J Intell Robot Syst 50(1):73–84

Niblack W (1985) An introduction to digital image processing. Strandberg Publishing Company Birkeroed, Denmark

Nilsson NJ (1971) Problem-solving methods in artificial intelligence. McGraw-Hill, New York

Nistér D (2003) An efficient solution to the five-point relative pose problem. In: IEEE Conference on Computer Vision and Pattern Recognition, vol 2. Madison, pp 195–202

Nistér D, Naroditsky O, Bergen J (2006) Visual odometry for ground vehicle applications. J Field Robotics 23(1):3–20

Nixon MS, Aguado AS (2012) Feature extraction and image processing, 3rd ed. Academic Press, London Oxford

Noble JA (1988) Finding corners. Image Vision Comput 6(2):121–128

Okutomi M, Kanade T (1993) A multiple-baseline stereo. IEEE T Pattern Anal 15(4):353–363

Ollis M, Herman H, Singh S (1999) Analysis and design of panoramic stereo vision using equi-angular pixel cameras. Robotics Institute, Carnegie Mellon University, CMU-RI-TR-99-04, Pittsburgh, PA

Olson E (2011) AprilTag: A robust and flexible visual fiducial system. In: Proceedings of the IEEE International Conference on Robotics and Automation (ICRA). pp 3400–3407

Orin DE, McGhee RB, Vukobratovic M, Hartoch G (1979) Kinematics and kinetic analysis of open-chain linkages utilizing Newton-Euler methods. Math Biosci 43(1/2):107–130

Ortega R, Spong MW (1989) Adaptive motion control of rigid robots: A tutorial. Automatica 25(6):877–888

Otsu N (1975) A threshold selection method from gray-level histograms. Automatica 11:285–296

Papanikolopoulos NP, Khosla PK (1993) Adaptive robot visual tracking: Theory and experiments. IEEE T Automat Contr 38(3):429–445

Papanikolopoulos NP, Khosla PK, Kanade T (1993) Visual tracking of a moving target by a camera mounted on a robot: A combination of vision and control. IEEE T Robotic Autom 9(1):14–35

Park FC (1994) Computational aspects of the product-of-exponentials formula for robot kinematics. IEEE T Automat Contr 39(3):643–647

Paul R (1972) Modelling, trajectory calculation and servoing of a computer controlled arm. Ph.D. thesis, technical report AIM-177, Stanford University

Paul R (1979) Manipulator Cartesian path control. IEEE T Syst Man Cyb 9:702–711

Paul RP (1981) Robot manipulators: Mathematics, programming, and control. MIT Press, Cambridge, Massachusetts

Paul RP, Shimano B (1978) Kinematic control equations for simple manipulators. In: IEEE Conference on Decision and Control, vol 17. pp 1 398–1 406

Paul RP, Zhang H (1986) Computationally efficient kinematics for manipulators with spherical wrists based on the homogeneous transformation representation. Int J Robot Res 5(2):32–44

Piepmeier JA, McMurray G, Lipkin H (1999) A dynamic quasi-Newton method for uncalibrated visual servoing. In: Proceedings of the IEEE International Conference on Robotics and Automation (ICRA). Detroit, pp 1 595–1 600

Pilu M (1997) A direct method for stereo correspondence based on singular value decomposition. In: Proceedings of the Computer Vision and Pattern Recognition, IEEE Computer Society, San Juan, pp 261–266

Pivtoraiko M, Knepper RA, Kelly A (2009) Differentially constrained mobile robot motion planning in state lattices. J Field Robotics 26(3):308–333

Pock T (2008) Fast total variation for computer vision. Ph.D. thesis, Graz University of Technology

Pollefeys M, Nistér D, Frahm JM, Akbarzadeh A, Mordohai P, Clipp B, Engels C, Gallup D, Kim SJ, Merrell P, et al. (2008) Detailed real-time urban 3D reconstruction from video. Int J Comput Vision 78(2):143–167, Jul

Pomerleau D, Jochem T (1995) No hands across America Journal. http://www.cs.cmu.edu/~tjochem/nhaa/Journal.html

Pomerleau D, Jochem T (1996) Rapidly adapting machine vision for automated vehicle steering. IEEE Expert 11(1):19–27

Posner I, Corke P, Newman P (2010) Using text-spotting to query the world. In: IEEE/RSJ International Conference on Intelligent Robots and Systems (IROS). IEEE, pp 3181–3186

Pounds P (2007) Design, construction and control of a large quadrotor micro air vehicle. Ph.D. thesis, Australian National University

Pounds P, Mahony R, Gresham J, Corke PI, Roberts J (2004) Towards dynamically-favourable quad-rotor aerial robots. In: Proceedings of the Australasian Conference on Robotics and Automation. Canberra

Pounds P, Mahony R, Corke PI (2006) A practical quad-rotor robot. In: Proceedings of the Australasian Conference on Robotics and Automation. Auckland

Pounds P, Mahony R, Corke PI (2007) System identification and control of an aerobot drive system. In: Information, Decision and Control. IEEE, pp 154–159

Poynton CA (2003) Digital video and HDTV: Algorithms and interfaces. Morgan Kaufmann, San Francisco

Poynton CA (2012) Digital video and HD algorithms and interfaces. Morgan Kaufmann, Burlington

Press WH, Teukolsky SA, Vetterling WT, Flannery BP (2007) Numerical recipes, 3rd ed. Cambridge University Press, New York

Prince SJ (2012) Computer vision: Models, learning, and inference. Cambridge University Press, New York

Prouty RW (2002) Helicopter performance, stability, and control. Krieger, Malabar FL

Pynchon T (2006) Against the day. Jonathan Cape, London

Rekleitis IM (2004) A particle filter tutorial for mobile robot localization. Technical report (TR-CIM-04-02), Centre for Intelligent Machines, McGill University

Rives P, Chaumette F, Espiau B (1989) Positioning of a robot with respect to an object, tracking it and estimating its velocity by visual servoing. In: Hayward V, Khatib O (eds) Lecture notes in control and information sciences. Experimental robotics I, vol 139. Springer-Verlag, Berlin Heidelberg, pp 412–428

Rizzi AA, Koditschek DE (1991) Preliminary experiments in spatial robot juggling. In: Chatila R, Hirzinger G (eds) Lecture notes in control and information sciences. Experimental robotics II, vol 190. Springer-Verlag, Berlin Heidelberg, pp 282–298

Roberts LG (1963) Machine perception of three-dimensional solids. MIT Lincoln Laboratory, TR 315, http://www.packet.cc/files/mach-per-3D-solids.html

Rosenfield GH (1959) The problem of exterior orientation in photogrammetry. Photogramm Eng 25(4):536–553

Rosten E, Porter R, Drummond T (2010) FASTER and better: A machine learning approach to corner detection. IEEE T Pattern Anal 32:105–119

Russell S, Norvig P (2009) Artificial intelligence: A modern approach, 3rd ed. Prentice Hall Press, Upper Saddle River, NJ

Sakaguchi T, Fujita M, Watanabe H, Miyazaki F (1993) Motion planning and control for a robot performer. In: Proceedings of the IEEE International Conference on Robotics and Automation (ICRA). Atlanta, May, pp 925–931

Salvi J, Matabosch C, Fofi D, Forest J (2007) A review of recent range image registration methods with accuracy evaluation. Image Vision Comput 25(5):578–596

Samson C, Espiau B, Le Borgne M (1990) Robot control: The task function approach. Oxford University Press, Oxford

Sanderson AC, Weiss LE, Neuman CP (1987) Dynamic sensor-based control of robots with visual feedback. IEEE T Robotic Autom RA-3(5):404–417

Scaramuzza D, Fraundorfer F (2011) Visual odometry [tutorial]. IEEE Robot Autom Mag 18(4):80–92

Scharstein D, Pal C (2007) Learning conditional random fields for stereo. In: IEEE Computer Society Conference on Computer Vision and Pattern Recognition (CVPR 2007). Minneapolis, MN

Scharstein D, Szeliski R (2002) A taxonomy and evaluation of dense two-frame stereo correspondence algorithms. Int J Comput Vision 47(1):7–42

Selig JM (2005) Goemetric fundamentals of robotics. Springer-Verlag, Berlin Heidelberg

Sharp A (1896) Bicycles & tricycles: An elementary treatise on their design an construction; With examples and tables. Longmans, Green and Co., London New York Bombay

Sheridan TB (2003) Telerobotics, automation, and human supervisory control. MIT Press, Cambridge, Massachusetts, 415 p

Shi J, Tomasi C (1994) Good features to track. In: Proceedings of the Computer Vision and Pattern Recognition. IEEE Computer Society, Seattle, pp 593–593

Shih FY (2009) Image processing and mathematical morphology: Fundamentals and applications, CRC Press, Boca Raton

Shirai Y (1987) Three-dimensional computer vision. Springer-Verlag, New York

Shirai Y, Inoue H (1973) Guiding a robot by visual feedback in assembling tasks. Pattern Recogn 5(2):99–106

Shoemake K (1985) Animating rotation with quaternion curves. In: Proceedings of ACM SIGGRAPH, San Francisco, pp 245–254

Siciliano B, Khatib O (eds) (2016) Springer handbook of robotics, 2nd ed. Springer-Verlag, New York

Siciliano B, Sciavicco L, Villani L, Oriolo G (2009) Robotics: Modelling, planning and control. Springer-Verlag, Berlin Heidelberg

Siegwart R, Nourbakhsh IR, Scaramuzza D (2011) Introduction to autonomous mobile robots, 2nd ed. MIT Press, Cambridge, Massachusetts

Silver WM (1982) On the equivalance of Lagrangian and Newton-Euler dynamics for manipulators. Int J Robot Res 1(2):60–70

Sivic J, Zisserman A (2003) Video Google: A text retrieval approach to object matching in videos. In: Proceedings of the Ninth IEEE International Conference on Computer Vision. pp 1 470–1 477

Skaar SB, Brockman WH, Hanson R (1987) Camera-space manipulation. Int J Robot Res 6(4):20–32

Skofteland G, Hirzinger G (1991) Computing position and orientation of a freeflying polyhedron from 3D data. In: Proceedings of the IEEE International Conference on Robotics and Automation (ICRA). Seoul, pp 150–155

Slama CC (ed) (1980) Manual of photogrammetry, 4th ed. American Society of Photogrammetry

Smith R (2007) An overview of the Tesseract OCR engine. In: 9th International Conference on Document Analysis and Recognition (ICDAR). pp 629–633

Sobel D (1996) Longitude: The true story of a lone genius who solved the greatest scientific problem of his time. Fourth Estate, London

Soille P (2003) Morphological image analysis: Principles and applications. Springer-Verlag, Berlin Heidelberg

Spong MW (1989) Adaptive control of flexible joint manipulators. Syst Control Lett 13(1):15–21

Spong MW, Hutchinson S, Vidyasagar M (2006) Robot modeling and control, 2nd ed. John Wiley & Sons, Inc., Chichester

Srinivasan VV, Venkatesh S (1997) From living eyes to seeing machines. Oxford University Press, Oxford

Stachniss C, Burgard W (2014) Particle filters for robot navigation. Foundations and Trends in Robotics 3(4):211–282

Steinvall A (2002) English colour terms in context. Ph.D. thesis, Ume Universitet

Stentz A (1994) The D* algorithm for real-time planning of optimal traverses. The Robotics Institute, Carnegie-Mellon University, CMU-RI-TR-94-37

Stewart A (2014) Localisation using the appearance of prior structure. Ph.D. thesis, University of Oxford

Stone JV (2012) Vision and brain: How we perceive the world. MIT Press, Cambridge, Massachusetts

Strasdat H (2012) Local accuracy and global consistency for efficient visual SLAM. Ph.D. thesis, Imperial College London

Strelow D, Singh S (2004) Motion estimation from image and inertial measurements. Int J Robot Res 23(12):1 157–1 195

Sünderhauf N (2012) Robust optimization for simultaneous localization and mapping. Ph.D. thesis, Technische Universität Chemnitz

Sussman GJ, Wisdom J, Mayer ME (2001) Structure and interpretation of classical mechanics. MIT Press, Cambridge, Massachusetts

Sutherland IE (1974) Three-dimensional data input by tablet. P IEEE 62(4):453–461

Svoboda T, Pajdla T (2002) Epipolar geometry for central catadioptric cameras. Int J Comput Vision 49(1):23–37

Szeliski R (2011) Computer vision: Algorithms and applications. Springer-Verlag, Berlin Heidelberg

Tahri O, Chaumette F (2005) Point-based and region-based image moments for visual servoing of planar objects. IEEE T Robotic Autom 21(6):1 116–1 127

Tahri O, Mezouar Y, Chaumette F, Corke PI (2009) Generic decoupled image-based visual servoing for cameras obeying the unified projection model. In: Proceedings of the IEEE International Conference on Robotics and Automation (ICRA). Kobe, pp 1 116–1 121

Taylor RA (1979) Planning and execution of straight line manipulator trajectories. IBM J Res Dev 23(4):424–436

ter Haar Romeny BM (1996) Introduction to scale-space theory: Multiscale geometric image analysis. Utrecht University

Thrun S, Burgard W, Fox D (2005) Probabilistic robotics. MIT Press, Cambridge, Massachusetts

Tissainayagam P, Suter D (2004) Assessing the performance of corner detectors for point feature tracking applications. Image Vision Comput 22(8):663–679

Titterton DH, Weston JL (2005) Strapdown inertial navigation technology. IEE Radar, Sonar, Navigation and Avionics Series, vol 17, The Institution of Engineering and Technology (IET), 576 p

Tomasi C, Kanade T (1991) Detection and tracking of point features. Carnegie Mellon University, CMU-CS-91-132

Triggs B, McLauchlan P, Hartley R, Fitzgibbon A (2000) Bundle adjustment – A modern synthesis. Lecture notes in computer science. Vision algorithms: theory and practice, vol 1 883. Springer-Verlag, Berlin Heidelberg, pp 153–177

Tsakiris D, Rives P, Samson C (1998) Extending visual servoing techniques to nonholonomic mobile robots. In: Kriegman DJ, Hager GD, Morse AS (eds) Lecture notes in control and information sciences. The confluence of vision and control, vol 237. Springer-Verlag, Berlin Heidelberg, pp 106–117

Uicker JJ (1965) On the dynamic analysis of spatial linkages using 4 by 4 matrices. Dept. Mechanical Engineering and Astronautical Sciences, NorthWestern University

Usher K (2005) Visual homing for a car-like vehicle. Ph.D. thesis, Queensland University of Technology

Usher K, Ridley P, Corke PI (2003) Visual servoing of a car-like vehicle – An application of omnidirectional vision. In: Proceedings of the IEEE International Conference on Robotics and Automation (ICRA). Taipai, Sep, pp 4 288–4 293

Valgren C, Lilienthal AJ (2010) SIFT, SURF & seasons: Appearance-based long-term localization in outdoor environments. Robot Auton Syst 58(2):149–156

Vanderborght B, Sugar T, Lefeber D (2008) Adaptable compliance or variable stiffness for robotic applications. IEEE Robot Autom Mag 15(3):8–9

Vedaldi A, Fulkerson B (2008) VLFeat: An open and portable library of computer vision algorithms. http://www.vlfeat.org

Wade NJ (2007) Image, eye, and retina. J Opt Soc Am A 24(5):1229–1249

Walker MW, Orin DE (1982) Efficient dynamic computer simulation of robotic mechanisms. J Dyn Syst-T ASME 104(3):205–211

Walter WG (1950) An imitation of life. Sci Am 182(5):42–45

Walter WG (1951) A machine that learns. Sci Am 185(2):60–63

Walter WG (1953) The living brain. Duckworth, London

Warren M (2015) Long-range stereo visual odometry for unmanned aerial vehicles. Ph.D. thesis, Queensland University of Technology

Weiss LE (1984) Dynamic visual servo control of robots: An adaptive image-based approach. Ph.D. thesis, technical report CMU-RI-TR-84-16, Carnegie-Mellon University

Weiss L, Sanderson AC, Neuman CP (1987) Dynamic sensor-based control of robots with visual feedback. IEEE T Robotic Autom 3(1):404–417

Westmore DB, Wilson WJ (1991) Direct dynamic control of a robot using an end-point mounted camera and Kalman filter position estimation. In: Proceedings of the IEEE International Conference on Robotics and Automation (ICRA). Seoul, Apr, pp 2 376–2 384

Whitney DE (1969) Resolved motion rate control of manipulators and human prostheses. IEEE T Man Machine 10(2):47–53

Wiener N (1965) Cybernetics or control and communication in the animal and the machine. MIT Press, Cambridge, Massachusetts

Wilburn B, Joshi N, Vaish V, Talvala E-V, Antunez E, Barth A, Adams A, Horowitz M, Levoy M (2005) High performance imaging using large camera arrays. ACM Transactions on Graphics (TOG) – Proceedings of ACM SIGGRAPH 2005 24(3):765–776

Wolf PR (1974) Elements of photogrammetry. McGraw-Hill, New York

Woodfill J, Von Herzen B (1997) Real-time stereo vision on the PARTS reconfigurable computer. In: Proceedings of the IEEE Symposium on FPGAs for Custom Computing Machines, Grenoble. pp 201–210

Xu G, Zhang Z (1996) Epipolar geometry in stereo, motion, and object recognition: A unified approach. Springer-Verlag, Berlin Heidelberg

Ying X, Hu Z (2004) Can we consider central catiodioptric cameras and fisheye cameras within a unified imaging model. In: Pajdla T, Matas J (eds) Lecture notes in computer science. Computer vision – ECCV 2004, vol 3 021. Springer-Verlag, Berlin Heidelberg, pp 442–455

Yoshikawa T (1984) Analysis and control of robot manipulators with redundancy. In: Brady M, Paul R (eds) Robotics research: The first international symposium. MIT Press, Cambridge, Massachusetts, pp 735–747

Zabih R, Woodfill J (1994) Non-parametric local transforms for computing visual correspondence. In: Ecklundh J-O (ed) Lecture notes in computer science. Computer Vision – ECCV 1994, vol 800. Springer-Verlag, Berlin Heidelberg, pp 151–158

Zarchan P, Musoff H (2005) Fundamentals of Kalman filtering: A practical approach. Progress in Astronautics and Aeronautics, vol 208. American Institute of Aeronautics and Astronautics

Zhang Z, Faugeras O, Kohonen T, Hunag TS, Schroeder MR (1992) Three D-dynamic scene analysis: A stereo based approach. Springer-Verlag, New York

Ziegler J, Bender P, Schreiber M, Lategahn H, Strauss T, Stiller C, Thao Dang, Franke U, Appenrodt N, Keller CG, Kaus E, Herrtwich RG, Rabe C, Pfeiffer D, Lindner F, Stein F, Erbs F, Enzweiler M, Knöppel C, Hipp J, Haueis M, Trepte M, Brenk C, Tamke A, Ghanaat M, Braun M, Joos A, Fritz H, Mock H, Hein M, Zeeb E (2014) Making Bertha drive – An autonomous journey on a historic route. IEEE Intelligent Transportation Systems Magazine 6(2):8–20

Index

Index of Functions, Classes and Methods

Classes are shown in **bold**, Simulink® models in *italics*, and methods are prefixed by a dot. All others are Toolbox functions.

General Index

Symbols

T